A. KOROLEV

Cell and Molecular Biology of Wood Formation

EXPERIMENTAL BIOLOGY REVIEWS

Series advisors:

D.W. Lawlor
AFRC Institute of Arable Crops Research, Rothamsted Experimental Station, Harpenden, Hertfordshire AL5 2JQ, UK

M. Thorndyke
School of Biological Sciences, Royal Holloway, University of London, Egham, Surrey TW20 0EX, UK

Environmental Stress and Gene Regulation
Sex Determination in Plants
Plant Carbohydrate Biochemistry
Programmed Cell Death in Animals and Plants
Biomechanics in Animal Behaviour
Cell and Molecular Biology of Wood Formation

Cell and Molecular Biology of Wood Formation

R.A. SAVIDGE

Faculty of Forestry and Environmental Management, University of New Brunswick, Canada

J.R. BARNETT

Department of Botany, University of Reading, UK

R. NAPIER

HRI Wellesbourne, Warwick, UK

BIOS

First published in 2000

A CIP catalogue record for this book is available from the British Library.

ISBN 1 85996 123 1

BIOS Scientific Publishers Ltd
9 Newtec Place, Magdalen Road, Oxford OX4 1RE, UK.
Tel. +44 (0) 1865 726286. Fax. +44 (0) 1865 246823
World Wide Web home page: http://www.bios.co.uk/

Production Editor: Andrea Bosher.
Typeset by Saxon Graphics Ltd, Derby, UK.
Printed by Biddles Ltd, Guildford, UK.

Contents

Contributors

Abe, Hisashi, Forestry and Forest Products Research Institute, Matsunosato, Kukizaki-Cho, Inashiki-gun, Ibaraki 305–8687, Japan

Abbott, James, University of Dundee, Department of Biological Sciences, Dundee DD1 4HN, UK

Aloni, Roni, Tel Aviv University, Department of Plant Sciences, Tel Aviv 69978, Israel

Altamura, Maria Maddalena, Universita Degli Studi Di Roma La Sapienza, Dipartimento di Biologia Vegetale, P le Aldo Moro 5, Rome, 00185, Italy

Andersson, Seppo, University of Helsinki, Department of Physics, PO Box 9, Helsinki, FIN-00014, Finland

André, Jean Pierre, I.N.R.A., Centre de Recherches d' Antibes, Station d' Agronomie et de Physiologie Végétale, 45 boulevard du Cap, BP 2078, 06606 Antibes, Cedex, France

Arend, Matthias, Universität München, Institut für Holzforschung, Winzerstrasse 45, 80797 München, Germany

Asante, Anning K, Crops Research Institute, P.O. Box 3785, Kumasi-Ashanti, Ghana

Baima, Simona, Instituto Nazionale Della Nutrizione, Unita-Di Nutrizione Sperimentale, Via Ardeatina 546, 00178 Rome, Italy

Barakate, Abdellah, University of Dundee, Department of Biological Sciences, Dundee DD1 4HN, UK

Barnett, John, The University of Reading, Department of Botany, Whiteknights, Reading, RG6 6AS, UK

Beritognolo, Isacco, I.N.R.A., Station d'Amerlioration des Arbes Forestiers, Ardon, F-45160 Olivet, Cedex, France

Bevan, Michael, John Innes Centre, Department of Cell Biology, Norwich Research Park, Colney, Norwich, NR4 7UH, UK

Bordenave, Marianne, Institut Jaques Monod, 2 place Jussieu, 75251 Paris, Cedex 05, France

Bossinger, Gerd, The University of Melbourne, School of Forestry, Creswick, Victoria, 3363, Australia

Boudet, Alain-M, UMR 5546, Signaux et Messages Cellulaires ches les Vegetaux, Pôle de Biotechnologie Végétale, 24, chemin de Borde-Rouge, 31326 Castanet, Tolosan, France

Bowtell, Richard W., University of Nottingham, Department of Physics, Magnetic Resonance Centre, University Park, Nottingham, NG7 2RD, UK

Breton, Christian, I.N.R.A., Station d'Amelioration des Arbes Forestiers, Ardon, F-45160 Olivet, Cedex, France

Burtin, Patricia, I.N.R.A., Station d'Amelioration des Arbes Forestiers, Ardon, F-45160 Olivet, Cedex, France

Campbell, Malcolm M, University of Oxford, Department of Plant Sciences, South Parks Road, Oxford, OX1 3RB, UK

Cãno-Delgado, Ana, John Innes Centre, Department of Molecular Genetics, Norwich Research Park, Colney, Norwich, NR4 7UH, UK

Carter, Carol E, The University of Reading, Department of Botany, Whiteknights, Reading, Berkshire, RG6 6AS, UK

Catesson, Anne Marie, Université P et M Curie, Physiologie Végétale, 4 place Jussieu, 75252 Paris, Cedex 05, France

Chaffey, Nigel, University of Bristol, Developmental Cell Biology, IACR-Long Ashton Research Station, Long Ashton, Bristol, BS41 9AF, UK

Charpentier, Jean-Paul, I.N.R.A., Station d'Amelioration des Arbes Forestiers, Ardon, Olivet, F-45160, France

Couée, Ivan, I.N.R.A., Station d'Amelioration des Arbes Forestiers, Ardon, Olivet, F-45160, France

Crow, Emma, Royal Botanic Gardens Kew, Jodrell Laboratory, Richmond, Surrey, TW9 3AB, UK

Cutler, David F., Royal Botanic Gardens Kew, Jodrell Laboratory, Richmond, Surrey, TW9 3AB, UK

Dehon, Laurence, Université des Sciences et Techniques du Languedoc, Laboratoire de Biotechnologie et Physiologie Végétales, Montpellier II, Place Eugène Bataillon, F-34095 Montpellier, Cedex 05, France

Domingo, Concha, Universidad Politécnica, Instituto de Biolog°a Molecular y Celular de Plantas, Camino de Vera s/n, 46022 Valencia, Spain

Downes, Geoff M, CSIRO, Forestry and Forest Products, GPO Box 252–12, Hobart, Tasmania 7001, Australia

Ehlers, Katrin, Justus-Liebig-Universität, Institut fur allgemeine Botanik und Pflanzenphysiologie, Senckenbergstrasse 17, D-35390 Giessen, Germany

Eklund, Leif, Växjö Universitet, Department of Engineering and Natural Sciences, Växjö, S-351 95, Sweden

Ermel, Fabienne F, Université P et M Curie, PCMP, UMR 7632, case 154, 4 place Jussieu, 75252 Paris, Cedex 05, France

Evans, Robert, CSIRO Forestry and Forest Products, Private Bag 10, Clayton South MDC, Victoria 3169, Australia

Farrar, John F, University of Wales, School of Biological Sciences, Bangor, Gwynedd, LL57 2UW, UK

Feigenbaum, Pua, Tel Aviv University, Department of Plant Sciences, Tel Aviv 69978, Israel

Fink, Siegfried, Albert-Ludwigs-Universität Freiburg, Institut fur Forstbotanik und Baumphysiologie, Bertoldstrasse 17, 79085 Freiburg, Germany

Follet-Gueye, Marie-Laure, Université de Rouen, SCUEOR, Faculté des Sciences, BP 118, 76134 Mont Saint Aignan, Cedex, France

Forneris, Natascha, University of New Brunswick, Faculty of Forestry and Environmental Management, Fredericton NB, E3B 6C2, Canada

Förster, Hartmut, Friedrich-Schiller-Universität Jena, Institut für Allgemeine Botanik, Bereich Pflanzenbiochemie, Kühnhäuser Strasse 101, D-99189 Erfurt-Kühnhausen, Germany

Fritts, Harold C., DendroPower, 5703 North Lady Lane, Tucson, Arizona 85704, USA

Fromm, Jörg, Universität München, Institut für Holzforschung, Winzerstrasse 45, 80797 München, Germany

Funada, Ryo, Hokkaido University, Faculty of Agriculture, Department of Forest Science, Kita 9, Nishi-9, Kita-ku, Sapporo 060–8589, Japan

Furusawa, Osamu, Hokkaido University, Department of Forest Science, Faculty of Agriculture, Kita-9, Nishi-9, Sapporo 060–8589, Japan

Goffner, Deborah, UMR 5546, Signaux et Messages Cellulaires ches les Vegetaux, Pôle de Biotechnologie Végétale, 24, chemin de Borde-Rouge, 31326 Castanet, Tolosan, France

Goldberg, Renee, Université P et M Curie, PCMP, UMR 7632, case 154, 4 place Jussieu, 75252 Paris, Cedex 05, France

Goodenough, Peter, The University of Reading, Department of Botany, Whiteknights, Reading, Berkshire, RG6 6AS, UK

Hagqvist, Risto, Finnish Forest Research Institute, Haapastensyrja Breeding Centre, Karkkilantie 247, Layliainen, Fin-12600, Finland

Hall, Madeleine, University of Manchester, School of Biological Sciences, 3.614 Stopford Building, Oxford Road, Manchester, M13 9PT, UK

Halpin, Claire, University of Dundee, Department of Biological Sciences, Dundee DD1 4HN, UK

Hodson, Jane, Albert-Ludwigs-Universität Freiburg, Institut fur Forstbotanik und Baumphysiologie, Bertoldstrasse 17, 79085 Freiburg, Germany

Höll, Wolfgang, Universität München, Lehrstuhl für Botanik der Technische, Am Hochanger 4, 85350 Freising, Germany

Iliev, Iliyan D, Bulgarian Academy of Sciences, Institute of Organic Chemistry and Centre of Phytochemistry, G. Bonchev Street, Bld. 9, Sofia 1113, Bulgaria

Jay-Allemand, Christian, I.N.R.A., Station d'Amelioration des Arbes Forestiers, Ardon, F-45160 Olivet, Cedex, France

Jones, Louise, University of Manchester, School of Biological Science, 3.614 Stopford Building, Oxford Road, Manchester M13 9PT, UK

Kalev, Natalie, Tel Aviv University, Department of Plant Sciences, Tel Aviv 69978, Israel

Klintborg, Alla, Växjö Universitet, Department of Engineering and Natural Sciences, Växjö, S-351 95, Sweden

Köckenberger, Walter, University of Nottingham, Department of Physics, Magnetic Resonance Centre, University Park, Nottingham, NG7 2RD, UK

Korolev, Andrey, Coleg y Brifysgol, Ysgol Gwyddorau Bioleg, Bangor, Gwynedd, LL57 2UW, UK

Krabel, Doris, Technische Universität Dresden, FR Forstwissenschaften, Institut für Forstbotanik und Forstzoologie, Pienner Strasse 7, 01737 Tharandt, Germany

Label, Phillippe, I.N.R.A., Centre de Recherches d'Orleans, Unite d'Amerlioration de Genetique et de Physiologie Forestieres, BP 20619 Ardon, F-45166 Olivet, Cedex, France

Lehto, Jouko, Finnish Forest Research Institute, Punkaharju Research Station, Finlandiantie 18, Punkaharju, Fin-58450, Finland

Leitch, Mathew A, University of Melbourne, School of Forestry, Water Street, Creswick, Victoria 3363, Australia

Lev-Yadun, Simcha, Department of Natural Resources, Agricultural Research Organisation, The Volcani Center, PO Box 6, Bet Dagan 50 - 250, Israel

Magel, Elisabeth A, Eberhard-Karls-Universität, Institut für Botanik, Lehrstuhl Physiologische Ökologie der Pflanzen, Auf der Morgenstelle 1, 72076 Tübingen, Germany

Matteucci, Antonella, Instituto Nazionale Della Nutrizione, Unita' di Nutrizione Sperimentale, Vi Ardeatina 546, 00178 Rome, Italy

McCann, Maureen, John Innes Centre, Department of Cell Biology, Colney Lane, Norwich, NR4 7UH, UK

McDougall, Gordon J., Scottish Crop Research Institute, Unit of Plant Biochemistry, Cell and Environmental Physiology Department, Mylnefield, Invergowrie, Dundee, DD2 5DA, UK

Metzlaff, Karin, University of Gent, Department of Plant Genetics, Ledeganckstraat 35, 9000 Gent, Belgium

Micheli, Fabienne, Institute Jacques Monod, Laboratorie d'Enzymologie en Milieu Structuri, 2 Place Jussieu, 75251 Paris, Cedex 05, France

Milioni, Dimitra, John Innes Centre, Department of Cell Biology, Norwich Research Park, Colney, Norwich, NR4 7UH, UK

Miura, Hiroto, Hokkaido Forest Products Research Institute, Nishikagura 1–10, Asahikawa 071–0198, Japan

Miura, Tatsuhiko, Hokkaido University, Faculty of Agriculture, Department of Forest Science, Kita-9 Nishi-9, Sapporo 060–8589, Japan

Morelli, Georgio, Instituto Nazionale della Nutrizione, UNS, Via Ardeatina 546, I-00178 Rome, Italy

Murphy, Richard J., Imperial College, Department of Biology, London, SW7 2AZ, UK

Napier, Richard, Horticulture Research International, Wellesbourne, Warwick, CV35 9EF, UK

Newman, Lisa, University of Oxford, Department of Plant Sciences, South Parks Road, Oxford, OX1 3RB, UK

Nicolay, Klaas, Universiteit Utrecht, Bijvoet Center for Biomolecular Research, Padulaan 8, NL-3584 CH Utrecht, The Netherlands

Ohtani, Jun, Hokkaido University, Faculty of Agriculture, Department of Forest Science, Kita-9, Nishi-9, Sapporo 060–8589, Japan

Paakkari, Timo, University of Helsinki, Department of Physics, PO Box 9, Helsinki, FIN – 00014, Finland

Pesonen, Erkki, Finnish Forest Research Institute METLA, PO Box 18, Vantaa, FIN-01301, Vantaa, Finland

Pommer, Ulrich, Friedrich-Schiller-Universität Jena, Institut für Allgemeine Botanik, Bereich Pflanzenbiochemie, Kühnhäuser Strasse 101, D-99189 Erfurt-Kühnhausen, Germany

Puech, Laurence, Albert-Ludwigs-Universität Freiburg, Institute fur Forstbotanik und Baumphysiologie, Bertoldstrasse 17, 79085 Freiburg, Germany

Ranocha, Philippe, UMR 5546, Signaux et Messages Cellulaires ches les Vegetaux, Pôle de Biotechnologie Végétale, 24, chemin de Borde-Rouge, 31326 Castanet, Tolosan, France

Richard, Luc, Institut Jacques Monod, 2 Place Jussieu, 75251 Paris, Cedex 05, France

Roberts, Keith, John Innes Centre, Department of Cell Biology, Norwich Research Park, Colney, Norwich, NR4 7UH, UK

Rozovsky, Sharon, Tel Aviv University, Department of Plant Sciences, Tel Aviv 69978, Israel

Ruberti, Ida, Universita Degli Studi Di Roma La Sapienza, Centro di Studio per Gli acidi Nucleici, c/o Dio di Genetics e Biologia Moleculare, Ple Aldo Moro 5, 00185 Rome, Italy

Saranpää, Pekka S, Finnish Forest Research Institute METLA, PO Box 18, FIN-01301 Vantaa, Finland

Sarén, Matti, University of Helsinki, Department of Physics, PO Box 9, Helsinki, FIN-00014, Finland

Sauter, Jörg J, Universität Kiel, Botanisches Institut und Botanischer Garten, Olshauserstrasse 40, D-24098 Kiel, Germany

Savidge, Rodney A, University of New Brunswick, Faculty of Forestry and Environmental Management, Fredericton, New Brunswick, E3B 6C2, Canada

Schimleck, Laurence R., CSIRO Forestry and Forest Products, Private Bag 10, Clayton South MDC, Victoria, Australia 3169

Serimaa, Ritva, University of Helsinki, Department of Physics, PO Box 9, Helsinki, FIN-00014, Finland

Shibagki, Masaki, Hokkaido University, Faculty of Agriculture, Department of Forest Science, Kita-9, Nishi-9, Sapporo 606–8589, Japan

Siiriä, Simo, University of Helsinki, Department of Physics, PO Box 9, FIN-00014 Helsinki, Finland

Stacey, Nicola, John Innes Centre, Department of Cell Biology, Norwich Research Park, Colney, Norwich, NR4 7UH, UK

Steeves, Valerie, University of New Brunswick, Faculty of Forestry & Environmental Management, Fredericton, E3B 6C2, Canada

Sundberg, Björn, SLU, Department of Forest Genetics and Plant Physiology, 901 83 Umeå, Sweden

Taylor, Neil, University of Manchester, School of Biological Science, 3.614 Stopford Building, Oxford Road, Manchester, M13 9PT, UK

Tomassi, Marisa, Universita Degli Studi Di Roma La Sapienza, Dipartimento di Biologia Vegetale, Ple Aldo Moro 5, 00185 Rome, Italy

Tomos, Deri, Coleg y Brifysgol, Ysgol Gwyddorau Bioleg, Bangor, Gwynedd, LL57 2UW, UK

Tuominen, Hannele, University of Helsinki, Institute of Biotechnology, PO Box 56, 00014 Helsinki, Finland

Türk, Stefan, Deutsche Sporthochschule Köln, Lehrstuhl Natursport und Ökologie, Institut für Schwimm-, Wasser-, und Kampfsport, 50927 Köln, Germany

Turner, Simon, University of Manchester, School of Biological Sciences, 3.614 Stopford Building, Oxford Road, Manchester, M13 9PT, UK

Uggla, Claes, Swedish University of Agricultural Sciences, BioAgri AB/Plant Pathology and Biocontrol Unit, P O Box 7035, SE-750 07 Uppsala, Sweden

van Bel, Aart J E, Justus Leibig Universität, Institut für allgemeine Botanik und Pflanzenphysiologie, Senckenbergstrasse 17, 35390 Giessen, Germany

Velling, Pirkko, Finnish Forest Research Institute, Vantaa Research Centre, PO Box 18, Vantaa 01301, Finland

Vian, Brigitte, I.N.R.A./I.N.A.P.G., Pathologie Végétale, 16 rue Claude Bernard, 75231 Paris, Cedex 05, France

Viherä-Aarnio, Anneli, Finnish Forest Research Institute, Vantaa Research Centre, PO Box 18, FIN-01301 Vantaa, Finland

Preface

One of the major events in the history of life on earth was the emergence of land plants, among which trees have a unique place. They represent the main CO_2-sinks in the biosphere and their wood is a major source of renewable raw material and energy. The acquisition in the Devonian Period of a lateral meristem, the cambium, following the building of the lignin biosynthetic pathway earlier in the primary era, allowed the 'making of wood' and the emergence of trees as a life form. The economic importance of wood has prompted extensive investigations into its physical properties, chemical composition and lignin biosynthesis. In contract, the study of the cambium has remained a kind of backwater, probably because of the reputed fragility of the meristem. For a long time, our understanding of the process by which xylem was produced by the cambium and how it is controlled progressed at a very slow pace, as can be judged from the various books and meeting reports published during the past 30 or 40 years.

In this context, it can be said that the recent meeting organized jointly by the SEB and EU COST ACTION E6 heralded a new era in cambial studies. It demonstrated the large interest in the cambium among scientists working in various areas of plant biology and illustrated the beneficial effects of combining viewpoints from a whole range of disciplines. This book is based on the contributions presented at the Heriot Watt University meeting in March 1999. It reflects the wide range of aspects discussed during the meeting and demonstrates the coming of age of the cambium as an integrated field of research. One can say that the 19th century saw the recognition of the cambium as a tissue by pioneers in the field. During the 20th century, in the wake of Bailey and Wareing, interest was mostly focused on the structure of living cambial cells and on the role of the growth substances in cambial activity. Now, at the beginning of the 21st century, research is accelerating in the hope of understanding at last how the cambium 'makes' the wood.

It is a special pleasure for me to acknowledge this major advance in a domain I always found fascinating. I am sure that readers will also find fascination in the prospects for understanding *Cell and Molecular Biology of Wood Formation* which are opened up by integration of results from molecular, biochemical, physiological, cytological, and anatomical approaches.

Our colleagues who had the difficult tasks of organizing the meeting and editing the many contributions from the presenters at the meeting must be warmly thanked for their initiative. The book they have produced will be of great value for many scientists and stimulate further advances in the field.

Anne-Marie Catesson

Acknowledgements

The editors wish to thank the Society for Experimental Biology, the European Commission COST ACTION E6 'Eurosilva' and Shell International Renewables, Forestry Research for their support. The Society provided the organisational framework for the meeting at the Heriot-Watt University, Edinburgh, from which this book developed. It also provided support for many speakers through the Plant Development Group. We are also grateful to the European Commission DGXII and CSIRO Forestry and Forest Products (CRC-HFPS), Australia for sponsoring the colour plates. Many scientists agreed to referee the chapters and we thank them for their suggestions, diligence and time. The assistance of Sabah Lamlom in helping to prepare the subject index is gratefully acknowledged.

COST

1

Biochemistry of seasonal cambial growth and wood formation – an overview of the challenges

R.A. Savidge

1. Introduction

Knowledge about the physiology of seasonal cambial activity and wood formation, otherwise known as secondary growth, remains rudimentary. Surprisingly little biochemistry data on the cambium and its derivative cells have yet been reported, and this in large measure undoubtedly explains the current state. This is not to suggest, however, that wood-formation biochemistry is in its infancy. Important progress has been made over the last century by a small number of dedicated researchers, and many useful treatises addressing various aspects of the topic are available (e.g., Bernard-Dagan, 1988; Brown *et al.*, 1996; Gahan, 1981; Hillis, 1987; Higuchi, 1985, 1997; Lewis and Sarkanen, 1998; Northcote, 1985; and this book). Nevertheless, the effort really has not begun to scratch the surface. Woods are not only immensely varied, but collectively they represent the most abundant form of biomass on earth. The number of investigations into their formative processes has by no means been in proportion to either their variability or abundance. Fortuitously, the cambium is much less variable than woods themselves (Larson, 1994) and, as detailed below, the cambial zone can be obtained more readily and abundantly in pure form than any other plant meristem. Cambial derivatives at different stages of becoming xylem elements can also be readily collected for molecular physiology studies, probably in greater quantity than any other developing tissue of eukaryotic biology. Thus, it seems ironic that there has not been major progress in this area. It appears that the explanation for the rudimentary understanding of wood-formation biochemistry resides primarily in a lack of appreciation of the opportunities.

There has been no lack of appreciation for woods themselves. There exists a massive, perhaps daunting, scientific literature addressing wood anatomy, wood chemistry and

Cell & Molecular Biology of Wood Formation, edited by R. Savidge, J. Barnett and R. Napier.

wood physical properties (see, for example, Carlquist, 1988; Hale, 1933; Mattheck and Kubler, 1995; Megraw, 1985; Meier, 1985; Panshin and de Zeeuw, 1980; Rowe, 1989; Shafizadeh and DeGroot, 1976; Sjöström 1993; Stamm and Sanders, 1966; Wheeler and Baas, 1998; Zobel and Sprague, 1998; Zobel and van Buijtenen, 1989). This literature tends to be purely descriptive and does not begin to explain the underlying physiological mechanisms of cambial activity or wood formation.

In most woody species, cellulose, hemi-celluloses and lignins constitute 90% or more of the dry matter in the xylem, but it should be recognized from the outset that there is much more to the chemistry of wood than those three classes of macromolecule. If there were not, all woods would be nearly white in colour, would tend to smell the same (i.e., have very little odour at all), and all would likely be as highly vulnerable to decay as are the woods of *Populus* spp. The chemical variation in microconstituents existing among woods is, in fact, immense. Thus, while each of the major phenomena of cellulose, hemi-cellulose and lignin biosynthesis still remains an open question, there is also a rich diversity of poorly known or still-not-at-all-explored ancillary biochemical reactions of wood formation awaiting elucidation.

The principal aim of this chapter is to facilitate ongoing research by developing perspective, removing unnecessary barriers and suggesting ways, some old and some new, in which cambial cells and their derivatives can be distinguished, isolated and manipulated for biochemistry investigations. I have not attempted to provide exhaustive coverage of the literature on the cambium and wood formation, knowing that much of it resides in other chapters of this volume. Instead, the reader is referred to some useful reference texts. Aspects of post-cambial metabolism are addressed in the chapters by Höll, Magel, Eklund and Klintborg and Lobel *et al.*

2. Practical significance of biochemistry research into cambial growth

Timber harvesting, wood processing and the utilization of wood/fibre products by humanity have immense socio-economical and ecological impacts. With a growing world population there is a rising demand for these products, hence for increased harvesting of timber from our forests. Global warming has also become a concern; thus, the role of forests in removing atmospheric CO_2 has gained new appreciation. Because of the immense variation in wood chemistry and the very many different kinds of wood in existence, an accurate guess of exactly how much carbon is tied up in woody biomass would be difficult to make, even if the biomass could be accurately estimated. Nevertheless, there can be no doubt that wood formation at the global level has a major role in checking unrestrained atmospheric CO_2 increases. In addition, once wood has been formed, its durability and inertness enable it, and therefore its carbon content, to accumulate over very long periods. In the extreme, the organic matter of wood can still be found in imperfectly petrified trees more than 200 million years old (*Figures 1a, 1b*). In-depth scientific knowledge of how trees make wood will obviously serve to meet the needs of humanity and, at the same time, ensure that the rate of CO_2 sequestration within woody biomass is optimal.

Timber is a *biosynthetic* end product and, therefore, the making of wood is a function of both gene expression and the catalytic rates of structural enzymes. Thus, to achieve a full understanding of wood formation, each component of the full set of intrinsic processes essential for diameter growth (i.e. chemical reactions and physical changes)

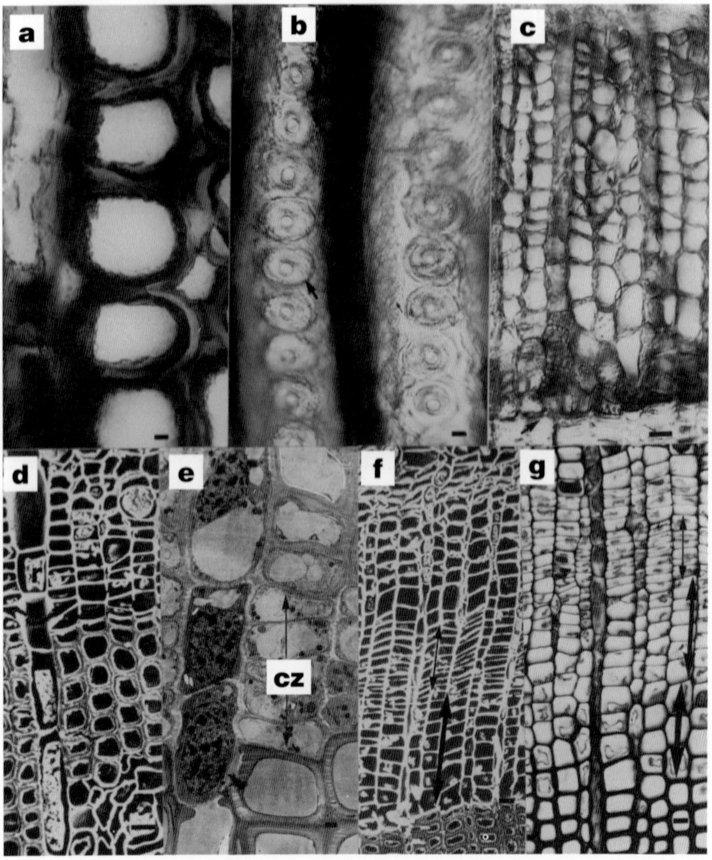

Figure 1. *(a) Cross section (30 μm) showing a ray (the 'A' is within it) and a radial file of tracheids (with residual organic matter) in a 225 million year old petrified conifer (Araucarioxylon); bar = 2 μm. (b) Same material as A, but a radial section (30 μm) showing bordered pits (large arrow) with residual organic matter. Note the inclined microfibrils (small arrow) still apparent in the secondary walls; bar = 2 μm. (c) Hand section showing tracheids produced* in vitro *in a cultured stem 'chip' with the in arbor latewood boundary (arrowed) at the bottom. The 'C' is near the cambial zone (8-year-old main-stem region,* Larix laricina*); bar = 10 μm. (d) Scanning electron micrograph of the transverse surface of a dormant* Pinus contorta *stem showing cambial zone in the region indicated by the two arrows, with phloem above and mature latewood below. In addition to the fusiform cambial cells, a cambial ray cell (r) is present and an axial resin canal is at the location of the bar; bar = 10 μm. (e) Transmission electron micrograph of a dormant* Pinus contorta *stem showing the width (between the arrows) of cambial zone with phloem at the top and mature latewood tracheids below. Note that cambial ray cell tangential walls have abundant plasmodesmata (small arrow) lacking in the walls of fusiform cells; bar = 1 μm. (f) Scanning electron micrograph of a transverse surface of a* Pinus contorta *stem with active cambium in late spring showing phloem (top), cambial zone (between the smaller arrow heads) and RE zone (between the larger arrow heads) with mature latewood at the bottom; bar = 8 μm. (g) Brightfield micrograph of the transverse surface (sectioned at 20 μm and stained with safranin–fast green) of a* Pinus contorta *stem in early summer showing phloem (top), cambial zone (between the smallest arrow heads), RE zone (between intermediate arrow heads) and SL zone (between large arrowheads) with fully autolysed 'mature' tracheids (transitional between early and late wood) at the bottom. A single ray traverses the image; bar = 8 μm.*

must be known and information on how each one of those components is affected by other processes, such as phytohormone and photosynthate production and export to the cambium, needs to be understood. Within each cell, both gene expression and catalysis are influenced by the surrounding microenvironment; thus, the roles in the cambial region of extrinsically derived factors as effectors of catalysis – such as water, light, CO_2, O_2, temperature, inorganic ions and gravitational force — also await in-depth study. Obviously, integrating all of the physical, molecular and themodynamic processes into a model explaining cell division, expansion and differentiation is going to require a major effort. However, given the attention it is due, it is unlikely to prove any more difficult than the work with microorganisms and animals which has led to such impressive progress in explaining the phenomena underlying growth and development of their cells, tissues and organs.

The world has no shortage of major problems to solve in terms of wise management of our biosphere. In the face of limitations in research funding and infrastructure, placing emphasis on the minutiae of cambial biochemistry could be construed as unnecessary or even irrelevant, research that is 'not able to see the forest for the trees.' Forestry in Canada, for example, has emphasized over the last century the need for highly visible research in order to know how to manage its forest ecosystems, certainly at the level of whole organisms and increasingly at the levels of stands and landscapes. Such holistic approaches, while readily comprehended by non-scientists, are in stark contrast to the reductionistic, highly esoteric non-visible approaches used by biochemists, biophysicists, cell and molecular biologists concerned with wood formation. It is not enough that those scientists who are active in areas of cambial research know that the effort will eventually make immense contributions to biosphere and forest management. It is essential that researchers find ways, and make time, to communicate effectively to non-biochemists about how molecular research can help.

Perhaps the most important point to be made is that biochemistry research is not about 'seeing' at all. Rather, it is a matter of going beyond the visible phenotype, whether it be forest, tree, log or individual fibre, and recognizing each for what it truly is, a biosynthetic end product, a manifestation of genotype – environment interactions that occurred in the past ($P = G \times E$). In living systems, it is not what we can see, rather what we cannot see, which drives growth and development and determines outcomes. The non-visible determinants of growth and development are what must be discovered, characterized and investigated in order to understand wood formation and provide a scientific basis for the management of wood supply and forest ecosystem health.

To date, the biochemistry basis for forecasting wood-supply in forestry has rarely extended beyond the concept that wood is somehow produced as a result of photosynthesis, with respiration being the downside to productivity (Cannell and Jackson, 1985; Landsberg, 1986). Although considerable research has been done with trees in those areas, the sobering reality is that measurements of photosynthesis and respiration have not been sufficient to explain photosynthate unloading from phloem to cambium, nor how unloaded photosynthate is metabolized once allocated to the cambium and its derivatives. In other words, whole-tree carbon-cycle research has provided almost no insight whatsoever into the factors determining how much wood is produced and how variation in wood quality arises. Silviculture research has demonstrated amply that the living crown has a major role in influencing both wood quality and quantity, and also that the environment in which a tree grows influences both growth rate and phenotype (Cannell and Jackson, 1985; Zobel and van Buijtenen, 1989). Nevertheless, knowing how to manipulate extrinsic factors influencing wood formation remains more of an art than a science because the underlying intrinsic processes remain speculative. For example, it is still not at all well understood when and which environmental factors (whether physical, chemical or biological in nature) exert their effects indirectly, by mediating biochemical processes in other tissues or organs subsequently to influence metabolism in the cambial region, *vis-à-vis* those acting directly on the cambium. Tree improvement research certainly has demonstrated that increased growth, and alterations in some aspects of wood quality, can be obtained through selection and breeding, thus confirming that there is a genetic basis for productivity (Zobel and Jett, 1995). However, outside the few loci addressed in this book, very little is yet known about the molecular genetics of wood formation *per se*.

Climate change is occurring, and it can no longer be assumed that history will repeat itself in terms of the environment forest trees grow in. In addition, the 'genotype' of stands has been funnelled from wild type to 'improved' and even clonal; i.e., from broad to much narrower genetic diversity. This narrowing has occurred with limited understanding of how the selected germplasms will react to, and cope with, the changing climate and other unforeseen stresses of next century. As expressed by Zobel and van Buijtenen (1989), 'anything that affects the growth of a tree can also affect its wood properties.' In other words, both G and E of the $P = G \times E$ expression are variables on the move, and the importance of maintaining a sufficient 'buffer' of genetic diversity is readily apparent. Even with such a buffer, the existing methods for predicting the productivity phenotype of our future forests can only be highly suspect. With climate change and limited information on tree physiology, the risk of inaccuracy in the estimates has become greater than ever. If framework information on the biochemistry of wood formation were in hand, the dynamics of $G \times E$ interactions would be much less problematic. In fact, with such information and modern-day computing capability, it should be possible to integrate the diverse kinetics and environmental switches into a working model to give, for the first time, forest managers a truly scientific tool for predicting productivity in the face of both environmental and genetic changes.

As biochemical knowledge on cambial growth increases, it will undoubtedly also facilitate useful biotechnological innovations, such as the generation of specialized genotypes for fibre-farm plantations. For example, many early wood anatomical investigations showed clearly that gene expression for polysaccharide and lignin biosynthesis were developmentally uncoupled, and present knowledge indicates that polysaccharides and lignin arise from a common precursor, glucose 6-phosphate (G6P). Thus, it was foreseen that any transformation reducing the allocation of G6P to lignification would simultaneously increase polysaccharide deposition during xylem development (Savidge 1985, 1986, 1996). Transgenic trees producing woods with reduced lignin content, higher pulp yield and less waste are now beginning to find application in forestry (Hu *et al.*, 1999), and there will undoubtedly be many related innovations springing from biochemistry investigations in this area. It is not inconceivable that biochemistry research into xylogenesis could also set the stage for industrial scale fibre production or molecular farming independently of the tree, for example from plant cells grown in vats on photosynthetically derived sugars (*Figure 1c*).

In view of the above, fundamental research to explain the biochemistry of cambial growth would seem to be more than amply justified. Indeed, in relation to maintaining life-supporting conditions in the terrestrial biosphere and meeting the needs of humanity it is difficult to think of a more important, or a more neglected, area of biochemistry.

3. Woods: chemically heterogeneous biosynthetic end products

3.1 *Anatomical similarities and differences among woods*

All woods are similar in being anisotropic, meaning that as materials their physical properties vary when tested in different planes (i.e., transverse, radial, tangential). Conifers (or 'softwoods') and *Gingko biloba* tend to be alike in basic structural makeup, as well as in lacking certain features – most conspicuously vessels – which are common traits in the angiosperms ('hardwoods'). However, even here there are exceptions. Woods of different species within any particular genus tend to be anatomically quite similar, but chemical analysis often reveals differences. All woods evidently have cellulose and lignin as their principal constituents. Beyond this, the commonality ends.

The variation among woods within and between species is immense. Every species tends to produce a unique kind of wood (Panshin and de Zeeuw, 1980; Wheeler and Baas, 1998), and this surely has its explanation in the genetic differences underlying speciation. Furthermore, in any species or, for that matter, any individual tree, the younger 'juvenile' wood produced in the crown has features which distinguish it from the older, more 'mature' wood of the bole (Zobel and Sprague, 1998). Branch wood and root wood tend to be distinct from the wood in the tree's main-stem axis. 'Normal' stem wood is very different in appearance and properties from reaction wood (i.e., compression or tension wood), and the same can apply to sap and heart woods, early and late woods etc. Even growing the same species in different geographical locations can result in astonishing differences in wood properties (Zobel and Sprague, 1998; Zobel and van Buijtenen, 1989).

The within-tree differences in wood arise in spite of the biological principal that the somatic genotype must be constant throughout the tree. If the principal of mitotic replication of the genome with perfect fidelity is correct, there can be no doubt that wood formation is not a singular, predetermined process of development, but is a very plastic expression of G × E interactions. There is evidence for DNA content varying as a function of both season and cambial age (Mellerowicz *et al.*, 1989); however, that variation is assumed to be strictly quantitative, not resulting in qualitative changes in the genotype. Thus, it appears that variation in the microenvironment experienced by cambial cells and their derivatives is the primary explanation for variability in woods.

Underlying the macro traits of every piece of wood are molecules determining wood colour, sheen, graininess, resistance to decay, etc. In addition, there are particular combinations of distinct cell types and sub-cellular features which, when viewed by skilled wood anatomists, permit keying out the wood to the source genus, even to the species (Carlquist, 1988; Panshin and de Zeeuw, 1980; Timell, 1986; Wheeler and Baas, 1998). Thus, the challenge for the biochemist interested in wood formation is threefold, to be able to explain how each of a number of cellular phenotypes is produced from cambial cells, to elucidate the biosynthetic pathways that engender these products and to integrate the two into a coherent picture.

3.2 *Wood chemistry*

Wood chemistry is concerned not only with discovering and characterizing particular molecules, but also in discovering compositional variation among known molecules at the micro- to nanometre scales within lamellae tediously dissected from individual woody elements (Meier, 1985). Wood chemistry also embodies bonding within polymers, both those that are monomerically homogeneous (such as cellulose) and those which can be heterogeneous (such as lignin and hemicelluloses). Intermolecular ionic and covalent bonding, topochemistry, hydrophobic and hydrophilic interfaces, the catalytic competence of wood (both enzymatic and non-enzymatic), stabilities and reactivities of organic compounds are all important in wood chemistry. Knowing the molecular structures in woods is not only essential for biochemistry, but also enables hypotheses to be formulated about how the molecules function.

Cellulose. Cellulose, considered as a singular linear strand of β-1,4-D-glucan, is uniform in comprising only cellobiose repeating units. However, cellulose microfibrils as found in cell walls are heterogeneous in a variety of ways. The cellulose content (percentage by dry weight) in wood varies with stem age, increasing from juvenile wood to mature wood, and it also varies within each annual ring, being lower in earlywood than in latewood. Length, number, degree of crystallinity versus amorphism and width of cellulose microfibrils all vary (Jones, 1971; Brown et al., 1996). At least seven distinct kinds of cellulose are known (Jones, 1971). Fortunately, cellulose I appears to be the only one present in wood. Production of cellulose I in cell-free systems has defied biochemists over the last century; however, putative cellulose synthase genes have been reported (see chapter by Turner *et al.*) and it may be that the field is on the verge of a breakthrough.

Lignin. By mass, lignin is a close second to cellulose in abundance, and lignin content in wood tends to vary inversely to that of cellulose (Panshin and de Zeeuw, 1980; Sjöström, 1993). Expressed as stored chemical (or, allocated photosynthetic) energy, lignin in conifers (but not in hardwoods, because they have approximately 10% lower lignin content) actually exceeds cellulose. Dry wood yields approximately 20 MJ per kilogram heat of combustion energy (Hale, 1933), but when a kilogram of lignin is completely oxidized in a pressure bomb it yields 25.1 MJ. The same mass of cellulose yields only 17.6 MJ (Shafizadeh and DeGroot, 1976). Lignin consists of aromatic monomers having higher bond energies than those of the aliphatic bonds in cellobiose.

Lignification unquestionably is the principal event resulting in tissues becoming 'woody.' Without lignin deposition, xylem cells probably would be considered by anatomists as collenchyma (no matter how thick and laminated their secondary walls) and the properties of non-lignified cells would be rather like those of the soft tissues derived from cambia of carrot, horseradish or beet roots. The strengthening and hydrophobicity imparted by lignin are, undoubtedly, essential for a tree to receive water for photosynthesis, elevate its crown, and withstand the loadings of snow, wind, ice and other forces of nature (Barghoorn, 1964). There is speculation in the literature that lignification is a form of excretion occurring to cleanse the plant of toxic or non-metabolizable waste products. However, given the roles of lignified tissues, lignin production during wood (and, in some species, also bark) formation can hardly be described as wasteful.

Progress has been made toward elucidating the structures of the several kinds of lignin and the overall biochemistry of lignification (Lewis and Sarkanen, 1998, and chapters in this volume). However, one of the primary conclusions so far is that lignification is highly complex and, therefore, needing much more research.

Hemicelluloses. In addition to cellulose and lignin, the third most abundant 'substance' in wood is actually a highly heterogeneous complex of polysaccharides traditionally referred to as the 'hemicelluloses' (Meier, 1985; Panshin and de Zeeuw, 1980; Sjöström, 1993). Although progress has been made in understanding biosynthesis of hemicelluloses (Northcote, 1985), their molecular structures for the most part remain uncertain. Wilkie (1985), who includes the pectins with the hemicelluloses, has provided useful and detailed insight into these diverse and highly complex structures and the methodologies needed to investigate them.

Other components. Cellulose, lignin and hemicelluloses account for at least 90% of the dry matter of wood in most species (Sjöström, 1993). In addition, woods contain a broad diversity of compounds extractable with organic solvents (Bernard-Dagan, 1988; Biggs, 1987; Higuchi, 1985, 1997; Hillis, 1987; Lewis and Sarkanen, 1998; Rowe, 1989; Sjöström, 1993). Polyphenols, including gallic acid derivatives, stilbenes, tannins, and phenylpropanoids, are present. The phenylpropanoids comprise coumarins, lignans, and an immense array of flavonoids, many in glycosidic form. Depending on the species, suberin may be present in abundance or not at all. There are also terpenoids, tropolones, fats, waxes, organic acids, alkaloids, diverse quinones, cinnamic acid and cinnamyl alcohol esters (with sugars, aromatic or aliphatic alcohols, hydroxy acids, sterols, terpenoids, alkaloids and other hydroxylated compounds serving as the alcohol moiety), cinnamoyl amides (with putrescine, spermidine, tyramine or spermine serving as amine moiety), starches and other water-miscible polysaccharides, sugars and related water-soluble carbohydrates, amino acids, nucleic acids and other heterocyclic compounds present. Both structural and catalytically active proteins are present within and on the cell walls of wood, and the catalytically active ones (though seldom studied) may be especially important for stabilization/transformation reactions occurring in mature woods, as well as in differentiating cambial derivatives (Pearce, 1996; Savidge *et al.*, 1998).

Wood density. Every wood has some ash content, typically 1–5% w/w but much higher in some species (Bailey and Reeve, 1994; Ellis, 1965; Lovestram *et al.*, 1990). Again, wood is neither uniform nor singular. As noted above, the chemistry of secondary walls is by no means constant, rather varies with the different lamellae of individual elements, across annual rings, and with stem age. Thus, even if the volume of secondary-wall material were constant at different positions in the tree (and it rarely is), wood density variations would be expected (Savidge, 1986). It becomes necessary, therefore, to consider density differences in terms of molecular composition as well as in terms of wall and void volumes (air having a density of zero, for practical purposes). Cellulose has a density of 1.60 g cm^{-3} (Jones, 1971), whereas the density of lignin is approximately 1.3 g cm^{-3} (Stamm and Sanders, 1966). Hemicelluloses, though highly heterogeneous, have been described as having densities near 1.65 g cm^{-3} (Stamm and Sanders, 1966). In addition, the densities of the numerous minor constituents can be integrated when known. Inorganic substances not uncommonly have densities

> 2 g cm^{-3}, and their content varies depending on the species and the location sampled in the tree (Ellis, 1965). 'Extractives' comprise diverse substances (Hillis, 1987; Lewis and Sarkanen, 1998; Rowe, 1989; Sjöström, 1993) and they vary considerably in density depending on their chemistry. Hydrocarbons and other non-polar constituents can have densities as low as 0.8 g cm^{-3}. In very high density woods, such as lignum vitae (*Guaiacum* spp.) and black ironwood (*Krugiodendron ferreum*), the enhancement of density evidently resides more in altered chemical composition (e.g., heightened content of inorganics) than in increased secondary wall thickening. In low density woods such as balsa (*Ochroma lagopus*) very thin secondary walls and a large lumen volume generally account for density reduction (Panshin and de Zeeuw, 1980).

3.3 *Some applied considerations of wood chemistry*

For the major wood-processing industries, the overall density of dry wood has been the most important quality consideration, the emphasis being on obtaining high yields of pulp and high strength in solid products, both of which usually vary directly in relation to changing wood density. Depending on the species and position within the tree, wood density can vary from as low as 0.04 to as high as 1.40 g cm^{-3} (Panshin and de Zeeuw, 1980). As already noted, density is determined to a large extent by the nature of wall material *vis-à-vis* the volume in the lumina of the wood elements. In general, latewood elements have less lumen space and thicker secondary walls than earlywood elements, particularly in mature wood. Thus, anything increasing the proportion of latewood present in each annual ring can significantly enhance wood density (Panshin and de Zeeuw, 1980; Zobel and van Buijtenen, 1989; Zobel and Jett, 1995).

For improving wood density, it is important to know which factors influence the extent of primary-wall radial expansion in cambial derivatives and which control how much secondary wall deposition occurs subsequently. Unquestionably, the availability of adequate soil water is conducive to maximizing both, but the extent of wall deposition during differentiation is influenced by photoperiod, temperature, light and nutrient status (Larson, 1962; Zobel and van Buijtenen, 1989). The older literature indicates that both processes are regulated by auxin (Savidge and Wareing, 1981; Larson, 1994). However, more recent observations indicate that radial expansion and secondary-wall formation are under separate forms of regulation, the extent of radial expansion being influenced by endogenous auxin content and the extent of secondary wall formation being under the control of a 'tracheid-differentiation factor' distinct from auxin (Savidge, 1994). The evidence to date suggests that activation, duration and synchronization of gene expression during secondary-wall deposition is regulated by one or only a few regulatory factors, possibly proteins produced from homeotic genes, operating to activate the families of structural genes needed for differentiation (Savidge, 1996). If this reasoning is correct, elucidating those regulatory mechanisms could well set the stage for increasing wood density and minimizing its variation within the tree.

In thermochemical pulping of natural (i.e. lacking recombinant DNA) woods, rarely is there more than a 50% yield of pulp. This is because the harsh physical and chemical treatments required to break down lignin and release individual fibres also hydrolyse large amounts of polysaccharide and other substances in woods. Thus, low pulp yield combined with high pulping and high-volume waste processing continue to be major efficiency challenges for industry (Sjöström, 1993). Anything that can be done to increase the mass of substances (such as cellulose) resistant to hydrolysis or sulfonation could have major impact on the industry. As noted above, reducing lignin content through genetic engineering is one approach to this problem. However, to date biotechnology has not focused specifically on the middle lamella and cell corners where lignin contents are highest. Those locations necessarily must be dissolved/ degraded in order for fibres to separate during pulping. How transgenic expression might be targeted to those locations remains unknown, but this knowledge could be linked to an explanation for lignification first being initiated at those locations.

Ecological biochemistry is in its infancy, and it is therefore difficult to predict what impact lignin transgenics will have on the biosphere. An alternative to biotechnology for achieving the goals of reduced lignin content and increased pulp yields may be to manage stands to sustain tree growth throughout the full growing season. For example, low soil water potential often limits late-summer growth. Harvesting schedules can be changed to permit trees to produce more mature wood. The lignin content in the juvenile cores of fast-grown plantation conifers is particularly high, whereas their cellulose content is relatively low (Zobel and Sprague, 1998; Zobel and van Buijtenen, 1989). Nevertheless, such juvenile-wood trees are what are being harvested as pulpwood with ever increasing frequency. Substantial increases in cellulose and corresponding decreases in lignin content occur following the transition from juvenile to mature wood production (Panshin and de Zeeuw, 1980). It is unknown whether longer rotation times using wild-type trees could compete with the gains offered by biotechnology, but the former might also address the attendant concerns about sustaining genetic diversity and health of forest ecosystems perhaps more effectively than can the biotechnology approach.

If improved long-term carbon sequestration were the objective, a case might be made for increasing, rather than for decreasing, lignin content in forest trees. Cellobiose contains 42% C, and monosaccharides common to the hemicelluloses for the most part contain 40% C. In contrast, coniferyl alcohol (the monomeric unit of the most abundant type of lignin) contains 67% C. Thus, woods with high overall lignin content are likely to have the higher C content. An even stronger argument could be made for increasing the suberin content in wood!

Compression wood production in the main stems of conifers is on the increase in plantation trees, and this has implications from the standpoint of utilization (Timell, 1986; Zobel and van Buijtenen, 1989). The core wood found in old-growth trees of natural forests is primarily of the juvenile type and contains little or no compression wood. The trend towards production of a compression wood juvenile core in plantation conifers is, evidently, a result of selection for accelerated height growth combined with intensive silvicultural practice. However, as noted above, the problem is exacerbated by the fact that plantations are being harvested at a younger age, resulting in little or no mature wood yield. Compression wood has not been found to be nearly as amenable to pulp and paper production, nor as dimensionally stable or safe in the form of solid wood products, as normal juvenile core wood (Timell, 1986). Thus, there is growing concern that wood quality has been compromised inadvertently in exchange for higher growth rates. Some consider that compression wood forms only under the influence of gravity (hence only in branches and leaning stems), but it should be clear that the anatomical characteristics of compression wood tracheids are entirely distinct from those of normal juvenile wood tracheids (Panshin and de Zeeuw,

1980; Timell, 1986). Timell (1986) reviewed numerous instances of natural compression wood formation occurring in upright stems. Moreover, in experimental systems where the stem is rigidly maintained in an orientation parallel to the gravity vector, compression wood can be induced in response to applications of high levels of exogenous auxin or, alternatively, by applying auxin-transport inhibitors (Savidge, 1983; Timell, 1986). Thus, at least part of the explanation for increased compression wood production in the juvenile core of fast growing trees may be that their cambial auxin content is elevated. Evidence supporting this hypothesis remains to be reported.

How gravitational force translates into a developmental signal in the cambial region remains unknown (Timell, 1986). As reported by Kennedy and Farrar (1965), manually tilting the stem of a conifer back and forth every few days results in the differentiation pathways of cambial derivatives being switched repeatedly from compression wood to normal (or opposite) wood formation. Kennedy and Farrar (1965) also rotated trees on a clinostat and found compression wood to form on the centrifugal sides. Although the physiological and biochemical mechanisms are poorly understood, there can be little doubt that gravitational force exerts a major effect on gene expression and biochemistry in the cambial region. Biologically, compression wood formation is a fascinating phenomenon and one of the best manifestations of the immense G × E plasticity existing in the cambium. This, combined with the many negative attributes displayed by compression wood in service to man, makes research into the biochemistry and biophysics of compression wood formation particularly attractive.

Another neglected research area having potential practical applications concerns the long-known ability of wood to accumulate irreversibly rare earth and transition metals. Even wood that has long been dead exhibits this remarkable capacity, as petrified woods show so beautifully (Daniels, 1998). In locations where radioactive mineral deposits have coincided with woods that had been permineralized (i.e. only partially petrified and still containing organic secondary wall materials), radionuclides became particularly enriched in the petrified logs (Baars, 1972). Woods within living trees also unquestionably bind rare earth, transition and other metals irreversibly (Aberg *et al.*, 1990; Bailey and Reeve, 1994; Bowers and Melhuish, 1988; Kennedy and Bergeron, 1991; Kohno *et al.*, 1988; Lovestram *et al.*, 1990). The metal-binding mechanism(s) of wood cell walls are far from understood, but clearly could be exploited.

4. General and specific pathways of development during wood formation

4.1 *Terminological pitfalls*

Scientific research into xylem and phloem formation in tree stems began with the discovery of the existence of the vascular cambium soon after the advent of the light microscope in the 1600s (Larson, 1994; Savidge, 1996). Descriptive anatomy grew strongly during the 18th century, and attempts were made during that and the next century to explain wood formation in mechanistic terms, despite minimal comprehension of the biochemical reactions underlying the formative processes. The early explanations for cambial growth embodied the implicit assumption that development was predetermined. The modern concept of developmental plasticity, a function of both epigenetic status and the physico-chemical environment within the cell, evidently had no meaning for those investigating morphogenesis at that time.

Mechanistic concepts and terminology arising out of early microscopy investigations into cambial growth and wood anatomy are still extant. Therefore, it is important for the biochemist to distinguish between terms describing what can be seen in the microscope and more imaginative terms embodying hypothetical still-unproven ideas.

Wilson *et al.* (1966) proposed that the term cambial zone be used to include cambial 'initials' and xylem and phloem 'mother cells,' based on historical precedent. However, as explained below, neither initials nor mother cells have yet been convincingly distinguished as entities having the epigenetic competencies ascribed to them in cell biology or biochemistry. Nor can they be distinguished unequivocally from other cells within the cambial zone. Therefore, one side of a 'research' community has either questioned or rejected the initial and mother cell concepts, whereas the other has insisted that they are the best possible explanation for cambial growth (Larson, 1994).

There are two quite different concepts, or definitions, of the cambial initial. The most commonly employed has one cell (some have argued that there could be two) within each radial file of the cambial zone being the source of, and having superior competence to, all other cells within its radial file. Thus, whether phloem or xylem is to develop is primarily controlled by events occurring within that initial, subsequent events following in a predetermined manner. In addition, this hypothetical cambial initial is the one cell that is indefinitely self-perpetuating. Such a hypothetical initial cell can be referred to as the 'periclinal-division initial' because the validity of the concept rests on the occurrence of periclinal divisions and consequent production of radial files of xylem and phloem elements. The second concept of the cambial initial can be referred to as the 'anticlinal-division initial' because it concerns the initiation of new radial files within the cambial zone. This concept is not at all hypothetical. Many studies have shown that new radial files arise in the cambial zone as the stem increases in circumference over the years, and that these new radial files are initiated through anticlinal divisions in the cambial zone. In *Pinus* spp., it has been shown unequivocally that more than one fusiform cell per radial file within the cambial zone can divide anticlinally to initiate new radial files, ruling out the possibility that there must be only one such cell per radial file (Savidge, 1985, 1990, 1996).

It is not uncommon at the fine structural level to see four fusiform cambial cells within a radial file sharing a common outer (primary) wall, and those four cells are known as Sanio's four (Larson, 1994). For those advocating the existence of periclinal-division initials, the presence of the common wall has been interpreted as evidence that the initial divided once in the periclinal plane to produce two 'mother' cells, and that the two mother cells each divided periclinally again yielding a foursome of 'daughter' cells. Based on such observations, the deduction that one cellular division is followed by two more is logical. However, as indicated above, the major difficulty with the periclinal-division 'initial' and 'mother cell' terminology has stemmed from the imposed developmental assumptions in the absence of supporting data. The major assumption is that the periclinal-division initial is the source of both xylem mother cells and phloem mother cells positioned on inner (centripetal) and outer (centrifugal) sides of the initial, respectively (Larson, 1994). When produced through periclinal division of the initial, the resulting xylem or phloem mother cell are their derivatives supposedly are committed to xylem or phloem development, respectively. Although plausible, there is no obvious reason why, following periclinal division of the hypothetical initial, one of the cellular progeny should retain the special 'initial' status and the other daughter cell relinquish it.

Elsewhere, it has been suggested that every cell within the cambial zone could be equally competent, in which case only anticlinal-division initials would exist and neither xylem nor phloem development would be predetermined (Savidge, 1985, 1990). Within this concept, the cambial zone arises and is maintained in response to basipetally transported auxin in conjunction with physical forces (Savidge and Wareing, 1981; Savidge, 1983, 1985, 1990, 1996). Differential behaviour of cambial cells within the cambial zone, and of cambial derivatives on the inner and outer peripheries of the cambial zone can be explained in terms of changes in the micro-environment experienced by the genome within each cell (Savidge, 1985, 1990, 1996). At least two sets of experimental data argue in favour of the above concept and against the existence of periclinal-division initials. Firstly, every fusiform cambial cell within a radial file of the cambial zone has been found equally competent for direct differentiation into a tracheary element, indicating that none of the cells in the cambial zone is better equipped to be self-perpetuating or exempt from terminal differentiation. Also, there is no obligatory requirement for preceding cell division (i.e., 'xylem mother cell' production) for xylogenesis to occur (Savidge, 1994). Secondly, it has been shown that every cell within the cambial zone will de-differentiate into axial parenchyma when the cambium's source of auxin is interrupted, again indicating that none of the cells in the cambial zone is programmed to be self-perpetuating (Savidge and Wareing, 1981; Savidge, 1983). Thus, in the absence of substantiating biochemical evidence for the existence of periclinal-division initials and xylem and phloem mother cells, there is no obvious need to embrace the notion of predetermined secondary growth in biochemistry research (Savidge, 1985, 1996).

Within the concept of G × E developmental plasticity at the cellular level, it is useful to recognize that just because definitive anatomical evidence for an aspect of growth or development (such as Sanio's four) has been obtained, it does not follow that such development necessarily always occurs.

Periclinal-division initials may be hypothetical, but the cambial zone of terrestrial plants is readily distinguished as a tissue between wood and bark, comprising a radially narrow (2–30 μm wide) region of adjoining cells lacking intercellular spaces, each cell having relatively small radial dimensions and competent to undergo cell division. Fusiform and ray cambial cells in the cambial zone are distinct entities readily distinguished by their contrasting protoplasma and extreme dimensional differences (*Figures 1d, 1e and 1f*).

4.2 *The central biochemistry themes underlying wood formation*

The central themes common to the formation of all woods can be summarized as follows. Firstly, woods always come into being through metabolic activity within living cells, activity which maintains, divides and enlarges the cells. Secondly, some proportion of the elements in secondary xylem of every woody perennial plant must be woody i.e., contain lignified secondary walls. Some of these must have walls containing bordered pits which, upon autolysis of the protoplasm, enable transmission of water from one tracheary element to another. Thirdly, the woody elements exhibit lamellae of secondary walls, normally consisting predominantly of cellulose. Lastly, both primary and secondary walls in woody elements normally contain lignin.

For all vascular plant species, the starting point for understanding the essentials of secondary xylem formation would seem to be to know the biochemistry underlying development of seven ubiquitous end products, viz.: (i) new cells, (ii) expanded cells, (iii) wall pits, (iv) cellulose, (v) lignin, (vi) secondary-wall lamellae and (vii) protoplasmically autolysed cells. As considered elsewhere (Savidge, 1996), regulatory proteins as well as structural enzymes are necessary considerations here.

Much that has been published on the regulation of the cell cycle, on cytokinesis and on cell-expansion activities as they occur in non-cambial plant cells is relevant to the cambium. It should be appreciated, however, that fusiform cells of the cambial region exhibit unusual phenomena during cell division and expansion, and very little is yet known about the underlying control mechanisms attending that remarkable meristem.

5. Distinguishing and isolating cells of the cambial region

5.1 *Microscopy, more an imaginary than a real bottleneck to progress*

Cambial fusiform cells are typically 1–3 μm and rarely as much as 6 μm in radial dimension. As shown in *Figures 1d* and *1e*, during dormancy there are very few, rarely more than six, cells per radial file in the cambial zone (CZ). The meristematically active CZ increases its number of cells per radial file (*Figures 1f and 1g*). Even then, however, the CZ *per se* remains very narrow in radial width. The characteristically narrow radial dimensions of fusiform cambial cells, combined with their being grouped together into a circumferential 'zone' sandwiched between wider cells of contrasting anatomy, readily distinguishes the CZ (*Figures 1d–g and 2a*). A dissecting microscope can provide some crude information about the cambial region, but if even coarse cellular resolution of the CZ is wanted, the researcher must resort to a compound microscope.

Depending on the developmental stage when the stem is sampled, the cambial region comprises, in addition to the CZ, one to several distinct sub-tissues, or zones, of xylo- and phloiogenesis, as shown for conifers in *Figures 1f, 1g and 2b–f*. Progress in the biochemistry of cambial growth hinges on knowing which stage of development the tissue is at. When sections are prepared using standard histochemical procedures for permanent mounts, microscopy is tedious and time-consuming and has undoubtedly been a bottleneck to biochemistry progress.

In general, hand-cut sections of fresh unfixed material provide a less distorted and otherwise less artifactual image of the dynamics of the living system. As shown in *Figure 2 (a–h)*, such sections can be adequate for many kinds of light microscopy investigations (e.g., using brightfield, interference contrast, and UV epifluorescence). Freshly harvested wood is naturally hydrated; thus, it tends to be soft and offers much less resistance to a sharp razor blade than does alcohol-dehydrated wood. Cross and radial sections can be produced with ease and mounted in distilled water, or glycerol.

Obtaining small bark-intact specimens from stems for sectioning can be more difficult than hand sectioning *per se*, particularly when cambial growth is actively underway, because at that time the bark splits away easily along the cambial region (Savidge *et al*., 1982). With small-diameter (i.e., young) stem or branch material, in order to minimize cambial damage the segment should be cut through the bark superficially into the wood around the full circumference, followed finally by 'walking' the razor blade back and forth through the wood and pith. For radial sectioning of the cambium, a short segment is first stood upright on a cutting board and split lengthwise through the pith. It is more difficult to obtain undamaged specimens for sectioning with older, larger diameter stems. The method is essentially the same as that used for

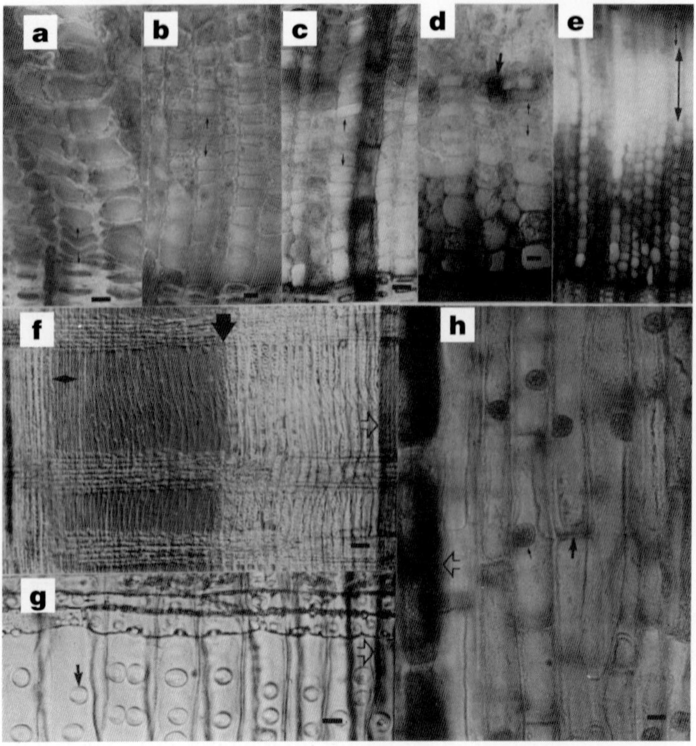

Figure 2. *(a) Fresh, hand-cut cross section of* Pinus resinosa *showing phloem (top), dormant CZ delimited by the arrows, and latewood tracheids; bar = 5 μm. (b) Fresh, hand-cut cross section of* Pinus strobus *showing phloem (top) active CZ delimited by the arrows, and RE zone between the CZ and latewood tracheids. Unstained section, brightfield optics; bar = 10 μm. (c) Fresh, hand-cut cross section of* Pinus strobus *showing phloem (top) active CZ delimited by the arrows, RE zone between the CZ and latewood tracheids. This section was briefly incubated in 20 mM guaiacol 1% H_2O_2 to reveal peroxidase activity concentrated in the ray and phloem; bar = 10 μm. (d) Fresh hand-cut cross section of* Picea rubens *showing phloem (top) active CZ delimited by the arrows, RE zone (2 cells per radial file), with SL zone and latewood tracheids below. This section was briefly incubated in 1.0 mM 6-diaminofluorene to reveal oxidative activity in both phloem (large arrow) and SL zone; bar = 5 μm. (e) Fresh, hand-cut cross section of* Pinus resinosa *showing phloem (top edge) active CZ delimited by the small arrows, RE zone delimited by the larger arrows, SL zone and fully autolysed earlywood tracheids, and latewood tracheids. This section was briefly incubated in 20 mM guaiacol 1% H_2O_2 to reveal peroxidase activity (at this time, present in the phloem, the rays, the SL zone, and the mature new tracheids; compare Figure 2,(c)). (f) Fresh, hand-cut radial section of* Pinus strobus *showing phloem (left), CZ delimited by the arrow heads, RE zone to the right of the CZ, and the boundary (large solid arrow) between the RE and SL zones. The latewood boundary is indicated by an open arrow. Three rays traverse the image. Note the inward curved cell tips (small arrow) of fusiform cells in RE and SL zones, the abundant bordered pits developing in the SL zone, and the absence of birefringence in cell walls of CZ and RE zones. Unstained, Nomarski image; bar = 50 μm. (g) Fresh hand-cut radial section of* Pinus banksiana *showing ray and fusiform cells in the SL zone, both actively differentiating into tracheids. Mature latewood is on the right. Note that the developing bordered pits in ray cells (small arrow) are of varied but much smaller diameters than those in axially oriented cells (large arrow), also that pitting can be in single or double files. Unstained, Nomarski image. The latewood boundary is indicated by the open arrow; bar = 10 μm. (h) Fresh, hand-cut radial section of* Pinus contorta *showing axial parenchyma that developed through transverse divisions (large arrow) from fusiform cambial cells in response to diminished auxin supply, resulting in disappearance of the CZ. Note that the nuclei (small arrow) do not exhibit the typical elongated character seen in fusiform cambial cells (see Figure 4(e)). Phloem axial parenchyma (open arrow) are on the left; bar = 5 μm.*

preparation of stem 'chips' for culturing (Savidge and Wareing, 1981; Savidge, 1993), except that surface sterilization is not done. Excess bark is first trimmed away in the region of interest, leaving several millimetres of crushed inner bark intact upon the cambium. Then, a 2–3 cm-wide incision perpendicular to the stem's long axis is forced carefully through the bark and several millimetres into the wood. A second oblique cut is made, beginning 2–3 cm along the axis from the first cut, angling the knife acutely in order to meet the first cut at its deepest position in the wood. The end result is a D-shaped specimen having cambial region sandwiched intact between phloem and xylem.

5.2 *Distinguishing the developmental zones by light microscopy*

During cambial growth, the several distinct zones can be distinguished and directly quantified by counting cell numbers per radial file in transverse sections (*Figures 1d–g and 2a–e*). The radial distances spanning the total number of cells making up each zone of course can also be measured; however, cell size varies depending on which zone is being measured. Moreover, it is the unit cell which is the fundamental unit of life, hence the most important feature to have quantitative data on. Counts of cell numbers per radial file in each developmental zone give essential information about current activity and preceding growth.

Temporal and spatial variations in zonal histogenesis occur commonly over the axes of tree stems, branches and roots (*Figure 3*). Therefore, microscopy data on the zones should be gathered using a specimen taken from the same position in the stem as the tissue for biochemical analysis. In stems producing reaction wood, neither the stage of growth nor the developmental zones will be at all representative of the tissue status 180° around the stem circumference.

For brevity's sake, the developmental zones are described here only in terms of how they arise in conifers growing in the temperate zones. Similar zones arise in hardwood stems; however, depending on the species, both the sequence of development and the manner of progression over the stem can differ.

The dormant CZ was described above. In late spring, before resumption of cell division, CZ fusiform cells become rehydrated and their protoplasm changes from having a dense gel-like appearance to, for the most part, appearing as one large clear vacuole with only a thin layer of parietal cytoplasm. The latter is clearly resolved only in the transmission electron microscope (*Figures 1e and 2a*). As vacuolation occurs, thickened radial walls of cambial fusiform cells become noticeably thinner, indicating that substances stored in these walls are remobilized, presumably through hydrolytic reactions. Concomitantly, cambial reactivation *per se* begins. The centripetal CZ cells — those adjoining the latewood — are frequently the first within a radial file to begin dividing as the cambium reactivates in spring (Larson 1994), but activity soon becomes more generalized throughout the CZ (*Figure 1g*). Following periclinal division, the daughter fusiform cells normally (but not invariably) regain their former radial dimensions by undergoing a small amount of primary-wall radial expansion growth. CZ fusiform cells also undergo tip or elongation growth; however usually only marginal increases occur over a growth season, making it difficult to detect such changes other than by retrospective analysis of mature xylem or phloem elements within serial sectioned radial files. CZ fusiform cells are also capable of shortening their lengths and changing their polarities using imperfect planes (in relation to the long axis of the fusiform cell) of periclinal or anticlinal division (Savidge and Farrar, 1984).

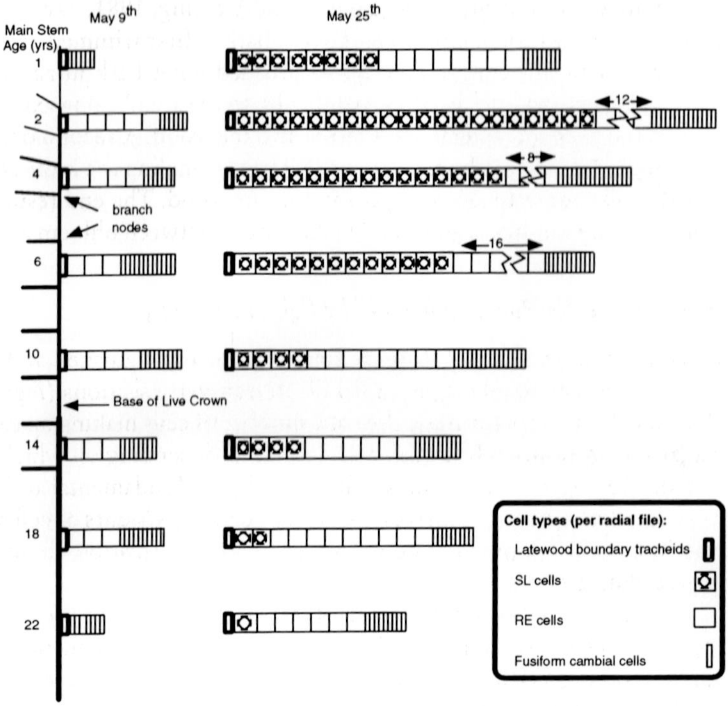

Figure 3. *Diagram to show the cell numbers per radial file and patterns of cambial reactivation observed by cutting hand sections from point locations over the main stems of two 25-year-old* Pinus strobus *trees felled on May 9th and May 25th. Note that on May 9th (before budbreak), the CZ was only beginning to reactivate in the 1-year and 22-year-old positions, whereas RE zones were already advanced at other locations. On May 25th, development of the SL zone was strongly advanced in the upper foliated crown region and tapered to less advanced states in progressively lower stem locations. Similar patterns of springtime reactivation have been observed in all conifers so far investigated in this way (Savidge and Wareing, 1981; Savidge, 1990, 1996; Savidge* et al., *1998). The observations indicate that initiation of development of the SL zone is spatially associated with foliage and under a form of regulation distinct from that controlling cambial reactivation and formation of the RE zone.*

Following the initial springtime CZ activity, rapid and massive radial enlargement begins, not uncommonly occurring in cells adjoining the latewood first (*Figures 1f, 2b and 2c*). In cells undergoing radial enlargment on the periphery of the CZ, no further cell-division occurs. Thus, a zone of nondividing, **r**adially **e**xpanded cells, the **RE** zone, arises between the CZ and latewood. The RE zone is > 90% water, and its thin-walled, enlarging and fully enlarged fusiform cells in no way resemble woody elements. The most massive radial expansion of cambial fusiform cells occurs in ring-porous hardwood species as the preliminary to earlywood vessel production. In addition to undergoing radial expansion, fusiform cells in the RE zone not uncommonly elongate through intrusive growth. The swollen and deformed tips of elongating fusiform cells can be most easily observed in radial sections (*Figure 2f*). (Note: The temporal sequence in conifers of a period of CZ cell-division activity before massive radial enlargement is one of the major differences in zonation development between conifers and at least some hardwood species. Particularly in ring-porous hardwoods native to the temperate zones, massive radial enlargement of latewood-adjoining CZ cells can precede cambial reactivation, as the first step in vessel-element differentiation.)

As the season progresses, cell division continues and cells of both cambial and RE zones accumulate, such that each increases in radial file cell number. Earliest CZ reactivation and maximal development (in terms of numbers of cells per radial file) of both cambial and RE zones occur in the stem's main axis at the base of the crown (*Figure 3*). In living branches, these events also occur first in the primary axis near the base of the foliated portion; however, reactivation in branches generally lags behind that in the main stem (Savidge and Wareing, 1981; Savidge *et al.*, 1998). CZ reactivation and RE-zone development spread both apically and basally over the axis, these changes occurring well before budbreak in conifers. (Note: In deciduous hardwoods, depending on the species, reactivation may commence first at bases of new shoots following bud break, or it may occur more or less simultaneously throughout the tree. In some species, it also occurs as described here for conifers.)

Depending on the presence and age of foliage, the enlarged RE-zone cells adjoining the latewood sooner or later commence bordered-pit development to initiate xylogenesis *per se*. As shown in *Figure 3*, the initiation of xylogenesis begins first in foliated regions, at about the time of budbreak (Savidge and Wareing, 1981; Savidge, 1990). Bordered-pit development in its earliest stages is not obvious in cross sections; however, pit development can be detected in radial sections using interference contrast optics. In those species producing ray tracheids, circular or elliptical modifications to the primary wall appear first in the radially expanded ray cells, heralding impending differentiation into a tracheary element. Before circular wall modifications become apparent, the previously straight tangential walls develop noticeable bulges at various, non-specific, locations along their lengths (*Figure 4a*). Spherical organelles of uncertain identity have been noted in the same cells (*Figure 4b*) beginning at about the same stage of development, and it is possible that they are associated with initiation of bordered-pit development. Wall bulging is followed by appearance of circular, or elliptical, modifications to the primary wall (*Figure 4c*). These rounded areas are not of constant diameter; those in fusiform RE-zone cells are generally much larger than those in ray cells. In species producing a torus, it appears centred in the circular or elliptical area of the primary wall soon after the rounded, primary-wall modification becomes detectable (*Figures 4c and 4d*). Following these early developments, newly deposited polysaccharide (commonly assumed, but never definitively proven, to be cellulose) appears around the circumference of the pit border as strongly birefringent rings (*Figure 4d*). The enlarged cell is otherwise still primary-walled at this stage, indicating that the rings of polysaccharidic material are produced independently of the production of the S_1 secondary wall layer that is to follow.

In conifers of the temperate zone, a protracted delay in initiation of xylogenesis attends the accumulation of RE-zone cells in stem regions beneath the base of the crown. (This delay is a developmental feature which clearly distinguishes conifers from hardwood species.) The RE zone can persist and increase in radial file cell-number near and below the base of the crown for two weeks or more before structural evidence of bordered-pit development becomes detectable (*Figures 1f, 2b, 2c and 3*). On the other hand, in younger foliated regions of the stem and branches, cells of the

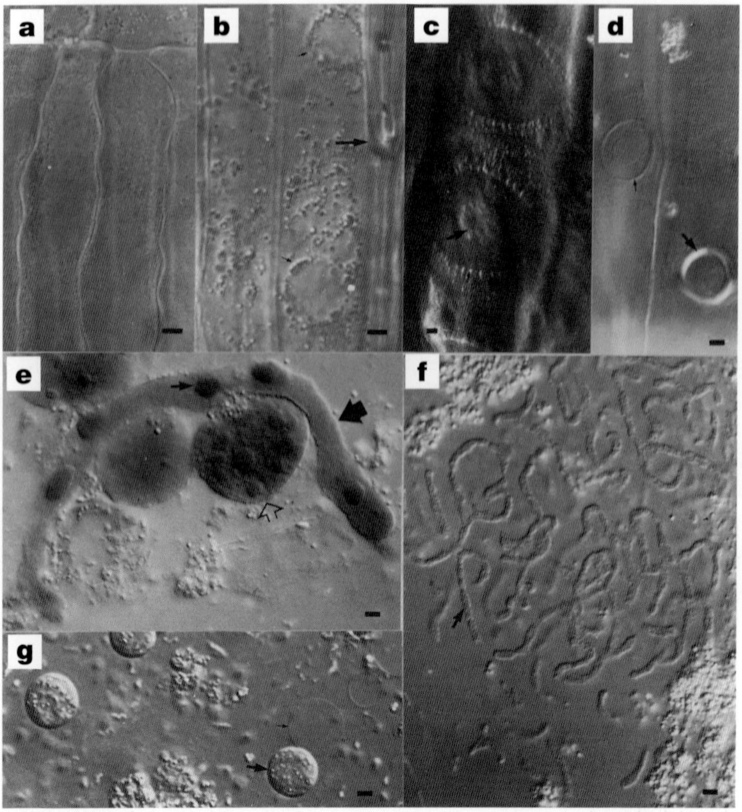

Figure 4. *(a) Fresh, hand-cut radial section of ray (top) and fusiform cells in the RE zone of* Pinus strobus *prior to appearance of bordered pits or the SL zone. Note that the tangential walls of the fusiform cells are bulging irregularly, the first indication of impending bordered-pit development; bar = 5 μm. (b) Fresh, hand-cut radial section of fusiform cells in the RE zone of* Pinus banksiana *prior to appearance of either bordered pits or the SL zone. The tangential wall of the previous year's latewood tracheid is arrowed. Note the two spherical organelles (small arrows) in the RE zone fusiform cell adjoining the latewood, and the beginning of waviness (far left) in the tangential walls of the other RE zone cell; bar = 10 μm. (c) Scanning electron micrograph of an RE-zone fusiform cell of* Pinus contorta *showing an early stage of bordered-pit development. The pit border (small arrow) is clearly demarcated, torus development (large arrow) is underway. Note that the primary wall material of the pit membrane interacts with the electron beam differently from the material outside of the pit borders; bar = 1 μm. (d) Fresh, hand-cut radial section of the RE zone in* Pinus strobus. *The fusiform cell on the left exhibits a developing bordered pit (small arrow) only slightly more advanced than that shown in Figure 4(c), whereas the overarching borders (large arrow) are clearly growing in the still-primary-walled cell at right; bar = 2 μm. (e) Nuclei from cambial cells of* Pinus strobus *obtained through bark peeling to release cellular protoplasma. Note that the fusiform cell nuclei (large solid arrow) are extremely elongated, whereas ray cell nuclei (open arrow) are more spherical. Variable numbers and sizes of nucleoli (small arrow) are present in both; bar = 1.0 μm. (f) Chromosomes (one is arrowed) from a cambial cell of* Pinus strobus *obtained through bark peeling; bar = 1.0 μm. (g) Organelles from cambial cells of* Pinus strobus *obtained through bark peeling to release the cellular protoplasma. In addition to abundant proplastids, mitochondria and other small organelles, at least two unidentified spherical organelles, one vacuole-like (small arrow) and the other with heterogeneous contents (large arrow), are present. Evidence (such as that shown in Figure 4(b)) indicates that the vacuole-like organelle transforms into the latter; bar = 1.0 μm.*

RE zone give rise to an RE zone producing bordered pits much more rapidly than in the lower, non-foliated regions. Once bordered-pit development has been initiated at any point in the stem, it continues and progresses radially outward (centrifugally) through the radial files of the RE zone. In addition to tracheary elements, other xylem-cell types lacking bordered pits develop from fusiform cells, particularly in hardwoods; thus, not all cells of the RE zone will necessarily initiate bordered-pit development.

Initiation of bordered-pit development is followed by formation of the '**SL** zone', or zone of **s**econdary-wall formation and **l**ignification. In the SL zone, S_1, S_2 and S_3 lamellae (S_1 and S_2 only, in the case of compression wood tracheids, or S_1, S_2 and G layers in the case of tension wood fibres) are generated internally to the primary wall (compare *Figures 2c* and *2d*). Histochemical evidence for the onset of lignification first becomes evident when the expanded cells are still quite thin-walled. Lignin becomes evident first in the apoplast at the cell corners, and also at locations where bordered pits are developing. When counter stains such as safranin–fast green are used on fixed specimens, the staining distinguishes RE-zone cells at a very early stage of S_1 deposition, but still lacking evidence for initiation of lignification, from both those which are still primary-walled RE cells and those which have initiated lignification. Thus, it appears that initiation of S_1 polysaccharide deposition always follows initiation of bordered-pit development and, in both time and space (i.e., position in the radial file), always precedes initiation of lignification.

The secondary-wall lamellae are produced in succession and are distinguished by their contrasting microfibrillar orientations. The possible role of microtubules in determining the orientations of microfibrils is addressed in other chapters of this book. The extent of delay occurring between ceasing to produce one lamella and beginning to produce the next layer is unknown, but it is probably correct to assume that a sequence of S_1, S_2 and S_3 sub-zones arises in the SL zone. Having a method to collect material from these successive stages of lamella deposition would certainly further progress.

Protoplasmic autolysis is initiated after secondary-wall formation and lignification are complete, however, a reliable method for diagnosing when protoplasmic autolysis is initiated remains wanting. Under the microscope, the open, transparent lumen of a fully autolysed woody element can be distinguished readily from one still having protoplasm (*Figure 2d*). In conifers, the SL cells adjoining latewood elements are the first to mature into earlywood tracheids through protoplasmic autolysis. Initiation of autolytic maturation also progresses centrifugally through the radial files. However, an observation that secondary-wall formation and lignification have been completed does not permit the deduction that autolysis must be underway, because a protracted and even permanent delay in initiation of the autolytic process is possible. Thus, with present knowledge, it is necessary to include autolysing as well as earlier stages of xylogenesis in the SL zone. Preceding, or in association with, protoplasmic autolysis, there may be deposited additional bands of secondary-wall material on the luminal side of the S_3 lamella, a warty layer, and other matter.

5.3 *Isolating zones for biochemistry investigations*

Each zone described above presumably becomes anatomically distinguishable in response to changing gene expression. Thus, for molecular-level research, combining two or more distinct zones of the cambial region into a single preparation would merge contrasting biochemistries. Therefore, when attempting such research it is necessary not only to recognize the zone, or stage of development in the microscope but also to achieve the highest possible resolution in separating the different zones. In the older literature, methodology statements such as the CZ being obtained from the xylem surface imply that incorrect zonations were assigned; thus, the conclusions from those reports may require reassessment.

The bark will not peel from the wood when the CZ is dormant except in autumn when xylogenesis is still occurring (see below). During the period of dormancy, the CZ can be obtained by carefully paring away bark with a sharp knife until only an extremely thin layer of translucent white tissue remains to be scraped from the mature wood for analysis. Even when great care in paring off the bark is taken, the residual tissue is likely to contain a larger mass of phloem than CZ, as the radial width of the CZ during dormancy rarely reaches 20 μm.

In early springtime, immediately before resumption of cambial cell-division activity, a narrow window of opportunity for obtaining dormant CZ arises. This occurs when the protoplasm of the fusiform cells is just becoming rehydrated. The bark peels cleanly, but with more difficulty than when cambial growth is fully underway, and the tissue tears through thin-walled, enlarged cells on the outer (centrifugal) edge of the CZ. Thus, following bark peeling, the dormant CZ is found adhering to and can be gently scraped from the wood surface as 'pure' CZ tissue, essentially free of phloem, xylem or cambial derivatives (Savidge *et al.*, 1982).

The tearing attending bark peeling during cambial activity occurs in the thin-walled cells, between the CZ and the RE zone (Savidge *et al.*, 1982). Thus, CZ tissue can be obtained by gently scraping the exposed inner bark surface. Microscopy indicates that the CZ scrapings do not contain mature phloem elements, rather only very thin-walled cambial cells. However, it should be noted that enlarged cambial derivatives on the centrifugal side of the CZ, when present, will also be part of the scrapings. Throughout the period of cambial growth, it is possible to obtain the CZ and each of the progressively developing xylogenic zones in massive amounts when large-diameter trees are available for felling.

By sampling the exposed tissue on the xylem surface on successive dates as spring changes into summer, it is possible to obtain 'pure' tissue of the RE zone, RE zone having early stages of bordered-pit development and RE combined with SL zones. By subtractive logic in conjunction with quantitative microscopy (recording cell numbers per radial file of CZ, RE and SL), biochemical changes attending development of each zone can, in principle, be deduced. One assumption attending this approach is that identical biochemical processes occur in the same zone of cellular differentiation no matter when it is sampled. While this is plausible, it remains to be substantiated.

Conifer cambium enters dormancy before latewood xylogenesis is complete. The RE zone gradually disappears when the cambium ceases producing derivatives at the end of the growing season. Bark peeling continues to be useful at the interface between the dormant CZ and the developing xylem; however, it becomes progressively more difficult as the last cambial derivatives complete their maturation into latewood tracheids. At this point in the annual cycle, dormant CZ can be scraped from the inner bark face and the SL zone, devoid of RE zone cells, can be obtained from the xylem surface. Thus, there is a narrow opportunity for investigating metabolism in dormant CZ and 'pure' SL zone cells at the end of each growing season.

Elsewhere in this book, the technique of serial tangential sectioning of frozen live stem material, followed by solvent or buffer extraction of the sections, is described. This method was first used by Grozdits and Ifju (1973) and, although very tedious, for some applications it clearly holds considerable promise as a refinement to the bark peeling approach. This is particularly true if the section thickness can be reduced to less than 3 μm (i.e., the commonly observed radial width of a fusiform cambial cell). When fractionating tissues by cryosectioning, the precision and repeatability of microtoming cannot be assumed to be constant. Therefore, it is critically important to conduct ancillary anatomical investigations (such as those described herein in the chapter by Micheli *et al.*) in order to know the stage of development of each cut section. Ideally, each section will contain only a single developmental stage. Achieving a perfectly parallel orientation between the microtome knife and the cambial zone is technically difficult, particularly when young (i.e., small diameter) stem pieces are investigated. Slanting sections containing several developmental stages or even distinct types of tissue cannot be considered more precise than bark peeling for fractionation. No matter what age of cambium is investigated, it is important to recognize that a single tangential section of 20 μm thickness may span the radial thickness of the entire cambial zone.

6. Manipulating cells of the cambial region

6.1 *Manipulating cambial growth: physical, chemical and genetic treatments*

A host of different treatments have been used to manipulate the extrinsic and intrinsic environments of the tree (*Table 1*). There is immense scope for these treatments to be used to enhance biochemistry understanding. Of course, untreated 'controls' must always be included in such investigations.

6.2 *Manipulating cambial cells in vitro*

As shown in *Figure 1c*, progress has been made toward simulating *in arbor* cambial growth *in vitro*. The reader interested in culturing the cambium *in vitro* is referred to chapters by Leitch and Savidge, and by Barnett and Asante, in this book.

Stem segments. Stem segments have long been used as an experimental system for investigating the metabolism underlying lignification. Radiolabelled or other precursors can be infiltrated as aqueous solutions into the segment bases and, following an incubation period, the cambial tissues can be investigated by bark peeling or used for sectioning and autoradiography to determine the effect of the treatment.

Most published research has employed leaf- or needle-bearing young stem material for infusion, perhaps because transpiration by the leaves or needles effectively lifts and distributes the supplied precursors throughout the stem. Potentially, there are several drawbacks to this approach. Preparing the stem segments requires cutting them, and even when done using inhibitors of callose synthase the wounding undoubtedly blocks phloem translocation at least partially, affecting the movement of the infused compound(s). Based on dye studies, uptake most likely occurs via the xylem. The available evidence indicates that the transpiration stream does not flow through cells in

Table 1. Some treatments for manipulating cambial growth

Physical	Chemical	Genetic
Light: –wavelength –intensity –photoperiod g force: –centrifugation –tilting warming/cooling electricity/magnetism external pressure: –compression –bending, looping stem wounding: –pinning, incising –re-orientation –girdling: –partial girdling –bridge girdling organ removal: –defoliation –debudding –derooting	Phytohormone application Defined media in culture: –cambial strips –stem chips –stem segments nutrient availability pH manipulation gas partial pressure: –O_2 –CO_2 –ethylene biochemicals: –effectors –toxins –herbicides/xenobiotics –substrates manipulation of water potential: –osmotica –irrigation –humidity	mutants/mutagenesis ploidy manipulation protoplast fusion applying 'other' genotypes: –grafting: –patch –stock-scion –root -unrelated species: –parasites –microbes recombinant DNA micromanipulatory implants –nuclei –mitochondria –plastids plasmids, viral DNA exogenous mRNA

the RE or SL zones. Therefore, in order for a test compound to reach the CZ/RE/SL regions, either it must be loaded into and transferred centrifugally by xylem ray cells or it must diffuse outwards in the apoplast. Foliar transpiration serves to lift the test compound into stem tissues, but it also introduces the complication of needing to determine whether the compound *per se* was utilized or if it was first metabolized in the leaves and one or more of its metabolic products then forwarded to the cambial region. Moreover, there is accumulating evidence that cambium of non-foliated stem segments responds distinctly from the cambium of foliated segments (Savidge, 1994). In these types of investigation, it would be informative if foliated and defoliated (but otherwise similarly treated) stem segments were compared.

A variation on the beaker-uptake procedure is the technique developed by Brown and Wodzicki (1969) where compounds or mixtures in aqueous solution are introduced by gravity flow into the end of a defoliated stem segment.

Tissue strips. Following bark peeling, very thin translucent strips of CZ can be obtained for live-cell investigations simply by slipping one point of needle-nosed forceps through to the phloem side of the CZ and lifting the strip away. The same approach can be used to lift living tissues from the woody surface. Mounted in water or glycerin, cyclosis remains active in these cells for several hours. Since the observations made by Thimann and Kaufman (1958) on cyclosis, evidently no further investigation has been done. There undoubtedly is scope for methodological as well as biochemistry advances here.

Protoplasts. The cambial zone is a complex tissue. Therefore, biochemical data obtained from the CZ or any of its derivative zones, although specific to a particular stage of growth or xylogenesis, cannot readily reveal what contributions were made by the distinct cell types in that zone. This difficulty can at least be partially circumvented by releasing protoplasts from cambial fusiform and ray cells, followed by separation of those protoplasts. Protoplasts from fusiform cells are several times larger in diameter than ray-cell protoplasts. The former appear primarily as a large vacuole with only a thin layer of parietal cyctoplasm; the latter present a dense protoplasm (Leinhos and Savidge, 1993).

Protoplasts are most readily and rapidly prepared by using the CZ and RE surface cells exposed by bark peeling. Many are released in this way without resort to cell-wall digesting enzymes. Using wall-digesting enzymes, protoplasts become increasingly more difficult to obtain as the differentiating cells produce progressively thicker lignified secondary walls (Leinhos and Savidge, 1993)

Organelles. Northcote (1985) and Barnett (1981) reviewed knowledge about organelles in relation to xylogenesis. However, very little if any cell fractionation research has yet been done on the cambium. Bark peeling is an exceptionally rapid and facile way to obtain large quantities of uncoagulated organelles suspended in their own cytosol in excellent condition. When the cell walls rupture during the bark-peeling process, the released protoplasma stream down the CZ- and RE-zone surfaces as a naturally buffered (pH 6.0–6.5, depending on the species) milky suspension ready for density gradient or other forms of separation. The nuclei of fusiform and ray cells in the protoplasma can readily be distinguished on the basis of their differing shapes (*Figure 4e*), and it is not uncommon to see chromosomes at various stages of mitosis (*Figure 4f*). In addition (*Figures 4b and 4g*), the membranes and cell walls can also be obtained in abundance and readily processed using standard procedures.

7. Some of the research challenges

As indicated above and in *Table 2*, cambial growth comprises a multitude of phenomena each presenting a research challenge. Many of these are addressed elsewhere in this book; thus, only an eclectic few are discussed briefly below.

7.1 *What controls formation and maintenance of the cambium?*

As discussed more fully in the chapters by Barnett and Asante, and Leitch and Savidge, formation of cambia can occur *de novo* in callus cultures, in parenchyma, in callus cells adjoining grafts, and in other unexpected locations. Thus, there appears to be no obligatory link between cambial formation and the development and/or activity of procambia, which arise in extending root and shoot apices. Tissue culture and other evidence have indicated that development of fusiform cells depends on the establishment of pressure relationships with the surrounding tissues (Savidge and Wareing,

Table 2. Some of the research challenges

Evolutionary relationships	Pit formation:
	–simple
Vascular cambium formation	–bordered
	–perforation plates
Annual dormancy cycle	
	Secondary wall lamellae:
Cambial sink strength:	–cellulose
–phloem unloading	–hemicelluloses
–xylem unloading	–lignins
–xylem/phloem interactions	–others
Perpetuation of the cambium	Reaction wood formation
Cell growth in the CZ	
	Metabolism in living cells
Fusiform cell polarity	–xylic
Planes of division	–phloic
Cell expansion	Catalytic activities of cell walls
	Resistance mechanisms in wood
Cell fate, determination	
Protoplasmic autolysis	

1981). However, additional factors also appear to be involved (Savidge, 1996). If a pine is grown in darkness, it produces long etiolated shoots with procambial strands. The vascular cambium does not form under these conditions despite extensive shoot elongation (*Figure 5a*). If a continuing supply of auxin from the apical end of a shoot is eliminated, the fusiform cells de-differentiate into axial parenchyma and the cambium disappears (*Figure 2b*).

7.2 *What biochemical mechanism underlies bordered-pit development?*

Following springtime cambial reactivation, once bordered-pit development has been initiated at any point in the stem, that initiation process progresses radially outward (centrifugally) rapidly through the radial files of the RE zone. Thus, it can be suggested that initiation of bordered-pit development signals a fundamental epigenetic change in the nature of the RE zone, one heralding the initiation of tracheid differentiation *per se.* The birefringent rings laid down will become the overarching borders of the bordered pits (Barnett, 1981). These rings can be isolated (*Figure 5b*), but their composition remains to be established. On the other hand, there is evidence that the overarching border can also be a continuous part of the developing secondary wall during differentiation of cells into primary-xylem tracheids (*Figure 5c*).

As shown in *Figure 4b*, spherical organelles of uncertain identity are associated with enlarged cells when they are initiating bordered-pit development. Two distinct spherical organelles, one vacuole-like and the other containing dense heterogeneous contents, and both of varied diameter, can be obtained (*Figures 4g and 5d*). One or both of these organelles appear to become appressed to the plasma membrane/primary

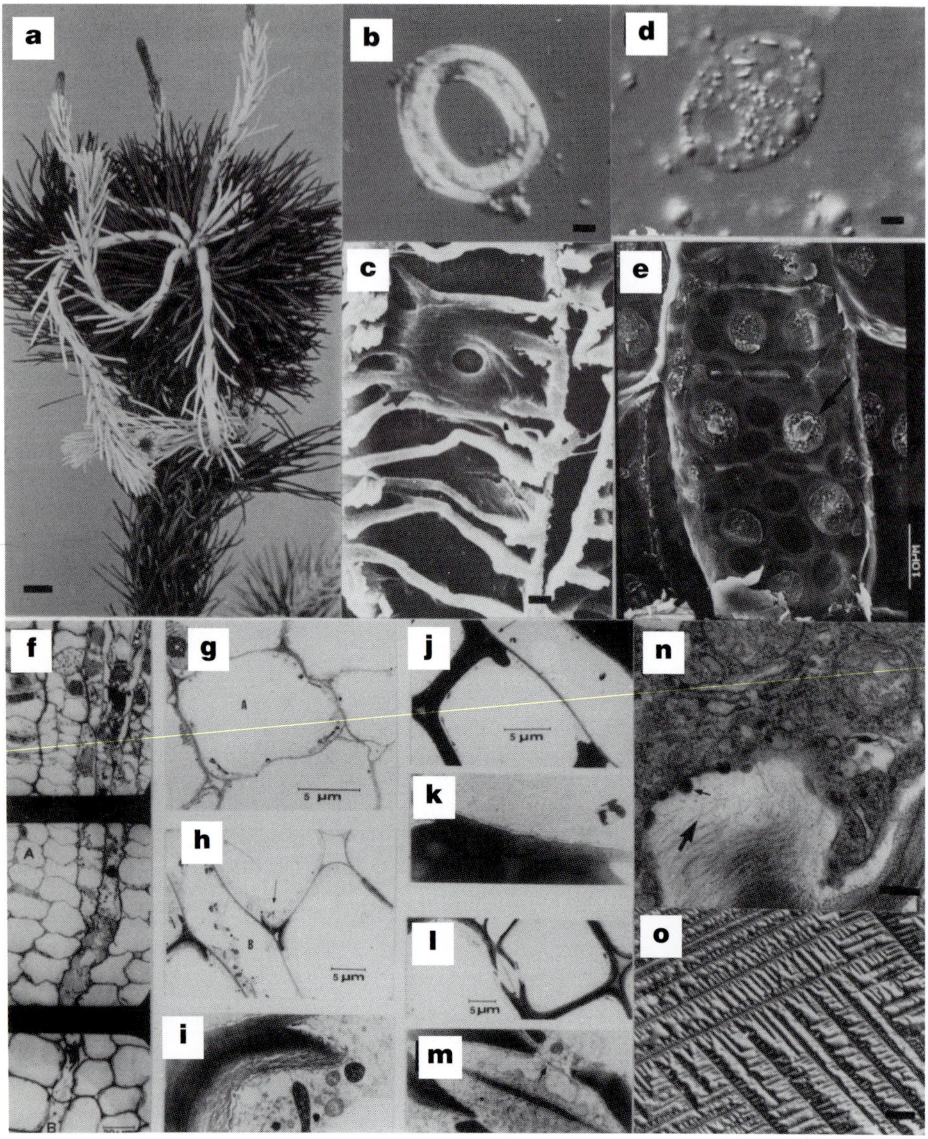

wall as disks in regions where both simple and bordered-pit development occur (*Figure 5e*). Thus, it is possible that these wall disks fulfil a determining role, perhaps inhibiting secondary-wall deposition, during pit development.

7.3 *How are the nature and number of secondary-wall lamellae determined?*

This topic is addressed at some depth by other chapters of this book, particularly in relation to the cytoskeleton. In transmission electron microscope images of the cambium and developing xylem (*Figures 5f–5n*), it is usual to see a space, of low electron density, between the membrane and the maturing secondary wall (*Figures 5f, 5j and 5n*). There has long been concern that this space may be merely a fixation

Figure 5. *(a) Etiolated shoots of* Pinus contorta *produced by growing the tree in continuous darkness. Primary xylem and procambial strands are present in such shoots; however, no evidence has been found for vascular cambium formation; bar = 1 cm. (b) A ring of Nomarski birefringent 'cellulose' isolated from a developing bordered pit of* Pinus strobus; *bar = 1.0 µm. (c) A bordered pit (arrowed) produced in apparent structural continuity with ribs of secondary-wall material; bar = 2 µm. (d) An unidentified organelle (compare Figure 4(g)) from cambial cells of* Pinus strobus *obtained through bark peeling to release cellular protoplasma. This type of organelle, containing highly heterogeneous contents, is distinct from proplastids in the cambium, is abundantly present in the SL zone and appears to be connected to pit development; bar = 1.0 µm. (e) Scanning electron micrograph showing disk-like organelles (arrowed) associated with pit development in* Pinus contorta; *bar = 10 µm. (f) Transmission electron micrograph of a* Pinus silvestris *stem showing the CZ (top), RE and SL zones essentially separated by the two grid bars, with phloem at the top of the section. The fusiform cambial derivative designated as 'A' is shown at higher magnification in Figure 5(g), and the ray cambial derivative designated as 'B' is shown at higher magnification in Figure 5(h), bar = 20 µm. (g) Fusiform cell (A) in the RE zone. Note the thin primary walls, the immense vacuole and the parietal cytoplasm. (h) A ray cell (B) and adjoining fusiform cells on either side at an early stage in SL development. Note the localized secondary-wall deposition (arrow) occurring in the fusiform but not in the ray cells. (i) Close up of the region arrowed in Figure 5(h). (j) and (k) A developing tracheid adjoining a ray cell. Note that secondary-wall formation has not occurred where its primary wall adjoins that of the ray cell, indicating that a cross-field pit will be there. The arrowed location is shown at higher magnification in Figure 5(k). (l) and (m) An example of advanced bordered-pit development, shown in detail in Figure 5(m). In the latter, note that there appears to be a membrane-bound heterogeneous organelle similar to that shown in Figure 5(d) lying upon the membrane of the developing torus. The overarching borders appear to be conforming to the organelle's membrane (arrow). (n) A developing rib of secondary-wall material in* Pinus contorta. *Note the electron-dense exocytotic blebs (small arrow) and fibrillar material (large arrow) in the outer regions of the developing rib. (o) Water-miscible polysaccharide obtained and crystallized from developing secondary walls of* Pinus *spp. Identically appearing material has been obtained from developing cotton (*Gossypium *sp.) boll fibres; bar = 0.1 µm.*

---

artifact, but cryo-fixation freeze-substitution work by Inomata *et al.* (1992) indicated that it does contain a non-fibrillar material 'which appeared to be composed of newly-deposited polysaccharides containing a great deal of water.' In my research, this water-miscible material (which can be obtained in massive quantity from developing xylem and has been confirmed to be polysaccharidic by molecular-weight fractionation, gas chromatography-mass spectroscopy of hydrolysed preparations, and nuclear magnetic resonance analysis of the polysaccharide in solution), when dried on the microscope slide, yields linear arrays suggestive of microfibrils (*Figure 5m*). Microscopically identical, or very similar, material has been obtained from developing fibres of cotton. Thus, it is possible that this polysaccharide is a nascent precursor of cellulose.

8. Conclusions

Understanding the process of wood formation is highly complex, requiring integration of biochemistry with wood anatomy, wood chemistry, wood physics and tree physiology. Each of the approximately 80 000 tree species on earth produces a more-or-less specific kind of wood. To make matters more complicated, several distinct types of wood develop from the cambium within each individual tree. In addition, heartwood formation and other types of post-cambial transformations also occur. Thus, there is no shortage of problems for the biochemist! There is a definite need in forestry to go beyond the visible phenotype, whether it be forest, tree, log or individual fibre, and recognize these for what they are, biosynthetic end products, manifestations of genotype–environment interactions that occurred in a living system. Once the non-visible determinants of growth, development and functioning have been discovered, characterized and investigated, a scientific basis for the management of wood supplies and forest ecosystem health will be in place.

Acknowledgements

I am grateful to the Natural Sciences and Engineering Research Council of Canada for its financial support. My thanks to the reviewers who commented on the first draft and to Hartmut Förster for his critical commentary on the penultimate draft.

References

Aberg, G., Jacks, G., Wickman, T. and Hamilton, P.J. (1990) Strontium isotopes in trees as an indicator for calcium availability. *CATENA* **17**: 1–11.

Baars, D.L. (1972) *Red Rock Country*. Doubleday/Natural History, Garden City, New York.

Bailey, J.H.E. and Reeve, D.W. (1994) Spatial distribution of trace elements in black spruce by imaging microprobe secondary ion mass spectrometry. *J. Pulp and Paper Sci.* **20**: J83–J86.

Barghoorn E.S. (1964) Evolution of cambium in geologic time. In: *The Formation of Wood in Forest Trees* (ed. MH Zimmermann). Academic Press, NY, pp 3–17.

Barnett, J.R. (1981) Secondary xylem development. In *Xylem Cell Development* (ed. J.R. Barnett). Castle House Publ., Tunbridge Wells, U.K., pp 47–95.

Bernard-Dagan, L. (1988) Biosynthesis of lower terpenoids: genetic and physiological controls in woody plants. In: *Genetic Manipulation of Woody Plants* (ed. J.W. Hanover and D.E. Keathley), Plenum Press, New York and London pp 329–351.

Biggs A.R. (1987) Occurrence and location of suberin in wound reaction zones in xylem of 17 trees species. *Phytopathology* **77**: 718–725.

Bowers, L.J. and Melhuish, J.H. (1988) Comparison of elemental concentrations in the wood of three tree species growing adjacent to an inactive chromium smelter, *Bull. Environ. Contamin. Toxicol.* **40**: 457–461.

Brown, C.L. and Wodzicki, T.J. (1969) A simple technique for investigating cambial activity and the differentiation of cambial derivatives. *Forest Science* **15**: 27–29.

Brown, R.M. Jr., Saxena, I.M. and Kudlicka (1996) Cellulose biosynthesis in higher plants. *Trends in Plant Sciences* **1**: 149–156.

Cannell, M.G.R. and Jackson, J.E., eds. (1985) *Attributes of Trees as Crop Plants.* ITE, NERC, Abbots Ripton, U.K.

Carlquist, S. (1988) *Comparative Wood Anatomy*. Springer-Verlag, Berlin, Heidelberg, New York, London, Paris, Tokyo.

Daniels, F. J. (1998) *Petrified Wood*. Western Colorado Publ., Grand Junction, CO.

Dute, R.R. (1994) Pit membrane structure and development in *Ginkgo biloba*. *IAWA J.* **15**: 75–90.

Ellis, E.L. (1965) Inorganic elements in wood. In *Cellular Ultrastructure of Woody Plants* (ed. W.A. Côté Jr.), Syracuse University Press, Syracuse, New York, pp 181–189.

Gahan, P.B. (1981) Biochemical changes during xylem element differentiation. In *Xylem Cell Development* (ed. J. R. Barnett), Castle House Publ, Tunbridge Wells, England, pp 168–191.

Grozdits, G.A. and Ifju, G. (1973) Nitrogen distribution in eastern hemlock and its relation to wood formation. *Wood Science* **6**: 1–8.

Hale, J.D. (1933) *Heating Value of Wood Fuels.* Dept. Interior, Forest Service, Forest Products Lab. Canada.

Higuchi, T., ed. (1985) *Biosynthesis and Biodegradation of Woody Components*. Academic Press, London.

Higuchi, T. (1997) *Biochemistry and Molecular Biology of Wood.* Springer-Verlag, Berlin, Heidelberg, New York, Tokyo.

Hillis, W.E. (1987) *Heartwood and Tree Exudates.* Springer-Verlag, Berlin, Heidelberg, New York, Tokyo.

Hu, W.-J., Harding, S.A., Lung, J., Popko, J.L., Ralph, J., Stokke, D.D., Tsai, C.-J., and Chiang, V.L. (1999) Repression of lignin biosynthesis promotes cellulose accumulation and growth in transgenic trees. *Nature Biotech.* **17**: 808–812.

Inomata, F., Takabe, K. and Saiki, H. (1992) Cell wall formation of conifer tracheid as revealed by rapid-freeze and substitution method. *J. Electron Microsc.* **41**: 369–374.

Jones, D.W. (1971) X-ray and electron diffraction. C.1. Structure studies. In *High Polymers, Vol. 5, Cellulose and Cellulose Derivatives, Part 4* (ed. N.M. Bikales and L. Segal), Wiley, New York, pp 117–150.

Kennedy, G. and Bergeron, S. (1991) Tree rings as monitors of heavy metal air pollution histories. *J. Radioanal. Nuc. Chem.* **151**: 337–343.

Kennedy, R.W. and Farrar, J.L. 1965 Tracheid development in tilted seedlings. In: *Cellular Ultrastructure of Woody Plants* (ed. W.A. Côté, Jr.), Syracuse University Press, Syracuse, New York, pp 419–453.

Kohno, M., Koiumi, Y., Okumura, K. and Mito, I. (1988) Distribution of environmental Cs-137 in tree rings. *J. Environ. Radioactivity* **8**: 15–19.

Landsberg, J.J. (1986) *Physiological Ecology of Forest Production.* Academic Press, London.

Larson, P.R. (1994) *The Vascular Cambium: Development and Structure.* Springer-Verlag, Berlin, Heidelberg, New York, London, Tokyo, Hong Kong, Barcelona, Budapest.

Larson, P.R. (1962) The indirect effect of photoperiod on tracheid diameter in *Pinus resinosa. Am. J. Bot.* **49**: 132–137.

Leinhos, V. and Savidge, R.A. (1993) Isolation of protoplasts from developing xylem of *Pinus banksiana* and *Pinus strobus. Can. J. For. Res.* **23**: 343–348.

Lewis, N.G. and Sarkanen, S., eds. (1998) *Lignin and Lignan Biosynthesis.* ACS Symp. Series 697, American Chemical Society, Washington, DC.

Lovestram, G., Johansson, E., Johansson, S. and Pallon, J. (1990) Elemental micro patterns in tree rings — a feasibility study using scanning microprobe analysis. *AMBIO* **19**: 87–93.

Mattheck, C. and Kubler, H. (1995) *Wood—The Internal Optimization of Trees.* Springer-Verlag, Berlin, Heidelberg, New York.

Megraw, R.A. (1985) *Wood Quality Factors in Loblolly Pine.* TAPPI Press, Technology Park, Atlanta, GA.

Meier, H. (1985) Localization of polysaccharides in wood cell walls. In *Biosynthesis and Biodegradation of Wood Polysaccharides* (ed. T. Higuchi), Academic Press, London, pp 43–50.

Mellerowicz, E.J., Riding, R.T. and Little, C.H.A. (1989) Genomic variability in the vascular cambium of *Abies balsamea. Can. J. Bot.* **67**: 990–996.

Northcote, D.H. (1985) Cell organelles and their function in biosynthesis of cell wall components: control of cell wall assembly during differentiation. In *Biosynthesis and Biodegradation of Woody Components* (ed. T. Higuchi), Academic Press, London, pp 87–108.

Panshin, A.J. and de Zeeuw, C. (1980) *Textbook of Wood Technology*, 4th ed. McGraw-Hill, New York, St. Louis, San Francisco.

Pearce, R.B. (1996) Antimicrobial defences in the wood of living trees. *New Phytol.* **132**: 203–233.

Rowe, J. W., ed. (1989) *Natural Products of Woody Plants.* Springer-Verlag, Berlin, Heidelberg, New York.

Savidge, R.A. (1983) The role of plant hormones in higher plant cellular differentiation. II. Experiments with the vascular cambium, and sclereid and tracheid differentiation in the pine, *Pinus contorta. Histochem J.* **15**: 447–466.

Savidge, R.A . (1985) Prospects for manipulating vascular cambium productivity and xylem cell differentiation. In *Attributes of Trees as Crop Plants* (eds. M.G.R. Cannell and J.E. Jackson), NERC Press, Abbots Ripton, England, pp 208–227.

Savidge, R.A. (1986) Prospects for genetic manipulation of wood quality. In *New Ways in Forest Genetics* (eds. F. Caron, A.G. Corriveau, T.J.B. Boyle), Proc. 20th CTIA mtg., Quebec, part 2, pp 159–165.

Savidge, R.A. (1990) Phytohormonal regulation of cambial growth in trees. In *Fast Growing Trees and Nitrogen Fixing Trees* (ed. D. Werner and P. Müller), Gustav Fischer, Jena, pp. 142–151.

Savidge, R.A. (1993) *In vitro* wood formation in 'chips' from merchantable stem regions of *Larix laricina. IAWA J.* **14**: 3–11.

Savidge, R.A. (1994) The tracheid-differentiation factor of conifer needles. *Int. J. Plant Sci.* **155**: 272–290.

Savidge, R.A. (1996) Xylogenesis, genetic and environmental regulation – a review. *IAWA J.* **17**: 269–310.

Savidge, R.A. and Farrar, J.L. (1984) Cellular adjustments in the vascular cambium leading to spiral-grain formation in conifers. *Can. J. Bot.* **62**: 2872–2879.

Savidge, R.A., Heald, J.K. and Wareing, P.F. (1982) Non-uniform distribution and seasonal variation of endogenous indol-3-ylacetic acid in the cambial region of *Pinus contorta* Dougl. *Planta* **155**: 89–92.

Savidge, R.A., Udagama-Randeniya, P.V., Xu, Y., Leinhos, V. and Förster, H. (1998) Coniferyl alcohol oxidase: a new enzyme spatio-temporally associated with lignifying tissues. In *Lignin and Lignan Biosynthesis* (eds. N.G. Lewis and S. Sarkanen), ACS Symp. Series 697, American Chemical Society, Washington, DC., pp. 109–130.

Savidge, R.A. and Wareing, P.F. (1981) Plant growth regulators and the differentiation of vascular elements. In *Xylem Cell Development* (ed. J R Barnett), Castle House Publ, Tunbridge Wells, England, pp 192–235.

Shafizadeh, F. and DeGroot, W.F. (1976) Combustion characteristics of cellulosic fuels. In: *Thermal Uses and Properties of Carbohydrates and Lignins* (eds. F. Shafizadeh, K.V. Sarkanen and D. A. Tillman), Interscience, N.Y.

Sjöström, E. (1993) *Wood Chemistry. Fundamentals and Applications*, 2nd ed. Academic Press, Inc. San Diego, New York, Boston, London, Sydney, Tokyo, Toronto.

Stamm, A.J. and Sanders, H.T. (1966) Specific gravity of the wood substance of loblolly pine as affected by chemical composition. *TAPPI* **49**: 397–400.

Thimann, K.V. and Kaufman, D. (1958) Cytoplasmic streaming in the cambium of white pine. In: *The Physiology of Forest Trees* (ed. K.V. Thimann), Ronald Press, New York, pp 479–492.

Timell, T.E. (1986) *Compression Wood in Gymnosperms*, Vols. 1–3. Springer-Verlag, Berlin, Heidelberg, New York, Tokyo.

Wheeler, E.A. and Baas, P. (1998) Wood identification — a review. *IAWA J.* **19**: 241–264.

Wilkie, K.C.B. (1985) New perspectives on non-cellulosic cell-wall polysaccharides (hemicelluloses and pectic substances) of land plants. In: *Biochemistry of Plant Cell Walls* (ed. C.T. Brett and J.R. Hillman), Cambridge University Press, Cambridge, London, New York, pp 1–37.

Wilson, B.F., Wodzicki, T.J. and Zahner, R. (1966) Differentiation of cambium derivatives: proposed terminology. *Forest Sci.* **12**: 438–440.

Zobel, B.J. and Jett, J.B. (1995) *Genetics of Wood Production.* Springer-Verlag, Berlin, Heidelberg, New York.

Zobel, B.J. and Sprague, J. R. (1998) *Juvenile Wood in Forest Trees*. Springer-Verlag, Berlin, Heidelberg, New York.

Zobel, B.J. and van Buijtenen, J.P. (1989) *Wood Variation – Its Causes and Control.* Springer-Verlag, Berlin, Heidelberg, New York.

2

Cytoskeleton, cell walls and cambium: new insights into secondary xylem differentiation

N. J. Chaffey

1. Introduction

Despite our modern, hi-tech world, there is still a great — and increasing — demand for wood and other tree products. Further, recent environmental and political interest has focused considerable attention on trees as CO_2-sinks in the context of 'global warming', and as a 'green energy' alternative to fossil fuels.

However, no two woods are identical. Variation in wood properties is recognized in the designation of gymnosperms as 'softwoods' and of angiosperms as 'hardwoods', and the identification of woods suitable for particular purposes, such as paper-production, or house-building. These differences in wood properties are a reflection both of the relative proportions of each cell type, and of their individual characteristics. The wide variety of wood (secondary xylem) cells is produced from just two basic cell types in the vascular cambium ('cambium'), the fusiform and ray initials. Consequently, during xylogenesis (wood formation), many changes take place to the cambial derivatives (see *Table 1*). Understanding the process of differentiation of each cell type is fundamental to an appreciation of these changes, and to the ability to 'design' wood for particular end-uses.

Notwithstanding a long history of the study of wood structure and cambial activity, there is still little understanding of how a cambial derivative differentiates as a wood-cell (Barnett, 1995), or *why* derivatives differ, although application of new techniques of cell biology (Chaffey, 1999) is now allowing us to answer those questions.

Cell & Molecular Biology of Wood Formation, edited by R. Savidge, J. Barnett and R. Napier.

Table 1. Some of the changes that take place during differentiation of axial secondary xylem elements in Aesculus hippocastanum

Feature	Change taking place	Cell type affected
Shape/size	Elongated	Fibres particularly
	Widened	All cells
	Shortened	Axial parenchyma cells
Cell walls	Thickened	All cells
	More-ordered	Secondary cf. primary walls of all cells
	Multi-layered	'S' layers of secondary wall
	Altered chemistry	e.g., all cells lignified
	Modified	e.g., pitted (all), perforated (vessel elements)

2. Role of the cytoskeleton in wood formation

2.1 *Why is the cytoskeleton of interest?*

Some of the most fundamental changes that take place during xylogenesis relate to cell size/shape and wall structure. Of great importance to, and a major constraint upon, plant-cell morphogenesis is the orientation of relatively inflexible cellulose microfibrils within their walls (e.g., Green, 1969). Ever since the first ultrastructural report of parallelism between microtubules (MT) and microfibrils (Ledbetter and Porter, 1963), MTs have been implicated in the orientation of cellulose (reviewed by Giddings and Staehelin, 1991). Although this hypothesis has yet to be proved to the satisfaction of all workers (see discussion in Barnett *et al.*, 1998), it has nevertheless focused attention on the possible role(s) of the cytoskeleton in plant-cell differentiation.

To date, the two most studied components of the plant cytoskeleton are the MTs and the microfilaments (MFs). Both are composed of proteins, tubulin and F-actin respectively, and can be several hundred μm long. However, the approximate diameter of an individual MT is 25 nm and 7 nm for a single MF, and although they are frequently aggregated into bundles, they are practically impossible to visualize directly without the aid of the transmission electron microscope (TEM). The comparatively recent development of antibodies against cytoskeletal proteins, and the application of indirect immunofluorescent (IIF) localization procedures for light microscopy, have increased greatly the ease with which the cytoskeleton can be studied (Lloyd, 1987). Investigation of the role of MTs in wood formation is justified because of their inferred role in cell-shaping and wall formation, and the importance of such events to xylogenesis. Furthermore, reports of co-localization of MFs and MTs (see review by Seagull, 1989), make MFs of interest in the context of xylogenesis.

2.2 *Historical background to cytoskeletal studies in wood formation*

Although the individual elements of the secondary vascular system (SVS) (the assemblage of secondary phloem, cambium, and secondary xylem) can be isolated relatively easily for biochemical analysis, study of their sub-cellular structure is compromised by the problems inherent in adequately fixing such material (Chaffey, 1997). Nevertheless, a number of TEM studies of the SVS have been performed in the last 30 years. Some of this work has dealt specifically with the cytoskeleton and shown co-orientation of cortical MTs and cellulose microfibrils in fibre walls (Cronshaw, 1965), and association of MTs with bordered pit formation (Robards and Humpherson, 1967). Unfortunately, use of the small area, ultrathin sections necessary for TEM does not lend itself readily to an appreciation of the 3-dimensional arrangement of the cytoskeleton within differentiating cells. Recent use of IIF on thick sections, however, has permitted a more in-depth study.

Credit for the first reported use of immunofluorescent visualization of the cytoskeleton within the tree SVS probably belongs to Uehara and Hogetsu (1993) for their IIF study of MTs in *Taxus* tracheids. Currently, only two groups are examining the role of the cytoskeleton in the tree SVS: Funada *et al.* primarily on gymnosperms (Abe *et al.*, 1995a, b; Prodhan *et al.*, 1995; Funada *et al.*, 1997; Furusawa *et al.*, 1998, see also their chapter in this book), and Chaffey *et al.*, working mainly with angiosperms. As with gymnosperms, studies in angiosperm trees have concentrated on MTs (Prodhan *et al.*; 1995; Chaffey *et al.*, 1996, 1997a, b, c, d, 1998, 1999), although IIF techniques for MFs have now been developed (Chaffey *et al.*, 1997d) and are being applied (Chaffey *et al.*, 2000).

The current vogue for IIF does not mean that more traditional approaches, like TEM, have no further contribution to make. Indeed, use of different techniques, such as combined TEM and IIF, is essential to a proper appreciation of cytoskeletal arrangements and interactions with other cell components, and to achieve a consensus view of their structure (Chaffey *et al.*, 1996). Extending studies of the cytoskeleton to the finer resolution achievable with the TEM significantly enhances the value of the IIF work. Epifluorescence IIF has been the method of financial necessity for many workers. As use of the confocal laser scanning microscope (CLSM) becomes more widespread, however (Hepler and Gunning, 1998), we should see a considerable advance in the type of IIF studies that are performed within the tree SVS (Chaffey, 1999).

2.3 *Scope of this review*

This review considers the role(s) of MTs and MFs during differentiation of the axial elements of secondary xylem within the root and shoot of the angiosperm tree, *Aesculus hippocastanum* L. (Hippocastanaceae) (horse-chestnut) during the first 5 years of growth. Except where indicated, results for both root and shoot are the same. The results summarized here were reported in Chaffey *et al.* (1996, 1997a, b, c, d, 1998, 1999, 2000).

2.4 *Anatomical background to the cambium and wood of* Aesculus

The SVS consists of secondary phloem to the outside of the organ, the cambial zone, and secondary xylem to the inside (*Figure 1d*). Because of difficulties in identifying the true cambial initials (Catesson, 1974), the term 'cambial zone' is here used to include those initial cells and their derivatives at the earliest stages of differentiation. Fusiform cambial cells are thin-walled, highly vacuolated, and elongated parallel to the organ axis, and their xylem derivatives differentiate as fibres, vessel elements, and axial parenchyma cells. All axial wood cells are significantly wider than their cambial precursors; while fibres are considerably longer, axial parenchyma cells significantly shorter, and vessel elements about the same length as fusiform cambial cells. At maturity, their walls are variously sculptured, modified at their ends, secondarily thickened and lignified. Ray cambial cells are also thin-walled and highly vacuolated, but approximately cuboid, while their derivatives elongate at right angles to the organ axis. Xylem ray cells are alive at maturity and develop thicker walls which lignify and become variously pitted. They are further differentiated into contact cells which have large-diameter contact pits in those walls which border vessel elements, and isolation cells which do not.

2.5 *Microtubules and wood formation*

The MTs of fusiform cambial cells are randomly arranged, and frequently overlap each other (*Figure 1a* and colour *Plate 1*). The loose meshwork of microfibrils described in the literature for the primary walls of wood cells (Desch and Dinwoodie, 1996) is consistent with this MT network, supporting the view that the latter may influence orientation of the former. As cambial derivatives differentiate, their radial width increases and their MTs adopt new arrangements (*Figure 1e*). At this stage of differentiation, the cell walls are unlignified and appear still to be primary.

During early stages of fibre differentiation, their MTs become progressively more ordered and more numerous as wall thickening proceeds, and eventually exhibit a dense, parallel-aligned, helical array (colour *Plate 1*). The MTs continue to be located parietally in the thin layer of cytoplasm as a single layer throughout the length of the fibre.

Elongation occurs at both ends of the fibre by a form of tip-growth (Wenham and Cusick, 1975) and, along with the increase in width, probably occurs within the cambial zone when such cells, and surrounding cells, are sufficiently 'plastic' to permit the extensive cellular rearrangements needed to accommodate such growth. Development of helically arranged MTs in incipient fibre cells is first detected in the cambial zone. Although the cell wall deposited during this period is considered to be primary, since it is laid down before cell growth has stopped, subsequent development of abundant parallel-associated, helically oriented MTs is more consistent with a phase of secondary cell wall thickening than with cell growth (Chaffey *et al.*, 1998). This pronounced increase in abundance of MTs, therefore, appears to relate to a switch from primary to secondary wall deposition.

At maturity, fibres have long-tapered ends and simple pits. The long axes of the pit aperture are oriented at a relatively acute angle to the cell's long axis, and are approximately parallel to the inferred pitch of the wall microfibrils, and to that of the MT helix. This apparent co-orientation of MTs and microfibrils is further circumstantial evidence for the idea that such cytoplasmic elements influence the arrangement of cellulose. MTs are associated with the edges of the developing pit, and with wall-thickening between the pits, but are absent from the area above the pit membrane. It is not known why or how these MT-free regions develop, but they probably reflect local differences in plasmalemma/cell wall properties at these sites. MT arrays associated with the development of axial parenchyma cells appear similar to those of fibres.

Vessel elements undergo a wide range of cell wall modifications in which the MTs appear to be intimately involved during their differentiation. Sites of bordered pits are initially marked out during the phase of cell expansion as MT-free regions within the

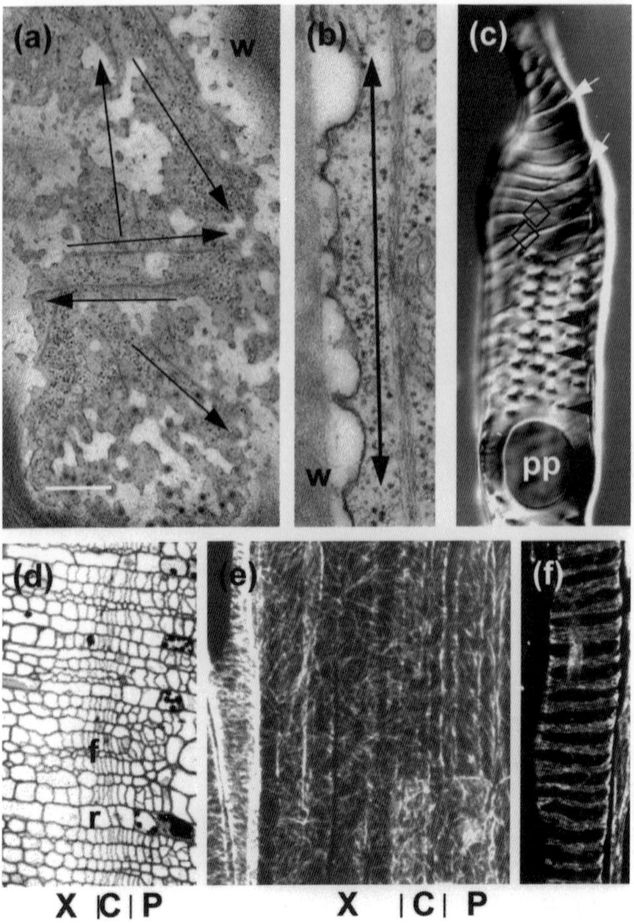

Figure 1. *Aspects of wood formation in* Aesculus hippocastanum *taproot ((a)–(d)) and stem ((e), (f)). (a) TEM of wall (w)/cytoplasm interface showing randomly oriented MTs (arrows), and (b) an axially oriented MF bundle (arrow) in fusiform cambial cells. (c) Isolated vessel element; note the simple perforation (pp), branching (◻) tertiary thickenings (arrows), and bordered pits (darts). (d) Transverse section; note the fusiform (f) and ray (r) cells of the cambial zone (C), and the radial cell files into secondary xylem (X) and secondary phloem (P). (e) CLSM of radial longitudinal section stained with anti-tubulin antibodies; note the change in MT orientation from random in cambium (C) to helical in secondary xylem (X) fibres and axial elements of the secondary phloem (P). (f) CLSM of anti-tubulin-stained MT bundles associated with tertiary-thickening in vessel element. Scale bar (in (a)) =0.5 μm.*

reticulate array of MTs present in fusiform cambial cells. Associated with this concentration of randomly oriented MTs at the periphery of the future bordered pit is the deposition of new cell wall material at this site. Subsequently, a ring of MTs develops at the periphery of the MT-free region (colour *Plate 1*). The random MTs at the pit-periphery eventually disappear, and the diameter of the MT ring becomes reduced as further wall deposition creates the over-arching pit-border. MTs between the MT rings eventually adopt a transverse/high-angled orientation (colour *Plate 1*) which is inferred to relate to a period of general wall-thickening in the cell. Contact pits develop in a manner similar to that of the inter-vessel bordered pits, but the MT ring does not decrease in diameter.

It should be emphasized that the periphery of the bordered pit is initially polygonal — usually hexagonal (colour *Plate 1*) — but oval or circular in the case of the contact pits. How the precise arrangements of MTs are initiated and maintained is as yet unknown, although one can speculate that MT-associated proteins (MAPs) may link MTs to the plasmalemma, and to each other, and that MT-organizing centres (MTOCs) might be involved. The reason(s) for the difference in behaviour of the MT rings between the two pit types is not understood, but it seems to be related to the fact that an over-arching border develops in bordered pits but not in contact pits, and to the behaviour of MFs at these sites (see section 2.6). It is, however, likely that the near-transverse MTs, which develop at late stages of formation of both pit-types, give rise to the MTs which appear to participate in tertiary-thickening.

Tertiary thickenings are broad-based ridges on the inner face of the vessel element secondary wall (*Figure 1c*). In early stages of tertiary-thickening, wide, near-transverse MT bands alternate with MT-free regions (*Figure 1f*). Later, the MT bands are replaced by, or rearranged as, narrower bands, then pairs of parallel MT bundles. It is conceivable that the MT bundles act as a 'funnel' through which dictyosome-derived secretory vesicles are channelled, thereby contributing to the development of the wall thickening. Each member of the MT bundle-pair is often seen to be connected to its neighbouring MT bundle in the adjacent MT bundle-pair by even finer threads of 'bridging MTs', which persist until late stages of tertiary-thickening. These bridging MTs possibly act as 'spacers'. They may help to keep the members of each bundle-pair close together so that the developing ridge of tertiary thickening is constrained to adopt the observed high profile, rather than to become a low-domed structure. Tertiary thickenings frequently branch (*Figure 1c*), as do the associated MT bundles.

The formation of MT bands associated with developing tertiary thickenings appears to take place *before* wall-thickening is detectable, suggesting that MTs may drive the process rather than follow in its wake. How the initial location of the tertiary thickenings is determined is unknown, but the fact that adjacent MT bundles and the corresponding thickenings appear to be regularly spaced suggests involvement of an organizing mechanism.

MT rings also accompany development of the simple perforations of vessel elements (colour *Plate 1*), suggesting similarities with the formation of bordered and contact pits, although on a much bigger scale. The formation of perforations at the ends of vessel elements implies that here, too, there is a distinctive plasmalemma/cell wall domain. As for bordered and contact pits, the MT ring appears to define a region from which MTs are excluded and which does not undergo secondary wall thickening.

It has not yet been possible to place a precise time-scale on the changes that occur during vessel element development, but it is clear that the various MT arrays can co-exist and the corresponding wall elaborations can take place concurrently. For example, *Figure 2b* in Chaffey *et al.* (1997d) shows a vessel element in which development of tertiary thickenings, bordered pits *and* the perforation are taking place at the same time, and on the same wall face.

Although the exact nature of the role of MTs in xylogenesis remains to be established, it is clear that it relates to fundamental characteristics of wall growth and elaboration, which distinguish the wood cells from each other.

2.6 *Microfilaments and wood formation*

The axially oriented bundles of MFs within fusiform cambial cells seen by TEM (*Figure 1b*), are revealed by IIF of thick longitudinal sections to be a helix, or a series of parallel helices, of extremely high pitch (colour *Plate 1*), apparently extending from one end of the cell to the other. Although the MF bundles may be branched, their overall net-axiality distinguishes them from the randomly arranged MTs in these cells, suggesting that each cytoskeletal component performs a different role.

A. hippocastanum cambium is non-storied, so increase in girth of the cambial cylinder is achieved by pseudotransverse division of fusiform cambial cells and the subsequent elongation of both daughter cells. It is believed that cell-elongation is accomplished by apical- [tip-] growth (Włoch and Polap, 1994). After periclinal division, incipient fibre cells also undergo a form of apical, intrusive tip-growth. MFs are implicated in cytoplasmic streaming (e.g., Collings *et al.*, 1996), which is a prominent feature of active fusiform cambial cells (e.g., Catesson, 1974), and it seems likely that the MF bundles may have such a role here. In view of the elongated, highly vacuolated nature of these cells, MF-mediated cyclosis will be of particular benefit in facilitating the distribution of metabolites etc. to support such processes as wall synthesis, especially in those cells undergoing tip-growth.

At early stages of differentiation of all axial secondary xylem cells, MFs retain their net-axial orientation. This essentially unchanged orientation contrasts with the behaviour of the MTs within these cells and implies that MFs are not involved in the change of MT orientation. Rather, the persistent axial orientation of the MFs supports the suggestion that they continue to be involved in delivery of secretory vesicles for cell wall synthesis at growing ends or, for example, of enzymes associated with wall modification which occurs during formation of vessel element perforations (reviewed by Butterfield, 1995). By contrast, the MTs are likely to be involved in directing the orientation/synthesis of cellulose microfibrils within the developing wall.

Although the net-axial orientation of MFs is retained in later stages of cell differentiation, additional MF arrangements occur. In fibres, ellipses of MFs are associated with sites of pits (colour *Plate 1*). The fact that both MTs and MFs appear to extend to the tips of these cells distinguishes fibres from the more widely known tip-growing systems of fungal hyphae, root hairs and pollen tubes (e.g., Kropf *et al.*, 1998). That, and the bipolar nature of their tip-growth makes wood-fibres a particularly instructive system for study, and comparison with the unipolar systems.

Vessel elements possess several distinct MF arrays:

(i) helically oriented MF bundles apparently extending the length of the cell;
(ii) MF rings associated with early stages of bordered pit formation, and the developing aperture (colour *Plate 1*);
(iii) circular MF arrays associated with development of contact pits;
(iv) linear MF arrays associated with tertiary-thickening, both parallel to the developing ridge of wall-thickening, and bridging the region between adjacent wall ridges;
(v) a circle of MFs at the periphery of the developing perforations; and,
(vi) a MF-meshwork at the perforation plate itself.

The MF ring associated with the developing border of the bordered pits also decreases in diameter as the over-arching pit border develops. This apparent co-localization of

MFs and MTs suggests that F-actin, possibly linked with myosin as actinomyosin, provides the motive-force for the ring contraction, in a manner akin to that of mammalian muscle (e.g., Jackson, 1982).

That MFs and MTs are co-oriented during later stages of tertiary-thickening suggests their co-localization, and the involvement of both cytoskeletal elements in this process. However, it is noteworthy that no broad bands of MFs have so far been seen in association with the broad bands of MTs which develop early in tertiary-thickening.

3. A cytoskeletal approach to designer wood

Mindful of the fact that any attempt to create 'designer wood' must not compromise the living system which produces it, manipulation of the cytoskeleton offers several possibilities for altering wood properties.

3.1 *Ratio of wood-cell types*

The shift in MT arrangement as cambial derivatives differentiate has been interpreted as a marker that cell-determination to a pathway of vascular differentiation has occurred (Chaffey *et al.*, 1997a), and that the cells are no longer meristematic. In other plant systems, multi-gene families for tubulin exist, tubulin isoforms are known, and there is evidence that different genes are expressed in different cell types (e.g., Uribe *et al.*, 1998). If different tubulin isoforms comprise the different MT arrays associated with different cell types in the tree SVS, overexpressing one or another isoform might promote the development of a particular cell type relative to another. Thus, it might be possible to produce fibres at the expense of vessel elements, or ray cells instead of axial elements. Although we have yet to examine whether different tubulin isoforms are present in the tree SVS, the technology exists for overexpressing tubulin genes in plant cells (e.g., Anthony and Hussey, 1998).

3.2 *Fibre qualities*

Microfibril angle. One of the major determinants of the 'usefulness' of wood-fibres is their microfibril angle (e.g., Desch and Dinwoodie, 1996). Largely because of this, fibres from outer mature wood (with a low microfibril angle) are more valuable than those with larger angles found in the inner, juvenile wood. Much published work links the angle of the MTs with that of the cellulose microfibrils; the work summarized here provides circumstantial evidence that this is also the case in hardwood trees. Consequently, alteration of the MT angle might have an impact on the microfibril angle, possibly leading to creation of 'mature wood fibres' in otherwise juvenile trees. Potentially, 'mature' wood could then be harvested earlier than at present.

Fibre dimensions. If the different MT densities associated with the two phases of fibre growth reflect differences in the associated tubulin isoforms, it might be possible to alter the duration of these phases by differential tubulin-gene expression. Thus, we might increase fibre length (useful for stronger paper) by prolonging the period when the low-density parallel-helical MT array exists, or reduce fibre length, and promote fibre-wall-thickening (hence calorific value), by advancing the onset of the phase associated with dense MT arrays.

3.3 *Vessel elements*

Pit-aperture diameter. There are a number of situations where alteration of the diameter of the inter-vessel bordered pit-apertures might be of value. *In vivo*, reduction in diameter might impede the spread of pathogenic fungi throughout the SVS. In industrial applications, increased pit-diameter should facilitate the penetration of preservatives, or, for example, of hemicellulolytic enzymes whose action enhances the bleachability of wood-pulp (Clark *et al.*, 1991). The possibility that an actino-myosin contractile system might be involved in development of bordered pits presents an obvious target for manipulation. The suggestion that the pits develop at specific plasmalemmal domains also presents a target for altering their density in a vessel element (and possibly the carbohydrate demand made on the cell, since more pits mean less wall needs to be made) by manipulation, for example, of location of MAPs and/or MTOCs.

Calorific value. Calorific value is highly relevant to use of wood as fuel. Since the heating value of lignin (28 MJ kg^{-1}) is almost twice that of cellulose (15 MJ kg^{-1} — El Bassam, 1996), increasing the amount of lignin in wood is of commercial concern. Hitherto, much work has been directed at reducing the quantity of lignin by 'molecular approaches' for the wood-pulping industry (e.g. Boudet, 1998). Using similar technology it should be possible to 'persuade' the tree to make more lignin. To be useful, though, any extra lignin needs to be retained within the tree, within the cell walls, so that its calorific value can subsequently be 'unlocked' upon combustion.

Manipulation of the amount of tertiary-thickening might help here. Although the role of these wall structures is not fully understood, and they are not found in all species, increase in their quantity might lead to an enhanced lignin content, as the amount of cellulose available for complexing with lignin is increased. Conversely, reducing the amount of tertiary-thickening might lessen the carbohydrate demand by individual vessel elements, potentially allowing such resources to be channelled into the development of more vessel elements or other wood-cells.

3.4 *Prospects*

The cytoskeleton is but one of several cell components which all play a part in wood-formation. As we discover more of the details of the process, other possibilities for manipulation of xylogenesis will become apparent. Although we do not yet have all the information needed to design wood, as long as the possibility — and the attendant economic benefits — exist, so should the impetus for further research in this area.

4. The future

Forests may disappear, but our demand for wood products — not least as 'green energy' — is likely to increase significantly in the near future. Thus, there is a real and pressing need to understand the cell biology of trees, particularly the process of wood-formation. To get the full benefit from such study, however, a multi-disciplinary approach is required, using modern techniques of both cell and molecular biology. To that end my current work is extending study of the cell biology of xylogenesis in angiosperm trees to hybrid aspen a widely accepted model hardwood tree (e.g. Chaffey, 1999a). Such work will benefit from exploitation of the recently generated expressed sequence tag (EST) library for hybrid aspen (Sterky *et al.*, 1998), and by comparison with wood-formation in *Arabidopsis* (Regan *et al.*, 1999).

Acknowledgements

I thank John Barnett and Peter Barlow for the excitement of the 'conker years', Sharon Regan for finally convincing me that *Arabidopsis* is not 'just a weed', Björn Sundberg for introducing me to hybrid aspen, and the anonymous reviewer for helpful suggestions. The *Aesculus* work was funded by the BBSRC under a LINK award to Reading University and IACR — Long Ashton Research Station. IACR receives grant-aided support from the BBSRC.

References

Abe, H., Funada, R., Ohtani, J. and Fukuzawa, K. (1995a) Changes in the arrangement of microtubules and microfibrils in differentiating conifer tracheids during the expansion of cells. *Ann. Bot.* **75**: 305–310.

Abe, H., Funada, R., Imaizumi, H., Ohtani, J. and Fukuzawa, K. (1995b) Dynamic changes in the arrangement of cortical microtubules in conifer tracheids during differentiation. *Planta* **197**: 418–421.

Anthony, R.G. and Hussey, P.J. (1998) Suppression of endogenous α and β tubulin synthesis in transgenic maize calli overexpressing α and β tubulins. *Plant J.* **16**: 297–304.

Barnett, J.R. (1995) Ultrastructural factors affecting xylem differentiation. In: *Vascular Cambial Derivatives* (ed. M. Iqbal). Gebrüder Borntraeger, Berlin, pp. 107–130.

Barnett, J.R., Chaffey, N.J. and Barlow, P.W. (1998) Cortical microtubules and microfibril angle. In: *Microfibril Angle in Wood* (ed. B.G. Butterfield). IAWA/IUFRO, Christchurch, New Zealand, pp. 253–271.

Boudet, A.-M. (1998) A new view of lignification. *Trends Plant Sci.* **3**: 67–71.

Butterfield, B.G. (1995) Vessel element differentiation. In: *Vascular Cambial Derivatives* (ed. M. Iqbal). Gebrüder Borntraeger, Berlin, pp. 93–106.

Catesson, A.-M. (1974) Cambial cells. In: *Dynamic Aspects of Plant Ultrastructure* (ed. A.W. Robards). McGraw-Hill Book Co., London, pp. 358–390.

Chaffey, N.J. (1997) Big *is* beautiful: science (and art) in wood microscopy. *Microsc. Today* **November:** 12–14.

Chaffey, N.J. (1999) Cambium: old challenges—new opportunities. *Trees* **13**: 138–151.

Chaffey, N.J., Barlow, P.W. and Barnett, J.R. (1996) Microtubular cytoskeleton of vascular cambium and its derivatives in roots of *Aesculus hippocastanum* L. (Hippocastanaceae). In: *Recent Advances in Wood Anatomy* (eds L.A. Donaldson, P.A. Singh, B.G. Butterfield and L.J. Whitehouse). New Zealand Forest Research Institute Ltd, Rotorua, New Zealand, pp. 171–183.

Chaffey, N.J., Barlow, P.W. and Barnett, J.R. (1997a) Cortical microtubules rearrange during differentiation of vascular cambial derivatives, microfilaments do not. *Trees* **11**: 333–341.

Chaffey, N.J., Barlow, P.W. and Barnett, J.R. (1998) A seasonal cycle of cell wall structure is accompanied by a cyclical rearrangement of cortical microtubules in fusiform cambial cells within taproots of *Aesculus hippocastanum* (Hippocastanaceae). *New Phytol.* **139**: 623–635.

Chaffey, N.J., Barlow, P.W. and Barnett, J.R. (2000) A cytoskeletal basis for wood formation in angiosperm trees: the involvement of microfilaments. *Planta* (in press).

Chaffey, N.J., Barnett, J.R. and Barlow, P.W. (1997b) Cortical microtubule involvement in bordered pit formation in secondary xylem vessel elements of *Aesculus hippocastanum* L. (Hippocastanaceae): a correlative study using electron microscopy and indirect immunofluorescence microscopy. *Protoplasma* **197**: 64–75.

Chaffey, N.J., Barnett, J.R. and Barlow, P.W. (1997c) Endomembranes, cytoskeleton and cell walls: aspects of the ultrastructure of the vascular cambium of taproots of *Aesculus hippocastanum* L. (Hippocastanaceae). *Int. J. Plant Sci.* **158**: 97–109.

Chaffey, N.J., Barnett, J.R. and Barlow, P.W. (1997d) Visualization of the cytoskeleton within the secondary vascular system of hardwood species. *J. Microsc.* **187**: 77–84.

Chaffey, N.J., Barnett, J.R. and Barlow, P.W. (1999) A cytoskeletal basis for wood formation in angiosperm trees: the involvement of cortical microtubules. *Planta* **208**: 19–30.

Clark, T., Steward, D., Bruce, M.E., McDonald, A.G., Singh, A.P. and Senior, D.J. (1991) Improved bleachability of radiata pine kraft pulps following treatment with hemicellulolytic enzymes. *Appita* **44**: 289–393 and 404.

Collings D.A., Wasteneys, G.O. and Williamson, R.E. (1996) Actin-microtubule interactions in the alga *Nitella*: analysis of the mechanism by which microtubule depolymerization potentiates cytochalasin's effects on streaming. *Protoplasma* **191**: 178–190.

Cronshaw, J. (1965) The organization of cytoplasmic components during the phase of cell wall thickening in differentiating cambial derivatives of *Acer rubrum*. *Can. J. Bot.* **43**: 1401–1407.

Desch, H.E. and Dinwoodie, J.M. (1996) *Timber Structure, Properties, Conversion and Use*. Macmillan Press Ltd, Basingstoke and London.

El Bassam, N. (1996) *Renewable Energy: Potential Energy Crops For Europe and The Mediterranean Region*. Food and Agricultural Organization of the United Nations, Rome.

Funada, R., Abe, H., Furusawa, O., Imaizumi, H., Fukuzawa, K. and Ohtani, J. (1997) The orientation and localization of cortical microtubules in differentiating conifer tracheids during cell expansion. *Plant Cell Physiol.* **38**: 210–212.

Furusawa, O., Funada, R., Murakami, Y. and Ohtani, J. (1998) The arrangement of cortical microtubules in compression wood tracheids of *Taxus cuspidata* visualized by confocal laser microscopy. *J. Wood Sci.* **44**: 230–233.

Giddings, T.H., Jr and Staehelin, L.A. (1991) Microtubule-mediated control of microfibril deposition: a re-examination of the hypothesis. In: *The Cytoskeletal Basis of Plant Growth and Form* (ed. C.W. Lloyd). Academic Press, London, pp. 85–99.

Green, P.B. (1969) Cell morphogenesis. *Annu. Rev. Plant Physiol.* **20**: 365–394.

Hepler P.K. and Gunning, B.E.S. (1998) Confocal fluorescence microscopy of plant cells. *Protoplasma* **201**: 121–157.

Jackson, W.T. (1982) Actomyosin. In: *The Cytoskeleton in Plant Growth and Development* (ed. C.W. Lloyd). Academic Press, London, pp. 3–29.

Kropf, D.L., Bisgrove, S.R. and Hable, W.E. (1998) Cytoskeletal control of polar growth in plant cells. *Curr. Opin. Cell Biol.* **10**: 117–122.

Ledbetter, M.C. and Porter, K.R. (1963) A 'microtubule' in plant cell fine structure. *J. Cell Biol.* **19**: 239–250.

Lloyd, C.W. (1987) The plant cytoskeleton: the impact of fluorescence microscopy. *Annu. Rev. Plant Physiol.* **38**: 119–136.

Prodhan, A.K.M.A., Funada, R., Ohtani, J., Abe, H. and Fukuzawa, K. (1995) Orientation of microfibrils and microtubules in developing tension-wood fibres of Japanese ash *(Fraxinus mandshurica* var. *japonica*). *Planta* **196**: 577–585.

Regan, S., Chaffey, N.J. and Sundberg, B. (1999) Exploring cambial growth with *Arabidopsis* and *Populus*. *J. Exp. Bot.* **50** (Suppl.): 33.

Robards, A.W. and Humpherson, P.G. (1967) Microtubules and angiosperm bordered pit formation. *Planta* **77**: 233–238.

Seagull, R.W. (1989) The plant cytoskeleton. *Crit. Rev. Plant Sci.* **8**: 131–167.

Sterky, F. *et al.* (1998) Gene discovery in the wood-forming tissues of poplar: analysis of 5,692 expressed sequence tags. *Proc. Natl Acad. Sci., USA* **95**: 13 330–13 335.

Uehara, K. and Hogetsu, T. (1993) Arrangement of cortical microtubules during formation of bordered pits in the tracheids of *Taxus*. *Protoplasma* **172**: 145–153.

Uribe, X., Torres, M.A., Capellades, M., Puigdomènech and Rigau, J. (1998) Maize α-tubulin genes are expressed according to specific patterns of cell differentiation. *Plant Molec. Biol.* **37**: 1069–1078.

Wenham, M.W. and Cusick, F. (1975) The growth of secondary wood fibres. *New Phytol.* **74**: 247–261.

Włoch, W. and Polap, E. (1994) The intrusive growth of initial cells in re-arrangement of cells in cambium of *Tilia cordata* Mill. *Acta Soc. Bot. Polon.* **63**: 109–116.

3

Ethylene, oxygen and carbon dioxide in woody stems during growth and dormancy

L. Eklund and A. Klintborg

1. Introduction

Higher oxygen and lower carbon dioxide levels are found in the wood during stem growth compared to levels found during dormancy in a number of woody species (Carrodus, 1971; Carrodus and Triffett, 1975; Eklund, 1990, 1993a, 2000; Eklund *et al.*, 1992, 1998; Hook *et al.*, 1972). The metabolism during growth that produces carbon dioxide and consumes oxygen and the lack of, or reduced, metabolism during dormancy may be the explanation. The lack of lateral gas transport through the cambial region and bark from the wood to the atmosphere or vice versa may also help to explain the altered levels of carbon dioxide and oxygen (Carrodus and Triffett, 1975; Eklund, 2000; Eklund and Lavigne, 1996; Hook and Brown, 1972; Hook *et al.*, 1972).

The level of the gaseous plant hormone ethylene is higher in the wood during growth compared to during dormancy when the level is low or undetectable. There is a direct relationship established between a high rate of growth, especially cambial growth, and ethylene levels measured in the sapwood and in the cambial region (Eklund, 1990, 1991a, 1993b; Eklund and Little, 1995; Eklund and Tiltu, 1999; Eklund *et al.*, 1992; Ingemarsson *et al.*, 1991a; Yamanaka, 1986). However, if the elevated ethylene level is the regulator of stem radial growth or just an effect of growth is still controversial.

In this chapter methods for measuring the levels of carbon dioxide, oxygen and ethylene, the regulation of synthesis or uptake of the gases and the physiological processes regulated by them will be discussed.

Cell & Molecular Biology of Wood Formation, edited by R. Savidge, J. Barnett and R. Napier.

2. Methods for sampling, identification and quantification of carbon dioxide, oxygen and ethylene

2.1 *Sampling methods*

One of the problems with analysing gases is their mobility, and thus it is more or less impossible to go out and collect a piece of wood, bring it back to the laboratory and start to extract the gas you want to measure. However, Carrodus (1971) essentially followed this procedure when he collected samples in the field and extracted carbon dioxide and oxygen back on the laboratory bench. He brought a piece of stem into the laboratory and through a series of steps made the wood piece smaller and smaller until the final 1 × 3 cm piece was placed in a sealed vial and the gases were allowed to evolve into the air in the vial. Carrudos (1971) assumed that only the outer portion of the wood exchanged gas with the surrounding air. Thus, taking that part of wood away the remaining piece of wood contained the original gas composition from the tree for at least a limited period of time. It is hard to evaluate the technique used by Carrodus (1971), as he did not try to verify the levels he found with some other technique. What was really a breakthrough with Carrodus (1971) work was the final identification and quantification of carbon dioxide and oxygen with the GC-MS technique. This brought the amount of gas necessary for quantification and identification down to μl-levels as compared to older methods requiring volumes in the range of litres of gas. It is easy to imagine the problems with homogeneity of the samples in the latter case. In an attempt to obtain on-line systems for sampling of small gas volumes from intact trees, two different sampling methods have been developed. In both cases the final identification and quantification is made with μl samples using a bench-top GC-MS.

Microdialysis. Within the cambial region the concentration of oxygen and ethylene of few-year-old seedlings (*Picea abies* (L.) Karst, *Pinus sylvestris* L.) and small twigs (*Pinus strobus* L.) as well as endogenous ethylene in apple fruits (*Malus domestica* L.) were collected by using microdialysis (Eklund, 1991a, 1993b; Eklund and Collin, 1991; Eklund and Lavigne, 1996; Eklund *et al.*, 1998). The microdialysis method was originally developed by a medical company, Carnegie medicin, Solna, Sweden (today: CMA Microdialysis AB, Stockholm, Sweden) and mostly used for sampling signal substances from rat brains, blood or other organs (Ungerstedt 1984). The dialysis membrane was placed at the end of a sewing needle-sized, eventually even smaller, probe (*Figure 1a*) that was inserted into the target organ. For sampling gas levels in the cambial region, the probe was inserted by hand to the wood/bark interface. The membrane that covered the probe tip had a length of 4 mm and a diameter of 1 mm. The inlet to the probe was connected to a pump enabling continuous perfusion of the probe with deionized water at a rate of 1–2 μl min^{-1}. The transfer rate of the gas of interest across the membrane was determined by measuring the decrease in concentration of gas dissolved in the perfusion liquid before and after passing the dialysis membrane (Eklund, 1991a; 1993b) and found to be around 25–30% depending on which gas was applied. A 100–500 ml aliquot of dialysate containing dissolved gases obtained from the cambial region was collected in a gas-tight vial from which a head-space sample was drawn for identification and quantification with GC-MS.

Gas-pocket-method. For sampling gases from big trees (*Picea abies* (L.) Karst, *Pinus sylvestris* L., *Quercus robur* L., *Acer platanoides* L.) a method has been used which is

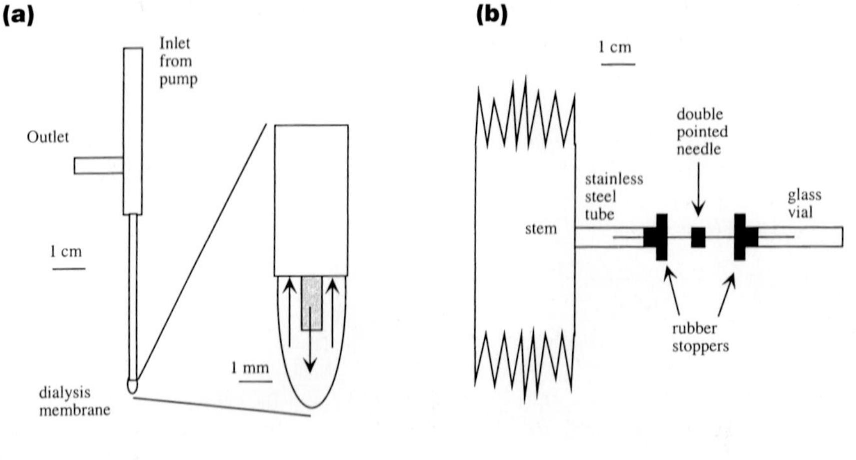

Figure 1. *Simple drawing showing the main characteristics of the (a) microdialysis probe and the (b) gas-pocket-method.*

described in detail in other works (Eklund, 1990, 1993a, 2000; Eklund and Tiltu, 1999; Eklund *et al.*, 1992; Ingemarsson *et al.*, 1991a). Normally, 10 mm wide holes were drilled through the bark and to a position 10 mm inside the cambium. Acid resistant stainless steel tubes were inserted into the holes and fixed in position leaving 1 cm^2 wood exposed to the gas in the tube. If the drilling and insertion of the steel tubes were made before the growing season, the effect of wounding was reduced and detected as increased ethylene compared to drilling made at the onset of growth.

The area around the tubes at the point of insertion was sealed with silicon grease to avoid leakage, and a gas tight rubber stopper was inserted into the end of each tube. The gas volume in the sealed tube was 10 ml. To each steel tube a glass vial sealed with a gas tight rubber stopper was connected through a double-pointed 20G needle that penetrated the rubber stopper in the steel tube and the vial (*Figure 1b*) thus, connecting the air in the sealed vial to the stem interior. On each sampling occasion, the glass vials were collected and new vials were connected. Before the first sampling occasion, the glass vials were left connected for 2 weeks and thereafter, between the sampling occasions, usually 2–3 weeks. A sample was then drawn from the vial with a syringe for identification and quantification.

The methods used for sampling gases from the stem interior have both advantages and disadvantages. The two obvious advantages are that firstly, the sampling can go on for a prolonged period of time in the same location inside the one tree and secondly, the sampled tree is intact. However, the previous statements are not perfectly true. When the tree grows the sampling location may be relocated by the newly formed cells. The more specific the sampling location, the greater the problem; sapwood is likely stable, while cambial zone may be a problem. The intactness of the tree can also be questioned. To avoid wound responses, inserting of the collection device has, when possible, been done during non-growing periods in order to allow the tree to heal the wound long before measurements start in the growing season. Fungal and/or bacterial infections in the wounded regions are also a potential problem, usually in parallel to the wounding effect. However, when the plant has healed its wound the likelihood of a fungal and/or bacterial impact on the data collected is less likely — although cannot be excluded. When insertion of the sampling devices have been done into actively growing plants, the problem of wound responses and microbial infection has to be considered thoroughly.

2.2 *Identification and quantification methods*

Carbon dioxide and oxygen. The concentrations of oxygen and carbon dioxide can be identified and quantified in several ways. In many cases there is a demand for portability, e.g. when measuring stem respiration in the field. Normally the technique involves use of portable infrared analysers. For very small samples, however, the use of a mass detector coupled to a gas chromatograph is an excellent tool enabling the injection of 1–5 μl gas samples. The gases in the atmosphere can be identified and quantified by focusing on the molecular ion of each gas, that is N_2+ (mw. = 28), O_2+ (mw. = 32) and Ar^+ (mw. = 40). For calculation of the concentrations, the concentrations of the ions in air samples can be used as a standard. Dilution of air with either Ar or N_2 showed that the mass detector response is linear between 4 and 79% N_2, 1 and 21% O_2 and 0.1 and 1% Ar. The method is fast; air passes through the GC column in less than a minute and the separation is made only on the mass detector.

Ethylene. The final identification and quantification of ethylene is readily made by gas chromatography using a flame ionization detector (FID). The use of FID detection for ethylene has been the standard since the technique became available in the 1960s and it is fast and accurate. The retention time can be less than a minute, which allows the possibility of 100s of daily measurements. Although use of a FID does not identify the molecule the use of mass detectors has not been common for ethylene measurement, because there is a limited risk to find another molecule co-chromatographing with ethylene. The FID responds to ethylene with a high degree of linearity. In my laboratory we have checked for, and found linearity from 10^{-1} to 10^3 ppm concentrations. By manipulating the incubation times, amount of tissue and vial volume, there is normally no problems over this range.

3. The source of carbon dioxide, oxygen and ethylene

3.1 *Carbon dioxide and oxygen*

The levels of oxygen and carbon dioxide in the sapwood are close to ambient, i.e. 20.8% and 0.035%, before growth starts in spring. However, as cambial growth commences carbon dioxide levels increase to 5–10% and oxygen levels decrease to 1–5%. When cambial growth slows and ceases, oxygen levels start to increase and carbon dioxide levels decrease, in both cases reaching spring time values in October (Eklund, 1990, 1993a, 1999; Eklund *et al.*, 1992). Such data do not leave much doubt that the decrease in oxygen and the increase in carbon dioxide levels depend on the respiration going on inside the cambium (Sprugel *et al.*, 1995). The main respiratory processes may be the differentiation of cambial derivatives into xylem and ray cells, heartwood formation and storage processes. However, there are other oxygen consuming processes as well, for example lignification, but the majority of oxygen consumption and carbon dioxide production comes from respiration. The question is, why is there such a dramatic build up of carbon dioxide and a decrease in oxygen? Two possible explanations are that the capacity to remove gases or the capacity for uptake of gases is reduced. The increase in carbon dioxide indicates a reduced capacity for removal while the low oxygen levels indicate a reduced capacity for uptake.

Hook and Brown (1972) found high resistance to lateral air diffusion through stems among species having no intercellular spaces in the cambium. On the other hand, lower resistance to lateral gas transfer was found among hydrophytic species with prominent intercellular spaces between cambial cells. Thus, in plants lacking intercellular spaces in the cambium lateral gas movement must take place with the gases dissolved in water. Gas diffusion in water is much slower, 3–4 orders of magnitude, than in air indicating an inhibition of lateral gas transport to and away from the xylem. Instead of being supplied laterally from the atmosphere, it has been hypothesized that the bulk of gas is supplied to, and removed from, the differentiating xylem through the transpiration stream (Eklund, 1990; Hook *et al.*, 1972).

In an experiment using cut *Pinus strobus* twigs, Eklund and Lavigne (1996) treated the cuttings basally and laterally with argon gas and measured its transport both in the transpiration stream and laterally through the cambium. The results showed clearly that gases are translocated acropetally in the transpiration stream but not laterally through the cambium from the xylem to the atmosphere or vice versa. The living portion of the xylem must therefore rely on the transpiration stream for gas-exchange. In the sapwood, 2–3 annual rings inwards from the cambium, the most rapid water transport takes place (Cermák *et al.*, 1991) and is assumed to provide oxygen to, and remove carbon dioxide from the growing cambial region and the differentiating xylem. The levels of internal oxygen in the outer sapwood in *Picea abies* trees decrease with decreasing amount of available water to the trees, indicating that increased drought lowers the amount of oxygen present in the sapwood (Eklund, 1999; Eklund *et al.*, 1992). Another indirect evidence that the transpiration stream is the main route for gas transfer to the interior of stems comes from data indicating that the sampling position on the tree determines the sapwood levels of oxygen. In droughted trees there is no statistically significant difference between the oxygen levels measured near or far from ground when the oxygen levels are the lowest, that is, in July and August (Eklund, 2000). A possible explanation for this may be that there are a large number of gas spaces available in the droughted tree (Hook and Brown, 1972; Zimmermann, 1981) thus facilitating the diffusion of oxygen vertically in the stem. However, when oxygen levels start to recover, the recovery is faster in the lower part of the stem in the droughted trees (Eklund, 2000). This may be explained by the fact that if oxygen is delivered to the stem in the transpiration stream, a decrease in oxygen demand should first be detected at the base. In the same experiment (Eklund, 2000), it was found that there was a statistically significant difference in oxygen levels in the sapwood between the lower and upper parts of the stem in irrigated trees. These results may be explained by the fact that a limited amount of oxygen is passing upwards in the stem and being consumed on the way. The latter results thus provide indirect evidence for a soil water dependent oxygen supply to the stem interior in trees.

3.2 *Ethylene*

Tree stems produce ethylene and the rate of production can be changed by stresses such as bending, shaking and flooding (Blake *et al.*, 1980; Brown and Leopold, 1973; Larson *et al.*, 1993; Robitaille and Leopold, 1974; Telewski 1990; Yamamoto and Kozlowski, 1987a, 1987b), and by application of the ethylene-generating compound Ethrel (2-chloroethylphosphonic acid) (Eklund and Little, 1995, 1996; Robitaille and Leopold, 1974; Telewski et al., 1983; Yamamoto and Kozlowski, 1987a, 1987b; Yamanaka, 1985). However, which parts of the shoot evolve ethylene is so far poorly investigated. We know that stems of *Abies balsamea* (L.) Mill. evolve more ethylene than needles do (Eklund and Little, 1995). Savidge (1988) reported that cambial tissue dissected from *Abies balsamea* stems evolved ethylene when provided with IAA and ACC, and Eklund (1991a, 1993b) demonstrated by microdialysis that ethylene occurred in the cambial region of nondissected *Picea abies* stems. The outer sapwood of *Picea abies*, *Quercus robur* and *Acer platanoides* stems evolved ethylene (Eklund, 1990, 1993a; Eklund and Tiltu, 1999; Eklund *et al.*, 1992) and ethylene evolution was greater from sapwood than heartwood in *Pinus sylvestris* (L.) stems (Ingemarsson *et al.*, 1991). In dissection experiments, ethylene evolution from *Chamaecyparis obtusa* and *Abies balsamea* stems was greater from inner cambial region than from outer phloem and cortex or xylem (Eklund and Little, 1998; Little and Eklund, 1999; Yamanaka, 1985), and it decreased inwards from outer sapwood to heartwood (Yamanaka, 1985). Similarly, ethylene evolution in *Pinus radiata* stems was highest from combined outer sapwood and cambial region and lowest from the heartwood. The highest evolution was found from the transition zone separating sapwood and heartwood (Shain and Hillis, 1972). If the shoot is separated into fractions, that is wounded, wound ethylene production is stimulated. Timing and amount of wound ethylene production vary with species, organ, tissue and severity of wounding (Abeles *et al.*, 1992; Ingemarsson, 1995; Telewski and Jaffe, 1986; Yamanaka, 1985). The basal rate of ethylene production and the production of wound ethylene vary among fractions. In *Abies balsamea* the basal rate was highest from the cambial region, followed by the combined cortex/periderm, wood and the needles. The basal rate was maintained for at least 4 h after excision but after 24 h the production of wound ethylene was high. Wound ethylene production was also delayed for several hours in experiments with nondissected stem segments of *Pinus taeda* (Telewski and Jaffe, 1986) and *Picea abies* (Ingemarsson, 1995), stem fractions of *Chamaecyparis obtusa* (Yamanaka, 1985), and a cambial region fraction excised from the stem of various conifers (Yamanaka, 1986). However, the capacity of the shoot fractions to evolve ethylene was similar before and after the production of wound ethylene started. Whether the basal rate or wound ethylene were measured, the cambial region was always the main site of production. The reason for the higher rate of ethylene evolution from the cambial region than from any other shoot fraction is not yet known. In *Pinus sylvestris* the major source for ethylene is the cambial region. In the same region was also found the highest activity of ACC-oxidase, the enzyme responsible for the conversion of the ethylene precursor 1-aminocyclopropane-1-carboxylic acid (ACC) to ethylene, as well as the highest activity of ACC-synthase, the enzyme responsible for the production of ACC (Tiltu *et al.*, unpublished manuscript).

4. Physiological effects of carbon dioxide, oxygen and ethylene

4.1 *Wood formation*

Ethylene plays a role in the regulation of xylem production in tree stems. Application of Ethrel increased tracheid production in *Pinus* spp. (Barker, 1979; Telewski and Jaffe, 1986; Yamamoto and Kozlowski, 1987a, 1987b), and xylem increment in woody angiosperms (Phelps *et al.*, 1980; Robitaille and Leopold, 1974; Yamamoto and Kozlowski, 1987c; Yamamoto *et al.*, 1987). Ethrel also altered the cell wall composition of *Picea abies* hypocotyls, increasing the content of cellulose and decreasing that of non-cellulosic polysaccharides (Eklund, 1991b; Ingemarsson *et al.*, 1991b) and promoted xylem differentiation in *Lactuca* pith explants (Miller *et al.*, 1984, 1985). Reinders-Gouwentak (1965) reported that cambial rest in *Fraxinus* shoots was overcome by ethylene chlorohydrin treatment. Ethylene evolution from the stem is higher during wood formation than when the cambium is dormant (Eklund, 1990, 1991a, 1993b; Eklund and Tiltu, 1999; Eklund *et al.*, 1992; Ingemarsson *et al.*, 1991b; Yamanaka, 1985). Mechanical perturbation increases both ethylene evolution and tracheid production in *Pinus taeda* (Telewski, 1990). In addition, there is evidence that cambial region cells synthesize ACC in *Pinus sylvestris* (Tiltu *et al.*, unpublished manuscript) and converts ACC to ethylene in *Abies balsamea* and *Pinus sylvestris* (Savidge, 1988; Tiltu *et al.*, unpublished manuscript).

Shoot radial growth depends on the import of carbohydrates for new cell synthesis and respiratory energy (Kozlowski, 1992). Ethylene treatment of *Abies balsamea* shoots increased the levels of glucose, fructose, sucrose and starch in the cambial region (Eklund and Little, 1998). This finding indicates that one possible mechanism for ethylene promoted wood formation is inceasing the sink activity in the cambium, which is commonly observed for fruits and other tissues (Abeles *et al.*, 1992). In the cambial region of *Pinus sylvestris,* starch synthase and sucrose synthase activities are increased and amylase and acid invertase activities decreased after ethylene treatment (Tiltu and Eklund, unpublished manuscript) indicating an important role for ethylene in the control of starch and sucrose levels in the cambial region (*Figure 2*).

4.2 *Early wood–late wood transition*

For a period after spring resumption of cambial cell-division, the cambial derivatives enlarge greatly in radial dimension to form a zone of tracheids. These wide-diameter early wood tracheids mature in spring and early summer, and tracheids having smaller diameters and thicker secondary walls, late wood tracheids, are produced later (Denne, 1989). This transition between the production of the two types of tracheids is gradual in some species and abrupt in others, and is accompanied or preceded by a reduced rate of cambial division (Creber and Chaloner, 1990; Savidge, 1993; Timell, 1986).

One possible explanation for the transition from early wood to late wood is that there is a decrease in the cambial region concentration of the growth promoter indole-3-acetic acid (IAA). Evidence indicates that IAA promotes both the production of tracheids and the enlargement of primary-walled cells (Cleland, 1995; Little and Pharis, 1995). There are conflicting reports as to whether the cambial region IAA concentration decreases during latewood formation. A decline in concentration was

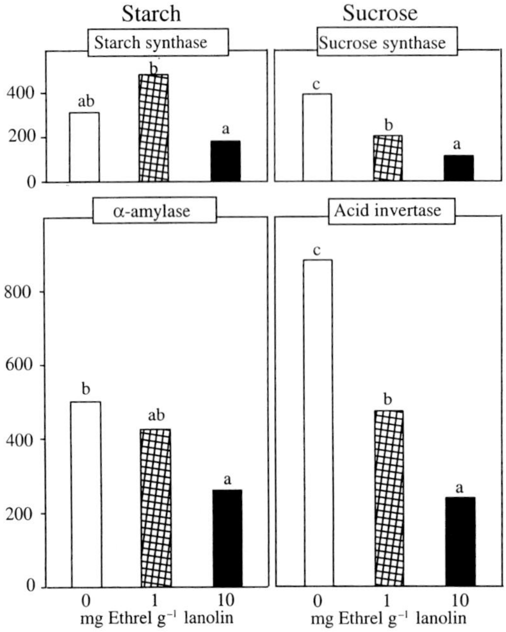

Figure 2. *The activities of starch synthase, sucrose synthase, acid invertase and amylase after ethylene treatment. Different letters indicate a statistically significant difference at $P \leq 0.05$ between the activities induced by the ethylene treatment.*

observed in *Pinus contorta* (Savidge *et al.*, 1982; Savidge and Wareing, 1984) and *Abies balsamea* (Sundberg *et al.*, 1987), but not in *Picea sitchensis* (Little and Wareing, 1981) and *Pinus sylvestris* (Sundberg *et al.*, 1991, 1993). It is possible that the onset of latewood formation is due to growth and maintenance respiration (Amthor, 1994; Sprugel *et al.*, 1995) decreasing the oxygen concentration in the cambial region to the extent that it becomes limiting for IAA action. Evidence for this possibility is that the oxygen concentration in the sapwood in *Picea abies* stems reaches a minimum well below 5% late during growth (Eklund, 1990; Eklund *et al.*, 1992). Such low oxygen concentrations inhibit IAA-induced proton secretion (Böttger and Hilgendorf, 1988; Evans and Vesper, 1980) and thus cell enlargement. However, so far such a mechanism has not been shown.

4.3 *Heartwood formation*

Heartwood is defined as 'the inner layers of wood which have ceased to contain living cells and in which the reserve materials have been removed or converted into heartwood substances', as defined by the IAWA (Anonymous, 1957). The inner part of a log, heartwood, is commonly darker than the outer part due to the incorporation of polyphenols into the xylem tissue. Heartwood formation occurs in a band that encircles the heartwood, an area called the transition zone.

Synthesis of heartwood compounds occurs in ray parenchyma cells or, in some species in axial parenchyma (Bamber and Fukazawa, 1985; Climent, 1997). Mitochondrial metabolism ceases close to the heartwood–sapwood boundary (Frey-Wyssling and Bosshard, 1959) as a consequence of the low oxygen content in the trunk (Carrodus and Triffett, 1975; Eklund, 1990, 1999). The lack of oxygen and high carbon dioxide concentration during the growing season (Eklund, 1990; Eklund *et al.*,

1992, 1998) promote the metabolism of less oxygen demanding reactions (Higuchi, 1976; Hillis, 1968). Such reactions are the pentosephosphate pathway resulting in cinnamic acid and finally different polyphenols, the synthesis of flavonoids from malonyl-CoA and 4-coumaryl-CoA (Carrodus, 1971) or the acetoacetyl pathway to isoprenoid substances. All these substances occur in the transition zone (Burtin *et al.*, 1998) and are components of heartwood. In order to test the hypothesis that carbon dioxide, oxygen or ethylene regulates heartwood formation, holes were drilled through the trunk at breast height and stainless steel tubes were inserted on each side into the holes. The end of the steel tube poking out from the stem was sealed with a rubber stopper so that a specific environment, depending on the treatment, was created in the hole drilled through the stem. Once a week from April to September gas, containing 2.5% ethylene, 99.5% carbon dioxide or 99.5% nitrogen, was injected into the holes in the stems. The results obtained in this experiment give strong evidence for ethylene being the major regulator of heartwood formation (*Figure 3*), thus confirming other indirect findings that ethylene is responsible for heartwood formation (Hillis, 1987; Shain and Hillis, 1973). In an experiment using walnut (*Juglans nigra*) Nelson and co-workers (1981) measured ethylene production in inner, middle and outer sapwood and found that more ethylene was produced in the outer than in the inner sapwood. These findings may reflect the cambial zone as the major region for ethylene synthesis and induction of heartwood formation by ethylene regulated from the cambial region. A problem is that Nelson and co-workers measured ethylene evolution from excised sapwood samples after a 20 h incubation period. Previous reports (Eklund and Little, 1998; Yamanaka, 1995) show that about 20–30 h after excision there is a peak in wound ethylene production; thus, the data of Nelson and co-workers do not necessarily represent the basal level of ethylene produced in the sapwood. So far we do not have access to data for the basal ethylene production in different parts of the sapwood — more research is needed.

The experiment also indicated that carbon dioxide and nitrogen treatment to create anoxia induce an increase in heartwood formation. For both gases the increase was similar, indicating that anoxia was the agent and not carbon dioxide *per se* as both the nitrogen and carbon dioxide treatments reduced the oxygen content to almost zero (Eklund, Wikman and Nilsson, unpublished manuscript).

4.4 *Compression wood*

Compression wood formation and tracheid production are promoted on the lower side, and not on the upper side of branches and tilted stems (Little and Eklund, 1999, Little *et al.*, 1990; Sundberg *et al.*, 1994; Timell, 1986; Yamaguchi *et al.*, 1980). Strong support for a causal relationship between ethylene and compression wood formation is increased ethylene evolution by the cambial region forming the compression wood (Little and Eklund, 1999). Other evidence for ethylene regulation of compression wood formation is that ethylene evolution from the lower half of *Cupressus arizonica* branches and tilted stems of *Abies balsamea* was greater than from the upper half of the stems (Blake *et al.*, 1980, Little and Eklund, 1999). In *Abies fraseri* and *Pinus taeda* stems mechanical shaking promoted both the emission of ethylene and the production of flexure wood, the latter having several characteristics similar to compression wood (Telewski, 1989, 1990; Telewski and Jaffe, 1986). Cambial region ACC was found only on the lower side of *Pinus contorta* branches (Savidge *et al.*, 1983). In contrast,

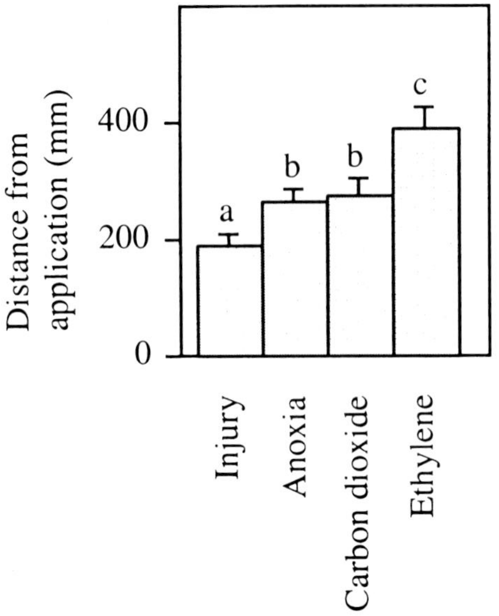

Figure 3. *The extension in acropetal direction of increased heartwood formation. Different letters indicate a statistically significant difference at $P \leq 0.05$ between the degrees of extension induced by the different treatments.*

Yamamoto and Kozlowski (1987a) found no difference in ethylene evolution from the upper and lower halves of tilted *Pinus densiflora* stems. However, those investigators performed their ethylene measurements after a 20 h incubation period, a period likely to induce production of stress ethylene (Eklund and Little, 1998; Morgan and Drew, 1997). Most information obtained so far indicates that ethylene is responsible for the regulation of compression wood formation.

5. Conclusions and perspectives for the future

Carbon dioxide and oxygen levels are different in growing and dormant tree stems. The reasons may be production and consumption by respiration, a gas tight cambial region not allowing lateral transport, and an inefficient vertical transport system. Physiological effects of the altered gas composition may be formation of heartwood components and thus the regulation of heartwood formation. There is also a possibility that the regulation of latewood formation is under control of altered gas composition although the results are both for and against. In the future we have to focus on accumulating further information about the spatial and temporal distribution of oxygen and carbon dioxide inside tree stems. Further, we have to identify processes in the formation of latewood, the programmed death of tracheids and heartwood formation that can possibly be enhanced or even induced by low oxygen levels or high carbon dioxide levels.

Ethylene is involved in the regulation of cambial activity, but we still do not know in detail how. What we know is that ethylene interacts with IAA, and it seems likely that it promotes accumulation of IAA to regions where ethylene levels are elevated. Ethylene also increases carbohydrate levels, especially sucrose and starch, in regions where ethylene levels are elevated. The activities of starch and sucrose synthesizing enzymes go up while activities of starch and sucrose degrading enzymes go down, thus explaining the increased levels of sucrose and starch. An exciting hypothesis is that a general effect of ethylene is to induce and/or accelerate the accumulation of (i) compounds necessary for regulation (IAA) and (ii) building blocks (carbohydrates) for cell growth. Data accumulated recently and still being processed does not counteract the hypothesis. Concerning qualitative regulation by ethylene, that is, compression wood and heartwood formation, another mechanism has to be found. Still we are in the dark in that case.

References

Abeles, F.B., Morgan, P.W. and Saltveit M.E. (1992) *Ethylene in Plant Biology*, 2nd edition. Academic Press, San Diego, California.

Amthor, J.S. (1994) Plant respiratory responses to the environment and their effects on the carbon balance. In: *Plant–Environment Interactions* (ed. Wilkinson, R.E.), Marcel Dekker, New York, pp. 501–554.

Anonymous (1957) International glossary of terms used in wood anatomy, prepared by the International Association of Wood Anatomists. *Trop. Woods* **107**: 1–36.

Bamber, R.K. and Fukazawa, K. (1985) Sapwood and heartwood, a review. *For. Prod. Abstr.* **8**: 265–278.

Barker, J.E. (1979) Growth and wood properties of *Pinus radiata* in relation to applied ethylene. *New Zealand J. For. Sci.* **9**: 15–19.

Blake, T.J., Pharis, R.P. and Reid, D.M. (1980) Ethylene, gibberellins, auxin and the apical control of branch angle in a conifer, *Cupressus arizonica*. *Planta* **148**: 64–68

Brown, K.M. and Leopold, A.C. (1973) Ethylene and the regulation of growth in pine. *Can. J. For. Res.* **3**: 143–145.

Burtin, P., Jay-Allemand, C., Charpentier, J.P. and Janin, G. (1998) Natural wood colouring process in *Juglans* sp. (*J. nigra*, *J. regia* and hybrid *J. nigra 23 x J. regia*) depends on native phenolic compounds accumulated in the transition zone between sapwood and heartwood. *Trees* **12**: 258–264.

Böttger, M. and Hilgendorf, F. (1988) Hormone action on trans membrane electron and H^+ transport. *Plant Physiol.* **86**: 1038–1043.

Carrodus, B.B. (1971) Carbon dioxide and the formation of heartwood. *New Phytol.* **70**: 939–943.

Carrodus, B.B. and Triffet, A.C.K. (1975) Analysis of respiratory gases in woody stems by mass spectrometry. *New Phytol.* **74**: 243–246.

Cermák, J., Cienciala, E., Kucera, J. and Hällgren, J.-E. (1991) Radial velocity profiles of water flow in trunks of spruce and oak and response of spruce tree to cut-off. *Tree Physiol.* **8**: 373–379.

Cleland, R.E. (1995) Auxin and cell elongation. In: *Plant Hormones: Physiology, Biochemistry and Molecular Biology* (Davies P.J., ed.), Dordrecht, The Netherlands: Kluwer Academic Publishers, pp. 214–227.

Climent, J., Gil, L. and Pardos, J.A. (1998) Xylem anatomical traits related to resinous heartwood formation in *Pinus canariensis* Sm. *Trees* **12**: 139–145.

Creber, G.T. and Chaloner, W.G. (1990) Environmental influences on cambial activity. In: *The Vascular Cambium* (Iqbal M., ed.), New York: John Wiley & Sons, pp. 159–199.

Denne, M.P. (1989) Definition of latewood according to Mork (1928). *IAWA Bull.* **10**: 59–62.

Eklund, L. (1990) Endogenous levels of oxygen, carbon dioxide and ethylene in stems of Norway spruce trees during one growing season. *Trees* **4**: 150–154.

Eklund, L. (1991a) Hormone levels in the cambial region of intact *Picea abies* during the onset of cambial activity. *Physiol. Plant.* **82**: 385–388.

Eklund, L. (1991b) Relations between indoleacetic acid, calcium ions and ethylene in the regulation of growth and cell wall composition in *Picea abies*. *J. Exp. Bot.* **42**: 785–789.

Eklund, L. (1993a) Movement and possible metabolism of ethylene in dormant *Picea abies*. *Plant Growth Regul.* **12**: 37–41.

Eklund, L. (1993b) Seasonal variation of O_2, CO_2 and ethylene in oak and maple stems. *Can. J. For. Res.* **23**: 2608–2610.

Eklund, L. (2000) Internal oxygen levels decrease during the growing season and with increasing stem height. *Trees*, 2(14): 177–180.

Eklund, L. and Collin, A.-K. (1991) Microdialysis, a new tool for sampling and manipulating internal ethylene concentration in apples. *J. Plant Physiol.* **137**: 155–157.

Eklund, L. and Lavigne, M.B. (1996) Restricted lateral gas movement in *Pinus strobus* branches. *Trees* **10**: 83–85.

Eklund, L. and Little, C.H.A. (1995) Interaction between indole-3-acetic acid and ethylene in the control of tracheid production in detached shoots of *Abies balsamea*. *Tree Physiol.* **15**: 27–34.

Eklund, L. and Little C.H.A. (1996) Laterally applied Ethrel causes local increases in radial growth and indole-3-acetic acid concentration in *Abies balsamea* shoots. *Tree Physiol.* **16**: 509–513.

Eklund, L. and Little, C.H.A. (1998) Ethylene evolution, radial growth and carbohydrate concentrations in *Abies balsamea* shoots ringed with Ethrel. *Tree Physiol* **18**: 383–392.

Eklund, L. and Tiltu, A. (1999) Cambial activity in 'normal' spruce *Picea abies* karst (L.) and snake spruce *Picea abies* (L.) Karst f. *virgata* (Jacq.) Rehd. in response to ethylene *J. Exp. Bot.*, **50**: 1489–1493.

Eklund, L., Cienciala, E. and Hällgren J.-E. (1992) No relation between drought stress and ethylene production in Norway spruce. *Physiol. Plant.* **87**: 297–300.

Eklund, L., Little, C.H.A. and Riding, R. (1998) Auxin and oxygen control of late wood formation in *Picea abies*. *J. Exp. Bot.* **49**: 205–211.

Evans, M.L. and Vesper, M.J. (1980) An improved method for detecting auxin-induced hydrogen ion efflux from corn coleoptile segments. *Plant Physiol.* **66**: 561–565.

Frey-Wyssling, A. and Bosshard, H.H. (1959) Cytology of the ray cells in sapwood and heartwood. *Holzforschung* **13**: 129–137.

Gref, R. and Ståhl, E. (1994) Lightwood induction in *Pinus sylvestris* by means of mechanical wounding. *Scand. For. Res.* **9**: 382–385.

Higuchi, T. (1976) Biochemical aspects of lignification and heartwood formation. *Wood Res.* **59/60**: 180–199.

Hillis, W.E. (1968) Chemical aspects of heartwood formation. *Wood Sci. Tech.* **2**: 241–259.

Hillis, W.E. (1987) *Heartwood and Tree Exudates*. Springer: Berlin, Heidelberg, New York.

Hook, D.D. and Brown, C.L. (1972) Permeability of the cambium to air in trees adapted to wet habitats. *Bot. Gaz.* **133**: 304–310.

Hook, D.D., Brown, C.L. and Wetmore, R.H. (1972) Aeration in trees. *Bot. Gaz.* **133**: 443–454.

Ingemarsson, S.M. (1995) Ethylene effects on peroxidases and cell growth patterns in *Picea abies* hypocotyl cuttings. *Physiol. Plant.* **94**: 211–218.

Ingemarsson, B.S.M., Eklund, L. and Eliasson, L. (1991a) Ethylene effects on cambial activity and cell wall formation in hypocotyls of *Picea abies* seedlings. *Physiol. Plant.* **82**: 219–224.

Ingermarsson, S.M., Lundquist, E. and Eliasson, L. (1991b) Seasonal variations in ethylene concentrations in the wood of *Pinus sylvestris* L. *Tree Physiol.* **8**: 273–279.

Kozlowski, T.T. (1992) Carbohydrate sources and sinks in woody plants. *Bot. Rev.* **58**: 107–222.

Larson, K.D., Schaffer, B. and Davies, F.S. (1993) Floodwater oxygen content, ethylene production and lenticel hypertrophy in flooded mango (*Mangifera indica* L.) trees. *J. Exp. Bot.* **44**: 665–671.

Little, C.H.A. and Eklund, L. (1999) Ethylene in relation to compression wood formation in *Abies balsamea* shoots. *Trees* **13**: 173–177.

Little, C.H.A. and Pharis, R.P. (1995) Hormonal control of radial and longitudinal growth in the tree stem. In: *Plant Stems: Physiology and Functional Morphology.* (Gartner B.L., ed.) San Diego: Academic Press, 281–319.

Little, C.H.A. and Wareing, P.F. (1981) Control of cambial activity and dormancy in *Picea sitchensis* by indol-3-y-lacetic and abscisic acids. *Can. J. Bot.* **59**: 1480–1493.

Little, C.H.A., Sundberg, B. and Ericsson, A. (1990) Induction of acropetal ^{14}C-photosynthate transport and radial growth by indole-3-acetic acid in *Pinus sylvestris* shoots. *Tree Physiol.* **6**: 177–189.

Miller, A.R., Crawford, D.L. and Roberts, L.W. (1985) Lignification and xylogenesis in *Lactuca* pith explants cultured in vitro in the presence of auxin and cytokinin: a role for endogenous ethylene. *J. Exp. Bot.* **36**: 110–118.

Miller, A.R., Pengelly, W.L. and Roberts, L.W. (1984) Introduction of xylem differentiation in *Lactuca* by ethylene. *Plant Physiol.* **75**: 1165–1166.

Morgan, P.W. and Drew, M.C. (1997) Ethylene and plant responses to stress. *Physiol. Plant.* **100**: 620–630.

Nelson, N.D., Rietveld, W.J. and Isebrands, J.G. (1981) Xylem ethylene production in five black walnut families in the early stages of heartwood formation. *For. Sci.* **27**: 537–543.

Phelps, J.E., McGinnes, E.A., Saniewski, J., Pieniazek, J. and Smolinski M. (1980) Some anatomical observations on the effect of morphactin IT 3456 and ethrel on wood formation in *Salix fragilis* L. *IAWA Bull.* **1**: 76–82.

Reinders-Gouwentak, C.A. (1965) Physiology of the cambium and other secondary meristems of the shoot. In: *Encyclopedia of Plant Physiology*. Vol. 15 (I). (ed. W. Ruhland). Springer Verlag: New York, pp 1077–1105.

Robitaille, H.A. and Leopold, A.C. (1974) Ethylene and the regulation of apple stem growth under stress. *Physiol. Plant.* **32**: 301–304.

Savidge, R.A. (1988) Auxin and ethylene regulation of diameter growth in trees. *Tree Physiol.* **4**: 401–414.

Savidge, R.A. (1993) Formation of annual rings in trees. In: *Oscillations and Morphogenesis*. (Rensing L, ed), New York: Marcel Dekker, pp. 343–363.

Savidge, R.A. and Wareing, P.F. (1984) Seasonal cambial activity and xylem development in *Pinus contorta* Dougl. in relation to endogenous indol-3-yl-acetic acid and (S)-abscisic acid levels. *Can. J. For. Res.* **14**: 676–682.

Savidge, R.A., Heald, J.K. and Wareing, P.F. (1982) Non-uniform distribution and seasonal variation of endogenous indol-3yl-acetic acid in the cambial region of *Pinus contorta* Dougl. *Planta* **155**: 89–92.

Savidge, R.A., Mutumba, G.M.C., Heald, J.K. and Wareing, P.F. (1983) Gas chromatography-mass spectroscopy identification of 1-aminocyclopropane-1-carboxylic acid in compressionwood vascular cambium of *Pinus contorta* Dougl. *Plant Physiol.* **71**: 434–436.

Shain, L. and Hillis, W.E. (1972) Ethylene production in xylem of *Pinus radiata* in relation to heartwood formation. *Can. J. Bot.* **51**: 1331–1335.

Sprugel, D.G., Ryan, M.G., Brooks, J.R., Vogt, K.A. and Martin, T.A. (1995) Respiration from the organ level to the stand. In: *Resource Physiology of Conifers: Acquisition, Allocation, and Utilization* (Smith WK, Hinkley TM, eds.), San Diego: Academic Press, pp. 255–299.

Sundberg, B., Tuominen, H. and Little, C.H.A. (1994) Effects of the indole-3-acetic acid (IAA) transport inhibitors N-1-naphthylphthalamic acid and morphactin on endogenous IAA dynamics in relation to compression wood formation in 1-year-old *Pinus sylvestris* (L.) shoots. *Plant Physiol.* **106**: 469–476.

Sundberg, B., Ericsson, A., Little, C.H.A., Näsholm, T. and Gref, R. (1993) The relationship between crown size and ring width in *Pinus sylvestris* L. stems: dependence on indole-3-acetic acid, carbohydrates and nitrogen in the cambial region. *Tree Physiol.* **12**: 347–362.

Sundberg, B., Little, C.H.A., Cui, K. and Sandberg, G. (1991) Level of endogenous indole-3-acetic acid in the stem of *Pinus sylvestris* in relation to the seasonal variation of cambial activity. *Plant Cell and Environ.* **14**: 241–246.

Sundberg, B., Little, C.H.A., Riding, R.T. and Sandberg, G. (1987) Levels of endogenous indol-3-acetic acid in the vascular cambium region of *Abies balsamea* trees during the activity-rest-quiescence transition. *Physiol. Plant.* **71**: 163–170.

Telewski, F.W. (1989) Structure and function of flexure wood in *Abies fraseri*. *Tree Physiol.* **5**: 113–121.

Telewski, F.W. (1990) Growth, wood density, and ethylene production in response to mechanical perturbation in *Pinus taeda*. *Can. J. For. Res.* **20**: 1277–1282.

Telewski, F.W. and Jaffe, M.J. (1986) Thigmomorphogenesis: the role of ethylene in the response of *Pinus taeda* and *Abies fraseri* to mechanical perturbation. *Physiol. Plant.* **66**: 227–233.

Telewski, F.W., Wakefield, A.H. and Jaffe, M.J. (1983) Computer-assisted image analysis of tissues of Ethrel-treated *Pinus taeda* seedlings. *Plant Physiol.* **72**: 177–181.

Timell, T.E. (1986) *Compression Wood in Gymnosperms, Vol. 2*. New York: Springer Verlag, pp. 1183–1262.

Ungerstedt, U. (1984) Measurement of neurotransmitter release by intercranial dialysis. In: *Measurement of Neurotransmittor Release in Vivo* (Marsden, C.A., ed.) New York: John Wiley, pp. 81–104.

Yamaguchi, K., Itoh, T. and Shimaji, K. (1980) Compression wood induced by 1-N-naphthylphthalamic acid (NPA), an IAA transport inhibitor. *Wood Sci. Technol.* **14**: 181–185.

Yamamoto, F. and Kozlowski, T.T. (1987a). Effects of flooding, tilting of stems, and Ethrel application on growth, stem anatomy and ethylene production of *Pinus densiflora* seedlings. *J. Exp. Bot.* **38**: 293–310.

Yamamoto, F. and Kozlowski, T.T. (1987b) Effect of Ethrel on growth and stem anatomy of *Pinus halepensis* seedlings. *IAWA Bull.* **8**: 11–19.

Yamamoto, F. and Kozlowski, T.T.. (1987c) Effects of flooding, tilting of stems, and ethrel application on growth, stem anatomy, and ethylene production of *Acer platanoides* seedlings. *Scand. J. For. Res.* **2**: 141–156.

Yamamoto, F., Angeles, G. and Kozlowski, T.T. (1987) Effect of ethrel on stem anatomy of *Ulmus americana* seedlings. *IAWA Bull.* **8**: 3–9.

Yamanaka, K. (1985) Ethylene production in *Chamaecyparis obtusa* phloem and xylem tissues in response to wounding. *Mokuzai Gakkaishi* **31**: 703–710.

Yamanaka, K. (1986) Wound ethylene production by phloem and cambium tissues in conifers. *Mokuzai Gakkaishi* **32**: 136–139.

Zimmermann, M.H. (1981) *Xylem Structure and the Ascent of Sap*. Springer: Berlin, Heidelberg, New York.

4

Cambial coniferin content as an indicator of the health status of conifers

V. J. Steeves and R.A. Savidge

1. Introduction

Much controversy surrounds the harvesting of old-growth forests around the world. Scientists, industry and environmental groups have yet to arrive at reliable criteria for determining the health status of trees, and hence stands and forests. Most methods for predicting forest health rely on macro features, such as the amount and efficiency of foliage (i.e., Waring, 1983; Blanche *et al.*, 1985), growth abnormalities such as departures from typical bole taper (e.g., Waring, 1985), radial growth and internode elongation (i.e., Kostka and Sherald, 1982; Lewis, 1995; Ducrey *et al.*, 1996), and foliar discoloration (i.e., Bernier *et al.*, 1989; Allen *et al.*, 1992; Kolb and McCormick, 1993; Wilmot *et al.*, 1995).

However, as Ducrey *et al.* (1996) state, 'measurement of loss of vigor is difficult because directly observable symptoms often do not develop.' Thus, micro features such as the chemistry and physiology governing wood formation (i.e., a healthy tree is one that is producing wood) in the cambium must also be considered as tools for predicting tree health.

As an example, cambial electrical resistance (ER) has been extensively studied as a possible indicator of tree health. Skutt *et al.* (1972) described a non-destructive method using resistance to pulsed electric current to detect discoloration and decay in living trees. Also in 1972, Tattar *et al.* determined that ER is a function of relative changes in concentrations of mobile ions, such as potassium. In 1974, Shigo and Shigo demonstrated application of an electrical meter, the Shigometer, for these measurements.

Cambial zones of healthy trees were shown to have lower resistance to pulsed electric current than stressed trees (Wargo and Skutt, 1975; Smith *et al.*, 1976), and Wargo and Skutt (1975) suggested that '...resistance is an indicator of tree vigor.' Resistance to pulsed electric current in the cambial zone was found to be inversely proportional to rate of tree growth (Shortle *et al.*, 1977) and this was subsequently used as a guide to thinning sugar maples (Shortle *et al.*, 1979). Measurements of ER

Cell & Molecular Biology of Wood Formation, edited by R. Savidge, J. Barnett and R. Napier.

were used successfully to predict Dutch elm disease in elm trees before visible symptoms were observed (Blanchard and Carter, 1980). Davis *et al.* (1980) found that balsam fir stands with increased levels of spruce budworm infestation had increased mean ER values. Kostka and Sherald (1982) described some limitations of ER values as indicators of tree health: 'Measurements of ER with the Shigometer were easily and quickly performed; however, ER readings did not reflect tree vigor as accurately as growth measurements.' Blanchard *et al.* (1983) found that lower ER values were related to higher growth rates in balsam fir, and Lindberg and Johansson (1989) corrected cambial ER of *Picea abies* to standard temperatures and found good correlation between growth rate and ER for both dominant and suppressed trees.

This chapter focuses on another possible indicator of tree health, with emphasis on conifers. Coniferin is a secondary metabolite found in abundance in active conifer cambia. An important physiological aspect of coniferin is that its presence varies seasonally. It has not been detected in the cambium during winter dormancy, nor in the early spring, prior to cambial reactivation. Following the dormancy period, coniferin appears in the cambial zone immediately before the resumption of cell-division (Savidge, 1988, 1989, 1991; Förster and Savidge, 1995; Fukushima *et al.*, 1997; Rowland and Arora, 1997; Savidge *et al.*, 1998) and thereafter is present throughout the period of cambial growth. Coniferin in the cambial zone becomes undetectable at the onset of cambial dormancy in autumn, before the last cambial derivatives have completed their maturation into latewood (Savidge, 1991; Förster and Savidge, 1995). Coniferin has never been detected in stems during cambial dormancy (Freudenberg, 1963; Terazawa and Miyake, 1984; Savidge, 1989, 1991; Förster and Savidge, 1995; Savidge *et al.*, 1998) and biochemical investigations have indicated that the presence or absence of UDPG:coniferyl alcohol glucosyltransferase activity is the explanation for the qualitative seasonal variation (Savidge and Förster, 1998).

Since coniferin is only present in the cambium of conifers during cambial activity, its presence serves as an indicator that cambial growth is occurring. This may be merely a coincidence of metabolism without physiological significance. On the other hand, it could be an indication that the gene expression underlying coniferin biosynthesis is linked to that controlling growth within the tree. If coniferin biosynthesis is obligately linked to primary metabolism underlying growth, as suggested by Savidge and Förster (1998), then coniferin content in the cambium should be affected when growth metabolism is experimentally altered. For example, Savidge (1989) found that girdling of tree stems, a treatment which accelerates cambial growth above and inhibits growth below the ring, had corresponding effects on cambial coniferin content.

To manipulate the vigour (i.e., growth metabolism) of conifers, different treatments can be used, including girdling, defoliation, debudding, and derooting. The coniferin concentration in the cambium of these trees can be readily monitored by quantitative nuclear magnetic resonance (NMR) spectroscopy, concomitantly observing the extent of cambial cell-division activity by microscopy (Savidge *et al.*, 1998).

2. Materials and methods

2.1 *Trees*

Twenty-four, 1 to 3 m tall, 8-year-old plantation *Pinus resinosa* Ait. were selected from the University of New Brunswick Woodlot for treatments and subsequent coniferin investigations in this study. Only those trees that appeared healthy, with no visible signs of decline or damage were chosen.

Fifteen, 4 to 6 m tall, 12-year-old plantation *Pinus strobus* L. were selected in late July from the University of New Brunswick woodlot for investigations into their coniferin content and cambial electrical resistance. The trees were deliberately selected to provide a range of health phenotypes, varying from completely dead snags to individuals having three ages (i.e., years) of dark green needles. The majority of the trees presented only first and second year foliage, with the latter more or less yellowed.

2.2 *Treatments*

Seven treatments to 'stress' the trees were performed: (i) debudding, (ii) defoliating, (iii) debudding and defoliating, (iv) derooting, (v) single ring in the 3-year-old main stem region (see Savidge, 1989), (vi) double ring in the 3-year-old main stem region, removing the needles from the stem between the two rings and (vii) double ring in the 3-year-old main stem region, with needles between the rings intact. An eighth set of untreated trees was used as the control. Trees chosen for treatments 1, 2 and 3 were only 1 m tall, while the rest of the trees, including controls, were 2 to 3 m tall. Treatments were performed in early-mid June, with each treatment being replicated three times.

Six weeks after the treatments were performed (late July), the 3-year-old stem region of each treated tree and control was cut from the tree, and brought back to the laboratory for analysis.

2.3 *Microscopy*

Small (1.0 to 1.5 cm × 1.0 cm × 0.5 cm) specimens containing cambium sandwiched between the bark and the mature wood were removed 5 mm above and below rings, and midway between the rings in double-ringed specimens. For all others, including controls, specimens were removed from the midpoint of the 3-year-old main stem region. A cross-section from each specimen was mounted in water and observed under a light microscope.

In the microscopy analysis, the numbers per radial file of cells in the cambial zone (CZ), the zone where cambial derivatives were primary walled and undergoing radial expansion (RE), the zone where secondary wall formation and lignification were occurring (SL), and the zone where the cells had completed protoplasmic autolysis to become new tracheids (NT) were all counted and recorded. The averages for the three replicates for each treatment and the controls were obtained and these results are depicted in *Figure 2*.

2.4 *Coniferin content by physicochemical methods*

The bark was peeled from the stem, and the exposed inner bark face and xylem surface were scraped together into a 2 ml cryogenic vial, which was then stored in liquid nitrogen. In the case of *P. resinosa*, the 3-year-old stem area was used for non-ringed trees, including controls. For double-ringed trees, 10 cm of stem, beginning 5 mm above the top ring or below the bottom ring were used. All of the area between rings was scraped, leaving a 5 mm buffer adjacent to each ring. For single-ringed trees,

10 cm of stem area were scraped both above and below the ring, beginning 5 mm from the ring. For *P. strobus*, the 8-year-old stem area was used following electrical resistance measurements (see below) in the same area.

Each of the samples was pulverized in liquid nitrogen followed by weighing of 100 mg of frozen powder. The powder was treated with 1.0 ml of extraction buffer containing 1.0 mg of methylphosphonic acid as an internal standard (Savidge *et al.*, 1998) and extracted for 15 minutes at 100°C. The suspension was then centrifuged (12 000 × g, 4°C) for 15 minutes and the supernatant pipetted into a 5 mm nuclear magnetic resonance (NMR) spectroscopy tube. Proton (^{1}H) NMR spectra were obtained at 400 MHz with a Unity 400 NMR spectrometer.

2.5 *Electrical resistance measurements*

Cambial electrical resistance was measured in the eight-year-old stem area of standing trees using a Keithley 614 electrometer. As electrodes, two sharp sterile needles (Becton-Dickinson 27G ½) were pushed through the bark to the point where resistance at the wood surface was encountered. The needles were spaced 1.00 cm apart, one axially above the other. Alligator clips were attached to the protruding stainless steel shafts and resistance values read immediately, taking each resistance measurement three times. The resistance over 1 cm between two points along the steel shaft was determined to be 0 KΩ, that is, lower than the sensitivity of the electrometer.

3. Results and discussion

3.1 *Results*

The combined integration of the aromatic signals for coniferin, δ 6.8–7.1 ppm, and the integration of the methylphosphonic acid (MPA) signal, δ 1.0 ppm, were obtained from ^{1}H NMR spectra (*Figure 1*). The integrated area for the coniferin signal was standardized with respect to the known mass of MPA. The concentration of coniferin in each sample ($mg_{coniferin}$ /ml_{buffer}) was determined by substituting the integrated coniferin signal into the calibration function (R^2 = 0.99993). Since each sample was extracted in 1.0 ml buffer, the amount of coniferin, in mg, is equal to the concentration ($mg_{coniferin}$ /ml_{buffer}). The calculated mass of coniferin was then divided by the mass of tissue extracted, giving coniferin content in ($mg_{coniferin}$ /g_{tissue}). The mean values for three replicates are depicted in *Figure 2*, with their standard errors. The minimum detection limit for coniferin was 12 μg.

Debudding had little effect on coniferin content in the cambial region of *P. resinosa*; however, defoliation resulted in a considerable decrease. A combination of debudding and defoliation resulted in an even greater decrease in coniferin content. Derooting had the greatest effect, with one of the three replicates having undetectable coniferin (*Figure 2*).

The amount of coniferin detected above single rings was not much different from controls, but the amount below rings was considerably reduced. Similar effects were seen in double-ringed specimens, with lower levels of coniferin below the bottom rings, and even lower levels between the rings, especially in the case where the needles were removed from the area between the rings (*Figure 2*).

The cell numbers per radial file found in the four anatomical zones of *P. resinosa* are shown in *Figure 2*. Microscopy indicated that the cambia of the controls were still

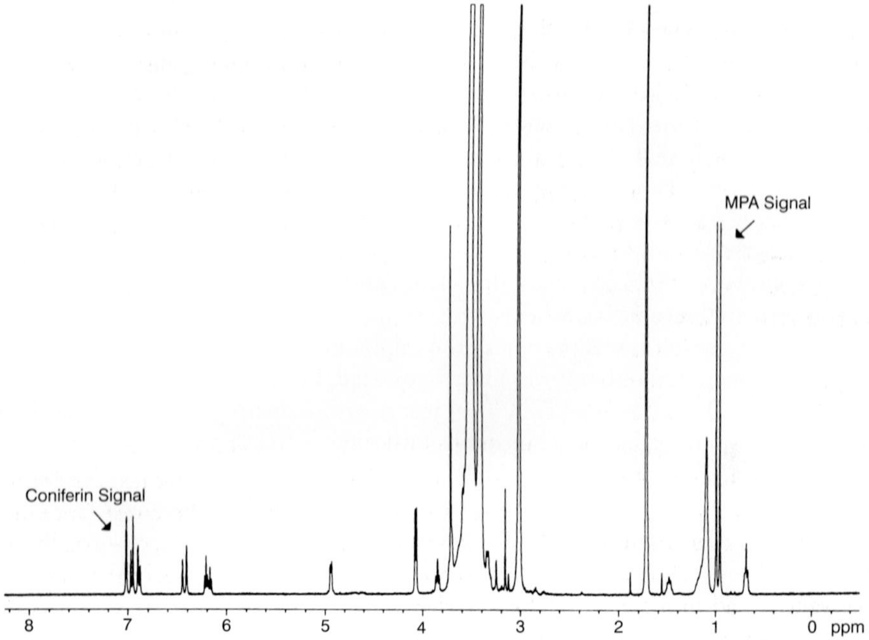

Figure 1. *NMR spectrum of coniferin with methylphosphonic acid (MPA) as an internal standard.*

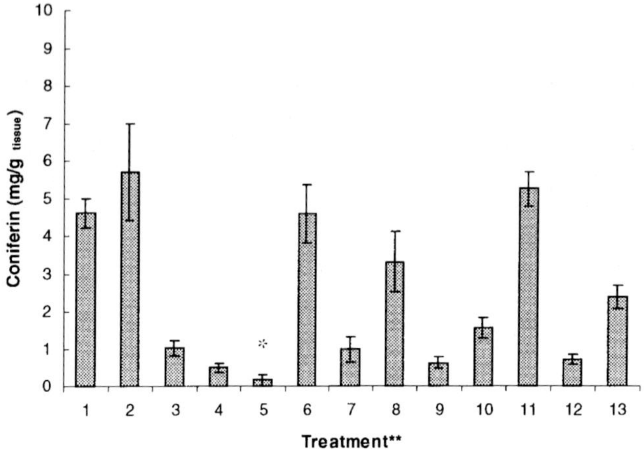

Tissue	Cell number per radial file												
CZ	9	5	0	6	14	10	0	8	11	0	6	0	0
RE	8	7	0	5	3	10	0	3	6	0	9	0	0
SL	10	7	4	5	0	11	4	5	3	8	11	6	6
NT	100+	60+	45	30	55	80+	70+	100+	80+	80+	65+	100	60+

Figure 2. *NMR and anatomy data for treated and control* P. resinosa *trees. * Coniferin was not detected in one of three replicates. ** 1, control; 2, debudded; 3, defoliated; 4, debudded + defoliated; 5, derooted; 6, above single ring; 7, below single ring; 8, above double ring (8–10 needles removed between rings); 9, between double ring; 10, below double ring; 11, above double ring (11–13 needles intact between rings); 12, between double ring; 13, below double ring*

active in July, and new tracheids were still forming in both RE and SL zones. The cambia in debudded trees, from above single rings, and from above double rings, were all still active and producing new tracheids. The cambia of defoliated trees, those from below all rings and those between the rings where needles were left intact, had all dedifferentiated such that the cambial zone consisted only of axial parenchyma cells (hence, CZ = 0 in *Figure 2*). At these locations there was a considerable decrease in the number of new tracheids formed, particularly in defoliated samples. In derooted trees, all cambial activity and radial expansion activity ceased when the roots were cut from the trees; however, SL-zone cells that had already begun differentiation into new tracheids matured into new tracheids.

In *P. strobus*, coniferin was present in the cambium when it was active and could not be detected when the cambium was inactive or dead. Electrometer values for electrical resistance, as measured immediately after inserting the electrodes, varied from 1.84 to 6.83 MΩ. Resistance values were unstable with time, increasing to a more stable value near 14 MΩ in all but one tree (8 MΩ) after elapse of several minutes. Correlation analysis (linear regression) of electrical resistance with coniferin yielded r = –0.00127, indicating no relationship (*Figure 3*).

3.2 *Discussion*

Sink strength of the cambium is weak, relative to shoot and root apices (Eis *et al.*, 1965; Gordon and Larson, 1968; Rangnekar and Forward, 1973). For example, in young trees more than 60% of photosynthate produced in the shoots can travel through the phloem to the roots without being unloaded to the stem's adjoining cambium (Bowen, 1985). The totality of factors regulating sink strength in general remain poorly understood, but trees presumably have feedback mechanisms for determining if there is sufficient nutriment available to permit allocation to the cambium. Within this logic, if strong cambial growth is occurring, it is likely that the whole tree is healthy.

Our data with *P. resinosa* indicate that the cambium's content of coniferin varies with changing physiological state, supporting the hypothesis that coniferin biosynthesis is obligately linked to cambial growth (Savidge, 1989; Savidge and Förster, 1998). The glucosyltransferase required for coniferin biosynthesis is unquestionably

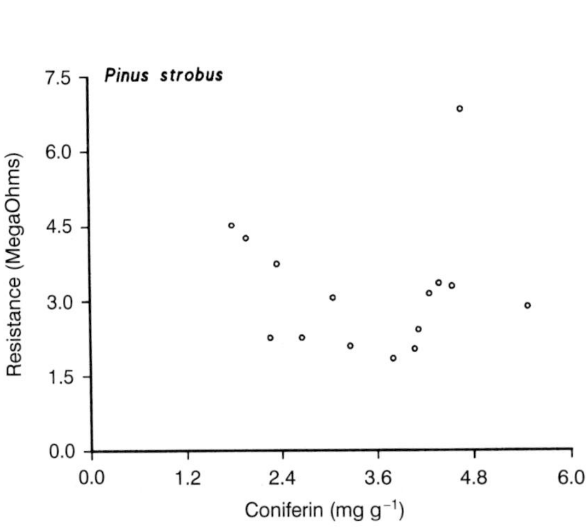

Figure 3. *Electrical resistance versus coniferin content in* P. strobus *stems.*

present in the cambium during active growth (Savidge and Förster, 1998); however, coniferin content of the cambium drops noticeably in areas below rings and is also greatly reduced in defoliated specimens. Thus, our data indicate that living foliage fulfills a crucial role in maintaining the coniferin content of the cambium. The physiological explanation for this is not clear at this time, although the absolute requirement of the crown in wood formation is well established. Coniferin content in the cambium was not found to be correlated with photosynthate nor auxin contents (Savidge, 1991), and an auxin-cytokinin combination was found to promote hydrolysis rather than biosynthesis of coniferin (Savidge, 1989). It appears, therefore, that coniferin biosynthesis is linked to an unknown factor(s) travelling to, or from, the needles. There is also evidence for coniferin biosynthesis in conifer needles (Slimestad and Hostettmann, 1996), raising the possibility that some coniferin may be exported from the needles to the cambium.

The lack of any correlation between electrical resistance (ER) and coniferin was not unexpected. Electrolyte concentration in solution is a function of the water potential of the tissue, and the electrical resistance of protoplasma is inversely related to its electrolyte concentration. Activity of the cambium in temperate-zone trees is invariably preceded by cambial rehydration and associated production of immense vacuoles in the fusiform cells. Moreover, many studies have shown that reduced water availability during summer can reduce or block the cambium's activity. It seems reasonable, therefore, to find relatively low ER values in association with active cambia and higher values with slow growing or inactive cambia. On the other hand, with current knowledge it would be difficult to argue that either the water content or the ER of the cambium was necessarily indicative of cambial activity. In spring, the living cambium rehydrates, but it does not necessarily reactivate and produce a new annual ring. In conifers, for example, diameter growth at the base of old branches and, in aged trees, at the base of the bole may not occur at all over many years, despite the continuing presence of the cambium at those positions (R.A. Savidge, unpublished observations). Evidently, conduction of water through xylem and sap translocation through phloem can continue without need for annual increments of new vascular tissue in those locations. Moreover, electrical assessment of tree health is complicated by the fact that insertion of electrodes into a tree stem does not yield measurements only as they exist in the cambium, rather the net effect generated by all tissues (rhytidome, phloem, CZ, RE, SL, mature wood) contacting the electrodes. The presence of coniferin in the cambium does appear to be specifically indicative of metabolism in support of cambial activity. The coniferin method is more invasive than the electrical resistance method; however, it need not be destructive. Current research is focusing on development of a facile non-destructive field method to diagnose cambial coniferin content.

Based on our observations to date, it appears that cambial coniferin content can serve as a diagnostically useful biochemical indicator of tree health status. We consider that it should be possible to develop a useful quantitative index of tree health for various applications, for example to decide whether an old-growth tree is ready for harvest, based on the content of coniferin in the cambium during the season of growth. If unhealthy trees (or parts thereof) could be detected and harvested before signs of decline occurred, secondary problems in declining forest stands such as insect infestations and decay could be reduced. Conversely, aged trees having continuing competence to be productive and contribute to overall ecosystem health would not have to be harvested prematurely.

Acknowledgements

We would like to thank Dr. L. Calhoun for his NMR expertise, as well as the Canadian Forest Service, Atlantic Division, for their interest and support. V. J. S. is grateful to NSERC Canada for a postgraduate scholarship.

References

Allen, D.C., Barnett, C.J., Millers, I. and Lachance, D. (1992) Temporal change (1988–1990) in sugar maple health, and factors associated with crown condition. *Can. J. For. Res.* **22**: 1776–1784.

Bernier, B., Paré, D. and Brazeau, M. (1989) Natural stresses, nutrient imbalances and forest decline in southeastern Quebec. *Water, Air, Soil Pollut.* **48**: 239–250.

Blanchard, R.O. and Carter, J.K. (1980) Electrical resistance measurements to detect Dutch elm disease prior to symptom expression. *Can. J. For. Res.* **10**: 111–114.

Blanchard, R.O., Shortle, W.C. and Davis, W. (1983) Mechanism relating cambial electrical resistance to periodic growth rate of balsam fir. *Can. J. For. Res.* **13**: 472–480.

Blanche, C.A., Hodges, J.D. and Nebeker, T.E. (1985) A leaf area–sapwood area ratio developed to rate loblolly pine tree vigor. *Can. J. For. Res.* **15**: 1181–1184.

Bowen, G.D. (1985) Roots as a component of tree productivity. In: *Attributes of Trees as Crop Plants* (eds M.G.R. Cannell and J.E. Jackson). ITE, NERC, pp. 303–315.

Davis, W., Shortle, W. and Shigo, A. (1980) Potential hazard rating system for fir stands infested with budworm using cambial electrical resistance. *Can. J. For. Res.* **10**: 541–544.

Ducrey, M., Duhoux, F., Huc, R. and Rigolot, E. (1996) The ecophysiological and growth responses of Aleppo pine (*Pinus halepensis*) to controlled heating applied to the base of the trunk. *Can. J. For. Res.* **26**: 1366–1374.

Eis, S., Garman, E.H. and Ebell, L.F. (1965) Relation between cone production and diameter increment of Douglas-fir (*Pseudotsuga menziesii* (Mirb.) Franco), grand fir (*Abies grandis* (Dougl.) Lindl.) and western white pine (*Pinus monticola* Dougl.). *Can. J. Bot.* **43**: 1553–1559.

Förster, H. and Savidge, R.A. (1995) Characterization and seasonal activity of UDPG: coniferyl alcohol glucosyl transferase linked to cambial growth in jackpine. In: *Proceedings of the Plant Growth Regulation Society of America. Twenty-second annual meeting.* (Fritz, C.D., ed.) Minneapolis, Minnesota.

Freudenberg, K. (1963) The glucosides of cambial sap of spruce. *Phytochemistry* **2**: 189–193.

Fukushima, K., Taguchi, S., Matsui, N. and Yasuda, S. (1997) Distribution and seasonal-changes of monolignol glucosides in *Pinus thunbergii*. *Mokuzai Gakkaishi* **43**: 254–259.

Gordon, J.C. and Larson, P.R. (1968) Seasonal course of photosynthesis, respiration and distribution of ^{14}C in young *Pinus resinosa* trees as related to wood formation. *Plant Physiol.* **43**: 1617–1624.

Kolb, T.E. and McCormick, L.H. (1993) Etiology of sugar maple decline in four Pennsylvania stands. *Can. J. For. Res.* **23**: 2395–2402.

Kostka, S.J. and Sherald, J.L. (1982) An evaluation of electrical resistance as a measure of vigor in eastern white pine. *Can. J. For. Res.* **12**: 463–467.

Lewis, T.E. (1995) *Tree Rings as Indicators of Ecosystem Health*. CRC Press, Boca Raton, Florida, USA.

Lindberg, M. and Johansson, M. (1989) The use of electrical resistance of cambium and phloem as a measure of tree vigor. *Scand. J. For. Res*. **4**: 175–185.

Rangnekar, P.V. and Forward, D.F. (1973) Foliar nutrition and wood growth in red pine: effects of darkening and defoliation on the distribution of ^{14}C-photosynthate in young trees. *Can. J. Bot.* **51**: 103–108.

Rowland, L.J. and Arora, R. (1997) Proteins related to endodormancy (rest) in woody perennials. *Plant Sci.* **126**: 119–144.

Savidge, R.A. (1988) A biochemical indicator of commitment to tracheid differentiation in *Pinus contorta*. *Can. J. Bot.* **66**: 2009–2012.

Savidge, R.A. (1989) Coniferin, a biochemical indicator of commitment to tracheid differentiation in conifers. *Can. J. Bot.* **67**: 2663–2668.

Savidge, R.A. (1991) Seasonal cambial activity in *Larix laricina* saplings in relation to endogenous indol-acetic acid, sucrose, and coniferin. *For. Sci.* **37**: 953–958.

Savidge, R.A. and Förster, H. (1998) Seasonal activity of uridine 5′-diphosphoglucose: coniferyl alcohol glucosyltransferase in relation to cambial growth and dormancy in conifers. *Can. J. Bot.* **76**: 486–493.

Savidge, R.A., Randeniya, P.U., Xu, Y., Leinhos, V. and Förster, H. (1998) Coniferyl alcohol oxidase: a new enzyme involved with lignification. In: *Lignin and Lignan Biosynthesis* (eds Lewis, N.G. and Sarkanen, S.), ACS Symposium Series **697**: 109–130.

Shigo, A.L. and Shigo, A. (1974) Detection of discoloration and decay in living trees and utility poles. *USDA For. Serv. Res. Pap.* NE-294. Northeast. For. Exp. Stn. Upper Darby, PA.

Shortle, W.S., Abusamra, J., Laing, F.M. and Morselli, M.F. (1979) Electrical resistance as a guide to thinning sugar maples. *Can. J. For. Res.* **9**: 436–437.

Shortle, W.S., Shigo, A.L., Berry, P. and Abusamra, J. (1977) Electrical resistance in tree cambium zone: relationship to rates of growth and wound closure. *For. Sci.* **23**: 326–329.

Skutt, H.R., Shigo, A.L. and Lessard, R.A. (1972) Detection of discolored and decayed wood in living trees using a pulsed electric current. *Can. J. For. Res.* **2**: 54–56.

Slimestad, R. and Hostettmann, K. (1996) Characterization of phenolic constituents from juvenile and mature needles of Norway spruce by means of high performance liquid chromatography-mass spectrometry. *Phytochemical Analysis* **7**: 42–48.

Smith, D.E., Shigo, A.L., Safford, L.O. and Blanchard, R. (1976) Resistances to a pulsed electrical current reveal differences between nonreleased, released, and released-fertilized paper birch trees. *For. Sci.* **22**: 471–472.

Tattar, T.A., Shigo, A.L. and Chase, T. (1972) Relationship between the degree of resistance to a pulsed electric current and wood in progressive stages of discoloration and decay in living trees. *Can. J. For. Res.* **2**: 236–243.

Terazawa, M. and Miyake, M. (1984) Phenolic compounds in living tissue of woods. II. Seasonal variations of phenolic glycosides in the cambial sap of woods. *Mokuzai Gakkaishi* **30**: 329–334.

Wargo, P.M. and Skutt, H.R. (1975) Resistance to pulsed electric current: an indicator of stress in forest trees. *Can. J. For. Res.* **5**: 557–561.

Waring, R.H. (1983) Estimating forest growth and efficiency in relation to canopy leaf area. *Adv. Ecol. Res.* **13**: 327–354.

Waring, R.H. (1985) Imbalanced forest ecosystems: assessments and consequences. *For. Ecol. Manage.* **12**: 93–112.

Wilmot, T.R., Ellsworth, D.S. and Tyree, M.T. (1995) Relationships among crown conditions, growth and plant and soil nutrition in seven northern Vermont sugarbushes. *Can. J. For. Res.* **25**: 386–397.

5

Seasonal variation in the K, Ca and P content and distribution of plasma membrane H^+-ATPase in the cambium of *Populus trichocarpa*

M. Arend and J. Fromm

1. Introduction

Seasonal changes in the ultrastructure of the vascular cambium were investigated in several hardwood species (Robards and Kidwai, 1969; Sennerby-Forsse, 1986; Farrar and Evert, 1997), showing that during dormancy fusiform cells are densely cytoplasmic with many small vacuoles while during cambial activity they are highly vacuolated. The purpose of the present study was to investigate seasonal variation in cambial ultrastructure in relation to changes in mineral content, because cell enlargement depends on the turgor pressure of the cells (Larson, 1994; Savidge, 1996). In addition, the distribution of the plasma membrane H^+-ATPase which might be involved in the uptake of potassium into enlarging cells was studied by immunolocalization. We chose poplar plants for the experiments because of the well-known advantages of the genus *Populus* as a model for study of the hardwood secondary vascular system (Chaffey, 1999) and, especially, for its ability to produce cuttings.

2. Seasonal variation in the fine structure of the vascular cambium

Branch tissues collected over a one-year period showed that during dormancy the cambial zone consists of only 3–4 layers of dense cytoplasmic cells with thick walls

Cell & Molecular Biology of Wood Formation, edited by R. Savidge, J. Barnett and R. Napier.

(*Figure 1*). Both fusiform and ray cells contain numerous small vacuoles and lipid droplets. Also nuclei, mitochondria, plastids, dictyosomes, ribosomes and endoplasmic reticulum occur frequently.

After cambial reactivation the cambium shows 6–7 cell layers per radial file with thin walls. The numerous small vacuoles are replaced by fewer larger vacuoles which appear electron-transparent (*Figure 2*). The cytoplasm is restricted to a narrow peripheral layer. In general, these results agree with previously published studies on seasonal changes in the ultrastucture of the vascular cambium.

3. Seasonal changes of the K, Ca and P content in the cambium

Recently it was shown that during cell enlargement a strong increase occurs in symplastic K content during tracheid differentiation in *Picea abies*, emphasizing a role of potassium in osmoregulation (Dünisch *et al.*, 1998). To investigate if increased vacuolation during cambial reactivation is correlated with changes in ion content we measured K, Ca and P by energy dispersive X-ray analysis (EDXA) in tissues such as are shown in *Figures 1* and *2*. The K content in the cambial zone showed a strong increase in April (*Figure 3*), during cambial reactivation when formation of larger vacuoles occurs in cambial cells. We interpret the data to mean that potassium plays a role in the generation of cambial turgor pressure. After the phase of cell enlargement, the K content decreased to less than 50% of that measured at the end of April. Such low K contents were detected on all other dates investigated during the seasonal cycle.

The Ca content increased by as much as 40% in April (*Figure 3*, 15.4.99). After this increase, Ca decreased immediately, indicating that this element is of importance especially for the onset of cambial reactivation. Since Ca is a major factor in the maintenance of structure and function of cell membranes and serves as a co-factor in several enzymes, its transient increase could be involved in the basic processes of the regulation of wood formation.

An increase in P content was detected during cambial reactivation although the P content is generally much lower than the K and Ca content (*Figure 3*, 15.4.99). The

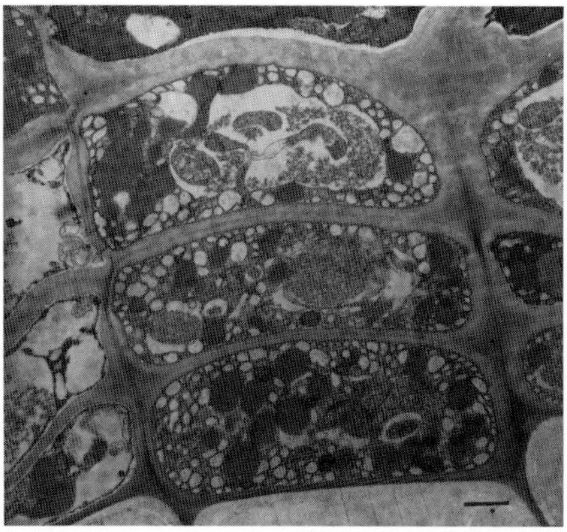

Figure 1. *Transverse TEM-section of the resting cambial zone of* Populus trichocarpa *collected in January (Bar: 1 μm).*

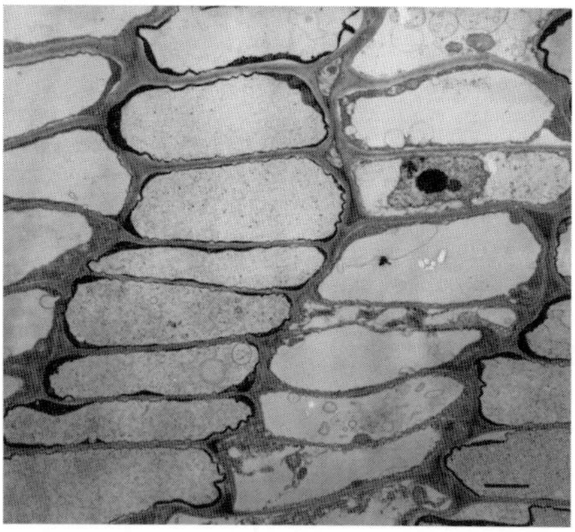

Figure 2. *Transverse TEM-section of the active cambial zone collected in April (Bar: 2 μm).*

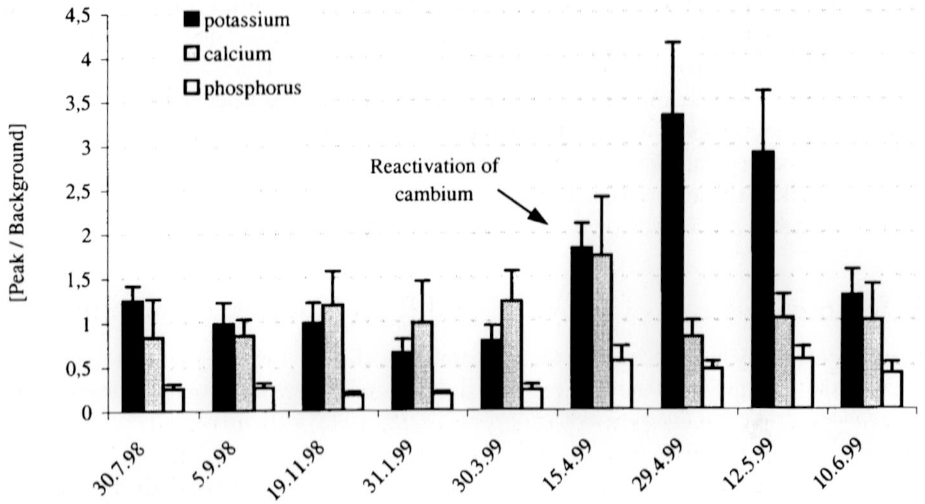

Figure 3. *Relative concentrations of K, Ca and P in the cambial zone during the period July 1998 to June 1999. The scale on the left side gives the peak to background ratio of recorded X-ray signals. Data are mean values obtained from cambial zones of 10 sections. The error bars represent standard deviations.*

concentration of adenine nucleotides increased in buds when they become swollen (Fromm and Eschrich, 1986), and it can be assumed that cambial growth also requires ATP for various processes (for example, phosphorylation of hexoses from starch hydrolysate). In addition, P is an important component of DNA and ribosomal RNA (Korn *et al.*, 1983).

4. Immunolocalization of PM-H$^+$-ATPase activity in the cambium

The plasmamembrane H$^+$-ATPase plays an important role in many functions such as cell nutrition and elongation. For instance, in rhizodermis cells the enzyme is known to be involved in active uptake of ions (Parets-Soler *et al.*, 1990). We studied PM-H$^+$-ATPase in 1-year-old poplar shoots during dormancy and cambial reactivation using a specific anti-PM-H$^+$-ATPase antibody on tissue sections, as reported by Baur *et al.* (1996). In winter, no evidence for the enzyme could be found in sections of dormant twigs, but during cambial reactivation the enzyme was found localized to almost all cells of the cambial zone (colour *Plate 2*, indicated by the antibody-specific yellow fluorescence). We saw no fluorescence in control sections of spring twigs when non-immune mouse antibody instead of primary antibody was used.

5. Conclusions

The results presented here suggest a key role for the PM-H$^+$-ATPase in the control of cambial activity and the regulation of vascular cell elongation. Since sugars can be taken up by cambial cells via symport with protons and potassium via antiport, a lowered wall pH can facilitate nutrient uptake and cell elongation. Apart from these results a decrease in apoplastic pH can activate wall-loosening processes which cause an increase in growth rate according to the acid growth theory (Rayle and Cleland, 1992). The experimental results support the view that ultrastructural changes like the formation of large vacuoles during cambial reactivation are caused by increasing K fluxes regulated by PM-H^+-ATPase activity. Thus, our work will focus on these aspects of the regulation of cambial activity in future. An important experimental approach will be the immunolocalization of various ion channels in the cambium using both fluorescence and electron microscopy.

References

Baur, M., Meyer, A.J., Heumann, H.G., Lützelschwab, M. and Michalke, W. (1996) Distribution of plasma membrane H^+-ATPase and polar current patterns in leaves and stems of *Elodea canadensis*. *Bot. Acta* **109**: 382–387.

Chaffey, N. (1999) Cambium: old challenges — new opportunities. *Trees* 13: 138–151.

Dünisch, O., Bauch, J., Müller, M. and Greis, O. (1998) Subcellular quantitative determination of K and Ca in phloem, cambium, and xylem cells of spruce (*Picea abies* [L.] Karst.) during earlywood and latewood formation. *Holzforschung* **52**: 582–588.

Farrar, J.J. and Evert, R.F. (1997) Seasonal changes in the ultrastructure of the vascular cambium of *Robinia pseudoacacia*. *Trees* **11**: 191–202.

Fromm, J. and Eschrich, W. (1986) Changes of adenine nucleotide and orthophosphate concentrations in buds of deciduous trees during spring reactivation. *Trees* **1**: 42–46.

Korn, A.P., Spitnik-Elson, P. and Elson, D. (1983) Specific visualization of ribosomal RNA in the intact ribosome by electron spectroscopic imaging. *Eur. J. Cell Biol.* **31**: 334–340.

Larson, P.R. (1994) *The Vascular Cambium: Development and Structure. Springer-Verlag Series in Wood Science*. Springer-Verlag, Berlin.

Parets-Soler, A., Pardo, J.M. and Serrano, R. (1990) Immunocytolocalization of plasma membrane H^+-ATPase. *Plant Physiol.* **93**: 1654–1658.

Rayle, D.L. and Cleland, R.E. (1992) The acid growth theory of auxin-induced cell elongation is alive and well. *Plant Physiol.* **99**: 1271–1274.

Robards, A.W. and Kidwai, P. (1969) A comparative study of the ultrastructure of resting and active cambium of *Salix fragilis* L. *Planta* **84**: 239–249.

Savidge, R.A. (1996) Xylogenesis, genetic and environmental regulation — a review. *IAWA Journal* **17(3)**: 269–310.

Sennerby-Forsse, L. (1986) Seasonal variation in the ultrastructure of the cambium in young stems of willow (*Salix viminalis*) in relation to phenology. *Plant Physiol.* **67**: 529–537.

6

Photosynthate allocation to the vascular cambium: facts and problems

J.J. Sauter

1. Introduction

With increasing interest in growing trees for biomass production it becomes important to know more about the vascular cambium which gives rise to the enormous mass of woody material. Details on the photosynthate allocation to this vascular cambium therefore is of basic interest. In contrast to herbaceous plants (Patrick, 1997), too little is known up to date, both, on the anatomical basis as well as on the physiological details of the transfer of assimilates from sieve elements to cambium cells, or, from the xylem to the cambium (e.g. van Bel, 1990). Whether an apoplastic or a symplastic route is chosen for the unloading of sieve elements along the axis of a woody stem before the photosynthates move radially towards the cambium is still a matter of debate (e.g., van Bel, 1993, Schulz, 1998). On the other hand, the remarkable evenness of growth in diameter of the large boles of trees leaves little doubt that a highly ordered supply of assimilates to the cambial cells must take place all along its length. In contrast to photosynthate allocation to cambial cells which has received little attention in the past, there is a great body of information on the hormonal stimuli which are needed to control cambial activity, i.e. its mitotic activity, as well as the future fate of the descendent cells (e.g., Aloni, 1987; Little and Pharis, 1995; Savidge, 1996). While these hormonal stimuli come primarily, though not exclusively, via the axial pathway in the cambium itself, the carbon skeletons needed in these cell division and cell differentiation processes are delivered predominantly via the radial route. Woody plants like trees have developed a highly differentiated radial tissue system, i.e. the rays, which in addition to their functioning in storage of organic and inorganic material serve as efficient radial translocation routes (Sauter, 1982; Sauter and Kloth, 1986; Sauter and Witt, 1997). One way to get information on the radial translocation capacity of this system is to study carefully the deposition and mobilisation events of carbohydrates

Cell & Molecular Biology of Wood Formation, edited by R. Savidge, J. Barnett and R. Napier.

and of nitrogenous compounds in this tissue. The nitrogenous compounds which are stored in the form of quite specific vegetative storage proteins, have been identified and characterized only during the last 10 to 15 years (van Cleve *et al.*, 1988; Wetzel *et al.*, 1989; Sauter and van Cleve, 1990; Harms and Sauter, 1991; Stepien *et al.*, 1994). The cambium, being passed continuously in its active stage by a prominent stream of sugars and many other substances important for growth and differentiation of its derivatives, imposes great difficulties, however, on the elucidation of its proper demand and its sink activities. Studying the deposition rates of carbohydrates and of nitrogenous compounds in the ray cells of the wood inside the cambium provides a way to estimate the minimal flow of assimilates through the cambium. In the present paper, therefore, an attempt is made to use numerous data collected on deposition events in the ray tissue in order to get an idea about the minimum flow rates of sugars and of nitrogenous compounds through the cambial cells themselves. In addition, results on the activity of sucrose synthase and its seasonal changes inside the cambium are reported which indicate that this enzyme in particular is involved in determining the sink activity of the cambium.

2. The structural relationship of rays and cambium

2.1 *Rays as radial routes of food supply*

The cambium is composed of two different kinds of cells: fusiform initials giving rise to axially elongated cells of the secondary xylem and phloem, and ray initials whose derivatives elongate normally more or less in the radial direction and form the main radial translocation path for organic and inorganic materials. The rays connect sieve elements with the cambium, on the one hand, and the cambium with the xylem elements on the other (*Figure 1*). There is little doubt that the vast amount of differentiating cells of the cambium area is supplied with carbohydrates, nitrogenous compounds and other substances mainly through this ray system which traverses the cambium at the ray initials (cf. van Bel, 1990, 1993; Sauter and Witt, 1997). Recent results using energy dispersive X-ray spectroscopy proved that the xylem rays serve also as an essential radial translocation route for minerals (e.g. Dünisch *et al.*, 1998). Therefore, more information on the translocation capacity of this tissue is urgently needed. However, up to now the exchange of materials from phloem to xylem via the rays, and vice versa, is still a widely undervalued route of transport (Van Bel, 1990). *Figure 1* illustrates schematically the histology of this route from phloem to xylem. It is recognized: (i) that the cambium is passed by the rays; more precisely, only the ray initials are traversed by the rays. Regarding the spacing of rays, it is of particular interest to note that the fusiform initials when beginning to suffer from shortage in food supply are evidently able to divide inequally. In this way the fusiform initials can give rise to new ray initials and thus to new radial routes for food supply (see Aloni, 1995 and literature cited). (ii) The phloem rays (PR) stay in contact with the sieve elements (SE). In case of gymnosperm trees, highly elaborated 'one-sided sieve areas' provide an intimate contact of the sieve cells with the Strasburger cells of a phloem ray (Sauter and Braun, 1968; Sauter, 1974; Sauter *et al.*, 1976). In case of angiosperm trees, the contact of the sieve tube/companion cell complexes is still a matter of debate due to conflicting results and a lack of information on woody stems (cf. van Bel, 1993; Patrick, 1997; Schulz, 1998). At present, therefore, we cannot but suspect that a symplastic route between sieve elements and ray cells also exists along the axial path in angiosperm trees trunks. (iii) Inside the cambium, in the wood, the rays are in close contact with the vessels by specialized ray cells, the so-called 'contact cells' (Sauter, 1972; Sauter and Witt, 1997) or vessel-associated cells (Czaninski, 1977). Experimental evidence suggests that uptake from and release into the vessels of sugars (Sauter, 1972, 1981, 1983), amino acids (Sauter, 1976; Sauter and van Cleve, 1992), and minerals (e.g. Dünisch *et al.*, 1998) proceeding via these cells, especially in spring. (iv) The basipetal flow of hormones in the cambial region itself is also indicated in the graph, though it is beyond the scope of this report. Nevertheless, its great significance for the regulation of the cambium's activity and for the differentiation of its derivatives deserves to be emphasized, as both events affect directly the sink strength of this area and thus the radial translocation processes in the rays.

2.2 *Factors involved in food supply by the rays*

The cambium in its active stage, i.e. from late spring until early fall, is passed continuously by a prominent flow of photoassimilates and of other compounds (see *Figure 3*). Such a passage of nutriment through or near a meristematic cell is quite unusual. It is reflected in the clear-cut differences in ultrastructural characteristics of these cells (e.g. Barnett, 1992; Catesson, 1994; Farrar and Evert, 1997). The supply of photoassimilates to the cambium depends in particular on: (i) The unloading of the assimilates of the sieve elements. In principle, this unloading can be apoplastic or symplastic (van Bel, 1993; Patrick 1997; Schulz, 1998). The results available for gymnosperm trees (e.g. Sauter, 1980; Sauter *et al.*, 1976) and for some herbaceous plants (see Patrick, 1997), however, are in favour of symplastic unloading. (ii) The radial, centripetal translocation in phloem rays to the cambium. Circumstantial information on the minimal flow rates are given below. (iii) The usage of the photoassimilates at the cambium site, which depends on cell division activity that in turn is under control of hormonal and

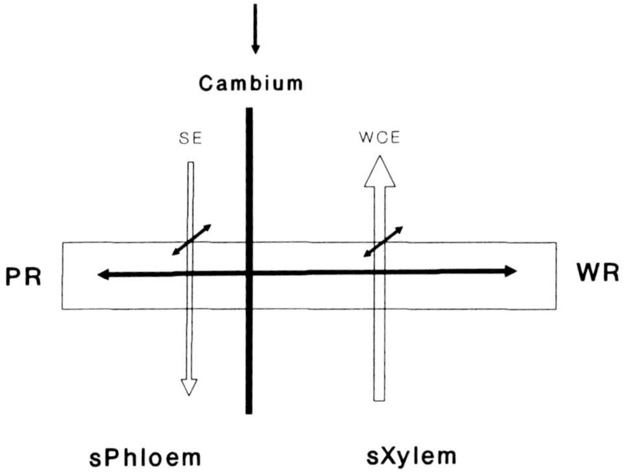

Figure 1. Diagram illustrating the main translocation pathways in a woody stem (for details see text). PR, WR = phloem ray, wood ray; SE = sieve elements; WCE = water conducting elements.

environmental stimuli (e.g. Savidge, 1996). (iv) Passage through the dividing cambial cells is important too, as it determines the possible import into the wood which in turn is related to the sink characteristics of cambial and xylem cells, e.g to the intensity of starch synthesis (cf. Sauter and Witt, 1997). (v) The centripetal translocation capacity of the xylem rays themselves and the subsequent storage in these cells might also contribute to the flow of photosynthates through the cambium. The mobilization of the storage materials in spring when these flow patterns are mostly reversed in direction, are not under the scope of this report.

With respect to the optimal supply of the cambium, some fundamental questions must be asked although most of them still await to be answered:

(i) What is the influence of photosynthate concentration in the sieve elements? There are noteworthy daily and seasonal fluctuations (see Zimmermann, 1971).
(ii) What regulates the unloading capacity at the sieve element/phloem ray intersection along a woody stem? What part is contributed by passive leakage in addition to symplastic unloading? The unloading capacity apparently is not a limiting factor, at least not for sugars, as will be shown below.
(iii) What is the radial translocation capacity of the ray cells? Can the flow rate be increased and what is its limit (for answers see sections 3.3 and 3.6)?
(iv) To what degree is the flow rate regulated by the sink activity of the cambium area and to what degree by the sink activity of the wood?
(v) What determines the maximum cell division activity of the cambium and in this way also its sink activity?
(vi) What regulates the sink activity of the storage tissues in the wood which in turn acts on the radial flow intensity of photoassimilate through the cambium?

From the growth rate at the cambium site, which can be measured quite exactly, we have a fairly good idea about the amount of photoassimilate which actually can be used for growth, but we don't know whether the cambial cells actually would have needed more, and, whether and to what extent the flow rates of materials from SE to the ray cells and within the rays towards the cambium are limiting the growth. On the other hand, the signals coming basipetally in the cambium cells themselves could play the pivotal role for the sink strength by controlling the cell division activity and the subsequent use of the assimilates during differentiation. Regarding the importance of the ray system as pathways for the supply of the photoassimilates to the cambium, the following answers are derived mainly on the basis of results obtained in poplar and willow.

3. Accumulation pattern of storage materials and flow rates of sugars and N-compounds

3.1 *The main storage materials*

The rays, in both the phloem and the xylem, are not only important as radial translocation pathways but also serves as the most prominent storage compartments of woody plants. In case of poplar and willow, ray cells are almost the only storage cells of the wood as axial xylem parenchyma are scanty. The main organic storage compounds are starch, sugars, fat, and specific vegetative storage proteins (e.g. Sauter and Witt, 1997, and literature cited). In *Figure 2*, the storage material of a wood ray cell during winter is shown, e.g. starch granules within amyloplasts, storage protein within distinct protein bodies, and fat within oleosomes. The microvesicles are suspected to be sites of accumulation of sucrose and/or its galactosides at this stage (Sauter *et al.*, 1996).

3.2 *Accumulation pattern of starch and deposition rates*

Figure 3 gives an example of starch deposition in the wood of poplar and willow. It is seen: (i) that starch content has its minimum in both species in early May, i.e. after budbreak and during leaf expansion, (ii) that starch deposition resumes in May, as soon as the leaves are fully grown, and continues until October when the leaves become senescent, (iii) that the speed of deposition can be quite different. This pattern is typical for many deciduous trees, although some remarkable differences also exist, especially in conifers (e.g. Hansen and Beck, 1994).

From this depostion pattern we can draw the following conclusions. The cambium in its active stage is passed continuously, i.e. from May until October, by a prominent stream of carbohydrates in the form of sugars. The carbohydrate supply during the vegetation period might change; however, the cambium clearly cannot use up all the incoming carbohydrates. Furthermore, the radial translocation capacity for sugars in the phloem ray cells is evidently higher than is demanded by the cambium. Finally, the unloading of sieve elements cannot be regulated by the cambium's activity alone because vigorous unloading continues in late summer when cambial divisions have already ceased while starch deposition continues in the wood.

Based on the starch deposition pattern in the branchwood of *Populus* and *Salix*, recorded for many individual years at the same tree stand, information on the

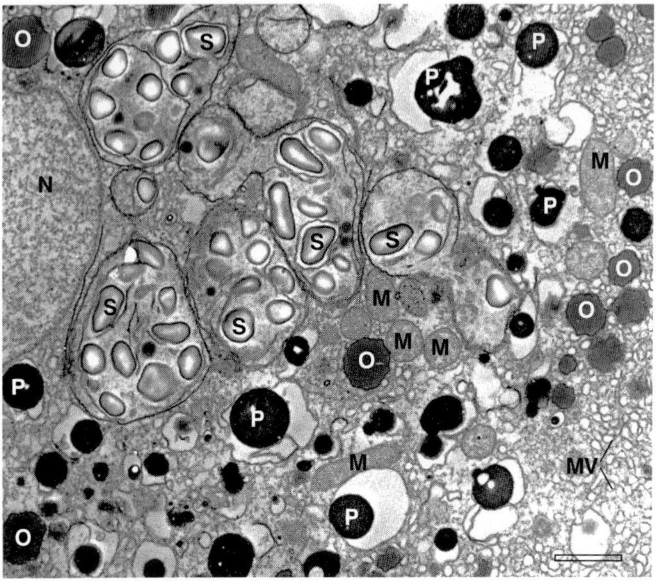

Figure 2. *Electron micrograph showing the main storage compounds of a ray cell in late fall. S = Starch in amyloplasts; P = storage protein in protein bodies; O = fat in oleosomes; N = nucleus; M = mitochondria; MV = vesicular endoplasmic reticulum. Scale bar = 1 μm.*

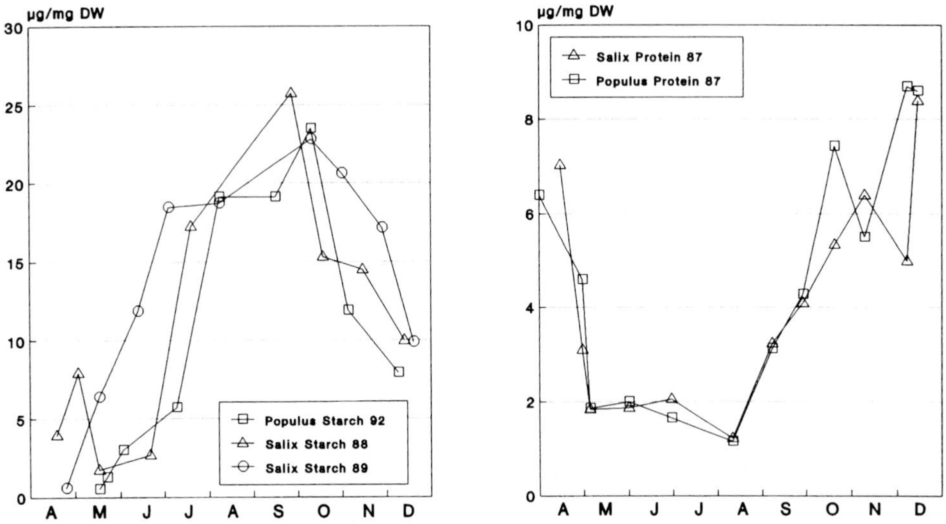

Figure 3. *Different deposition patterns of starch (left graph) and of protein (right graph) in branchwood of poplar and willow.*

deposition rates have been obtained, both on average rates over rather long periods as well as on rates during periods with intense starch deposition (*Table 1*). When a rather long period, for example, from May until September, is taken the starch deposition rate (SDR) becomes 0.136 µg starch mg^{-1} DW day^{-1} (*Table 1*). When periods with a rapid increase in starch content of the wood are selected this value can increase most significantly to values from 0.200 to 0.400 µg starch mg^{-1} DW day^{-1}. The maximum SDR of 0.490 µg starch mg^{-1} DW day^{-1} was recorded for poplars during a 3 week period in summer 1989. Interestingly, the SDR for willow are found to lie exactly in the same range. Again, the SDR did not increase over 0.5 µg mg^{-1}DW day^{-1}. These values give us not only an idea about the radial translocation capacity of the rays but also about the minimum sugar flux through the cambium which is necessary to enable these SDRs in the wood.

3.3 *Radial flux of sugars through ray cells*

The starch accumulation rates permit deduction of the minimum radial flux of sugars through the rays and thus through their initials in the cambium itself. Based on measurements of the tangential lumen area of ray cells (RCTA) and on a measured SDR of 0.18 µg mg^{-1}DW day^{-1}, a radial flux of 80.7 pmol glucose (Glc) cm^{-2} RCTA sec^{-1} had been reported earlier (Sauter and Kloth 1986). When the SDR of 0.491 µg mg^{-1}DW day^{-1} (the maximum deposition rate recorded in poplar branches) is taken, this value augments to 220 pmol Glc cm^{-2} RCTA sec^{-1}. For willow with a SDR of 0.3114 µg mg^{-1}DW day^{-1} (*Table 1*), the flux through the ray cells and thus through the ray initials lies in the same order of magnitude, e.g. at 141 pmol Glc cm^{-2} RCTA sec^{-1}. It might be justified, therefore, to put the average long-term flow rates of sugars through the cambium in poplar and willow somewhere within the range of 100 to 200 pmol Glc cm^{-2} RCTA sec^{-1}. However, these flow rates for sugars must be regarded to be minimum values, firstly, because they are calculated for rather long periods of starch deposition, and secondly, because no deduction has been made for the loss by respiration, by conversion into other constituents, by increasing the cell sugar levels, and by leakage into the apoplast.

3.4 *Transport capacity of a plasmodesma*

From these starch deposition rates we can also obtain a rather good estimate on the transport capacity for sugars of a plasmodesma in ray cells. We need to know the RCTA, the pit frequency in these tangential walls, and the number of plasmodesmata per pit field. These values have been gathered earlier in a painstaking study at the electron microscopical level for poplar (Sauter and Kloth, 1986).

For a SDR of 0.18 μg mg^{-1}DW day^{-1} and a radial sugar flux of 80.7 pmol Glc cm^{-2} RCTA sec^{-1}, a translocation capacity of a plasmodesma in poplar ray cells has been reported to be between 1.0 to 1.7 × 10^{-7} pmol Glc $plasmodesma^{-1}$ sec^{-1}, depending on whether the frequency of plasmodesmata isolation cells or of contact cells of a ray have been used (cf. Sauter and Kloth, 1986). For a measured SDR of 0.491 μg mg^{-1}DW day^{-1}, this value augments to 2.7 to 4.6 × 10^{-7} pmol Glc $plasmodesma^{-1}$ sec^{-1}. As evidenced by the SDR obtained for willow (*Table 1*), the transport capacity of a plasmodesma in willow lies in the same order of magnitude, namely, 1.7 to 3.0 × 10^{-7} pmol Glc $plasmodesma^{-1}$ sec^{-1}, provided the anatomical characteristics are similar to poplar, a closely related species. Again, these values are minimum values for the reasons given above. The validity of these values, however, becomes substantially supported by the values estimated for mesophyll cells, bundle sheath cells, and companion cells, e.g. 2–8 × 10^{-7} pmol $plasmodesma^{-1}$ sec^{-1} (see Gunning and Robards, 1976). The suspicion of Gunning and Robards (1976) that the limits of translocation capacity of plasmodesmata could lie approximately in this range thus receives additional weight. It would be interesting to know the number of plasmodesmata traversing the ray initials of the cambium in order to estimate the possible minimum plasmodesmatal sugar flow rate at this site.

3.5 *Accumulation pattern of vegetative storage proteins*

The ray cells in phloem and xylem are sites for prominent deposition of vegetative storage proteins. Such storage proteins are highly specific proteins (Stepien *et al.*, 1994) both in gymnosperm (Harms and Sauter, 1991) and angiosperm trees (van Cleve *et al.*, 1988; Wetzel *et al.*, 1989). By immuno-gold localization they were found to accumulate within membrane-bound organelles in cells, as protein bodies (see *Figure 2*; Sauter and van Cleve, 1989, 1990). *Figure 3* shows that their accumulation pattern in the wood differs substantially from that of starch. After their mobilization in spring, the protein content stays at a low level throughout the vegetative period. Poplar and willow behave similarly (*Figure 3*). The deposition starts usually in early fall, concomitantly with the yellowing of leaves when their major nitrogen-compounds are hydrolysed for retrieval and storage in the branches and in the stem.

This peculiar pattern of deposition poses some interesting questions: (i) Is the supply of N-compounds via the phloem rays during the growth period in summer too low to trigger synthesis and deposition of vegetative storage protein in the wood rays? (ii) Does cambial activity and the differentiation of its derivatives use up all the

Table 1. *Starch deposition rates (SDR) and estimated minimum radial sugar flux rates (RSFR) in glucose (Glc) units through the ray cell tangential area (RCTA) in the branchwood of poplar and willow.*

Period (year/month/ day)	Period length (days)	Changes in starch content (μg mg^{-1} DW)	SDR (μg mg^{-1} DW day^{-1})	RSFR (pmol Glc cm^{-2} RCTA sec^{-1})
Populus				
1984				
5/25 – 9/20	117	0 – 15.9	0.136	61
6/8 – 8/18	71		0.180	80.7
1987				
8/10 – 9/7	28	2.6 – 9.2	0.236	106
9/7 – 10/19	42	9.2 – 18.4	0.219	
9/28 – 10/19	21	10.5 – 18.4	0.376	
1988				
6/20 – 7/18	28	1.9 – 13.7	0.421	189
1989				
6/12 – 7/3	21	10.0 – 20.3	0.490	220
1990				
5/7 – 6/5	29	0.5 – 5.2	0.162	
9/3 – 10/1	27	12.8 – 21.8	0.333	
1991				
7/10 – 8/7	28	4.5 – 8.2	0.132	
8/7 – 9/16	40	8.2 – 21.0	0.320	143
Salix				
1987				
9/7 – 9/28	21	8.0 – 14.4	0.309	138
9/28 – 10/19	21	14.5 – 19.6	0.242	
1988				
8/15 – 9/26	42	12.5 – 25.7	0.314	141
1989				
6/12 – 7/3	21	11.8 – 18.4	0.314	

N-compounds unloaded from the SE and delivered radially through the rays? If so, the conclusion could be that N-compounds are limiting the cambium's growth in particular. This would be in sharp contrast with the carbohydrates, which appear to be always present in abundance. Indeed, the results of girdling experiments showed clearly that synthesis and deposition of typical vegetative storage proteins in the wood rays can be triggered even during summer, e.g. above a girdle (Sauter and Neumann, 1994; Stepien and Sauter, 1994). This is support for the view that an increased import of N-compounds is required. The same conclusion was drawn when storage protein synthesis was induced in poplar trees by increased nitrogen supply and low temperature under long-day conditions (van Cleve and Apel, 1993).

The deposition rate of storage proteins in fall permits derivation of an estimate on the radial translocation capacity of the rays for N-compounds and thus also on the possible fluxes of N-compounds through the cambium. The protein deposition rates

(PDR) listed in *Table 2* show that they lie almost in the same order of magnitude as the starch deposition rates. The potential of the rays to translocate a high amount of N-compounds thus is evident, at least for the fall period. High rates are not found during the growth period possibly because the cambium itself is using up the incoming N-compounds. Based on these PDR, the radial flow of N-compounds can be estimated in the same way as was done for the sugars. For a PDR of 0.150 μg protein mg^{-1} DW day^{-1}, the radial flow of N-compounds through the ray cells of poplar expressed in glutamine (GLN) units becomes 74.6 pmol GLN cm^{-2} RCTA sec^{-1}. This result shows that the ray initials of the cambium must be able to translocate at least such an amount of N-compounds. Whether and to what extent the cambial cells are supplied with N-compounds via the xylem rays in spring when the protein bodies become mobilized within a relatively short period (Sauter and van Cleve, 1992), cannot be answered yet. The most prominent release of glutamine and other amino acids into the vessels parallels the rapid decrease in protein content of the wood during this stage and makes it impossible to judge on the amount translocated in radial direction.

3.6 *Effect of girdling on deposition of storage compounds*

The girdling of branches enables study of the effects of an artificially created surplus of photosynthate on the accumulation rates of storage materials, as well as on their final accumulation levels in the storage tissue, compared with controls. In this way circumstantial information on the flow rate of sugars and of N-compounds through the cambium can also be gathered. Results obtained from poplar branches 6 and 10 weeks

Table 2. *Protein deposition rates (PDR) and estimated minimum radial flux rates of nitrogen compounds (RFRN) in glutamine (GLN) units through the ray cell tangential area (RCTA) in the branchwood of poplar and willow.*

Period (year/month/day)	Period length (days)	Changes in protein content (μg mg^{-1} DW)	PDR (μg mg^{-1} DW day^{-1})	RFRN (pmol GLN cm^{-2} RCTA sec^{-1})
Populus				
1987				
8/8 – 9/9	32	1.15 – 3.13	0.062	
9/9 – 9/28	19	3.13 – 4.27	0.060	30.8
9/28 – 10/19	21	4.27 – 7.42	0.150	74.6
1988				
9/26 – 10/17	21	2.83 – 3.88	0.050	
10/17 – 10/24	7	3.88 – 7.27	0.484[a]	240.7[a]
Salix				
1987				
8/10 – 9/7	28	1.21 – 3.24	0.072	
9/7 – 10/19	42	3.24 – 5.33	0.050	24.9
10/19 – 11/9	21	5.33 – 6.39	0.050	

[a] Period of rapid leaf senescence and abscission.

after girdling are shown in *Table 3* (for details see Sauter and Neumann, 1994). It is seen (i) that starch deposition above the ring increased from 11.9 to 26.5 μg starch mg^{-1} DW. This proves that starch deposition normally is not proceeding at its maximum rate and can be stimulated. (ii) Girdling also increased sugar levels in ray cells inside the cambium substantially, both total sugar content and content of sucrose. They were increased by as much as 50% while reducing sugars stayed at about the same level. Interestingly, the protein content increased significantly in the rays above the girdle, indicating that induction of storage protein synthesis in summer is possible. These results illustrate that the flow of sugars and N-compounds through the cambium which normally takes place (see control twigs) can be increased substantially under experimental conditions. The maximum deposition rates shown for starch and protein, have not been surpassed under the climatic conditions of the experiment.

4. Sucrose synthase activity of the cambial zone

Considering the fact that a prominent flow of sugars passes the cambium continuously during its active stage, some fundamental questions arise: What determines the amount of sugars taken out of the radial pathway by the cambium as carbon skeletons for growth processes? What inhibits significant accumulation of starch in the plastids of the cambial cells when they are steadily passed by a prominent flow of sugars during summer? These questions illustrate that much more information on the physiology of cambial cells themselves, and in particular on the enzymes involved in the metabolism of carbohydrates is needed. Sucrose synthase (SuSy) is an enzyme which has been found at much increased levels in many sink tissues, e.g. in potato tubers (Geigenberger and Stitt, 1993), in developing tomato fruits (Wang *et al.*, 1993), in carrot roots (Sebkova *et al.*, 1995), and in bean pods (Sung *et al.*, 1994) and is believed to be involved in the regulation of the sink activity of an organ. Moreover, the concentration of sucrose in the cytoplasm of a sink tissue was suggested to regulate sucrose cleavage by changing both SuSy activity and the level of that enzyme (Claussen *et al.*, 1986). Recent results also showed increased activity of SuSy in differentiating xylem of *Robinia* (Hauch and Magel, 1998), indicating a possible pivotal role of SuSy in

Table 3. *Influence of girdling on deposition of starch and protein in 2–3-year old poplar twigs. The content is shown 6 (Jul 17) and 10 weeks (Aug 14) after girdling. The sugar content is given for sucrose and reducing sugars in addition to total content (cf. Sauter and Neumann, 1994).*

Date	Twig	Starch (μg mg^{-1} DW)	Protein (μg mg^{-1} DW)	Sugars (μg mg^{-1} DW)		
				Total	Sucrose	Reducing
July 17	control	11.9	2.2	11.1	5.0	4.2
	above	26.5	5.1	17.3	7.0	5.3
	below	15.7	2.6	8.9	3.9	4.8
Aug. 14	control	22.6	2.3	6.9	2.4	2.5
	above	32.0	4.7	11.0	7.7	2.5
	below	27.7[a]	3.4	4.5	2.8	1.2

[a] new laterals developed below the girdled region after remission from apical dominance.

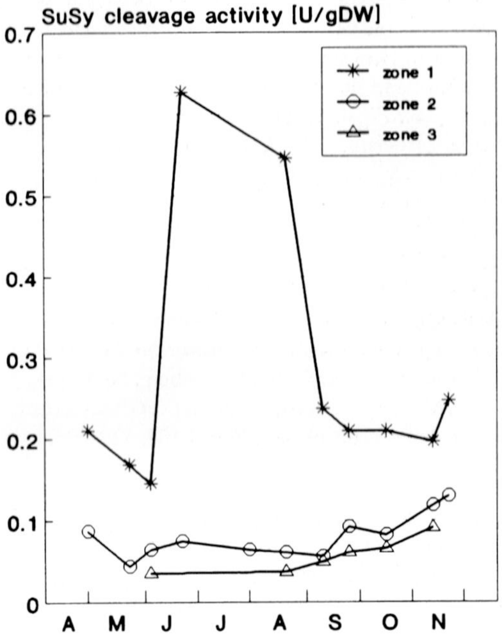

Figure 4. *Seasonal changes of sucrose synthase activity in a poplar stem. Zone 1, 2, 3 = zone directly inside the cambium, in the middle of sapwood, inner part of sapwood, respectively (after Schrader, 1999).*

supplying carbon skeletons for the cambial growth area. Interestingly, a membrane-associated form of SuSy was found by Amor *et al.* (1995) which is evidently involved in the synthesis of cellulose. The increased activity of SuSy in *Robinia* differentiation zone is now supported by almost identical results in a poplar stem. *Figure 4* shows the dramatic increase in SuSy activity in the differentiating zone interior to the cambium during early June; the activity of the middle and inner sapwood was much lower at the same time. Moreover, girdling experiments on poplar branches indicated that the sucrose concentration increased above the girdled part and also led to a concomitant increase in SuSy activity (Schrader, 1999). Cleavage of sucrose by SuSy in the cambium area thus seems to be a fundamental event intimately related both to the sink strength of this site and the potential in delivering the carbon skeletons necessary for well ordered cambial growth.

References

Aloni, R. (1987) Differentiation of vascular tissues. *Ann. Rev. Plant Physiol.* **38**: 179–204.

Aloni, R. (1995) Differentiation of the ray system in woody plants. *Bot. Rev.* **61**: 45–84.

Amor, Y., Haigler C.H., Johnson, S., Wainscott, M. and Delmer, D.P. (1995) A membrane-associated form of sucrose synthase and its potential role in synthesis of cellulose and callose in plants. *Proc. Natl. Acad. Sci. (USA)* **92**: 9353–9357.

Barnett, J.R. (1992) Reactivation of the cambium in *Aesculus hippocastanum* L.: a transmission electron microscope study. *Ann. Bot.* **70**: 169–177.

Catesson, A.-M. (1994) Cambial ultrastructure and biochemistry: changes in relation to vascular tissue differentiation and the seasonal cycle. *Int. J. Plant Sci.* **155**: 251–261.

Claussen, W., Loveys, B.R. and Hawker, J.S. (1986) Influence of sucrose and hormones on the activity of sucrose synthase and invertase in detached leaves and leaf sections of eggplants (*Solanum melongena*). *J. Plant Physiol.* **124**: 345–357.

Czaninski, Y. (1977) Vessel-associated cells. *IAWA Bull. n.s.* **3**: 51–55.

Dünisch, O., Bauch, J., Müller, M. and Greis, O. (1998) Subcellular quantitative determination of K and Ca in phloem, cambium, and xylem cells of spruce (*Picea abies* (L.) Karst.) during earlywood and latewood formation. *Holzforschung* **52**: 582–588.

Farrar, J.J. and Evert, R.F. (1997) Seasonal changes in the ultrastructure of the vascular cambium of *Robinia pseudoacacia*. *Trees* **11**: 191–201.

Gunning, B.E.S. and Robards, A.W. (1976) Plasmodesmata: current knowledge and outstanding problems. In: *Intercellular Communication in Plants: Studies on Plasmodesmata* (eds. B.E.S. Gunning and A.W. Robards). Springer, Berlin, Heidelberg, New York, pp. 297–311.

Geigenberger, P. and Stitt, M. (1993) Sucrose synthase catalyses a readily reversible reaction *in vivo* in potato tubers and other plant tissues. *Planta* **189**: 329–339.

Hansen, J. and Beck, E. (1994) Seasonal changes in the utilization and turnover of assimilation products in 8-year-old Scots pine (*Pinus sylvestris* L.) trees. *Trees* **8**: 172–182.

Harms, U. and Sauter, J.J. (1991) Storage proteins in the wood of *Taxodiaceae* and of *Taxus*. *J. Plant Physiol.* **138**: 497–499.

Hauch, S. and Magel, E. (1998) Extractable activities and protein content of sucrose-phosphate synthase, sucrose synthase and neutral invertase in trunk tissues of *Robinia pseudoacacia* L. are related to cambial wood production and heartwood formation. *Planta* **207**: 266–274.

Little, C.H.A and Pharis, R.P. (1995) Hormonal control of radial and longitudinal growth in the tree stem. In: *Plant Stems: Physiology and Functional Morphology* (ed. B.L. Gartner). Academic Press, San Diego, pp. 289–319.

Patrick, J.W. (1997) Phloem unloading: sieve element unloading and post-sieve element transport. *Annu. Rev. Plant Physiol. Plant Mol. Biol.* **48**: 191–222.

Sauter, J.J. (1972) Respiratory and phosphatase activities in contact cells of wood rays and their possible role in sugar secretion. *Z. Pflanzenphysiol.* **67**: 135–145.

Sauter, J.J. (1974) Structure and physiology of Strasburger cells. *Ber. Dtsch. Bot. Ges.* **87**: 327–336.

Sauter, J.J. (1976) Analysis of amino acids and and amides in the xylem sap of *Salix caprea* L. in early spring. *Z. Pflanzenphysiol.* **79**: 276–280.

Sauter, J.J. (1980) The Strasburger cells — equivalents of companion cells. *Ber. Dtsch. Bot. Ges.* **93**: 29–42.

Sauter, J.J. (1981) Evidence for sucrose efflux and hexose uptake in the xylem of *Salix*. *Z. Pflanzenphysiol.* **103**: 183–187.

Sauter, J.J. (1982) Transport in Markstrahlen. *Ber. Dtsch. Bot. Ges.* **95**: 593–618.

Sauter, J.J. (1983) Efflux and reabsorption of sugars in the xylem. II. Seasonal changes in sucrose uptake in *Salix*. *Z. Pflanzenphysiol.* **111**: 429–440.

Sauter, J.J. and Braun, H.J. (1968) Histologische und cytochemische Untersuchungen zur Funktion der Baststrahlen von *Larix decidua* Mill., unter besonderer Berücksichtigung der Strasburger Zellen. *Z. Pflanzenphysiol.* **59**: 420–438.

Sauter, J.J. and Kloth, S. (1986) Plasmodesmatal frequency and radial translocation rates in ray cells of poplar (*Populus* × *canadensis* Moench *'robusta'*). *Planta* **168**: 377–380.

Sauter, J.J. and van Cleve B. (1989) Immunochemical localization of a willow storage protein with a poplar storage protein antibody. *Protoplasma* **149**: 175–177.

Sauter, J.J. and van Cleve B. (1990) Biochemical, immunochemical and ultrastructural results on protein storage in poplar (*Populus* × *canadensis 'robusta'*) wood. *Planta* **183**: 92–100.

Sauter, J.J. and van Cleve B. (1992) Seasonal variation of amino acids in the tracheal sap of *Populus* × *canadensis* and its relation to protein body mobilization. *Trees* **7**: 26–32.

Sauter, J.J. and Neumann, U. (1994) The accumulation of storage material in ray cells of poplar wood (*Populus × canadensis 'robusta'*): Effect of ringing and defoliation. *J. Plant Physiol.* **143**: 21–26.

Sauter, J.J. and Witt, W. (1997) Structure and function of rays: storage, mobilization, transport. In: *Trees – Contributions to Modern Tree Physiology* (eds. H. Rennenberg, W. Eschrich and H. Ziegler). Backhuys Publishers, Leiden, pp. 177–195.

Sauter, J.J., Dörr, I. and Kollmann, R. (1976) The ultrastructure of Strasburger cells (= albuminous cells) in the secondary phloem of *Pinus nigra var. austriaca* (Hoess) Badoux. *Protoplasma* **88**: 31–49.

Sauter, J.J., Wisniewski, M. and Witt, W. (1996) Interrelationships between ultrastructure, sugar level, and frost hardiness of ray parenchyma cells during frost acclimation and deacclimation in poplar (*Populus × canadensis* Moench *'robusta'*) wood. *J. Plant Physiol.* **149**: 451–461.

Savidge, R.A. (1996) Xylogenesis, genetic and environmental regulation – a review. *IAWA Journal* **17**: 269–310.

Schrader, S. (1999) Enzyme des Saccharosestoffwechsels im Holz von *Populus* und ihre Rolle bei den jahresperiodischen Kohlenhydratveränd erungen Doctoral thesis, U. Kiel, 159 pp.

Schulz, A. (1998) Phloem. Structure related to function. *Progress in Botany* **54**: 429–475.

Sebkova, V., Unger, C., Hardegger, M. and Sturm, A. (1995) Biochemical, physiological and molecular characterization of sucrose synthase from *Daucus carota*. *Plant Physiol.* **108**: 75–83.

Stepien, V. and Sauter, J.J. (1994) Ringing induces the accumulation of vegetative storage proteins in poplar bark. *Trees* **9**: 88–92.

Stepien, V., Sauter, J.J. and Martin, F. (1994) Vegetative storage proteins in woody plants. *Plant Physiol. Biochem.* **32**: 185–192.

Sung, S.S., Sheih, W.J., Geiger, D.R. and Black, C.C. (1994) Growth, sucrose synthase, and invertase activities of developing *Phaseolus vulgaris* L. fruits. *Plant, Cell and Environment* **17**: 419–426.

van Bel, A.J.E. (1990) Xylem-phloem exchange *via* the rays: the undervalued route of transport. *J. Exp. Bot.* **41**: 631–644.

van Bel, A.J.E. (1993) The transport phloem. Specifics of its functioning. *Progress in Botany* **54**: 134–150.

van Cleve, B., Clausen, S. and Sauter, J.J. (1988) Immunochemical localization of a storage protein in poplar wood. *J. Plant Physiol.* **133**: 371–374.

van Cleve, B. and Apel, K. (1993) Induction by nitrogen and low temperature of storage-protein synthesis in poplar trees exposed to long days. *Planta* **189**: 157–160.

Wang, F., Sanz, A., Brenner, M.L. and Smith, A. (1993) Sucrose synthase, starch accumulation and tomato fruit sink strength. *Plant Physiol.* **101**: 321–327.

Wetzel, S., Demmers, C. and Greenwood, J.S. (1989) Seasonally fluctuating bark proteins are a potential form of nitrogen storage in three temperate hardwoods. *Planta* **178**: 275–281.

Zimmermann, M.H. (1971) Transport in the phloem. In: *Trees structure and function* (eds. M.H. Zimmermann and C.L. Brown). Springer, New York, Heidelberg, Berlin, pp. 221–279.

7

Symplasmic organization of the transport phloem and the implications for photosynthate transfer to the cambium

A.J.E. van Bel and K. Ehlers

1. Introduction

Based on the limited knowledge at present, an attempt is made to identify modes and pathways of photosynthate transfer from phloem to cambium. The analysis will show the importance of the symplasmic organization of cambium and cambium-associated vascular tissues, and the positioning of photosynthate carriers in symplast domains. The analysis will further raise speculations on the physiological implications of the potential pathways for photosynthate transfer and their impact on photosynthate allocation within the plant.

The cambium is an axial sink tissue, the occurrence of which is limited to a few ferns, the gymnosperms and the dicotyledons. It is nourished through lateral supply by the vascular tissues, both xylem and phloem. The cambium receives photoassimilates from the transport phloem. This notion prompts us to discriminate between the respective sections of the phloem in order to define the physiological environment of the cambium. Therefore, the sequential tasks of the phloem are reviewed briefly in the next section.

2. The different tasks of the phloem in seed plants

The phloem in seed plants comprises three functional sequential units each of which has a specific task (*Figure 1*). The phloem collects the photosynthate in the phloem loading zone of the mature leaves (collection phloem), distributes the photosynthate towards the centres of growth and storage (transport phloem), where the

Cell & Molecular Biology of Wood Formation, edited by R. Savidge, J. Barnett and R. Napier.

photosynthate is released into the proliferating, expanding or accumulating cells in the phloem unloading zone of the terminal sinks (release phloem).

As pointed out above, the cambium itself is nourished by the transport phloem, along which simultaneous release/retrieval processes are taking place (Minchin and Thorpe, 1987). The cambium and the other heterotrophic cells of the stem profit from the release of photoassimilates from the transport phloem. The conducting elements (sieve cells associated with Strasburger cells in gymnosperms, and sieve elements (SEs) or sieve-tube members associated with companion cells (CCs) in angiosperms) are arranged in longitudinal arrays (sieve tubes). The ultrastructure of the sieve elements is discussed extensively in '*Sieve Elements*' edited by Behnke and Sjolund (1990).

In dicotyledons, the structure of the conducting elements seems to reflect the function of each particular phloem unit. Whilst the CCs are much larger than the SEs in the phloem loading zone (Gamalei, 1989), the CCs cover only about 30% of the SE-surface in the transport phloem (at least, in evolutionarily advanced dicotyledons; van Bel 1996), and are smaller or even missing in the phloem unloading zone (Offler and Patrick, 1984; Schulz, 1994). The decreasing volume ratio between companion cells and sieve elements along the phloem pathway may be related to a decreasing photosynthate retention in collection, transport and release phloem, respectively. The CC/SE volume ratios thus suggest that transport phloem is a functional hybrid of collection and release phloem permitting a controlled and limited lateral escape of photosynthate along the transport path.

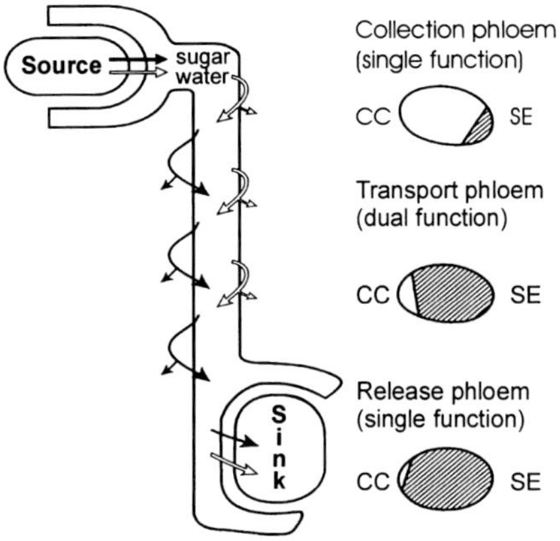

Figure 1. *A dynamic version of the Münch mass flow model (from van Bel, 1996). Photoassimilates are translocated via the phloem through essentially leaky instead of hermetically sealed pipes. The solute content, and implicitly the turgor, are controlled by release/retrieval mechanisms in the sieve element/companion cell complexes (SE/CC-complexes). Differential release/retrieval balances control the influx/efflux of sugars and water in the various phloem zones. In the collection phloem (phloem loading), the influx will dominate, in the release phloem (phloem unloading) the efflux. In the transport phloem having a dual task (nourishment of axial and terminal sinks), the balance between influx and efflux varies with the requirements of the plant. The gradual loss of solute and commensurate amounts of water towards the sink, where massive delivery of water and solutes takes place, has been ascribed to a proton-motive force gradient in the SE/CC-complexes from source to sink (van Bel, 1995). Alternatively, the relative size reduction of companion cells along the source-to-sink path may explain a decreasing retrieval capacity of the SE/CC-complexes in the direction of the sink.*

3. Ultrastructure of the companion cells in dicotyledons

The SEs are enucleate and possess a highly reduced cytoplasm (Behnke and Sjolund, 1990). As the SEs themselves are only capable of a very limited range of activities, the larger part of their physiological functions is taken over by the CCs. The CCs also play a major role in controlling the release/retrieval processes along the transport phloem (Van Bel, 1993; Stadler *et al.*, 1995; Stadler and Sauer, 1996; van Bel and Kempers, 1997; Kühn *et al.*, 1997). All CCs share a few ultrastructural traits; they possess a dense cytoplasm containing many mitochondria and ribosomes. They are connected symplasmically to the SEs by PPUs (pore-plasmodesma units). These specialized cell connections are branched at the side of the CC and single-channelled at the side of the SE. Probably, the PPUs are the key elements in the interaction between SE and CC (see for a review, Van Bel and Kempers, 1997) suggesting in turn that the SE/CC-complex operates as a functional unit.

CCs show structural peculiarities which may mirror functional specialization. For the CCs in the collection phloem, the structural adaptations are the most obvious. In the phloem loading zone, three characteristic types of companion cells are distinguished: transfer cells (TCs), smooth-walled cells or ordinary companion cells (SCs or OCCs) and intermediary cells (ICs). The TCs have virtually no plasmodesmata at the interface with the mesophyll (< 0.1 plasmodesmata per μm^2; Gamalei, 1989), possess large central vacuoles and cell wall protrusions commensurate with the transit of photosynthate (Gamalei, 1989; Wimmers and Turgeon, 1991; Gamalei *et al.*, 1992). The SCs (Turgeon, 1996; Turgeon and Medville, 1998) have an intermediate number of plasmodesmata at the interface with the mesophyll (0.1–10 plasmodesmata per μm^2; Gamalei, 1989), contain a few vacuoles and possess smooth cell walls (Turgeon, 1996). The ICs are connected symplasmically to the mesophyll by many plasmodesmata (> 10 plasmodesmata per μm^2; Gamalei, 1989), contain highly fragmented vacuoles or vesicles (Turgeon *et al.*, 1975; Gamalei, 1989) and have no cell wall protrusions. The occurrence of TCs and ICs is associated with distinctly different modes of phloem loading (Turgeon and Wimmers, 1988; van Bel *et al.*, 1992, 1994): an apoplasmic mode via carriers in the TC plasma membrane and a symplasmic mode using a plasmodesmal pathway (ICs). Different mechanisms of phloem loading may be correlated with carbohydrate metabolism (Gamalei, 1985; Turgeon *et al.*, 1993; Flora and Madore, 1996) and will vary from species to species. In apoplasmically loading species organic carbon is translocated in the form of sucrose, whereas complementary or prevailing carbon translocation as galactosyl-oligosaccharides takes place in symplasmically loading species.

In a recent study, the ultrastructure of CCs in the secondary transport phloem of four dicotyledons was investigated (Kempers *et al.*, 1998). All CCs shared the usual traits, but differed in the substructure of the cell wall. Whilst the CCs of *Zinnia* and *Vicia* showed cell wall invaginations, the companion cells of *Lythrum* and *Cucurbita* possessed smooth walls. The vacuoles in the latter two species were divided with the strongest vesiculation in *Cucurbita*.

The differences in CC-ultrastructure in the transport phloem appear to bear a weak resemblance to those in the collection phloem of the same species (Kempers *et al.*, 1998). The minor vein phloem of *Zinnia* and *Vicia* contains TCs, that of *Lythrum* probably SCs, and that of *Cucurbita* ICs (Gamalei, 1991). Taken together, there is some evidence in favour of a functional continuity between the collection and the transport phloem (Kempers *et al.*, 1998). Thus, symplasmic organization and carbohydrate management in the transport phloem, and, consequently, in the 'cambial environment' may differ between the species.

4. Translocation of photosynthate and lateral release from the transport phloem in dicotyledons

A matter of debate for a long time has been whether or not translocation in sieve tubes occurs by mass flow (see in Zimmermann and Milburn, 1975). Recently, confocal laser scanning images of fluorochromes moving in the sieve tubes of intact *Vicia* plants provided convincing support in favour of mass flow (Knoblauch and van Bel, 1998). The fluorescent front moved steadily; the phloem-mobile probes did not halt or accumulate at the sieve pores despite the presence of protein deposits on the sieve plate wall. Mass flow, however, does not imply that all solutes in the phloem stream are being translocated at the same speed (Eschrich *et al.*, 1972; van Bel, 1993). As in xylem vessels (van Bel, 1976), differential lateral escape of the solutes will result in a different net translocation rate for each single compound. The lateral escape rate depends on accumulation mechanisms along the pathway (van Bel, 1993). This behaviour is exemplified by the retention of fluorochromes that have escaped from the sieve tubes and become sequestered in the CC vacuoles along the translocation path.

Cambial activity and that of other axial sinks may be one of the factors which control lateral escape of photosynthate from the sieve tubes. In turn, the cambial capacity of accumulating photosynthate is correlated with the symplasmic organization of the transport phloem (*Figure 2*). If SE/CC-complexes, the eventual intervening phloem parenchyma cells, and the cambium are linked intimately by plasmodesmal connections, transfer to the cambium depends on the photosynthate gradient through the symplast (*Figure 3a*). However, should there be any symplasmic barrier on the transport pathway, carrier-mediated transport may be the limiting factor of photosynthate transfer to the cambium (*Figures 3b–d*).

Symplasmic organization is also a major determinant of the mechanism of unloading in terminal sinks (Patrick and Offler, 1996; Fisher and Oparka, 1996). Unloading from release phloem is supposed to be symplasmic in vegetative sinks such as root tips (Oparka *et al.*, 1994, 1995; Schulz, 1994) and young leaves (Turgeon, 1987; Schmalstig and Geiger, 1987), attributed to what appears to be a symplasmically continuous route from sieve element to sink cells. In elongating shoots, phloem unloading was concluded to be mainly symplasmic (McNeil, 1976; Schmalstig and Cosgrove, 1990), but an apoplasmic component may operate in parallel (Minchin and Thorpe, 1987; Wood *et al.*, 1998).

5. Experimental assessment of the symplasmic organization of the transport phloem

As outlined above, the cellular organization of the transport phloem indicates the pathways and mechanisms involved in the lateral photosynthate transfer to the cambium. The spatial positioning of the phloem and cambium cells and their plasmodesmal connectivity are key factors to be considered. Actually, our knowledge on the symplasmic organization of the transport phloem is restricted to the symplasmic

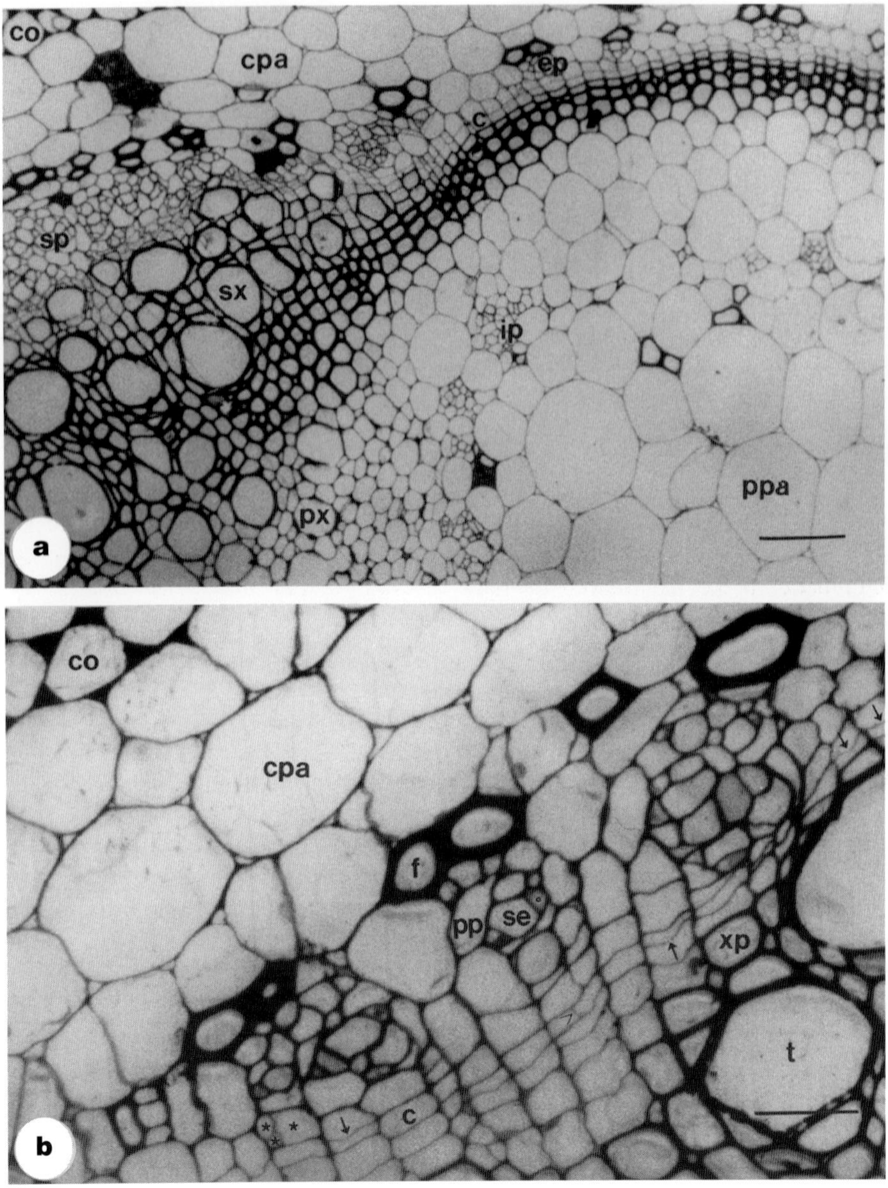

Figure 2. *Spatial position of the cells involved in photoassimilate supply of the cambium as exemplified by cross-sections of a vascular bundle (a) and the phloem/cambium zone (b) in the tomato stem. (a) Semithin cross-section of a vascular bundle (bar = 100 μm). co = collenchyma, cpa = cortex parenchyma, sp = secondary phloem, sx = secondary xylem, px = primary xylem, ep = external phloem, c = cambium, ip = internal phloem, ppa = pith parenchyma. (b) Semithin cross-section of the external phloem/cambium zone (bar = 50 μm). f = fibre, pp = phloem parenchyma, se = sieve element, ° = companion cell, * = differentiating SE/CC-complex, > = new cell wall (tangential), ↓ = new cell wall (radial), xp = xylem parenchyma, t = tracheary element.*

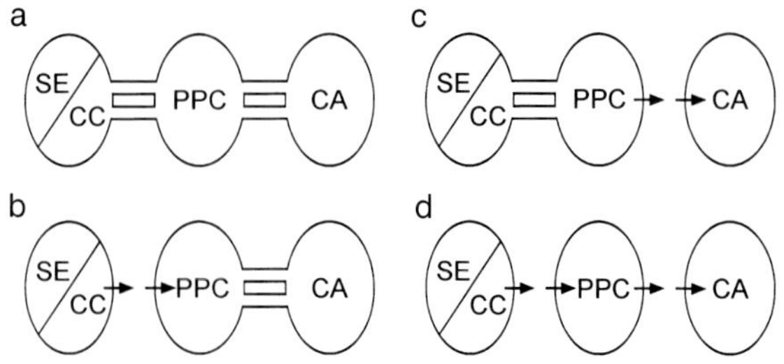

Figure 3. Potential pathways of photosynthate transfer from the sieve element/companion cell complex (SE/CC) to the cambium (CA) in transport phloem. (a) Symplasmic continuity from SE/CCs to the cambium. (b) Symplasmic discontinuity at the interface between SE/CC and the intervening phloem parenchyma cells (PPCs). (c) Symplasmic discontinuity at the PPC-CA interface. (d) Symplasmic discontinuity at the SE/CC-PPC and the PPC-CA interfaces. For reasons outlined in the text, configuration (b) is supposed to be the most common one. Arrows represent transmembrane movement of photosynthate.

connectivity between SE/CC-complexes and adjacent phloem parenchyma cells (PPCs) (Kempers *et al.*, 1998).

5.1 *Plasmodesmograms of the SE/CC-complexes in the transport phloem*

Plasmodesmograms visualizing the plasmodesmal frequencies at the interfaces between transport phloem cells (Kempers *et al.*, 1998) suggest that, above all, the interfaces between SE/CC-complexes and PPCs present the largest barrier to symplasmic transport. In the transport phloem tissues investigated, the plasmodesmal frequencies were by far the lowest at these interfaces. If symplasmic transport is engaged in photosynthate transport to the cambium, this interface may present a critical and limiting step.

The interspecific differences in symplasmic connectivity between CCs and adjacent PPCs are smaller in transport than in collection phloem. The ratio between the highest and the lowest plasmodesmal frequencies in the species investigated is about 10 (Kempers *et al.*, 1998). The interspecific differences in plasmodesmal frequency between the CCs in the collection phloem and the adjacent mesophyll tissues vary by a factor of 1000 (Gamalei, 1991). This emphasizes once again that the distinct differences in ultrastructure between the CCs in the collection phloem, observed in diverse species, are less evident in transport phloem.

5.2 *Membrane potential mapping*

Open plasmodesmata provide pathways for electrical currents (Lew, 1994, 1996; Holdaway-Clarke *et al.*, 1996), so that membrane potentials tend to equalize between cells which are linked symplasmically. Should cell groups be isolated symplasmically from each other, differences in membrane potential can be maintained. Thus clusters of differential membrane potentials are diagnostic for the position of symplasmic barriers.

Membrane potentials were recorded systematically in the vascular tissues, including the phloem, of tomato petioles (van der Schoot and van Bel, 1989, 1990). Clusters of

Table 1. *Membrane potentials of phloem parenchyma and the sieve element/companion cell complex (SE/CC-complex) in transport phloem. The membrane potentials of the upper group were measured by blind impalement or use of aphids. The lower group was measured using glass microelectrodes for impalement into tissue slices under the microscope. Separate data for sieve element (second number) and companion cell (first number) are given for Solanum (from van Bel, 1993).*

	Phloem parenchyma (mV)	SE/CC complex (mV)	Reference
Mimosa pudica	–60	–160	Sibaoka (1962)
Salix exigua		–160	Wright and Fisher (1981)
Mimosa pudica		–150	Fromm and Eschrich (1988b)
Cucurbita pepo		–170	Eschrich *et al.* (1988)
Cucurbita pepo		–150	Opritov and Pyatygin (1989)
Solanum lycopersicum	–70	–140/-150	van der Schoot and van Bel (1990)
Ricinus communis	–60	–90	van Bel and Kempers (1991)
Salix alba	–50	–130	van Bel and Kempers (1991)
Lupinus albus	–70	–105	van Bel and van Rijen (unpubl.)

similar membrane potentials showed up suggesting the existence of symplasmic domains in vascular tissues (van der Schoot and van Bel, 1990). In subsequent experiments, fluorochromes were injected into vascular cells of the same tissues. The distribution of fluorochromes after intracellular injection confirmed the existence of symplasmic domains which largely overlie the membrane potential clusters. The membrane potential maps indicated a symplasmic isolation between SE/CC-complexes and PPCs in the transport phloem (van der Schoot and van Bel, 1990). In tomato, the membrane potentials of the SE/CC-complexes were more negative than those of the neighbouring cells. Other studies (Sibaoka, 1962; van Bel and Kempers, 1991; van Bel and van Rijen, 1994) showed similar tendencies (*Table 1*). All these data suggest that assimilate transfer to the cambium resembles models (b) or (d) in *Figure 3*.

Later investigations revealed a more differentiated pattern (van Bel, 1996). In all species studied, membrane potentials of SE/CC-complexes and PPCs in the transport phloem were distinctly different. However, the ratios between the membrane potentials fell into two categories. In one group of species (*Lupinus, Senecio, Vicia*) belonging to the apoplasmic loaders, the membrane potentials of the SE/CC-complexes were more negative than that of the PPCs (*Table 2*). In the other group (*Epilobium, Lamium, Ocimum*) representing the symplasmic loaders, the situation was the reverse (*Table 2*).

5.3 *Intracellular injection and application of fluorochromes*

In initial intracellular injections into SE/CC complexes of the transport phloem (van der Schoot and van Bel, 1989; van Bel and Kempers, 1991; van Bel and van Rijen, 1994), membrane-impermeant fluorochromes remained restricted to the SE/CC-complexes. This phenomenon seems to confirm that the few plasmodesmata between SE/CC-

Table 2. The ratio ($\Delta \Psi_{SE/CC} / \Delta \Psi_{PP}$) between the membrane potentials ($\Delta \Psi$) of the sieve element/companion cell complex (SE/CC) and phloem parenchyma (PP) in the transport phloem of symplasmic (type 1) and apoplasmic species (type 2) (from van Bel, 1996).

	$\Delta \Psi_{SE/CC} / \Delta \Psi_{PP}$
Type 1	
Epilobium angustifolium	0.80
Lamium album	0.60
Ocimum basilicum	0.76
Type 2	
Lupinus luteus	1.26
Senecio viscosus	1.87
Senecio vulgaris	0.94
Vicia faba	1.16

complex and PPC (*Figure 4*) were non-functional or temporarily closed. Yet, it was suspected that the closure is due to the use of tissue strips. Stripping or excision of phloem tissue would have induced wound-triggered plasmodesmal closure at the interface between SE/CC-complex and PPCs.

However, fluorochrome application to or injection into intact plants produced an identical restriction to the SE/CC-complexes in the transport phloem (Oparka *et al.*, 1994, 1995; Rhodes *et al.*, 1996; Knoblauch and van Bel, 1998). The fluorescent dye 5,6-carboxyfluorescein diacetate applied to the leaves of *Arabidopsis* stayed within the SE/CC-complexes in the transport phloem of the root. The concomitant escape of the fluorochrome from the release phloem in the root tip (Oparka *et al.*, 1994, 1995) was a strong argument for a symplasmic isolation of the SE/CC-complexes in the transport phloem. Symplasmic discontinuity between SE/CC-complexes and PPCs in the transport phloem does not need to be permanent, though. Release of 5,6-carboxyfluorescein from transport phloem sieve elements under certain growing conditions indicates temporary gating of the plasmodesmata at the interface between SE/CC-complexes and PPC (Hayes *et al.*, 1987; Shen and Patrick, 1990; Patrick and Offler, 1996). This would allow symplasmic transport possibly in parallel with the apoplasmic transfer. Reviews of terminal sinks reveal a complexity in the post-sieve element pathways that may also apply to phloem unloading into the cambium (Fisher and Oparka, 1996; Patrick and Offler, 1996).

6. The organization of the transport phloem and the post-phloem pathways

The pathways of photosynthate transfer from the phloem to the cambium may also be multiple (*Figure 4*). Several potential pathways which may function in parallel are available for the lateral photosynthate transfer to the cambium. What we know with some certainty is that the photoassimilates are delivered by the SE/CC-complexes. What happens thereafter is speculative.

Experimental assessment indicates that most of the photosynthate is released into the apoplast (Oparka *et al.*, 1994, 1995; Rhodes *et al.*, 1996; Knoblauch and van Bel,

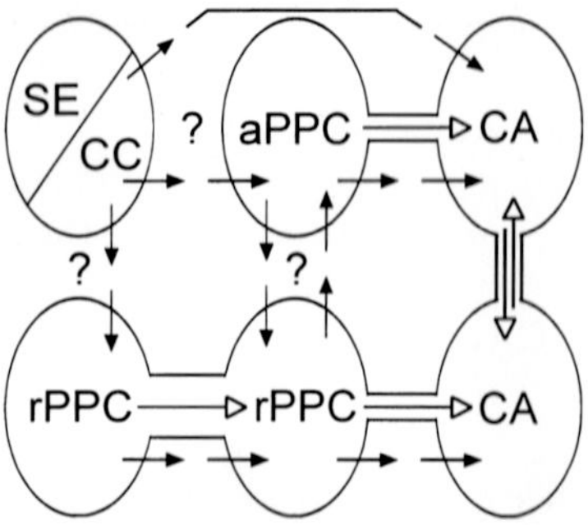

Figure 4. The spatial complexity of the structural frame for lateral photosynthate transfer in the transport phloem. Symplasmic connectivity is only known for the interfaces between SE/CC and PPC (Kempers et al., *1998) and between the PPCs (Sauter, 1982; Sauter and Kloth, 1986; van Bel, 1990; van der Schoot and van Bel, 1990). At interfaces showing symplasmic discontinuity, only transmembrane movement of photosynthate takes place. At any interface with symplasmic continuity, however, apoplasmic transfer may occur in parallel. Apoplasmic and symplasmic transfer is represented by arrows with closed and open arrow heads, respectively. The widths of the symplasmic corridors are estimates of the symplasmic connectivity. Question marks identify the sites where apoplasmic transfer may occur despite the presence of limited numbers of plasmodesmata. Intercellular transport of injected fluorochromes was not observed at these interfaces (van der Schoot and van Bel, 1989; van Bel and Kempers, 1991; van Bel and van Rijen, 1994; Rhodes* et al., *1996). The same applies to carboxyfluorescein diacetate application to intact plants (Oparka* et al., *1994, 1995; Knoblauch and van Bel, 1998), albeit not in all cases (Patrick and Offler, 1996). SE/CC = sieve element/companion cell complex, aPPPC = axial phloem parenchyma cell, rPPC = radial phloem parenchyma cell, CA = cambial cell.*

1998). Limited experimental evidence obtained by intracellular injection of fluorochromes (van der Schoot and van Bel, 1990) suggests symplasmic connectivity between phloem parenchyma and cambium. According to this concept, photoassimilates are taken up from the apoplast either directly by the cambium or by the PPCs followed by symplasmic transfer to the cambium (*Figure 4*).

Under certain circumstances, a symplasmic pathway is available in parallel (Hayes *et al.*, 1987; Shen and Patrick, 1990; Patrick and Offler, 1996). In this instance, photoassimilates are transferred from SE/CC-complex to the cambium through a symplasmic continuum (*Figure 4*).

An important point to note is the behaviour of differentiating cells in the cambium. It is known that cambial cells are connected symplasmically among themselves (van der Schoot and Van Bel, 1990; van Bel and van Rijen, 1994). Cambial cells differentiating into SE/CC-complexes tend to shut off all their plasmodesmal connections (van Bel and van Rijen, 1994)

7. Conclusions on the symplasmic organization of transport phloem

(i) Low plasmodesmal frequencies at the interface between SE/CC-complex and PPCs in the transport phloem (Kempers *et al.*, 1998) suggest a bottleneck for symplasmic transport as illustrated in *Figures 3b* or *3d*.

(ii) Despite the presence of the plasmodesmal connections between SE/CC-complex and PPC, fluorochromes stay in the SE/CC-complexes, with the exception of plants grown under some specific conditions.
(iii) The distinctly different membrane potentials also indicate closure of the plasmodesmata between SE/CC-complexes and PPCs in the transport phloem.
(iv) Closure of the plasmodesmata at the interface implies an apoplasmic photoassimilate transit from SE/CC-complex to PPC. Potential gating of these plasmodesmata would allow symplasmic photoassimilate transfer from the transport phloem.
(v) If photoassimilates are released into the apoplast, the energization and number of carrier species involved in transport over the plasma membrane of the SE/CC-complex will be decisive for the rate of release. The same applies for the uptake by the receiver cells.
(vi) With the exception of the symplasmic connectivity between SE/CC-complexes and the adjacent PPCs, little is known on the symplasmic organization of the SE-to-cambium pathway.

8. The battle for photosynthates in the apoplast and the consequences for photosynthate supply to the cambium

As a rough assessment of the symplasmic organization, we assume that SE/CC-complexes in the transport phloem are largely isolated and the majority of the photo-

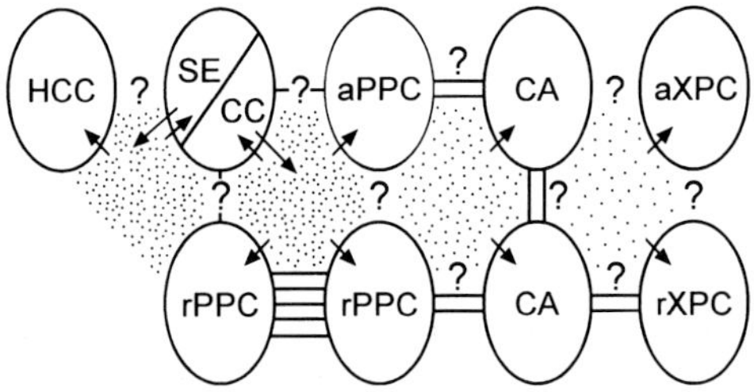

Figure 5. The competition for photosynthate in the apoplast on the basis of the presumptive symplasmic organization of transport phloem and adjoining tissues. Competitiveness in retrieving photosynthate from the apoplast (represented by the inwardly directed arrows) is determined by the thermodynamic forces, the number and affinity of carriers on the respective plasma membrane surfaces and, possibly, the metabolic activity. Functional symplasmic continuity between SE/CC-complex and the surrounding cells is presumably low (see section 4), as well as the continuity between axial and radial parenchyma (van der Schoot and van Bel, 1990). The radial parenchyma probably possesses abundant symplasmic continuity (e.g. Sauter and Kloth, 1986). The striping represents the estimated functional symplasmic continuity. Question marks indicate different degrees of uncertainty concerning symplasmic continuity: no data on the plasmodesmal frequency and symplasmic transfer (free question mark), plasmodesmal frequencies known, but no symplasmic transfer of fluorochromes observed (question marks within the plasmodesmal striping), plasmodesmal frequencies unknown, but symplasmic transfer of fluorochromes observed (question marks outside the plasmodesmal striping). Stippling represents the approximate apoplasmic concentration of photosynthate. SE/CC = sieve element/companion cell complex, HCC = heterotrophic cortex cell, aPPC = axial phloem parenchyma cell, rPPc = radial phloem parenchyma cell, CA = cambial cell, aXPC = axial xylem parenchyma cell, rXPC = radial xylem parenchyma cell.

synthate is released into the apoplast. Part of the photosynthate is retrieved by the SE/CC-complexes, the rest is absorbed by vascular and non-vascular cells (*Figure 5*). Much of the apoplasmic photosynthate will be taken up by PPCs which are presumed to be symplasmically connected to the cambial cells.

This concept conforms with the existence of an apoplasmic pool built up by permanent leakage from the sieve elements of the transport phloem (Minchin and Thorpe, 1984; Minchin *et al.*, 1984). The net photosynthate release from the SE/CC-complexes into the apoplast results from the balance between release and retrieval (Aloni *et al.*, 1986; Minchin and Thorpe, 1987; Grimm *et al.*, 1990). In contrast to this view, PCMBS (an inhibitor of sucrose – H^+ co-transport) did not affect photosynthate translocation (Grimm *et al.*, 1997). The mechanism of release is unknown, but may be facilitated diffusion driven by the outward-directed gradient of photosynthates.

The cells surrounding the phloem apoplast are competing for the photosynthate (*Figure 5*). Among the SE/CC-complexes, the cambium-associated PPCs and the cambium cells, the cells kinetically best fitted for the job will win this competition. The strength of the thermodynamic sink depends on the transport machineries (carriers + driving forces) of the plasma membranes involved (*Figure 5*).

Little is known about the carrier systems and driving forces of the PPCs (van Bel, 1996). As for sucrose retrieval by SE/CC-complexes, carrier systems were found in the phloem of celery petioles (Daie, 1986). Later, sucrose carriers in the transport phloem were identified and localized in transgenic plants. The carriers were found to be located on the CC plasma membrane in *Plantago* (Stadler *et al.*, 1995) and *Arabidopsis* (Stadler and Sauer, 1996) and on the SE plasma membrane in Solanaceae (Kühn *et al.*, 1997), respectively. It is thought that the proton-driven retrieval by the SE/CC-complex is favoured by a steep pH-gradient (van Bel, 1993, 1996).

9. Speculations on the consequences of the phloem design for photosynthate allocation between axial and terminal sinks

For reasons of convenience, a simplified concept of the symplasmic framework in which lateral transport to the cambium occurs is adopted. The considerations here are restricted to three compartments bordering the transport phloem apoplast: the sieve element/companion cell complexes, the phloem parenchyma cells and the cambial cells (*Figure 6*).

As has been set out before, the ratios between the membrane potentials of SE/CC-complexes and PPCs in the transport phloem fall into two categories (*Table 2*). There seems to be a correlation between the membrane potential ratios and the configuration of the collection phloem. Given the structural continuity of the phloem in various species (Kempers *et al.*, 1998), the ultrastructure of the CCs in the transport phloem may be correlated with their ability to retrieve apoplasmic photosynthate. Species with TCs in the minor veins ('apoplasmic species') and transport phloem show more negative SE membrane potentials in the transport phloem than species with ICs in the minor veins ('symplasmic species').

The differences in cell (ultra)structure and the associated cell physiology of the phloem may result in different partitioning of photosynthates between axial and terminal sinks. Based on the membrane potential ratios, the SE/CC-complexes in the transport phloem of apoplasmic species may have a better chance to retain photosyn-

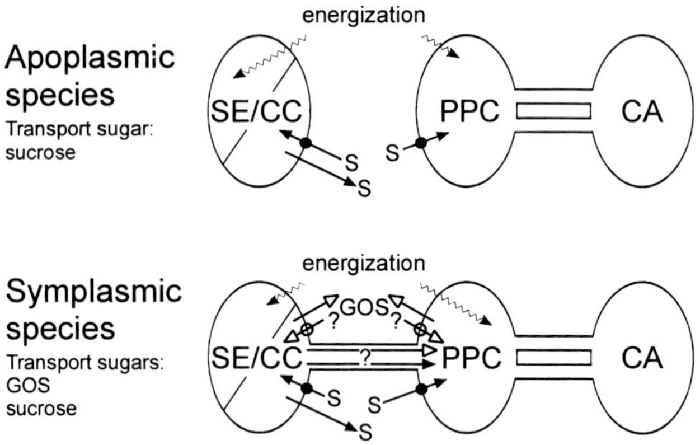

Figure 6. *Speculative impact of the differential uptake capacity (= driving force + number of carriers) in SE/CCs and PPCs in the transport phloem of symplasmic and apoplasmic phloem loaders (van Bel, 1996). This is a simplified presentation of the symplasmic connectivity and the associated pathway of lateral transfer of photoassimilates from the sieve elements to the cambium. The coiled arrows represent the relative driving forces (=energization) for uptake, the arrows with circles the carriers for transport sugars (sucrose -S- or galactosyl-oligosaccharides -GOS-). SE/CC = sieve element/companion cell complex, PPC = phloem parenchyma, CA = cambium.*

thate. In contrast, the PPCs in transport phloem may have an advantage in accumulating the photosynthate from the apoplast in symplasmic species. It should be noted, that the membrane potential alone is not decisive for the ability to retrieve. The proton gradient and the number of carriers (which are unknown) are just as important.

Provided that SE/CC-complexes in apoplasmic species have a higher retrieval potential than SE/CC-complexes in symplasmic species along the whole transport phloem, the cambium in apoplasmic species will be less competitive and deplete less of the passing assimilate pool (*Figure 6*). Consequently, cambial growth is reduced in comparison to symplasmic species and more photosynthate is directed to the terminal sinks. The input of more energy-carrying substances into the terminal sinks is speculated to lead to higher relative growth rates (van Bel, 1996).

Obviously, these speculations do not take into account that the transport to the cambium may be less dependent on carrier systems in symplasmic species than in apoplasmic species due to a higher capacity of symplasmic transport (*Figure 6*). Higher plasmodesmal frequencies were observed at the interface between SE/CC-complex and PPC in *Cucurbita* than in the other species investigated (Kempers *et al.*, 1998).

10. Concluding remarks

In conclusion, we know very little about the structural and physiological environment of the photosynthate transfer in the cambial zone. Vital information is missing concerning the parameters involved in photosynthate supply of the cambium. How is the symplasmic organization of the zone extending from the sieve elements to the cambium? Nature, the frequencies and gating of plasmodesmata all determine the fluxes of materials and messages towards the phloem. Where do permanent or temporary symplasmic bottlenecks occur? The positioning and number of uptake/release carriers together with the driving forces across the respective membranes are decisive for the flux of photoassimilate. These crucial questions must be answered before any picture of the photosynthate supply of the cambium can be drawn.

References

Aloni, B., Wyse, R.E. and Griffith, S. (1986) Sucrose transport and phloem unloading in stem of *Vicia faba*: possible involvement of a sucrose carrier and osmotic regulation. *Plant Physiol.* **81**: 482–486.

Behnke, H.D. and Sjolund, R.D. (1990) *Sieve Elements. Comparative Structure, Induction and Development*. Springer Verlag: Berlin, Heidelberg, New York.

Daie, J. (1986) Kinetics of sugar transport in isolated vascular bundles and phloem tissue of celery. *J. Amer. Hort. Sci.* **111**: 216–220.

Eschrich, W., Evert, R.F. and Young, J.H. (1972) Solution flow in tubular semipermeable membranes. *Planta* **107**: 279–300.

Fisher, D.B. and Oparka, K.J. (1996) Post-phloem transport: principles and problems. *J. Exp. Bot.* **47**: 1141–1154.

Flora, L.L. and Madore, M.A. (1996) Significance of minor-vein anatomy to carbohydrate transport. *Planta* **198**: 171–178.

Gamalei, Y.V. (1985) Characteristics of phloem loading in woody and herbaceous plants. *Fiziol. Rast.* **32**: 866–875.

Gamalei, Y.V. (1989) Structure and function of leaf minor veins in trees and herbs. A taxonomic review. *Trees* **3**: 96–110.

Gamalei, Y.V. (1991) Phloem loading and its development related to plant evolution from trees to herbs. *Trees* **5**: 60–64.

Gamalei, Y.V., Pakhomova, M.V. and Sjutkina, A.V. (1992) Ecological aspects of phloem transport. I. Temperature. *Fiziol. Rast.* **39**: 1068–1078.

Grimm, E., Bernhardt, G., Rothe, K. and Jacob, F. (1990) Mechanism of sucrose retrieval along the phloem path — a kinetic approach. *Planta* **182**: 480–485.

Grimm, E., Jahnke, S. and Rothe, K. (1997) Photoassimilate translocation in the petiole of *Cyclamen* and *Primula* is independent of lateral retrieval. *J. Exp. Bot.* **48**: 1087–1094.

Hayes, P.M., Patrick, J.W. and Offler, C.E. (1987) The cellular pathway of radial transfer of photosynthates in stems of *Phaseolus vulgaris* L.: effects of cellular plasmolysis and p-chloromercuribenzenesulphonic acid. *Ann. Bot.* **59**: 635–642.

Holdaway-Clarke, T.L., Walker, N.A. and Overall, R.L. (1996) Measurement of the electrical resistance of plasmodesmata and membranes of corn suspension culture cells. *Planta* **199**: 537–544.

Kempers, R., Ammerlaan, A. and van Bel, A.J.E. (1998) Symplasmic constriction and ultrastructural features of the sieve element/companion cell complex in the transport phloem of apoplasmically and symplasmically phloem-loading species. *Plant Physiol.* **116**: 271–278.

Knoblauch, M. and van Bel, A.J.E. (1998) Sieve tubes in action. *Plant Cell* **10**: 35–50.

Kühn, C., Franceschi, V.R., Schulz, A., Lemoine, R. and Frommer, W.B. (1997) Macromolecular trafficking indicated by localization and turnover of sucrose transporters in enucleate sieve elements. *Science* **275**: 1298–1300.

Lew, R.R. (1994) Regulation of the electric coupling between *Arabidopsis* root hairs. *Planta* **193**: 67–73.

Lew, R.R. (1996) Pressure regulation of the electrical properties of growing *Arabidopsis thaliana* L. root hairs. *Plant Physiol.* **112**: 1089–1100.

McNeil, D. (1976) The basis of osmotic pressure maintenance during expansion growth in *Helianthus annuus* hypocotyls. *Aust. J. Plant Physiol.* **3**: 311–324.

Minchin, P.E.H. and Thorpe, M.R. (1984) Apoplastic phloem unloading in the stem of bean. *J. Exp. Bot.* **35**: 538–550.

Minchin, P.E.H. and Thorpe, M.R. (1987) Measurement of unloading and reloading of photo-assimilate within the stem of bean. *J. Exp. Bot.* **38**: 211–220.

Minchin, P.E.H., Ryan, K.G. and Thorpe, M.R. (1984) Further evidence of apoplastic unloading into the stem of bean: identification of the phloem buffering pool. *J. Exp. Bot.* **35**: 1744–1753.

Offler, C.E. and Patrick, J.W. (1984) Cellular structures, plasma membrane surface areas and plasmodesmatal frequencies of seed coats of *Phaseolus vulgaris* L. in relation to photosynthate transfer. *Aust. J. Plant Physiol.* **11**: 79–90.

Oparka, K.J., Duckett, C.M., Prior, D.A.M. and Fisher, D.B. (1994) Real-time imaging of phloem unloading in the root tip of *Arabidopsis. Plant J.* **6**: 759–766.

Oparka, K.J., Prior, D.A.M. and Wright, K.M. (1995) Symplastic communication between primary and developing lateral roots of *Arabidopsis thaliana. J. Exp. Bot.* **46**: 187–198.

Patrick, J.W. and Offler, C.E. (1996) Post-sieve element transport of photoassimilates in sink regions. *J. Exp. Bot.* **47**: 1165–1177.

Rhodes, J., Thain, J.F. and Wildon, D.C. (1996) The pathway for systemic electrical signal conduction in the wounded tomato plant. *Planta* **200**: 50–57.

Sauter, J.J. (1982) Transport in Markstrahlen. *Ber. deutsch. Bot. Ges.* **95**: 593–618.

Sauter, J.J. and Kloth, S. (1986) Plasmodesmatal frequency and radial translocation rates in ray cells of poplar (*Populus* × *canadensis* Moench 'robusta'). *Planta* **168**: 377–380.

Schmalstig, J.G. and Cosgrove, D.J. (1990) Coupling of solute transport and cell expansion in pea stems. *Plant Physiol.* **94**: 1625–1634.

Schmalstig, J.G. and Geiger, D.R. (1987) Phloem unloading in developing leaves of sugar beet. I. Evidence for pathway through the symplast. *Plant Physiol.* **79**: 49–52.

Schulz, A. (1994) Phloem transport and differential unloading in pea seedlings after source and sink manipulations. *Planta* **192**: 239–249.

Schulz, A. (1995) Plasmodesmal widening accompanies the short-term increase in symplasmic phloem unloading of pea root tips under osmotic stress. *Protoplasma* **188**: 22–37.

Sibaoka, T. (1962) Excitable cells in *Mimosa. Science* **137**: 226.

Shen, Y.-W. and Patrick J.W. (1990) Cellular pathway and hormonal promotion of phloem unloading in stems of *Phaseolus vulgaris* L. *Acta Phytophysiol. Sin.* **16**: 115–124.

Stadler, R. and Sauer, N. (1996) The *Arabidopsis thaliana* AtSUC2 gene is specifically expressed in companion cells. *Bot. Acta* **109**: 299–306.

Stadler, R., Brandner, J., Schulz, A., Gahrtz, M. and Sauer, N. (1995) Phloem loading by the PmSUC2 sucrose carrier from *Plantago major* occurs into companion cells. *Plant Cell* **7**: 1545–1554.

Turgeon, R. (1987) Phloem unloading in tobacco sink leaves: insensitivity to anoxia indicates a symplastic pathway. *Planta* **171**: 73–81.

Turgeon, R. (1996) Phloem loading and plasmodesmata. *Trends Plant Sci.* **1**: 418–423.

Turgeon, R. and Medville, R. (1998) The absence of phloem loading in willow leaves. *Proc. Natl. Acad. Sci.*, USA 95: 12 055–12 060.

Turgeon, R. and Wimmers, L.E. (1988) Different patterns of vein loading of exogenous [^{14}C]-sucrose in leaves of *Pisum sativum* and *Coleus blumei*. *Plant Physiol.* **87**: 179–182.

Turgeon, R., Beebe, D.U. and Gowan, E. (1993) The intermediary cell: minor-vein anatomy and raffinose oligosaccharide synthesis in the Scrophulariaceae. *Planta* **191**: 446–456.

Turgeon, R., Webb, J.A. and Evert, R.F. (1975) Ultrastructure of minor veins in *Cucurbita pepo* leaves. *Protoplasma* **83**: 217–232.

van Bel, A.J.E. (1976) Different mass transfer rates of labeled sugars and tritiated water in xylem vessels and their dependency on metabolism. *Plant Physiol.* **57**: 911–914.

van Bel, A.J.E. (1990) Xylem-phloem exchange via the rays: the undervalued route of transport. *J. Exp. Bot.* **41**: 631–644.

van Bel, A.J.E. (1993) The transport phloem. Specifics of its functioning. *Prog. Bot.* **54**: 134–150.

van Bel, A.J.E. (1995) The low-profile directors of C- and N-economy: the parenchyma cells associated with the translocation channels. In: *Plant Stems: Physiology and Functional Morphology* (ed. B. Gartner). Academic Press: San Diego, pp. 205–222.

van Bel, A.J.E. (1996) Interaction between sieve element and companion cell and the consequences for assimilate distribution. Two structural hardware frames with associated physiological software packages in dicotyledons? *J. Exp. Bot.* **47**: 1129–1140.

van Bel, A.J.E. and Kempers, R. (1991) Symplasmic isolation of the sieve element–companion cell complex in the phloem of *Ricinus communis* and *Salix alba* stems. *Planta* **183**: 69–76.

van Bel, A.J.E. and Kempers, R. (1997) The pore/plasmodesma unit; key element in the interplay between sieve element and companion cell. *Progr. Bot.* **58**: 278–291.

van Bel, A.J.E. and van Rijen, H.V.M. (1994) Microelectrode-recorded development of the symplasmic autonomy of the sieve element/companion cell in the stem phloem of *Lupinus luteus* L. *Planta* **192**: 165–175.

van Bel, A.J.E., Ammerlaan, A. and van Dijk, A.A. (1992) A three-step screening procedure to identify the mode of phloem loading in intact leaves. Evidence for symplasmic and apoplasmic phloem loading associated with the type of companion cell. *Planta* **182**: 31–39.

van Bel, A.J.E., Gamalei, Y.V., Ammerlaan, A. and Bik, L.P.M. (1994) Dissimilar phloem loading in leaves with symplasmic or apoplasmic minor-vein configurations. *Planta* **186**: 518–525.

van der Schoot, C. and van Bel, A.J.E. (1989) Glass microelectrode measurements of sieve tube membrane potentials in internodes and petioles of tomato (*Solanum lycopersicum*). *Protoplasma* **149**: 144–154.

van der Schoot, C. and van Bel, A.J.E. (1990) Mapping menbrane potential differences and dye coupling in internodal tissues of tomato (*Lycopersicum esculentum* L.). *Planta* **182**: 9–21.

Wimmers, L. and Turgeon, R. (1991) Transfer cells and solute uptake in minor veins of *Pisum sativum* leaves. *Planta* **186**: 2–12.

Wood, R.M., Patrick J.W. and Offler, C.E. (1998) The cellular pathway of short-distance transfer of photosynthates and potassium in the elongating stem of *Phaseolus vulgaris* L. A physiological assessment. *Ann. Bot.* **82**: 337–345.

Zimmermann, M.H. and Milburn, J.A. (1975) *Transport in Plants. I. Phloem Transport. Encyclopedia of Plant Physiology*. Springer Verlag: Berlin, Heidelberg, New York.

8

Water and solute relations of the carrot cambium studied at single-cell resolution

D. Tomos, A. Korolev, J. Farrar, K. Nicolay, R. Bowtell and W. Köckenberger

1. Introduction

1.1 *Hydraulic machines*

Plants are hydraulic machines, driven by membrane transport and transpiration. In some cases organs actually move mechanically as a result of such processes (Irving *et al.*, 1997), but more generally hydraulics are harnessed to circulate metabolites and other solutes between different parts of the plant. The short and long distance transport of solutes and of water are intimately linked at several levels. Nowhere is this more evident than in the vascular cambium. As yet little work has been done on these aspects in tree cambia, but here we present information on a homologous situation in the carrot (*Daucus carota*) storage taproot. Within this single organ must occur upward water transport in the xylem, from absorptive roots to crown, and downward flow of photoassimilate solution in the phloem. There must also be horizontal movement from the phloem to the storage tissues, net flow due to cell extension growth and, possibly, lateral flow from the xylem to the tissues.

The focus for many of these processes is the vascular cambium (*Figure 1*). The cells of this tissue undergo expansion and act as a pathway for phloem contents moving in a centripetal (inward) direction. They may also, at times, act as a path for xylem water in a centrifugal (outward) direction. To understand the 'plumbing' of such a system, it is essential to be able to analyse the behaviour of individual cells within the organ. Analysing integrated solute and water transport in any tissue is not easy; in the case of the carrot it is made more difficult by having to study living tissue up to 10 mm below the surface. We have used single-cell sampling and analysis (SiCSA) techniques (Tomos and Leigh, 1999) to provide some information about this, and have begun to

Cell & Molecular Biology of Wood Formation, edited by R. Savidge, J. Barnett and R. Napier.

apply **m**agnetic **r**esonance **i**maging techniques (MRI; Verscht *et al.*, 1998) to provide further information in a truly non-invasive fashion.

1.2 *Tissue partition chromatography*

The passage of water and solutes across tissues is governed by a small number of physical parameters that describe the electrochemical driving forces, the conductivity coefficients and the interlinking of solute and water flow. In the case of metabolites, metabolic parameters are also involved (e.g., to maintain solute gradients). At

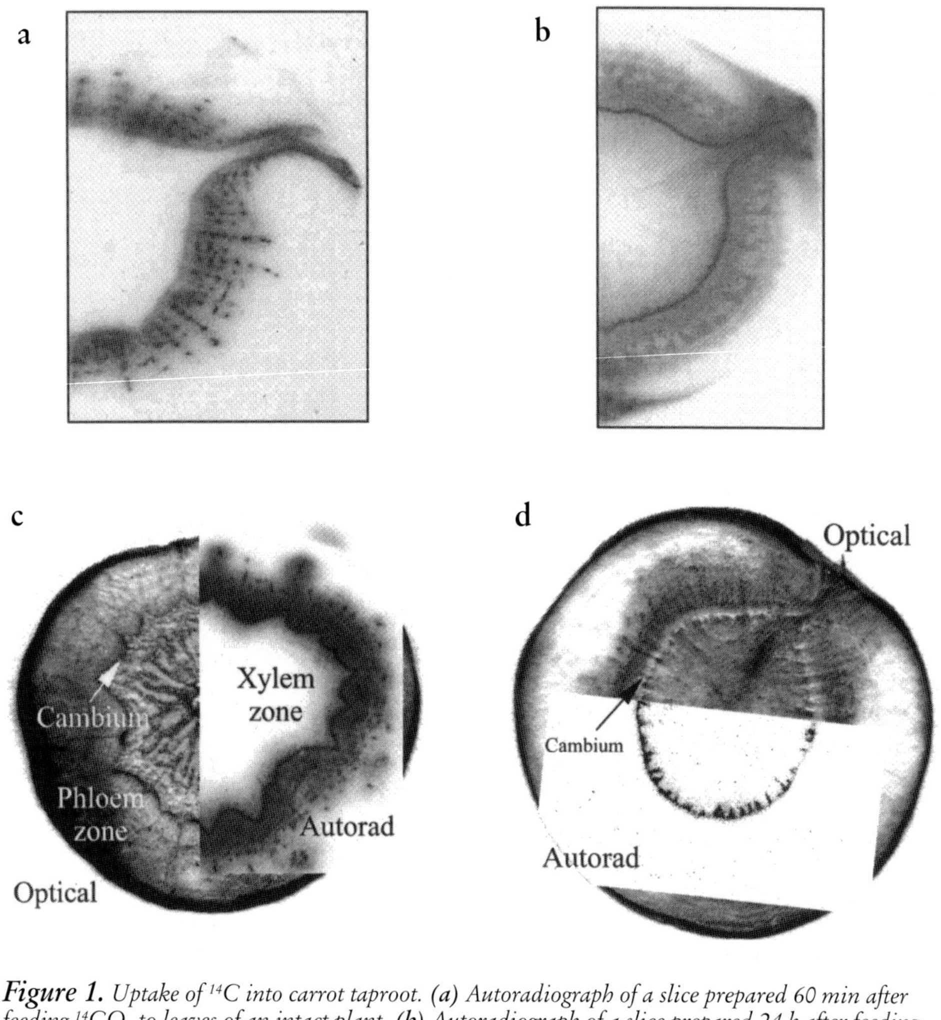

Figure 1. *Uptake of ^{14}C into carrot taproot. (**a**) Autoradiograph of a slice prepared 60 min after feeding $^{14}CO_2$ to leaves of an intact plant. (**b**) Autoradiograph of a slice prepared 24 h after feeding $^{14}CO_2$ to leaves. (**c**) (Right sector) Autoradiograph of a slice prepared 4 h after feeding $^{14}CO_2$ to leaves; (Left sector) optical image of the same freeze-dried slice after autoradiography, superimposed to identify the labelled regions of the root. (**d**) Autoradiograph of an excised slice of taproot after incubation of the slice for 4 h in ^{14}C-sucrose, superimposed on an optical image of the same slice to locate the tissue types. The label is largely restricted to the cambium and the outermost cells of the xylem. Each root is approximately 2 cm in diameter.*

dimensions of the order of cells or less, diffusion will be important. At a larger scale, tissues can be compared to a partition chromatography system, with bulk mobile and stationary aqueous phases (Tomos and Wyn Jones, 1988). The distribution of solutes between these phases is determined by 'partition coefficients' that are generally determined by membrane transport properties. These are usually highly selective and can be both passive (channels) and active (pumps and transporters). The mobile phases, represented by the apoplast and the symplast, will have varying conductances and may be interrupted by permeability barriers. Such barriers are the Casparian strip for the apoplast and closed or absent plasmodesmata for the symplast. The distribution of the 'partition coefficients' is a function of the membrane transport properties of individual cells along the pathways. There is much interest in identifying such cell-type specific activities (Leigh and Tomos, 1993).

Although their water content will not be static, the vacuolar compartments that dominate most tissue volumes may be considered as the major stationary phase of the system. In some cases the fixed charges of the apoplast will also provide a genuine (ion-exchange) stationary phase.

Attempts have been made to express the simpler case of water relations in terms of electrical circuit analogues (Molz and Ferrier, 1982). In the specific case of the radius of absorptive (fibrous) roots this has led to significant advances in our comprehension of the integrated function of such a complex organ (Steudle and Peterson, 1998). Extending these water models to solute transport presents considerable challenges — it is surprising how little is known about the pathways of solute movement and their driving forces across any tissues other than long distance movement in xylem and phloem (Leigh and Tomos, 1993). To understand the workings of a tissue at this level, solute and water relations cannot be studied in isolation. As noted above, the carrot taproot cambium presents a valuable model system of uncomplicated, linear, geometry. Similar information is also being collected from the barley leaf (Fricke *et al.*, 1995; Koroleva *et al.*, 1998; Cuin *et al.*, 1995).

2. Techniques of study

2.1 *Carbon-14 labelling and autoradiography*

Carbon-14 labelling has been used widely to study accumulation, transport and partitioning of assimilate in plants (e.g., Farrar, 1993). We have used the techniques in two ways. First, to visualize routes of transport and sites of photoassimilate accumulation, 30 minute pulse-feeding of $^{14}CO_2$ was given to illuminated leaves. Second, to study sugar uptake by taproot tissues, taproot slices were exposed to solutions of uniformly ^{14}C-labelled sugar, freeze-dried and autoradiographed to analyse ^{14}C distribution (Farrar, 1993). In some experiments the taproot slices were autoradiographed both before and after extraction by ethanol and water to evaluate the incorporation of ^{14}C into non-soluble material. Autoradiographs and light images of taproot slices were superimposed to identify the label distributions in different tissues (*Figure 1c*).

2.2 *Pressure probe and SiCSA*

The cell pressure probe (Hüsken *et al.*, 1978; Tomos and Leigh, 1999) enables the turgor pressure of individual cells to be measured in intact plants. It is comprised of a fine oil-filled microcapillary mounted in such a way that the pressure in the oil can be both altered and measured. On inserting the tip of the capillary (typically 1–5 μm diameter) into a target cell using a micromanipulator, the turgor pressure of the cell forces cell sap some way into the capillary. This results in a visible boundary between the two fluids. By increasing the pressure in the oil, this boundary can be forced back to, or near to, the point of insertion into the cell. Once stationary, the boundary then acts as a datum point for the technique. To maintain the stationary boundary, the pressure in the oil — which is continuously measured with a solid state pressure transducer — must equal that of the cell sap (i.e., the cell turgor). It has been used for a wide range of tissues, and can probe cells down to a depth of about 1 mm (e.g., Pritchard *et al.*, 1989, who used it to measure the pressure profile across the diameter of a wheat root). It is possible to measure the pressures of deeper cells by using excised tissues (e.g., red beet; Leigh and Tomos, 1983). However, the turgor pressures of such cells are likely to differ from those of the intact root due to changes within the apoplast. Following excision, desiccation of the excised tissue must be avoided and the simplest way to achieve this is to bathe the tissue in an aqueous solution. However, the apoplastic space will then be in equilibrium with the medium used to bathe the tissue. This may be very different from the situation *in planta*. The alternative, of maintaining the tissue in a water saturated atmosphere, is equally ineffective as the apoplast will be contaminated to an unknown degree with solutes from the cut cells of the surface of the tissue.

For *Beta vulgaris* taproot and *Suaeda maritima* leaves we have started to develop a procedure that appears to reveal the turgor pressure of cells of intact plants albeit by measuring those of sections (Lawrence and Tomos, 1998). The cut tissue is bathed in oil immediately after excision. Initially, the cell turgor pressure is low due to the broken cell contents contaminating the apoplast. Within an hour, however, it recovers. In both species, the new equilibrium pressure is considerably lower than the cell osmotic pressure. It is, however, indistinguishable from cells of the same organs measured from the surface of intact plants. This indicates an osmotic adjustment back to the initial turgor pressure. Moreover, it shows that this turgor is lowered *in planta* by regulation of the concentration of solutes in the apoplast.

It is possible to use the probe capillary to sample the contents of the cells. By quickly releasing the pressure in the oil reservoir, an undiluted sample of cell sap is forced into the capillary tip. This can be harvested, generally under paraffin oil to avoid evaporation (Tomos and Leigh, 1999). This permits the measurement of osmotic pressure by picolitre osmometry (Malone *et al.*, 1989), inorganic ions by microdroplet X-ray analysis (Hinde *et al.*, 1998) and organic solute concentrations by enzyme-linked fluorescence microscopy (Fricke *et al.*, 1994; Koroleva *et al.*, 1998). Solute concentrations can be measured to a resolution of 5–10 mM.

2.3 *Magnetic resonance imaging*

Magnetic resonance imaging is a non-invasive and non-destructive technique which can be used to measure the distribution of water (Bentrup, 1996) and sugars (Verscht *et al.*, 1998) in biological samples. Carrot taproot is a suitable model for MRI because of its large diameter, cylindrical shape and relatively simple and well-studied anatomy. In our case the carrot roots were inserted either into the 400 mm bore of a 200 MHz NMR microscope (Utrecht) or into the 89 mm bore of a 500 MHz NMR microscope

(Nottingham) for 1–4 h to image the water distribution inside the intact taproot. The proton NMR image (*Figure 2*) shows a non-uniform distribution of image intensity in the taproot. The low signal (dark) corresponds to lower water concentration and/or a higher density of air spaces. After imaging, the same taproots were harvested for histological study. The MR and light images of the same taproot slices were superimposed to identify the position of different tissues (*Figure 2*).

3. Results

3.1 *Path of photoassimilate to the cambium*

Labelled material can be detected in the carrot taproot within 60 min of feeding $^{14}CO_2$ to the leaf (*Figure 1a*). It is at first limited to points just centrifugal to the cambium. These points presumably represent the active phloem elements responsible for long range transport of photoassimilates. Longitudinal autoradiographs (not shown) indicated that axial transport is very rapid and by 60 minutes, when the first autoradiograph was made, this tissue is uniformly labelled. Over the following 24 hours, labelled material moves both centrifugally and centripetally (*Figure 1b*). Away from the cambium, the outer phloem parenchyma becomes increasingly labelled during this period. The patterns of its movement towards the cambium show more detailed features. The active phloem remains strongly (if more diffusely) labelled, but now a second region of accumulation appears. This corresponds to the cambium. Further centripetal movement appears to be limited with the inner xylem parenchyma and pith being significantly less labelled than the outer phloem tissue. Extracting soluble components before autoradiography indicated that much of the label associated with the phloem was soluble, whereas that associated with the cambium was insoluble — presumably polymeric polysaccharide and protein.

More information of the pathway of organic solutes to the cambium was gained by incubating excised tissue discs directly in labelled sucrose, glucose and fructose. Significantly, very little label was accumulated by the cells on the phloem-side of the

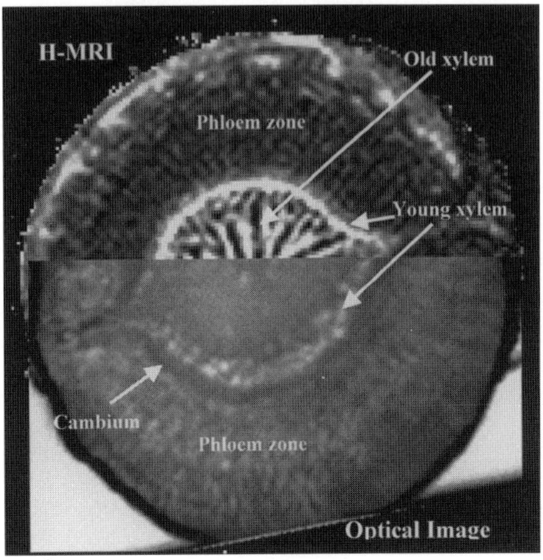

Figure 2. *Superimposition of the proton MRI and optical images of the same taproot slice. For the nmr (upper) sector, the darker parts of the image represent lower water concentrations or air spaces. (Section approx. 2 cm diameter.)*

cambium (*Figure 1d*) despite these cells containing high concentrations of sugars (see below). In contrast label accumulated on the xylem side. Supporting data has been obtained following quantification of the uptake of radioactive sugars into excised sections of separated xylem and phloem (data not shown).

3.2 *Distribution of specific solutes at the cambium*

Use of SiCSA provided information about the distribution of solutes within the mature taproot at the resolution of individual vacuoles. Across the entire root a characteristic pattern was seen (*Figure 3a and b*). The vacuoles of tissues either side of the cambium — the inner phloem and outer xylem — are dominated by sugars (sucrose and hexose). The solutes of the cells of the root periphery (periderm) and core (pith), on the other hand, are predominantly electrolytes.

The situation within the cambium, however, was unexpected. Cells sampled as close to the meristem as possible yielded contents with very low concentrations of both K^+ and sugars. Their osmotic pressures, although lower than those of surrounding cells, were higher than the sum of these accounted solutes (*Figure 3c*). We currently assume the shortfall to be due to amino acids — on the basis that the nascent cells may resemble the cells of the young carrot. The low concentration of K^+ is consistent with that of the surrounding cells. The low sugar, however, is not. A suggestion is that the sugar is rapidly polymerized or metabolized by the growing cell. As this happens a steep sugar gradient is maintained by which these solutes will be driven along the symplastic pathway from the phloem.

3.3 *Water relations of the cambium*

As noted above, it is not a trivial matter to measure the water relations of an intact large organ, such as carrot taproot, at single-cell resolution. Furthermore, the indirect techniques successfully applied to *Suaeda* and *Beta* did not provide unequivocal results for *Daucus*. The turgor pressures of peridermal cells of mature, hydroponically-grown, intact taproots were very similar to their osmotic pressure (data not shown). This indicates that the wall space (apoplast) of peridermal tissue *in planta* has a water potential close to zero. When these same cells were probed in excised tissues under paraffin oil there was a recovery of turgor pressure from the initial low level found immediately following excision. It did not, however, reach the values of the intact tissue. Various interpretations of this are possible, but it is clear that a technique developed for *Beta* and *Suaeda* cannot be assumed to work for other plants. We did find that the cambial, and all other taproot, tissues undergo osmotic adjustment to an equilibrium pressure (*Figure 3d*). Indeed this equilibrium is at a similar value for all tissues (approximately 5 bar, 200 mMolal), despite a wide range of cell osmotic pressures (*Figure 3c*). While this indicates a degree of turgor regulation in every tissue, there is no way of knowing how complete regulation might be in this system at present.

We used global water relations information to attempt to estimate the bulk flow of water in the region of the cambium. First we have assumed that water movement is predominantly radial in the region between the phloem sieve elements and the functional xylem vessels. Active phloem was located by autoradiography (*Figure 1a*). An attempt to identify active xylem was made by feeding the basal part of the taproot in basic Fuchsin. On sectioning the root, a ring of discrete points of dye is seen just inside

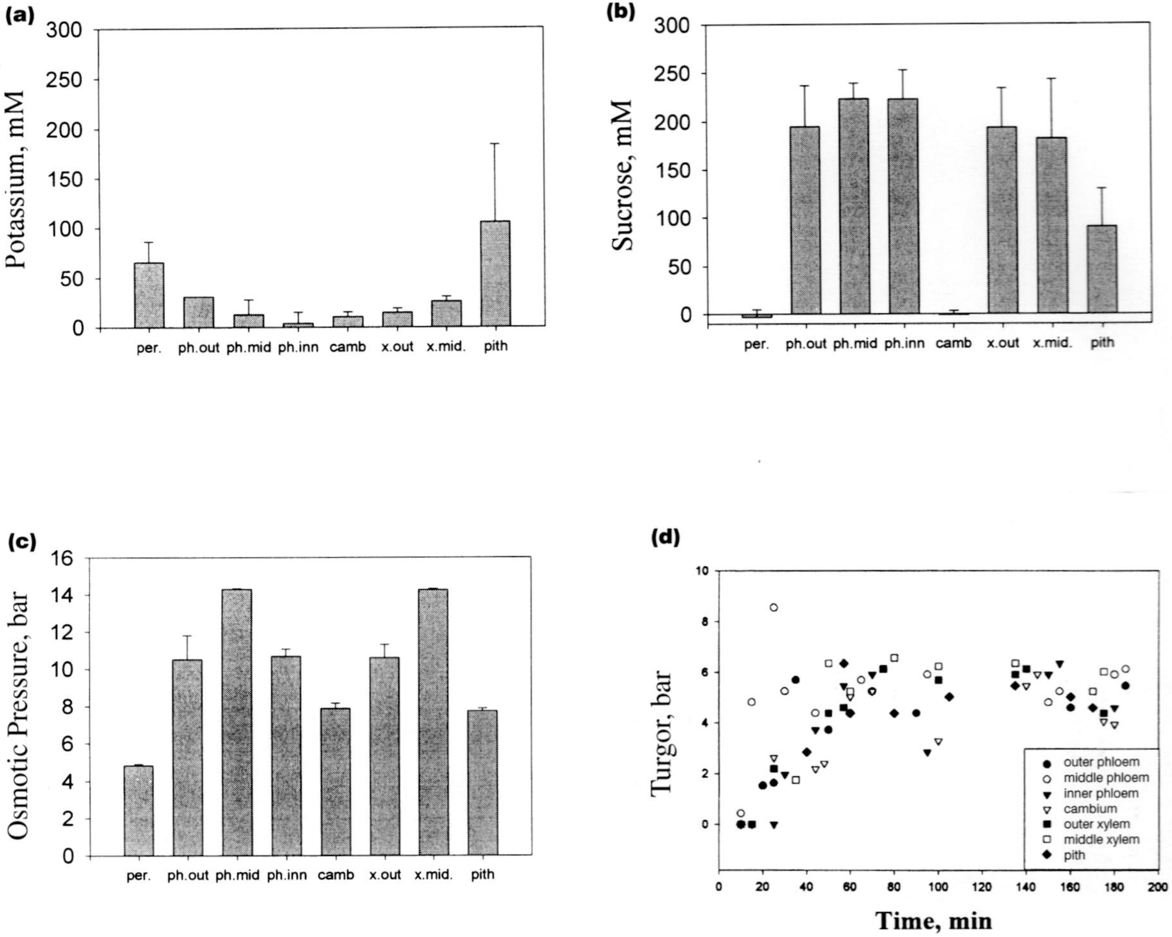

Figure 3. *Solutes in a cross section of mature carrot taproot measured using SiCSA. (a) Distribution of potassium. (b) Distribution of sucrose. (c) Distribution of osmotic pressure. (d) Turgor recovery of cells in a slice cut from an intact taproot and immediately bathed in paraffin oil. per – periderm, ph – phloem, camb – cambium, χ – xylem, inn–inner, mid – middle, out – outer.*

the cambium (not shown). As with the outer phloem, the inner xylem and pith do not seem to be involved in bulk axial water transport — although we concede that dyes are not always good indicators of water flow (Canny, 1990). The use of MRI also indicates mobile water in the same outer xylem region (*Figure 2*). A high proton MRI signal was detected in the young (outer) xylem zone and in an unidentified net of cells in the inner xylem. In some cases it appeared possible to discriminate between water movement in the radial and axial directions in phloem and xylem parenchyma (not shown).

At 50 days after germination, the taproot has a dry weight of some 2 g and a wet weight of 16 g and a relative growth rate of 0.19 day^{-1}. If the dry weight increase were to be largely due to derivatives of sucrose, it would imply a sucrose delivery of 1.1 mMoles day^{-1}. Meanwhile the net water accumulation is 2.6 g day^{-1}. For this to be entirely supplied to the taproot via the phloem would imply a phloem sucrose concentration of 1.1 mMoles/2.6 g H_2O, or 423 mMolal. If the true phloem concentration were higher, this would indicate a net influx of water into the tissue from both phloem and xylem (*Figure 4*). If the phloem concentration were less it would indicate a net flow of water from phloem to xylem. Current estimates of sucrose concentration in the phloem from other taproot tissues are, indeed, somewhat higher than this (Winzer *et al.*, 1996), indicating the former, although this may well show diurnal changes that we are at present unable to estimate. We have not analysed the nature of the sugars transported by carrot tissue, but assume that it resembes the majority of the Apiaceae (Baker and Milburn, 1989).

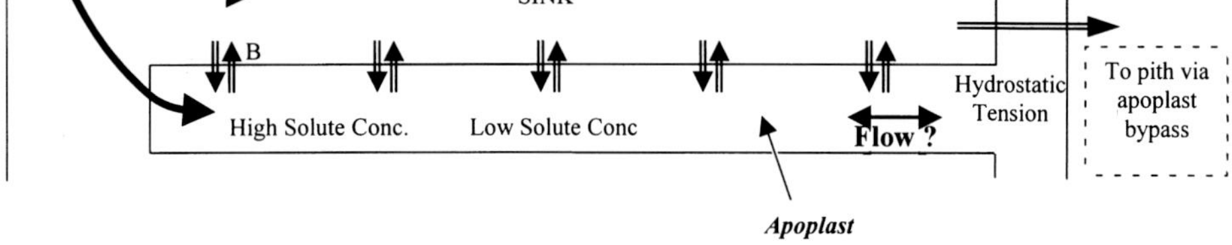

Figure 4. *Model for the tissue chromatogram associated with carrot cambium. The symplast and apoplast are mobile aqueous phases moving over the 'stationary' phase of the vacuoles. Water and solutes enter the system from phloem elements. Some water may enter from the xylem, especially under low transpiration tension. Solutes are distributed between the individual cells due to the activity of a series of partition events at both the tonoplast (A) and plasmamembrane (B). There appears to be a blockage to sugar flow at the cambium due either to symplastic discontinuity or to the strength of the metabolic sink. A symplastic bypass of the xylem must occur in order for sugars to reach the pith and inner xylem. Indeed there appears to be a very active retrieval of apoplastic sugar by the cells around the functional xylem. Note that the hydrostatic tension of the xylem is transmitted directly to the adjacent apoplast.*

If the phloem does supply the bulk of the net accumulation of water into the taproot of 2.6 g day^{-1}, an estimate of radial flow rates from the phloem can be made. The carrot at 50 days has a typical length of 17 cm and a mean cambial diameter of 7 mm, giving a surface area for the 'cylinder' of active phloem of 38 cm^2. The mean radial water flow rate for this is 700 μm day^{-1}. Furthermore, since the phloem and xylem parenchyma can be readily separated due to a line of weakness at the cambium, the relative growth rates of each tissue can be measured. At 50 days, the tissue outside the cambium is expanding at 0.19 day^{-1} and that inside the cambium at 0.23 day^{-1}. From this it can be calculated that the inward flow from the phloem is 390 μm day^{-1} (0.27μm min^{-1}) and the outward 310 μm day^{-1} (0.22 μm min^{-1}).

As far as the cambium is concerned it is the first figure that is relevant. As it is unlikely that the phloem sucrose content is much lower than 400 mMolal and more than 1.2 Molal, this indicates a very slow bulk flow across the cambium — one against which diffusion would be more than enough to move solutes. It does, however, raise the possibility of significant hydrostatic pressure gradients in the apoplast — set up by the hydrostatic tension of the xylem. In *Beta vulgaris* this tension is sufficient to dehydrate the tissue so that a measurable diurnal shrinkage of the entire organ can be measured (Palta *et al.*, 1987).

4. Discussion

Analysis of the carrot cambium has revealed a complex pattern of solute and water relations. Two observations are of particular interest. These are the very weak uptake of sugar from the apoplast by the cells of the innermost layer of phloem and the low sugar content of the cambium.

The weak uptake of sugar by excised tissue slices from the apoplast of the phloem parenchyma between the sieve elements and the cambium has several implications. First, since the vacuoles of this tissue contain the highest sugar concentrations of any part of the root, it indicates that sugar transport to it from the phloem sieve tubes is symplastic. Second it suggests that the osmotic adjustment of these cells in excised tissue slices under oil (*Figure 3d*), and assumed to occur *in planta*, was due to membrane transport of solutes other than sugars.

At the cambium there appears to be a limit to further symplastic flow, shown by both the distribution of label fed via the leaves of intact plants (*Figure 1a–c*) and by sucrose uptake by excised tissue (*Figure 1d*). This barrier to solutes may be due either to the absence of an open plasmodesmatal continuum at this point, or to the sink strength of the cambium itself. Note that the cambium is unique in its ability to take up sugars from both symplastic and apoplastic routes. This would imply the barrier is due to sink strength rather than anatomy.

The pathway of sugars in tissues on the xylem side of the cambium raises more questions. Clearly sugars must reach the xylem parenchyma — and even, at lower concentrations, the pith. Apoplastic pathways for sugars to pass from cambium to xylem parenchyma seem unlikely unless the conducting xylem vessels could be isolated from such a pathway. Lack of isolation would result in the apoplast solutes being dragged back to the leaf in the transpiration stream. The preferential uptake of apoplast solutes by the outer xylem may indeed be a response to this requirement. Sugars 'leaking' into the apoplast would be salvaged at this point for symplastic passage past the zone of active xylem long-range flow. This would effectively separate the two processes while retaining the transpiration-derived driving force for bulk flow within the apoplast. If indeed there is a discontinuity of the symplast within the cambium, this salvage would be the only route of sugars to the xylem and pith — and may be the role of the rays that cross the cambium. The path from phloem to xylem is an explicit example of the tissue chromatography model. The cells between phloem sieve tubes and xylem vessels remove solutes from the mobile apoplastic phase, cafeteria-style, to suit the situation (*Figure 4*). This could change with diurnal rhythm and development. We note, however, that it does not seem that water flow from phloem to xylem is very extensive.

The low sugar content of the cambium was also unexpected. These cells contain negligible sugar and inorganic electrolytes. They are very different from their neighbouring cells in this respect. They do, however, accumulate sugars from the apoplast — but these are rapidly converted to both insoluble polymers and, we believe, amino acids that act as the major protoplast osmoticum. Their osmotic pressure is lower than the surrounding tissues. This would appear to suggest that their turgor pressure might also be lower. If this were true, although it would usefully drive bulk symplastic flow towards the cambium, it is difficult to imagine how the cambium would be able to mechanically grow against the overlying tissue. We believe that this paradox is solved by the turgor pressures of the outer cells being regulated at a lower level by the presence of significant solutes in their apoplastic space. The apoplastic solute gradient between inner phloem and cambium could be maintained by uptake by the cambium followed by immediate metabolism. If this were the case it suggests that the apoplast osmotica involved are sugars. This would be in contrast to *Beta vulgaris*, where the apoplast solutes appear to be potassium salts (Leigh and Tomos, 1983). We have discussed the general role of cell wall solutes in turgor adjustment elsewhere (Clipson *et al.*, 1985; Wyse *et al.*, 1986; Tomos, 1988; Tomos *et al.*, 1992; Tomos and Leigh, 1999). The solutes of the phloem apoplast would have roles both in transport and turgor regulation. We show here that the various tissues of the taproot are all capable of at least a certain level of turgor adjustment. To date, however, we have been frustrated in determining the completeness of this.

Unlike long distance transport of water and solutes, the short-range inter-cellular pathways are poorly described. Our current knowledge resembles that of the movement of the blood in animals before Harvey linked together the two halves of the circulation system. By using techniques capable of supplying quantitative values to solute and water relations at the resolution of the individual cell, however, we may be able to provide some answers. Tissues such as the vascular cambium are certainly on this interface.

Acknowledgement

We thank the BBSRC for grant support for much of this work and the EU large scale facility for NMR for the use of facilities.

References

Baker D.A. and Milburn J.A. (1989) *Transport of Photoassimilates.* Longman, London, pp. 349.

Bentrup, F.-W. (1996) NMR-microscopy: Observing xylem and phloem conduits at work. *Bot. Acta* **109**: 177–179.

Canny, M.J. (1990) What becomes of the transpiration stream? *New Phytol.* **114**: 341–368.

Clipson, N.J.W., Tomos, A.D., Flowers, T.J. and Wyn Jones, R.G. (1985) Salt tolerance in the halophyte *Suaeda maritima* L. Dum. The maintenance of turgor pressure and water-potential gradients. *Planta* **165**: 392–396.

Cuin, T.A., Ashenden, T. and Tomos, A.D. (1995) The influence of ozone on the distribution of solutes between individual cells of barley leaves. *J. Exp. Bot.* **46** (suppl): 16.

Farrar, J.F. (1993) Carbon partitioning. In: *Photosynthesis and Production in a Changing Environment. A Field and Laboratory Manual* (eds D.O. Hall, J.M.O. Scurlock, H.R. Bolhar-Nordenkampf, R.C. Leegood and S.P. Long.). Chapman & Hall, London, pp. 233–246.

Fricke, W., Hinde, P.S., Leigh, R.A. and Tomos, A.D. (1995) Vacuolar solutes in the upper epidermis of barley leaves. Intercellular differences follow patterns. *Planta* **196**: 40–49.

Fricke, W., Leigh, R.A. and Tomos, A.D. (1994) Concentrations of inorganic and organic solutes in extracts from individual epidermal, mesophyll and bundle-sheath cells of barley leaves. *Planta* **192:** 310–316.

Hinde, P., Richardson, P., Koyro, H.W. and Tomos, A.D. (1998) Quantitative X-ray microanalysis of solutes in individual plant cells: a comparison of microdroplet and *in situ* frozen-hydrated data. *J. Microscopy — Oxford* **191**: 303–310.

Hüsken, D., Steudle, E. and Zimmermann, U. (1978) Pressure probe technique for measuring water relations of cells in higher plants. *Plant Physiol.* **61**: 158–163.

Irving, M.S., Ritter S., Tomos, A.D. and Koller, D. (1997) Phototropic response of the bean pulvinus: Movement of water and ions. *Botanica Acta* **110**, 118–126.

Koroleva, O.A., Farrar, J.F., Tomos, A.D. and Pollock, C.J. (1998) Carbohydrates in individual cells of epidermis, mesophyll and bundle sheath in barley leaves with changed export or photosynthetic rate. *Plant Physiol.* **118:** 1525–1532.

Lawrence, R.A and Tomos, A.D. (1998) What is the role of the apoplast in the maintenance of cell turgor pressure in water stressed plants. *J. Exp. Bot.* **49**: suppl 6.

Leigh, R.A. and Tomos, A.D. (1983) An attempt to use isolated vacuoles to determine the distribution of sodium and potassium in cells of storage roots of red beet (*Beta vulgaris* L). *Planta* **159:** 469–475.

Leigh, R.A. and Tomos, A.D. (1993) Ion distribution in cereal leaves: pathways and mechanisms. *Phil. Trans. Roy. Soc. London B*, **341:** 75–86.

Malone, M., Leigh, R.A. and Tomos, A.D. (1989) Extraction and analysis of sap from individual wheat leaf cells: the effect of sampling speed on the osmotic pressure of extracted sap. *Plant Cell Environ.* **12:** 919–926.

Molz, F.J. and Ferrier, J.M. (1982) Mathematical treatment of water movement in plant cells and tissues: a review. *Plant Cell Environ.* **5:** 191–206.

Palta, J., Wyn Jones, R.G. and Tomos A.D. (1987) Leaf diffusive conductance and tap root cell turgor pressure in sugarbeet. *Plant Cell Environ.* **10**: 735–740.

Pritchard, J., Williams, G., Wyn Jones, R.G. and Tomos, A.D. (1989) Radial turgor pressure profiles in growing and mature zones of wheat roots — A modification of the pressure probe. *J. Exp. Bot.* **40:** 567–571.

Steudle, E. and Peterson, C.A. (1998) How does water get through roots? *J. Exp. Bot.* **49:** 775–788.

Tomos, A.D. (1988) Cellular water relations in plants. In: *Water Science Reviews 3*. (ed. F. Franks). Cambridge University Press, pp. 186–277.

Tomos, A.D. and Leigh, R.A. (1999) The pressure probe: A versatile tool in plant cell physiology. *Ann. Rev. Plant Phys. Plant Mol. Biol.* **50**: 447–472.

Tomos, A.D., Leigh, R.A., Palta, J. and Williams, J.H.H. (1992) Sucrose and cell water relations. In: *Assimilate Partitioning Within and Between Organisms* (eds C.J. Pollock and J.F. Farrar). BIOS Scientific Publishers, pp. 71–89.

Tomos, A.D. and Pritchard, J. (1994) Biophysical and biochemical control of cell expansion in roots and leaves. *J. Exp. Bot.* **45:** 1721–1731.

Tomos, A.D. and Wyn Jones, R.G. (1988) Some transport properties of cells within tissues. In: *Solute Transport in Plant Cells and Tissues* (eds D.A. Baker and J.L. Hall). Longman, pp. 220–250.

Verscht, J., Kalusche, B., Köhler, J., Köckenberger, W., Metzler, A., Haase, A. and Komor, E. (1998) The kinetics of sucrose concentration in the phloem of individual vascular bundles of *Ricinus communis* seedlings measured by nuclear magnetic microimaging. *Planta* **205**: 132–139.

Winzer, T., Lohaus, G. and Heldt, H.-W. (1996) Influence of phloem transport, N-fertilization and ion acumulation on sucrose storage in the taproots of fodder beet and sugar beet. *J. Exp. Bot.* **47**: 863–870

Wyse, R.E., Zamski, E. and Tomos, A.D. (1986) Turgor regulation of sucrose transport in sugar beet taproot tissue. *Plant Physiol.* **81**: 478–481.

9

Influence of sucrose on cambial activity

D. Krabel

1. Introduction

1.1 *Historical perspective*

The vascular cambium is responsible for the radial growth of a tree and produces the elements of secondary phloem and xylem which constitute the predominant portion of the tree. Due to its multiple functions, viz. the renewal of secondary phloem and xylem, the maintenance of the meristem itself, the translocation of nutrients between phloem and xylem and finally the transmission of plant growth regulators in axial and radial direction (Catesson *et al.*, 1994), this meristem has been the subject of investigations by plant physiologists and ecologists for more than a century.

Much work concerning the cambium covers anatomical, cytological and morphological aspects of cambial activity, the differentiation of xylem and phloem and the concept of cambial rest (Iqbal and Ghouse, 1990; Catesson, 1980, 1981, 1984; Philipson *et al.*, 1971; Little and Savidge, 1987; Roberts *et al.*, 1988; Catesson and Lachaud, 1993; Larson, 1994). Other investigations have attempted to answer questions about physiological aspects of cambial activity and the influence of growth stimulators such as phytohormones (see Bannan, 1955; Sheldrake and Northcote, 1968; Little and Bonga, 1974; Riding and Little, 1984, 1986; Sundberg and Little, 1987; Little and Savidge, 1987). A few articles have been published about genomic variability (e.g., Mellerowicz *et al.*, 1989), the uptake of water into and cold hardiness of cambial cells (Wort, 1962; Ladefoged, 1952; Essiamah and Eschrich, 1985; Wisniewski and Ashworth, 1986; Catesson, 1989) and glucanase activity associated with cambial reactivation (Krabel *et al.*, 1994). The site-related processes of storage and mobilization of carbohydrates have been discussed (e.g. by Kramer and Kozlowski, 1960, 1962; Kozlowski and Keller, 1966; Kidwai and Robards, 1969a, 1969b; Barnett 1973; Catesson, 1989; Lachaud, 1989).

Since growth, cell division and differentiation processes obviously depend on the availability of photoassimilates, the translocation of sugars within a source-sink

Cell & Molecular Biology of Wood Formation, edited by R. Savidge, J. Barnett and R. Napier.

system must be regarded as the main physiological process in the cambial cells. Sucrose is the main sugar being transported in trees (Shiroya *et al.*, 1962; Zimmermann and Brown, 1971; Zimmermann and Ziegler, 1975; for exceptions see Zimmermann and Ziegler, 1975) is directly provided by photosynthesis or produced from hydrolysis of starch stored in parenchyma cells. According to the sink strength of the tissue or individual cells, sucrose, amino acids and other phloem-mobile substances are transported by mass flow but unloaded independently (Blechschmidt-Schneider *et al.*, 1997). Due to the high number of plasmodesmata, symplastic diffusion of sucrose to provide cells with carbon skeletons has been suggested to occur during the period of mitotic activity (Catesson *et al.*, 1994), whereas during the resting season the apoplastic compartment might be the pathway for nutrient allocation. To a large part carbon skeletons are needed for the formation of structural elements like cell walls, but additionally the formation of amino acids as a result of nitrogen assimilation depends on carbon skeletons too.

Based on a source-sink model, this paper suggests that during phases of increased cell division activity, the conversion of assimilates in the cambium is enhanced. This would mean that at that time the cambium and its derivatives represent a stronger 'sink', that is a place of increased consumption of sucrose, than during the period of rest, which starts in autumn upon conclusion of cell division activity.

Results obtained from a number of studies of ligneous perennials show that the cambial zone has to be interpreted as a strong sink for sucrose, the major carbohydrate being translocated (Zimmermann and Brown, 1971; Giaquinta, 1980). Studies on the breakdown of sucrose conducted by Xu *et al.* (1989) with cambial tissue from loblolly pine roots show that the enzymes of the sucrose synthase pathway play a major role for the entry of sucrose into plant cellular metabolism. The results of Sung *et al.* (1993a, 1993b) provided evidence for a periodicity related to the seasons of sucrose synthase activity for the cambium (root and stem of the same species). Since sucrose synthase together with acid invertase is responsible for the cleavage of sucrose, increases or decreases of the activities of those enzymes permit conclusions to be drawn about the sink strength of the tissue concerned (Claussen, 1983; Hubbard *et al.*, 1989).

1.2 *Obstacles to progress*

The difficulty which arises when working in the 'cambial field' is, first of all, the identification of cambial initials in a tissue. Reliable molecular and biochemical markers needed to answer the question whether all cells of one radial file go back to a single initial, or to more than one, are still missing (Eschrich, 1994). Moreover, it is hardly possible to harvest cambial material consisting of cells (e.g., phloem or xylem 'mother cells') of the same cell type and degree of differentiation (i.e., age), which one can use for biochemical or molecular investigations.

The studies by Catesson *et al.* (1994) were preliminary attempts to find biochemical and histochemical markers of the distinction between cambial initials and cambial derivatives. Using immunological techniques, they were able to localize pectin methylesterase in the walls of cambial derivatives but not in those of cambial initials. Additional evidence for tissue-specific pathways of phloem and xylem differentiation was obtained from their microanalytical work on individual pyridine nucleotides within the cambial zone of *Robinia pseudoaccacia.* Catesson *et al.* (1994) reported

NAD and NADH to be dominant in phloem-side cambial derivatives, and they concluded that their corresponding reducing equivalents provide substrates and energy for growth processes and for the synthesis of parenchymatous tissue. In xylem-side cambial derivatives, they measured high amounts of NADP and interpreted it as important reduction equivalents for the biosynthesis of cell wall material.

To investigate these questions further, biochemical aspects of cambial growth at the cell level were investigated in *Thuja occidentalis* and *Fagus sylvatica* by applying micro-dissection and enzymatic cyling techniques in conjunction with light and electron microscopy and ^{14}C-labelling techniques.

2. Species investigated and general approach

In order to find a marker for seasonal cambial activity and differentiation of cambial cells into phloem and xylem, the application of micromethods seemed to be suitable. The search in my investigations focused on sucrose as the starting point of glycolysis and the termination of gluconeogenesis. Combined microdissection with enzymatic cycling techniques (Lowry and Passonneau, 1972) was applied to stems of five-year-old *Thuja occidentalis* (cambium anatomy described in detail by Bannan, 1955) and 15-year-old *Fagus sylvatica* (Krabel *et al.*, 1993, 1994; Krabel, 1998). The anatomy of *F. sylvatica* has been carefully investigated by Farooqui and Robards (1979). In contrast to *Thuja*, the phloem and the xylem structures are much more complicated. This approach provided information about seasonal fluctuations of sucrose concentration in selected cambial derivatives numbering as small as five to ten cells. *Thuja occidentalis* was used because of its regular phloem and xylem anatomy, which facilitates the characterization of young undifferentiated cells (for the method in detail, see Krabel *et al.*, 1994). *Fagus sylvatica*, as the most widely distributed dicotyledonous tree species in Germany, seemed to be a suitable hardwood species for comparison purposes.

For the anatomical and biochemical studies small stem sections of both tree species were sampled during the phases of rest, reactivation of cell division and of increased activity. The tissue samples were quickly transferred to nearly solidified nitrogen. Before freeze-drying the material was dissected into radial and tangential sections with a freezing microtome. Until dissection into tissue fragments and determination of sucrose content, the samples were stored under vacuum at –25°C.

3. Dynamics of sucrose concentration in cambial tissue

Contradicting the hypothesis mentioned above, in both tree species the values of sucrose in the cambial zone cells were at a maximum in winter (December), compared to spring and summer values (*Figures 1 and 2*). The sucrose content in the phloem was also high in December.

The high December sucrose concentrations may have been due to intensified hydrolysis of reserve materials at that time. The existence of a negative correlation between starch and sucrose contents has long been known, as has the fact that both compounds can be readily transformed into the other (see e.g., Fekete, 1969). Thus, an increase in soluble sugars during natural or artificially induced frost hardiness is often

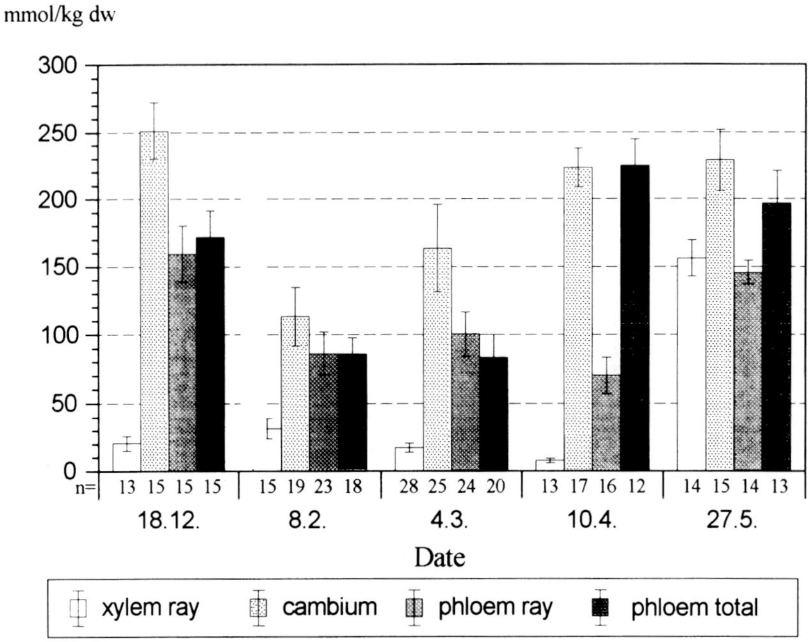

Figure 1. Sucrose concentration (mmol/kg dw) in different cambial derivatives (n=number of collections per three trees) of Thuja occidentalis *collected in February, March, April, May and December.*

described and mostly discussed in connection with the hydrolysis of starch (see e.g., Essiamah and Eschrich, 1985; Fischer and Höll, 1992; Höll, 1981, 1985; Sauter, 1988).

The sucrose values in phloem rays of both tree species, being likewise high, suggest that starch, primarily stored in ray parenchyma, is hydrolysed and symplastically unloaded into the cambial cells (*Figures 1 and 2*). The same reasoning applies to the phloem parenchyma cells. Occurrence of starch hydrolysis became obvious from the sucrose measurements of young 'phloem total' cells (phloem parenchyma, sieve cells/elements and companion cells), particularly in *Fagus* in December. According to the hypothesis of Jeremias (1964, 1968), the sucrose accumulation occurring at that time may be explained by decreasing temperatures leading to restricted respiration; thus, the acidity of the tissues is diminished. The optimum of the starch-degrading and sucrose-forming enzymes, near pH 7.0, is approached with the increasing pH. Consequently, the osmotic value in the cells is increased and the freezing point lowered. Contrary to this, Sauter (1967) demonstrated that starch degradation can also take place at high temperatures, making clear that the emptying and filling of wood ray cells with reserve materials is a phenomenon controlled by multiple factors (Höll, 1975).

The hydrolysis of starch presupposes enzyme activity. Hence, the hypothesis of Farooqui and Robards (1979), Rao and Catesson (1987) and Barnett (1992), asserting that dormancy observed using light-microscopy is not a stage of complete metabolic inactivity, is supported. Also, the vesicle-producing dictyosomes present in beech in December (Krabel, 1998; *Figures 3 and 4*) and the presence of some amino acids especially in winter in *Thuja* (D. Krabel *et al.*, unpublished data) permit the conclusion that

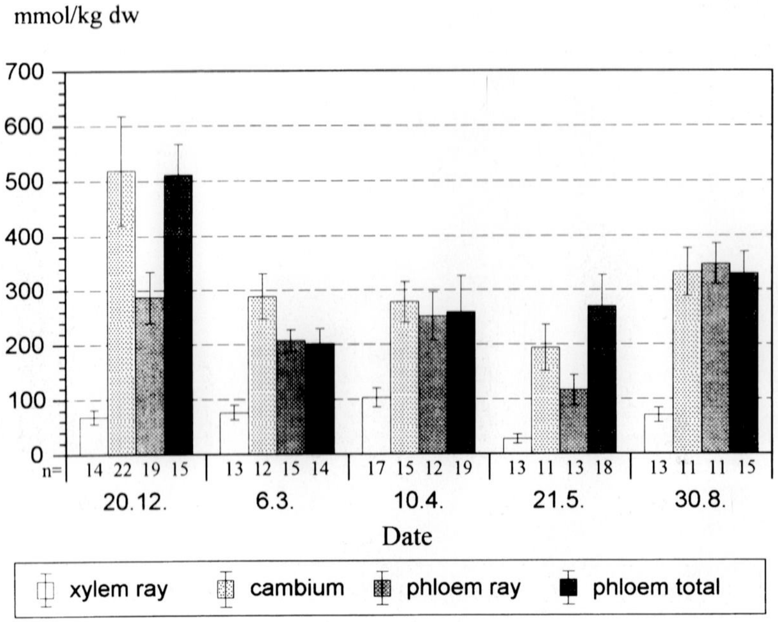

Figure 2. Sucrose concentration (mmol/kg dw) in different cambial derivatives (n=sum of collections per three trees) of Fagus sylvatica *collected in March, April, May, August and December (Krabel, 1998).*

cold-adapted metabolism takes place in the plant. The structural changes observed in the cell, for instance the formation of numerous small vacuoles instead of a few big ones as is the case during the phase of activity, is another adaptation strategy of the plant to situations of frost (see, e.g., Timell, 1974; Sauter and van Cleve, 1991).

The relatively high concentrations of sugar measured in the cambium during the visibly active phases (*Figures 1 and 2*) are supposedly attributable to its stronger 'sink' effect compared to the adjacent tissues and, therefore, to enhanced provision with assimilates (Ap Rees, 1974). As becomes obvious from microautoradiographic studies of stem cross-sections of beech, in springtime primarily the cambium and its youngest derivatives (i.e., the young xylem elements) are supplied with ^{14}C-labelled assimilates (*Figure 5*). At that time, the phloem of beech consists merely of the previous-year finally differentiated sieve elements, and they show ^{14}C assimilate translocation. This result supports the concept that they over-winter in a state retaining their full functional ability and the following spring are capable of transporting assimilates. Thus, the translocational ability of beech sieve elements does not cease after one growing season.

Similar data on the translocation of assimilates were obtained by Glerum and Balatinecz (1980) with *Pinus banksiana* and Fischer and Höll (1992) with *Pinus sylvestris*. These results clearly demonstrate that the formation of the youngest xylem elements is supported by products of recent photosynthesis. Moreover, hydrolysis products of reserve materials do not seem to play any noteworthy role in the formation of earlywood elements. In contrast to this, the results of the investigations conducted by Hansen and Beck (1990) on *Pinus sylvestris* indicated that the formation of earlywood clearly correlates with a decrease of reserve products.

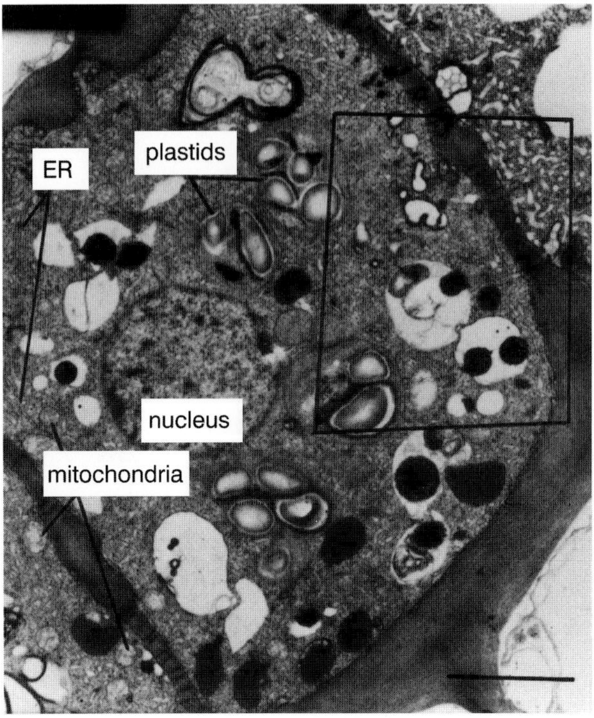

Figure 3. Fagus sylvatica *(mid-February) – ray initials during rest showing young plastids filled with starch. Scale 2.7 μm.*

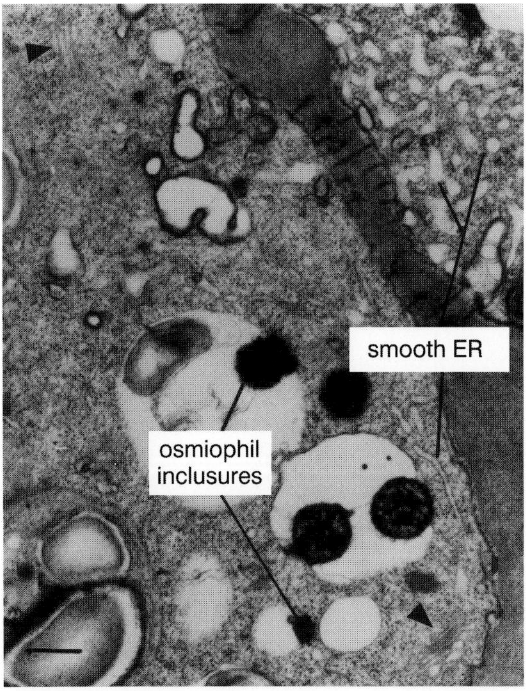

Figure 4. Vesicle producing dictyosomes (triangle) and an intensively tubular compartment forming smooth ER can be observed. Scale bar 0.4 μm.

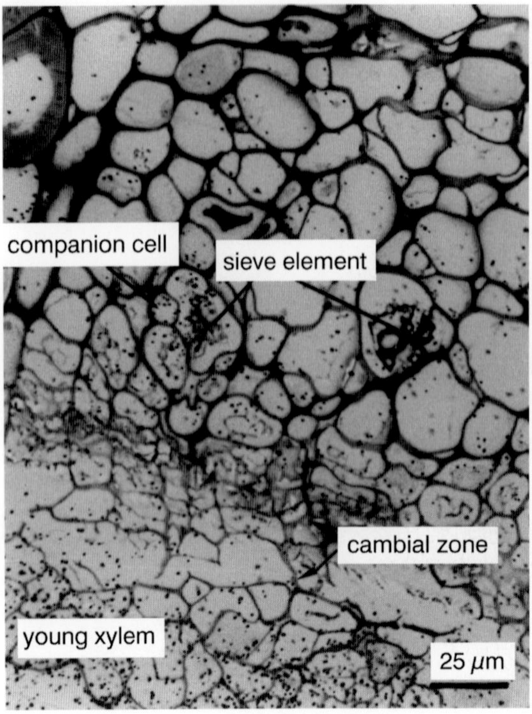

Figure 5. Fagus sylvatica *(beginning of June). -^{14}C assimilates are homogenously distributed over the cambial zone. In the zone of young phloem elements mainly sieve elements formed in the previous year are labelled (Krabel, 1998).*

A polar distribution of phosphatase activity in the phloem ray cells was described by Sauter and Braun (1968) for *Pseudotsuga*. Phosphatase activity intensified towards the cambial zone, so that a distinct sugar gradient developed, and the highest values were measured invariably at the cambial zone. However, the generally very high sugar concentration in the cambial zone suggests that the requirement of this highly active meristem at that time, though essentially supported by recent assimilates, was also supplemented by products of starch hydrolysis. This hypothesis is supported by strong ^{14}C labelling in the cambial zone and by dissolution of plastid starch as observed by electron microscopy in phloem rays of beech. Ziegler (1964) assumed that the emerging buds need the major part of the mobilized reserve materials for their development, with cambial metabolism receiving the 'surplus' of these reserves, implying that the cambium is not a primary sink.

In April and May with spring reactivation, sucrose concentrations increase in the wood rays and in the phloem cells of *Thuja occidentalis* and *Fagus sylvatica* (*Figure 2* and Krabel *et al*., 1993). This suggests enhanced radial transport at those times. The increased content in the ray cells may result from the hydrolysis of starch and, additionally, from the transport via the youngest fully differentiated sieve cells/elements capable of transporting sucrose at that early date. As shown by electron microscope investigations, cells in the cambial region are linked in radial and tangential directions *via* numerous plasmodesmata (*Figure 6*). The symplastic link to the rays and sieve elements or sieve cells, is likewise conspicuously developed. It is suggested, therefore, that sucrose and other metabolites are unloaded into this meristematic zone *via* symplastic connections.

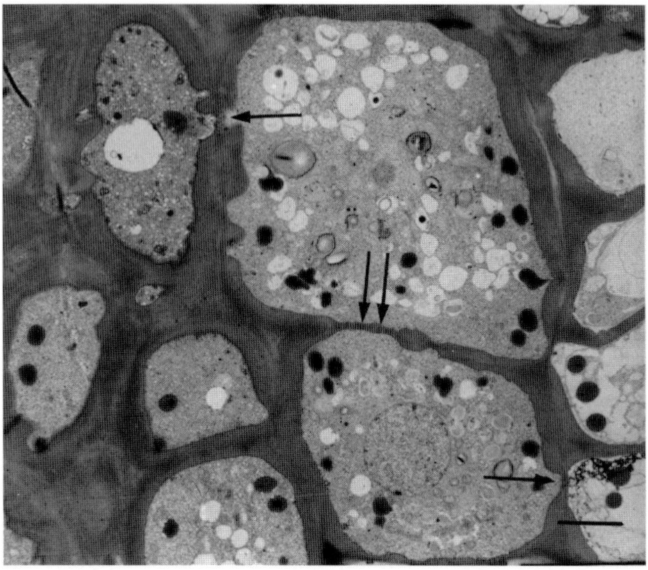

Figure 6. *Multiseriate ray cells of* Fagus sylvatica *showing numerous plasmodesmata in tangential and radial cell walls (arrows) (Krabel, 1998). Scale bar 2.7 μm.*

The temporarily decreased sugar concentration measured in *Thuja occidentalis* and *Fagus sylvatica*, in the phloem rays in *Thuja* during April and in the phloem and xylem rays during May in *Fagus* (*Figures 1 and 2*, respectively) permit the conclusion that sucrose consumption during strong cell division activity exceeds the supply. High sucrose phosphate synthase activity occurred in May in beech in the tissue isolates investigated here (Krabel, 1998); thus, sucrose synthesis evidently takes place and is necessary for the above given reasons. At that time, competing 'sinks' lead to a decrease of sucrose content in the cambial region (Larson, 1969; Sung *et al.*, 1989). Barnett (1981) interpreted the conspicuously thin walls in early wood vessels to be a consequence of the competition for assimilates between longitudinal and radial growth, with the longitudinal growth embodying the stronger 'sink'.

Supposedly, an increase of sucrose content and thus of the osmotic value in the cells will occur shortly before the beginning of the so-called 'slipping of bark' (Esau, 1977). However, such an increase could not be ascertained in the present investigation. Possibly, the hydration of the tissue can be explained in terms of other osmotica, such as potassium ions. X-ray microanalytical investigations showed that in the lumen of the cambial initials, distinctly higher potassium contents were present in beech during May, and in *Thuja* during June, than were ascertainable in winter samples (Krabel, 1998). The osmo-regulatory effect of increased [K^+] would be a possible explanation for the 'slipping of bark'. Moreover, meristematic tissues such as the cambium are preferentially supplied with potassium (Cassagnes *et al.*, 1969; Schulte-Baukloh, 1992). A proper potassium supply is essential for meristematic growth processes. For example, potassium is involved in loosening of cell walls—a process necessary for extension growth (Mengel and Kirkby, 1987). Although the connection of potassium with carbon metabolism is complex, the hydrolysis and synthesis of reserves taking place in autumn and winter, especially in the phloem and the rays, presupposes the activity of certain enzymes activated by potassium. The high concentrations of potassium occurring in the phloem cells may also exist because of the important role potassium has in sucrose loading and unloading processes.

4. Identification and characterization of cambial derivatives

As became obvious from the studies of seasonal sucrose distribution in the various cell types, the tree is well adapted to respond to changes of 'external' environmental conditions, such as the change of seasons. A possible adaptation to changes taking place inside the plant, for instance within single cells at a certain time, would likewise be conceivable. One may expect adaptation to be expressed also in terms of different metabolite contents. In this vein, the composition of the cell milieu of a young fibre cell may be regarded as different from that of a young sieve cell, and this difference may become measurably apparent as changes in metabolite composition and/or concentration. One can expect that those cambial derivatives which are different in function will also possess different magnitudes of sink strength. Detecting such differences could contribute to an exact identification and characterization of young cambial derivatives.

Attempts to characterize cambial initials by biochemical methods were made by Catesson (1992), Catesson *et al.* (1994) and Gugleilmino *et al.* (1997). They showed that by the occurrence of cellulose microfibrils and the presence of some enzymes, such as xylan synthases or methyl esterases, one can differentiate between young phloem and young xylem elements. Furthermore, the radial wall of the phloem derivatives could be characterized by early superimposition of a microfibrillar skeleton and of a low content of xylans. In contrast, the radial walls of the xylem elements were distinguished by their lower numbers of microfibrils as well as by an abundance of xylans. Additionally, there was an obviously rising trend in content of pectin acids and calcium ions from the phloem towards the xylem (Catesson,1992). Data by Catesson *et al.* (1994) suggest pectin methylesterase activity will be a helpful tool to discriminate between a cambial initial and its daughter cells. These studies were based on immunological techniques showing the presence of the enzyme in the walls of cambial derivatives, but not in those of cambial initials.

Sucrose data obtained from microanalytical investigations revealed that closely adjoining tissue zones had significant differences (Krabel *et al.*, 1993, 1994). Thus, these cell isolates manifested specific metabolism. The latter is also confirmed by the investigations by Sung *et al.* (1993a, 1993b) into the enzymes of the sucrose synthase pathway of loblolly pine roots and stem.

First indications for tissue-specific sucrose metabolism within the cambial zone became obvious from experiments by Sung *et al.* (1993a, 1993b). Her data indicated that in spring the phloem side of the cambium became active earlier than the cells of the cambial zone near the xylem. About two weeks later, when both sides were active, she also measured differences in sucrose synthase activity between the phloem- and xylem sides of the cambium.

It is supposed that the morphogenetic effect of sucrose gradients is very large (Siebers and Ladage, 1973; Bodson and Outlaw, 1985; Gamalei, 1991). At lower concentrations, cell division and differentiation are stimulated, whereas at high concentrations such processes are suppressed *in situ* in the cambial zone. For example, investigations on roots revealed that the availability of carbohydrate controls cell development via mitosis (van't Hof and Kovacs, 1972). However, the optimal sucrose concentrations for induction of cambial cell division and for the differentiation of derivatives have been found to be very similar (Gamalei, 1991).

5. Conclusions

Since growth processes of woody plants depend on the availability of photoassimilates, the translocation of sugars has to be considered as one of the main physiological processes occurring in trees. For *Thuja occidentalis* as well as *Fagus sylvatica*, the highest sucrose concentrations were determined in the youngest cambial derivatives including cambial initials, whereas the lowest content was in xylem rays. The increase in sucrose concentration during cambial rest gives strong evidence to the hypothesis that an anatomically dormant cambium is not necessarily a metabolically dormant one. Thus, those cells are able to react directly to changing environmental conditions.

The results on sucrose concentration in selected cell types show significant differences between cells of different type and function. Each cell can be characterized by its individual metabolism, due to the season and the ontogenetic stage. Future studies should concentrate on storage materials such as starch, and the corresponding enzyme activities should be investigated, in order to improve knowledge about transport and storage of photoassimilates in cambial zone cells.

Acknowledgements

This investigation was supported by Deutsche Forschungsgemeinschaft and in part by a grant from IRSIA to M. Bodson. Skilfull technical assistance of R. Burchardt and C. Förster is gratefully acknowledged.

References

Ap Rees, (1974) Pathways of carbohydrate breakdown in higher plants. In: *International Review of Biochemistry* (ed. D.H. Northcote), Buttersworth: London, 90–127.

Bannan, M.W. (1955) The vascular cambium and radial growth in *Thuja occidentalis* (L.). *Can. J. Bot.* **33**: 113–138.

Barnett, J.R. (1981) Secondary xylem cell development. In: *Xylem Cell Development* (ed. J.R. Barnett), Castle House Publications: Tunbridge Wells, 47–95.

Barnett, J.R. (1992) Reactivation of the cambium in *Aesculus hippocastanum* L. *Ann. Bot.* **70**: 169–177.

Blechschmidt-Schneider, S., Eschrich, W. and Jahnke, S. (1997) Phloem loading and unloading processes. In: *Trees–Contributions to Modern Tree Physiology* (eds. H. Rennenberg, W. Eschrich, and H. Ziegler), Backhuys Publishers: Leiden, pp. 139–163.

Bodson, M. and Outlaw, W.H. Jr. (1985) Elevation in the sucrose content of the shoot apical meristem of *Sinapis alba* at floral evocation. *Plant Physiol.* **79**: 420–424.

Catesson, A.M. (1980) The vascular cambium. In: *Control of Shoot Growth in Trees* (ed. C.H.A. Little), Proc. IUFRO Workshop: Fredericton, NB Canada, pp. 12–40.

Catesson, A.M. (1981) Le cycle saisonnier des cellules cambiales chez quelques feuilles. *Bull. Soc. Bot. Fr. Actualités Botaniques* **128**: 43–51.

Catesson, A.M. (1984) La dynamique cambiale. *Annales des Sciences Naturelles, Botanique* **13**: 23–43.

Catesson, A. M. (1989) Cambial cytology and biochemistry. In: *Radial Growth in Plants. I. Vascular Cambium* (ed. M. Iqbal), Research Studies Press: Taunton.

Catesson, A.M. (1992) Ultrastructural and biochemical changes in the cambial zone in relationship with cell differentiation and seasonal activity. *IAWA Bull.* **13**: 251–267.

Catesson, A.M. and Lachaud, S. (1993) Le cambium, structure, functionnement et contrôle de l'activité saisonnière. *Acta Bot. Gallica* **140**: 337–350.

Catesson, A.M., Bonnemain, J.L., Eschrich, W. and Magel, E. (1994) The cambium and its derivative tissues: biochemical changes in relation to cell differentiation and seasonal activity. In: *EUROSILVA—Contribution to Forest Tree Physiology* (eds. H. Sandermann, Jr., M. Bonnet), Masimbert, INRA: Paris, pp. 57–77.

Cassagnes, P.J., Magny, J., Carles, P.A. and Carles, J. (1969) Contributions to investigations into the accumulation of minerals during growth of apple fruit (Reine des Reinettes). *C. R. Acad. Sci.* **269**, 708–711.

Claussen, W. (1983) Untersuchungen über den Zusammenhang zwischen der Verteilung der Assimilate und der Saccharose-Syntethase-Aktiyvtät in *Solanum melongena* L. 1. Charakterisierung und Verteilung der Saccharose-Syntethase. *Z. Pflanzenphysiologie* **110**: 165–173.

Claussen, W., Loveys, B.R. and Hawker, J.S. (1986) Influence of sucrose and hormones on the activity of sucrose synthase. *J. Plant Physiol.* **124**: 345–375.

Esau, K. (1977) *Anatomy of Seed Plants*. John Wiley & Sons: New York, Chichester, Brisbane, Toronto, Singapore.

Eschrich, W. (1995) *Funktionelle Pflanzenanatomie*. Springer-Verlag: Heidelberg.

Essiamah, S. and Eschrich, W. (1985) Changes of starch content in the storage tissues of deciduous trees during winter and spring. *IAWA Bull.* **6**: 97–106.

Farooqui, P., Robards, A.W. (1979) Seasonal changes in the ultrastructure of cambium of *Fagus sylvatica* L. *Proc. Indian Acad. Sci.* **88B**: 463–472.

Farrar, J.J. and Evert, R.F. (1997) Seasonal changes in the ultrastructure of the vascular cambium of *Robinia pseudoacacia*. *Trees* **11**: 191–202.

Fekete, de, M.A.R. (1969) Zum Stoffwechsel der Stärke. I. Die Umwandlung von Saccharose in Stärke in den Kotyledonen von *Vicia faba*. *Planta* **87**: 311–323.

Fischer, C. and Höll, W. (1992) Food reserves of scots pine (*Pinus sylvestris* L.). II. Seasonal changes and radial distribution of carbohydrate and fat reserves in pine wood. *Trees* **6**: 147–155.

Gamalei, Y. (1991) Phloem loading and its development related to plant evolution from trees to herbs. *Trees* **5**: 50–64.

Giaquinta, R.T. (1980) Translocation of sucrose and oligosaccharides. In: *Biochemistry of Plants*, Vol. 3 (eds. P. K. Stumpf and E.E. Conn), New York: Academic Press, pp. 271–320.

Glerum, C. and Balatinecz, J.J. (1980) Formation and distribution of food reserves during autumn and their subsequent utilization in jack pine. *Can. J. Bot.* **58**: 40–54.

Guglielmino, N., Liberman, M., Catesson, A.M., Mareck, A., Prat, R., Mutafschiev, S. and Goldberg, R. (1997) Pectin methylesterases from poplar cambium and inner bark: localization, properties and seasonal changes. *Planta* **202**: 70–75.

Hansen, J. and Beck, E. (1990) The fate and path of assimilation products in the stem of 8-year-old Scots pine (*Pinus sylvestris* L.) trees. *Trees* **4**: 16–21.

Höll, W. (1975) Radial transport in rays. In: *Transport in Plants. I. Phloem Transport. Encyclopedia of Plant Physiology New Series 1* (eds. M.H. Zimmermann and J.A. Milburn), Springer-Verlag: Berlin, Heidelberg, New York, 432–450.

Höll, W. (1981) Eine dünnschichtchromatographische Darstellung des Jahresganges löslicher Zucker im Stammholz von drei Angiospermen und einer Gymnosperme. *Holzforschung* **35**: 173–175.

Höll, W. (1985) Seasonal fluctuation of reserve materials in trunkwood of spruce (*Picea abies* [L.] Karst.). *Plant Physiol.* **117**: 355–362.

Hubbard, N.L., Huber, S.C. and Pharr, D.M. (1989) Sucrose phosphate synthase and acid invertase as determinants of sucrose concentration in developing muskmelon (*Cucumis melo* L.) fruits. *Plant Physiol.* **91**: 1527–1534.

Iqbal, M. and Ghouse, A.K.M. (1990) Cambial concept and organisation. In: *The Vascular Cambium* (ed. M. Iqbal), Research Study Press: Somerset, pp. 1–36.

Jeremias, K. (1964) Über die jahresperiodisch bedingten Veränderungen der Ablagerungsform der Kohlenhydrate in vegetativen Pflanzenteilen. *Botan. Stud.* **15**: 1–96.

Jeremias, K. (1968) Zum Verhalten einiger Kohlenhydrate in Blättern und Rinden der Pappelsorten Oxford, Rochester und Androscoggin. *Mitt. Ver. Forstl. Standortskd. Forstpflanzenzüchtung* **18**: 89–94.

Krabel, D., Eschrich, W. and Bodson, M. (1993) Seasonal changes of sucrose concentration in trees. *Acta Bot. Gallica* **140**: 360–362.

Krabel, D., Bodson, M. and Eschrich, W. (1994) Seasonal changes in the cambium of trees. I. Sucrose content in *Thuja occidentalis*. *Bot. Acta* **107**: 54–59.

Krabel, D. (1998) Mikroanalytische Untersuchungen zur Physiologie des Baumkambiums am Beispiel von *Thuja occidentalis* L. und *Fagus sylvatica* L. *Cont. Forest Sci.* **2**: 1–97.

Lachaud, S. (1989) Participation of auxin and abscisic acid in the regulation of seasonal variations in cambial activity and xylogenesis. *Trees* **3**: 125–137.

Ladefoged, K. (1952) The periodicity of wood formation. *Kgl. Danske Vidensk. Selsk. Biol. Skr.* **7**: 1–98.

Larson, P.R. (1969) Wood formation and the concept of wood quality. *Yale Uni. Bull.* **74**: 24–54.

Larson, P.R. (1994) *The Vascular Cambium: Development and Structure*. Springer-Verlag: Heidelberg, Berlin, New York.

Little, C.H.A. and Bonga, J.M. (1974) Rest in the cambium of *Abies balsamea*. *Can. J. Bot.* **52**: 1723–1730.

Little, C.H.A. and Savidge, R.A. (1987) The role of plant growth regulators in forest tree cambial growth. *Plant Growth Reg.* **6**: 137–169.

Lowry, O.H. and Passonneau, J.V. (1972) *A Flexible Method of Enzymatic Cycling Analysis*. Academic Press: New York.

Mellerowicz, E.J., Coleman, W.K., Riding, R.T., Little, C.H.A. (1992) Periodicity of cambial activity in *Abies balsamea*. I. Effects of temperature and photoperiod on cambial dormancy and frost hardiness. *Physiologia Plantarum* **85**: 515–525.

Mengel, K. and Kirkby, E.A. (1987) *Principles of Plant Nutrition*. International Potash Institute: Bern.

Philipson, W.R., Ward, J.M. and Butterfield, B.G. (1971) *The Vascular Cambium*. Chapman and Hall: London.

Rao, K.S., Catesson, A.M. (1987) Changes in the membrane components of nondividing cambial cells. *Can. J. Bot.* **65**: 246–254.

Riding, R.T. and Little, C.H.A. (1984) Anatomy and histochemistry of *Abies balsamea* cambial zone cells during the onset and breaking of dormancy. *Can. J. Bot.* **62**: 2570–2579.

Riding, R.T. and Little, C.H.A (1986) Histochemistry of the dormant vascular cambium of *Abies balsamea*: changes associated with tree age and crown position. *Can.J. Bot.* **62**: 2082–2087.

Roberts, L.W., Gahan, P.B., Aloni, R. (1988) *Vascular Differentiation and Plant Growth Regulators*. Springer Series in Wood Science. Springer-Verlag: Berlin.

Sauter, J.J. (1967) Der Einfluß verschiedener Temperaturen auf die Reservestärke in parenchymatischen Geweben von Baumsproßachsen. *Z. Pflanzenphysiol.* **56**: 340–357.

Sauter, J.J., Braun, H.J. (1968) Enzymatic polarity in ray parenchyma cells of conifers in spring. *Z. Pflanzenphysiol.* **58**: 378–381.

Sauter, J.J. (1988) Temperature induced changes in starch and sugars in the stem of *Populus* × *canadensis* (robusta). *J. Plant Physiol.* **132**: 608–612.

Sauter, J.J. and van Cleve, B. (1991) Biochemical and ultrastructural results during starch-sugar-conversion in ray parenchyma cells of *Populus* during cold adaption. *J. Plant Physiol.* **139**, 19–26.

Sheldrake, A. R. and Northcote, D.H. (1968) The production of auxin by tobacco internode tissues. *New Phytol.* **67**: 1–13.

Shiroya, T., Lister, G.R., Slankis, V., Krotkov, G., Nelson, C.D. (1962) Translocation of the products of photosynthesis to roots of pine seedlings. *Can. J. Bot.* **40**: 1125–1135.

Schulte-Baukloh, C. and Fromm, J. (1993) The effect of calcium starvation on assimilate partitioning and mineral distribution of the phloem. *J. Exp. Bot.* **44**: 1703–1707.

Siebers, A.M. and Ladage, C.A. (1973) Factors controlling cambial development in the hypocotyl of *Ricinus communis* L. *Acta Bot. Neerl.* **22**: 416–432.

Sundberg, B. and Little, C.H.A. (1987) Effects of defoliation on tracheid production and the level of indol-3-acetic acid in *Abies balsamea* shoots. *Physiol. Plantarum* **94**: 1721–1727.

Sung S.-J.S., Xu, D.-P., Galloway, C.M. and Black, C.C. (1988) A reassessment of glycolysis and gluconeogenesis in higher plants. *Physiologia Plantarum* **72**: 650–654.

Sung, S.-J.S., Xu, D.-P. and Black, C.C. (1989) Identification of actively filling sucrose sinks. *Plant Physiol.* **89**: 1117–1121.

Sung, S.-J.S. Kormanik, P.P. and Black, C.C. (1993a) Vascular cambial sucrose metabolism and growth in loblolly pine (*Pinus taeda* L.) in relation to transplanting stress. *Tree Physiol.* **12**, 243–258.

Sung, S.-J.S., Kormanik, P.P. and Black, C.C. (1993b) Understanding sucrose metabolism and growth in a developing sweetgum plantation. In: *Proceedings of the 22nd Southern Forest Tree Improvement Conference,* Atlanta, pp. 114–119.

Timell, T.E. (1974) *Ultrastructure of the Dormant and Active Cambial Zone and the Dormant Phloem Associated with the Formation of Normal and Compression Wood in* Picea abies *(L.) Karst*. State University of New York, College of Environmental Science and Forestry, Syracruse.

Van't Hof, J. and Kovacs, C.J. (1972) Mitotic cycle regulation in the meristem of cultured roots: the principal control point hypothesis. In: *The Dynamics of Meristem Cell Populations* (M.W. Miller and C. Kuehnert, eds.), Plenum Press: New York, pp. 5–33.

Wiesniewski, M.E. and Ashworth, E.N. (1986). A comparison of seasonal ultrastructural changes in stem tissues of peach (*Prunus persica*) that exhibit contrasting mechanisms of cold hardiness. *Bot. Gaz.* **147**: 407–417.

Wort, D.J. (1962) Physiology of cambial activity. In: *Tree Growth* (ed. T.T. Kozlowski), Ronald Press: New York, pp. 89–96.

Xu, D.-P., Sung, S.-J.S., Loboda, T., Kormanik, P.P. and Black, C.C. (1989) Characterization of sucrolysis *via* uridine diphosphate and pyrophosphate-dependent sucrose synthase pathway. *Plant Physiol.* **90**: 635–642.

Ziegler, H. (1964) Storage, mobilization and distribution of reserve material in trees. In: *The Formation of Wood in Forest Trees* (ed. M.H. Zimmermann), Academic Press: New York, London, pp. 303–320.

Zimmermann, M.H. and Brown, C.L. (1971) *Trees: Structure and Function*. Springer Verlag: New York.

Zimmermann, M.H. and Ziegler, H. (1975) List of sugars and sugar alcohols in sieve-tube exudates. In: *Encyclopedia of Plant Physiology (New Series). Transport in Plants I. Vol. 1* (M.H. Zimmermann and J.A. Milburn, eds.), Springer-Verlag: Berlin, Heidelberg, New York.

10

Dark fixation by the cambium and its derivatives

N.A. Forneris and R.A. Savidge

1. Introduction

Cambium and living xylem cells in trees do not contain chloroplasts, except in young shoots, and thus, they cannot photosynthesize. They nevertheless require photosynthates for metabolism in support of cambial growth and maintenance of the living state. Carbon fixed during photosynthesis is loaded into the phloem and translocated from the leaves as organic compounds (sucrose, in most species) to the heterotrophic stem and the root system. These compounds may be unloaded directly to the cambium from the phloem, or may return in the xylem transpiration stream from the root system to the cambium as modified carbon-containing substances, notably glutamine and asparagine. In addition, inorganic carbon generated through respiratory activities moves in both phloem and xylem transport streams.

During the dark reaction of photosynthesis, the enzyme ribulose bisphosphate carboxylase captures CO_2 and catalyses the carboxylation of ribulose 1,5-*bis*phosphate, followed by reduction of the fixed carbon to the level of carbohydrate and regeneration of ribulose 1,5-*bis*phosphate. The 'dark reaction' (synonyms: C3 pathway, Calvin-Benson cycle) is so named because its discovery involved investigations performed in the absence of light, and to distinguish it from the light reaction of photosynthesis. However, it is well established that the ATP and NADPH required to drive the C_3 pathway are generated during the photosynthetic light reaction, hence that the dark reaction is actually a light-dependent reaction.

In addition to the dark reaction of photosynthesis, there are other reactions in plants which are not light dependent but which nevertheless incorporate inorganic carbon into organic molecules. These reactions are collectively referred to as 'dark fixation' because carbon fixation occurs under conditions of continuous darkness, whether chloroplasts are present or absent in the tissue. Thus, dark fixation is distinct from and should not be confused with the dark reaction of photosynthesis.

Photosynthesis does not work in isolation and is not the only component of plant productivity (Penning DeVries, 1974). Dark fixation, respiration and other

Cell & Molecular Biology of Wood Formation, edited by R. Savidge, J. Barnett and R. Napier.

biosynthetic reactions are the other part of the carbon balance equation, and they are no less important than photosynthesis to either plant productivity or plant health (*Figure 1*). As will become evident below, components of dark fixation are vitally important in primary metabolism of all living plant cells, and dark fixation also serves as a mechanism for scavenging respired CO_2, hence increasing the amount of photo-assimilated carbon retained by the plant. At present, almost nothing is known about dark fixation in relation to cambial growth.

1.1 *Examples of dark fixation*

The process of dark fixation was first reported to occur in bacteria (Wood and Werkman, 1935). Most organisms, including animals, have some ability to assimilate CO_2 or bicarbonate (HCO_3^-) into organic molecules (Ting, 1971). Dark fixation has two suggested purposes in C_3 plants: (i) an anapleurotic role in the replenishment of tricarboxylic acid cycle intermediates, in particular through the activity of phospho*enol*pyruvate carboxylase, and (ii) recapture of respired CO_2 (Andreo *et al.*, 1987; Latzko and Kelly, 1983; Melzer and O'Leary, 1987; Splittstoesser, 1966).

As indicated in *Table 1*, dark fixation comprises several possible types of catalysis each of which requires energy, an enzyme and one or more carbon compounds as substrates in order for inorganic carbon to be assimilated into organic molecules (Ting, 1971). One of the metabolically more important dark fixation reactions utilizes HCO_3^-, NH_4^+, and ATP as precursors to generate carbamoyl phosphate, an intermediate in both the urea cycle and pyrimidine biosynthesis, and hence an essential compound for both protein and nucleic acid synthesis. Conversion of pyruvic acid to oxaloacetic acid in the mitochondrion is another essential aspect of dark fixation in support of primary metabolism.

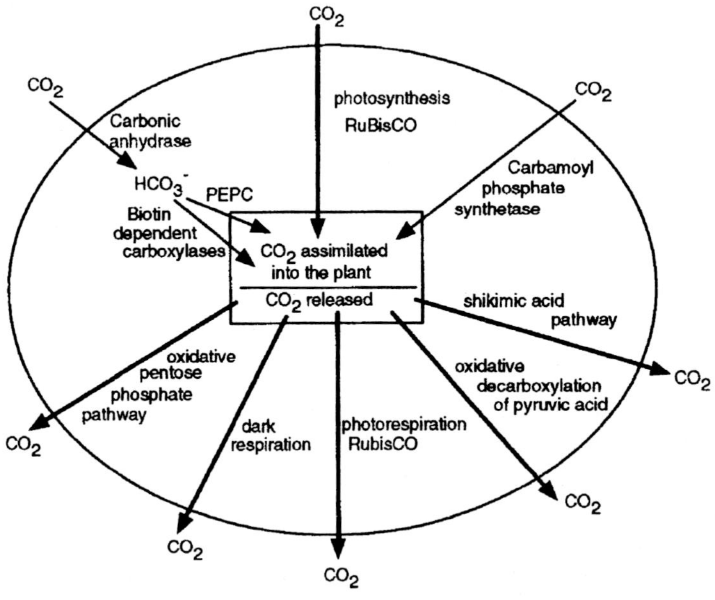

Figure 1. Possible paths of CO_2 fixation and release.

Table 1. Some enzymes of dark CO_2 fixation

Enzyme	Reaction	Reference
phospho*enol*pyruvate carboxylase (EC 4.1.1.31)	phosph*enol*pyruvate + HCO_3^- ⇒ oxaloacetate + Pi	Wingler *et al.* (1994)
acetyl CoA carboxylase (EC 6.4.1.2)	acetyl CoA + HCO_3^- + ATP ⇒ malonyl CoA + ADP + Pi	Konishi *et al.* (1996)
Carbamoyl phosphate synthetase (EC 6.3.4.16)	NH_4 + CO_2 + ATP + H_2O ⇒ carbamoyl phosphate + ADP + Pi	Taylor and Stewart (1981)
glycine synthase (EC 2.1.2.10)	NH_4 + HCO_3^- + methylenetetra-hydrofolate ⇒ glycine	Stryer (1988)
pyruvate carboxylase (EC 6.4.1.1)	pyruvate + HCO_3^- + ATP ⇒ oxaloacetate + ADP + Pi	Wurtele and Nikolau (1990)
propionyl CoA carboxylase (EC 6.3.4.10)	propionyl CoA +HCO_3^- + ATP ⇒ D-methylmalonyl CoA + ADP + Pi	Wurtele and Nikolau (1990)
3-methylcrotonyl CoA carboxylase (EC 6.3.4.11)	3-methylcrotonyl CoA +HCO_3^- + ATP ⇒3-methylglutaconyl CoA + ADP + Pi	Alban *et al.* (1993)

The C_4 pathway of photosynthesis, prevalent in monocots and also associated with crassulacean acid metabolism (CAM) in many xerophytes, involves an initial carboxylation event via phospho*enol*pyruvate carboxylase, followed by subsequent decarboxylation of C_4 acids to provide CO_2 in support of the C_3 photosynthetic pathway. However, phospho*enol*pyruvate carboxylase is not dependent on light for its activity. In fact, through the activity of this enzyme together with carbonic anhydrase (an enzyme catalysing conversion of CO_2 into bicarbonate) CAM plants assimilate the bulk of their CO_2 in darkness. Although carbonic anhydrase and phospho*enol*pyruvate carboxylase are frequently associated with photosynthesis in the plant physiology literature, their activities have also been reported in non-green plant tissues and could be important to the cambium.

1.2 *Carbon reduction during dark fixation*

The reduction of carbon into carbohydrates in the C_3 photosynthetic pathway is driven by NADPH generated during the light reaction of photosynthesis. Dark-fixed carbon is also reduced to generate fatty acids, amino acids, nucleotides, steroids and other compounds. The oxidative pentose phosphate pathway is the source of NADPH needed for reductions of dark-fixed carbon. The oxidative pentose phosphate pathway is undoubtedly highly active during cambial activity and wood formation, it being the same pathway that generates ribose 5-phosphate in support of nucleic acid/nucleotide biosynthesis and erythrose 4-phosphate in support of phenylalanine and lignin biosynthesis.

1.3 *Energy driving dark fixation*

The biosynthesis of all cell components is unquestionably coupled to the hydrolysis of ATP, whether that ATP be generated during the light reaction of photosynthesis or subsequently, from catabolism of carbohydrates or other organic substances. The exergonic energy released during hydrolysis of ATP (or other energy-rich organic compounds) depends on the availability of water, as well as on having the energy-rich compound present. Biochemistry has tended to focus on reactions occurring in living systems where limiting water is not a consideration, but for the heterotrophic tissues of perennial woody species a different perspective is needed. Water content in the cambial region of temperate-zone trees can vary markedly from one season to the next, the cambial cells exhibiting profound structural changes in association with changing water content. How these changes affect energy relations remains unknown.

In the case of the cambium, roots and other non-green tissues, it is also important to recognize that energy generated other than through ATP hydrolysis is undoubtedly important to metabolism and growth, although very little research has yet been done to quantify other contributions. For example, biochemistry theory has the rate of enzymatic catalysis essentially doubling for every 10°C rise in temperature. On this basis, soil water at 30°C entering the root system will energize cambial metabolism in the roots eight times more than soil water at 0°C. This is simplistic, however, because the rate of respiration at 20°C is higher in plants native to cold climates than in those of warm climates (Forward, 1960). Infrared radiation impinging directly on tree stems and the heat energy in the air also affect the activation energy of enzymes, which varies depending on what climate a plant species is adapted to (Wareing, 1985).

2. Sources of carbon dioxide

A large proportion of the gas present in woody tissues is CO_2 (Carrodus and Trifett, 1975; Jenson, 1969) and has been shown to be between 2.8% and 4.2% higher than in the atmosphere (Eklund, 1990, 1993; Jensen, 1969). Carbon dioxide exists in an equilibrium between three molecular species, carbon dioxide, bicarbonate and carbonic acid, depending on the pH. Bicarbonate is more predominant at alkaline or neutral pH, being twenty times as abundant as CO_2 at pH 7.4 (Rubio, 1986), and thus is likely to be the main source of CO_2 in cambium and developing xylem cells at physiological pH. In addition to equilibria achieved in aqueous solution as a function of pH, conversion between CO_2 and HCO_3^- also occurs through the activity of the enzyme carbonic anhydrase, an enzyme described as ubiquitous in plants (Sühltemeyer *et al.*, 1993).

There are three possible sources for the CO_2 /HCO_3^- found in woody stems: (i) the atmosphere, (ii) uptake of dissolved inorganic carbon through the roots, and (iii) CO_2 generating enzymes such as those associated with respiration. Carbon dioxide was found to enter young stems of *Fraxinus excelsior* L. and *Picea abies* (L.) Karst. through lenticels present in the bark (Langenfeld-Heyser et al., 1996; Langenfeld-Heyser, 1989). Radioactive label from $^{14}CO_2$ that entered the stem through the lenticels was found in both photosynthetic and non-photosynthetic cells including the cambium. Export of dark fixation products to the cambium was ruled out; thus, the label present in the cambium was attributed to the dark reaction of photosynthesis (Langenfeld-Heyser, 1989). The possibility of entry of CO_2 through the bark may apply more to younger stems and may not be possible in all stems. The trees that were studied were one to three years of age and possessed photosynthetic bark. As the tree ages the photosynthetic layer of bark is replaced with a deep, non-photosynthetic periderm layer (Gartner, 1996).

Carbon dioxide has been shown to be transported in the transpiration stream (Vapaavuori and Pelkonen, 1985; Vuorinen *et al.*, 1989; Zelawski *et al.*, 1970). Vuorinen *et al.* (1989) found uptake of water soluble inorganic carbon occurred in both light and darkness. However, carbon fixation in the dark after 24 hours was only 50% of that occurring in the light. This transportation ability allows for the uptake of CO_2 from the soil surrounding the roots. The soil can also have a higher concentration of CO_2 than the atmosphere as a result of the decomposition of organic matter (Larcher, 1975), but the uptake of CO_2 from the soil is not considered to be very high (Vapaavuori and Pelkonen, 1985; Vuorinen *et al.*, 1989; Zelawski *et al.*, 1970). While movement of HCO_3^- within the stem could be demonstrated it could not be determined whether the incorporation of label was a result of carboxylation of HCO_3^- transported directly to the site of fixation or from CO_2 produced through the decarboxylation of malate that had been generated through phospho*enol*pyruvate carboxylase activity in the roots.

Eklund and Lavigne (1995) found that there was a restriction of lateral gas movement in *Pinus strobus* L. branches between the woody tissue and the air. They hypothesized that respiratory CO_2 and O_2 in support of respiration are removed from and brought into the stem by the transpiration stream (Eklund and Lavigne, 1995). Transportation of CO_2 / HCO_3^- has been thought to occur to convey them to locations within the plant capable of photosynthetic carbon fixation (Stringer and Kimmerer, 1993; Zelawski *et al.*, 1970). The above ^{14}C labelling studies found evidence not only of photosynthetic CO_2 fixation but also of dark fixation.

Respiration and related enzymatic activity is probably the greatest source of internal CO_2. The stem of a tree has a large proportion of living cells requiring growth and maintenance energy. Respiration in parenchyma cells in support of growth and maintenance may be partially responsible for the high CO_2 concentration within woody stems (Höll and Lendzian, 1973; Shain and Mackay, 1973). Besides the tricarboxylic acid cycle, CO_2 is also released through the oxidative pentose phosphate cycle, the shikimic acid pathway and other pathways (*Figure 1*). The oxidative pentose phosphate pathway is not only critically important for RNA, DNA and NADPH production, it also produces the starting reactant for the shikimic acid pathway. This produces the aromatic amino acids needed for protein synthesis and aromatic-based secondary metabolism, including lignification. The amount of CO_2 released during the production of lignin is assumed to be substantial as lignin is the second most abundant organic compound on earth (Lewis and Sarkanen, 1998).

Although carbon dioxide released during metabolism by non-photosynthesizing tissue has been presumed to escape into the atmosphere (Stringer and Kimmerer, 1993), dark fixation provides a mechanism(s) to recapture it. The ability to relocate respired CO_2 may be of particular importance to trees, given that they have a high proportion of respiring to photosynthetic tissue (Hari *et al*, 1991; Ryan, 1990; Stringer and Kimmerer, 1990; Waring and Schlesinger, 1985) and that much of the respiring tissue is in contact with the transpiration stream (Stringer and Kimmerer, 1993). Due to the high solubility of both CO_2 and HCO_3^- in water, it is possible that the respired CO_2 from non-photosynthetic tissue could be moved in the phloem, xylem or ray conducting streams to other parts of the plant for reassimilation (Stringer and

Kimmerer, 1993). Although the tendency has been to assume that refixation of CO_2 from non-photosynthetic cells takes place in photosynthetic cells (Billings and Godfrey, 1967; Langenfeld-Heyser, 1989; Stringer and Kimmerer, 1993; Zelawski, 1969), the possibility also exists that a portion of this CO_2 /HCO_3^- could be reassimilated in the non-photosynthetic cells of the cambium and xylem, through dark fixation. The refixation of respiratory CO_2 may perform an important metabolite replenishing function for the cambium, xylem and other non-photosynthetic organs.

3. Enzymes of dark fixation

3.1 *Phospho*enol*pyruvate carboxylase*

Phosphoe*nol*pyruvate carboxylase (PEPC) catalyses the carboxylation of phospho*enol*pyruvate to oxaloacetate which is then further converted to malate or aspartate (*Table 1*). Bicarbonate is the form of CO_2 utilized by this enzyme. PEPC has been quite extensively studied in a number of plant and tree species, particularly C_4 and CAM species. PEPC was first discovered in the chloroplasts of a C_3 plant in 1953 (Bandurski and Greiner, 1953) and was later shown to perform the preliminary carboxylation reaction in what are now labelled the C_4 and CAM photosynthetic pathways. PEPC activity has been shown to be fairly widespread in almost all plant species and tissues, including C_3 species (Mazelis and Vennesland, 1957). Although PEPC does not appear to have been investigated in the cambium, its activity has been investigated in other heterotrophic tissues.

The reaction catalysed by PEPC may be one of the most important dark-fixation reactions occurring in plant tissue through its fulfilment of a fundamental anapleurotic, or metabolite replenishing role in metabolism (Melzer and O'Leary, 1987). The focus of PEPC research in C_3 plants has tended towards activity within photosynthetic tissue. The role of PEPC in C_3 plant metabolism, particularly in non-photosynthetic tissues still requires investigation. PEP carboxylase found in C_3 plants is not considered to be a part of the plant's photosynthetic pathway, as it is in C_4 and CAM plants.

The replenishment of tricarboxylic acid cycle intermediates and formation of four carbon skeletons for malate and aspartate production are two of the main proposed anapleurotic functions of PEPC. Activity has been shown to occur in heterotrophic tissues to support these proposed functions. A prime example of this is Indian pine (*Monotropa uniflora* Linn.), a non-chlorophyllous vascular plant. *In vitro*, PEPC activity was shown to occur in both the flower and stem of this plant (Malik *et al*., 1983). Malik *et al.* (1983) proposed that the fixation of CO_2 by PEPC may be an important way for this xylem producing plant to generate organic acids, amino acids and NADPH needed for biosynthesis. PEPC may have a similar function in the cambium. Activity of PEPC has also been demonstrated in other heterotrophic plant tissues such as pollen and flowers which are traditionally large sinks for photosynthate (Schnable and Mayer, 1976; Stanley *et al.* 1958). The conversion of malate, for example, into various tricarboxylic acid cycle products may be an important anapleurotic means for replenishing those molecules needed in support of primary metabolism.

There has been only limited investigation into the activity of PEPC in the xylem. Höll (1974) and Höll and Meyer (1977) found xylem from *Robinia pseudoacacia*, *Tilia cordata* Mill. and *Fraxinus excelsior* L. trees, 18, 32 and 45 years of age, respectively, to have PEPC activity. The largest proportion of assimilated CO_2 was in aspartate. There was high malate dehydrogenase activity but no appreciable accumulation of malate (Höll, 1974). The lack of accumulation of malate may be of physiological significance as malate can have an inhibitory effect on PEPC. The activity of PEPC was not found to be rate limiting; instead, the enzymatic reactions that follow carboxylation of phosph*enol*pyruvate and convert oxaloacetate into other products appeared to be the rate limiting steps. Activity of PEPC was reported to be highest in the youngest annual ring and to decrease towards the centre of the stem, which may help to explain the higher CO_2 concentration generally found in the centre of the stem (Höll and Meyer, 1977).

Another purpose for PEPC in plant metabolism is linked to the assimilation of nitrogen and production of amino acids. This can be seen in particular through the activity of PEPC in roots in relation to varied nitrogen supply. Nitrogen is frequently a limiting nutrient, particularly in temperate coniferous forests (Ingestad *et al.*, 1986). PEPC may be the enzyme that connects carbohydrate metabolism to amino acid metabolism (Paul *et al*, 1978). Increased PEPC activity in response to increased nitrogen supply has been found in plants both with and without nodulated roots (Anderson *et al.*, 1987; Cramer *et al.*, 1993; Lawrie and Wheeler, 1975; Wallenda *et al.*, 1996; Wingler et al, 1994). Wallenda *et al.* (1996) found an increase in PEPC activity and a decrease in starch content in roots after an increased supply of nitrogen to three year old Norway spruce for seven months. When a plant experiences a surplus of nitrogen, photosynthate is reappropriated from the production of carbohydrates, such as sucrose or starch, towards the production of amino acid and protein (Paul *et al*, 1978; Wallenda *et al.*, 1996). Amino acids, protein and organic acids that are produced in the roots may then be transported out of the roots to other parts of the plant in support of growth.

For a more comprehensive view of the mechanisms of the enzyme PEPC consult one of the following reviews (Andreo *et al.*, 1987; Davies, 1979; Latzko and Kelly, 1983; O'Leary, 1982).

3.2 *Other carbon dioxide fixation enzymes*

Although PEPC is commonly referred to as the enzyme of dark fixation, a number of other enzymes have the capacity for carbon fixation. These enzymes include acetyl CoA carboxylase, carbamoyl phosphate synthetase, glycine synthase, pyruvate carboxylase, 3-methylcrotonyl CoA carboxylase, propionyl CoA carboxylase and any other biotin dependent carboxylase (*Table 1*). Propionyl CoA carboxylase and 3-methylcrotonyl CoA carboxylase were only discovered in plant tissue in the past decade, so it is not unrealistic to assume that there may be currently uncharacterized biotin-dependent carboxylases or other carboxylases present in the cambium. Do any of these other carbon fixation enzymes have an impact on carbon acquisition, the carbon balance of the plant, or play a fundamental role in the metabolism of the cambium? This question has yet to be answered, as the activities of these enzymes in the cambium have not been investigated.

Acetyl CoA carboxylase catalyses the first committed step in the synthesis of fatty acids. Fixation of CO_2 by this enzyme may not represent a net assimilation of CO_2, but more of a cyclic exchange of CO_2 as the assimilated carbon is decarboxylated after a few steps in the production of fatty acids. Even though the activity of acetyl CoA carboxylase may not represent a gain of carbon its activity in the cambium and xylem, it may be very important in conifer species which tend to store fat (Höll and Goller, 1982).

Carbamoyl phosphate synthetase and glycine synthase are both involved in nitrogen metabolism, specifically the carboxylation of CO_2 with ammonia and ammonium (Taylor and Stewart, 1981). These reactions are important for the incorporation of nitrogen and the formation of amino acids, thus are essential for primary metabolism. Propionyl CoA carboxylase and 3-methylcrotonyl CoA carboxylase have both been found to occur in plants. Their role in plant metabolism is still unknown and under investigation (Wurtele and Nikolau, 1992; Alban *et al.*, 1993). None of these enzymes has been studied in the cambium or developing xylem, so their importance in the carbon balance of the stem is unknown.

The $\delta^{13}C$ value of plant tissue can provide some clues about the enzyme involved in the carboxylation of CO_2. The $\delta^{13}C$ value indicates to what extent the ^{13}C content of a product differs from that of the substrate. Phospho*enol*pyruvate carboxylase does not discriminate against $\delta^{13}C$ to the same extent as does ribulose bisphosphate carboxylase (rubisco), the carboxylating enzyme of photosynthesis. The effect of this discrimination can be seen when comparing $\delta^{13}C$ contents of C_3 plants, whose first carboxylating enzyme is rubisco, and CAM/C_4 plants, whose first carboxylation occurs through PEPC. The $\delta^{13}C$ values for C_3 plants is -28 ‰ while for CAM/C_4 plants it is -14 ‰ due to decreased $\delta^{13}C$ discrimination by PEPC. Based on $\delta^{13}C$ studies conducted on tobacco leaves, Melzer and O'Leary (1987) stated that the main function of PEPC is anapleurotic and not the refixation of respiratory CO_2. However, this may not be true of heterotrophic tissues such as the cambium or xylem. The $\delta^{13}C$ value of non-foliar tissue tends to be less negative than that of leaves (Brown *et al.*, 1996). This implies that to some extent incorporated carbon is being supplied by an enzyme other than rubisco in larger quantities in non-foliar tissue than foliar tissue.

4. Dark fixation in the pine cambium

4.1 *Methods*

To date, evidence of dark CO_2 fixation activity in the cambium is very limited. The cambia of coniferous species do not appear to have been previously investigated. We have obtained evidence of dark fixation activity in the cambium and developing xylem of *Pinus banksiana* Lamb. and *Pinus strobus* L. Cambium and developing xylem tissue samples were collected from 15 to 20 year old *P. strobus* and *P. banksiana* at different times during the growing season. Trees were harvested from the University of New Brunswick Woodlot (Fredericton, New Brunswick, Canada). Samples were collected by peeling the bark, scraping off the cambium and xylem and quick-freezing the tissue in liquid nitrogen until ready for use. Microscopy was performed using hand sections to determine the phenology of the tissue.

Protein extracts were prepared by grinding tissue in liquid nitrogen and extracting the protein in a buffer containing 50 mM Tris-HCl (pH 8.0), 20 mM mercaptoethanol, 10% glycerol and 10 mM $MgCl_2$. Following centrifugation, the supernatant was brought to 80% ammonium sulfate saturation. The precipitated protein was collected by centrifugation and desalted using a Sephadex G-25 column. The resulting protein fractions were assayed for activity. ^{14}C-labelled HCO_3^- was used to investigate dark fixation activity. The assay consisted of 200 µl protein extract, 1 mM PEP, 0.5 mM $MgCl_2$, 0.2 mM HCO_3^- (0.09 µCi), 50 mM Tris-HCl (pH 8.0). The reaction was terminated after one hour by the addition of acetic acid. Aliquots were dried on filter paper and counted using liquid scintillation spectroscopy. Thin layer chromatography of the reaction products was also conducted using two solvent systems (7:3, ethanol: H_2O; 45:8:8, benzene:methanol:acetic acid). TLC plates were analysed using a TLC plate scanner (Bioscan System 200 Imaging Scanner). Malate and aspartate standards were co-chromatographed with the products to aid in product identification. ^{13}C-labelled bicarbonate was also used in assays and the products were analysed using gas chromatography/mass spectroscopy (GC/MS).

4.2 *Results*

Results indicated dark fixation activity was occurring in the cambium and developing xylem of *P. strobus* and *P. banksiana*. Fixation of CO_2 into acid stable products was found to occur in samples harvested from April 24th to August 8th (*Figure 2*). Microscopy of tissue harvested on these dates show the cambium to be active. There was no activity in protein extracted from dormant cambium (October 10th). Thin layer chromatography showed one main product of CO_2 fixation. The Rf value for the fixation product differed from that of malate and aspartate. GC/MS did not confirm the incorporation of ^{13}C into aspartate, either in scanning or more sensitive selected-ion monitoring modes. Thus the product could not be deduced to be malate or aspartate; however, this may in part be due to analytical sensitivities. Incubations conducted without PEP as a co-substrate did not yield any appreciable incorporation. This tends to imply that dark fixation is the result of PEPC activity even though malate and aspartate have not been identified as the reaction products to this point. Further investigations are underway to identify the fixation product and to look at kinetic activity and gene expression of both PEPC and acetyl CoA carboxylase, another important carboxylating enzyme in the cambium.

5. Implications of dark fixation

The fixation of inorganic carbon dioxide into organic matter by plants is essential to the survival of all life. The fixed carbon is used by both plants and animals as a source of molecular building blocks and energy. Fixation of carbon dioxide in the stems of woody plants may be of major physiological importance to the plant. High levels of CO_2 in the stem could conceivably lead to interruptions of the transpiration stream (Höll, 1974), resulting in the disruption of transpiration and other metabolic processes in the plant. Even small amounts of fixation by a source other than photosynthesis could have a significant effect on the productivity of the plant over the long term. Vuorinen *et al.* (1989) found that even though the percentage of carbon taken up by roots was very minor, one to two percent, in comparison to photosynthesis, *Salix* plants grown with elevated CO_2 medium for 1 month showed a 30% increase in dry weight over control plants. If CO_2 is fixed in the cambium through the process of dark fixation, even if the percentage is small in comparison to photosynthesis, it still may have an effect on long term productivity. The anapleurotic role of dark fixation is also a critically important physiological process. Without the replenishment of intermediary metabolites,

	24-Apr	5-May	21-May	10-Jun	14-Aug	28-Aug	10-Oct
CZ	4	12	14	14	10	8	6
RE	0	0	10	10	1	5	0
SL	0	0	2	2	16	8	26
NT	0	0	0	0	20	20	++

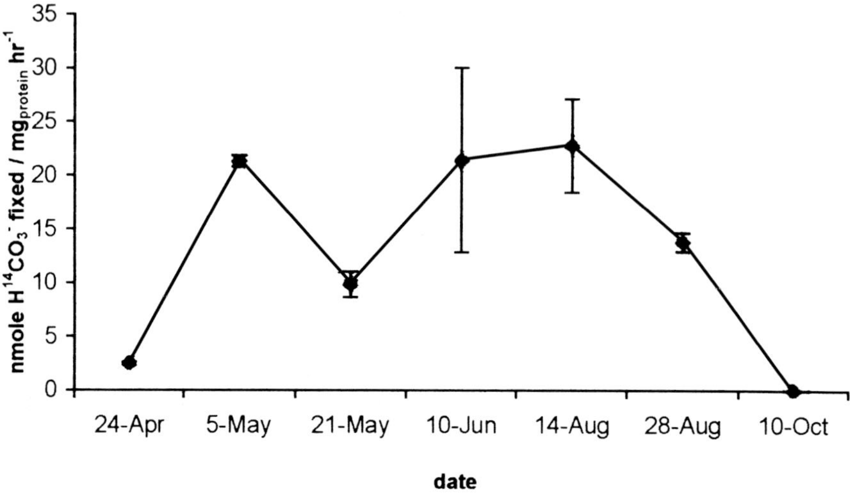

Figure 2. *HCO_3^- fixation at different times during the growing season (standard error of the mean, n = 3). Phenological data on cells per radial file also shown (CZ = cambial zone; RE = radially expanding cells; SL = cells with lignifying secondary walls; NT = new tracheids (++ represents greater than 25).*

respiration and biosynthesis would cease to function. Respired CO_2 converted to HCO_3^- by carbonic anhydrase or physico-chemical processes may function as the source of substrate for anapleurotic dark fixation reactions.

The carbon balance of the forest is an important factor when considering global warming and the increase in atmospheric CO_2. Carbon assimilation affects the global environment. With an ever changing environment, it is becoming increasingly important to understand the carbon balance of plants and trees. As a result of the Kyoto protocol on global warming, many countries are currently looking for ways to reduce carbon emissions. A possible route for combating increasing atmospheric CO_2 is to increase carbon storage through forest management. Trees provide a large mechanism for carbon storage. Respiratory carbon loss from the stems of trees can range from sixteen to twenty-three percent of annual net photosynthesis and is an important factor in the carbon balance of both trees and the forest (Ryan, 1989). Trees competent to recycle and re-fix CO_2 efficiently may have increased productivity and, may therefore store more carbon and release less into the atmosphere through respiration. Dark fixation may affect gas exchange measurements. Even if dark CO_2 fixation does not contribute significantly to plant productivity it must be understood in order to properly interpret gas exchange measurements (Zelawski *et al.*, 1970). In order to model the effect of trees on the ecosystem carbon balance, the processes of carbon fixation and release within the tree must be better understood.

Dark fixation cannot be construed to be a substitute for photosynthetic CO_2 fixation and it may be argued that if models use net photosynthesis, any anapleurotic fixation will already be accounted for in this measurement. However, this assumes that the only form of energy available to the tree in support of metabolism is that captured from photosynthetically active radiation (400–700 nm) through the light reaction of photosynthesis. This of course is naive (see Section 1.3), as a tree exposed to photosynthetically active radiation, but under sub-zero temperatures will not grow. Thermal energy applied directly on the stem of a tree is critically important to support cambial growth (Savidge and Wareing, 1981). The energy contained in liquid water molecules is also available, particularly through the innumerable exergonic hydrolysis reactions occurring in support of growth and development. The reactions of dark fixation may not respond to environmental change in the same way that photosynthesis does. Thus, the use of only the changes in net photosynthesis to model vegetation responses to environmental change will not necessarily reflect what is actually happening within the tree.

6. Conclusion

The activity of dark CO_2 fixation may be significant in the metabolism of plants and in particular woody plants. It cannot be forgotten that there is more to trees than just leaves. At present there has been insufficient research on the ability of the cambium and its derivatives to fix CO_2 in order to fully understand the role of dark fixation in the metabolism and growth of woody plants. Dark fixation activity has been found in the cambium, developing xylem and mature xylem. Whether dark fixation has a purely anapleurotic function or serves a larger purpose has not been completely investigated. Dark CO_2 fixation has an effect on the carbon balance of the tree, as well as on forest ecosystems and the global environment. The effects of internal and external environmental conditions, such as hormones or elevated atmospheric CO_2, on dark fixation must be understood to model and predict the response of vegetation to climate change accurately.

Acknowledgement

N.A.F. is grateful to NSERC Canada for a postgraduate scholarship.

References

Alban, C., Baldet, P., Axiotis, S. and Douce, R. (1993) Purification and characterization of 3-methylcrotonyl-coenzyme A carboxylase from higher plant mitochondria. *Plant Physiol.* **102**: 957–965.

Anderson, M.P., Heichel, G.H. and Vance, C.P. (1987) Nonphotosynthetic CO_2 fixation by alfalfa (*Medicago sativa* L.) root nodules. *Plant Physiol.* **85**: 283–289.

Andreo, C.S., Gonzalez, D.H. and Iglesias, A.A. (1987) Higher plant phosphoenolpyruvate carboxylase, structure and regulation. *FEBS Letters* **213(1)**: 1–8.

Billings, W.D. and Godfrey, A.J. (1967) Photosynthetic utilization of internal carbon dioxide by hollow-stemmed plants. *Science* **158**: 121–123.

Bandurski, R.S. and Greiner, C.M. (1953) The enzymatic synthesis of oxalacetate from phosphoryl-enolpyruvate and carbon dioxide. *J. Biol. Chem.* **204**: 781–786.

Brown, K.R., Thompson, W.A., Camm, E.L., Hawkins, B.J. and Guy, R.D. (1996) Effects of N addition rates on the productivity of *Picea sitchensis, Thuja plicata,* and *Tsuga heterophylla* seedlings. *Trees* **10**: 198–205.

Carrodus, B.B. and Triffett, A.C.K. (1975) Analysis of composition of respiratory gases in woody stems by mass spectroscopy. *New Phytol.* **74**: 243–246.

Cramer, M.D., Lewis, O.A.M. and Lips, S.H. (1993) Inorganic carbon fixation and metabolism in maize roots as affected by nitrate and ammonium nutrition. *Physiologia Plantarum* **89**: 632–639.

Davies, D. (1979) The central role of phosphoenolpyruvate carboxylase in plant metabolism. *Ann. Rev. Plant Physiol.* **30**: 131–158.

Eklund, L. (1990) Endogenous levels of O_2, CO_2 and ethylene in stems of Norway spruce trees throughout a growth period. *Trees* **4**: 150–154.

Eklund, L. (1993) Seasonal variations of O_2, CO_2 and ethylene in oak and maple stems. *Can. J. For. Res.* **23**: 2608–2610.

Eklund, L. and Lavigne, M.B. (1995) Restricted lateral gas movement in *Pinus strobus* branches. *Trees* **10**: 83–85.

Forward, D.F. (1960) Effect of temperature on respiration. In: *Encyclopedia of Plant Physiology* (ed. W. Ruhland), Springer: Berlin. Vol. **12(2)**: 234–258.

Gartner, B.L. (1996) Does photosynthetic bark have a role in the production of core vs. outer wood? *Wood and Fiber Sci.* **28**: 53–61.

Hari, P., Nygren, P. and Korpilahti, E. (1991) Internal circulation of carbon within a tree. *Can. J. For. Res.* **21(4)**: 514–515.

Höll, W. (1974) Dark CO_2 fixation by cell-free preparations of the wood of *Robinia pseudoacacia*. *Can. J. Bot.* **52**: 727–734.

Höll, W. and Goller, I. (1982) Free sterols and steryl esters in the trunkwood of *Picea abies* (L.) Karst. *Z. Pflanzenphysiol. Bd.* **106**: 409–418.

Höll, W. and Lendzian, K. (1973) Respiration in the sapwood and heartwood of *Robinia pseudoacacia. Phytochemistry* **12**: 975–977.

Höll, W. and Meyer, H. (1977) Enzyme catalysed CO_2 fixation by wood preparations of *Tilia cordata* Mill. and *Fraxinus excelsior* L. *Holzforschung* **6**: 184–187.

Jensen, K.F. (1969) Oxygen and carbon dioxide concentrations in sound and decaying red oak trees. *Forest Sci.* **15**: 246–251.

Konishi, T., Shinohara, K., Yamada, K. and Sasaki, Y. (1996) Acetyl-CoA carboxylase in higher plants: Most plants other than Gramineae have both the prokaryotic and the eukaryotic forms of this enzyme. *Plant and Cell Physiol.* **37(2)**: 117–122.

Langenfeld-Heyser, R. (1989) CO_2 fixation in stem slices of *Picea abies* (L.) Karst: Microautoradiographic studies. *Trees* **3**: 24–32.

Langenfeld-Heyser, R., Schella, B., Buschmann, K. and Speck, F. (1996) Microautoradiographic detection of CO_2 fixation in lenticel chlorenchyma of young *Fraxinus excelsior* L. stems in early spring. *Trees* **10**: 225–260.

Larcher, W. (1975) *Physiological Plant Ecology.* Springer-Verlag: Berlin.

Latzko, E. and Kelly, G.J. (1983) The many-faceted function of phosphoenolpyruvate carboxylase in C_3 plants. *Physiol. Veg.* **21(5)**: 805–815.

Lawrie, A.C. and Wheeler, C.T. (1975) Nitrogen fixation in the root nodules of *Vicia faba* L. in relation to the assimilation of carbon. *New Phytologist* **74**: 437–445.

Lewis, N.G., and Sarkanen, S. (eds.) (1998) *Lignin and Lignan Biosynthesis.* ACS Symposium Series 697, Am. Chem. Soc., Washington, D.C.

Liedvogel, B. (1986) Acetyl coenzyme A and isopentenylpyrophosphate as lipid precursors in plant cells — biosynthesis and compartmentation. *J. Plant Physiol.* **124**: 211–222.

Malik, C.P., Gill, R.K. and Singh, M.B. (1983) Nonphotosynthetic CO_2 fixation in *Monotropa uniflora* Linn. — a colourless angiosperm. *Indian J. Exp. Bot.* **21**: 405–407.

Mazelis, M. and Vennesland, B. (1957) Carbon dioxide fixation into oxalacetate in higher plants. *Plant Physiol.* **32**: 591–600.

Melzer, E. and O'Leary, M.H. (1987) Anapleurotic CO_2 fixation by phosphoenolpyruvate carboxylase in C_3 plants. *Plant Physiol.* **84**: 58–60.

Paul, J.S., Cornwell, K.L. and Bassham, J.A. (1978) Effects of ammonia on carbon metabolism in photosynthesizing isolated mesophyll cells from *Papaver somniferum* L. *Planta* **142**: 49–54.

Penning de Vries, F.W.T. (1974) Products, requirements and efficiency of biosynthesis: a quantitative approach. *J. Theor. Biol.* **45**: 339–377.

O'Leary, M.H. (1982) Phosphoenolpyruvate carboxylase: An enzymologist's view. *Ann. Rev. Plant Physiol.* **33**: 297–315.

Rubio, V. (1986) Enzymatic HCO_3^- fixation: a common mechanism for all enzymes involved? *Bioscience Reports* **6(4)**: 335–347.

Ryan, M.G. (1989) Growth and maintenance respiration in stems of *Pinus contorta* and *Picea engelmannii*. *Can. J. For. Res.* **20**: 48–57.

Salisbury, F.B. and Ross, C.W. (1992) *Plant Physiology*. Wadsworth Publishing Co.: Bilmont, California.

Savidge, R.A. and Wareing, P.F. (1981) Plant growth regulators and the differentiation of vascular elements. In: *Xylem Cell Development* (ed. Barnett, J. R.), pp. 192–235

Schnabl, H. and Mayer, I. (1976) Dark fixation of CO_2 by flowers of cut roses. *Planta* **131**: 51–55.

Shain L. and Mackay J.F.G. (1973) Seasonal fluctuation in respiration of aging xylem in relation to heartwood formation in *Pinus radiata. Can J. Bot.* **51**: 737–741.

Splittstoesser, W.E. (1966) Dark CO_2 fixation and its role in the growth of plant tissue. *Plant Physiol.* **41**: 755–759.

Stanley, R.G., Young, L.C.T. and Graham, J.S.D. (1958) Carbon dioxide fixation in germinating pine pollen (*Pinus ponderosa*). *Nature* **182**: 1462–1463.

Stringer, J.W. and Kimmerer, T.W. (1993) Refixation of xylem sap CO_2 in *Populus deltoides. Physiologia Plantarum* **89**: 243–251.

Stryer, L. (1988) *Biochemistry.* W.H. Freeman and Co. New York.

Sültemeyer, D., Schmidt, C. and Fock, H.P. (1993) Carbonic anhydrases in higher plants and aquatic microorganisms. *Physiologia Plantarum* **88**: 179–190.

Taylor, A.A. and Stewart, G.R. (1981) Tissue and subcellular localization of enzymes of arginine metabolism in *Pisum sativum*. *Biochem. Biophys. Res. Commun.* **101**: 1281–1289.

Ting, I.P. (1971) Nonautotrophic CO_2 fixation and crassulacean acid metabolism. In: *Photosynthesis and Photorespiration* (eds M.D. Hatch, C.B. Osmond and R.O. Slayter), Wiley-Interscience: New York, pp. 169–188.

Vapaavuori, E.M. and Pelkonen, P. (1985) HCO_3^- uptake through the roots and its effect on the productivity of willow cuttings. *Plant and Cell Environment* **8**: 531–534.

Vuorinen, A.H., Vapaavuori, E.M. and Lapinjoki, S. (1989) Time-course of uptake of dissolved inorganic carbon through willow roots in light and in darkness. *Physiologia Plantarum* **77**: 33–38.

Wallenda, T., Schaeffer, C., Einig, W., Wingler, A., Hampp, R., Seith, B., George, E. and Marschner, H. (1996) Effects of varied soil nitrogen supply on Norway spruce (*Picea abies* [L.] Kars). *Plant and Soil* **186**: 361–369.

Wareing, P.F. (1985) Tree growth at cool temperatures and prospects for improvement by breeding. In: *Attributes of Trees as Crop Plants* (eds M.G.R. Cannell and J.E. Jackson), ITE, NERC: Abbots Ripton, England, pp 80–88.

Waring, R.H. and Schlesinger, W.H. (1985) *Forest Ecosystems Concepts and Management.* Academic Press: New York, New York.

Wingler, A., Einig, W., Schaeffer, C., Wallenda, T., Hampp, R., Wallander, H. and Nylund, J. (1994) Influence of different nutrient regimes on the regulation of carbon metabolism in Norway spruce (*Picea abies* (L.) Karst.) seedlings. *New Phytologist* **128**: 323–330.

Wood, H.G. and Werkman, C.H. (1935) The utilization of CO_2 in the dissimilation of glycerol by the propionic acid bacteria. *J. Bacteriology* **30(1)**: 48–53.

Wurtele, E.S. and Nikolau, B.J. (1990) Plants contain multiple biotin enzymes: discovery of 3-methylcrotonyl-CoA carboxylase, propionyl carboxylase and pyruvate carboxylase in the plant kingdom. *Archives of Biochemistry and Biophysics* **278:** 179–186.

Wurtele, E.S. and Nikolau, B.J. (1992) Differential accumulation of biotin enzymes during carrot somatic embryogenesis. *Plant Physiol.* **99**: 1699–1703.

Zelawski, W, Riech, F.P. and Stanley, R.G. (1970) Assimilation and release of internal carbon dioxide by woody plant shoots. *Can. J. Bot.* **48**: 1351–1355.

11

Wood formation in hybrid aspen (*Populus tremula* L. × *Populus tremuloides* Michx.) grown under different nitrogen regimes

L. Puech, S. Türk, J. Hodson and S. Fink

1. Introduction

The middle of this century saw a growing scientific interest in the structural, chemical and mechanical properties of poplar wood which were generally assessed using light microscopy (Liese and Ammer, 1957; Meyer-Uhlenried, 1957). The pulp and paper industries required basic information about the composition of poplar wood and how to influence it and increase growth. Special interest was therefore accorded to the effects of climatic factors (Necesany and Cetlová, 1969) and fertilization (Safford and Czapowskyj, 1986). The positive effects of fertilization on growth were mainly described at a phenological level, that is using the stem diameter, the total height and the crown size (Safford and Czapowskyj, 1986; Nilsson and Wasielewski, 1987), and little work was carried out on the effect of nitrogen on the anatomical properties of the wood. In recent decades, the problem of forest decline has stimulated research on nitrogen fertilization, in poplar and other species, in relation to other stresses such as mineral deficiencies and air pollution. Anatomical studies have also proved useful as complements to physiological studies (Pell *et al.*, 1995; Günthardt-Goerg *et al.*, 1996; Puech and Mehne-Jakobs, 1997).

It is well-known that the vascular cambium is responsible for wood formation and that influencing the activity of cambium may modify the properties of the wood. There are several consecutive phases involved in the development of xylem elements. These include division in the cambial zone, differentiation, cell expansion, secondary wall formation and autolysis of xylem elements destined for water transport and

Cell & Molecular Biology of Wood Formation, edited by R. Savidge, J. Barnett and R. Napier.

mechanical support (Larson, 1994). Xylem development is coordinated in space and cross-sectional analysis reveals a succession of discrete zones corresponding to the different developmental states. The rate of xylem cell production, that is the number of divisions in the cambial zone per unit of time, is determined by the rate of cell cycling of the individual cambial zone cells and by the number of dividing cells. Similarly, the rate and duration of expansion and secondary wall formation affect cell size and wall thickness of the xylem elements, thus determining the final properties of the wood (Wilson and Howard, 1968; Gregory, 1971; Dodd and Fox, 1990; Ridoutt and Sands, 1994). A correlation reportedly exists between the radial width of the expansion zone and the final cross-sectional lumen area of fibres in the wood of hybrid aspen (Tuominen *et al.*, 1997). Although most efforts have been directed at elucidating the effect of growth regulators on cambial activity in order to improve or at least influence wood formation, some work has been done on the effect of nitrogen fertilization, but mainly in gymnosperms (Brix and Mitchell, 1980; Sundberg *et al.*, 1993).

We have used modern light microscopy techniques, such as computer-assisted image analysis, to investigate the role of different nitrogen treatments on the wood formation processes in hybrid aspen (*Populus tremula* L. × *P. tremuloides* Michx.) grown under controlled conditions. We intended to establish the effect of nitrogen on the relationship between the histological properties of the cambium and its derivative tissues and the final properties of the wood, in particular the dimensions and density of different xylem cell types.

2. Materials and methods

2.1 *Plant cultivation*

Hybrid aspen (*Populus tremula* L. × *P. tremuloides* Michx.) was vegetatively propagated in sterile cultures. Rooted plants were then potted in mineral wool and cultivated in a controlled-environment chamber with a day/night temperature of 20°C/15°C, an 18 h-photoperiod, a photon flux density of 300 μE m^{-2} s^{-1} provided by HQI-TS 400 W DH^{-1} metal halogen lamps (Osram, München, Germany) and a relative humidity of at least 70%. The plants were freely watered with a complete Wallco 51–10–4 + micronutrient solution (Cederoth International AB, Falun, Sweden) daily. Four different nitrogen treatments were applied by supplying the plants with different diluted nutrient solutions with an N-content corresponding to 100 mg N l^{-1}, 50 mg N l^{-1}, 25 mg N l^{-1} and 12.5 mg N l^{-1}, provided as NH_4NO_3.

2.2 *Phenotypic characterization and sampling procedure*

After a growth period of 2 months (June–July 1998) the trees had reached a height of approximately 1 m. During the growth of the trees, the lateral buds were removed. The stem was sampled by cutting at approximately 10 cm from the ground (further defined as the 'base') and divided into appropriate segments for further analysis: two consecutive 1 cm segments were taken from the base for microscopical observation and wood maceration. During the growth of the trees, the lateral buds were removed. The following characteristics were measured at the time of sampling: height, basal stem diameter, number of leaves longer than 1 cm and finally, the width and length of fully expanded leaves between the 20th and the 25th internode from the 'top' (defined as the position of the 12th leaf from the apex).

2.3 *Anatomical characterization*

Specimens from the base of the stems, consisting of mature xylem and extraxylary tissues, were fixed in cold 4% glutaraldehyde in 25 mM phosphate buffer, pH 7.2, for at least one day. The samples were then dehydrated in an ascending series of acetone concentrations and embedded in a methacrylate resin (Fink, unpublished). Transverse and longitudinal sections, 3 μm and 5 μm thick, were cut with a Leica-2065-microtome (Leica Inst. GmbH, Nussloch, Germany) using a diamond knife. The sections were stained polychromatically with successive incubations in 0.1% acriflavin/3% safranin O, 1% auramin O and 2% methylene blue. Parallel sections were also stained for polysaccharides with periodic acid-Schiffs reagent (PAS) as described by Gerlach (1977) and for proteins with 0.02% Coomassie blue R250 (Gahan, 1984). The sections were mounted in Eukitt (Thoma, Freiburg, Germany) and observed using a Zeiss-Axiophot light microscope (Carl Zeiss, Oberkochen, Germany).

The widths of the different developmental zones, and the number and size of fusiform cells in the cambial zone were measured for ten radial files on one section per tree using an ocular measuring scale. The developmental zones measured on transverse sections comprised:

(i) the cambial zone, consisting of dividing cells,
(ii) the expansion zone, defined by the presence of primary-walled expanding xylem fibres and vessels,
(iii) the maturation zone, defined by the presence of secondary thickening in the fibre walls.

The distinction between the different phases was made according to the radial dimensions of the cells, the thickness of tangential walls, the presence of birefringence and the presence or absence of cytoplasm (Skene, 1969; Wodzicki, 1971). The tangential diameter of cambial fusiform cells, the number of fusiform initials per ray initial and the density of rays in the cambium were determined in ten randomly selected intervals between rays on transverse sections. The widths of the phloem and of the remaining bark (cortex and periderm) were measured on radial sections where the transition from phloem to cortex was recognized by the disappearance of sieve tubes and phloem fibres. The densities of vessels and rays were determined on photographs of transverse sections by measuring the number of vessels and rays crossed by lines drawn parallel to the cambium at 400 μm and 800 μm from the cambial zone. The length of the line was approximately 2 mm.

The cross-sectional lumen area of fibres and vessels was measured on transverse sections using a computer-assisted image analysing system. These sections were stained with PAS in order to intensify the contrast of the cell walls and to reduce staining of the cell contents. Before analysis, fibres and vessels were distinguished using UV-fluorescence in combination with the PAS staining (Puech and Mehne-Jacobs, 1997).

For the measurement of the fibre lumen area, five 225 μm × 300 μm pictures were taken with a 3CCD colour video camera (Sony DXC-930P) both at the inner limit of the expansion zone and in the recently matured xylem on each section. For the measurement of vessel lumen area, five 600 μm × 450 μm pictures were taken in the same way in the recently differentiated xylem. The images were improved by removing cell content, restoring broken cell walls, and enhancing contrast and sharpness with Adobe-Photoshop D1–2.5.1. (Adobe Systems Inc., Mountain View, CA). Image analysis was carried out using Image-Pro Plus 1.1 for Windows (Media Cybernetics, Silver Spring, MD). In the digital pictures, the lumen area of fibres and vessels was measured automatically. For this purpose, threshold values were set to allow the whole cell lumen to be counted as an object. Border objects were excluded automatically while ray cells were removed manually. A lower size limit was set at 50 μm^2 to exclude fibre ends.

To measure fibre and vessel lengths, a 1-cm-long sample was collected from the base of the stem. The xylem was chopped into thin sticks, and macerated in acidified sodium chlorite solution (Spearin and Isenberg, 1947). The length of 150 fibres and 50 vessels (including tips) in each sample was measured manually after staining with toluidine blue O 0.5% (Jensen, 1962) using Image-Pro Plus 1.1 for Windows (Media Cybernetics, Silver Spring, MD).

2.4 *Statistical analysis*

Analysis of variance (ANOVA) and the Tukey test were used to analyse differences among the treatments within each measurement. Student's t-test was also applied to assess the statistical significance of the difference between the means of the two extreme nitrogen treatments ($p < 5\%$).

3. Results

3.1 *Phenology*

Reducing N-supply below 100 mg N l^{-1} induced a decrease in apical growth, whereas a reduction in radial growth, reflected by the radial diameter at stem base, was observed only at levels below 50 mg N l^{-1} (*Table 1*). Further reductions in radial growth were achieved when the N-levels were lowered from 25 mg N l^{-1} to 12.5 mg N l^{-1}. The number of leaves exceeding 1 cm also correlated positively with the nitrogen level (*Table 1*).

3.2 *Radial width of stem tissues*

The secondary tissues of the stem are composed of cells in various phases of division, expansion, and maturation, forming the cambial zone and the juvenile xylem and phloem, as well as mature xylem and phloem. These tissues are embedded in the primary body of the stem, which includes the cortex, the primary phloem and xylem, and the pith (*Figure 3a, b, e and f*). Radial widths of most of those tissues were measured at the base of the stem. A correlation between the above-mentioned reduction in radial diameter and a reduction in xylem width (significantly decreased under 50 mg N l^{-1} and further decreased under 25 mg N l^{-1}) was observed (*Figure 1*). The width of the bark remained uninfluenced by the N-supply (*Figure 1*). The width of the phloem, however, increased with the reduction in N-supply (observed below

Table 1. Phenotypical characteristics of hybrid aspen grown at different nitrogen regimes. (100 mg Nl[-1], 50 mg Nl[-1], 25 mg Nl[-1] and 12.5 mg Nl[-1])[a]

Growth parameters	N100		N50		N25		N12,5	
	Mean value	Std-error	Mean value	Std-error	Mean value	Std error	Mean value	Std error
Total length (cm)	120.5	±1.6a	101.7	±1.6b	94.8	±4.1b	95.3	±2.4b
Diameter at stem base (cm)	1.27	±0.04a	1.27	±0.03a	0.97	±0.02b	0.75	±0.03c
Number of leaves (>1 cm)	70.6	±2.4a	68.8	±2.8a	58.4	±1.3b	45.2	±0.7c
Ratio xylem/phloem	13.9	±0.5a	12.5	±0.4a	7.2	±1.1b	4.7	±0.4b

[a] Mean and standard error of measurements made on four trees per treatment. Radial width of xylem and phloem were measured at ten randomly selected positions on radial longitudinal sections sampled from the base of the trees. Different letters indicate statistically significant differences between the means as determined by an ANOVA and a Tukey test ($p < 5\%$).

50 mg N l[-1]) (*Figures 2, 3a and b*). Concomitantly, the width of the remaining bark was reduced (*Figures 2, 3a and b*). Consequently, a decrease in the xylem/phloem ratio was observed below 50 mg N l[-1] and further accentuated below 25 mg N l[-1] (*Table 1*).

3.3 *Developmental zones and xylogenesis*

Radial widths of developmental zones involved in xylogenesis (for definition see Section 2.3) were measured at the base of the stem. The cambium was significantly thinner in the plants supplied with 12.5 mg N l[-1] (*Figures 2 and 3a–d*) and was constituted of fewer cells than in the other treatments (*Table 2, Figures 3c and d*). However, the radial and tangential diameters of the fusiform cambial cells were constant in all

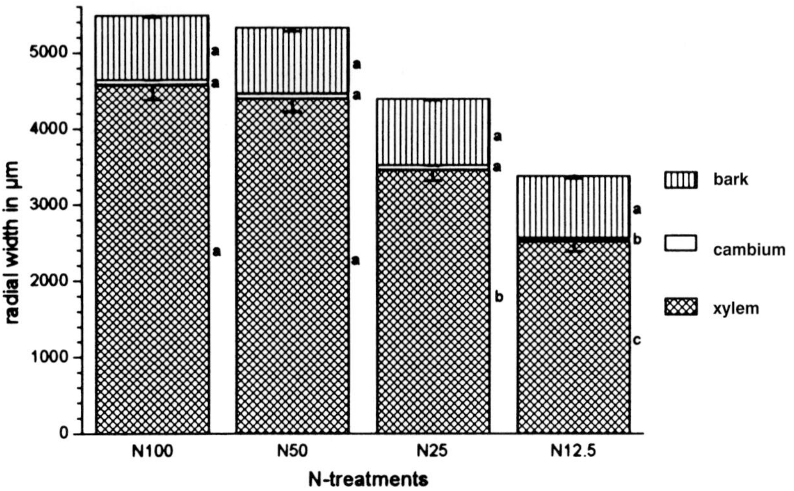

Figure 1. Radial width of stem regions of hybrid aspen grown under different nitrogen regimes (100 mg N l[-1], 50 mg N l[-1], 25 mg N l[-1] and 12.5 mg N l[-1]). The measurements were made in ten randomly selected rows on a transverse section sampled from the base of four trees per treatment. Different letters indicate statistically significant differences between the means as determined by an ANOVA and a Tukey test ($p < 5\%$). The bars indicate standard error.

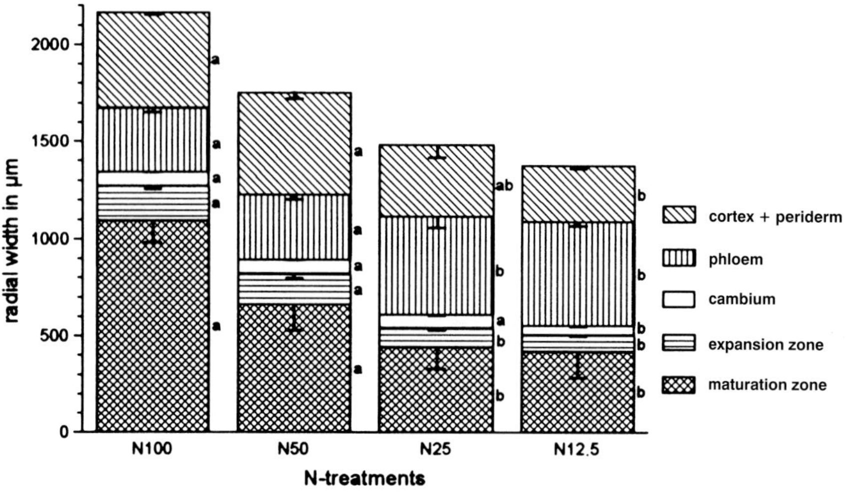

Figure 2. *Radial width of different stem tissues and developmental zones of hybrid aspen grown under different nitrogen regimes (100 mg N l^{-1}, 50 mg N l^{-1}, 25 mg N l^{-1} and 12.5 mg N l^{-1}). The measurements were made at ten randomly selected positions on a radial longitudinal section (phloem and remaining bark) and in ten randomly selected rows on a transverse section (cambium, expansion zone and maturation zone) sampled from the base of four trees per treatment. Different letters indicate statistically significant differences between the means as determined by an ANOVA and a Tukey test ($p < 5\%$). The bars indicate standard error.*

N-treatments applied (*Table 2*). The length of the fusiform cambial cells at the base, deduced from the length of the vessel elements measured after wood maceration, remained unaffected by the N-supply (*Table 2*).

Reductions of the radial width of the zone of expanding xylem elements ('expansion zone') and of the zone of xylem maturation ('maturation zone') were observed when N-supply was maintained below 50 mg N l^{-1} (*Figures 2, 3c and d*).

3.4 *Wood anatomy*

The wood anatomy at the base of the stem was characterized by the lumen areas of the fibres (before and after secondary wall completion) and of the vessels, by the length of the fibres in macerated wood, thus giving an evaluation of fibre elongation, and by the density of rays and vessels taken at 400 µm and 800 µm from the cambium.

Figure 3. *Anatomical appearance of the stem base of hybrid aspen grown with 100 mg N l^{-1} (a, c, e, g) and 12.5 mg N l^{-1} (b, d, f, h). (a) and (b) Bark, cambial zone and first cambial derivatives on the xylem side; bar = 100 µm. (c) and (d) Young phloem, cambial zone and cambial derivatives on the xylem side; (d), starch grains in the phloem parenchyma and in the rays on the phloem side (arrow head); bar = 50 µm. (e) and (f) Immature and recently matured xylem; (f), starch accumulation in the rays of the mature xylem (arrow head); bar = 100 µm. (g) and (h) Mature xylem; (h), accumulation of starch in all the ray cells except the contact cells; bar = 50 µm. (a), (b), (c), (e), (g) Polychromatical staining; (d), (f), (h) PAS staining. Cx: cortex + periderm; Ph: phloem; CZ: cambial zone; EZ: expansion zone; MZ: maturation zone; F: fibre; V: vessel; R: ray; CC: contact cell.*

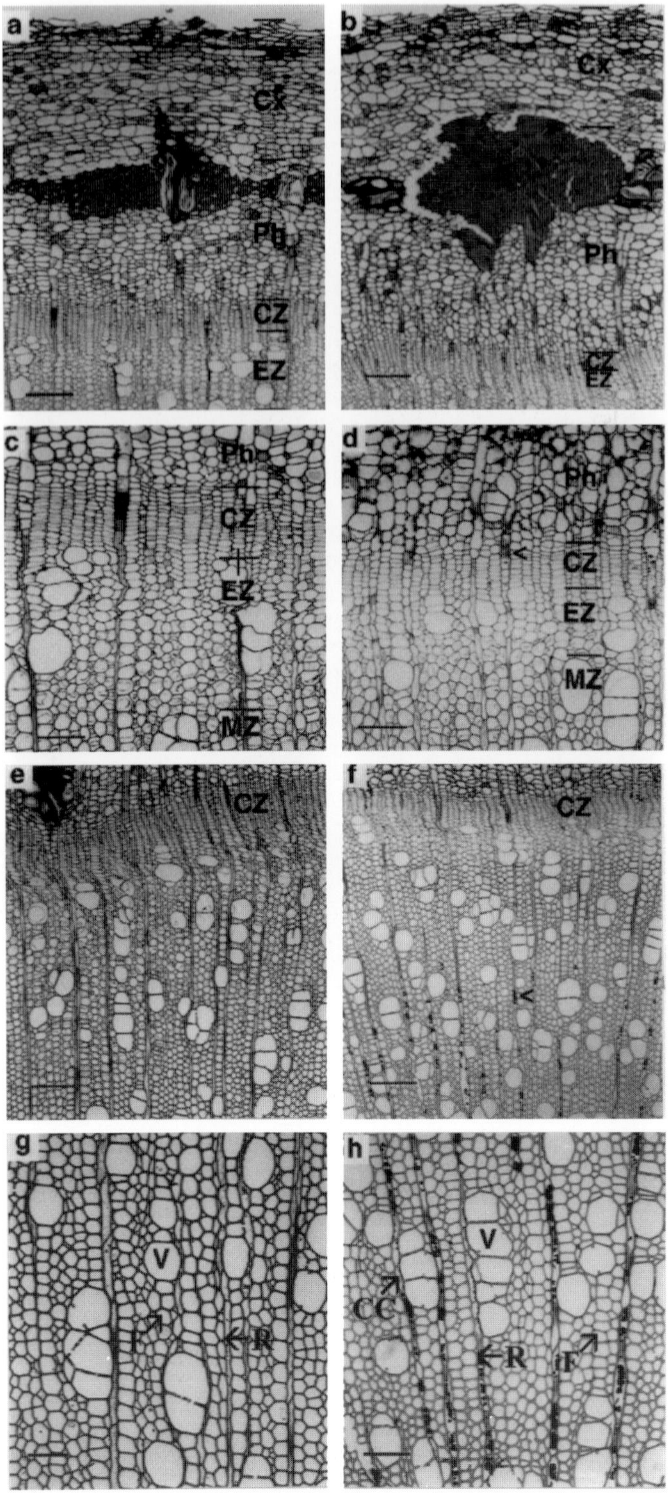
a
Cx
Ph
CZ
EZ
b
Cx
Ph
CZ
EZ
c
Ph
CZ
EZ
MZ
d
Ph
CZ
EZ
MZ
e
CZ
f
CZ
g
V
F
R
h
V
CC
R
F

Table 2. *Anatomical properties of developing and mature xylem in hybrid aspen grown at different nitrogen regimes. (100 mg N l^{-1}, 50 mg N l^{-1}, 25 mg N l^{-1} and 12.5 mg N l^{-1})*[a]

Cell type		N100		N50		N25		N12,5	
		Mean value	Std-error	Mean value	Std-error	Mean value	Std-error	Mean value	Std-error
Cambial cells									
Radial diameter (μm)		5.8	±0.1a	5.9	±0.2a	6.0	±0.1a	5.7	±0.2a
Number per radial file		13.2	±0.3a	12.7	±0.5a	11.5	±0.5a	9.1	±0.7b
Tangential diameter (μm)		13.3	±0.2a					12.1	±0.4a
Number of fusiform cells/ray initial		5.72	±0.41a					4.7	±0.15a
Ray density in the cambium (*n*/mm)		13.38	±1.03a					17.7	±1.03b
Fibres									
Length (μm)		425	±13a	442	±8a	439	±9a	459	±7a
Lumen area, young (μm)		168	±5ab	169	±6a	147	±7bc	136	±0.4c
Lumen area, old (μm)		145	±9a	148	±3a	121	±6b	115	±0.3b
Ratio young/old		1.17	±0.10a	1.14	±0.03a	1.23	±0.09a	1.19	±0.01a
Vessels									
Length (μm)		289	±6a	307	±6a	302	±6a	283	±24a
Lumen area (μm2)		1512	±69A	1547	±89	1393	±110	1213	±29B
Density (*n*/cm2)	400 μm	73	±8a	85	±3ab	106	±9b	96	±7ab
	800 μm	85	±4A	84	±6	101	±16	120	±8B
Elongation									
Fibres/vessels		1.47	±0.04a	1.44	±0.03a	1.45	±0.01a	1.66	±0.20a
Rays	cambium								
Density (n/mm)	400 μm	123	±5A	117	±12	138	±8	147	±7B
	800 μm	113	±6A	108	±5	135	±9	137	±7B

[a] Mean and standard error of measurements all made on transverse sections sampled from the base of the stem of four trees per treatment except fibre and vessel length. The radial diameter and the number of cambial zone fusiform cells were measured on ten randomly selected rows per section. The tangential diameter of the fusiform cambial cells, the number of fusiform cambial cells per ray initial, and the density of rays in the cambium were measured at ten different positions per section in the two extreme treatments. The fibre and vessel lumen areas were measured in five images (0.067 mm^2 and 0.24 mm^2, respectively) taken in the mature xylem ('old' fibres and vessels) or at the limit of the expansion zone ('young' fibres). The density of rays and vessels in the xylem was determined by measuring the number of vessels and rays crossed by a line drawn parallel to the cambium at a distance of 400 μm or 800 μm from the cambial zone. The length of vessels and fibres was measured from a macerated xylem sample taken at the base of the stem. Different lower case letters indicate statistically significant differences between the means as determined by an ANOVA and a Tukey test ($p < 5\%$). The capital letters indicate a statistical significance at 5% of the difference between the means of the two extreme treatments using a Student's t-test.

The lumen areas of both recently expanded and mature fibres were significantly reduced when nitrogen was supplied at levels below 50 mg N l^{-1} (*Table 2, Figures 3g and h*). The ratio of the lumen areas of recently expanded fibres to mature fibres, a quantification of the rate of cell wall thickening during maturation, remained uninfluenced by N-supply (*Table 2*). Fibre length and elongation (fibre length/vessel length) were not affected (*Table 2*).

A Student t-test between the treatments 100 mg N l^{-1} and 12.5 mg N l^{-1} showed that such a dramatic decrease in N-supply induced a significant reduction in vessel lumen area (*Table 2, Figures 3e–h*).

The density of the vessels was not clearly affected by the different nitrogen treatments. However, a tendency to an increase was observed (there was a significant difference between the 100 mg N l^{-1} and the 12.5 mg N l^{-1} treatments with a t-test) at 800 μm from the cambial zone (*Table 2, Figures 3e–h*).

Furthermore, a student's t-test also revealed an increase in the density of the rays in the 12.5 mg N l^{-1} treatment both at 400 μm and at 800 μm from the cambium (*Table 2, Figure 3e–h*). An increased density of ray initials was also observed in the cambium but the ratio of the number of fusiform cells to ray initials remained unaffected by the N-regime (*Table 2*).

3.5 *Histochemistry*

A staining of polysaccharides with PAS revealed an accumulation of starch in the stem of aspen grown under the low N-regime. The starch content of the stem at the base was strongly dependent on the N-supply. Well-supplied plants stored almost no starch in the stem (only a few starch grains were observed in the cortex and in the phloem parenchyma) (data not shown). The starch content increased dramatically when N-supply was reduced below 50 mg N l^{-1}, reaching a maximum at 12.5 mg N l^{-1}. At the lowest N-level, large amounts of starch were observed in the cortex, the phloem parenchyma, the pith and the rays (*Figures 3d, f and h*). The distribution of starch in the rays was heterogeneous, decreasing from the phloem to the cambium (*Figure 3d*). Starch was then absent from the cambium (*Figure 3d*) and the living xylem (*Figure 3f*). Starch storage increased with xylem maturation (*Figure 3f*). In the mature xylem, only the ray contact cells were nearly starch-free (*Figure 3h*).

A staining of proteins with Coomassie blue revealed little variation between the treatments. Almost no storage protein could be observed. The cell contents of physiologically active cells, especially the companion cells in the phloem, all ray cells in the immature xylem, the young fibres and vessels, and the ray contact cells in the mature xylem, were well stained (data not shown).

4. Discussion

Nitrogen fertilization is known to increase the allocation of photosynthates in the shoots to the disadvantage of the roots (Davidson, 1969; Ingestad, 1980; Ingestad and Ågren, 1988; Pregitzer *et al.*, 1990). As a result, the growth of the crown and the number of leaves are increased (Pell *et al.*, 1995; Günthardt-Goerg *et al.*, 1996; Ibrahim *et al.*, 1997). This was confirmed in the present experiment. Surprisingly, there was no gradual decrease in plant height when the N-supply was decreased although the plants supplied with the highest N-level were significantly taller than all the others. Most of the other phenological and histological parameters selected to characterize growth and wood formation were affected only at nitrogen levels below 50 or 25 mg N l^{-1}.

The observed reduction in stem diameter as an effect of a reduced N-supply has been described by other studies (Ibrahim *et al.*, 1997; Günthardt-Goerg *et al.*, 1996). This phenomenon has been correlated to a reduced width of the xylem, the bark remaining unaffected (Brix and Mitchell, 1980; Günthardt-Goerg *et al.*, 1996). The observed increase in the width of the phloem in our experiment, in addition to a reduced width of the xylem, is in accordance with these studies. This suggests that nitrogen not only stimulates the production of xylem cells by the cambium, but also influences the behaviour of the cambial initials that divide towards the phloem or the xylem. However, the width of the phloem may not only be affected by the rate of cell production of the cambium, but may also depend considerably on further division of the differentiated cells. For instance, the distribution pattern and size of the phloem fibres may strongly influence the width of the whole tissue (*Figure 3*). The phloem as a whole has not been adequately investigated in the stem until now (Fink, 1986; Vollenweider *et al.*, 1994). This is mainly due to the complexity of the tissue, and more work is needed to characterize the rates of phloem cell production and further division patterns of the different phloem cells.

The size and number of fusiform cambium cells may be used to characterize the activity of the cambium. For example, larger cells in older trees reveal a reduced rate of periclinal cell division, both in the initials and in the mother cells (Iqbal and Ghouse, 1990; Vollenweider *et al.*, 1994). On the other hand, the number of cells across the cambial zone reflects a balance between the rates of cell division of the cambial initials and cambial mother cells, and the differentiation of the new derivatives at the limit of the cambium (Fahn and Werker, 1990). In a previous experiment, transgenic hybrid aspen, over-expressing genes involved in auxin synthesis, showed a broader cambial zone made up of wider cells than were found in the wild-type plants. The number of fusiform cambial cells remained unchanged. As a result of the observed decrease in cambial division activity, as shown by the presence of wider cells, the transgenic plants developed less xylem (Tuominen *et al.*, 1997). In the present study, reducing the N-supply also decreased the xylem production. However, this phenomenon was correlated to a narrower cambial zone made up of fewer fusiform cambial cells. Since the size of these cells remained constant, it can be assumed that the reduced width of the cambium was due to a reduced division activity of the fusiform initials rather than a reduced division activity of the mother cells. It is, of course, also conceivable that the differentiation of the derivatives started earlier.

Separating the phases of cell expansion and wall thickening in hybrid aspen is artificial since they comprise a continuous sequence. Nevertheless, the phases, or developmental zones, proved to be important individually since there is some evidence that the final form and size of the xylem cells depend on the length of time spent in each phase (Denne and Dodd, 1981; Tuominen *et al.*, 1997). Reducing N-supply induced a reduction in the width of the zone of expanding xylem elements. This correlated with a reduction in the cross-sectional area of the fibres. Such a correlation between the duration of the expansion phase and the size of the fibres has already been observed (Tuominen *et al.*, 1997). Furthermore, we were able to show that the reduction in the cross-sectional lumen area of the mature fibres was already complete at the end of the expansion phase: the N-supply had no influence on the further thickening of the fibre cell walls. In addition, the lumen area of the vessels was influenced in a similar way in this experiment. The length of the vessels and fibres and the elongation process of the fibres were not affected by the N-treatments. Variations in the length of xylem elements have been reported to occur naturally along the stem (depending on the age of the tissues) or across the stem (Liese and Ammer, 1957; Meyer-Uhlenried, 1957) rather than under the effect of external factors. N-supply affected not only the rates of expansion but also the production of different xylem elements. Vessel and ray density increased in the low-N treatment. Increased ray density was already observed in the cambium where the fusiform cell to ray initial ratio and, to a stronger degree, the tangential diameter of the fusiform cambial cells, tended to decrease when the N-supply was reduced. This ratio is believed to be stable and to depend on the species more than on external factors (Iqbal and Ghouse, 1990). Our observations indicate that the general pattern of the wood was conserved but reduced in space.

Stem parenchymatic tissues are known to play an important role in carbon and nitrogen allocation (Ibrahim *et al.*, 1997; Pregitzer *et al.*, 1990) and rays seem to be affected by variations in N-supply. Staining transverse sections for polysaccharides and proteins revealed interesting results. Hybrid aspen supplied with the highest nitrogen level (100 mg N l^{-1}) grew actively throughout the whole experiment. This situation is comparable to spring activity (Sauter, 1965; Witt and Sauter, 1994; Sauter and Van Cleve, 1994). At that time of the year, carbon and nitrogen are both required for growth and practically no storage is observed in the stem (no starch accumulation). Kozlowski (1962) reported that 45% of the carbohydrate production of a tree may be used in wood production. Reducing N-supply both reduces growth (demand for carbon) and N availability for N compounds such as amino-acids and proteins (Pretziger *et al.*, 1990). Since the carbon is not needed, it is stored, mainly in the form of starch. Large amounts of starch are observed in the parenchymatic cells of the stem in the low N-treatment, except in the living, thus carbon-demanding, xylem. Contact cells without a storage role are also starch-free. Since it has been shown that sucrose is involved in cambial activity, it is not surprising that no starch was observed in the cambium and in the living xylem (Catesson *et al.*, 1995).

The influence of nitrogen on wood formation observed in this study was similar in some aspects to the effects of modified endogenous auxin levels observed in hybrid aspen in previous studies (Tuominen *et al.*, 1995, 1997). This could be the result of a direct influence on the cambium or of a secondary effect on shoot growth. A relationship also exists between nitrogen supply and the hormonal status of the plants (Michniewizc *et al.*, 1976; Sundberg *et al.*, 1993; Marschner, 1995) and it would be interesting to take a closer look at the auxin distribution at different nitrogen levels.

Acknowledgements

The authors wish to thank Dr Björn Sundberg from the Swedish University of Agricultural Sciences in Umeå, Sweden, for cultivating and providing the plant material, and Mrs Karin Waldmann and Mr Kjell Olofsson for technical assistance. The present study has been carried out with financial support from the Commission of the European Communities, Agriculture and Fisheries (FAIR) specific RTD programme, CT 98–3972, 'Wood formation processes: the key to improvement of the raw material'.

It does not necessarily reflect its views and in no way anticipates the Commission's future policy in this area.

References

Brix, H. and Mitchell, A.K. (1980) Effect of thinning and nitrogen fertilization on xylem development in Douglas-fir. *Can. J. For. Res.* **10**: 121–128.

Catesson, A.-M., Bonnemain, J.-L., Eschrich, W. and Magel, E. (1995) The cambium and its derivative tissues: biochemical changes in relation to cell differentiation and seasonal activity. In: *EUROSILVA—Contribution to Forestry Physiology.* Les Colloques de l'INRA 76.

Davidson, R.L. (1969) Effects of soil nutrients and moisture on root/shoot ratios in *Lolium perenne* L. and *Trifolium repens* L. *Ann. Bot.* **33**: 571–577.

Denne, M.P. and Dodd, R.S. (1981) The environmental control of xylem differentiation. In: *Xylem Cell Development* (ed. Barnett, J.R.), Castle House Publications: Tunbridge Wells, pp. 236–255.

Dodd, R.S. and Fox, P. (1990) Kinetics of tracheid differentiation in Douglas-fir. *Ann. Bot.* **65**: 649–657.

Fahn, A. and Werker, E. (1990) Seasonal cambial activity. In: *The Vascular Cambium* (ed. Iqbal, M.), Wiley: New York, pp. 139–157.

Fink, S. (1986) Microscopical investigations on wood formation and function in diseased trees. *IAWA Bull. n.s.* **7**: 351–355.

Gahan, P.B. (1984) *Plant Histochemistry and Cytochemistry.* Academic Press, Inc.: London.

Gerlach, D. (1977) *Botanische Mikrotechnik.* Georg Thieme Verlag: Stuttgart, Germany.

Gregory, R.A. (1971) Cambial activity in Alaskan white spruce. *Am. J. Bot.* **58**: 160–171.

Günthardt-Goerg, M.S., Schmutz, P., Matyssek, R., Bucher, J.B., Percy, K.E., Cox, R.M. and Jensen, K.F. (1996) Leaf and stem structure of poplar (*Populus × euramericana*) as influenced by O_3, NO_2, their combination and different soil N supplies. *Can. J. For. Res.* **26**: 4, 649–657.

Ibrahim, L., Proe, M.F. and Cameron, A.D. (1997) Main effects of nitrogen supply and drought stress upon whole-plant carbon allocation in poplar. *Can. J. For. Res.* **27**: 1413–1419.

Iqbal, M. and Ghouse, A.K.M. (1990) Cambial concept and organisation. In: *The Vascular Cambium* (ed. Iqbal, M.), pp. 1–36. Wiley: New York.

Ingestad, T. (1980) Nitrogen stress in birch seedlings. II. N, K, P, Ca and Mg nutrition. *Physiol. Plant.* **45**: 149–157.

Ingestad, T. and Ågren, G.I. (1988) Nutrient uptake and allocation at steady-state nutrition. *Physiol. Plant.* **72**: 450–457.

Jensen, W.A. (1962) *Botanical Histochemistry*. W.H. Freeman and Co.: San Francisco, London.

Kozlowski, T.T. (1962) Photosynthesis, climate and tree growth. In: *Tree Growth* (ed. Kozlowski, T.T.), Ronald Press: New York, pp. 149–164.

Larson, P.R. (1994) *The Vascular Cambium.* Springer Verlag: Berlin.

Liese, W. and Ammer U. (1957) Untersuchungen über die Länge der Holzfaser bei der Pappel. *Holzforschung* **11**: 169–174.

Marschner, H. (1995) *Mineral Nutrition of Higher Plants* (2nd ed.) Academic Press: London, New York.

Meyer-Uhlenried, K.-H. (1957) Holzanatomische Untersuchungen an der Pappel. *Holzforschung* **11**: 150–157.

Michniewicz, M., Rozej, B. and Stopinska, J. (1976) The influence of nitrogen nutrition on the dynamics of growth and metabolism of endogenous growth regulators in Scotch pine (*Pinus sylvestris*, L.) seedlings. *Acta Soc. Bot. Pol.* **45**: 495–510.

Necesany, V. and Cetlová, J. (1969) Über den Einfluß von äußeren Wuchsbedingungen auf die Struktur des Pappelholzes (*Populus molinifers*). *Holzforschung* **23**: 73–82.

Nilsson, L.O. and Wasielewski, D. (1987) Influence of fertilization in a natural *Populus tremula* stand. *Scand. J. For. Res.* **2**: 343–348.

Pell, E.J, Sinn, J.P. and Johansen, C.V. (1995) Nitrogen supply as a limiting factor determining the sensitivity of *Populus tremuloides* Michx. to ozone stress. *New Phytologist* **130**: 437–446.

Pregitzer, K.S., Dickmann, D.I., Hendrick, R. and Nguyen, P.V. (1990) Whole-tree carbon and nitrogen partitioning in young hybrid poplars. *Tree Physiol.* **7**: 79–93.

Puech, L. and Mehne-Jacobs, B. (1997) Histology of magnesium-deficient Norway spruce needles influenced by nitrogen source. *Tree Physiol.* **17**: 301–310.

Ridoutt, B.G. and Sands R. (1994) Quantification of the processes of secondary xylem fibre development in *Eucalyptus globulus* at two height levels. *IAWA J.* **15**: 417–424.

Safford, L.O. and Czapowskyj, M.M. (1986) Fertilizer stimulates growth and mortality in a young *Populus-Betula* stand: 10-year results. *Can. J. For. Res.* **16**: 807–813.

Sauter, J.J. (1965) Untersuchung zur Physiologie der Pappalholzstrahlen. Jahresperiodischer Verlauf der Stärkespeicherung im Holzstrahlparenchym. *Z. Pflanzenphysiol.* **55**: 246–258.

Sauter, J.J. and Van Cleve, B. (1994) Storage, mobilisation and interrelations of starch, sugars, protein and fat in the ray storage tissue of poplar trees. *Trees* **8**: 297–304.

Skene, D.S. (1969) The period of time taken by cambial derivatives to grow and differentiate into tracheids in *Pinus radiata* D.Don. *Ann.Bot.* **33**: 253–262.

Spearen, W.E. and Isenberg, I.H. (1947) The maceration of woody tissues with acetic acid and sodium chlorite. *Science* **105**: 214.

Sundberg, B., Ericsson, A., Little, C.H.A., Näsholm, T. and Gref, R. (1993) The relationship between crown size and ring width in *Pinus sylvestris* L. stems: dependence on indole-3-acetic acid, carbohydrates and nitrogen in the cambial region. *Tree Physiol.* **12**: 347–362.

Tuominen, H., Sitbon, F., Jakobsson, C., Sundberg, G., Olsson, O. and Sundberg, B. (1995) Altered growth and wood characteristics in transgenic hybrid aspen expressing *Agrobacterium tumefaeciens* T-DNA indole-acetic-acid-biosynthetic genes. *Plant Physiol.* **109**: 1179–1189.

Tuominen, H., Puech, L., Fink, S. and Sundberg, B. (1997) A radial concentration gradient of indole-3-acetic acid is related to secondary xylem development in hybrid aspen. *Plant Physiol.* **115**: 577–585.

Velten, J., Velten L., Hain, R. and Schell, J. (1984) Isolation of a dual plant promoter fragment from the Ti plasmid of *Agrobacterium tumefaeciens*. *EMBO J.* **3**: 2723–2730.

Vollenweider, P., Dustin, I., Hofer, R.-M., Vittoz, P. and Hainard, P. (1994) A study of the cambial zone and conductive phloem of common beech (*Fagus sylvatica* L.) using an image analysis method. I. Influence of tree age on the structure. *Trees* **9**: 106–112.

Wilson, B.F. and Howard, R.A. (1968) A computer model for cambial activity. *For. Sci.* **14**: 77–90.

Witt, W. and Sauter, J.J. (1994) Starch metabolism in poplar wood ray cells during spring mobilisation and summer deposition. *Physiol. Plant.* **92**: 9–16.

Wodzicki, T.J. (1971) Mechanism of xylem differentiation in *Pinus silvestris* L. *J. Exp. Bot.* **22**: 670–687.

12

The formation of cambium from callus in grafts of woody species

J.R. Barnett and A.K. Asante

1. Introduction

Study of the development and behaviour of cambial cells is hindered by the fact that they cannot be maintained *in vitro* because they dedifferentiate to callus on being removed from the stem. It is well known that callus can be induced to differentiate into 'tracheary elements' similar to primary xylem elements. However, no instance of direct differentiation of callus into cells with the characteristics of secondary xylem elements *in vitro* has been reported. Those instances in which secondary xylem (and phloem) elements have been produced from callus *in vitro*, have involved prior formation of a cambium (Harris and Barnett, 1975; Washer *et al.*, 1977). Formation of cambium from callus occurs readily *in vivo* during wounding responses including graft union formation in woody stems. The graft union therefore offers a system for the study of the formation of cambium and the maintenance of its meristematic activity once formed.

This chapter will examine the suggestion that two factors are involved in the formation of cambium: positional information provided by flow of chemical signals from pre-existing cambia, and pressure to trigger the formation of cambium from callus.

2. The anatomy of the developing graft union

Graft formation occurs readily in many woody species, both in stems and branches, and in roots. In natural woodlands and forests a careful observer will find many instances of branches or stems which have joined as a result of being forced into contact with each other during growth. Skilled horticulturists can create bizarre ornamental effects using this propensity of woody plants to graft together (*Figure 1a and b*).

Cell & Molecular Biology of Wood Formation, edited by R. Savidge, J. Barnett and R. Napier.

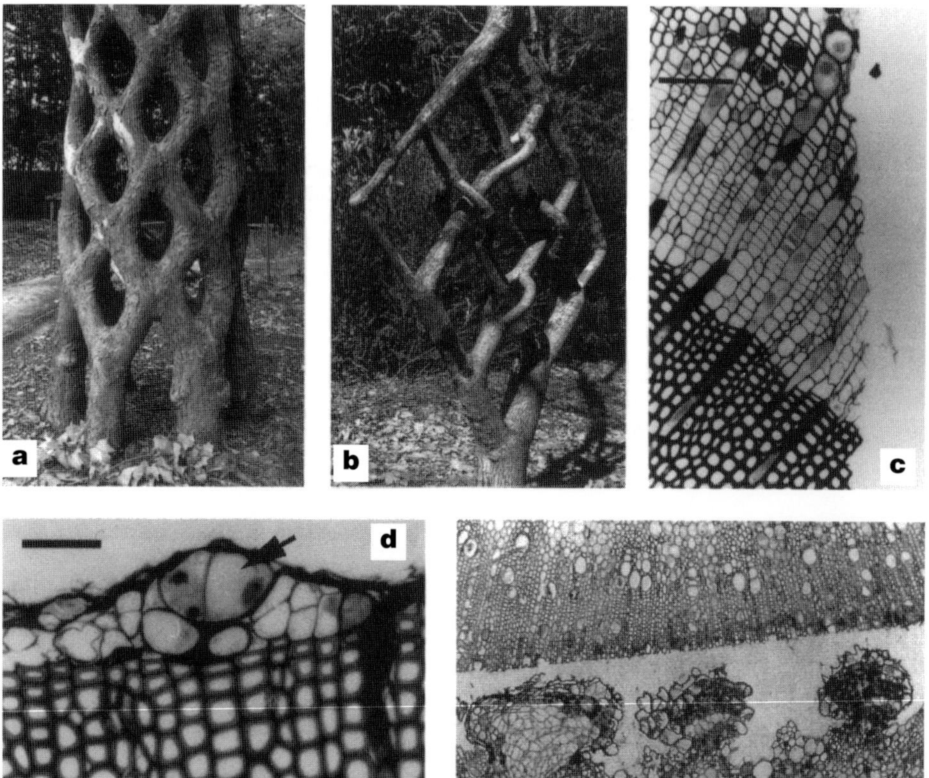

Graft union formation is related to the normal wound response in woody stems (Barker, 1954). Removing a sliver of bark from a stem induces dedifferentiation of undamaged living cells to form callus at and near the surface of the wound. Reunion of the cambia occurs only after a callus mass has been produced in contact with the severed cambial cells (Sharples and Gunnery, 1933).

During preparation of a graft, the grafter's knife cuts through bark, cambium and secondary xylem of the scion and rootstock to an extent which varies according to the type of graft being prepared (Garner, 1990) (*Figure 1c*). In a side veneer graft, for example, the prepared scion and stock are bound together, with scion and stock diameters carefully matched, and cambia aligned as illustrated in *Figure 3*. This type of graft was used for the Sitka spruce, mango and cashew grafts illustrated in this chapter. Apple grafts were prepared by chip budding (*Figure 4*). In this procedure a shield-shaped piece of tissue containing xylem, cambium, phloem and outer bark, and enclosing a bud, is cut from a shoot. A similar piece of tissue is cut from the stem of the scion, leaving a cavity into which the bud-containing chip is bound (*Figure 4*).

Callus develops at the cut surfaces of the scion and rootstock as it does in wound repair. The new cambial cells formed from the callus link the cambia of the partners resulting in a single cambium encircling the graft union.

The sequence of events in the formation of a successful union is essentially the same, with only minor variations, for all species described in the literature, and may be divided into three main stages: cohesion of the scion and rootstock; proliferation of callus, and vascular differentiation (Moore, 1984).

2.1 *Cohesion*

The first observable tissue response following assembly of the graft is resin secretion (in those species with resin canals), and the formation of a so-called necrotic layer or plate (Stoddard and McCully, 1979, 1980; Moore and Walker, 1981a, b) made up of dead and damaged cells (Miller and Barnett, 1993; Asante and Barnett, 1997) (Figure 1d). The necrotic layer has also been called a barrier zone (Sass, 1932), insulating layer (Mendel, 1936), isolation layer (Copes, 1969), and contact layer (Dormling, 1963; Copes, 1969; Mergen, 1954). With the exception of 'necrotic', these names imply a function for the layer. Although its role in graft formation is unknown, there is no evidence that its presence inhibits graft formation (Copes, 1969).

The appearance of the necrotic layer suggests it is formed as a result of cells being crushed when the graft partners are bound together. However, Artschwager (1951) found that application of pressure to micrografts of *Parthenium argentatum* and *Helianthus annuus* with smooth well-matched surfaces led to the formation of a thin necrotic layer. If no pressure was applied, and the cut surfaces were uneven, a thick and irregular layer was produced. Cline and Neely (1983) described a similar layer formed at the cut end of geranium cuttings one day after insertion into a soil medium.

Figure 1. *(a) and (b). Bizarre ornamental grafted sycamore trees at the late Alex Erlandson's 'tree circus' near Santa Cruz, California. These illustrate the ease with which woody species can be grafted successfully by a skilled horticulturist. (c) Transverse section through a freshly prepared active rootstock of* Picea sitchensis. *In this view it can be seen that the grafter's knife has cut through (from top to bottom) secondary phloem, cambium, differentiating xylem and mature secondary xylem. Scale bar = 0.1 mm. (d) Transverse section through a rootstock of* Picea sitchensis *showing the beginnings of formation of callus (arrow). The cut surface is covered by a dark staining necrotic layer. Scale bar = 0.1 mm. (e) Transverse section through a developing graft (prepared by the apical side-veneer method) in* Mangifera indica. *The scion is at the top, the rootstock below in this illustration. The rootstock is producing callus to fill the space between the badly matched partners. Scale bar = 0.1 mm. (f) Transverse section through a graft in cashew (*Anarcardium occidentale L.*) in which callus cells are being produced in rows originating in the pith of the rootstock (above) and the scion (below). This growth has pushed the graft partners apart. Scale bar = 0.1 mm.*

The detection of such a layer in cuttings suggests a possible role in the prevention of entry of pathogens. Failure of grafts is usually attributed to physiological factors such as water stress (Barnett and Weatherhead, 1989) rather than to pathogens. Fungal hyphae were observed in tracheid lumina in graft unions of Sitka spruce, without apparently affecting the health of the plant (I. Weatherhead, unpublished observations).

2.2 *Callus formation*

There are conflicting views on the origin of callus, possibly a result of the fact that different authors described species-specific reactions. Barker (1954), Buck (1954), Braun (1959) and Fujii and Nito (1972) stated that the cambium is a major source of callus, while Sass (1932), Juliano (1941), Sharples and Gunnery (1933) suggested that most callus came from tissues external to it. Brown and Sax (1962) found that callus formation was mainly from severed ray cells, with a small proportion from immature xylem elements in *Populus trichocarpa* and *Pinus strobus*. Dormling (1963) emphasized the importance of severed needle traces in the production of callus, a view supported by Miller and Barnett (1993).

Microscopy of developing grafts in Sitka spruce (Miller and Barnett, 1993), apple (Soumelidou *et al.*, 1994) and mango (Asante and Barnett, 1997) showed that callus arose from cortical and ray parenchyma, resin canal epithelium, cambial cells and developing tracheids at the enlargement stage. It seems probable that any living cell near the wounded surface has the potential to form callus, given suitable conditions.

Callus production has been reported to begin first in the scion then later in the rootstock in *Picea abies* and *Pinus sylvestris* (Dormling, 1963), *Pseudotsuga menziesii* (Copes, 1969) and *Picea sitchensis* (Miller, 1991). Weatherhead (1986) observed that grafts of *Picea sitchensis* which failed to form a union were characterized by the failure of either component to form callus. Graft failure was attributed to low water potential in the scion, resulting in no turgor for cell expansion and callus formation. This suggests scion activity is a prerequisite for callus formation by the rootstock (where turgor should not be lacking under experimental conditions).

Callus formation is crucial to the survival of the scion, since it represents a way for water to bypass vessels and tracheids damaged during preparation of the graft partners to reach undamaged conducting tissues below the cut surfaces. Callus formed by the scion and rootstock links directly to ray parenchyma and other living cells and merges to form a single tissue during the critical period before a contiguous vascular system can be established between the two components. At the same time plasmodesmata form *de novo* between cells originating from scion and rootstock (Kollman and Glockmann, 1985; Kollmann *et al.*, 1985; Shi-jie and Cheng-hou, 1988).

Cell proliferation appears initially to be without order, cells enlarging and dividing randomly to fill any gaps between the partners left as a result of irregularities in the mating surfaces (*Figure 1e*). Later, however, divisions become more ordered, and rows of cells are produced which clearly have a common ancestry. These rows are usually aligned perpendicularly to the graft interface (*Figure 1f*). The formation of rows of callus cells in this way can be interpreted as evidence supporting the view that the favoured division plane is at right angles to the direction of pressure exerted on the cells, or to the principal direction of cell enlargement.

The proliferation of callus is such that it forces the scion and rootstock apart (*Figure 1f*). In doing so, it breaks up the necrotic layer, which is apparently digested by enzymes secreted by the callus cells which then absorb the products (Artschwager, 1951; Kostoff, 1928; Krenke, 1933; Mergen, 1954; Shi-jie and Cheng-hou, 1988; Stoddard and McCully, 1979, 1980; Soumelidou *et al.*, 1994).

2.3 *The formation of a cambium*

The formation of cambial cells is often preceded by the formation of tracheary elements within the callus (Weatherhead and Barnett, 1986; Soumelidou *et al.*, 1994) (*Figure 2b*). In *P. sitchensis*, the first tracheids produced have a combination of a primary xylem character (helical wall thickening) with a secondary one (bordered pits) (Weatherhead and Barnett, 1986). This phenomenon has been postulated as being part of a differentiation continuum in which families of structural genes are 'repeatedly co-expressed during cellular differentiation to generate increasingly more complex sub-types' (Savidge, 1996). In *Malus*, vessel elements are formed apparently randomly at a distance from the cambium (Soumelidou *et al.*, 1994).

Cambium cells appear after a period of callus formation which varies according to species (*Figure 2a*). Cambial contact has been reported as occurring between seven (Soumelidou *et al.*, 1994) and 21 days (Sass, 1932) in *Malus*, 14 days in *Nicotiana* (Crafts, 1934) and peach (Fletcher, 1964), 15 days in citrus (Mendel, 1936), 22 days in pear/quince (Ermel *et al.*, 1993), 17 days in *Pseudotsuga menziesii* (Copes, 1969), between 21 days and 42 days in *Picea abies* and *Pinus sylvestris* (Dormling, 1963), between 50 and 90 days in *Picea sitchensis* (Barnett and Weatherhead, 1988) and 90 days in poplar (Braun, 1959).

The process begins with the cells closest to the existing cambia, and spreads until the stock and scion cambia are joined by new cambium cells (*Figure 2a* and *2b*). After transverse and vertical cambial continuity has been established, normal secondary tissues are produced by the newly differentiated cambium. Cambial union is effected even if the two cambia are not well aligned when the graft is assembled. Severe lateral misalignment may result in complete failure of the scion and rootstock cambia to join. If this happens on one side only of the graft, while the cambia on the other side join, then the graft can survive. The resulting plant will develop a groove down one side of the stem in the position of the graft as secondary tissues fail to form. A perfect match at the time of assembly may not necessarily lead to a straightforward union. The action of callus production in forcing the partners apart will means that cambium differentiation has to follow a sinuous S or Z pathway through the callus to link the edges of the cambia in the scion and rootstock (*Figure 2b*). In some cases this may be a very long path with the new cambia forming from each severed edge initially in a direction diametrically opposite to that required for union. Randomly distributed tracheary elements occur between the xylem-producing faces of the loop. A similar phenomenon has been described in pear/quince grafts by Ermel *et al.* (1993), who related the severity of the sinuosity to the degree of incompatibility between the graft partners.

Where the pathway of differentiation is long, it is common to find fusiform cells with their long axis parallel to the line of new cambium, i.e. they lie at right angles to the axis of the stem, rather than parallel to it (*Figure 2b*). Even where the pathway is relatively short and direct, there is a tendency for cells to be aligned obliquely to the stem axis.

A striking feature found in grafts made on dwarfing and semi-dwarfing rootstocks in *Malus* is the presence of swirling patterns of cells in the callus in the rootstock above

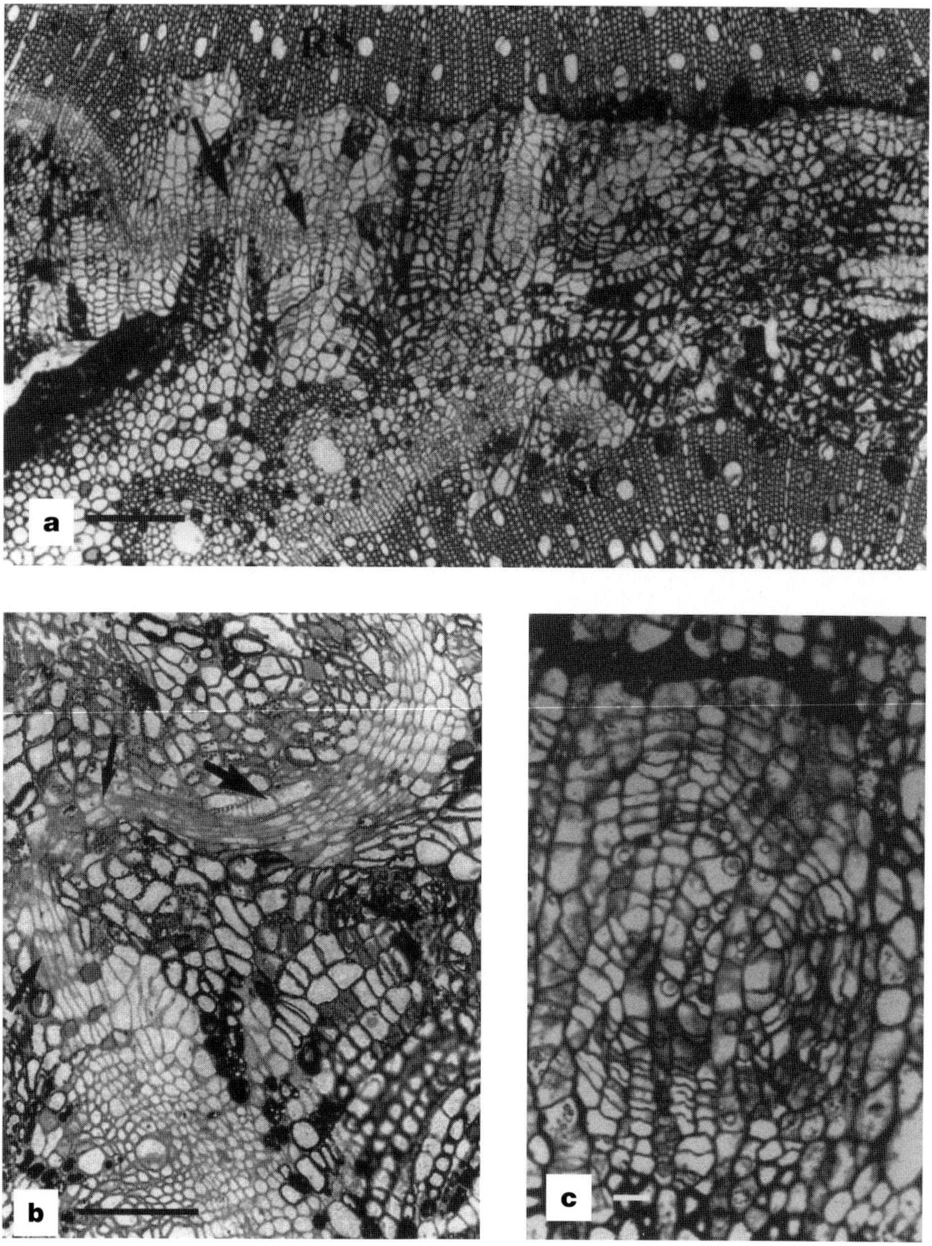

Figure 2. *(a) Transverse section through a developing graft of* Mangifera indica. *Cambial cell differentiation is spreading through the callus from the edge of the rootstock cambium (arrows). No differentiation is yet visible in the callus associated with the scion (bottom of the figure). Scale bar = 0.1 mm. (b) Transverse section through a developing graft (prepared by the chip budding technique) in apple. Note the sinuous path followed by the linking cambium (small arrows), and the fact that the long axes of the cambial cells, and the xylem elements that have been produced from them (large arrow), are oriented at right angles to the stem axis. Scale bar = 0.1 mm. (c) A swirl of cambium-like cells which have differentiated from callus above the bud in an apple graft in which a dwarfing rootstock was used. Scale bar = 10 μm*

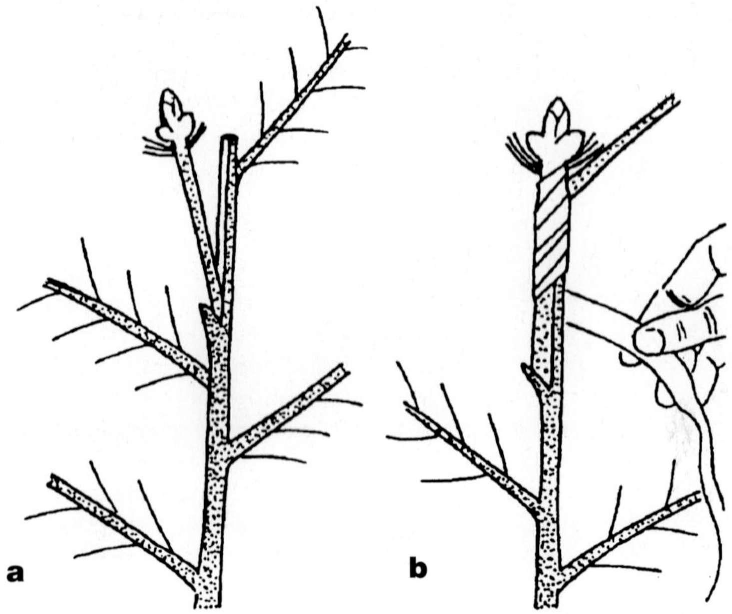

***Figure 3**. Diagramatic representation of side-veneer graft assembly in Sitka spruce. (a) Introducing the scion to the rootstock. (b) Binding scion and rootstock together. From Weatherhead (1986).*

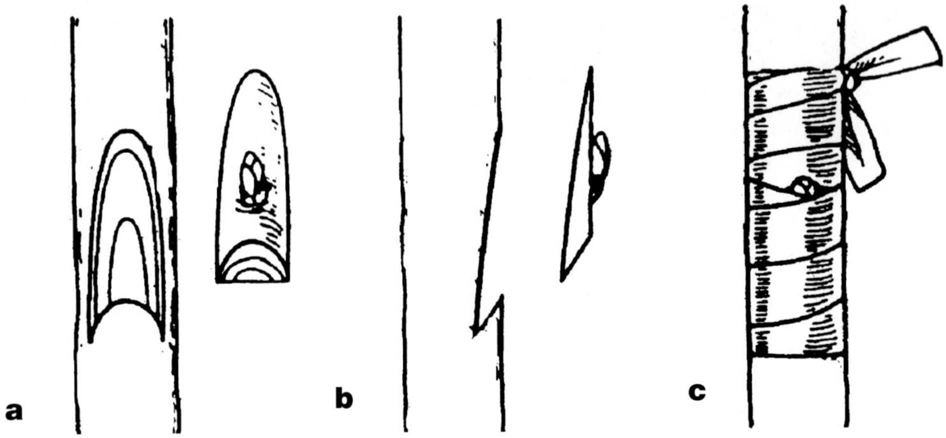

***Figure 4.** Chip-budding. (a) Showing the prepared rootstock (left) and scion, front view. (b) Side view. (c) Graft assembly completed with the partners bound together with tape. From Soumelidou (1993).*

the grafted bud (*Figure 2c*). These patterns consist of parenchyma cells in the centre, surrounded by cambium-like cells which divide in a circular pattern. No differentiation into xylem or phloem of any of these cells was observed. This feature has also been reported in pear/quince grafts by Ermel *et al.* (1993) who suggest it may be related to congruity or compatibility problems associated with interspecific grafts, or grafts between normal scions and dwarfing rootstocks.

3. Factors involved in cambium formation

The differentiation of cambium from callus, so difficult to induce *in vitro*, readily occurs in properly assembled grafts, meaning that the factors responsible must be sought in differences between the systems. The most obvious differences between the two systems are that in the graft, differentiated plant tissues are in contact with the callus, and the callus is confined under pressure between the scion and rootstock.

The fact that only those callus cells on the shortest path between the severed cambia of the graft partners differentiate indicates that the severed cambia themselves provide information which determines which callus cells will differentiate to form cambium. The observation that grafts do not normally take unless the scion and rootstock are tightly appressed during graft assembly suggests that pressure is in some way involved. A combination of pressure and a chemical signal may be what is required to produce a successful graft.

3.1 *The role of pressure in cambial maintenance and development*

There is good evidence that pressure is involved in the maintenance of the cambial state in normal stems. Brown and Sax (1962) demonstrated that when bark strips were separated from woody stems of *Populus trichocarpa* and *Pinus strobus.* Leaving the cambium exposed, there was a proliferation of callus at the exposed surface. When the bark strips were tied firmly back in place, but separated from the cambium by a plastic film, the cambium did not dedifferentiate into callus, but produced derivatives which differentiated normally. External pressures of 0.25 and 0.5 atmospheres applied to the stripped surface ensured that the cambium continued to produce derivatives which differentiated normally. External pressure of 1.0 atmospheres significantly reduced cambial activity. The authors considered that a callus mass was essential for cambial repair and that as it developed, pressure produced between the cells resulted in distinct rows of radially aligned cells. Brown (1964) also showed that confining callus cells induced them to form pitted, lignified secondary walls. Bloch (1941) stated that the first cell divisions from a wounded meristem always occur parallel to the wound surface. This has been attributed to a loss of pressure in the plane perpendicular to the new walls (Copes, 1969) or to diffusion of a wound hormone from the cut surface (Bloch, 1941).

Makino *et al.* (1983) and Lintilhac and Vesecky (1981) showed that the application of mechanical force to the surface of a callus mass had a direct affect on new cell wall orientation. Lintilhac and Vesecky devised an apparatus which enabled them to control the pressure applied to callus cultures. Compression forces between 5 and 8 atmospheres induced an ordering of cell division, which resulted in the formation of cambium-like areas.

Instances have also been reported of the differentiation of a cambium in callus formed *in vivo* and *in vitro*. Harris and Barnett (1975) found that an active cambium had formed within old callus nodules produced by dedifferentiation of ray cells in internal cavities of *Pinus ponderosa* wood. These nodules were spherical, up to 1 cm in diameter, with a bark-like surface enclosing concentric layers of tissues, with secondary xylem in the centre, surrounded by a cambium, which in turn was enclosed by secondary phloem. The longitudinal axis of the cells of these layers were arranged parallel to the nodule surface, but showed no consistent preferred orientation. Smaller hard spherical nodules, with a similar concentric internal structure of secondary-type tissues, were observed to develop at the surface of *Pinus radiata* callus which had been left to grow on unchanged medium for six weeks (Washer *et al.*, 1977). Similar observations were made by Gautheret (1959) and Steward *et al.* (1958). The hard surface layer of the nodules probably resists outward growth and causes an accumulation of pressure inducing differentiation of cambium, which produces secondary xylem and phloem elements.

If pressure helps to induce cambial development from callus in the above cases, then it should also be involved in cambium formation in grafts. In support of this, Braun (1959) stressed the importance of pressure during union formation, stating that union in *Populus* occurred only after a sufficient pressure level had been achieved by vigorous callus formation from the rootstock.

Pressure accumulation in a graft depends on resistance to expansion of callus cells which provides the necessary physical environment for the development of cambium. Unless the scion and rootstock are bound together, turgor pressure generated by callus formation and growth will be dissipated as a result of the scion and stock moving apart. The need for pressure explains why sufficient callus is formed to push the scion and stock apart against the resistance of the binding tape. The elastic tape is stretched as it is applied tightly to the graft, and considerable force is needed to stretch it further. It is possible that the difference in time taken by different species to achieve cambial union may be related to the rates of callus production in those species, and hence the time taken to build up the necessary pressure.

3.2 *Positional information*

Pressure alone cannot account for the observed events at the graft interface, otherwise many more callus cells would form cambium. It is apparent that some signal must pass through the callus, marking the line to be followed by the differentiation process. As has been seen, this line is often obliged to follow a tortuous path, a fact which raises interesting questions about the nature of the signal.

Warren Wilson and Warren Wilson (1961) reviewed early ideas on the positional control of regenerating cambia. One theory they considered was called by them the 'free surface' theory, under which a cambium forms beneath and parallel to a natural or artificial tissue surface. Another was named by them the 'cambial ring' theory in which regeneration of cambium is caused by interruption of the old cambial ring, the regenerating cambia extending from the cut edges of the ring in such a way as to form a closed ring. Following a series of experiments, they concluded that neither theory could explain their observations. Instead they proposed their 'gradient induction' hypothesis under which a gradient of some factor 'as yet unknown', arises perpendicular to exposed surfaces following wounding, and that new cambium arises at a particular position on the gradient, which also determines the direction of subsequent development of xylem and phloem. Warren Wilson (1978) proposed that cambial development is induced at a critical position along an auxin–sucrose gradient. Such gradients have been invoked to account for the control of secondary xylem and phloem development (see Savidge (1996) for review).

Aloni (1987), however, suggested that it was the flow of inductive stimuli rather than static gradients which is important. Indeed, Aloni and co-workers have demonstrated in a number of systems that wound induced changes in direction of flow of growth substances can be invoked to explain differentiation of parenchyma to xylem and phloem (see Aloni, 1995 for review).

It is possible that both a gradient and a flow may be required. This is suggested by the observation that where the path followed by cambial development is long, the new cambial cells near the centre of the path have their long axis parallel to it. The axial alignment of new cambial cells formed close to the pre-existing cambia may reflect the predominantly basipetal flow of a signal in this region, while cells further away are aligned perpendicular to the stem axis as a result of lateral diffusion of signal through the callus.

The signal received by callus cells located near severed cambium cells is probably an auxin flux which typically results in vascular redifferentiation across a graft interface (Moore, 1982). It has been suggested that there is an absolute requirement for auxin for vascular redifferentiation across a graft union between isolated internodes (Parkinson and Yeoman, 1982). In woody species auxin is believed to move primarily in the cambial region and to be involved in the regulation of cambial activity and xylem development (Little and Savidge, 1987). Diffusion of a signal into the callus from the severed cambia would also provide the type of concentration gradient required by Warren Wilson's hypothesis. Tuominen *et al.* (1997) demonstrated a radial concentration gradient in the cambial region of hybrid aspen and found that the peak level was in the cambial zone, with levels reducing towards the zone of secondary wall formation in developing xylem. This suggests that the correct auxin level, coupled with polar flow, may be involved in the formation of cambium, while lower levels in the callus may be associated with direct differentiation of tracheary elements deeper within the callus.

The importance of a polar signal is emphasized by observations of what happens in its absence, as for example in callus nodules (Washer *et al.*, 1977), where there is localized graft incompatibility (Ermel *et al.*, 1993) and in dwarfing apple grafts (Soumelidou *et al.*, 1994). The swirling patterns of cambium-like cells observed in rootstock callus are formed above the bud trace, presumably because there is no polar auxin flow in that region. The lack of polarity results in the observed swirling pattern of growth, while the cambium-like cells formed did not produce vascular elements, indicating that the stimulus for this process was absent.

4. Consequences for understanding of the cambium

The developing graft union is a system in which cambial cells develop from callus in an analogous way to their formation from interfascicular parenchyma in stems. It therefore provides a model system for the study of those factors which are responsible for the development of the cambium in a normal stem, and for maintaining it in a meristematic state. It is apparent that differentiation of callus to cambium requires both a physical (pressure) and biochemical stimulus, and it perhaps is not unreasonable to suppose that normal cambium requires both to maintain its integrity in a stem. In a normal stem, the cambium is growing against an immovable object, the xylem it has already produced, and a strong boundary, the cortex and bark. Wounding, or preparing the grafting surfaces, releases this pressure and causes the cambial cells (and other parenchyma cells) to lose the positional signal they were given by the radial pressure gradient.

During extension growth of the primary stem, which is driven by turgor pressure, the pressure is constantly eased by expansion of the tissues against the low resistance of the epidermis (which itself expands to accommodate the growth). It can be postulated that the cessation of extension growth, and the formation of a less-yielding periderm, coupled with increase in turgor to drive radial expansion, creates the pressure conditions for the formation of interfascicular cambium. In this case, the cambium in the vascular bundles provides the biochemical signal providing positional information, and controlling which parenchyma cells differentiate. Cambium in the vascular bundles is formed from a procambium (O'Brien, 1981), and it could be postulated that the cohesiveness and enlargement of developing primary phloem and xylem cells in the bundle is the source of resistive pressure to maintain these cells in a meristematic state. A study of pressure changes within stems and roots is currently underway to test these hypotheses.

References

Aloni, R. (1987) Differentiation of vascular tissues. *Annu. Rev. Plant Physiol.* **38**: 179–204.

Aloni, R. (1995) The induction of vascular tissues by auxin. In: *Plant Hormones: Physiology, Biochemistry and Molecular Biology.* 2nd edition. (ed. P.J. Davis), Kluwer Academic Publishers, Dordrecht, London, pp. 531–546.

Artschwager, E. (1951) Anatomical studies on graft unions between guayule and sunflower (in interspecific and intergeneric grafts with special reference to formation of rubber in guayule). *U.S. Department of Agriculture Technical Bulletin*, **1040**.

Asante, A. and Barnett, J.R. (1997). Graft union formation in mango (*Mangifera indica* L.). *J. Hort. Sci.* **72**: 781–790.

Barker, W.G. (1954) A contribution to the concept of wound repair in woody stems. *Can. J. Bot.* **32**: 486–490.

Barnett, J.R. and Weatherhead, I. (1988). Graft union formation in Sitka spruce: a scanning electron microscope study. *Ann. Bot.* **61**: 581–587.

Barnett, J.R. and Weatherhead, I. (1989). The effect of scion water potential on graft success in Sitka spruce (*Picea sitchensis*). *Ann. Bot.* **64**: 9–12.

Bloch, R. (1941). Wound healing in higher plants. *Bot. Rev.* **7**: 110–146.

Braun, H.J. (1959) Die normalen Verwachsungsvorgänge nach Propfung on Laubbäumen. II. Die Verfahren des seitlichen Anplattens und der Kopulation. *Zeitschrift für Botanik* **47**: 145–166.

Brown, C.L. (1964) The influence of external pressure on the differentiation of cells and tissues cultured *in vitro*. In: *The Formation of Wood in Forest Trees* (ed. M.H. Zimmermann), Academic Press, New York, pp. 389–404.

Brown, C.L. and Sax, K. (1962) The influence of pressure on the differentiation of secondary tissues. *Amer. J. Bot.* **49**: 683–691.

Buck, G.J. (1954) The histology of the bud graft union in roses. *Iowa State College Journal of Science* **28**: 587–602.

Cline, M.N. and Neely, D. (1983). The histology and histochemistry of the wound-healing process in geranium cuttings. *J. Amer. Soc. Hort. Sci.* **108**: 496–502.

Copes, D.L. (1969) Graft union formation in Douglas fir. *Amer. J. Bot.* **56**: 285–289.

Dormling, I. (1963) Anatomical and histological examination of the union of scion and stock in grafts of Scots pine (*Pinus sylvestris* L.) and Norway spruce (*Picea abies* (L.) Karst.). Stockholm. *Studia Forestalia Suecia* **13**.

Ermel, F.E., Catesson, A.M. and Poësselo, J.-L. (1993) Néoformation cambiale et incompatabilité de greffe chez les associations poirier/cognassier. *Acta Botanica Gallica* **140**: 363–370.

Fujii, T. and Nito, N. (1972) Studies on the compatibility of grafting fruit trees. I. Callus fusion between the rootstock and scion. *J. Jap. Soc. Hort. Sci.* **41**: 1–10.

Garner, R.J. (1990) *The Grafter's Handbook*, 5th Edition. Cassel, London.

Gautheret, R.J. (1959) *La Culture des Tissus Végétaux.* Masson, Paris.

Harris, J.M. and Barnett, J.R. (1975) Differentiated callus nodules in resin pockets of *Pinus ponderosa* Laws. *New Zealand J. Forestry Sci.* **5**: 226–229.

Juliano, J.B. (1941) Callus development in graft unions. *Philippine J. Sci.* **75**: 245–254.

Kollmann, R. and Glockmann, C. (1985) Studies on graft unions. I. Plasmodesmata between cells of plants belonging to different unrelated taxa. *Protoplasma* **124**: 224–235.

Kollmann, R., Yang, S. and Glockmann, C. (1985) Studies on graft unions. II. Continuous and half plasmodesmata in different regions of the graft interface. *Protoplasma* **125**: 19–29.

Kostoff, D. (1928) Studies on callus tissue. *Amer. J. Bot.* **15(10)**: 565–579.

Krenke, N.P. (1933) *Wundkompensation, Transplantation und Chimären bei Pflanzen.* Verlag-Springer, Berlin.

Lintilhac, P.M. and Vesecky, T.B. (1981) Mechanical stress and cell wall orientation in plants. II. The application of controlled directional stress to growing plants; with a discussion on the nature of the wound reaction. *Amer. J. Bot.* **68**: 1222–1230.

Little, C.H.A. and Savidge, R.A. (1987). The role of plant growth regulators in forest tree cambial growth. *Plant Growth Reg.* **6**: 137–169.

Makino, R., Kuroda, K. and Shimaji, K. (1983) Callus formation and effects of applied pressure to the cultured cambial explant of Sugi (*Cryptomeria japonica* D. Don). *Bulletin of the Wood Research Institute Kyoto University No* **69**.

Mendel, K. (1936) The anatomy and histology of the bud-union in citrus. *Palestine J. Bot. Hort. Sci.* **1**: 13–46.

Mergen, F. (1954) Anatomical study of slash pine graft unions. *J. Florida Acad. Sci.* **17**: 237–245.

Miller, H. (1991) *The Effect of Heat on Graft Union Formation in Dormant Sitka Spruce (Picea sitchensis (Bong.) Carr.).* Ph.D. Thesis, University of Reading.

Miller, H. and Barnett, J.R. (1993) The formation of callus at the graft interface in Sitka spruce. *IAWA J.* **14**: 13–21.

Moore, R. (1982) Graft formation in *Kalanchoe blossfeldiana. J. Exp. Bot.* **33**: 533–540.

Moore, R. (1984) The role of direct cellular contact in the formation of compatible autografts in *Sedum telephoides. Ann. Bot.* **54**: 127–133.

Moore, R. and Walker, D.B. (1981a) Studies for vegetative compatibility–incompatibility in higher plants. I. A structural study of a compatible autograft in *Sedum telephoides* (Crassulaceae). *Amer. J. Bot.* **68**: 820–830.

Moore, R. and Walker, D.B. (1981b) Studies for vegetative compatibility–incompatibility in higher plants. II. A structural study of an incompatible heterograft between *Sedum telephoides* (Crassulaceae) and *Solanum pennellii* (Solanaceae). *Amer. J. Bot.* **68**: 831–842.

O'Brien, T.P. (1981) The primary xylem development. In: *Xylem Cell Development* (ed. J.R. Barnett), Castle House Publications. Tunbridge Wells, U.K., pp. 14–46.

Parkinson, M. and Yeoman, M.M. (1982) Graft union formation in cultured, explanted internodes. *New Phytol.* **91**: 711–719.

Sass, J.E. (1932) Formation of callus knots on apple grafts as related to the histology of the graft union. *Bot. Gazette* **94**: 364–380.

Savidge, R.A. (1996) Xylogenesis, genetic and environmental regulation — a review. *Int. Assoc. Wood Anatomists J.* **17**: 269–310.

Sharples, A. and Gunnery, H. (1933) Callus formation in *Hibiscus rosa-sinensis* and *Hevea brasiliensis* Mull-Arg. *Ann. Bot.* **47**: 829–839.

Shi-Jie, Y.J. and Cheng-Hou, L. (1988) Paramural bodies in callus beside the isolation layer of graft union. *Acta Bot. Sin.* **30**: 480–484.

Steward, F.C., Mapes, M.O. and Mears, K. (1958) Growth and organised development of cultured cells. II. Organisation in cultures grown from freely suspended cells. *Amer. J. Bot.* **45**: 705–708.

Stoddard, F.L. and McCully, M.E. (1979) Histology of the development of the graft union in pea roots. *Can. J. Bot.* **57**: 1486–1501.

Stoddard, F.L. and McCully, M.E. (1980) Effects of excision of stock and scion organs on the formation of the graft union in *Coleus*: a histological study. *Bot. Gazette* **141**: 401–412.

Soumelidou, K. (1993) Investigations of vascular differentiation and the dwarfing effects of apple rootstocks. PhD Thesis. The University of Reading.

Soumelidou, K., Battey, N.H., John, P. and Barnett, J.R. (1994) The anatomy of the developing bud union and its relationship to dwarfing in apple. *Ann. Bot.* **74**: 605–611.

Tuominen, H., Puech, L., Fink, S. and Sundberg, B. (1997) A radial concentration gradient of indole-3-acetic acid is related to secondary xylem development in hybrid aspen. *Plant Physiol.* **115**: 577–585.

Warren Wilson, J. (1978) The position of regenerating cambia: auxin/sucrose ratio and the gradient induction hypothesis. *Proc. Royal Soc. London, series B* **203**: 153–176.

Warren Wilson, J. and Warren Wilson, P.M. (1961) The position of regenerating cambia — a new hypothesis. *New Phytol.* **60**: 63–73.

Washer, J., Reilly, K.J. and Barnett, J.R. (1977) Differentiation in *Pinus radiata* callus culture: The effect of nutrients. *New Zealand J. Forestry Sci.* **7**: 321–328.

Weatherhead, I. (1986) *Causes of Graft Failure in Sitka Spruce,* Picea sitchensis *(Bong.) Carr.* PhD Thesis. The University of Reading.

Weatherhead, I. and Barnett, J.R. (1986). Development and structure of unusual xylem elements during graft union formation in *Picea sitchensis*. *Ann. Bot.* **57**: 593–598.

13

Cambial growth and auxin gradients

B. Sundberg, C. Uggla and H. Tuominen

1. Introduction

Wood is formed by cambial growth, which includes cell divisions in the cambial meristem, followed by expansion and maturation of the cambial derivatives (Larson, 1994). A strict regulation of these developmental processes in time and space produces typical patterns of wood variation along the stem (Zobel and Buijtinen, 1989). The plant hormone indole-3-acetic acid (IAA) is an endogenous auxin that has been demonstrated to affect many aspects of secondary xylem (wood) formation when applied to cambial tissues. Consequently, IAA has long been considered to be an important regulator of cambial growth (for reviews see Wareing *et al.*, 1964; Larson, 1969; Brown, 1970; Savidge and Wareing, 1981; Little and Savidge, 1987; Lachaud, 1989; Aloni, 1991; Little and Pharis, 1995). However, there has been a failure to demonstrate a consistent relationship between endogenous IAA and cambial growth processes, and to develop a unifying concept for the role of IAA in cambial growth. This is due to limited knowledge, not only of mechanisms of IAA perception and signal transduction, but also of its metabolism, transport and resulting concentration and distribution in plant tissues.

In this chapter, we will first summarize briefly the evidence for a role of IAA in different developmental processes of cambial growth. Thereafter, recent data on the IAA balance in cambial tissues will be presented, providing the basis for a new concept of the role of IAA in controlling wood formation in time and space.

2. IAA and cambial growth

2.1 *Role in cambial cell division*

The observation that an exogenous source of IAA could replace the expanding apical shoot in inducing cambial cell division and xylem differentiation first indicated a role for IAA in cambial growth (Snow, 1935; Gouwentak, 1941). Since then, many experiments with exogenous IAA have been performed in both conifers and angiosperm

Cell & Molecular Biology of Wood Formation, edited by R. Savidge, J. Barnett and R. Napier.

trees to elucidate its role in wood formation. It became clear that apically applied IAA is transported polarly in stem tissues (Hollis and Tepper, 1971; Odani, 1975; Zamski and Wareing, 1974; Little, 1981), and induces secondary xylem and phloem formation in a dose-dependent manner in both young shoots (Digby and Wareing, 1966; Doley and Leyton, 1968; Zajaczkowski, 1973; Little and Bonga, 1974; Sheriff, 1983; Zakrzewski, 1983, 1991; Cui and Little, 1993) and mature stems (Little and Sundberg, 1991; Leitch and Savidge, 1995). It was also demonstrated that cambial tissues that were isolated from apical IAA sources by debudding or girdling responded by dedifferentiation of cambial zone cells (cambial zone = cambial meristem, see Wilson *et al.*, 1966 for terminology) to axial parenchyma cells by transverse septation (Evert and Kozlowski, 1967; Evert *et al.*, 1972; Savidge, 1983).

With the development of accurate techniques for IAA measurement in plant tissues, the physiological relevance of auxin supplementation and depletion experiments could be evaluated. Debudding and defoliation of young shoots in conifer and angiosperm species was demonstrated to result in a depletion of endogenous IAA and an inhibition of cambial growth in the subjacent stem (Savidge and Wareing, 1981; Sundberg and Little, 1987; Rinne *et al.*, 1993; Wang *et al.*, 1997). Apical application of different concentrations of IAA to debudded conifer shoots raised internal IAA levels comparable to those in intact shoots, and induced a corresponding increase in cambial growth (Little and Wareing, 1981; Sundberg and Little, 1987; 1990; Wang *et al.*, 1997; Sundberg and Uggla, 1998). Inhibitors of polar IAA transport applied around the circumference of intact *Pinus sylvestris* shoots mimicked the effects of debudding and resulted in both IAA depletion and an inhibition of cambial growth below the point of treatment (Sundberg *et al.*, 1994). Moreover, IAA applied to *intact Pinus sylvestris* shoots supplemented the endogenous supply of polarly transported IAA to produce a physiological relevant increase in IAA that was associated with a stimulation of xylem production (Sundberg and Little, 1990). Taken together, it can be concluded that: (i) the expanding shoot is a major supplier of polarly transported IAA, (ii) polarly transported IAA maintains the structure and activity of the cambial meristem and (iii) the amount of IAA supplied to the cambial tissues by polar transport controls the extent of cell divisions in the cambial meristem and therefore xylem production.

2.2 *Role in tracheary element differentiation*

Polarly transported IAA also induces the differentiation of primary vascular strands along its path of transport in a quantitative manner (Sachs, 1981; Jacobs, 1984). Moreover, there is a voluminous literature on the xylogenic effects of IAA applied to isolated tissues, callus and cell cultures of both gymnosperms and angiosperms (Roberts, 1988). Often the response to IAA is preceded by cell division (Fosket, 1968; Fosket and Torrey, 1969; Sussex *et al.*, 1972; Behnke and Schulz, 1980; Phillips and Arnott, 1983), but parenchyma cells can also be induced by auxin to differentiate directly into tracheary elements (Fukuda and Komamine, 1980, 1981; Warren Wilson *et al.*, 1994a). Thus, auxin is not only required for cell division but also for differentiation of tracheary elements. However, experimental evidence for the latter function in ontogenesis of primary and secondary vascular tissues of intact plants is complicated by the fact that auxin is also needed for cell divisions which are required to produce the derivatives that will differentiate into tracheary elements.

2.3 *Role in other processes of cambial growth*

IAA has also been implicated in the control of radial diameter of xylem elements, and less decisively, in the control of size and frequency of rays, frequency of vessels, thickness of secondary walls and the formation of reaction wood (Aloni, 1987; Little and Savidge, 1987; Lev Yadun and Aloni, 1995; Little and Pharis, 1995). However, these ideas are mainly based on auxin supplementation and depletion experiments where the resulting internal IAA level has not been measured. Conclusions drawn from such studies must therefore be viewed with caution, because the probability that the response is due to unphysiological IAA levels in the observed tissues, or that assumptions about the resulting internal IAA level are incorrect, cannot be excluded. For example, IAA (0.1% in lanolin) supplied apically to decapitated conifer shoots can restore the IAA level to that present in intact shoots (Little and Wareing, 1981; Sundberg and Little, 1990). But the same treatment applied to *Populus tremula* × *tremuloides* shoots was found to result in internal levels that were only 10–20% of those in intact shoots (Sundberg, unpublished). Such differences in uptake and transport of exogenous IAA between experimental systems may explain the conflicting views about the role of IAA in radial expansion of xylem derivatives. In most species, including *Populus*, a stimulation of radial expansion of xylem elements has been observed below a point application of auxin (Larson, 1960; Digby and Wareing, 1966; Doley and Leyton, 1968; Hejnowicz and Tomaszewski, 1969; Denne and Wilson, 1977; Meichenheimer and Larson, 1985). On the other hand, an inhibitory effect of auxin on vessel size was reported in *Phaseolus* and *Melia azedarach* (a ring-porous tree) (Aloni and Zimmermann, 1983; Aloni, 1991). In experiments where IAA was supplied in liquid solution through isolated stem segments (hence was not limited by uptake and polar transport) a positive correlation between auxin concentration and radial diameter of tracheids (Sheriff, 1983) and vessels (Zakrzewski, 1983, 1991) was observed up to a certain threshold level, beyond which auxin became inhibitory. Thus, applied IAA will stimulate xylem cell size as long as the resulting internal concentration is below optimum for cell expansion, as in *Populus*, and inhibit xylem cell size when resulting internal concentration becomes supraoptimal, which was probably the case in *Phaseolus* and *Melia*.

In addition to traditional approaches in which auxin levels were manipulated, transgenic herbaceous plants with increased or decreased IAA concentration have been produced (Klee *et al.*, 1987; Romano *et al.*, 1991; Sitbon *et al.*, 1992). In general, these plants exhibit similar phenotypic effects to those described in plants treated with exogenous IAA. However, evaluation of phenotypic responses in these transgenic plants is complicated by their much altered phenotype and developmental pattern.

The results from experiments with exogenous IAA and transgenic IAA-overproducing plants have demonstrated that IAA has great potential for affecting a multitude of processes controlling developmental patterns and morphology of xylem elements. A requirement for IAA in cell expansion can be concluded from extensive experiments in many plant species (Kutshera, 1994), but for the other observed IAA effects on xylem structure the physiological relevance remains to be demonstrated. The challenge is to understand the mode of action of IAA, and to what extent processes that can be modulated by auxin under experimental conditions actually are under IAA control in the intact plant.

3. Earlier models explaining the role of IAA in seasonal and spatial control of cambial growth

The early finding that expanding shoots were a major source of polarly transported IAA gave rise to the concept of a decreasing auxin concentration gradient from shoot to root. It was also generally believed that endogenous auxin was absent in the dormant cambium and that the concentration peaked during the growing season (Larson, 1962). On the basis of these ideas about spatial and seasonal variation of endogenous IAA, and the knowledge that IAA stimulated cambial cell division and expansion of cambial derivatives, some general hypotheses on its regulation of cambial growth patterns were formulated (Wareing, 1958; Larson, 1962, 1969). It was proposed that cambial reactivation in the spring was induced by increasing auxin concentration, initiated by auxin synthesized in expanding buds and transported down the stem by polar transport. Formation of radially narrow latewood cells and the eventual cessation of cambial growth were found to be associated with cessation of apical growth, and therefore considered to be induced by decreasing auxin concentrations. The concept of a basipetally decreasing gradient in auxin concentration was used to explain variations in annual ring width along the stem, as well as the acropetal spread of latewood induction (Larson, 1962). In a conflicting hypothesis launched in 1983 (Aloni and Zimmermann, 1983), the presumptive decreasing axial auxin gradient was invoked to explain the often observed pattern of increasing xylem cell diameter and decreasing density of vessels down the stem.

Although these hypotheses have persisted up to date, they suffer from the fact that the proposed distribution patterns of endogenous IAA in time and space were never convincingly demonstrated. In fact, later measurements of seasonal and spatial variation of endogenous IAA in the cambial region of trees have yielded conflicting results, and no consistent relationship between endogenous IAA concentration and cambial growth patterns has been established (Little and Pharis, 1995). This is due to problems associated with analysing multi-tissue samples (i.e. phloem, cambium and developing xylem), as the tissues differ widely in their concentrations of IAA (see Uggla *et al.*, 1998). The proportions of different tissues included in the sample, which depends on the activity state of the cambium, will greatly affect the estimated IAA concentration. This is obvious from the data presented in *Figure 1*, which show that estimates of IAA concentration in combined cambial region tissues do not reflect the seasonal variations in IAA supply to these tissues. This conclusion highlighted the need for more tissue-specific IAA measurements to evaluate its role in the different developmental phases of cambial growth. The recent development of more sensitive techniques for IAA analyses (Edlund *et al.*, 1995) has made such an approach possible.

4. Towards a new concept of the role of IAA in cambial growth

4.1 *Cambial growth is determined by rate and duration of developmental phases and requires positional information*

Following cell division in the cambial zone, cambial derivatives develop into wood cells by expansion, secondary wall formation and finally autolysis. These phases are separated in space, resulting in a radial pattern of discrete developmental zones (Larson, 1994). Each cell divides, expands and forms a secondary wall at a certain rate over a certain period of time. These two components, rate and duration, together

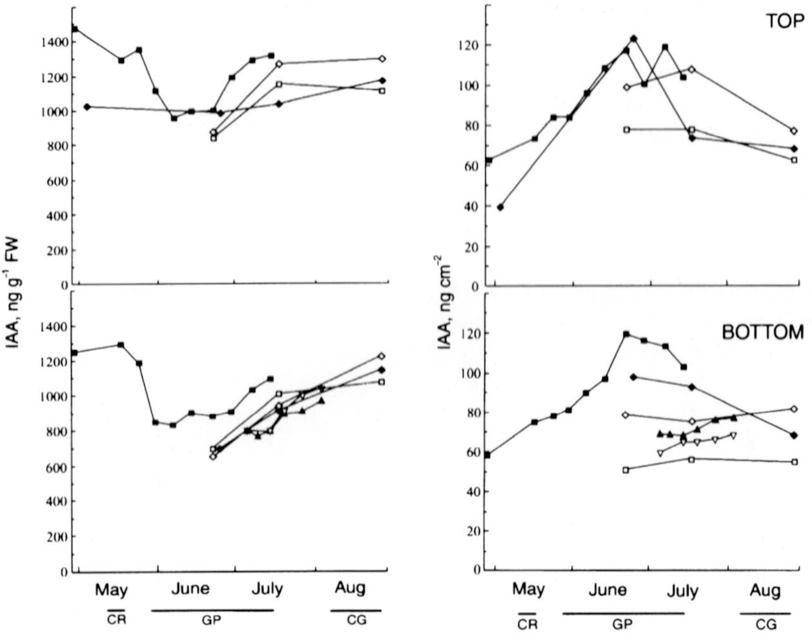

***Figure 1**. Seasonal variation of endogenous IAA in cambial region tissues (i.e. differentiating xylem cells, cambial zone cells, and differentiating, mature and one year-old phloem cells) measured within the crown (top) and at breast height (bottom) in mature* Pinus sylvestris *trees. The figure summarizes data from several investigations performed in different stands and years (data from Sundberg* et al. *1990 (◆), 1991(■), 1993 (□,◇) and Sundberg unpublished (▲,▽). The data are presented as a concentration (ng g^{-1} fresh weight) and as an amount (ng per cm^{-2} stem area). The calculated concentration of IAA peaked in the dormant cambium, and dropped during the growing season. In contrast, the amount of IAA present in the cambial region tissues was greatest during the growing season. This is because concentration is calculated by dividing the amount of IAA in the sample by the weight of the sample, and the increase in sample weight during cambial reactivation was greater than the increase in IAA amount (see Sundberg* et al.*, 1991). From this example it can be concluded that estimates of IAA concentration in combined cambial region tissues are not suitable to describe variations in IAA supply to these tissues if samples with different tissue compositions are compared. CR=cambial reactivation; GP=grand period of tracheid production; CG=production of latewood tracheids and cessation of cambial growth.*

determinine the overall growth rate of the cambium and the final morphology of the wood cell (Whitmore and Zahner, 1966; Wilson and Howard, 1968; Skene, 1969, 1972; Wodzicki, 1971; Dodd and Fox, 1990; Ridoutt and Sands, 1994). It is therefore essential to understand the mechanism that controls both of these aspects of cambial growth.

The rates of cell cycling (measured as mitotic frequency), cell expansion and secondary wall formation (measured as the increase in radial width and wall thickness of cells along a radial file) are not uniform within each developmental zone, but decrease from early to late stages of the developmental phase (Bannan, 1955; Wilson, 1964; Imagawa and Ishida, 1972; Abe *et al.*, 1995; Yoshinaga *et al.*, 1997). This suggests that the mechanism(s) controlling the rate of the cambial growth processes is not uniform within each developmental zone.

The duration of meristematic activity for each xylem mother cell is reflected in the radial population of dividing cambial zone cells (assuming that the majority of cambial zone cells will develop into xylem elements), and therefore also by the radial width of the zone of cell division. The time that each cambial derivative spends in the zones of expansion and wall formation, on the other hand, depends both on the rate of cell production in the cambial meristem and on the radial width of the developmental zone. It follows that the width of each developmental zone is important for the duration of developmental events, and therefore involved in the control of cambial growth.

It is conceivable that the width of the zones is under a developmental control that gives positional information according to which cambial derivatives switch between the different phases of division, expansion, secondary wall formation and final autolysis. The developmental pattern is co-ordinated between radial cell files, indicating that such positional information is acting on a supracellular level. Traditionally, IAA has been considered to regulate the rate of different processes, such as cell cycling and cell expansion. However, in view of the morphogenetic properties of IAA, it may also function as a developmental signal that regulates the width of the developmental zones. An effect of IAA on both rate and duration of developmental phases has been claimed (Wodzicki and Zajaczkowski, 1974; Denne and Wilson, 1977; Porandowski *et al.*, 1982; Nix and Brown, 1987) but, in general, little attention has been paid to this issue.

4.2 *Auxin as a positional signal in cambial growth*

Positional information conveyed by concentration gradients of morphogens has been shown to be of fundamental importance for developmental regulation in animal systems (Wolpert, 1981; 1996). Implicit in this hypothesis is the idea that cells develop according to their position along a concentration gradient of a morphogen, the morphogenetic field. Similar mechanisms are believed to co-ordinate plant development (Holder, 1979; Warren Wilson and Warren Wilson, 1984; Meinhardt, 1996), but morphogens creating such fields were not demonstrated.

Highly sensitive mass-spectrometry measurements of IAA in tangential 30-μm cryosections obtained across the developing secondary vascular tissues of *Pinus sylvestris* L. and *Populus tremula* L. × *tremuloides* Michx. revealed concentration gradients which share features of the morphogenetic fields found in animal systems (*Figure 2 and* colour *Plate 3*) (Uggla *et al.*, 1996; Tuominen *et al.*, 1997). Both species exhibited a high concentration of IAA in the cambial zone and its most recent derivatives. From the peak value, the concentration decreased towards phloem and xylem, reaching low and relatively stable levels close to the transition between expanding and secondary-wall forming cells. Further characterization, involving measurements of IAA specifically in ray and fusiform cell types dissected out from dividing and maturing tissues, revealed that IAA distribution was homogenous across axial and radial cell systems. It was also demonstrated that the amount of IAA in the cambial region did not fluctuate within substantial (cm scale) areas of the stem (Uggla, 1998). Cambial derivatives are therefore exposed to a stable supra-cellular field of decreasing IAA concentration during their development into xylem and phloem cells. The requirement of IAA for cell division, expansion and differentiation implies that the radial IAA gradient functions as a morphogenetic field with the potential to give positional signalling in cambial growth.

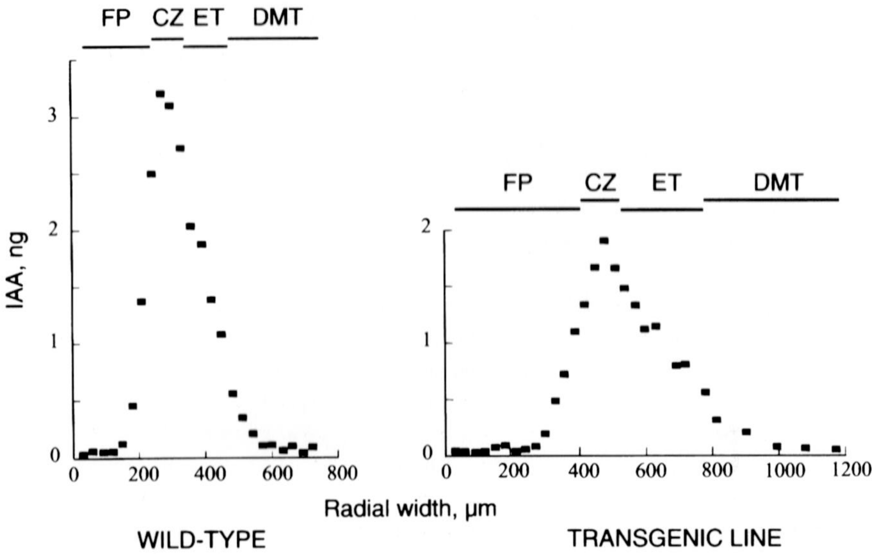

Figure 2. *Radial distribution of IAA across the cambial region of a wild-type and a transgenic* Populus tremula × tremuloides *expressing the IAA biosynthetic genes from* Agrobacterium tumefaciens. *Each data point represents the IAA content in a* cm^{-2}*, 30-µm thick, tangential section. Horizontal bars indicate the approximate position of developmental zones. FP=zone of differentiating and functional phloem; CZ=cambial zone; ET=zone of expanding differentiating tracheids; DMT=zone of differentiating tracheids forming secondary walls as well as mature dead tracheids. Figure redrawn from Tuominen* et al. *(1997).*

In particular, it is suggested that cell division is maintained at high auxin levels, and that critical concentration thresholds define the radial position where cambial derivatives switch from dividing to expanding, and from expanding to secondary-wall forming. However, it is clear from our data that the cambial zone is not positioned within a certain concentration window of IAA, emphasizing the interaction of IAA with other signals. Evidence for a spatially co-ordinated developmental pattern along an IAA gradient was obtained in an experiment where IAA was applied to one end of a lettuce pith explant (Warren Wilson *et al.*, 1994a). Mitotic activity, cell enlargement and tracheary element differentiation were restricted to tissues within a distance of 3 mm from the IAA donor. Mitotic activity was most frequent at a distance of around 0.5 mm from the donor block, whereas cell enlargement was most abundant at distances of around 1 to 1.5 mm. Differentiation to tracheary elements without prior cell division was observed close to the IAA source, whereas strands of tracheary elements preceded by cell division activity were formed at a greater distance from the IAA source. This suggests that IAA was eventually canalized to transport pathways along the explant (Sachs, 1981). The results support the idea that different developmental processes predominate at different IAA concentrations, which is consistent with the view that the IAA gradient provides positional information to cells during vascular development.

That IAA provides positional information in cambial growth does not rule out the widespread view that IAA concentration is involved in regulating the rate of cell cycling and expansion. In fact, this idea is supported by the observation that cambial derivatives exhibit decreasing rates of cell cycling and expansion across different developmental zones, which is associated with their exposure to decreasing auxin concentrations in each zone.

4.3 *Observations supporting a role for IAA as a positional signal in cambial growth*

The idea that IAA provides positional information during cambial growth was evaluated by visualizing the radial IAA distribution across cambial tissues with different growth rates in *Pinus sylvestris* trees (Uggla *et al.*, 1998). A close correlation was found between the radial width of the cambial zone and the radial width of the IAA distribution, supporting the idea that IAA provides positional information for determining the width of the cambial meristem. The width of the cambial zone was closely related to the number of dividing cambial zone cells and therefore to growth rate. Furthermore, the width of the IAA distribution was closely correlated with the total amount of IAA per stem area unit. Together, these correlations suggest a chain of causal relationships; the amount of polarly transported IAA determines the radial width of the IAA distribution pattern, which in turn controls cambial growth rate by defining the radial width of the cambial zone. The importance of the radial population of cambial zone cells rather than cell cycling rate (measured as mitotic index) in determining the rate of xylem production has also been demonstrated for both *Pinus strobus* and *Picea glauca* when comparing dominant and suppressed trees in the same stand (Wilson, 1964; Gregory and Wilson, 1968). However, when comparing trees originating from different latitudes and therefore with different genetic background, rates of cell cycling were found to be important for rates of xylem formation (Gregory, 1971).

It was also observed that the radial IAA distribution approximates the radial width of the entire zone of expanding cells, i.e. the combined width of dividing and expanding cells (Uggla *et al.*, 1996; 1998; Tuominen *et al.*, 1997). This suggests that cambial derivatives will continue to expand as long as they are positioned within the field of appropriate IAA concentrations. It follows that the width of the IAA gradient modulates the radial width of xylem elements by affecting the duration of their expansion. In Scots pine, radially wider tracheids were associated with wider IAA gradients (Uggla *et al.*, 1998). However, an increased cambial growth rate will influence the duration of cell expansion due to an increased flow of cells through the zone of cell expansion. Thus, IAA may affect the duration of cell expansion by influencing both the rate of production of cambial derivatives and the radial width of the expansion zone.

Controlled perturbation of the IAA distribution pattern across the cambial region tissues would be helpful in evaluating its role as a positional signal in cambial growth. This was achieved by ectopic expression of the *Agrobacterium tumefaciens* IAA biosynthetic genes in transgenic hybrid aspen (Tuominen *et al.*, 1997). The transgenic trees did not differ in the amount of IAA in the cambial region tissues compared to wild type trees, but displayed a wider distribution pattern of IAA with a lower peak level in the cambial zone (*Figure 2*). This was associated with wider developmental zones of both division and expansion, providing further evidence for the idea that cambial derivatives continue their expansion as long as they are positioned within a field of appropriate IAA levels. However, the wider zone of cell division was not associated with a larger number of dividing cells. This was due to the lower rate of cell cycling in

the transgenic trees, which could be attributed to the decreased IAA concentration in the cambial zone. Furthermore, the transgenic trees exhibited a larger cross-sectional area of xylem fibres. This increase in fibre diameter was apparently a result of a longer period of cell expansion, as judged from the increase in the width of the zone of cell expansion together with the slower rate of cell production Taken together, the results give additional support to the role of IAA in regulating the width of the different developmental zones.

The observation that tracheary element differentiation is induced by critical auxin concentration in cell and tissue cultures suggests that the auxin gradient is involved in determining the radial position at which developing cambial derivatives become committed to tracheary differentiation. This idea can be extended to incorporate the view that the auxin field gives information for determining the position of the cambial initial and of xylem and phloem mother cells on each side of the initial. It has been found that low concentrations of applied auxin induce phloem only, whereas high concentrations induce both xylem and phloem (Aloni, 1980). On the other hand, it has been suggested that sucrose is a controlling factor in phloem and xylem differentiation, phloem being induced by auxin and high sucrose levels, and xylem by auxin together with low sucrose levels (Wetmore and Rier, 1963; Jeffs and Northcote, 1966; Warren Wilson *et al.*, 1994b). In secondary growth, it is not known at what developmental stage (or radial position) cambial derivatives become committed to xylem and phloem cells. But from the visualization of endogenous IAA distribution, it is clear that cambial derivatives on both sides of the cambial initial are exposed to the same range of IAA concentration. This rules out IAA level as the only determinant for xylem and phloem differentiation. The Warren Wilsons (1984) hypothesized that opposing gradients of sucrose and auxin provided positional information for xylem and phloem development. Interestingly, we have documented the presence of a steep concentration gradient of sucrose across the cambial zone of *Pinus sylvestris* (*Figure 3*) (Uggla, 1998). Consequently, there are large changes in the ratios of sucrose to auxin across the cambial zone, which may have the spatial resolution required for defining phloem and xylem mother cells.

5. Spatial and seasonal variation of IAA and its control of cambial growth patterns — a new model

A new concept for the role of IAA in controlling seasonal and spatial patterns of wood formation in temperate-zone tree species is presented. The concept is based on the visualization of IAA distribution across cambial region tissues of *Pinus sylvestris* (*Figure 4*) (Uggla *et al.*, 1996, 1998; Uggla, 1998), and the idea that IAA controls both rate and duration of cell cycling and expansion of developing xylem cells. Although the data come from a conifer, we suggest that this concept is valid also for angiosperm trees.

The dormant cambium of both conifers and angiosperms stems contains significant levels of IAA (Little and Wareing, 1981; Savidge and Wareing, 1981, 1982; Wignall and Browning, 1988; Sundberg *et al.*, 1990). In fact, measurements of IAA specifically in cambial zone cells of slow-growing Scots pine trees showed that the concentration in the dormant cambium in the mature stem was similar to that in the active cambial meristem (Uggla *et al.*, 1996). IAA transport seems to be maintained as long as temperature is favourable, but with reduced efficiency in the dormant cambium of

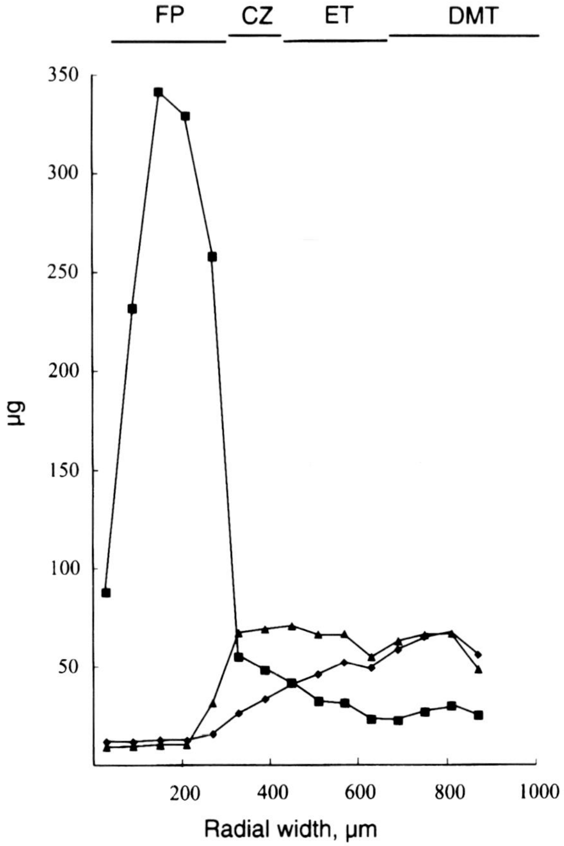

Figure 3. *Radial distribution of sucrose (■), fructose (▲) and glucose (◆) across the cambial region of* Pinus sylvestris. *The sugars were measured in tangential cryosections (obtained as described in Uggla* et al., *1996) by gas chromatography-mass spectrometry using phenyl-β-D-glucoside as an internal standard. Each data point represents the sugar content in a cm^{-2}, 30-µm wide, tangential section. Horizontal bars indicate the approximate position of developmental zones. FP=zone of differentiating and functional phloem; CZ=cambial zone; ET=zone of expanding differentiating tracheids; DMT=zone of differentiating tracheids forming secondary walls as well as mature dead tracheids.*

angiosperm trees (Hollis and Tepper, 1971; Zamski and Wareing, 1974; Odani, 1975; Little, 1981; Odani, 1985; Lachaud, 1989). Apparently auxin for polar transport can also be synthesized in trees during dormancy (Savidge and Wareing, 1982).

To induce cambial reactivation, a critical auxin level has to be reached. This can be deduced from detached dormant shoots placed in favourable environmental conditions where cambial reactivation is observed at the base of the shoot, or above a girdle, where the endogenous IAA will accumulate in the dormant cambium as a result of polar transport (Brown and Cormack, 1937; Little *et al.*, 1990). Thus, intact young shoots are dependent on a *de novo* supply of IAA from expanding buds in order to reactivate (Sundberg and Little, 1990). The endogenous auxin present in the dormant cambium of mature stems is sometimes sufficient for cambial reactivation as indicated

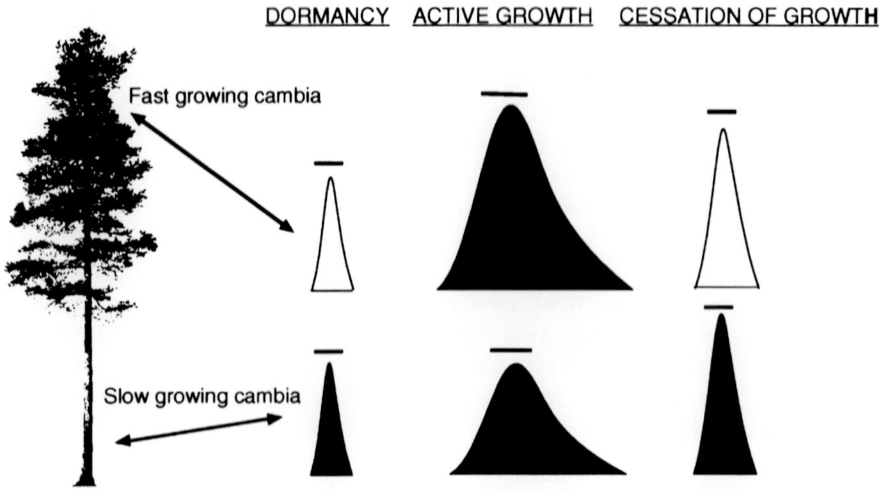

Figure 4. *Schematic model of seasonal variation of the radial distribution of IAA across cambial region tissues in slow-growing and fast-growing cambia. Solid figures represent IAA distributions based on actual data (Uggla* et al.*, 1996, 1998a, b), whereas open figures are hypothesized distributions. Horizontal bar indicates the positions of the cambial zone.*

by experiments with girdling or local heating (Evert and Kozlowski, 1967; Evert *et al.*, 1972; Egierszdorff, 1981; Savidge and Wareing, 1981; Little and Wareing, 1981; Oribe and Kubo, 1997), whereas in other cases it is not and additional IAA is needed to induce cambial reactivation (Leitch and Savidge, 1995; Oribe and Kubo, 1997). But this does not imply that the auxin needed for cambial reactivation in mature stems originates from *de novo* synthesis in reactivating buds. Auxin levels in the cambium may increase due to *de novo* synthesis in cambial tissues (see below) or possibly due to accumulation of polarly transported auxin as a result of decreasing transport rates in older cambia (Suttle, 1991). Therefore, the original idea that cambial reactivation is induced by a basipetal auxin wave from expanding buds is not necessarily valid along the whole length of the tree stem, and it is clear from cambial reactivation patterns that at least in conifers this is not the case (Savidge and Wareing, 1981)

Once cambial reactivation is initiated, maintenance of cambial growth is dependent on a continuous supply of IAA through the polar transport system. The amount of IAA that is supplied to the cambial tissues will increase in association with active shoot growth. Our data indicate that this increase results primarily in an increased radial distribution of IAA, and only when the supply is large enough will the concentration of IAA in the dividing and expanding cells also increase (*Figure 4*) (Uggla *et al.*, 1998). This implies that auxin stimulates cambial growth primarily by acting as a positional signal, increasing the population of dividing cambial zone cells through its radial distribution pattern, and secondarily by stimulating the rate of cambial cell cycling through an increased concentration in the meristem.

The transition from the formation of earlywood to narrow-diameter and thick-walled latewood tracheids is under developmental control. It involves a decreased duration of cell expansion due to narrowing of the expansion zone, in combination with an increased duration of wall formation due to a widening of the zone of secondary wall formation and a reduced rate of cell production in the cambial zone (Whitmore and Zahner, 1966; Wodzicki, 1971; Dodd and Fox, 1990). Induction of latewood tracheids and the eventual cessation of cambial activity have been ascribed to decreased IAA concentration (Larson, 1960, 1969). However, in slow-growing cambia of Scots pine trees, the amount of IAA in cambial region tissues did not change during the transition from earlywood to latewood formation (Sundberg *et al.*, 1990, 1991, 1993). More detailed analysis revealed that the radial width of the auxin gradient decreased and, consequently, the peak concentration of IAA in the cambial zone increased (*Figure 4*) (Uggla, 1998). The decreasing width of the IAA gradient was closely associated with a decreasing width of the zone of dividing and expanding cells. This is in line with the role of IAA as a positional signal, but it excludes the idea that induction of latewood formation is due to decreased IAA concentration. The data rather suggest that IAA has a function in the earlywood to latewood transition by determining the decrease in the width of the zone of cell expansion, the major factor behind the narrow diameter of latewood cells. Because the amount of IAA did not change during this period it seems likely that this control by IAA is not directly mediated via an altered supply of IAA to the cambial tissues. It is rather due to the yet unknown mechanism(s) determining the radial width of the IAA concentration gradient.

Our data demonstrate clearly that cessation of cambial cell division is not due to decreasing IAA concentration in the meristem (*Figure 4*) (Uggla, 1998). This can also be concluded from earlier experiments with exogenous IAA, which demonstrate that the response to IAA in stimulating cambial cell division decreases towards the autumn to reach a minimum during the period of rest (Zajaczkowski, 1973; Little and Bonga, 1974; Sundberg at al., 1987; Mellerowicz *et al.*, 1992). After a period of chilling the cambium regains its ability to respond to IAA (Little and Bonga, 1974; Odani, 1980).

6. Some aspects on IAA homeostasis in cambial region tissues

6.1 *Origin of IAA*

The IAA present in cambial region tissues originates by polar transport, as indicated by the almost complete depletion of IAA below the site of treatment with IAA transport inhibitors (Sundberg *et al.*, 1994). Without doubt, the expanding apical shoot is the major auxin source of polarly transported IAA in subjacent internode(s) (Sundberg and Uggla, 1998). However, in view of the slow rate of polar transport it would take months for an auxin molecule to cover the distance from shoot to root in a large tree, which suggests that there are additional sources of IAA. Early investigators proposed *de novo* synthesis of IAA in the cambial meristem (see Berlyn, 1982), and recent data provide some evidence for this idea (Sundberg and Uggla, 1998). Mature leaves are another putative source of polarly-transported IAA, and, at least for angiosperm trees, there is polar transport of IAA down the petiole, which is involved in the control of leaf abscission (Morris and Small, 1990). But compared to the expanding shoot, mature leaves are minor sources of IAA and experiments in conifers demonstrated that removal of mature leaves on the 1-year-old internode did not affect the IAA level in the internode (Sundberg *et al.*, 1987; Sundberg and Uggla, 1998). On the other hand, it has been shown that extensive defoliation of trees induces decreased IAA levels in the stem (Savidge and Wareing, 1982). Such decrease in IAA level may merely reflect an overall decrease of tree vigour, which is an unavoidable result of such treatment. However, considering the large amount of leaf biomass on a whole-tree basis, the idea that mature leaves make a significant contribution to the pool of IAA that is polarly transported in the stem cannot be ruled out. This may be particularly important towards the end of the cambial growing season after the period of shoot expansion. Another putative source of IAA is hydrolysis of auxin stored as amide and ester conjugates. IAA conjugates are of great importance, for example during seed germination (Bandurski *et al.*, 1995). But they are not likely to be an important source of free IAA in cambial region tissues, as hydrolysable IAA conjugates occur only in trace amounts in extraxylary tissues of Scots pine and hybrid aspen (Sundberg *et al.*, 1990; Tuominen *et al.*, 1995).

Apart from the pool of IAA in polar transport, endogenous IAA has been detected in xylem and phloem sap (Crozier *et al.*, 1980; Hoad, 1995). This IAA may partly originate from polarly-transported IAA diffusing laterally from the high concentrations present in the cambial zone. In support of this hypothesis, lateral transport of radiolabelled IAA from cambial tissues to sieve tubes in *Salix* has been demonstrated (Hoad *et al.*, 1971). In contrast, radiolabelled IAA fed to the polar transport system was not found in phloem sap of pea plants (Morris and Kadir, 1972). It has also been suggested that IAA in the phloem sap originates from synthesis in mature leaves (Borkovec *et al.*, 1994; Hoad, 1995; Cambridge and Morris, 1996). Cambridge and Morris (1996) further showed that radiolabelled IAA fed to mature pea leaves was transported to the immature tissues of the shoot apex where it entered the basipetal polar transport pathway. Based on this evidence they suggested that mature leaves were the major source of polarly-transported IAA. However, this idea is not supported by experiments where mature leaves were removed (Sundberg and Little, 1987), and it is also unclear if the fate of IAA fed to mature leaves reflects the situation in the intact plant. With present knowledge, no firm conclusion can be drawn about the origin and fate of endogenous IAA in xylem and phloem sap, and the role of this IAA pool in cambial growth, if any, remains to be demonstrated.

6.2 *Radial and longitudinal distribution of IAA*

The cambial zone and its most recent derivatives form the main pathway for polar transport in stem tissues. This view is supported by the observation that the peak concentration of endogenous IAA occurs in these tissues, and by autoradiographic studies of radiolabelled IAA in polar transport (Nix and Wodzicki, 1974; Morris and Thomas, 1978; Lachaud and Bonnemain, 1984). The exact cellular compartment for this transport remains to be determined. The radial distribution pattern of IAA across the cambial region tissues is determined by: (i) the supply of polarly transported IAA, (ii) the amount of IAA that is laterally transported and (iii) the catabolism and/or removal of laterally transported IAA. The latter is essential for the creation of the concentration gradient. IAA removal may be a result of IAA reaching the stream of mass flow in xylem and phloem. However, a more controlled catabolism of IAA also seems likely. The relative importance of these factors and the mechanism(s) of radial transport/diffusion are not known.

The large concentration difference of IAA across cambial region tissues complicates the concept of a decreasing concentration gradient of auxin down the stem. The cambial derivatives will be exposed to a great range of IAA concentrations during their development, and longitudinal concentration gradients can only be discussed when comparing tissues in the same developmental stage. Moreover, possible IAA biosynthesis in stem tissues, unequal transport rates in internodes of different ages, and the changing size of the transport channel within the stem will all affect the longitudinal pattern of IAA distribution. When the auxin distribution across the cambial region was visualized at different positions along the stem of slow-growing Scots pine trees, it was evident that the peak concentration of IAA in the cambial zone cells did not decrease down the stem (Uggla *et al.*, 1998). Thus, the concept of decreasing longitudinal concentration gradients of IAA is not always valid.

7. Summary

A major role for polarly transported IAA in plants is to initiate primary vascular strands, to induce the vascular cambium, and maintain and organize cambial growth. This is achieved through its morphogenic properties, which are required for cell division, cell expansion and xylem- and phloem-element differentiation. Radial concentration gradients of IAA across the tissues developing into xylem and phloem are inferred to provide positional information. We suggest that this information controls the radial width of the developmental zones of division and expansion. The radial width of these zones is critical for determining the time each cambial derivative spends in each developmental phase.

We have presented evidence that the amount of IAA supplied by polar transport will influence the rate of xylem production through changes in IAA concentration and distribution pattern across the cambial tissues. We suggest that this control operates primarily by affecting the radial population of dividing cells (i.e., the size of the meristem) via changes in radial width of the IAA distribution, and secondarily, when the supply is large, by affecting the rate of cell cycling via changes in the IAA concentration in cambial zone cells. The potential of exogenous IAA applied to cambial tissues to modify many other aspects of cambial growth suggests that it has additional roles in the control of natural patterns in wood variation in trees. This remains, however, to be evaluated. The recent advance in isolating and characterizing components in the perception and signal transduction pathway of IAA (Abel and Guilfoyle, 1998; Hobbe, 1998; Leyser, 1998) has opened the prospect for progress in this area.

Acknowledgement

We wish to thank Dr C.H.A. Little for valuable comments.

References

Abe, H., Funada, R., Othani, J and Fukazawa K. Changes in the arrangement of microtubules and microfibrils in differentiating conifer tracheids during the expansion of cells. *Ann. Bot.* **75**: 305–310.

Abel, S. and Guilfoyle T.J. (1998) Aux/IAA proteins and auxin signal transduction. *Trends Plant Sci.* **3**: 205–207.

Aloni, R. (1980) Role of auxin and sucrose in the differentiation of sieve and tracheary elements in plant tissue cultures. *Planta* **150**: 255–263.

Aloni, R. (1987) Differentiation of vascular tissues. *Annu. Rev. Plant. Physiol.* **38**: 179–204.

Aloni, R. (1991) Wood formation in deciduous hardwood trees. In: *Physiology of Trees* (ed A.S. Raghavendra). John Wiley & Sons, Inc., pp.175–197.

Aloni, R. and Zimmermann, M.H. (1983) The control of vessel size and density along the plant axis. A new hypothesis. *Differentiation* **24**: 203–208.

Bandurski, R.S., Cohen, J.D., Slovin, J.P., and Reinecke, D.M. (1995) Auxin biosynthesis and metabolism. In: *Plant Hormones, Physiology, Biochemistry and Molecular Biology* (ed P.J. Davies). Kluwer Academic Publishers: Dordrecht, pp. 39–65.

Bannan, M.W. (1955) The vascular cambium and radial growth in *Thuja occidentalis* L. *Can. J. Bot.* **33**: 113–138.

Behnke, H.-D. and Schulz, A. (1980) Fine structure, pattern of division, and course of wound phloem in *Coleus blumei*. *Planta* **150**: 357–365.

Berlyn, G.P. (1982) Morphogenetic factors in wood formation and differentiation. In: *New Perspectives in Wood Anatomy* (ed P. Baas). Mertinus Nijhoff/Dr W Junk Publishers: The Haague, pp. 123–150.

Borkovec, V., Didehvar, F. and Baker, D.A. (1994) The biosynthesis and translocation of ^{14}C-IAA in *Ricinus communis*. *Plant Growth Regul.* **15**: 137–141.

Brown, A.B. and Cormack, R.G.H. (1937) Stimulation of cambial activity, locally in the region of application and at a distance in relation to a wound, by means of heteroauxin. *Can. J. Res. Sect. C* **15**: 433–441.

Brown, C.L. (1970) Physiology of wood formation in conifers. *Wood Sci.* **3**: 8–22.

Cambridge, A.P. and Morris, D.A. (1996) Transfer of exogenous auxin from the phloem to the polar auxin pathway in pea (*Pisum sativum* L.). *Planta* **199**: 583–588.

Crozier, A., Loferski, K., Zaerr, J.B. and Morris, R.O. (1980) Analysis of picogram quantities of indole-3-acetic acid by high performance liquid chromatography-fluorescence procedures. *Planta* **150**: 366–370.

Cui, K.-M. and Little, C.H.A. (1993) The effects of exogenous IAA and GAs on phloem and xylem production in *Pinus sylvestris* and *Picea abies*. *Chinese J. Bot.* **5**: 145–153.

Denne, M.P. and Wilson, J.E. (1977) Some quantitative effects of indoleacetic acid on the wood production and tracheid dimensions of *Picea*. *Planta* **134**: 223–228.

Digby, J. and Wareing, P.F. (1966) The effect of applied growth hormones on cambial division and the differentiation of the cambial derivatives. *Ann. Bot.* **30**: 539–548.

Dodd, R.S. and Fox, P. (1990) Kinetics of tracheid differentiation in Douglas-fir. *Ann. Bot.* **65**: 649–657.

Doley, D. and Leyton, L. (1968) Effects of growth regulating substances and water potential on the development of secondary xylem in *Fraxinus*. *New Phytol.* **67**: 579–594.

Edlund, A., Eklöf, S., Sundberg, B., Moritz, T., and Sandberg, G. (1995) A microscale technique for gas chromatography-mass spectrometry measurements of picogram amounts of indole-3-acetic acid in plant tissues. *Plant Physiol.* **108**: 1043–1047.

Egierszdorff, S. (1981) The role of auxin stored in Scots pine trunk during spring activation of cambial activity. *Biol. Plant.* **23**: 110–115.

Evert, R.F. and Kozlowski, T.T. (1967) Effect of isolation of bark on cambial activity and development of xylem and phloem in trembling aspen. *Amer. J. Bot.* **54**: 1045–1054.

Evert, R.F., Kozlowski, T.T. and Davis, J.D. (1972) Influence of phloem blockage on cambial growth of sugar maple. *Amer. J. Bot.* **59**: 632–641.

Fosket, D.E. (1968) Cell division and the differentiation of wound-vessel members in cultured stem segments of *Coleus*. *Proc. Natl. Acad. Sci.* **59**: 1089–1096.

Fosket, D.E. and Torrey, J.G. (1969) Hormonal control of cell proliferation and xylem differentiation in cultured tissues of *Glycine max.* var. *Biloxi*. *Plant Physiol.* **44**: 871–880.

Fukuda, H. and Komamine, A. (1980) Direct evidence for cytodifferentiation to tracheary elements without intervening mitosis in a culture of single cells isolated from the mesophyll of *Zinnea elegans*. *Plant Physiol.* **65**: 61–64.

Fukuda, H. and Komamine, A. (1981) Relationship between tracheary element differentiation and the cell cycle in single cells isolated from the mesophyll of *Zinnea elegans*. *Physiol. Plant.* **52**: 423–430.

Gouwentak, C.A. (1941) Cambial activity as dependent on the presence of growth hormone and the presence of non-resting conditions of stems. *Proc. Amsterdam Acad. Sci.* **44**: 654–663.

Gregory, R.A. (1971) Cambial activity in Alaskan white spruce. *Am. J. Bot.* **58**: 160–171.

Gregory, R.A. and Wilson, B.F. (1968) A comparison of cambial activity of white spruce in Alaska and New England. *Can. J. Bot.* **46**: 733–734.

Hejnowicz, A. and Tomaszewski, M. (1969) Growth regulators and wood formation in *Pinus silvestris*. *Physiol. Plant.* **22**: 984–992.

Hoad, G.V. (1995) Transport of hormones in the phloem of higher plants. *Plant Growth Regul.* **16**: 173–182.

Hoad, G.V., Hillman, S.K. and Wareing, P.F. (1971) Studies on the movement of indole auxins in willow. *Planta* **99**: 73–88.

Hobbey, L.J. (1998) Auxin: Molecular genetic approaches in *Arabidopsis*. *Plant Physiol. Biochem*. **36**: 91–102.

Holder, N. (1979) Positional information and pattern formation in plant morphogenesis and a mechanism for the involvement of plant hormones. *J. Theor. Biol.* **77**: 195–212.

Hollis, C.A. and Tepper, H.B. (1971) Auxin transport within intact dormant and active white ash shoots. *Plant Physiol.* **48**: 146–149.

Imagawa, H. and Ishida, S. (1972) Study on the wood formation in trees. Report III. Occurrence of the overwintering cells in cambial zone in several ring-porous trees. *Res. Bull. Hokkaido Univ. For.* **29**: 207–221.

Jacobs, W.P. (1984) Functions of hormones at tissue level of organization. In: *Hormonal Regulation of Development II, The Function of Hormones from the Level of the Cell to the Whole Plant, Encyclopedia of Plant Physiology, (New Series), Vol 2* (ed T.K. Scott). Springer: Berlin, pp. 149–171.

Jeffs, R.A. and Northcote, D.H. (1967) The influence of indol-3yl acetic acid and sugar on the pattern of induced differentiation in plant tissue culture. *J. Cell. Sci.* **2**: 77–88.

Klee, H.J., Horsch, R.B., Hinchee, M.A., Hein, M.B. and Hoffman, N.L. (1987) The effects of overproduction of two *Agrobacterium tumefaciens* T-DNA auxin biosynthetic gene products in transgenic petunia plants. *Genes Dev.* **1**: 86–96.

Kutshera, U. (1994) The current status of the acid-growth hypothesis. *New Phytol.* **126**: 549–569.

Lachaud, S. (1989) Participation of auxin and abscisic acid in the regulation of seasonal variation in cambial activity and xylogenesis. *Trees* **3**: 125–137.

Lachaud, S. and Bonnemain, J.L. (1984) Seasonal variations in the polar-transport pathways and retention sites of [^{3}H]indole-3-acetic acid in young branches of *Fagus sylvatica* L. *Planta* **161**: 207–215.

Larson, P.R. (1960) A physiological consideration of the springwood-summerwood transition in red pine. *For. Sci.* **6**: 110–122.

Larson, P.R. (1962) Auxin gradients and the regulation of cambial activity. In: *Tree Growth* (ed T.T. Kozlowski). The Ronald Press Company, New York, pp. 97–117.

Larson, P.R. (1969) Wood formation and the concept of wood quality. *Yale Univ.: Sch. For. Bull.* **74**: 1–54.

Larson, P.R. (1994) *The Vascular Cambium, Development and Structure*. Springer-Verlag: Berlin.

Leitch, M.A. and Savidge, R.A. (1995) Evidence for auxin regulation of bordered-pit positioning during tracheid differentiation in *Larix laricina*. *IAWA* **16**: 289–297.

Lev-Yadun, S. and Aloni, R. (1995) Differentiation of the ray system in woody plants. *Bot. Rev.* **61**: 49–88.

Leyser, O. (1998) Auxin signalling: protein stability as a versatile control target. *Curr. Biol.* **8**: R305–R307.

Little, C.H.A. (1981) Effect of cambial dormancy state on the transport of [1-^{14}C]indol-3-ylacetic acid in *Abies balsamea* shoots. *Can. J. Bot.* **59**: 342–348.

Little, C.H.A and Bonga, J.M. (1974) Rest in the cambium of *Abies balsamea*. *Can. J. Bot.* **52**: 1723–1730.

Little, C.H.A. and Wareing, P.F. (1981) Control of cambial activity and dormancy in *Picea sitchensis* by indole-3-ylacetic and abscisic acids. *Can. J. Bot.* **59**: 1480–1493.

Little, C.H.A. and Savidge, R.A. (1987) The role of plant growth regulators in forest tree cambial growth. *Plant Growth Regul.* **6**: 137–169.

Little, C.H.A. and Sundberg, B. (1991) Tracheid production in response to indole-3-acetic acid varies with internode age in *Pinus sylvestris* stems. *Trees* **5**: 101–106.

Little, C.H.A. and Pharis, R.P. (1995) Hormonal control of radial and longitudinal growth in the tree stem. In: *Plant Stems: Physiology and Functional Morphology* (ed Gartner, B.). Academic Press: San Diego, pp. 281–319.

Little, C.H.A., Sundberg, B. and Ericsson, A. (1990) Induction of acropetal ^{14}C-photosynthate transport and radial growth by indole-3-acetic acid in *Pinus sylvestris* shoots. *Tree Physiol.* **6**: 177–189.

Meichenheimer, R.D. and Larson, P.R. (1985) Exogenous auxin and N-1-naphthylphthalamic acid effects on *Populus deltoides* xylogenesis. *J. Exp. Bot.* **163**: 320–329.

Meinhardt, H. (1996) Models of biological pattern formation: common mechanism in plant and animal development. *Int. J. Dev. Biol.* **40**: 123–134.

Mellerowicz, E.J., Coleman, W.K., Riding, R.T. and Little, C.H.A. (1992) Periodicity of cambial activity in *Abies balsamea*. I. Effects of temperature and photoperiod on cambial dormancy and frost hardiness. *Physiol. Plant.* **85**: 515–525.

Morris, D.A. and Kadir, G.O. (1972) Pathway of auxin transport in the intact pea seedling (*Pisum sativum* L.). *Planta* **107**: 171–182.

Morris, D.A. and Thomas, A.G. (1978) A microautoradiographic study of auxin transport in the stem of intact pea seedling. *J. Exp. Bot.* **29**: 147–157.

Morris, D.A. and Small, D.K. (1990) Auxin transport and the regulation of petiole abscission in cotton (*Gossypium hirsutum* L.): A comparison of indol-3yl-acetic acid and phenylacetic acid. *Plant Growth Regul.* **9**: 201–214.

Nix, L.E. and Wodzicki, T.J. (1974) The radial distribution and metabolism of IAA-^{14}C in *Pinus echinata* stems in relation to wood formation. *Can. J. Bot.* **52**: 1349–1355.

Nix, L.E. and Brown, C.L. (1987) Cellular kinetics of compression wood formation in slash pine. *Wood and Fiber Science*. **19**: 126–134.

Odani, K. (1975) Transport of indoleacetic acid in dormant shoots of *Populus deltoides*. *J. Jap. For. Soc.* **57**: 144–147.

Odani, K. (1980) Effects of chilling on exogenous indole-3-acetic acid-mediated wood formation and vessel diameter of white ash and poplar. *Mokuzai Gakkaishi* **26**: 127–131.

Odani, K. (1985) Indole-3-acetic acid transport in pine shoots under the stage of true dormancy. *J. Jap. For. Soc.* **67**: 332–334.

Oribe, Y. and Kubo, T. (1997) Effect of heat on cambial reactivation during winter dormancy in evergreen and deciduous conifers. *Tree Physiol.* **17**: 81–87.

Phillips, R. and Arnott, S.M. (1983) Studies on induced tracheary element differentiation in cultured tissues of tubers of the Jerusalem artichoke, *Helianthus tuberosus*. *Histochemical J.* **15**: 427–436.

Porandowski, J., Rakowski, K., and Wodzicki, T.J. (1982) Apical control of xylem formation in the pine stem. II. Response of differentiating tracheids. *Acta Soc. Bot. Pol.* **51**: 203–214.

Ridoutt, B.G. and Sands, R. (1994) Quantification of the processes of secondary xylem fibre development in *Eucalyptus globulus* at two height levels. *IAWA J.* **15**: 417–424.

Rinne, P., Tuominen, H. and Sundberg, B. (1993) Growth patterns and endogenous indole-3-acetic acid concentrations in current-year coppice shoots and seedlings of two *Betula* species. *Physiol. Plant.* **88**: 403–412.

Roberts, L.W. (1988) Evidence from wound responses and tissue cultures In: *Vascular Differentiation and Plant Growth Regulators*. Springer, Berlin, pp. 63–88.

Romano, C.P., Hein, M.B. and Klee, H.J. (1991) Inactivation of auxin in tobacco transformed with the indoleacetic acid-lysine synthetase gene of *Pseudomonas savastanoi*. *Genes Dev*. **5**: 438–446.

Sachs, T. (1981) The controls of the patterned differentiation of vascular tissues. *Adv. Bot. Res.* **9**: 151–162.

Savidge, R.A. (1983) The role of plant hormones in higher plant cellular differentiation. II. Experiments with the vascular cambium, and sclereid and tracheid differentiation in the pine, *Pinus contorta*. *Histochemical J*. **15**: 447–466.

Savidge, R.A. and Wareing, P.F. (1981) Plant growth regulators and the differentiation of vascular elements. In: *Xylem Cell Development* (ed Barnett, J.R.). Tunbridge Wells: Castle House Publishing Ltd., pp. 192–235.

Savidge, R.A. and Wareing, P.F. (1982) Apparent auxin production and transport during winter in nongrowing pine tree. *Can. J. Bot*. **60**: 681–691.

Sheriff, D.W. (1983) Control by indole-3-acetic acid of wood production in *Pinus radiata* D. Don segments in culture. *Aust. J. Plant. Physiol*. **10**: 131–135.

Sitbon, F., Hennion, S., Sundberg, B., Little, C.H.A., Olsson, O. and Sandberg, G. (1992). Transgenic tobacco plants coexpressing the *Agrobacterium tumefaciens iaaM* and *iaaH* genes display altered growth and indoleacetic acid metabolism. *Plant Physiol.* **99**: 1062–1069.

Skene, D.S. (1969) The period of time taken by cambial derivatives to grow and differentiate into tracheids in *Pinus radiata* D. Don. *Ann. Bot*. **33**: 253–262.

Skene, D.S. (1972) The kinetics of tracheid development in *Tsuga canadensis* Carr. and its relation to tree vigour. *Ann. Bot*. **36**: 179–187.

Snow, R. (1935) Activation of cambial growth by pure hormones. *New Phytologist* **34**: 347–360.

Sundberg, B. and Little, C.H.A. (1987) Effect of defoliation on tracheid production and the level of indole-3-acetic acid in *Abies balsamea* shoots. *Physiol. Plant.* **71**: 430–435.

Sundberg, B. and Little, C.H.A. (1990) Tracheid production in response to changes in the internal level of indole-3-acetic acid in 1-year-old shoots of Scots pine. *Plant Physiol*. **94**: 1721–1727.

Sundberg, B. and Uggla, C. (1998) Origin and dynamics of indoleacetic acid under polar transport in *Pinus sylvestris*. *Physiol. Plant.* **104**: 22–29.

Sundberg, B., Little, C.H.A. and Cui, K. (1990) Distribution of indole-3-acetic acid and the occurrence of its alkali-labile conjugates in the extraxylary region of *Pinus sylvestris* stems. *Plant Physiol*. **93**: 1295–1302.

Sundberg, B., Tuominen, H. and Little, C.H.A. (1994) Effects of the indole-3-acetic acid (IAA) transport inhibitors *N*-1-naphthylphthalamic acid and morphactin on endogenous IAA dynamics in relation to compression wood formation in 1-year-old *Pinus sylvestris* (L.) shoots. *Plant Physiol.* **106**: 469–476.

Sundberg, B., Little, C.H.A., Riding, R.T. and Sandberg, G. (1987) Levels of endogenous indole-3-acetic acid in the vascular cambium region of *Abies balsamea* trees during the activity–rest–quiescence transition. *Physiol. Plant.* **71**: 163–170.

Sundberg, B., Little, C.H.A., Cui, K. and Sandberg, G. (1991) Level of endogenous indole-3-acetic acid in the stem of *Pinus sylvestris* in relation to the seasonal variation of cambial activity. *Plant, Cell and Environment* **14**: 241–246.

Sundberg, B., Ericsson, A., Little, C.H.A., Näsholm, T. and Gref, R. (1993) The relationship between crown size and ring width in *Pinus sylvestris* L. stems: dependence on indole-3-acetic acid, carbohydrates and nitrogen in the cambial region. *Tree Physiol*. **12**: 347–362.

Sussex, I.M., Clutter, M.E. and Goldsmith, M.H. (1972) Wound recovery by pith cell redifferentiation: structural changes. *Amer. J. Bot*. **59**: 797–804.

Suttle, J.C. (1991) Biochemical basis for the loss of basipetal IAA transport with advancing physiological age in etiolated *Helianthus* hypocotyls. *Plant Physiol*. **96**: 875–880.

Tuominen, H., Sitbon, F., Jacobsson, C., Sandberg, G., Olsson, O. and Sundberg, B. (1995) Altered growth and wood characteristics in transgenic hybrid aspen expressing *Agrobacterium tumefaciens* T-DNA indoleacetic acid-biosynthetic genes. *Plant Physiol.* **109**: 1179–1189.

Tuominen, H., Puech, L., Fink, S. and Sundberg, B. (1997) A radial gradient of indole-3-acetic acid is related to secondary xylem development in hybrid aspen. *Plant Physiol.* **115**: 577–585.

Uggla, C. (1998) *New Perspectives on the Role of Auxin in Wood Formation*. Ph.D. thesis. Acta Universitatis Agriculturae Sueciae, Silvestria 58, Umeå.

Uggla, C., Mellerowicz, E.J. and Sundberg, B. (1998) Indole-3-acetic acid controls cambial growth in Scots pine by positional signaling. *Plant Physiol.* **117**: 113–121.

Uggla, C., Moritz, T., Sandberg, G. and Sundberg, B. (1996) Auxin as a positional signal in pattern formation in plants. *Proc. Natl. Acad. Sci. USA* **93**: 9282–9286.

Wang, Q., Little, C.H.A. and Oden, P.C. (1997) Control of longitudinal and cambial growth by gibberellins and indole-3-acetic acid in current-year shoots of *Pinus sylvestris*. *Tree Physiol.* **17**: 715–721.

Wareing, P.F. (1958) The physiology of cambial activity. *J. Inst .Wood Sci.* **1**: 34–42.

Wareing, P.F., Hanney, C.E.A. and Digby, J. (1964) The role of endogenous hormones in cambial activity and xylem differentiation. In*: The Formation of Wood in Forest Trees* (ed. Zimmermann M.H.). Academic Press: New York, pp. 323–344.

Warren Wilson, J. and Warren Wilson, P.M. (1984) Control of tissue patterns in normal development and in regeneration. In*: Positional Controls in Plant Development* (eds Barlow, P. and Carr, D). Cambridge University Press: Cambridge, pp. 225–280.

Warren Wilson, J., Keys, W.M.S., Warren Wilson, P.M. and Roberts, L.W. (1994a) Effects of auxin on the spatial distribution of cell division and xylogenesis in lettuce pith explants. *Protoplasma* **183**: 162–181.

Warren Wilson, J., Roberts, L.W., Warren Wilson, P.M. and Gresshoff, P.M. (1994b) Stimulatory and inhibitory effects of sucrose concentration on xylogenesis in lettuce pith explants; possible mediation by ethylene biosynthesis. *Ann. Bot.* **73**: 65–73.

Wetmore, R.H. and Rier, J.P. (1963) Experimental induction of vascular tissues in callus of angiosperms. *Am. J. Bot.* **50**: 418–430.

Whitmore, F.W. and Zahner, R. (1966) Development of the xylem ring in stems of young red pine trees. *For. Sci.* **12**: 198–210.

Wignall, T.A. and Browning, G. (1988) Epicormic bud development in *Quercus rubor* L. Studies of endogenous IAA, ABA, IAA polar transport and water potential in cambial tissues and effects of exogenous hormones on bud outgrowth from stem explants. *J. Exp. Bot.* **39**: 1667–1678.

Wilson, B.F. (1964) A model for cell production by the cambium of conifers. In: *The Formation of Wood in Forest Trees* (ed. Zimmermann, M.H.). Academic Press: New York, pp. 19–36.

Wilson, B.F. and Howard, R.A. (1968) A computer model for cambial activity. *For. Sci.* **14**: 77–90.

Wilson, B.F., Wodzicki, T.J. and Zahner, R. (1966) Differentiation of cambial derivatives: proposed terminology. *For. Sci.* **12**: 438–440.

Wodzicki, T.J. (1971) Mechanisms of xylem differentiation in *Pinus silvestris* L. *J. Exp. Bot.* **22**: 670–687.

Wodzicki, T.J. and Zajaczkowski, S. (1974) Effect of auxin on xylem tracheid differentiation in decapitated stems of *Pinus sylvestris* L. and its interaction with some vitamins and growth regulators. *Acta Soc. Bot. Pol.* **43**: 129–148.

Wolpert, L. (1981) Positional information and pattern formation. *Phil. Trans. R. Soc. Lond. B Biol. Sci.* **295**: 441–450.

Wolpert, L. (1996) One hundred years of positional information. *Trends Genet.* **12**: 359–364.

Yoshinaga, A., Fujita, M and Saiki, H. (1997) Secondary wall thickening and lignification of oak xylem components during latewood formation. *Mokuzai Gakkaishi* **43**: 377–383.

Zajaczkowski, S. (1973) Auxin stimulation of cambial activity in *Pinus sylvestris* L. The differential cambial response. *Physiol. Plant.* **29**: 281–287.

Zakrzewski, J. (1983) Hormonal control of cambial activity and vessel differentiation in *Quercus rubur*. *Physiol. Plant.* **57**: 537–542.

Zakrzewski, J. (1991) Effect of indole-3-acetic acid (IAA) and sucrose on vessel size and density in isolated stem segments of oak (*Quercus robur*). *Physiol Plant.* **81**: 234–238.

Zamski, E. and Wareing, P.F. (1974) Vertical and radial movement of auxin in young sycamore plants. *New Phytol.* **73**: 61–69.

Zobel, B.J. and Buijtenen, J.P. (1989). *Wood Variation: Its Causes and Control*. Springer-Verlag: Berlin.

14

UDPG:coniferyl alcohol glucosyltransferase and coniferin biosynthesis — a regulatory link to seasonal cambial growth in conifers

H. Förster, V. Steeves, U. Pommer and R.A. Savidge

1. Introduction

E-coniferin, the 'coniferin' present in the cambium of conifers, is the 4-*O*-β-D-glucopyranoside of *trans*-coniferyl alcohol. Coniferin has long been known to be a major cambial metabolite in the Coniferophyta, and it has also been reported in woody angiosperms (e.g., Terazawa *et al.*, 1984) including Liliopsida (i.e., monocot) species (Nan *et al.*, 1997). Early research in Germany demonstrated that when the coniferyl alcohol moiety of coniferin was radiolabelled, the radioactivity was readily incorporated into lignin of spruce (Freudenberg, 1964). Freudenberg (1964) considered that coniferin served as a 'reservoir' of lignin precursor (i.e., of coniferyl alcohol) because, unlike coniferyl alcohol, the glucoside was found not to be susceptible to oxidation by enzymes of lignification. In a similar vein, Neish (1964) speculated that 'if for some reason there is an overproduction such as may occur in spruce cambium, I would imagine that coniferyl alcohol is stabilized as a glucoside, perhaps by reaction with UDP-glucose.'

In his initial investigations into the biosynthesis of coniferin, Ibrahim (1977) again emphasized that during coniferin formation 'lignin precursors' were being glucosylated. There is no question today that coniferyl alcohol as the aglycone can be utilized as a lignin precursor (Lewis and Sarkanen, 1998), and when coniferin is hydrolysed by a coniferin-specific β-glucosidase, the resulting aglycone has been shown to be coniferyl alcohol (Leinhos *et al.*, 1994). Thus, it is not surprising that the physiological role of coniferin has been considered almost exclusively in relation to the known role

Cell & Molecular Biology of Wood Formation, edited by R. Savidge, J. Barnett and R. Napier.

of coniferyl alcohol in lignification. There have been many reviews which address that concept, including those by Grima-Pettenati and Goffner (1999), Baucher *et al.* (1998) and Whetten *et al.* (1998).

Acknowledging that coniferin does have a role in supplying coniferyl alcohol for lignification, the primary focus of this chapter nevertheless is on what coniferin biosynthesis can tell us about the regulation of the seasonal cycle of growth and dormancy in conifers, of which lignification is only one aspect.

2. Coniferin biosynthesis and metabolism

2.1 *Background on CAGT*

Coniferin is only one of a great many glucosides which are produced in plants as both aliphatic and aromatic compounds (Harborne, 1964; Hopkinson, 1969). Plant glucosides arise through the activities of various glucosyltransferases (E.C. 2.4.1), enzymes which catalyse transfer of glucose from a sugar nucleoside diphosphate to the corresponding aglycone. The only enzyme known to catalyse *E*-coniferin biosynthesis is uridine 5′-diphosphoglucose:coniferyl alcohol glucosyl transferase (CAGT). Only very modest CAGT activities have been detected in plants low on the evolutionary scale (e.g., Bryophyta and Polypodiophyta), and much higher activities have been found in most Pinophyta and a few woody Magnoliophyta (Ibrahim, 1977).

Ibrahim and Grisebach (1976) published the first report on purification and characterization of a CAGT. Despite the fact that CAGT is most readily available from conifers, the enzyme was first isolated from suspension cultures of Paul's scarlet rose and found to catalyse the transfer of glucose from UDP glucose (UDPG) to coniferyl alcohol, yielding the highest substrate specificity reported to date for a CAGT ($K_m = 3.3 \times 10^{-9}$ M for coniferyl alcohol; $K_m = 2 \times 10^{-5}$ M for UDPG).

Schmid and Grisebach (1982) subsequently purified and characterized CAGT from *Picea abies*. Immunofluorescence detection in young *P. abies* seedlings showed the enzyme being localized to, predominantly, the epidermis, subepidermis, and vascular bundles. Intracellularly, the enzyme was found in the parietal cytoplasmic layer (Schmid *et al.*, 1982). CAGT was subsequently investigated in several conifer species in greater detail (Förster and Savidge, 1995; Savidge and Förster, 1998). Despite the investigations of Ibrahim and subsequent reports, CAGT evidently has yet to be assigned a formal E.C. number.

As indicated in *Figure 1*, a coniferin-specific β-glucosidase exists in the cambial region, delivering oxidizable monolignols by hydrolysing cinnamyl glucosides (Marcinowski and Grisebach, 1978; Burmeister and Hösel, 1981; Leinhos *et al.*, 1994; Dharmawardhana *et al.*, 1995, 1999). This coniferin-specific glucosidase isolated from *Pinus banksiana* (Leinhos *et al.*, 1994) and *P. contorta* (Dharmawardhana *et al.*, 1995) was localized to developing xylem. Nevertheless, coniferin-specific β-glucosidase isolated from lignifying tissues was found only in low abundance (Leinhos *et al.*, 1994) and its catalytic activity evidently is not competitive with that of CAGT found in the same tissues, at the same developmental stage (Savidge and Förster, 1998). A cDNA encoding a glucosidase of *Pinus contorta* has recently been isolated and proposed to be a target gene for overall reduction in lignin deposition (Dharmawardhana *et al.*, 1999).

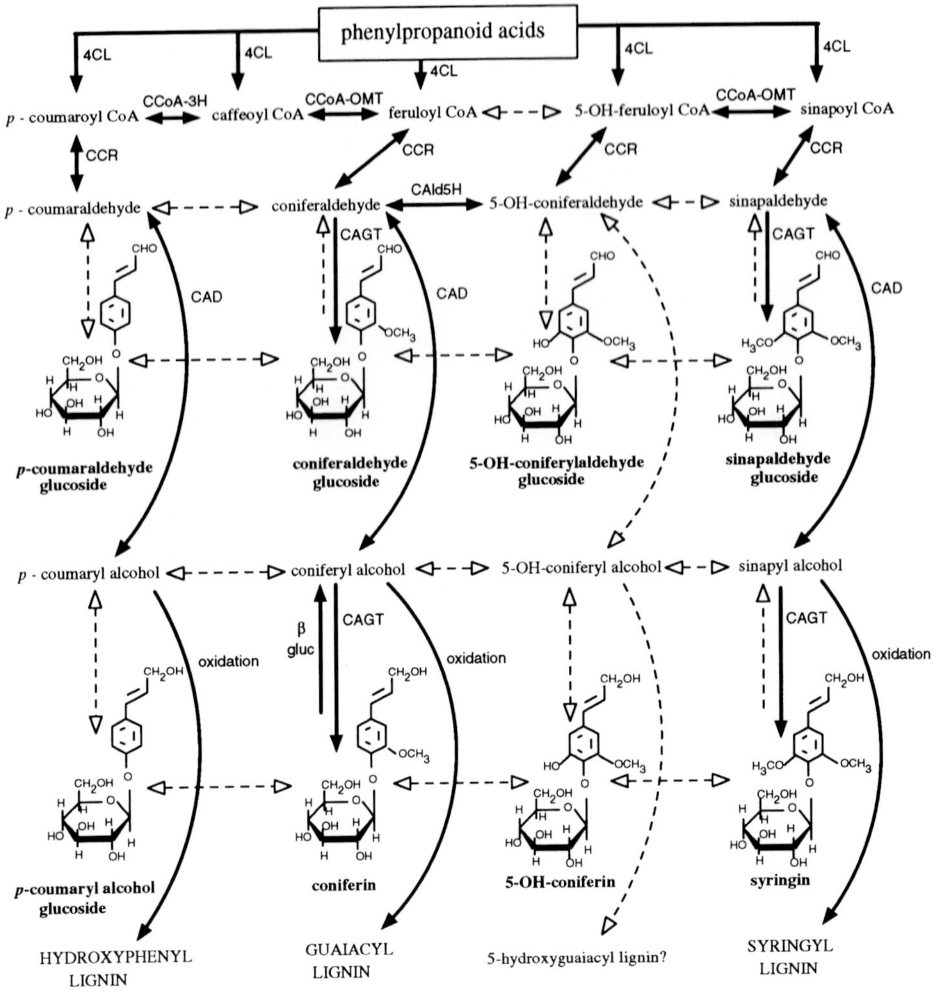

Figure 1. *Pre-lignification metabolism of monolignol and monolignol glucosides (indicated by bold-faced labels). Solid arrows: catalysis has been demonstrated in vitro. Dashed arrows: circumstantial evidence exists for the indicated metabolic step but catalysis remains to be demonstrated in vitro. 4CL = 4-coumarate coenzyme A ligase; CCoA-3H=coumaroyl coenzyme A 3-hydroxylase; CCoA-OMT=coumaroyl coenzyme A O-methyltransferase; CCR=cinnamoyl coenzyme A reductase; CAld5H=coniferyl aldehyde 5-hydroxylase; CAD=coniferyl alcohol dehydrogenase; CAGT=uridine 5′-diphosphoglucose:coniferyl alcohol glucosyltransferase; β-gluc=β-glucosidase; oxidation=oxidative enzymes, for example, peroxidase, laccase, coniferyl alcohol oxidases.*

2.2 *Seasonal variation in endogenous coniferin*

In conifer species, formation of the active cambial zone (CZ) followed by a zone interior to the CZ of radially expanded primary-walled cambial derivatives (RE zone) requires several weeks, and only after both active CZ and RE zones have been generated does the initiation of wood formation and attending onset of lignification occur (Savidge and Wareing, 1981, 1984; Savidge *et al.*, 1998).

In a phenological study of cambial reactivation, Savidge (1988) found HPLC evidence for a biochemical indicator of impending tracheid differentiation in cambial extracts of *Pinus contorta*. The indicator metabolite was isolated and characterized as *E*-coniferin (Savidge, 1989). Quantitative investigation of endogenous indol-3-y lacetic acid, sucrose and coniferin in the cambium of *Larix laricina* throughout the course of a year revealed that, of the three, only coniferin content varied parallel to seasonal cambial activity (Savidge, 1991). In both CZ and developing xylem of *Pinus banksiana*, endogenous coniferin content oscillated in relation to the annual cycle of growth and dormancy (Savidge and Förster, 1998; Savidge *et al.*, 1998).

In springtime, coniferin first became detectable in the CZ in association with resumption of cell-division, indicating that coniferin biosynthesis occurred much earlier than the onset of lignification (Savidge, 1988, 1989, 1991; Fukushima *et al.*, 1997). Coniferin in the CZ (as isolated by bark peeling) remained present only during the period of active cambial growth, disappearing from the CZ in autumn prior to completion of lignification in the latewood of the annual ring (Savidge, 1988, 1989, 1991). In *Larix, Picea and Pinus* spp., coniferin was not detectable in dormant cambium by HPLC, gas chromatography, or selected-ion-monitoring gas chromatography-mass spectroscopy (Savidge, 1988, 1989, 1991).

2.3 *Seasonal variation in CAGT activity*

Seasonal changes in metabolic processes are frequently observed (e.g., Terazawa and Miyake, 1984; Savidge, 1991), but poorly understood. Our springtime investigations into *Pinus banksiana* indicated that both CAGT activity and the accumulation within the CZ of a detectable amount of endogenous coniferin are associated — both temporally and spatially — with resumption of cell-division, rather than with lignification (Förster and Savidge, 1995; Savidge and Förster, 1998; Savidge *et al.*, 1998). Similarly, late summer and autumn investigations indicated that CAGT activity is somehow connected to ongoing mitotic/cytokinetic activity, as CAGT activity in the CZ disappeared with the onset of dormancy despite ongoing lignification and maturation of the latewood elements (Förster and Savidge, 1995; Savidge and Förster, 1998). Thus, the annual cycle of cambial growth and dormancy in conifers appears to be linked to the production of the metabolite coniferin, and the evidence indicates that this qualitative relationship is due to the expression of the corresponding glucosyltransferase. To our knowledge, CAGT is the first enzyme to be qualitatively linked to the annual cycle of growth and dormancy in perennial woody plants (Savidge and Förster, 1998). The fact that both coniferin and CAGT activity are present in the CZ only when the cambium is active points to a probable control mechanism for seasonal cambial activity at the level of gene expression (Savidge, 1996).

2.4 *Physiological roles of cinnamyl alcohol glucosides*

Cinnamyl alcohol glucosides (*p*-coumaryl alcohol glucoside, coniferin and syringin) have frequently been isolated from different plants (for a survey, see Förster *et al.*, 1999). Despite the fact that lignin is the second most abundant organic substance on earth, after cellulose, monolignols and their corresponding glucosides have only been found in small quantities in the Magnoliophyta. Cell suspensions of *Linum flavum* are a noteworthy exception, yielding coniferin at 5.5 mmol g^{-1} (van Uden *et al.*, 1990). In conifers, concentrations of endogenous coniferin as high as 10 mmol g^{-1} and 9 mmol g^{-1} were reported in *P. banksiana* (Savidge and Förster, 1998) and *Larix laricina* saplings, respectively (Savidge, 1991). Large quantities (3.7 mmol g^{-1}) were also obtained from cambial sap of spruce (Schmid and Grisebach, 1982). Coniferin content in the cambium of conifers varies depending on when during the growing season the cambium is harvested (Savidge, 1988, 1989, 1991; Förster and Savidge, 1995; Savidge and Förster, 1998).

Compounds in addition to lignin can be synthesized from monolignol glucosides and may have biological activity in plants. Addition of coniferin stimulated the production of podophyllotoxin, a cytotoxic lignan (Woerdenbag *et al.*, 1990) and coniferin served as a precursor to lignans in the cambial region of *Tsuga heterophylla* (Krahmer *et al.*, 1970). A correlative behaviour between the levels of lignans and cinnamyl alcohol glucosides was observed before and after hatching of gall larvae in *Picea glauca* (Kraus and Spiteller, 1997). The promotion of cell division by dihydroconiferyl alcohol glucosides to initiate rapidly-growing crown gall tumours in *Vinca rosea* (Binns *et al.*, 1987; Lynn *et al.*, 1987) is especially remarkable, since the stimulation of cell growth and differentiation is usually attributed to phytohormones. A putative defence reaction is illustrated by the adaptive oscillation of Z-coniferin in *Fagus sylvatica* in response to attack by beech scale (Dübeler *et al.*, 1997). Z-Isoconiferin and z-syringin were reported to induce wall shedding in ascospores of xylariaceae (Chapela *et al.*, 1991) Coniferin was also described as an inducer of *vir*-gene activities in *Agrobacterium tumefaciens* causing crown gall disease (Morris and Morris, 1990; Delay *et al.*, 1994; Nan *et al.*, 1997).

CAGT, in catalysing formation of *E*-coniferin and other cinnamyl alcohol glucosides (see below) evidently competes with coniferyl alcohol dehydrogenase and monolignol oxidizing enzymes for coniferyl alcohol (*Figure 1*). Coniferin is effective in inhibiting coumarate-CoA-ligase (Voo *et al.*, 1995) and cinnamyl alcohol dehydrogenase (CAD; O'Malley *et al.*, 1992), the latter being considered a core enzyme of lignification. Our data (see below) indicate that the glucosyltransferase must also compete with CAD for available cinnamyl aldehydes by forming the corresponding aldehydic glucosides (Savidge and Förster, 1998; Förster *et al.*, 1999). As illustrated in *Figure 1*, the substrate competition possible between CAGT and other enzymes in the lignification pathway very probably diminishes precursor flow into lignification. Thus, it is apparent that monolignol glucosides could fulfil a role in regulating lignin biosynthesis.

Despite the focus on lignification in regard to coniferin, CAGT has not been considered as an enzyme capable of influencing either content or composition of lignin. There is evidence that syringyl lignin can arise from coniferyl aldehyde (Osakabe *et al.*, 1999); however, Matsui *et al.* (1996) also reported interconversion of labelled coniferin and syringin (the 4-*O*-β-D-glucoside of sinapyl alcohol) in *Magnolia kobus* by transformation of the corresponding aglycones. As *Figure 1* indicates, the guaiacyl:syringyl ratio of lignin can be modified through the activities of enzymes forming and hydrolysing coniferin and syringin (Piumi *et al.*, 1998; Förster *et al.*, 1999). Repeating the ideas of Freudenberg (1964), Grima-Pettenati and Goffner (1999) suggested that coordinated transport, storage and mobilization of the glucosidic precursors may be rate limiting steps in lignification. CAGT activity may also be part of the explanation for the aldehyde:alcohol ratio in guaicyl lignin, because CAGT competes with CAD for both coniferyl aldehyde and coniferyl alcohol (Savidge and Förster, 1998; *see also Figures 2* and *3*).

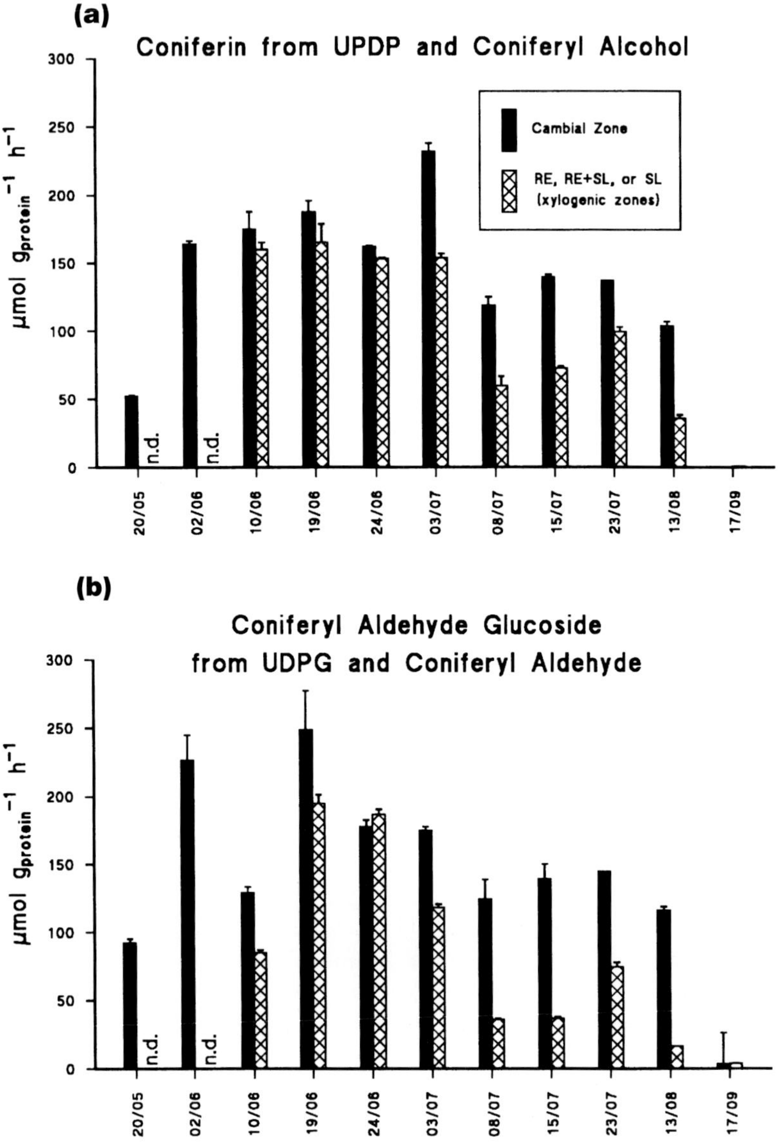

Figure 2. In vitro *production (1 h, 36°C) of coniferin (a) and the glucoside of coniferyl aldehyde (b) by partially purified enzyme preparations obtained from the cambial zone and zones of xylogenesis. The 13–14-year-old main-stem region of 20-year-old* Pinus strobus *was assayed on the several successive dates (day/month) indicated under the X axis. Product quantification was by reversed-phase HPLC (Savidge, 1988). n.d.=activity was not determined. Error bars are standard errors of mean* (n=3).

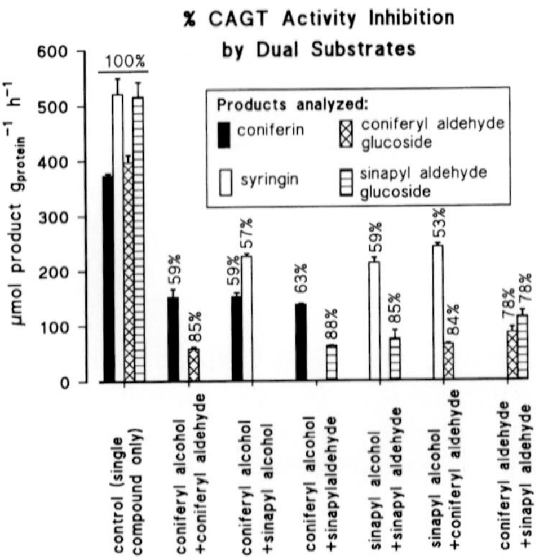

Figure 3. *Per cent reduction of* in vitro *CAGT activity by dual substrates, relative to controls receiving only a single substrate. All of the substrates were provided at a concentration of 1 mmol* g^{-1}protein. *Incubations (1 h, 36°C) were done using partially purified CAGT from the 14-year-old cambial zone of a 20-year-old* Pinus strobus *harvested on June 19th. Error bars are standard errors of mean* (n=*3*).

3. UDPG:coniferyl alcohol glucosyltransferase activity in *Pinus strobus*

3.1 *Material and methods*

Enzyme. *P. strobus* trees were harvested on dates throughout the growth season, and CAGT was extracted from tissues (cambial zone, CZ; zone of radial expansion, RE; and zone of secondary wall formation and lignification, SL) separated by bark peeling. The CZ and xylogenic zones (RE, RE+SL, or SL) were scraped individually into liquid nitrogen. Enzyme isolation, purification and enzymatic assays were carried out according to Savidge and Förster (1998).

Product analysis. Aliquots of the enzyme assay were chromatographed by HPLC on a C18 ODS Ultrasphere column (Savidge, 1988; Förster and Savidge, 1995; Savidge and Förster, 1998). Coniferin peaks were detected at 262 nm and confirmed by co-chromatography using authentic *E*-coniferin crystallized from *P. strobus* (Savidge, 1989). Coniferyl aldehyde glucoside and sinapyl aldehyde glucoside peaks were quantified at 332 nm and 320 nm, respectively, and co-chromatographed with authentic references isolated and confirmed by ^{1}H NMR, as described by Förster *et al.* (1999). Coniferin and syringin were collected from the HPLC, lyophilized, derivatized with Sylon TP and analysed in scanning mode by GC-MS, using conditions as previously described by Savidge (1989).

Microscopy. Before extraction and HPLC analysis, handcut sections of cambium and developing xylem were examined by microscopy to ascertain the phenological stage of the trees being investigated.

3.2 *Results*

Seasonal coniferin content and CAGT activity. In agreement with the previous finding of a direct correlation between seasonal CAGT activity and cambial growth in *P.*

banksiana (Savidge and Förster, 1998), a similar relationship was found in *P. strobus* (*Figure 2a*). CAGT activity appeared in the CZ at the time when cell division activity was commencing for another growing season, about three weeks before the first produced cambial derivatives began developing into earlywood xylem. The major differences in coniferin content between developing xylem and CZ observed in *P. banksiana* (Savidge and Förster, 1998) were not found in *P. strobus*. Whereas more than 15 times higher CAGT activity was present in developing xylem compared to CZ of *P. banksiana*, the same two zones extracted from *P. strobus* exhibited more or less comparable catalytic rates of CAGT over the growing period (*Figure 2a*).

Specificity of CAGT. The CAGT isolated from *P. strobus* exhibited a pronounced specificity for uridine-5-diphosphate glucose (*Table 1*). None of the other glucose nucleotides tested could be efficiently substituted for UDPG, although trace activities were detected with GDPG and CDPG in the case of coniferin formation, and with TDPG for coniferyl aldehyde glucoside formation. In agreement with results reported previously (Ibrahim and Grisebach, 1976; Ibrahim, 1977; Schmid and Grisebach, 1982), UDP-glucose was the preferred sugar donor for the glucosylation of cinnamyl alcohols. It should be mentioned, however, that in cambial sap of spruce TDP-glucose exhibited 17 % activity in comparison to UDPG (Schmid and Grisebach, 1982).

Unexpectedly, CAGT purified from *P. strobus* expressed high activity when UDP-galactose was provided to the enzyme (*Table 1*). Cinnamyl alcohols showed better turnover rates than cinnamyl aldehydes. It remains to be clarified whether the sugar moiety was inserted as glucose or galactosel; UDP-glucose-4-epimerase activity in the enzyme preparation cannot be excluded (Dalessandro and Northcote, 1977). This possibility may be resolvable using UDP-xylose, a highly efficient inhibitor of the epimerase (Pazchowski and Wojciechowski, 1994).

Cinnamyl aldehydes were reported as suitable substrates for CAGT (Ibrahim and Grisebach, 1976; Schmid and Grisebach, 1982) and *in vitro* led to the formation of the corresponding 4-*O*-β-D-glucosides (*Table 2*, see also Förster *et al.*, 1999). Unexpectedly, CAGT from *P. strobus* produced higher amounts of cinnamyl aldehyde glucosides than monolignol glucosides although the specific turnover rate (i.e., ratio between administered substrate and obtained product) was about 8 times lower (*Figure 2b*). Nevertheless,

Table 1. *Relative contributions of sugar nucleotides to coniferin biosynthesis* in vitro

Relative activity[a] (%)	Coniferin	Syringin	Coniferyl aldehyde glucoside	Sinapyl aldehyde glucoside
		Sugar nucleotides		
UDP-glucose	100.00	100.00	100.00	100.00
GDP-glucose	1.83	0.00	0.00	0.00
TDP-glucose	0.00	0.00	2.57	0.00
CDP-glucose	1.81	0.00	0.00	0.00
UDP-galactose	90.00	89.00	48.00	48.00

[a] Activities were relative to 100% for UDP-glucose.

the *in vitro* generation of coniferyl aldehyde glucoside revealed a similar seasonal pattern to that observed for coniferin (compare *Figures 2a* and *2b*).

The substrate specificity of CAGT is summarized in *Table 2*. The observed substrate specificities of these partially purified enzymes generally agree, but there are also some exceptional differences. Notably, there are remarkably higher rates of glucosylation of cinnamic acids in angiosperms (*Forsythia* and *Rosa* spp.) compared to conifers. In general, aldehydes are suitable substrates for CAGT, evidently even in the case of syringaldehyde (however, the structure of the glucosidic product has not yet been investigated). Sinapyl alcohol was utilized by CAGT in *P. strobus* cambium to an even higher extent than coniferyl alcohol, but it was used much less effectively in *P. abies* (*Table 2*).

Inhibition of CAGT activity. Cinnamyl alcohols and cinnamyl aldehydes were in competition with each other for CAGT from the cambial zone (*Figure 3*). All of the substrates depicted in *Figure 3* were applied as 1 mmol g^{-1} solutions. Coniferyl alcohol and sinapyl alcohol administered together diminished the formation of coniferin and syringin to 41% and 43%, respectively. Only 22% of the original turnover was reached when the cinnamyl aldehydes were incubated together. In a mixture of cinnamyl alcohols and cinnamyl aldehydes, formation of cinnamyl aldehyde glucosides was

Table 2. *CAGT activity* in vitro *with a selection of aglycones*

Substrate	Activity[a] (%) relative to coniferyl alcohol				
	Pinus strobus	*Pinus*[b] *banksiana*	*Forsythia*[c] *ovata*	*Paul's*[d] *scarlet rose*	*Picea*[e] *abies*
Cinnamyl alcohols and -aldehydes					
coniferyl alcohol	100.00	100.00	100.00	100.00	100.00
sinapyl alcohol	139.43	96.00	82.00	74.00	14.00
p-coumaryl alcohol	n.d.	0.00	0.00	<1	7.00
coniferaldehyde	106.26	n.d.	n.d.	46.00	63.00
sinapaldehyde	138.02	n.d.	n.d.	n.d.	48.00
Phenolics					
luteolin	3.14	n.d.	n.d.	<1	2.00
apigenin	2.05	n.d.	n.d.	n.d.	4.00
naringenin	1.08	n.d.	n.d.	n.d.	8.00
trans-cinnamic acid	1.41	0.00	n.d.	n.d.	n.d.
p-coumaric acid	0.27	0.00	0.00	0.00	3.00
caffeic acid	0.49	0.00	n.d.	18.00	2.00
ferulic acid	2.38	0.00	18.00	20.00	4.00
sinapic acid	1.35	0.00	14.00	21.00	3.00
2-OH-benzoic acid	5.08	n.d.	0.00	<1	n.d.
syringaldehyde	11.41	n.d.	n.d.	n.d.	n.d.
mandelic acid	0.00	0.00	n.d.	n.d.	n.d.
vanillin	2.76	0.00	n.d.	n.d.	n.d.
vanillic acid	0.00	0.00	0.00	<1	14.00

[a]Activities were determined by calculation of the amount of UDPG incorporation into each substrate.
[b] Savidge and Förster (1998). [c] Ibrahim (1977). [d] Ibrahim and Grisebach (1976). [e] Schmid and Grisebach (1982). n.d. = not determined

more strongly affected (between 12% and 16% of the control) than that of cinnamyl alcohol glucosides (between 37% and 41%). However, it made almost no difference when coniferyl aldehyde or sinapyl aldehyde were incubated together with coniferyl or sinapyl alcohol (*Figure 3*).

Coniferin itself was reported to inhibit CAGT in a noncompetitive manner with product inhibition starting at about 10 mmol g^{-1} (Schmid and Grisebach, 1982). However, coniferin added to CAGT of *P. banksiana* showed no inhibitory effect up to 10 mmol g^{-1} coniferin (Förster and Savidge, 1995). Thus, strong inhibition of CAGT by the monolignols and their corresponding aldehydes at substrate concentrations ten times lower indicates efficient interference. Our *in vitro* results clearly demonstrate the possibility for *P. strobus* CAGT to catalyse 'sidestep' reactions along the lignification pathway.

4. Conclusions and outlook

CAGT, catalysing formation of glucosides of both cinnamyl alcohols and aldehydes, has previously been considered merely as an enzyme catalysing production of a storage reserve of monolignols in support of lignification. To date, there is no evidence that glucosides such as coniferin are susceptible to oxidative polymerization or other kinds of metabolism, excepting hydrolysis; thus, if coniferin is only a storage reserve, there is no obvious reason for it not continuing to reside in the stem during dormancy. However, both coniferin and the activity of CAGT vanish from the cambial zone upon its entry into dormancy, which occurs well before the completion of lignification in the finally maturing latewood. Both coniferin accumulation and CAGT activity reappear in the cambial zone in springtime well before either lignification or any other aspect of earlywood formation are initiated. Thus, a growing body of evidence indicates that coniferin and the gene expression underlying CAGT activity, although capable of contributing to lignification, must be more fundamentally linked to the overall control of seasonal cambial growth in conifers. The physiological regulation of seasonal cambial growth is a longstanding, unanswered question and, to our knowledge, no other metabolites nor products of gene expression have been observed to have the qualitatively perfect spatio-temporal relation to the annual cycle of growth and dormancy that are displayed by coniferin and CAGT, respectively. Thus, continuing research into coniferin and CAGT would appear to be justified as a means of providing insight into the control of seasonal cambial growth.

Acknowledgements

This research was supported by the Ministry of Science, Research and Culture of Thuringia. R.A.S. acknowledges NSERC research funding. We would like to thank Ms Ingrid Beuthel and Ms Simone Burger for excellent technical support.

References

Baucher, M., Monties, B., VanMontagu, M. and Boerjan, W. (1998) Biosynthesis and genetic engineering of lignin. *Crit. Rev. Plant Sci.* **17:** 125–197.

Burmeister, G. and Hösel, W. (1981) Immunohistochemical localization of β-glucosidases in lignin and isoflavone metabolism in *Cicer arietinum* L. seedlings. *Planta* **152:** 578–586.

Binns, A.N., Chen, R.H., Wood, H.N. and Lynn, D.G. (1987) Cell division promoting activity of naturally occurring dehydrodiconiferyl glucosides: Do cell wall components control cell division? *Proc. Natl. Acad. Sci. USA* **84**: 980–984.

Chapela, I.H., Petrini, O. and Hagmann (1991) Monolignol glucosides as specific recognition messengers in fungus–plant symbioses. *Phys. Mol. Plant Pathol.* **39**: 289–298.

Dalessandro, G. and Northcote, D.H. (1977) Changes in enzymic activities of nucleoside diphosphate sugar interconversions during differentiation of cambium to xylem in sycamore and poplar. *Biochem. J.* **162**: 267–279.

Delay, D., Dyé, F., Wisniewski, J-P. and Delmotte, F. (1994) Synthesis and *Agrobacterium vir*-inducing activities of coniferyl alcohol β-glucosides. *Phytochemistry* **36:** 289–298.

Dharmawardhana, D.P., Ellis, B.E. and Carlson, J.E. (1995) A β-glucosidase from lodgepole pine xylem specific for the lignin precursor coniferin. *Plant Physiol.* **107:** 331–339.

Dharmawardhana, D.P., Ellis, B.E. and Carlson, J.E. (1999) cDNA cloning and heterologous expression of coniferin β-glucosidase. *Plant Molec. Biol.* **40**: 365–372.

Dübeler, A., Voltmer, G., Gora, V., Lunderstädt, J. and Zeeck, A. (1997) Phenols from *Fagus sylvatica* and their role in defence against *Cryptococcus fagisuda*. *Phytochemistry* **45:** 51–57.

Förster, H. and Savidge, R.A. (1995) Characterization and seasonal activity of UDPG: coniferyl alcohol glucosyltransferase linked to cambial growth in jackpine. *Proc. Plant Growth Reg. Soc. Am.* **22**: 402–407.

Förster, H., Pommer, U. and Savidge, R.A. (1999) Metabolic activity of uridine-5′-diphospho-glucose: cinnamyl alcohol glucosyltransferase as an intrinsic indicator of cambial growth in conifers. In: *Plant Polyphenols 2: Chemistry, Biology, Pharmacology, Ecology* (eds. G.G. Gross, R.W. Hemingway and T. Yoshida). Kluwer Academic/Plenum Publ.: New York, Boston, Dordrecht, London, Moscow, pp. 371–392.

Freudenberg, K. (1964) The formation of lignin in the tissue and *in vitro*. In: *The Formation of Wood in Forest Trees* (ed. M.H. Zimmermann). Academic Press: New York, pp. 203–218.

Fukushima, K., Taguchi, S., Matsui, N. and Yasuda, S. (1997) Distribution and seasonal changes of monolignol glucosides in *Pinus thunbergii. Mokuzai Gakkaishi* **43**: 254–259.

Grima-Pettenati, J. and Goffner, D. (1999) Lignin genetic engineering revisited. *Plant Sci.* **145**: 61–65.

Harborne, J.B., ed. (1964) *Biochemistry of Phenolic Compounds*. Academic Press: N.Y.

Hopkinson, S.M. (1969) The chemistry and biochemistry of phenolic glycosides. *Q. Rev. Chem. Soc.* **23**: 98–124.

Ibrahim, R.K. (1977) Glucosylation of lignin precursors by uridine diphosphate glucose: coniferyl alcohol glucosyltransferase in higher plants. *Z. Pflanzenphysiol.* **85:** 253–262.

Ibrahim, R.K. and Grisebach, H. (1976) Purification and properties of UDP-glucose: coniferyl alcohol glucosyltransferase from suspension cultures of Paul's scarlet rose. *Arch. Biochem. Biophys.* **176:** 700–708.

Krahmer, R.L., Hemingway, R.W. and Hillis, W.E. (1970) The cellular distribution of lignans in *Tsuga heterophylla* wood. *Wood Sci. Tech.* **4**: 122–139.

Kraus, C. and Spiteller, G. (1997) Comparison of phenolic compounds from galls and shoots of *Picea glauca*. *Phytochemistry* **44:** 59–67.

Leinhos, V., Udagama-Randeniya, P.V. and Savidge, R.A. (1994) Purification of an acidic coniferin-hydrolysing β-glucosidase from developing xylem of *Pinus banksiana. Phytochemistry* **37:** 311–315.

Lewis, N.G. and Sarkanen, S. (1998) *Lignin and Lignan Biosynthesis*. ACS Symposium Series 697, American Chemical Society, Washington, DC.

Lynn, D.G., Chen, R.H., Manning, K.S. and Wood, H.N. (1987) The structural characterization of endogenous factors from *Vinca rosea* crown gall tumors that promote cell division of tobacco cells. *Proc. Natl. Acad. Sci. USA* **84:** 615–619.

Marcinowski, S. and Grisebach, H. (1978) Enzymology of lignification. Cell-wall-bound β-glucosidase for coniferin from spruce (*Picea abies*) seedlings. *Eur. J. Biochem.* **87:** 37–44.

Matsui, N., Fukushima, K., Yasuda, S. and Terashima, N. (1996) On the behavior of monolignol glucosides in lignin biosynthesis. III. Synthesis of variously labelled coniferin and incorporation of the label into syringin in the shoot of *Magnolia kobus*. *Holzforschung* **50:** 408–412.

Morris, J.W. and Morris, R.O. (1990) Identification of an *Agrobacterium tumefaciens* virulence gene inducer from the pinaceous gymnosperm *Pseudotsuga menziesii*. *Proc. Natl. Acad. Sci. USA* **87:** 3614–3618.

Nan, G.L., Tang, C.S., Kuehnle, A.R. and Kado, C.I. (1997) *Dendrobium* orchids contain an inducer of *Agrobacterium* virulence genes. *Physiol. Molec. Pl. Path.* **51:** 391–399.

Neish, A.C. (1964) Cinnamic acid derivatives as intermediates in the biosynthesis of lignin and related compounds. In: *The Formation of Wood in Forest Trees* (ed. M.H. Zimmermann). Academic Press: New York, pp. 219–242.

Osakabe, K., Tsao, C.C., Laigeng, L., Popko, J.L., Umezawa, T., Carraway, D.T., Smeltzer, R.H., Joshi, C.P. and Chiang, V.L. (1999) Coniferyl aldehyde 5-hydroxylation and methylation direct syringyl lignin biosynthesis in angiosperms. *Proc. Natl. Acad. Sci. USA* **96:** 8955–8960.

O'Malley, D.M., Porter, S. and Sederoff, R.R. (1992) Purification, characterization, and cloning of cinnamyl alcohol dehydrogenase in loblolly pine (*Pinus taeda* L.). *Plant Physiol.* **98:** 1364–1371.

Paczkowski, C. and Wojciechowski, Z.A. (1994) Glucosylation and galactosylation of diosgenin and solasodine by soluble glucosyltransferase(s) from *Solanum melongena* leaves. *Phytochemistry* **35:** 1429–1434.

Piumi, L., Daubresse, N., Francesch, Ch., Rolando, Ch., Jouanin, L., Lapierre, C. and Pollet, B. (1998) A new synthesis of labelled coniferin and analogues. Application to metabolism studies in plants. In: *Polyphenols Communication 98* (eds. F. Charbonnier, J.-M. Delacotte and C. Rolando). Lille, pp. 435–436.

Savidge, R.A. (1988) A biochemical indicator of commitment to tracheid differentiation in *Pinus contorta*. *Can. J. Bot.* **66:** 2009–2012.

Savidge, R.A. (1989) Coniferin, a biochemical indicator of commitment to tracheid differentiation in conifers. *Can. J. Bot.* **67:** 2663–2668.

Savidge, R.A. (1991) Seasonal cambial activity in *Larix laricina* saplings in relation to endogenous indol-3-ylacetic acid, sucrose and coniferin. *Forest Sci.* **37:** 953–958.

Savidge, R.A. (1996) Xylogenesis, genetic and environmental regulation — a review. *IAWA Journal* **17:** 269–310.

Savidge, R.A., Udagama-Randeniya, P.V., Xu, Y., Leinhos, V. and Förster, H. (1998) Coniferyl alcohol oxidase: a new enzyme involved with lignification. In: *Lignin and Lignan Biosynthesis* (eds. Lewis, N.G. and Sarkanen, S.). ACS Symposium Series 697, American Chemical Society, Washington, DC, pp. 109–130.

Savidge, R.A. and Förster, H. (1998) Seasonal activity of uridine-5′-diphosphoglucose: coniferyl alcohol glucosyltransferase in relation to cambial growth and dormancy in conifers. *Can. J. Bot.* **76:** 486–493.

Savidge, R.A. and Wareing, P.F. (1981) Plant growth regulators and the differentiation of vascular elements. In: *Xylem Cell Development* (ed. J.R. Barnett). Castle House Publ.: UK, pp. 192–235.

Savidge, R.A. and Wareing, P.F. (1984) Seasonal cambial activity and xylem development in *Pinus contorta* in relation to endogenous indol-3-ylacetic and (*S*)-abscisic acid levels. *Can. J. Forest Res.* **14:** 676–682.

Schmid, G. and Grisebach, H. (1982) Enzymatic synthesis of lignin precursors. Purification and properties of UDPglucose: coniferyl alcohol glucosyltransferase from cambial sap of spruce (*Picea abies* L.). *Eur. J. Biochem.* **123:** 363–370.

Schmid, G., Hammer, D.K., Ritterbusch, A. and Grisebach, H. (1982) Appearance and immunohistochemical localization of UDP-glucose: coniferyl alcohol glucosyltransferase in spruce (*Picea abies* (L.) Karst.) seedlings. *Planta.* **156:** 207–212.

Terazawa, M. and Miyake, M. (1984) Phenolic compounds in living tissue of woods. II. Seasonal variations of phenolic glycosides in the cambial sap of woods. *Mokuzai Gakkaishi* **30:** 329–334.

Terazawa, M., Okuyama, H. and Miyake, M. (1984) Isolation of coniferin and syringin from the cambial tissue and inner-bark of some angiospermous woods. *Mokuzai Gakkaishi* **30:** 409–412.

van Uden, W., Pras, N., Batterman, S., Visser, J.F. and Malingré, T.M. (1990) The accumulation and isolation of coniferin from a high-producing cell suspension of *Linum flavum* L. *Planta* **183:** 25–30.

Voo, K.S., Whetten, R.W., O'Malley, D.M. and Sederoff, R.R. (1995) 4-Coumarate-coenzyme-A ligase from loblolly pine xylem — isolation, characterization, and complementary–DNA cloning. *Plant Physiol.* **108:** 85–97.

Whetten, R.W., MacKay, J.J. and Sederoff, R.R. (1998) Recent advances in understanding lignin biosynthesis. *Annu. Rev. Plant Physiol. Plant Mol. Biol.* **49**: 585–609.

Woerdenbag, H.J., van Uden, W., Frijlink, H.W., Lerk, C.F., Pras, N. and Malingré, T.M. (1990) Increased podophyllotoxin production in *Podophyllum hexandrum* cell suspension cultures after feeding coniferyl alcohol as a β-cyclodextrin complex. *Plant Cell Rep.* **9:** 97–100.

15

Isolation of cambium-specific genes from *Eucalyptus globulus* Labill

G. Bossinger and M.A. Leitch

1. Introduction

The structure of the plant body is dependent on meristem cell division and on the differentiation of tissues that derive from such meristems. We have developed systems which appear to be particularly promising for the analysis of genes involved in xylogenesis, featuring wood being produced *in vitro* (see also Leitch and Savidge, this book). Our general approach to wood-gene and gene construct analysis *in vitro* is depicted in *Figure 1*. In this *in vitro* system with *Eucalyptus globulus* Labill., wood formation proceeds from cambial cells through radial expansion to fibre, vessel and ray-cell differentiation, lignification and secondary wall formation. On average more than 80 fibre cells are produced per radial file within a period of two to three months in culture. The system is responsive to *Agrobacterium*-mediated gene transfer (data not shown) and can be used for the analysis of genes involved in xylogenesis such as those whose identification we describe here.

2. Methods and results

2.1 *Targeting cambium specific sequences*

Random cDNA sequencing approaches have recently been used to analyse the molecular basis of xylem formation in both pine and poplar (Allona *et al.*, 1998; Sterky *et al.*, 1998), yielding 1097 and 5692 expressed sequence tags (ESTs) respectively. Based on the determination of putative gene function by sequence homology, and through the identification of sequence clusters, some deductions can be made as to the level of expression of certain genes. However, the distinction between housekeeping genes and others which specifically control cambial differentiation and wood formation and

Cell & Molecular Biology of Wood Formation, edited by R. Savidge, J. Barnett and R. Napier.

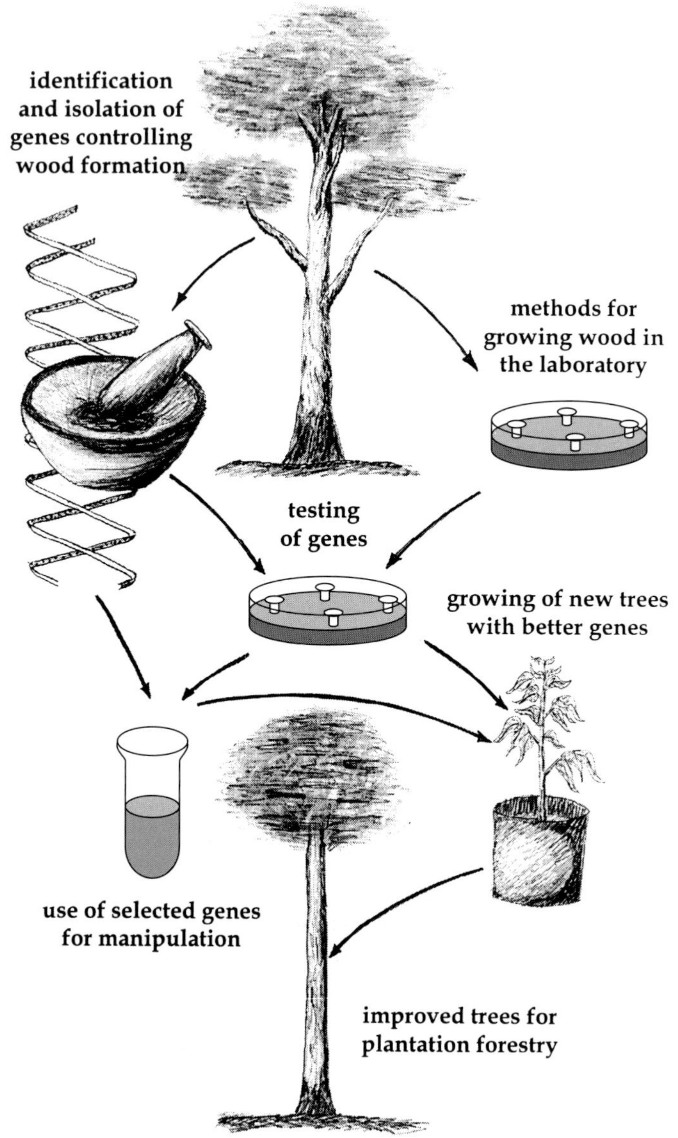

Figure 1. *Wood-gene and gene construct analysis* in vitro.

which are differentially expressed in the relevant tissue becomes more blurred the higher the number of ESTs that have to be analysed.

In an attempt to identify genes that specifically control cambial differentiation and wood formation, we isolated mRNA and prepared cDNAs from cambial tissue of 3-year-old eucalypt trees (*Eucalyptus globulus* Labill.) at 40% stem height. The bark was removed and scrapings were taken from both the exposed xylogenic tissue and from the inside of the bark to ensure representation of cambial initials while avoiding contamination with phloem tissue. To minimize the more ubiquitously expressed housekeeping genes, we subtracted from our cambial expressed sequences cDNAs prepared from mature leaves of the same trees, using a cDNA subtraction system (CLONTECH PCR-Select™) with subsequent suppression PCR (Siebert *et al.*, 1995). The resulting preparation was considerably enriched for gene fragments differentially expressed in woody tissue; therefore, they are likely candidates for identifying the control of cambial differentiation and wood formation. With the same method, rare and abundant fragments will have been equalized in their level of representation.

2.2 *Analysis of cambial cDNA fragments*

Subtracted and amplified cDNA fragments were cloned and 117 cambial cDNA inserts were sequenced (Table 1). After removal of ribosomal sequences and sequences smaller than 50 bases, our analysis following BLASTX (Altschul *et al.*, 1990) and BLASTN searches of GenBank non-redundant databases revealed that most cDNA fragments (50; 54%) were novel sequences with no significant similarity to any protein or DNA sequence in the databases. This 54% compares to 12% found in non-subtracted poplar EST experiments (Sterky *et al.*, 1998). Forty-three (46%) of our sequences showed homology to other ESTs or known genes with potential involvement in cell elongation and cell wall construction, lignin biosynthesis, carbon supply for cellulose, cell proliferation, amino-acid biosynthesis and others (*Table 1*). Details on some genes to which homology was demonstrated can be found in Okazawa *et al.* (1993), Nishitani (1995), Lee *et al.* (1997), Liu *et al.* (1994), Kiefermeyer *et al.* (1996), Craig *et al.* (1999), Rasmussen and Means (1987), Van Doorsselaere *et al.* (1993), Zegzouti *et al.* (1999) and many others. All sequences were submitted to GenBank and have been allocated accession numbers AW191301 to AW191393.

Respective patterns of expression were determined by dot-blot hybridization, identifying strongly and differentially expressed cDNA fragments as compared to cDNAs containing poly(A)-tails (as controls) and those representing ribosomal RNA and, therefore, giving signals with both cambial and leaf probes (*Figure 2*). The presumably cambium-specific cDNA fragments are now being used for the isolation of full length cDNAs and genomic clones, as well as for the identification of cambium-specific promoter sequences. The actual function of these genes, however, in many cases still has to be elucidated by overexpression and anti-sense experiments in suitable model systems.

3. Summary

To better understand the molecular basis of wood formation we isolated tissue-specific cambial sequences using a subtractive approach with subsequent suppression PCR. This approach yielded a number of novel cambial expressed cDNA fragments, as well

Table 1. *Subtracted cambium-minus-leaf cDNAs*

Inserts representing	No.	Homology[a] Strong	Homology[a] Marginal	Strong diff. expression
Known genes	43	37	6	5
Novel sequences	50			3
Ribosomal RNA	16			
<50 bases	4			
Primer only	2			
Others	2			
Total	117			8

[a] Homology	P-value	Score	%-Identity
Strong	0.005	100	40
Marginal	0.010	80	30

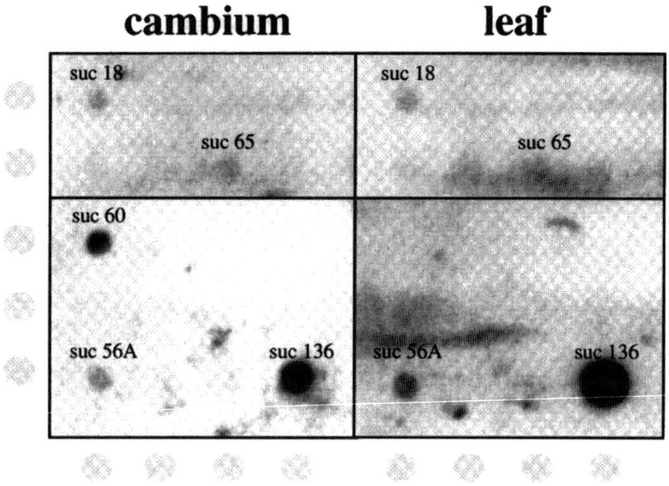

Figure 2. *Dot-blot analysis of subtracted cambial cDNAs. Individual cDNAs were hybridized with ^{32}P-dATP-labelled total ss-cDNA prepared from cambial mRNA ('cambium') and mature leaf mRNA ('leaf') respectively. Strong cambium-specific expression can be observed for Suc60. Suc56A and Suc136 contain Poly(A)-tails and were used as controls. Suc18 apparently has escaped successful subtraction while Suc65 represents a ribosomal RNA fragment.*

as fragments of many known genes that are involved at various stages in the wood formation process. For some of those, strong cambium-specific expression could be established.

References

Allona, I., Quinn, M., Shoop, E., Swope, K., St. Cyr, S., Carlis, J., Riedl, J., Retzel, E., Campbell, M.M., Sederoff, R. and Whetten, R.W. (1998) Analysis of xylem formation in pine by cDNA sequencing. *Proc. Natl. Acad. Sci.* **95**: 9693–9698.

Altschul, S.F., Gish, W., Miller, W., Myers, E.W. and Lipman, D.J. (1990) Basic local alignment search tool. *J. Mol. Biol.* **215**: 403–410.

Craig, J., Barratt, P., Tatge, H., Dejardin, A., Handley, L., Gardner, C.D., Barber, L., Wang, T., Hedley, C., Martin, C. and Smith, A.M. (1999) Mutations at the *rug4* locus alter the carbon and nitrogen metabolism of pea plants through an effect on sucrose synthase. *Plant J.* **17**: 353–362.

Kiefermeyer, M.C., Gomord, V., O'Connell, A., Halpin, C. and Faye, L. (1996) Cloning and sequence analysis of laccase-encoding cDNA clones from tobacco. *Gene* **178**: 205–207.

Lee, D., Meyer, K., Chapple, C. and Douglas, C.J. (1997) Antisense suppression of 4-coumarate:coenzyme A ligase activity in *Arabidopsis* leads to altered lignin subunit composition. *Plant Cell* **9**: 1985–1988.

Liu, L., Dean, J.F.D., Friedman, W.E. and Eriksson, K.E.L. (1994) A laccase-like phenoloxidase is correlated with lignin biosynthesis in *Zinnia elegans* stem tissues. *Plant J.* **6**: 213–224.

Nishitani, K. (1995) Endo-xyloglucan transferase, a new class of transferase involved in cell wall construction. *J. Plant Res.* **108**: 137–148.

Okazawa, K., Sato, Y., Nakagawa, T., Asada, K., Kato, I., Tomita, E. and Nishitani, K. (1993) Molecular cloning and cDNA sequencing of endoxyloglucan transferase, a novel class of glycosyltransferase that mediates molecular grafting between matrix polysaccharides in plant cell walls. *J. Biol. Chem.* **268**: 25364–25368.

Rasmussen, C.D. and Means, A.R. (1987) Calmodulin-regulation of cell proliferation. *EMBO J.* **6**: 3961–3968.

Siebert, P.D., Chenchik, A., Kellogg, D.E., Lukyanov, K.A. and Lukyanov, S.A. (1995) An improved method for walking in uncloned genomic DNA. *Nucleic Acids Res.* **23**: 1087–1088.

Sterky, F., Regan, S., Karlsson, J., Hertzberg, M., Rohde, A., Holmberg, A., Amini, B., Bhalerao, R., Larsson, M., Villarroel, R., Van Montagu, M., Sandberg, G., Olsson, O., Teeri, T.T., Boerjan, W., Gustafsson, P., Uhlén, M., Sundberg, B. and Lundeberg, J. (1998) Gene discovery in the wood-forming tissues of poplar: analysis of 5692 Expressed Sequence Tags. *Proc. Natl. Acad. Sci. USA* **95**: 13 330–13 335.

Van Doorsselaere, J., Gielen, J., Van Montagu, M. and Inze, D. (1993) A cDNA encoding S-adenosyl-L-methionine synthetase from poplar. *Plant Physiol.* **102**: 1365–1366.

Zegzouti, H., Jones, B., Frasse, P., Marty, C., Maitre, B., Latche, A., Pech, J.C. and Bouzayen, M. (1999) Ethylene-regulated gene expression in tomato fruit: characterization of novel ethylene-responsive and ripening-related genes isolated by differential display. *Plant J.* **18**: 589–600.

16

Cambial activity and xylem differentiation in walnut (*Juglans* sp.)

P. Label, I. Beritognolo, P. Burtin, L. Dehon, I. Couée, C. Breton, J.-P. Charpentier and C. Jay-Allemand

1. Introduction

Wood has many uses in construction, furnishing or pulp and paper manufacture and 'wood quality' remains a big issue for the wood market. Wood quality can be defined as those wood properties of importance to industry. One major difficulty of research in this field is the considerable difference in time-scales between industrial needs and tree development. At any given time, industry may adapt rapidly to some characteristics of the wood resource, but improving those characteristics is possible only over the long period of time which is necessary to grow and produce new trees (Larson, 1969). For industrial uses such as construction and furniture, general requirements might include accelerated tree radial growth; homogenization of wood in terms of annual growth ring increment and wood density; and increased natural wood durability (avoiding chemical treatments) are important. Translating these expectations into biological terms enables the drawing up of basic research guidelines for forest science research. Walnut wood has potential in this way, because of its industrial importance as a high value-added product.

On one hand, radial growth is directly related to cambial activity and increased radial growth could be achieved by modifying cambial activity (Larson, 1994). On the other hand, wood mechanical properties are directly related to wood anatomy and xylem structure. The key point would be then to understand the control of xylem differentiation into different cell types: fibres, vessels, parenchyma (Roberts, 1976; Barnett, 1981; Roberts *et al.*, 1988). Finally, natural wood durability and colour are obviously related to heartwood formation involving tyloses and deposition of secondary metabolites such as phenolic compounds (Hillis, 1987).

Cell & Molecular Biology of Wood Formation, edited by R. Savidge, J. Barnett and R. Napier.

Wood formation is a physiological continuum where cambial cells divide, elongate and differentiate into young xylem to constitute sapwood tisues. After some years, sapwood differentiates into heartwood. Gaining basic knowledge on these aspects of wood formation is a primary goal of research at INRA-Orléans, and started early in 1997. Up to now the work has been focused on biochemistry and molecular physiology of wood formation in walnut (*Juglans* sp.), and on cambial activity, xylem differentiation and heartwood formation. Our aim is to increase the understanding of the making of walnut wood and to gain information for the wood-chain industry.

2. Cambial activity

Cambium is the site of cambial activity producing phloem outwards and xylem inwards (Butterfield, 1975). Here, we will consider only xylem cell production which involves cell division and the beginning of cell elongation (Larson, 1994). Cell division is a final stage of cell cycle control which is active during interphase (John, 1996). After cell division, primary cell wall synthesis occurs before secondary wall formation (Barnett, 1981). At the same time, determination events commit cells to differentiation to fibres, vessels and parenchyma cells (Roberts, 1976; Higuchi, 1997).

To gain understanding of the xylem cell production process, we have developed two different approaches. The aim of both strategies is to identify genes involved in the control of radial growth. A neutral strategy consisted in analysing the global gene expression during meristematic activity of the cambium, and involved the construction of a cDNA library from cambial tissues. Another strategy is being developed, where known biological functions are targeted and genes involved in these functions studied. This is illustrated by the cloning of the *cdc2a* gene from hybrid walnut (*Juglans nigra* × *J. regia*) and the study of its expression pattern during radial growth.

3. Construction of a cDNA library from active cambium

Considering the cambium as a meristem leads us to address a simple question: what are the differences of gene expression between cambium and other meristems in the tree? In a first approach, the construction of a cDNA library from the cambial zone of hybrid walnut is a neutral strategy by which we intend to screen differentially expressed genes. The analysis of a cDNA library obtained from hybrid walnut cambial zone is under way and general characteristics of the library can already be described.

Tissue was sampled during radial growth two weeks after full leaf expansion. Stem pieces were frozen in liquid nitrogen and lyophilized under vacuum at –70°C. After 'bark' peeling, the 'bark' inner face was scraped producing a white pulverulent powder containing cambial zone enriched tissue (*Figure 1*), carefully avoiding collection of coloured phloem tissues. Polyadenylated RNA (poly(A+) RNA) was purified from 0.2 g DW of cambial zone tissue by extraction and isolation using magnetic Dynabeads (L. Duroux and C. Breton, pers. comm.). Synthesis of first strand and second strand complementary DNA was done using Stratagene's ZAP cDNA Synthesis Kit. Library construction into Uni-ZAP XR vector was performed using Stratagene's ZAP-cDNA Gigapack III Gold Cloning Kit. Titration of the primary library and amplification of the Uni-ZAP XR library were developped following Stratagene's recommendations. Primary library titre was 1.2×10^5 pfu ml^{-1} at 99% insert efficiency and amplified library titre was 4.0×10^{10} pfu ml^{-1} at 93% insert efficiency. To estimate insert size a specific amplification was done by PCR using T3–T7 primers (sequences of which border the insert position in the vector) (*Figure* 2). Taking into account the 145 bp size increase due to vector DNA amplification within T3–T7 primers, the estimated mean insert size of the library is 800 bp and 90% of inserts range from 300 bp to 1500 bp. This library is now to be screened against cDNAs from other active meristems of the tree in order to select genes differentially expressed in the cambium.

For other trees species cDNA libraries, obtained from cambial zone or differentiating xylem, were used as Expressed Sequence Tags (EST) databases (http://www.ncbi.nlm.nih.gov/dbEST/dbESTsummary.html): *Populus tremula* × *Populus tremuloides* (4809 ESTs), *Pinus taeda* (1953 ESTs), *Populus balsamifera* subsp. *trichocarpa* (883 ESTs), *Malus domestica* (430 ESTs), *Pseudotsuga menziesii* (87 ESTs), *Pinus radiata* (67 ESTs), *Picea abies* (38 ESTs). An EST project is a development intended for walnut cambium cDNA library.

4. Cloning of genes involved in cell cycle control

Target genes involved in control of the cell cycle (Riabowol *et al.*, 1989; Nurse, 1990; De Veylder *et al.*, 1998; Dudits *et al.*, 1998) can be identified from published data.

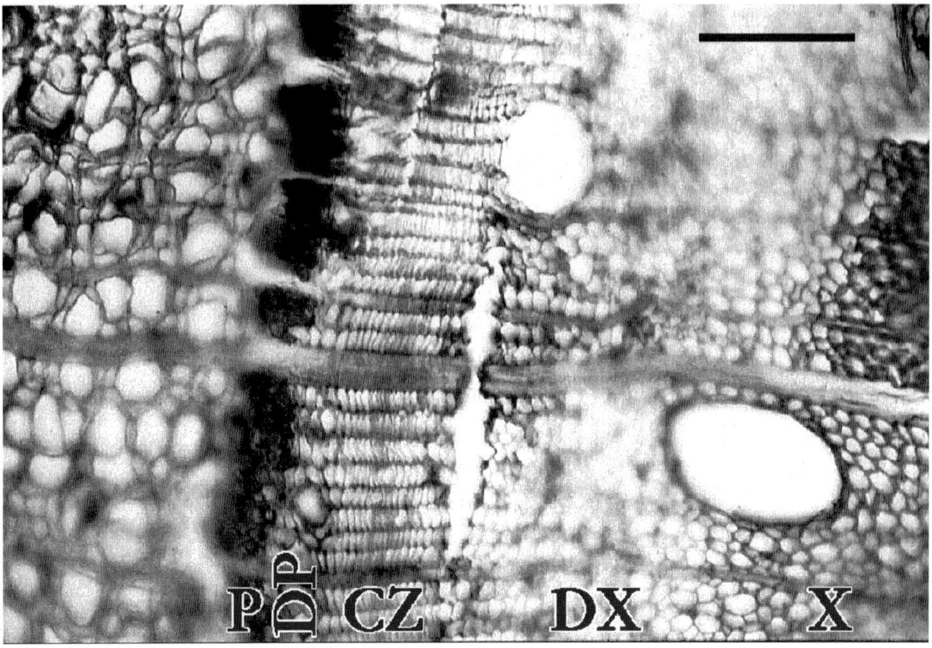

Figure 1. *Transverse section of a* Juglans nigra *stem. Sectioning at a thickness of 25 μm was performed on fresh stem tissues collected in July. Sections were observed without staining under a light microscope. P, phloem; DP, differentiating phloem; CZ, cambial zone; DX, differentiating xylem; X, xylem. The picture shows the separation zone between CZ and DX during 'bark' peeling procedure. Scraping inside the 'bark' until coloured phloem (P) appears enables the collection of cambial zone enriched tissues. Bar represents 0.05 mm.*

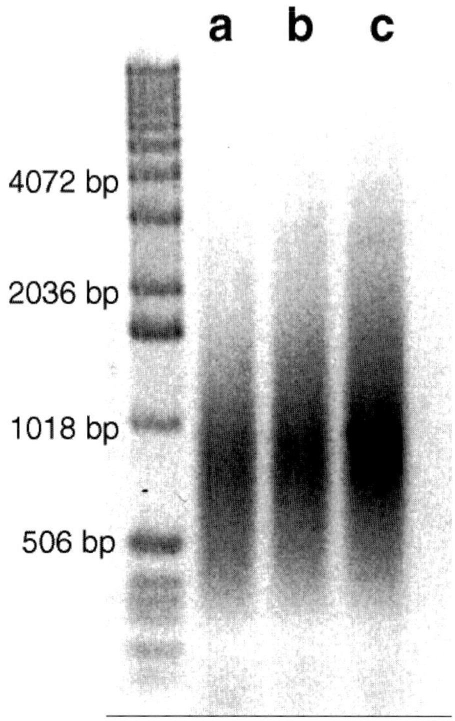

Figure 2. Electrophoretic analysis of Uni-ZAP ZR cambial zone hybrid walnut cDNA library amplified by PCR using T3–T7 primers. Electrophoresis was performed on 1.2% agarose gel at 10 V cm^{-1}. Lanes a, b and c are respective dilutions by 1/100, 1/10 and no dilution of the library. The gel was stained using ethidium bromide and scanned on Hitachi's FMBIO II analyser.

Amongst those genes, one coding for a cyclin-dependent kinase (*cdc2a*) has been isolated in several species (Dudits et al., 1998). We postulate that this gene could be a marker of cell division rate and thus could be used to determine the period of cambial activity at the molecular level. Our first goal is to isolate homologous sequences of such molecular markers of cambial activity from hybrid walnut. A 597 bp sequence of *cdc2a* gene has been obtained through reverse transcriptase-PCR (RT-PCR), and its expression pattern during radial growth of hybrid walnut was studied. *Figure 3* illustrates the first results on the study of the relationship between radial growth and *cdc2a* purified mRNA levels for hybrid walnut during the growing season.

Results show increasing levels of *cdc2a* mRNA concomitant with radial growth that suggests *cdc2a* could be a marker of cambial activity in hybrid walnut. From this point, the work now aims at determining the full sequence of *cdc2a* cDNA. In a second phase, the genome organization of cdc2 genes will be approached by analysing a hybrid walnut together with its parental genotypes.

5. Xylem differentiation

After xylem mother cell division, xylem cells enter differentiation leading to three main cell types: fibres, vessels and parenchyma. Fibres are formed by a cell enlargement phase and a phase of secondary wall formation (cellulose deposition and lignification) (Boudet et al., 1995) followed by cell autolysis (Mittler and Lam, 1995; Wang *et al.*, 1996). Vessel elements are also differentiated by cell autolysis and lignification. Parenchyma cells are the only longitudinal xylem cells remaining alive for

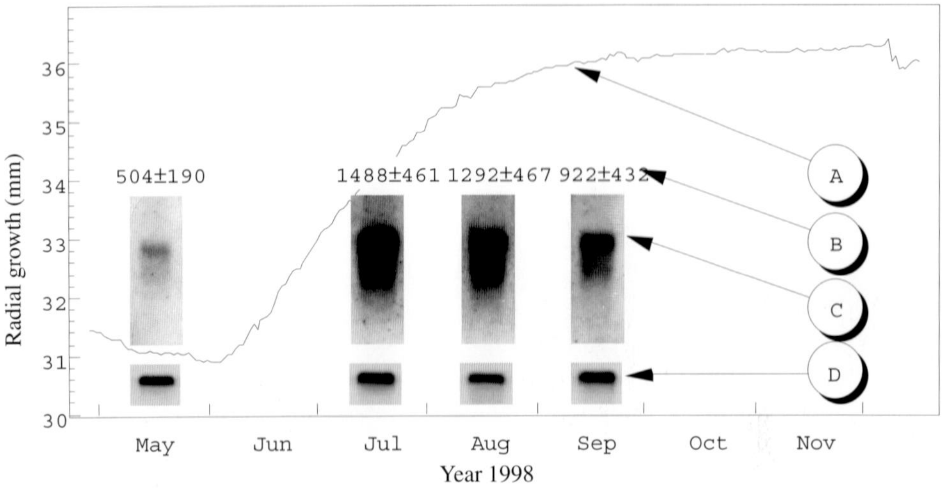

Figure 3. *Expression of the* cdc2a *gene during radial growth of hybrid walnut.* **A**: *Daily measurements of radial growth in mm during year 1998;* **B**: *yield in total RNA extracted from the cambial zone reported in micrograms per gram dry weight of tissue;* **C**: *expression pattern of* cdc2a *mRNA;* **D** *internal control of mRNA amounts by constitutive aquaporin expression level. Radial growth* (**A**) *was measured by an automated system mounted on the tree so the stem's increasing diameter mechanically impacts voltage of a variable resistor connected as a voltage divider. The voltage tension was digitized by an analogue-to-digital converter plugged into a PC-based data recording system. Total RNA* (**B**) *was measured spectrophotometrically by the absorbance at 260 nm and purity of the nucleic acid samples was checked by the 260 nm/280 nm absorbance ratio (Sambrook* et al., *1989). mRNAs were purified from 10µg total RNA using magnetic Dynabeads according to Dynal's recommendations. Poly(A+)-RNA (including mRNA) were run on 1% agarose gel in denaturing conditions (Sambrook* et al., *1989), analysed by northern blot and revealed using digoxigenin-labelled homologous probes (Boehringer Mannheim) either using a homologous probe against cdc2a* (**C**) *or probe against an aquaporin constitutively expressed gene* (**D**) *as an internal control of mRNA amounts .*

several years. Radially, ray parenchyma are differentiated from ray initials and constitute rays which remain alive until heartwood formation occurs (Hillis, 1987; Larson, 1994). To understand xylem cell differentiation we need to integrate at the physiological level biochemical data describing enzyme and substrate interactions and molecular data describing gene expression patterns related to key enzymes involved in such interactions. Up to now xylem differentiation in hybrid walnut has been studied only on a biochemical basis.

Walnut is a diffuse-porous wood and water storage in the stem could play an important role for day/night growth variations in relation to transpiration processes in the crown (Kozlowski, 1972; Garnier and Berger, 1986). Thus, this specific water status should be taken into account in biochemical studies of xylem formation during tree growth. Relative water content has been measured across stem sections and throughout the growing season in order to determine the relationship between fresh weight and dry weight for black walnut (*Juglans nigra*) wood samples. *Figure 4* illustrates water content measured during the growing season (July). The most striking observation is that there is relatively lower water content in sapwood compared with bark and heartwood. This radial pattern has already been reported for walnut and other species (Stewart, 1967).

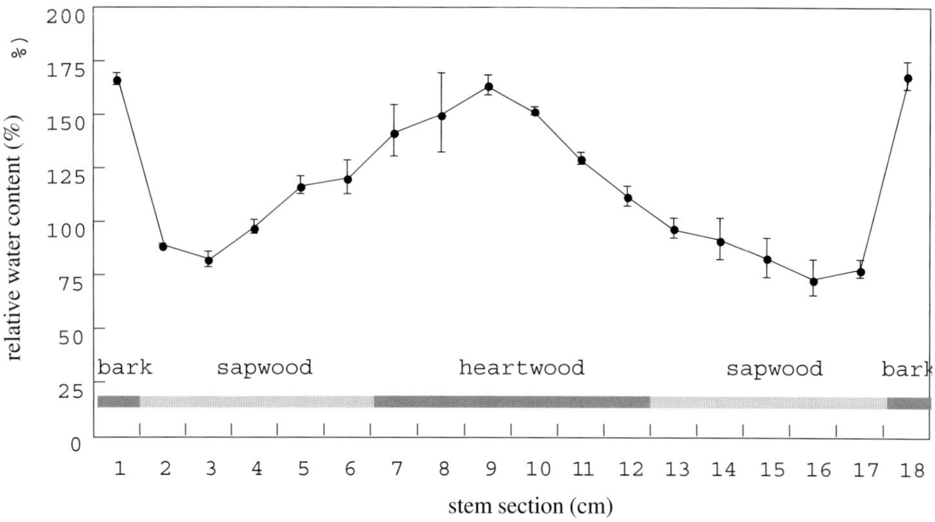

Figure 4*. Variation of relative water content across a black walnut stem section sampled in July 1998. Percent water content was relative moisture content according to the following calculation: (fresh weight – dry weight) × dry weight^{-1} × 100. Pieces of wood (1 cm^3) were analysed concurrently. Bars represent standard error on three repetitions.*

As a first biochemical approach to the study of xylem differentiation in hybrid walnut, we measured cinnamyl-alcohol dehydrogenase (CAD) activities related to lignification processes (Whetten and Sederoff, 1995) and aminopeptidase (AP) activities related to protein degradation and autolysis (Fukuda, 1996). All biochemical analyses have been performed in differentiating phloem, young xylem (formed during the year of collection) and xylem (formed the year before collection) of hybrid walnut. *Table 1* summarizes data on CAD activities and AP activities. Different phenolic compounds known to be involved in wood colouring (Burtin *et al.*, 1998) have also been analysed.

As already reported by numerous authors for other trees, results of CAD activity suggest that CAD is a marker of young xylem and outer sapwood in hybrid walnut (Boudet *et al.*, 1995). AP activity is higher in phloem tissues and shows a poor correlation with xylem differentiation (Beers and Freeman, 1997). Naphthoquinone levels are high in phloem and show a decreasing trend during xylem differentiation. Phenolics which have been analysed are substrates for enzymes involved in heartwood colouring and their accumulation in the sapwood is probably crucial for later stages of xylem differentiation during ageing and tissue maturation (Burtin *et al.*, 1998).

6. Heartwood formation

Heartwood formation is characterized by two striking processes: formation and deposition of heartwood compounds (extractives) and death of parenchyma cells. Parenchyma cells remain alive in old sapwood and are responsible for the changes occurring during transformation of sapwood into heartwood (Bosshard, 1965; Fengel, 1970; Nobuchi et al., 1984). Heartwood is usually defined as the inner part of the stem,

Table 1. *Biochemical characterization of xylem differentiation in hybrid walnut* (Juglans nigra × J. regia). *An extraction of total soluble proteins was performed on walnut tissues by pooling three successive extracts for each tissue using polyethylene glycol to prevent protein denaturation. Total protein content was measured by Coomassie staining (Bradford, 1976). Cinnamyl-alcohol dehydrogenase (CAD) activities were measured by the formation of coniferyl aldehyde from coniferyl alcohol at 400 nm (Hawkins and Boudet, 1994) and aminopeptidase (AP) activities were measured by the formation of naphthylamine from leucine naphthylamine at 335 nm (Cook and Adams, 1997). Naphthoquinones (juglone + hydrojuglone glucoside) and flavonols (myricitrine + quercitrine) were measured by acetone-water (1:1, v:v) extraction and HPLC analysis (Claudot* et al., *1992) and are expressed as 6-methoxyflavone equivalent (6-methoxyflavone was added at 10^{-4} M in the phenolic extract as internal standard). Measurements were performed in phloem, young xylem (differentiating xylem) and xylem collected in June 1997.*

	Phloem	Young xylem (1997)	Xylem (1996)
Proteins (mg g^{-1} fresh weight)	0.4	0.5	0.5
CAD (μmol min^{-1} mg^{-1} protein)	3.1	6.8	5.1
AP activity (μmol min^{-1} mg^{-1} protein)	0.4	0.2	0.2
Naphthoquinones (μmol g^{-1} fresh weight)	317.2	23.3	11.7
Flavonols (μmol g^{-1} fresh weight)	12.3	1.0	0.2

which, in the growing tree, has ceased to contain living cells, and in which the reserve materials have been removed or converted into heartwood substances (Hillis, 1987). When compared to surrounding sapwood, heartwood is frequently darker in colour and richer in phenolic compounds (extractives). A narrow transition zone (one or two growth rings) is present between sapwood and heartwood. This zone can be sometimes distinguished by a paler or intermediate colour and is considered as a primary stage of the conversion of sapwood into heartwood (Hillis, 1987). The increased content of phenolic substances is the most important feature of heartwood. It directly affects parameters of wood quality evaluation, such as colour and durability. For this reason the physiological processes leading to extractives accumulation are of great scientific and industrial interest.

7. Extractives accumulation

Despite their importance, the physiology of extractives formation is not well known. In many broadleaves and conifers the extractives and their precursors are specifically accumulated in the inner part of the sapwood (transition zone) (Gardner and Barton, 1960; Hillis, 1966; Magel *et al.*, 1994; Burtin *et al.*, 1998). Our results show a specific accumulation of coloured compounds in which gallic and ellagic acid derivatives are major compounds. Besides this accumulation, a major phenolic precursor in walnut (hydrojuglone glucoside) accumulates in the sapwood and rapidly disappears in heartwood (*Figure 5*). Two main hypotheses have been proposed to explain this accumulation: translocation from outer sapwood through ray parenchyma cells, and *in situ* formation. Although it is not possible to exclude the translocation of extractives (Stewart, 1966), experimental evidences suggest that these substances would be formed *in situ* in the inner sapwood, as well as in the transition zone. Many workers have reported a significant increase in metabolic activity in the transition zone, related to the synthesis of heartwood extractives (Hillis, 1987). Cytological observations suggested that the death of the parenchyma cells is preceded by a period of increased cellular activity (Chattaway, 1952). An increase of respiration in the transition zone has been observed in many species including *Pinus radiata* (Shain and Mackay, 1973), and *Robinia pseudoacacia* L. (Höll and Lendzian, 1973). Moreover, an increase in the activity of specific enzymes controlling extractives synthesis has been observed in the transition zone of several species. In *Robinia pseudoacacia* an increase of phenylalanine ammonia-lyase (PAL) and chalcone synthase (CHS) activities was observed in the transition zone as compared to outer sapwood. In addition, some CHS isoforms were specifically detected only in the transition zone, suggesting an activation of specific CHS genes (Magel *et al*., 1991; Magel and Hübner, 1997).

As far as *in situ* synthesis is concerned, the carbon skeleton and the energy needed for extractives formation can be provided by primary metabolites, such as lipids, starch and sugars. Generally starch and lipids are accumulated and stored in sapwood parenchyma cells. They could be used, together with soluble sugars transported from

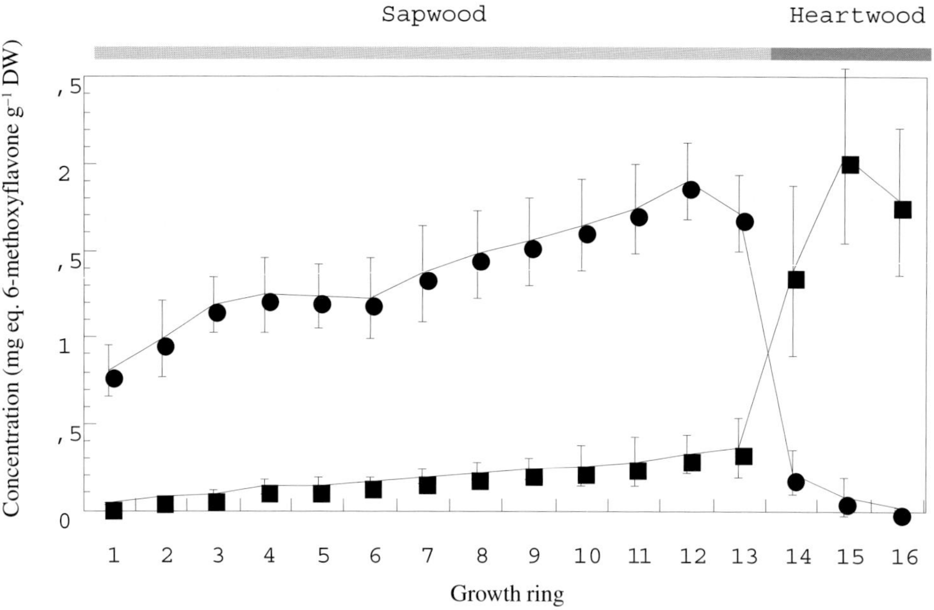

Figure 5. *Hydrojuglone glucoside and ellagic acid derivative concentrations across hybrid walnut stem section sampled in July 1998. Hydrojuglone glucoside (●) and ellagic acid derivative contents (■) were measured by spectrophotometric absorbance after HPLC fractionation. The internal standard was 6-methoxyflavone.*

the leaves, for the synthesis of heartwood extractives (Higuchi, 1997). Labelled glucose administered to the cambial region has been converted, after a few days, to labelled polyphenols at the heartwood periphery in *Eucalyptus sieberi* (Hillis and Hasegawa, 1963) and *Prunus yedoensis* (Hasegawa and Shiroya, 1965). In *Robinia pseudoacacia* L., the activity of sucrose degrading enzymes (sucrose synthase, neutral invertase) was higher in the differentiating xylem during the cambial growth season (spring-summer) and in the transition zone during heartwood formation in autumn. The highest activities of sucrose synthase were in accordance with highest levels of gene expression detected by competitive RT-PCR. These data suggest that sucrose is used for xylogenesis at the cambium level and for biosynthesis of extractives in the inner sapwood (Magel *et al.*, 1994; Hauch and Magel, 1998).

Higuchi (1997) proposed a scheme based on three main steps to describe the physiological processes leading to the transition from sapwood to heartwood: (i) alterations in the metabolism of living parenchyma cells activated by hormonal (ethylene) and physical factors such as reduction of water content, (ii) increased metabolic activity, synthesis of heartwood extractives, (iii) cellular death, liberation of extractives stored in the vacuole, oxidation and polymerization of extractives by chemical and/or enzymatic processes. Even though some chemical and biochemical aspects of heartwood formation have been elucidated by experimental work, the physiology of this complex process remains poorly understood, especially at the molecular level. Very little or no information is available on gene expression in ageing wood tissues.

8. Gene expression during heartwood formation

One objective of our work is to study the molecular basis of heartwood formation in walnut, through the analysis of gene expression in the sapwood and transition zone of mature trees. Our aim is to look in the transition zone for differential expression of genes that could be linked to heartwood formation. The experimental strategy is based on the expression analysis of candidate genes. The chemical composition of walnut extractives includes many polyphenols which could partly originate from the flavonoids pathway (Burtin et al., 1998). Key enzymes of phenylpropanoid pathway including (PAL, cinnamate 4-hydroxylase (C4H), 4-coumarate:CoA ligase (4CL)), and of flavonoid biosynthesis (CHS, flavanone 3-hydroxylase (F3H), dihydroflavonol-4-reductase (DFR) have been targeted as candidate genes potentially controlling extractives formation and wood quality.

Some difficulties were encountered in applying molecular biology techniques to old xylem. Such problems are related to the small proportion of living cells, the low metabolic activity, and the chemical composition of wood tissues. A protocol based on CTAB buffer extraction from freeze-dried wood (Chang *et al.*, 1993) was chosen for the extraction of total RNA. The amount of RNA which it is possible to extract from the walnut transition zone may be as low as 0.2–0.4 $\mu g\ g^{-1}$ dry weight (1500 to 3000 times less than what it is possible to obtain from young differentiating xylem). Reverse transcriptase-PCR (RT-PCR) technology offers an efficient tool to detect gene expression from small amounts of transcripts (Larrick and Siebert, 1995). As a first step, a RT-PCR approach was applied to amplify and clone genes expressed in young xylem. Degenerated primers were prepared according to multiple alignments of amino acid sequences and used to amplify cDNA sequences corresponding to the target genes. Degenerated primers for CHS were prepared according to El Euch (El Euch *et al.*, 1998). Fifty nanograms of total RNA were used for each RT-PCR reaction. Messenger RNAs (mRNA) were purified from total RNA using Dynabeads Oligo $(dT)_{25}$ (Dynal A.S., Oslo, Norway) and used for RT-PCR reactions. The expression of genes coding for PAL, 4CL, CHS, F3H, and DFR was detected in the sapwood and transition zones of *Juglans nigra* L. trees. To our knowledge, this is the first report showing the activation of phenylpropanoid metabolism at the transcription level in the transition zone.

The amplified cDNA fragments were cloned and sequenced, thus providing homologous probes for subsequent expression studies. The RT-PCR approach was used to study the expression of CHS, F3H and DFR genes across radial sections of trunks from the youngest sapwood to the transition zone. The first results show that these genes are expressed in all sapwood tissues (*Figure 6*). For the CHS gene, stronger RT-PCR signals were observed in differentiating xylem and older tissues (inner sapwood and transition zone) as compared with outer sapwood. For F3H and DFR genes, only slight differences between tissues could be observed. RT-PCR does not provide a quantitative measure of gene expression. Nevertheless the marked differences observed for the CHS gene suggest that a differential expression could exist between young and old sapwood. Quantitative analysis will be carried out to detect differential levels of gene expression using northern blot and dot-blot analyses.

A crucial point in studying heartwood formation is to select the periods characterized by high activities in extractives accumulation and heartwood expansion. In *Robinia pseudoacacia* L. these processes appear to be seasonally regulated and linked to tree phenology. The accumulation of extractives and the radial expansion of heartwood

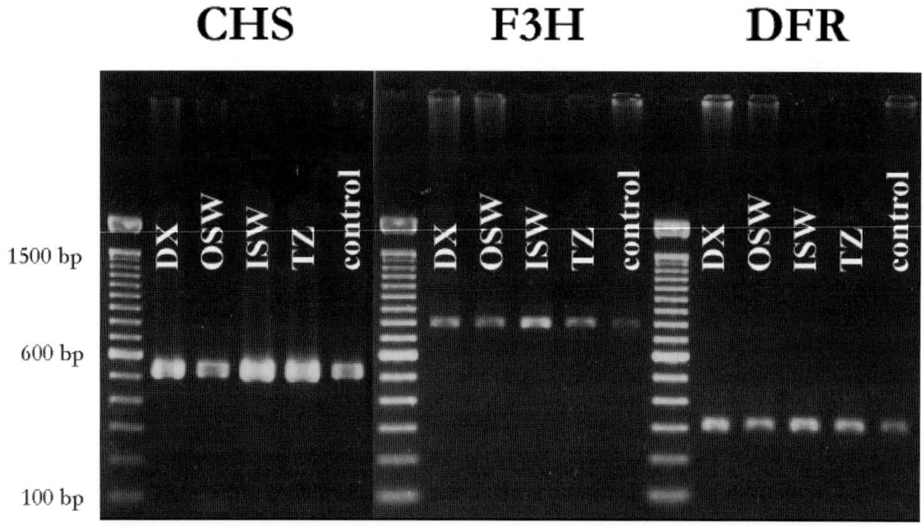

Figure 6. *Gene expression detected by RT-PCR in* Juglans nigra L. *wood tissues sampled in July 1997. RT-PCR reactions were performed with 50 ng mRNA. Two-fifths of the PCR product were separated by electrophoresis on 1.5% agarose gel and visualized after ethidium bromide staining. CHS: Chalcone synthase; F3H: flavanone-3-hydroxylase; DFR: dihydroflavonol-4-reductase; Ladder 100 bp 300 ng; DX: differentiating xylem; OSW: outer sapwood; ISW: inner sapwood; TZ: transition zone; control: cloned cDNA reference.*

follow the period of cambial activity and are more marked in late summer and autumn (Nobuchi et al., 1984; Magel *et al.*, 1994; Hauch and Magel, 1998). A year-round sampling of walnut trees was carried out in order to monitor the temporal dynamics of heartwood formation. Aged black and hybrid (*Juglans major* × *J. regia*) walnut trees were collected every 6 weeks between June 1997 and May 1998. Analyses of extractives are being carried out on these samples. These data will provide a temporal distribution of extractive accumulation. On this base, key periods will be located and used for detailed analyses of gene expression. The results of molecular analyses will be related to chemical analyses of extractives content to give an integrated description of extractives accumulation.

Acknowledgements

The authors thank the European Community for funding our work by the FAIR CT96–1887 Walnut-BRAINS research programme.

References

Barnett, J.R. (ed.) (1981) *Xylem Cell Development*. Castle House Publications Ltd., Tunbridge Wells, pp. 307.

Beers, E.P. and Freeman, T.B. (1997) Proteinase activity during tracheary element differentiation in *Zinnia* mesophyll cultures. *Plant Physiol.* **113**: 873–880.

Bosshard, H.H. (1965) Aspects of the aging process in cambium and xylem. *Holzforsch.* **19**: 65–69.

Boudet, A.M., Lapierre, C. and Grima-Pettenati, J. (1995) Tansley Review No. 80. Biochemistry and molecular biology of lignification. *New Phytol.* **129**: 203–236.

Bradford, M.M. (1976) A rapid and sensitive method for the quantitation of microgram quantities of protein utilizing the principle of protein-dye binding. *Anal. Biochem.* **72**: 248–254.

Burtin, P., Jay-Allemand, C., Charpentier, J.P. and Janin, G. (1998) Natural wood colouring process in *Juglans* sp. (*J. nigra*, *J. regia* and hybrid *J. nigra* 23 × *J. regia*) depends on native phenolic compounds accumulated in the transition zone between sapwood and heartwood. *Trees — Structure and Function* **12**: 258–264.

Butterfield, B.G. (1975) Terminology used for describing the cambium. *IAWA Bulletin 1975* **1**: 13–14.

Chang, S., Puryear, J. and Cairney, J. (1993) A simple and efficient method for isolating RNA from pine trees. *Plant Mol. Biol. Rep.* **11**: 113–116.

Chattaway, M.M. (1952) The sapwood-heartwood transition. *Aust. For.* **26**: 25–34.

Claudot, A.C., Drouet, A. and Jay-Allemand, C. (1992) Tissue distribution of phenolic compounds in annual shoots from adult and rejuvenated hybrid walnut trees. *Plant Physiol. Biochem.* **30**: 565–572.

Cook, M. and Adams, Z. (1997) Purification and characterization of an arginyl peptidase from the chloroplast stroma of pea seedlings. *Plant Physiol. Biochem.* **35**: 163–168.

De Veylder, L., Van Montagu, M. and Inzé, D. (1998) Cell cycle control in *Arabidopsis*. In: *Plant Cell Division* (ed. Francis, D., Dudits, D. and Inzé, D.) Portland Press: London, pp. 21–45.

Dudits, D., Magyar, Z., Deak, M., Meszaros, T., Miskolczi, P., Feher, A., Brown, S., Kondorosi, E., Athanasiadis, A., Pongor, S., Bako, L., Koncz, C. and Gyorgyey, J. (1998) Cyclin-dependent and calcium-dependent kinase families: response of cell division cycle to hormone and stress signal. In: *Plant Cell Division* (ed. Francis, D., Dudits, D. and Inzé, D.), Portland Press: London, pp. 21–45.

El Euch, C., Jay-Allemand, C., Pastuglia, M., Doumas, P., Charpentier, J.P., Capelli, P. and Jouanin, L. (1998) Expression of antisense chalcone synthase RNA in transgenic hybrid walnut microcuttings. Effect on flavonoid content and rooting ability. *Plant Mol. Biol.* **38**: 467–479.

Fengel, D. (1970) Ultrastructural changes during aging of wood cells. *Wood Sci. Technol.* **4**: 176–188.

Fukuda, H. (1996) Xylogenesis: initiation, progression, and cell death. *Annu. Rev. Plant Physiol. Plant Mol. Biol.* **47**: 299–325.

Gardner, K.H. and Barton, G.M. (1960) The distribution of dihydroquercetin in Douglas-fir and western larch. *For. Prod. J.* **10**: 171–173.

Garnier, E. and Berger, A. (1986) Effect of water stress on stem diameter changes of peach trees growing in the field. *J. Appl. Ecol.* **23**: 193–209.

Hasegawa, M. and Shiroya, M. (1965) The formation of phenolic compounds at the sapwood-heartwood boundary. In: *Proc. Mtg. Sect. 41* (ed. IUFRO), pp. 10.

Hauch, S. and Magel, E. (1998) Extractable activities and protein content of sucrose-phosphate synthase, sucrose synthase and neutral invertase in trunk tissues of *Robinia pseudoacacia* L. are related to cambial wood production and heartwood formation. *Planta* **207**: 266–274.

Hawkins, S.W. and Boudet, A.M. (1994) Purification and characterisation of cinnamyl alcohol dehydrogenase isoforms from the periderm of *Eucalyptus gunii* Hook. *Plant Physiol.* **104**: 75–84.

Higuchi, T. (1997) *Biochemistry and Molecular Biology of Wood*. Springer Verlag: Berlin.

Hillis, W.E. (1966) Variation in polyphenol composition within species of *Eucalyptus. Phytochem.* **5**: 541–556.

Hillis, W.E. (1987) *Heartwood and Tree Exudates*. 1st Edn. Springer Verlag: Berlin.

Hillis, W.E. and Hasegawa, M. (1963) The formation of polyphenols in trees. 1. Administration of ^{14}C glucose and subsequent distribution of radioactivity. *Phytochem.* **2**: 195–199.

Höll, W. and Lendzian, K. (1973) Respiration in the sapwood and heartwood of *Robinia pseudoacacia. Phytochem.* **12**: 975–977.

John, P.C.L. (1996) The plant cell cycle: conserved and unique features in mitotic control. In: *Progress in Cell Cycle Research* (ed. Meijer, L., Guidet, S. and Vogel, L.). Plenum Press: New York, pp. 59–72.

Kozlowski, T.T. (1972) Shrinking and swelling of plant tissues. In: *Water Deficit and Plant Growth. Plant Responses and Control of Water Balance* (ed. Kozloswki, T.T.). Academic Press: New York, pp. 1–64.

Larson, P. (1969) *Wood Formation and the Concept of Wood Quality*. Yale University, New Haven.

Larson, P.R. (1994) *The Vascular Cambium: Development and Structure*. 1st Edn. Springer Verlag: Berlin.

Magel, E. and Hübner, B. (1997) Distribution of phenylalanine ammonia lyase and chalcone synthase within trunks of *Robinia pseudoacacia* L. *Bot. Acta* **110**: 314–322.

Magel, E., Jay-Allemand, C. and Ziegler, H. (1994) Formation of heartwood substances in the stemwood of *Robinia pseudoacacia* L. II. Distribution of nonstructural carbohydrates and wood extractives across the trunk. *Trees* **8**: 165–171.

Magel, E.A., Drouet, A., Claudot, A.C. and Ziegler, H. (1991) Formation of heartwood substances in the stem of *Robinia pseudoacacia* L. I. Distribution of phenylalanine ammonium lyase and chalcone synthase across the trunk. *Trees* **5**: 203–207.

Mittler, R. and Lam, E. (1995) *In situ* detection of nDNA fragmentation during the differentiation of tracheary elements in higher plants. *Plant Physiol.* **108**: 189–193.

Nobuchi, T., Sato, T., Iwata, R. and Harada, H. (1984) Season of heartwood formation and the related cytological structure of ray parenchyma cells in *Robinia pseudoacacia* L. *Mozukai Gakkaishi* **30**: 628–636.

Nurse, P. (1990) Universal control mechanism regulating onset of M phase. *Nature* **344**: 503–508.

Riabowol, K., Draetta, G., Brizuela, L., Vandre, D. and Beach, D. (1989) The cdc2 kinase is a nuclear protein that is essential for mitosis in mammalian cells. *Cell* **57**: 393–401.

Roberts, L.W. (1976) *Cytodifferentiation in plants: Xylogenesis as a Model System*. Cambridge University Press: Cambridge.

Roberts, L.W., Gahan, P.B. and Aloni, R. (1988) *Vascular Differentiation and Plant Growth Regulators*, Springer Verlag: Berlin.

Sambrook, J., Fritsch, E.F. and Maniatis, T. (1989) *Molecular Cloning: a Laboratory Manual:* 2nd Edn. Cold Spring Harbor Laboratory Press: Cold Spring Harbor, N.Y.

Shain, J. and Mackay, J.F.G. (1973) Seasonal fluctuations in respiration of aging xylem in relation to heartwood formation in *Pinus radiata. Can. J. Bot.* **51**: 737–741.

Stewart, C.M. (1966) Excretion and heartwood formation in living trees. *Science* **153**: 1068–1074.

Stewart, C.M. (1967) Moisture content of living trees. *Nature* **214**: 138–140.

Wang, H., Li, J., Bostock, R.M. and Gilchrist, D.G. (1996) Apoptosis: a functional paradigm for programmed plant cell death induced by a host-selective phytotoxin and invoked during development. *Plant Cell* 8: 375–391.

Whetten, R. and Sederoff, R. (1995) Lignin biosynthesis. *Plant Cell* 7: 1001–1013.

17

Hormonal control of vascular differentiation in plants: the physiological basis of cambium ontogeny and xylem evolution

R. Aloni, P. Feigenbaum, N. Kalev and S. Rozovsky

1. Introduction

This chapter presents current thoughts and new evidence on the role of auxin, ethylene, gibberellin and cytokinin in controlling vascular differentiation in plants. The chapter deals with mechanisms that control the following: the transition from procambium to cambium, differentiation of perforations and vessel endings, the evolution from tracheids to vessels and fibres, the control of wood formation in ring-porous vs. diffuse-porous trees, and the role of roots in vascular differentiation.

Evidence is provided that the complexity of vascular tissues is a result of specialization of the vascular components to specific signals, their level of stimulation and the sensitivity of the vascular meristem to hormonal signals. These have shaped vascular differentiation during plant evolution.

2. Control of vascular differentiation by a flow of signals

Due to the complexity of vascular tissues, comprising several types of cells, they are induced by a number of developmental signals which flow in the axial and radial directions. Auxin is the major controlling signal and is involved in all aspects of vascular differentiation. The continuity of vascular tissues along the plant axis is a result of the steady polar flow of auxin from leaves to roots (Aloni, 1987).

Cell & Molecular Biology of Wood Formation, edited by R. Savidge, J. Barnett and R. Napier.

Developing buds and young growing leaves induce vascular differentiation below them. In spring the young leaves stimulate cambial reactivation and the formation of new phloem and xylem which extend downwards from the developing buds toward the root tips (Reinders-Gouwentak, 1965; Aloni *et al.*, 1997a, 1997b). The removal of young leaves from *Coleus* stems reduces or prevents vascular differentiation below the excised leaves (Jacobs, 1952; LaMotte and Jacobs, 1963). In *Arabidopsis* leaves, vascular tissues become progressively confined towards the leaf margin as the concentration of auxin transport inhibitors is increased, suggesting that the leaf vascular system depends on inductive signals from the margin of young leaves (Mattsson *et al.*, 1999). Furthermore, vessel differentiation starts in the lobes of young leaf primordia in *Arabidopsis*, where hydathodes develop, indicating major auxin synthesis sites in the growing lobes (Aloni, R., Ivanir-Zachut L. and Feldman, L.J., unpublished).

It is well known that the major signal produced by the young leaves is auxin, which moves in a polar fashion towards the roots. The pioneering work of Jacobs (1952) demonstrated that indole-3-acetic acid (IAA) is the limiting and controlling factor in xylem regeneration around a wound in *Coleus*. A source of auxin replaces the effect of the young leaves in promoting vascular regeneration (Jacobs, 1952, 1998; Thompson and Jacobs, 1966; Bruck and Paolillo, 1984). Low auxin levels induce phloem and phloem anastomoses between the bundles (Aloni, 1980; Aloni and Peterson, 1990; Aloni and Barnett, 1996), while there is a need for higher auxin stimulation for xylem differentiation (Aloni 1980, 1987, 1995).

In the canalization hypothesis Sachs (1981) suggested that auxin flow, which starts by diffusion, induces a polar auxin transport system which promotes auxin movement and leads to canalization of auxin flow along a narrow file of cells. The continuous polar transport of auxin through these cells induces a further complex series of events which finally result in the formation of a vascular bundle. The canalization of auxin flux determines the orderly pattern of vascular differentiation from leaves to roots (Sachs, 1981).

3. Control of the vascular meristem: from procambium to cambium

In the plant body the primary and secondary vascular tissues originate from embryonic tissues, called the vascular meristems, whose cells retain the ability to divide and multiply. Two meristematic stages are distinguished: the procambium and the cambium. The procambium is the apical meristem that produces primary phloem and primary xylem in young elongating organs, while the cambium is the lateral meristem that produces secondary phloem and secondary xylem in gymnosperms and dicotyledons (Larson, 1994). Both vascular meristems are induced and controlled by the steady polar flow of auxin from leaves to roots (Sachs, 1981; Aloni, 1995; Uggla *et al.*, 1996, 1998). Cambial activity is also promoted by other growth regulators, like gibberellin (Digby and Wareing, 1966) and cytokinin (Aloni, 1993).

There are difficulties in determining where the transition lies between the meristems because there is no clear-cut criterion by which the procambium and the cambium can be separated. Therefore, the procambium and the cambium are usually regarded as a continuum representing two developmental stages of a single meristem (Larson, 1982). The cambium is first detected in the cotyledonary node of seedlings (Saks and Aloni, 1985) and in young nodes (Bruck and Paolillo, 1984) and the appearance of the cambium in the nodal regions is ascribed to higher levels of auxin in these regions.

Wounding experiments led Fahn *et al.* (1972) to suggest that the difference between the procambium and the cambium may be a result of stimulus concentration and the age of the reacting tissue. Catesson (1964) claimed that the formation of the cambium can be clearly defined by the appearance of vascular rays in tangential sections. The cambium of most plants produces vascular rays (Carlquist, 1988; Larson, 1994) which are probably induced by a radial signal flow (Carmi et al., 1972; Sachs, 1981; Lev-Yadun and Aloni, 1995). According to the *ethylene aeration hypothesis* (Lev-Yadun and Aloni, 1995) ethylene produced in the xylem moves via the vascular meristem in the centrifugal direction and induces ray initials. However, this hypothesis has not yet been confirmed experimentally. In a recent unpublished study on *Ricinus communis* and *Hibiscus canabinus* we have found that inhibitors of ethylene synthesis (cobalt ions) and action (silver ions) prevented ray differentiation (results are not shown), whereas treatments that elevated ethylene production (wounding, flooding, grafting and ethrel application) stimulated ray initiation, ray enlargement (*Figure 1b*) and the fusion of small rays to large ones (*Figure 1c*). Therefore, we propose that ethylene plays an important role in the transition from procambium to cambium. We suggest that there is a need for a threshold level of ethylene for inducing the first ray initials, and these rays do characterize the transition from procambium to cambium. Ethylene, which promotes and regulates ray differentiation in the cambium (Lev-Yadun and Aloni, 1995), has been also suggested to control the differentiation of radial tracheids in the rays of pine trees (Kalev and Aloni, 1999).

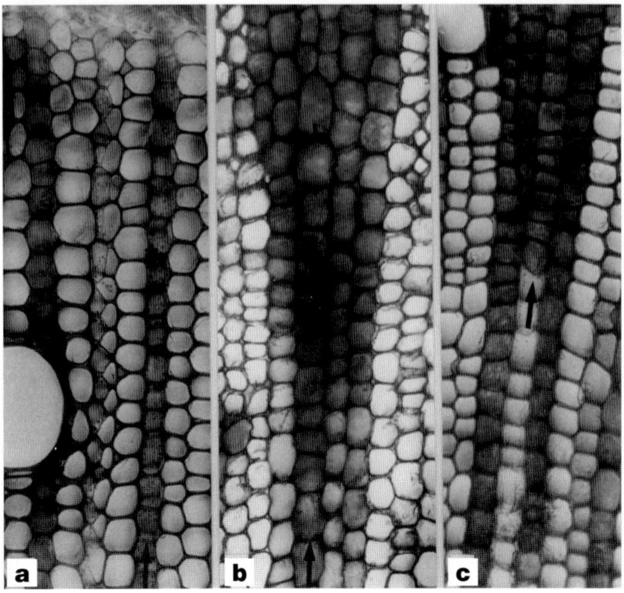

Figure 1. *Photomicrographs of cross sections in stems of* Hibiscus canabinus, *showing: a typical narrow vascular ray in a control stem (a), and the effects of ethylene (originating from 1% ethrel in lanolin) inducing ray enlargement (b), and integration of rays (c). The cytoplasm in the ray cells is darker and the rays are marked by arrows. The oldest xylem cells are located in the lower part. All photographs are at the same magnification (× 150).*

4. Tracheids with perforations

Two basic types of conducting elements are distinguished in the xylem: the *tracheid* is an imperforated cell with bordered pits, while the *vessel element* has perforations in pits at the end walls. The tracheid is a more primitive type of cell than the vessel element (Bailey and Tupper, 1918; Bailey, 1953). Vessel elements are more efficient conductors of water than tracheids, since the water in vessel elements flows through the perforations rather than diffusing through the cell wall (Zimmermann, 1983). Tracheids are the only type of water conducting cells in most vascular plants other than angiosperms. The earliest tracheids in the vascular tissues of ancient land plants were recorded from about 430 million years ago, in the early Silurian plant *Pinnatiramosus quianensis* (Cai *et al.*, 1996) and in the late Silurian–early Devonian plant *Cooksonia pertoni* (Edwards *et al.*, 1992). Vessel elements were recorded by the Late Permian (Li *et al.*, 1996), about 140 million years ago, and became dominant in angiosperms, although many dicotyledonous genera contain both tracheids and vessels (e.g. *Quercus, Rosa* etc.). Vessel elements have evolved independently from tracheids in several diverse groups of vascular plants. The vessel element is thus an excellent example of parallel evolution (Bailey, 1944). Vessels were found in a giganopteric fossil (Li *et al.*, 1996), in *Marsilea*, *Selaginella* (of the Lycophyta) (Duerden, 1940), *Equisetum* (of the Sphenophyta) (Bierhorst, 1958), some gymnosperms (*Pteridium*, *Gnetum*, *Ephedra* and *Welwitschia*) and in angiosperms (dicotyledons and monocotyledons).

Application of a high auxin concentration to the hypocotyl of young pine seedlings (Kalev and Aloni, 1998) induced a limited number of tracheids with perforations (*Figure 2*). These tracheids with perforations originated from cortex parenchyma. A low auxin level (0.05% w/w naphthaleneacetic acid, NAA, in lanolin) induced only tracheids. A high auxin concentration (0.5% NAA) induced a few tracheids with perforations (less than 0.1% of the tracheary elements) among the numerous tracheids. The high auxin concentration (0.5% NAA) induced tracheids with perforations bearing one or two (*Figure 2*) perforations on opposite side walls, thus forming short files of tracheids with perforations, like a vessel of two or three elements each. The diameter of the openings in a perforation of pine seedlings ranges from 7–35 μm and was characterized by a border around the perforation (*Figure 2*).

Evidence for the potential to eliminate primary cell wall in conifers could be detected in their bordered pit ultrastructure, where the primary wall is lysed in the pit margo forming a net-like margo around the central torus (see Sano *et al.*, 1999). Therefore, it is possible that the perforations in the auxin-induced tracheids in pine originated from bordered pits induced to further development by the increase in auxin stimulation.

Naturally occurring perforated tracheids are very rare and were previously observed in two conifers: *Thuja occidentalis* (Jane, 1956) and *Sequoia sempervirens* (Bannan, 1958). In both species the perforations were large and simple and continued through the growth of more than one year. We therefore suggest that this pattern of radial files of perforated tracheids was induced by a stimulus, possibly the polar flow of auxin in the cambium (Uggla *et al.*, 1996, 1998), which probably moved continuously and preferably through specific files of fusiform initials from which the perforated tracheids differentiated. For *Thuja*, the origin of the wood from which the section was cut is not known, but it had features of compression wood (Jane, 1956). In *Sequoia* the

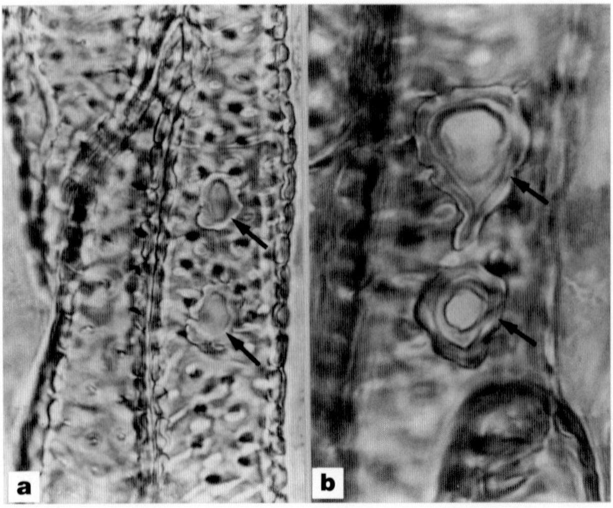

Figure 2. *Longitudinal views of tracheids with perforations, induced by auxin (0.5% NAA in lanolin) in the hypocotyl of young* Pinus pinea *seedlings. (a) shows a tracheid with two openings (arrows) on a side wall (×600). (b) shows a magnified view of two perforations (arrows) with a relatively thick rim around the openings (×700).*

origin of the xylem was a lower branch from a mature tree (Bannan, 1958). Reaction wood formation is generally hypothesized to develop in branches in response to gravity, which induces a lateral gradient of auxin across the branch. Compression wood develops in the lower side of the branch in the region of high auxin concentration (see Timell, 1986 for a review). Thus, our findings suggest a possible explanation for the earlier reports on the formation of perforations in tracheids, indicating that a relatively high auxin stimulation was probably the factor which induced perforation formation in both the naturally occurring (Jane, 1956; Bannan, 1958) and the experimentally-induced tracheids with perforations in pine (*Figure 2*).

5. From tracheids to vessels and fibres

Vessels and fibres in the xylem of angiosperms originated from tracheids of more primitive plants (see Esau, 1965; Fahn, 1990). Auxin is well known to induce tracheids (see Savidge, 1996). Auxin induced short tracheids which differentiated from parenchyma cells in the hypocotyl of young pine seedlings, while combinations of auxin and gibberellin promoted the differentiation of long tracheids (Kalev and Aloni, 1998). Gibberellin, in the presence of auxin, promoted tracheid elongation by stimulating intrusive growth of both the upper and lower ends of the differentiating tracheids. Kalev and Aloni (1998) suggested that in the course of xylem evolution, the original hormonal mechanism for tracheid differentiation in conifers had become more specific. From the original mechanism for tracheid differentiation in conifers (a combination of auxin and gibberellin) the signals for each xylem element in angiosperms, has become specific: auxin by itself induces vessel elements (Jacobs, 1952; Sachs, 1981; Aloni, 1995), whereas gibberellin, in the presence of auxin, has become the specific signal for fibre differentiation (Aloni, 1979, 1987; Roberts *et al.*,

1988). Thus, the structural changes from tracheids to vessel elements and fibres represent physiological changes which have occurred during evolution in the hormonal mechanisms regulating these cells (Kalev and Aloni, 1998).

For further clarification of the evolution of vascular differentiation there is a need to use a genetic approach to uncover the genes that control the differentiation of fibres (Zhong *et al.*, 1997), vessels (Fukuda, 1996; Hardtke and Berleth, 1998; Nelson, 1998), the arrangement of phloem and xylem in vascular bundles and the evolution of the stele (Zhong *et al.* 1999).

6. The control of vessel endings

The operating units of the xylem in angiosperms are the vessels and not vessel elements (Zimmermann, 1983). The vessel, which is made up of numerous vessel elements connected end-to-end by perforation plates, is limited in length by vessel endings of imperforate walls at both extremities. The frequency of vessel endings is significantly higher in nodes of young stems of both trees and herbaceous species (Saleo *et al.*, 1984; Wiebe *et al.*, 1984). Therefore node regions are considered 'safety zones' because gaseous emboli and fungal spores fail to pass through the endings. A hydraulic 'safety zone' was also found at the base of root-shoot junctions in some cereal species (Luxová, 1986; Aloni and Griffith, 1991). At the root-shoot junctions, short tracheids and vessel endings occur and they protect the vessels of the root from embolism originating in the shoot. Cereals grown under stress usually develop 'safety zones', while those in mesic conditions normally do not (Aloni and Griffith, 1991). Vessel endings and tracheids occur also at the base of leaves of the palm *Rhapis excelsa* (Zimmermann and Sperry, 1983), thus separating the wide vessels of the stem from a possible vapour-block which may originate in old leaves. The occurrence of vessel endings and tracheids in organ junctions results in the hydraulic segmentation of the plant (see Zimmermann, 1983).

In spite of the physiological significance of vessel endings in water conduction no information is available on their differentiation, probably because of the technical difficulty of locating vessel endings in nodes or organ junctions. However, vessel endings can be induced experimentally along a node, where their location is determined by a cut (Aloni and Inding, 1989). With this method we have analysed the effect of ethylene, gibberellin and auxin on the formation of the secondary wall in vessel endings in earlywood *Melia azedarach* (*Figure 3*). The hormones in lanolin were applied in spring time, a few weeks before bud break, after removing the apical bud. The study was conducted over two successive years (from March to May). In control branches (treated with lanolin) vessel endings were characterized by rounded pits in various patterns (*Figure 3a*). Ethylene (originating from 0.02% ethrel) disrupted pit formation usually in the centre of the vessel ending (*Figure 3b*), achieving maximal expression in vessel endings with completely pitless areas (not shown). Gibberellin (0.02% gibberellic acid in lanolin) induced thread-like patterns with long and narrow pits (*Figure 3c*). Intermediate patterns were also found in the various hormonal treatments. An auxin treatment (0.02% NAA) usually prevented the formation of vessel endings, resulting in open vessels with incomplete vessel endings (not shown) near the site of auxin application. This indicates that a stream of auxin induces perforations and only where its pathway is laterally redirected a vessel ending can be formed. We suggest that the applied hormones affected pit formation by affecting cortical microtubule orientation (see Shibaoka, 1994; Kalev and Aloni, 1999). Gibberellin probably

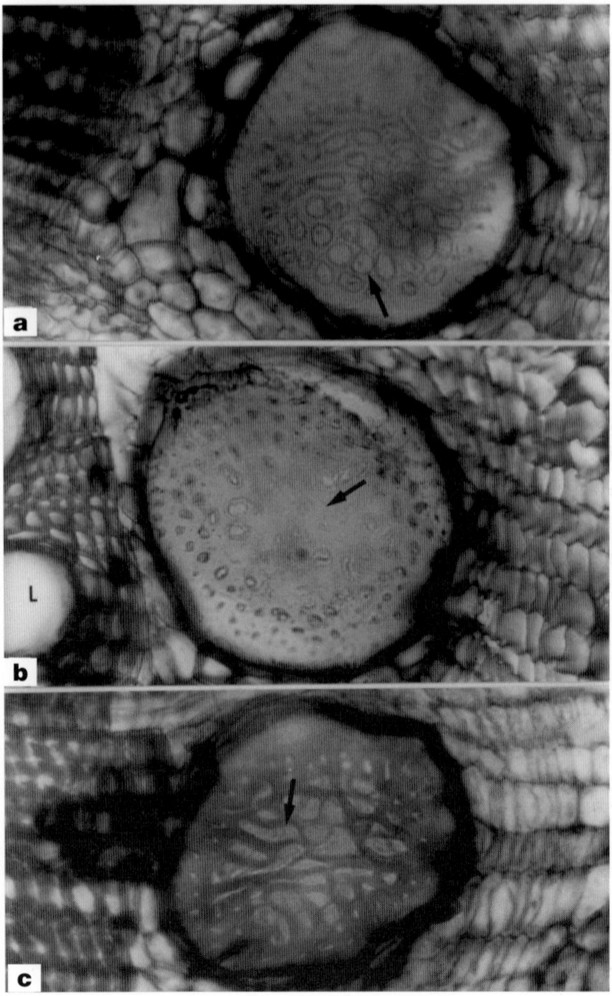

Figure 3. *Photomicrographs of cross sections in stems of* Melia azedarach, *showing magnified views of vessel endings in a control stem (a), in a stem treated with 0.02% ethrel in lanolin (b), and with 0.02% gibberellic acid (c). The arrows mark in the vessel endings a rounded pit in the control (a), a pitless area cause by ethylene (b) and a narrow elongated pit induced by gibberellin (c). All the vessel endings shown are of earlywood vessels. 'L' marks a latewood vessel in (b). All photographs are in the same orientation (the latewood is at the left side) and are at the same magnification.*

promoted longitudinal patterns of cortical microtubules resulting in longitudinal patterns of pits, while ethylene disturbed an organized orientation of the microtubules as suggested by Shibaoka (1994) and Warren Wilson and Warren Wilson (1993) and therefore results in pitless vessel endings.

Circular vessels are a special case of vessels which do not have vessel endings. These non-functional circular vessels occur normally in branch junctions, usually in the upper side of the branch, and their size and frequency increase continuously with age and stem width (Lev-Yadun and Aloni, 1990). It has been suggested that the differentiation of circular vessels is induced by movement of auxin in circular patterns (Sachs and Cohen, 1982; Aloni and Wolf, 1984). Circular vessels, which also occur in tumour tissues (Aloni *et al.*, 1998), can be induced experimentally by treatments that promote circular flow of auxin (Sachs and Cohen, 1982; Aloni *et al.*, 1997a).

7. Control of wood formation in diffuse-porous vs. ring-porous trees

In temperate deciduous hardwoods, the size differences of the vessels, or pores, in the early- and latewoods are quite marked and two categories of deciduous trees are determined: diffuse-porous species and ring-porous species. In ring-porous wood, the vessels produced at the beginning of a growth season are significantly wider than those produced at the end of the season, while in diffuse-porous wood, the vessels are more or less uniform (see Aloni, 1991). Along the plant axis there is a general increase in vessel size from leaves to roots (Aloni, 1987). The six-point hypothesis (Aloni and Zimmermann, 1983) ascribes this general increase in vessel size to a gradient of decreasing auxin concentrations from leaves to roots. High auxin levels near the young leaves induce narrow vessels because of their rapid differentiation, while low auxin concentrations further down result in slow differentiation, which permits more cell expansion before secondary wall deposition, and therefore results in wide vessels (Aloni and Zimmermann, 1983).

The pattern and size of vessels in the secondary xylem might be modified in the very same tree species by environmental conditions. Under favourable environmental conditions, Liphscitz (1995) found that in both a diffuse-porous poplar (*Populus euphratica*) and a ring-porous oak (*Quercus ithaburensis*), when extension growth is intensive wide rings with diffuse-porous wood are produced, while under stress conditions (where extension growth is suppressed) narrow rings characterized by ring-porous wood are formed.

Forest trees in which leaf emergence consistently occurs in early spring are diffuse-porous species, while ring-porous species are late-leafing trees (Lechowicz, 1984). Diffuse-porous trees produce young leaves for a longer period (Aloni *et al.* 1997a) and possess greater growth intensity than ring-porous trees. Diffuse-porous trees enjoy a relatively long season of extension growth under good conditions, while species in the ring-porous group are adapted to grow in less favorable conditions (Gutherie, 1989; Aloni, 1991) and have short season of extension growth (Aloni *et al.* 1997a) .

Aloni (1991) and Wheeler and Baas (1991) suggested that ring-porous trees have originated from diffuse-porous plants. According to the limited growth hypothesis (Aloni, 1991) continuous selective pressures in limiting environments finally resulted in the development of the specialized ring-porous wood that maximized the efficiency of water conduction. Evidence which supports the limiting-growth hypothesis (see Aloni, 1991) shows that the selection for ring-porous wood has led to a decrease in the intensity of vegetative growth, accompanied by reduced levels of growth regulators. The latter was followed by an increase in the sensitivity of the cambium to relatively low auxin stimulation. These physiological changes created the special internal conditions that enable the differentiation of very wide and long earlywood vessels during spring in the ring-porous trees by very low-level streams of auxin.

In ring-porous species, the first vessels (which are very wide) are initiated two to six weeks prior to the onset of leaf expansion and secondary wall deposition on the vessel elements was completed from one week before to three weeks after leaf expansion. Conversely, in diffuse-porous species, the first vessel element formation was initiated two to seven weeks after the onset of leaf expansion and secondary wall deposition was completed four to nine weeks after leaf expansion (Suzuki *et al.*, 1996).

It has been shown that low-level streams of auxin control the differentiation of wide earlywood vessels in ring-porous trees. Moderate or high auxin levels applied to ring-porous trees at the time of bud break limit the size of the earlywood vessels and result in a diffuse-porous type of wood (Aloni, 1991). The wide earlywood vessels in ring-porous species are induced just prior to the breaking of buds by very low-level streams of auxin descending from the swelling buds (Aloni, 1991; Aloni *et al.*, 1997a, 1997b). Only because the cambium is very sensitive in ring-porous trees can it respond to the low auxin stimulation, and the reactivation in spring occurs very rapidly throughout the entire tree. Conversely, in diffuse-porous trees the cambium is less sensitive and there is a need for higher auxin levels from young growing leaves for reactivation. Therefore, in diffuse-porous trees the reactivation of the cambium commences at the base of swelling buds and proceeds gradually downward toward the base of the tree. Priestley and Scott (1936) found that it might take several weeks for the 'wave' of cambial activity to extend from the twigs of a large tree to the base of its trunk.

During winter, the phloem of both the diffuse-porous tree magnolia (*Magnolia kobus*) and the ring-porous oak (*Quercus robur*) is dormant and is characterized by heavy deposits of dormancy callose. In spring, just before bud break the sieve tubes of a ring-porous oak become almost callose-free, while those of the diffuse-porous tree retain the dormancy callose. Application of the same auxin concentration (1% NAA in lanolin) to excised dormant branches of both species just before bud break resulted in the removal of the dormancy callose from the sieve tubes of magnolia (Aloni *et al.*, 1997b), similar to the effect obtained in *Vitis vinifera* (Aloni *et al.*, 1991), and an opposite effect in the oak — an accumulation of callose on the most recently formed sieve tubes (Aloni *et al.*, 1997b). These results suggest that the phloem in the ring-porous tree is extremely sensitive and reacts to very low levels of stimuli from folded leaves while they are still in dormant-looking buds (Aloni, 1991, 1995; Aloni *et al.*, 1997a, 1997b). The production of callose on the sieve tubes in the ring-porous branches as a result of auxin application demonstrates that the applied auxin (1% NAA) constitutes an overdose for the sensitive ring-porous oak tissue.

The early development of phloem and xylem (before bud break) in ring-porous species is an ecological adaptation which prepares the vascular system of these trees to function immediately at the beginning of their growing season which is relatively short. In spring, the phloem system in diffuse-porous trees begins to function after bud break, while the phloem system and the xylem system are ready to function in ring-porous species before leaf enlargement. The key to the difference in seasonal vascular system reactivation is the heightened sensitivity of the system to growth regulators and possibly other developmental stimuli in ring-porous species. By having an early regeneration of their vascular systems, ring-porous trees have become adapted to make the most efficient use of the relatively short growing season available to them (Aloni *et al.*, 1997b).

8. Role of roots in vascular differentiation

Roots do not induce vascular differentiation nor must they be present in order to obtain vascular tissues in stems. The roots, however, have two major functions regarding vascular differentiation, namely: (i) the root orients the pattern of vascular differentiation towards the root tip by acting as a sink for the continuous flow of auxin deriving from young leaves (Sachs, 1968) and (ii) the root apices are sources of inductive stimuli that promote vascular development (Aloni, 1993). The major developmental signals of the root are cytokinins, which increase the sensitivity of the vascular meristems and conductive tissues to developmental signals originating in the leaves (Aloni, 1991, 1995) and promote the differentiation of both phloem and xylem (Aloni *et al.*, 1990; Baum *et al.*, 1991; Aloni, 1993, 1995). However, there is doubt if plants produce cytokinins by themselves. It has been recently suggested that cytokinins are produced by microbial symbionts of plants and not by the plants themselves (Holland, 1997).

Experiments with excised internodes of *Coleus*, receiving IAA in lanolin at their apical ends and cytokinins in aqueous solution at their bases, revealed that cytokinins are controlling factors in the regeneration of sieve-tubes (Aloni *et al.*, 1990) and vessels (Baum *et al.*, 1991; Aloni, 1993). Cytokinins also promote fibre differentiation (Aloni, 1982; Saks *et al.*, 1984). However, cytokinins alone, or root apices in the absence of an auxin source, do not induce vascular differentiation in stem tissues. There is evidence that along the stem axis of *Coleus* the internodes may show different responses to cytokinin (Baum *et al.*, 1991). This probably indicates that different sites in the plant body may operate at different levels of cytokinin. In addition, cytokinins may be subjected to diverse metabolic pathways in different plant organs (Bayley *et al.*, 1989). Studies of transgenic plants with altered levels of cytokinin confirm the involvement of cytokinin as a controlling factor in xylem differentiation (Medford *et al.*, 1989; Li *et al.*, 1992).

9. Summary

This chapter considers the role of ethylene in controlling the transition from procambium to cambium. It also provides a physiological basis for analysing xylem evolution (Carlquist, 1988) and stelar architecture (Stein, 1993; Zhong *et al.*, 1999) in land plants. We have detected three major hormonal trends which have shaped xylem evolution: (i) specialization of the signals for each vascular element, (ii) modification of the level of stimulation and (iii) a general increase in cambium sensitivity to hormonal signals. Typical evidence for these trends are presented: (i) during evolution, the original mechanism for tracheid differentiation, a combination of auxin and gibberellin, has become more specific for each xylem element: auxin by itself induces vessels, whereas gibberellin, in the presence of auxin, induces fibres, (ii) a high auxin concentration promotes perforation development in tracheids of pine, while removal of the apical bud and young leaves (major auxin sources) results in the formation of vessel endings in angiosperms and (iii) cambia of conifers are less responsive to exogenous auxin than cambia of hardwoods (Savidge and Wareing, 1981). In temperate deciduous hardwoods, an increase in cambium sensitivity occurred in limiting environments resulting in the development of ring-porous trees from diffuse-porous species. In ring-porous trees, the sensitive cambium enables wide earlywood vessel differentiation by low-level streams of auxin just before bud break. The promoting role of cytokinins in increasing the sensitivity of the cambium to developmental signals originating in leaves is also discussed.

There is a clear need to apply molecular genetics to uncover the genes involved in the control of vascular evolution in plants. For further progress in this field there is a need to combine developmental, physiological and genetic approaches for solving open questions of vascular differentiation. Efforts should be made to reveal the synthesis sites of plant hormones, their metabolic pathways and transport in the tissues, to understand the interactions between different plant hormones in controlling cambium activity and vascular differentiation, and to uncover the genetic basis for tissue sensitivity.

References

Aloni, R. (1979) Role of auxin and gibberellin in differentiation of primary phloem fibers. *Plant Physiol.* **63**: 609–614.

Aloni, R. (1980) Role of auxin and sucrose in the differentiation of sieve and tracheary elements in plant tissue cultures. *Planta* **150**: 255–263.

Aloni, R. (1982) Role of cytokinin in differentiation of secondary xylem fibers. *Plant Physiol.* **70**: 1631–1633.

Aloni, R. (1987) Differentiation of vascular tissues. *Annu. Rev. Plant Physiol.* **38**: 179–204.

Aloni, R. (1991) Wood formation in deciduous hardwood trees. In: *Physiology of Trees* (ed. A.S. Raghavendra). Wiley and Sons, New York, pp. 175–197.

Aloni, R. (1993) The role of cytokinin in organized differentiation of vascular tissues. *Aust. J. Plant Physiol.* **20**: 601–608.

Aloni, R. (1995) The induction of vascular tissues by auxin and cytokinin. In: *Plant Hormones: Physiology, Biochemistry and Molecular Biology*, 2nd edn. (ed. P.J. Davis), Kluwer Academic Publishers, Dordrecht, pp. 531–546.

Aloni, R., Alexander, J.A. and Tyree, M.T. (1997a) Natural and experimentally altered hydraulic architecture of branch junctions in *Acer saccharum* Marsh. and *Quercus velutina* Lam. trees. *Trees* **11**: 255–264.

Aloni, R. and Barnett, J.R. (1996) The development of phloem anastomoses between vascular bundles and their role in xylem regeneration after wounding in *Cucurbita* and *Dahlia*. *Planta* **198**: 595–603.

Aloni, R., Baum, S.F. and Peterson, C.A. (1990) The role of cytokinin in sieve tube regeneration and callose production in wounded *Coleus* internodes. *Plant Physiol.* **93**: 982–989.

Aloni, R. and Griffith, M. (1991) Functional xylem anatomy in root-shoot junctions of six cereal species. *Planta* **184**: 123–129.

Aloni, R. and Inding, F.E. (1989) An experimental method for studying the differentiation of vessel ending. *Ann. Bot.* **64**: 589–592.

Aloni, R. and Peterson, C.A. (1990) The functional significance of phloem anastomoses in stems of *Dahlia pinnata* Cav. *Planta* **182**: 583–590.

Aloni, R. and Peterson, C.A. (1997b) Auxin promotes dormancy callose removal from the phloem of *Magnolia kobus* and callose accumulation and earlywood vessel differentiation in *Quercus robur*. *J. Plant Res.* **110**: 34–44.

Aloni, R., Raviv, A. and Peterson, C.A. (1991) The role of auxin in the removal of dormancy callose and resumption of phloem activity in *Vitis vinifera*. *Can. J. Bot.* **69**: 1825–1832.

Aloni, R. and Wolf, A. (1984) Suppressed buds embedded in the bark across the bole and the occurrence of their circular vessels in *Ficus religiosa*. *Am. J. Bot.* **71**: 1060–1066.

Aloni, R., Wolf, A., Feigenbaum, P., Avni, A. and Klee, H.J. (1998) The *Never ripe* mutant provides evidence that tumor-induced ethylene controls the morphogenesis of *Agrobacterium tumefaciens*-induced crown galls on tomato stems. *Plant Physiol.* **117**: 841–847.

Aloni, R. and Zimmermann, M.H. (1983) The control of vessel size and density along the plant axis — a new hypothesis. *Differentiation* **24**: 203–208.

Bailey, I.W. (1944) The development of vessels in angiosperms in morphological research. *Am. J. Bot.* **31**: 421–428.

Bailey, I.W. (1953) Evolution of the tracheary tissue of land plants. *Am. J. Bot.* **40**: 4–8.

Bailey, I.W. and Tupper, W.W. (1918) Size variation in tracheary cells. I. A comparison between the secondary xylems of vascular cryptograms, gymnosperms and angiosperms. *Amer. Acad. Art Sci. Proc.* **54**: 149–204.

Bannan, M.W. (1958) An occurrence of perforated tracheids in *Thuja occidentalis* L. *New Phytol.* **57**: 132–134.

Baum, S.F., Aloni, R. and Peterson, C.A. (1991) The role of cytokinin in vessel regeneration in wounded *Coleus* internodes. *Ann. Bot.* **67**: 543–548.

Bayley, A.D., Van Staden, J., Mallett, J.A. and Drewes, S.E. (1989) The *in vitro* metabolism of [8-^{14}C]benzyladenine by excised organs of tomato plants. *Plant Growth Regul.* **8**: 193–204.

Bierhorst, D.W. (1958) Vessels in *Equisetum*. *Am. J. Bot.* **45**: 534–537.

Bradford, K.J. and Trewavas, A.J. (1994) Sensitivity thresholds and variable time scales in plant hormone action. *Plant Physiol.* **105**: 1029–1036.

Bruck, D.K. and Paolillo, Jr., D.J. (1984) Replacement of leaf primordia with IAA in the induction of vascular differentiation in the stem of *Coleus*. *New Phytol.* **96**: 353–370.

Cai, C., Ouyang, S., Wang, Y., Fang, Z., Rong, J., Geng, L. and Li, X. (1996) An early Silurian vascular plant. *Nature* **379**: 592.

Carlquist, S. (1988) *Comparative Wood Anatomy, Systematic, Ecological, and Evolutionary Aspects of Dicotyledon Wood.* Springer-Verlag, Berlin.

Carmi, A., Sachs, T. and Fahn, A. (1972) The relation of ray spacing to cambial growth. *New Phytol.* **71**: 349–353.

Catesson, A.M. (1964) Origine, functionnement et variations cytologiques saisonières du cambium de l'*Acer pseudoplatanus* L. (Acèracèes). *Ann. Sci. Nat. Bot., 12e ser.*, **5**: 229–503.

Digby, J. and Wareing, P.F. (1966) The effect of applied growth hormones on cambial division and the differentiation of the cambial derivatives. *Ann. Bot.* **30**: 539–548.

Duerden, H. (1940) On the xylem elements of certain ferns. *Ann. Bot.* **4**: 523–531.

Edwards, D., Davies, K.L. and Axe, L. (1992) A vascular conducting strand in the early land plant *Cooksonia*. *Nature* **357**: 683–683.

Esau, K. (1965) *Plant Anatomy*, 2nd edn. Wiley and Sons, New York.

Fahn, A. (1990) *Plant Anatomy*. Pergamon Press, Oxford.

Fahn, A., Ben-Sasson, R. and Sachs, T. (1972) The relation between the procambium and the cambium. In: *Research Trends in Plant Anatomy* (ed. A.K.M. Ghouse), Tata McGraw-Hill, New Delhi, pp. 161–170.

Fukuda, H. (1996) Xylogenesis: initiation, progression, and cell death. *Annu. Rev. Plant Physiol. Plant Mol. Biol.* **47**: 299–325.

Gutherie, R.L. (1989) Xylem structure and ecological dominance in a forest community. *Am. J. Bot.* **76**: 1216–1228.

Hardtke, C.S., and Berleth, T. (1998) The *Arabidopsis* gene *MONOPTEROS* encodes a transcription factor mediating embryo axis formation and vascular development. *EMBO J.* **17**: 1405–1410.

Holland, M.A. (1997) Are cytokinins produced by plants? *Plant Physiol.* **115**: 865–868.

Jacobs, W.P. (1952) The role of auxin in differentiation of xylem around a wound. *Am. J. Bot.* **39**: 301–309.

Jacobs, W.P. (1998) Hormone transport and action in the green shoot: long-term studies of a clonal stock of *Coleus blumei* (Labiatae). *Am. J. Bot.* **85**: 1353–1363.

Jane, F.W. (1956) Perforated vertical tracheids in *Sequoia sempervirens* Endl. *New Phytol.* **55**: 367–369.

Kalev, N. and Aloni, R. (1998) Role of auxin and gibberellin in regenerative differentiation of tracheids in *Pinus pinea* L. seedlings. *New Phytol.* **138**: 461–468.

Kalev, N. and Aloni, R. (1999) Role of ethylene and auxin in regeneration and orientation of tracheids in *Pinus pinea* L. seedlings. *New Phytol.* **142**: 307–313.

LaMotte, C.E. and Jacobs, W.P. (1963) A role of auxin in phloem regeneration in *Coleus* internodes. *Dev. Biol.* **8**: 80–98.

Larson, P.R. (1982) The concept of cambium. In: *New Perspective in Wood Anatomy* (ed. P. Baas), Martinus Nijhoff Publishers, The Hague, pp. 88–121.

Larson, P.R. (1994) *The Vascular Cambium, Development and Structure*. Springer-Verlag: Berlin.

Lechowicz, M.J. (1984) Why do temperate deciduous trees leaf out at different times? Adaptation and ecology of forest communities. *Am. Nat.* **124**: 821–842.

Lev-Yadun, S. and Aloni, R. (1990) Vascular differentiation in branch junctions of trees: circular patterns and functional significance. *Trees* **4**: 49–54.

Lev-Yadun, S. and Aloni, R. (1995) Differentiation of the ray system in woody plants. *Bot. Rev.* **61**: 45–84.

Li, Y., Hagen, G. and Guilfoyle, T.J. (1992) Altered morphology in transgenic tobacco plants that overproduce cytokinins in specific tissues and organs. *Dev. Biol.* **153**: 386–395.

Li, H., Taylor, E.L. and Taylor, N.T. (1996) Permian vessel elements. *Science* **271**: 188–189.

Liphscitz, N. (1995) Ecological wood anatomy: changes in xylem structure in Israeli trees. In: *Wood Anatomy Research 1995*, (ed. W. Shuming), Proc. Inter. Symp. Tree Anatomy and Wood Function, Tianjin, China. International Academic Publishers, Beijing, pp. 12–15.

Luxová, M. (1986) The hydraulic safety zone at the base of barley roots. *Planta* **169**: 465–470.

Mattsson, J., Sung, Z.R. and Berleth, T. (1999) Responses of plant vascular systems to auxin transport inhibition. *Development* **126**: 2979–2991.

Medford, J.L., Horgan, R., El-Sawi, Z. and Klee, H.J. (1989) Alteration of endogenous cytokinins in transgenic plants using chimeric isopentenyl transferase gene. *Plant Cell* **1**: 403–413.

Nelson, T. (1998) Polarity, vascularization and auxin. *Trends Plant Sci.* **3**: 245–246.

Priestley, J.H. and Scott, L.I. (1936) A note upon summer wood production in the trees. *Proc. Leeds. Phil. Soc.* **3**: 235–248.

Reinders-Gouwentak, C.A. (1965) Physiology of the cambium and other secondary meristems of the shoot. In: *Handbuch der Pflanzenphysiologie XV/1* (ed. W. Ruhland), pp. 1077–1105. Springer, Berlin.

Roberts, L.W., Gahan, P.B. and Aloni, R. (1988) *Vascular Differentiation and Plant Growth Regulators.* Springer-Verlag, Berlin.

Sachs, T. (1968) The role of the root in the induction of xylem differentiation in peas. *Ann. Bot.* **32**: 391–399.

Sachs, T. (1981) The control of the patterned differentiation of vascular tissues. *Adv. Bot. Res.* **9**: 151–162.

Sachs, T. and Cohen, D. (1982) Circular vessels and the control of vascular differentiation in plants. *Differentiation* **21**: 22–26.

Saks, Y. and Aloni, R. (1985) Polar gradients of tracheid number and diameter during primary and secondary xylem development in young seedlings of *Pinus pinea* L. *Ann. Bot.* **56**: 771–778.

Saks, Y., Feigenbaum, P. and Aloni, R. (1984) Regulatory effect of cytokinin on secondary xylem fiber formation in an *in vivo* system. *Plant Physiol.* **76**: 638–642.

Saleo, S., LoGullo, M.A. and Siracusano, L. (1984) Distribution of vessel ends in stems of some diffuse and ring-porous trees: the nodal region as 'safety zones' of the water conducting system. *Ann. Bot.* **54**: 543–552.

Sano, Y., Kawakami, Y. and Ohtani (1999) Variation in the structure of intertracheary pit membranes in *Abies sachalinensis*, as observed by field-emission scanning electron microscopy. *IAWA J.* **20**: 375–388.

Savidge, R.A. (1996) Xylogenesis, genetic and environmental regulation. *IAWA J.* **17**: 269–310.

Savidge, R.A. and Wareing, P.F. (1981) Plant growth regulators and the differentiation of vascular elements. In: *Xylem Cell Development* (ed. J.R. Barnett) Castle House Publications. Tunbridge Wells, UK, pp. 192–235.

Shibaoka, H. (1994) Plant hormone-induced changes in orientation of cortical microtubule: alteration in the cross-linking between microtubule and the plasma membrane. *Annu. Rev. Plant Physiol.* **45**: 527–544.

Stein, W. (1993) Modeling the evolution of stelar architecture in vascular plants. *Int. J. Plant Sci.* **154**: 229–263.

Suzuki, M., Yoda, K. and Suzuki, H. (1996) Phenological comparison of the onset of vessel formation between ring-porous and diffuse-porous deciduous trees in a Japanese temperate forest. *IAWA J.* **17**: 431–444.

Thompson, N.P. and Jacobs, W.P. (1966) Polarity of IAA effect on sieve-tube and xylem regeneration in *Coleus* and tomato stems. *Plant Physiol.* **41**: 673–682.

Timell, T.E. (1986) *Compression Wood in Gymnosperms*. Vol. 2. Springer-Verlag, Berlin.

Uggla, C., Mellerowicz, E.J. and Sundberg, B. (1998) Indole-3-acetic acid controls cambial growth in Scots pine by positional signaling. *Plant Physiol.* **117**: 113–121.

Uggla, C., Moritz, T., Sandberg, G. and Sundberg, B. (1996) Auxin as a positional signal in pattern formation in plants. *Proc. Natl. Acad. Sci. USA* **93**: 9282–9286.

Warren Wilson, J. and Warren Wilson, P.M. (1993) Mechanisms of auxin regulation of structural and physiological polarity in plants, tissues, cells and embryos. *Aust. J. Plant Physiol.* **20**: 555–571.

Wheeler, E.A. and Baas, P. (1991) A survey of the fossil record for dicotyledonous wood and its significance for evolutionary and ecological wood anatomy. *IAWA Bull. n.s.* **12**: 275–332.

Wiebe, H.H., Greer, R.L. and Van Alfen, N.K. (1984) Frequency and grouping of vessel endings in alfalfa (*Medicago sativa*) shoots. *New Phytol.* **97**: 583–590.

Zhong, R., Taylor, J.T. and Ye, Z.-H. (1997) Disruption of interfascicular fiber differentiation in an *Arabidopsis* mutant. *Plant Cell* **9**: 2159–2170.

Zhong, R., Taylor, J.J. and Ye Z.-H. (1999) Transformation of the collateral vascular bundles into amphivasal vascular bundles in an *Arabidopsis* mutant. *Plant Physiol.* **120**: 53–64.

Zimmermann, M.H. (1983) *Xylem Structure and the Ascent of Sap*. Springer-Verlag, Berlin.

Zimmermann, M.H. and Sperry, J.S. (1983) Anatomy of the palm *Rhapis excelsa*. 9: Xylem structure of the leaf insertion. *J. Arnold Arbor.* **64**: 599–609.

18

Inductive effects of conifer needles on xylogenesis — auxin not the explanation

R.A. Savidge

1. Introduction

1.1 *The overall response*

Under the experimental conditions shown in *Figure 1a,* mature living conifer needles on isolated stem segments induce cells of the cambial zone to differentiate directly into tracheary elements (TEs) (Savidge and Wareing, 1981a; Savidge, 1994). The location of the differentiation response is restricted to the junction of the needle's vascular trace with the vascular cambium. The response does not occur at the trace-cambium junction if its associated needle, or needle cluster, has been either removed or, in older stems, has naturally abscised, despite the continuing presence at the junction of a short-shoot or axillary apical meristem. Thus, the factor triggering differentiation at the trace-cambium junction evidently is either something exported from the needle, something taken up by the needle (e.g., through the transpiration stream), or both.

The overall differentiation response results in the conversion of the cambium into a woody fusiform-shaped area, axially elongated but rarely extending more than two millimetres above and below the trace-cambium junction. To left and right of the junction, the woody area includes only those cambial cells within close proximity to the egression point of the vascular trace from the stem (*Figure 1b*). The differentiation response is invariably strongest immediately at the junction, resulting in a closed arc of new xylem which sometimes is sufficiently large to be visible with the unaided eye (*Figure 1b*). With increasing distance from the junction, a decreasing proportion of the cell population within the cambial zone becomes TEs. As explained below, under certain conditions the overall response domain can be strongly enhanced.

Cell & Molecular Biology of Wood Formation, edited by R. Savidge, J. Barnett and R. Napier.

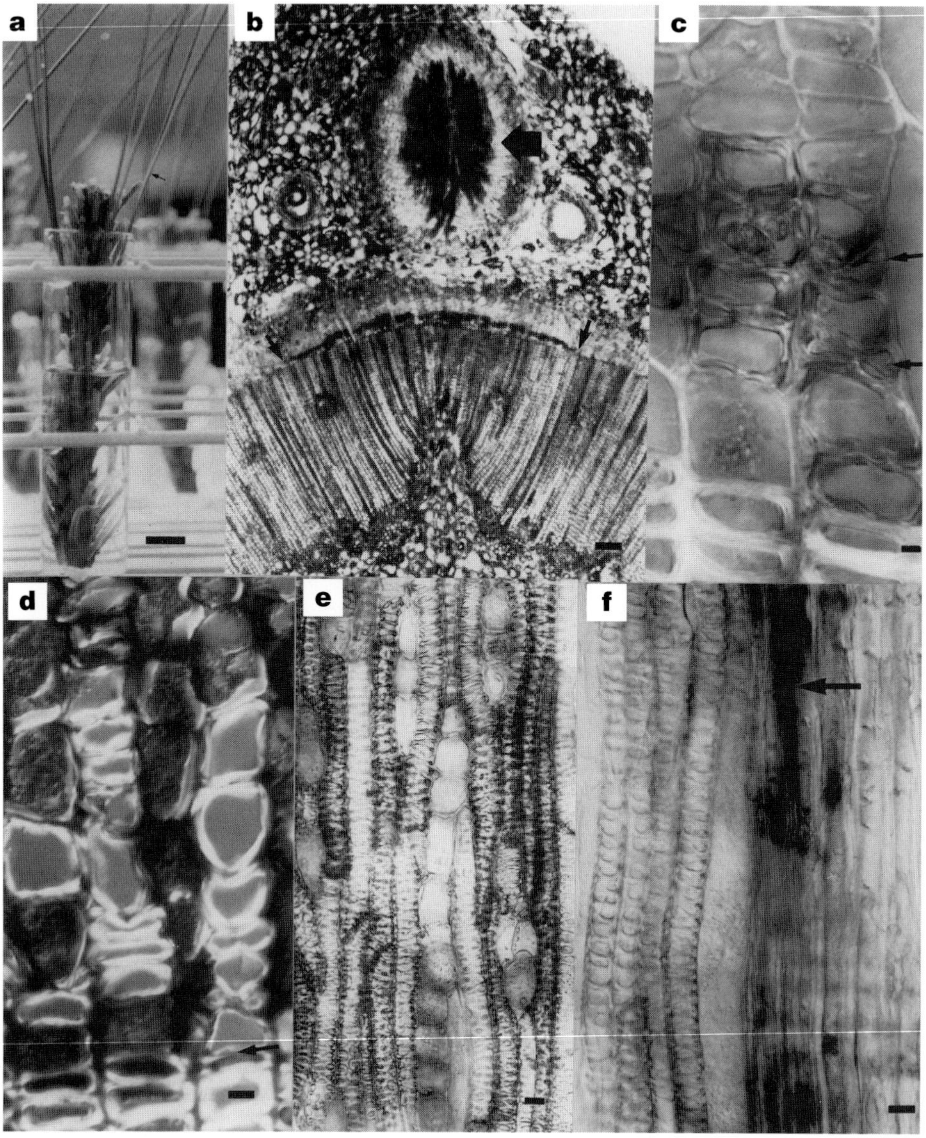

1.2 *The differentiation response is novel*

The induced differentiation occurring at the trace-cambium junction evidently is without precedent in the plant science literature in that 'dormant' cells convert into TEs before either vacuolation or resumption of cell-division activity has occurred. Vacuolation, cell-division activity and radial expansion of primary-walled cambial derivatives normally always occur in the cambium in springtime prior to the initiation of earlywood formation. Because differentiation is induced to occur in dormant cells, a localized region of the cambial zone *per se* differentiates into TEs (*Figure 1b*). In the immediate region of the trace-cambium junction, it is not uncommon for every fusiform cell of a radial file to be transformed, in effect eliminating entire radial files

Figure 1. *(a) A 2-year-old dormant* Pinus resinosa *cutting standing in water in a tube, to show the experimental setup. Three to four pairs of needles (arrow) can be spaced equidistantly around the circumference to enable genotypic replication. Bar = 1 cm. (b) Fresh hand-cut cross section (unstained, brightfield) of 'dormant'* P. resinosa *showing the short-shoot vascular trace (solid arrow) in the cortex which connects the stem vasculature with that of the short-shoot and associated pine needles. The arc of tracheids (between the two small arrows) induced in the cambial zone a short distance apical to the trace-cambium junction is also clearly visible. Note the opacity of the xylem within both the arc and the trace and how it contrasts with the secondary xylem in the stem. Bar = 100 μm. (c) Fresh hand-cut cross section (unstained, Nomarski optics) of* P. resinosa *showing two radial files of the 'dormant' cambial zone that were induced to differentiate into tracheids. Note that the arrowed tracheids have the radially narrow dimensions of fusiform cambial cells, an indication that they underwent no radial expansion before differentiating. Normal late wood tracheids are at the bottom. Bar = 2 μm. (d) A 12 μm thick safranin–fast green stained cross section of FAA-fixed 'dormant' cambial zone of* P. resinosa *showing radial files of fusiform cells that differentiated into tracheids at the trace-cambium junction. Phloem is at the top; the late wood boundary is arrowed. Nomarski optics. Bar = 2 μm. (e) Fresh hand-cut tangential section through the 'dormant' cambium of* P. resinosa *in the region of the trace-cambium junction showing the primary-xylem-like character of the tracheids induced to differentiate in the cambial zone. Bar = 10 μm. (f) Fresh hand-cut radial section of* P. resinosa *in the vicinity of the trace-cambium junction showing induced tracheids on the left that differentiated in the 'dormant' stem, and normal late wood tracheids (right). The darker regions (arrowed) of normal late wood tracheids were produced in response to the treatment and are due to the presence of an intense orange to rust-coloured chromophore in the walls of those elements. Bar = 10 μm.*

---

from the cambium (*Figures 1c and 1d*). As is evident in *Figures 1c* and *1d,* radial enlargement of fusiform cambial cells sometimes does, and sometimes does not occur in response to the single needle treatment. Not uncommonly, the radially compressed fusiform cells of the cambial zone differentiate without any radial enlargement occurring. Thus, it appears that cell expansion also is not an essential preliminary for differentiation to occur. Cell-division and expansion are predictable responses of the cambium to exogenous auxin, again an indication that the inductive mechanism exerted by needles cannot be explained in terms of auxin effects.

The response at the trace-cambium junction is all the more novel in that the cambial cells are induced to transform into primary-xylem TEs (*Figures 1e and 1f*). Bordered-pitted secondary-xylem tracheids normally arise from the conifer cambium, whereas primary xylem TEs usually are associated with activity of the procambium during shoot, root and needle elongation. Thus, the single-needle treatment seemingly causes the vascular cambium to revert to procambium. With increasing distance from the trace-cambium junction (*Figure 2a*), only isolated fusiform cells within the cambial zone are found to have differentiated, indicating that some cells are more readily induced than their neighbours. Cambial ray cells occasionally also differentiate and, in addition, approximately 10% of the population of living ray parenchyma in the xylem (a lower proportion in the phloem) near the trace-cambium junction are induced to differentiate into TEs in response to the presence of a single intact needle pair. Differentiation of ray cells appears never to be preceded by either cell-division or cell-expansion activity.

In addition to primary TEs, secondary-xylem tracheids sometimes also are produced in dormant cuttings in response to the single needle treatment, but primary types are the more conspicuous. The secondary-xylem TEs can be distinguished by the presence of bordered pits, or bordered-pit-like structures, in their otherwise massive

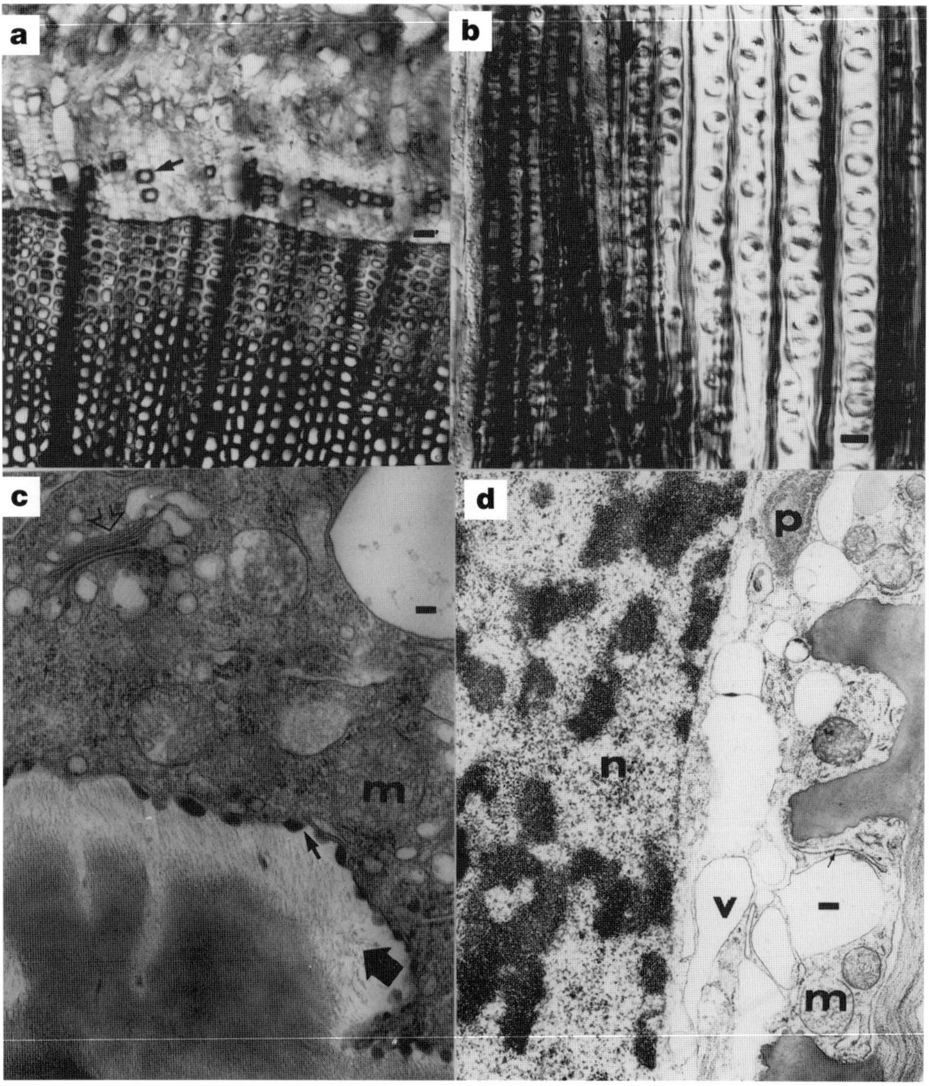

secondary-wall lamellae. In single-needle experiments with *Pinus contorta*, secondary xylem TEs were produced in the vicinity of the immediate junction of the needle trace with the stem cambium (Savidge and Wareing, 1981a). However, with increasing distance above and below that junction, a gradation in the extent of secondary-wall deposition occurred, varying from scalariform wall thickenings to simple annular ribs and, beyond that, into cells exhibiting no evidence for differentiation (Savidge and Wareing, 1981a).

If *Pinus resinosa* stem segments having single needle pairs are prepared in the illustrated manner (*Figure 1a*) after active earlywood formation has already commenced, the differentiation processes occurring in the cambial region at the trace-cambium junction abruptly alter, leading to production of radially narrow rather than radially wide tracheids (*Figure 2b*). Thus, the induced TEs arising at the junction during cambial

Figure 2. *(a) Fresh hand-cut cross section of* Pinus resinosa *showing scattered tracheids (arrowed) that were produced in the 'dormant' cambial zone at about 1 mm from the trace-cambium junction. Bar = 20 μm. (b) Fresh hand-cut radial section of* P. resinosa *showing dark late wood tracheids (extreme right), five light-coloured earlywood tracheids, and radially narrow tracheids with darkened walls. The latter were induced to differentiate by treating the cutting with active cambium in the way shown in* Figure 1a. *The point where an abrupt cessation of normal early wood formation was followed by needle-pair induced TE differentiation is indicated by the arrow at top. Bar = 10 μm. (c) Transmission electron micrograph (glutaraldehyde–OsO_4 fixed, uranyl acetate–lead citrate stained) of a 'dormant' fusiform cambial cell of* Pinus contorta *actively differentiating into a tracheid near the trace-cambium junction. A dense coarse protoplasm with abundant dictyosomes (open arrow), mitochondria (m), rough and smooth ER and free polyribosomes (small arrow) is characteristic of induced differentiation. The presence of electron-dense exocytotic blebs (large arrow) derived from the ER is highly diagnostic of active differentiation in this system (see Savidge and Barnett 1993). Note the non-coalesced fibrillar material (large solid arrow) on the extremities of the developing rib of secondary wall. Bar (within a vacuole) = 0.1 μm. (d) Same trace-cambium junction as shown in* Figure. 2c, *but showing an induced tracheid at a more advanced stage of differentiation. Secondary-wall formation (right side) is complete and digestion of the protoplasm is accompanied by increased vacuolation (v). A nucleus (n) with densely staining chromatin occupies the left half of this figure. Mitochondria (m), proplastids (p) and rough ER (small arrow) are also evident. Bar (within a vacuole) = 0.25 μm.*

activity are distinct from those produced during dormancy in being mostly secondary xylem TEs whereas those induced when the cambium is dormant are, for the most part, primary xylem TEs. The abrupt transition from radially enlarged to radially narrow TEs in response to the single needle pair treatment during earlywood formation againsuggests that auxin is not the inductive factor derived from the needle.

1.3 *Fine structure of differentiating cells*

Transmission electron microscopy of active differentiation of needle-induced 'dormant' cambial-zone cells yielded a number of intriguing observations (Savidge and Barnett, 1993). The protoplasm during differentiation does not become highly vacuolated, as normally occurs during wood formation; rather, it increases in electron density (in glutaraldehyde–osmium tetroxide fixed and lead citrate–uranyl acetate stained cells), becomes coarsely granular and appears to be strongly enriched in polyribosomes, endoplasmic reticulum, Golgi and mitochondria (note that quantitative studies on organellar changes have yet to be done). Electron-dense exocytotic blebs are conspicuously associated with and highly diagnostic of cells actively differentiating into TEs (*Figure 2c*), those blebs being not at all associated with either cells that have completed their secondary-wall deposition (*Figure 2d*) nor with non-differentiating cells. Remarkably, during induced differentiation of dormant cambial cells into TEs, no evidence for microtubules has yet been seen to be associated with the developing ribs of secondary wall material (*Figure 2c*), although microtubules have commonly been observed to be associated with secondary wall formation during normal secondary xylem development. We are continuing to search for microtubules in association with the needle-induced differentiation response; however, if in fact microtubules are not present, the conclusion would seem to be inescapable that microtubules are unnecessary for secondary-wall formation (see Savidge and Barnett, 1993).

When deposition of secondary-wall material during induced differentiation is completed, the protoplasm begins to autolyse in association with ballooning of cisternae of endoplasmic reticulum (ER) into small vacuoles. As those small vacuoles increase in size they fuse to create one large vacuole and, concomitantly, the plasma membrane retracts from the walls. Nuclei, rough ER, mitochondria and proplastids persist until late in the autolytic process (*Figure 2d*), when they deteriorate and eventually disappear (Savidge and Barnett, 1993).

1.4 *Auxin from needles not the explanation*

Soon after the needle-induced differentiation response at the trace-cambium junction was discovered, the explanation for it was sought within the concept of auxin (i.e., indol-3-ylacetic acid, IAA) production within and export to the cambium from needles (Savidge and Wareing, 1981a, 1981b, 1982, 1984). There is no doubt that conifer needles contain endogenous IAA nor that removal of needles results in strongly reduced IAA content in the cambium, even during dormancy (Savidge and Wareing, 1981b, 1982, 1984). In addition, exogenous auxin when applied to young (1-year-old or less) dormant stem cuttings has, for various species of the Coniferophyta and Magnoliophyta, frequently been reported to be effective in inducing cambial cell division, radial expansion of cambial derivatives, and differentiation of cambial cells into TEs. For conifer stem segments, however, the evidence indicates that living intact needles also are essential if exogenous auxin is to promote wood formation. That is, numerous instances have been seen where only cambial cell division and primary-wall radial expansion occurred in response to exogenous auxin applications to artificially defoliated young stem segments (Savidge and Wareing, 1981a; Savidge, 1983). Moreover, in defoliated stem segments having cambia aged more than two years, or in stem segments lacking needles through natural abscission, the cambium has not been found to produce tracheids in response to exogenous auxin applications (Savidge, 1983). Thus, it appears that the effectiveness of exogenous auxin in inducing cambial activity and wood formation in young foliated conifer cuttings has part of its explanation in the juvenile physiology of that material, possibly because it contains storage reserves or has epigenetic competence not present in older stem regions.

Repeated investigations using the experimental design indicated in *Figure 1a* provided no support for the hypothesis that auxin issuing from the needles could be the explanation for the differentiation response occurring at the trace-cambium junction (e.g., Savidge and Wareing, 1981a; Savidge, 1983, 1994). In fact, the observations resulted in a distinction between the roles of IAA and the needle factor in controlling cambial growth. As explained above, the single needle pair induced differentiation of cambial cells into TEs in the absence of normal cambial reactivation. When exogenous IAA was applied to the apical ends of young stem segments having single needle pairs, as shown in *Figure 1a*, the cambial fusiform cells cleared their protoplasm (i.e., became highly vacuolated) and divided periclinally to produce radially enlarged derivatives as normally occurs during springtime cambial reactivation. However, differentiation of the enlarged cambial derivatives into TEs was restricted to the region of the trace-cambium junction having an intact needle pair (Savidge and Wareing, 1981a).

The possibility that a combination of auxin and one or more additional phytohormones could induce the differentiation response was also extensively investigated. Cytokinins were particularly of interest within the hypothesis that the intact needle pair could be drawing them to the trace-cambium junction in the transpiration stream.

Moreover, findings with *Zinnia* mesophyll and other experimental *in vitro* systems have pointed to an auxin-cytokinin combination being essential for induction of differentiation into TEs. However, experiments with various cytokinins and at several auxin-cytokinin molar ratios yielded no evidence that the response at the trace-cambium junction can be explained in this way (Savidge and Wareing, 1981a, 1981b; Savidge, 1983; Savidge and Barnett, 1993; Savidge, 1994). It should be emphasized, however, that not all cytokinins have been investigated, nor have all auxin conjugates or all possible ratios of those auxins and cytokinins.

1.5 *A 'tracheid-differentiation factor'*

In view of the above and additional findings (Savidge and Wareing, 1981a, 1981b), it was proposed that a tracheid-differentiation factor (tdf) distinct from phytohormones or other known regulators of growth and development was produced in and exported from conifer needles. Thus, in effect it was proposed that the cambium is not determined, or epigenetically committed, to producing tracheids, rather is merely competent to produce cells which, upon receiving the appropriate differentiation signal, are induced to differentiate as TEs or other cell types (Savidge, 1994, 1996). This reasoning is in line with the concept of the cellular phenotype arising as an interaction between the genotype and the environment, but it does not exclude the possibility that some epigenetic changes also attend cambial growth. It was also suggested that the tdf controlled how much secondary-wall deposition occurred during differentiation, presumably by tdf concentration in the cambial zone becoming diluted with increasing distance from the trace-cambium junction in the described experimental system (Savidge and Wareing, 1981a). More recently, this hypothesis was restated to the effect that 'the activation, duration and synchronization of structural gene expression is regulated by one or only a few regulatory factors — possibly proteins produced from homeotic genes — which activate families of structural genes in general' (Savidg, 1996).

Wood density is an important quality consideration for wood-processing industries, and density increases as the extent of secondary wall thickening increases during wood formation. Thus, the existence of a needle-generated transmissible tdf controlling both the initiation and the extent of secondary-wall formation during differentiation clearly is a matter of fundamental importance to wood production and utilization. This chapter reviews and updates the state of knowledge about the tdf and its associated effects.

2. Recent developments

2.1 *D-myo-inositol-1,4,5-*tris*-phosphate (IP_3)*

It was observed that D-*myo*-inositol 1,4,5-*tris*-phosphate (IP_3) from bovine brain (Sigma Chemical Company, I 4009) when applied as a one-off dose (2 μg in water) to the apical surfaces of pine-needle stumps elicited unusual tracheid-differentiation in the cambial zone, similar to that induced by intact needles (Savidge, 1994). The differentiation response was actually somewhat unusual from that induced by intact needles in that the 'tracheids' developing in the cambial zone exhibited secondary wall features even more extraordinary than those induced by an intact single needle pair (*Figures 3a, 3b*). Investigations into ^3H–IP_3 movement after applying the radiolabelled compound as a droplet to the apical surface of a needle stump indicated that the ^3H moved downward through the needle tissue into the cambium and, in the stem cambium, moved apically as well as basally, in agreement with tracheid differentiation being induced both above and below the trace-cambium junction by intact conifer needles (Savidge, 1994). A large number of additional compounds – phytohormones, sugars, vitamins, amino acids and other substances – when applied to needle stumps in the same manner were unable to reproduce the inductive effects seen with intact needle pairs or exogenous bovine IP_3; thus, it was considered that IP_3 could be the endogenous tdf (Savidge, 1994).

The observations of IP_3 induction of tracheid differentiation were based on several well-replicated and controlled experiments and, as illustrated (*Figures 3a, 3b*), extraordinary phenotypes of tracheary elements were produced in response to the IP_3 treatment, whereas no differentiation occurred in any of the control cuttings. However, just as intact needles are not always effective in inducing differentiation at

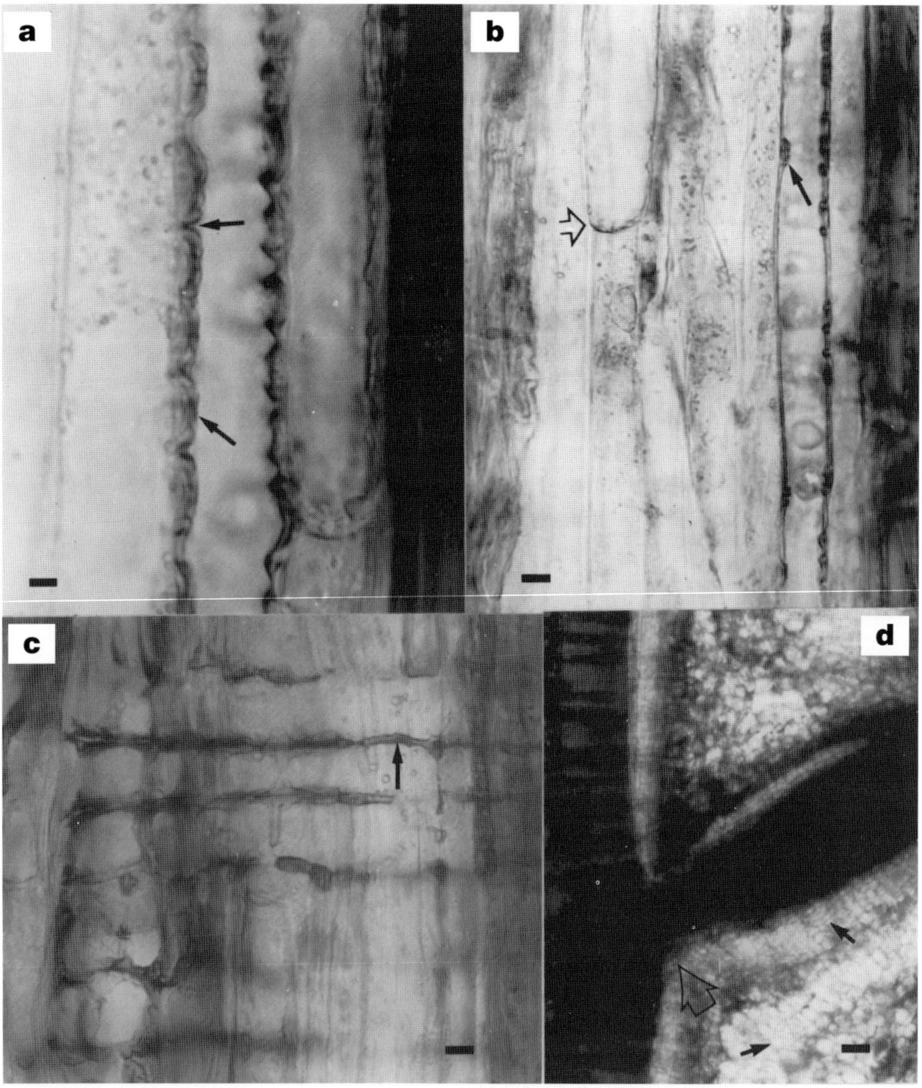

the trace-cambium junction, following the report of IP_3 induction (Savidge, 1994) subsequent investigations yielded variable, sometimes nil, results with IP_3 (R.A. Savidge, unpublished). The reason for this is not at all clear; however, as noted previously (Savidge, 1994), IP_3 as purchased commercially is far from a pure preparation. Moreover, the high cost of the commercial preparation has been prohibitive to large scale investigations.

There can be considerable variation among stem segments in the tdf bioassay when evaluated qualitatively (i.e., based on the presence or absence of new tracheids in the cambial zone), and the reason for this remains unclear (Savidge, 1994). Neither cutting size (length, diameter) nor the stage of dormancy appear to be factors influencing the qualitative response. Nevertheless, the variation in the response could be explainable in terms of varied genotypes or other physiological factors (Savidge, 1994).

2.2 *Wounding induction of differentiation*

Wounding is known to result in ethylene production, and there is some evidence that ethylene has a role in differentiation of cambial derivatives into TEs (Savidge, 1988). However, neither direct wounding of pine cuttings nor applications of ethylene precursors (Ethrel = 2-chloroethylphosphonic acid; ACC = 1-aminocyclopropane-1-carboxylic acid) alone or in combination with exogenous auxin have been found to induce tracheid differentiation in dormant cambial zone cells. Moreover, other research indicated that cambial wounding inhibited both cambial activity and tracheid differentiation during cambial growth (Savidge and Farrar, 1984).

In experiments involving injection of microlitre volumes of water-soluble compounds (individual phytohormones, combinations of phytohormones and other suspected regulators) into needle-stump bases, it was noted that the controls (receiving only distilled H_2O injected into the base of needle stumps) produced new tracheids (R.A. Savidge, unpublished). That observation led to experiments to determine the effect that wounding of the base of needle stumps had on differentiation at the trace-cambium junction. Pine needle pairs (*P. resinosa*) intact on cuttings, as per *Figure 1a*, were cut through the prophyll sheath such that only five-millimetre-long

Figure 3. *(a) Fresh hand-cut radial section of 'dormant' cambial zone of* Pinus resinosa *in the region of the trace-cambium junction showing the unusual secondary wall formation (arrowed) induced in fusiform cambial cells by application of* D-myo-*inositol 1,4,5-*trisphosphate *(IP_3) as a droplet to the apical surface of a five-millimetre long needle-pair stump. The dark elements on the right are late wood tracheids. Bar = 1 μm. (b) Same conditions as* Figure 3a, *showing other examples of highly unusual secondary-walled formation (arrows) induced by application of IP_3. Latewood is on the right, phloem on the left. Bar = 2 μm. (c) Fresh hand-cut radial section of 'dormant' cambial zone of* P. resinosa *in the region of the trace-cambium junction showing unusual wall thickenings ('wall bars' – see text) on ray cell walls (arrowed) that were induced to form in response to the treatment shown in* Figure 1a. *Latewood is on the right and phloem on the left. Bar = 1 μm. (d) Fresh hand-cut radial section showing the trace-cambium junction (large open arrow) of* P. resinosa *after being provided a combination of 1-naphthalene acetic acid and zeatin riboside (each at 1 mg ml^{-1}) in 5 μl water by injecting the solution into a waxed well (see* Figure 4b*). Tracheid differentiation was not induced at the trace-cambium junction; however, the treatment promoted cell division and expansion hence proliferation of parenchyma cells within the xylem of the short-shoot trace. Proliferation and enlargement of cambial and cortical cells around the short shoot (solid arrows) also occurred, resulting in the unusual swollen appearance of the short-shoot region. Stem latewood xylem is on the left. Bar = 0.1 mm.*

basal stumps remained attached to the stem segment. Wounding of those residual needle-pair bases was done by inserting a sterile metal needle (27G ½, Becton Dickinson no. 305109) once through the prophyll sheath into the white to yellow needle tissue at the base of the prophyll sheath. The depth of penetration was determined, by observation of hand-cut radial sections, to extend to the vicinity of the short-shoot apical meristem supporting the needle pair. Control stumps were not wounded. As indicated in *Figure 4a*, following a month of incubation of the cuttings at room temperature, more than 50% of the wounded stumps responded by producing new tracheids at the trace-cambium junction, whereas control stumps produced no new tracheids.

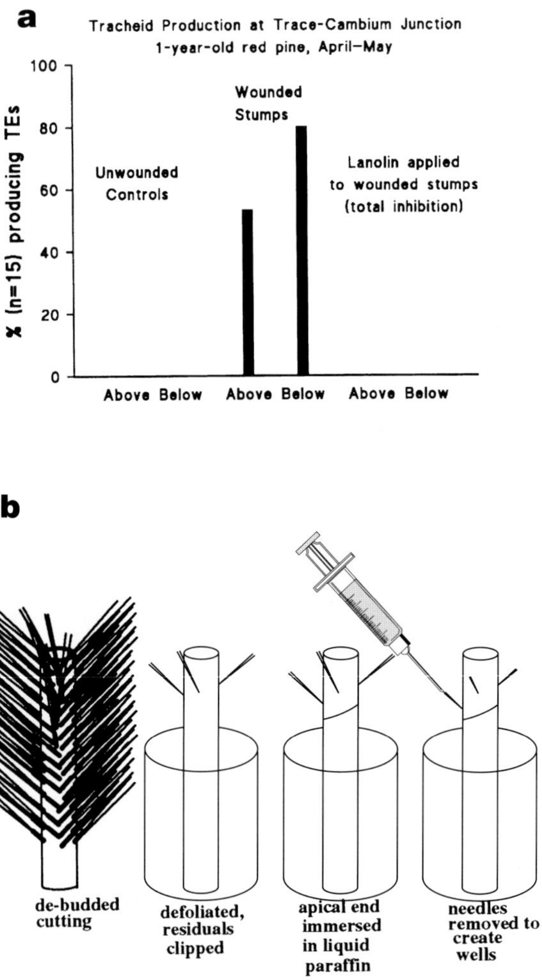

Figure 4. *(a) Percentage of trace-cambium junctions induced to produce new tracheids above and below the junction in dormant* P. resinosa *cuttings having only 5-mm-long needle pair stumps, using 15 replicates per treatment. No TEs were produced in the controls nor in wounded needle-pair stumps to which lanolin was applied. (b) Sequential procedure (left to right) for producing miniature waxed (therefore resistant to water loss) wells for testing the effect of applied aqueous solutions on differentiation at the trace-cambium junction.*

The nature of tracheid differentiation occurring at the trace-cambium junction in response to wounding of needle-pair stumps was similar to that occurring in response to intact needle pairs, except that all stem ages (1, 2 and 3 years in *P. resinosa*) responded more strongly. Not uncommonly, the wound-induced response extended axially 10 mm or more above and below the junction of the needle trace with the cambium (R.A. Savidge, unpublished). These observations appear to contradict the primary report on the tracheid-differentiation factor from pine needles (Savidge and Wareing, 1981a), where it was reported that the number of TEs differentiating at the trace-cambium junction was proportional to the length of residual intact needle, with no response occurring when only a needle pair stump remained. However, wounding was not investigated in that seminal work, and in the present experiments the absence of any induced differentiation in non-wounded control stumps corroborates the original observations. The wound response data appear to indicate that the tdf does not actually originate in the green parts of the needles, rather in their chlorotic bases or associated apical meristem; however, it nevertheless could be that a factor (e.g., IP_3) travelling basipetally out of the needles could be eliciting production of the tdf down-stream, in effect a signal cascade.

Wounding of intact whole needle pairs in the same way at the same location as just described for wounded needle stumps enhanced the differentiation response at the trace-cambium junction. In early spring, wounding increased the qualitative frequency of the differentiation response from 50% to 100% of the dormant cuttings. On the other hand, when the same experiment was repeated in late spring, just before cambial reactivation, the wounding treatment made no difference to the outcome. Stem regions from which pine needle pairs had been fully removed were also wounded by inserting a sterile needle into the residual apical meristem regions. These 'controls' yielded no new tracheids at their trace-cambium junctions (R.A. Savidge, unpublished).

The observations on wounding induction of tracheid differentiation may be related to those on *Zinnia* mesophyll indicating that wounding is an essential trigger for induction of TE differentiation in that system (Roberts *et al.*, 1997).

2.3 *Lanolin inhibition of differentiation*

Concomitantly with the above wounding experiments, similarly wounded pine-needle stumps were also provided lanolin paste (Fisher, L7–500). The paste was applied to the apical ends of the stumps. As indicated in *Figure 4a*, lanolin effectively inhibited wound induced TE differentiation at the trace-cambium junction. Innumerable past investigations into hormonal regulation of vascular development have used lanolin as the medium in which to mix and apply exogenous hormones. The strong inhibition by lanolin (*Figure 4a*) of tracheid differentiation possibly is related to the observations by Gersani (1987) on an auxin-responsive system; 100 times more auxin in lanolin paste than in agar was required to induce xylogenesis. Lanolin is a complex mixture of esters and polyesters of 33 high-molecular-weight alcohols and 36 fatty acids obtained from wool (Budavari, 1989). Thus, although the inhibitory effect of lanolin on differentiation was clearly apparent, it will require some effort to determine which constituent(s) in lanolin blocks differentiation and exactly what its mechanism is.

The observation of inhibition by lanolin prompted the development of an alternative method for delivering hypothetical regulators to the trace-cambium junction, as shown in *Figure 4b*. After preparing the cuttings as shown in *Figure 1a*, the residual needles are clipped to be even with the top of the cutting. The top end of the cutting and clipped needles are then briefly immersed in melted paraffin to a depth sufficient to coat the prophyll sheaths of the residual needles fully with the wax. The cutting is withdrawn from the paraffin, and when the wax has solidified the pine needle pairs can be extracted by exerting a quick upward tug on the needles with forceps. Removal of the needle pair leaves a water-impermeable 'waxed well' within the paraffin-coated prophyll sheath. Compounds in aqueous solution can be pipetted (5 – 10 μl) into the well and left to diffuse the short distance (< 1 mm) into the residual short-shoot apex at the base of the well, and from there into the trace-cambium junction.

This waxed-well method shown in *Figure 4b* has now been used extensively. When auxins or auxin–cytokinin combinations are pipetted into such wells, tracheid differentiation has not been promoted; however, as shown in *Figure 3d*, cell division (i.e., proliferation) and enlargement of cells does occur in response to those treatments. The waxed-well method also permits re-insertion of an intact needle pair into the well, and it has been observed that a transposed needle pair does result in tracheid differentiation at the trace-cambium junction, provided that water is pipetted into the well before inserting the needle-pair base (R.A. Savidge, unpublished).

2.4 *Evidence for a novel type of wall deposition*

As shown in *Figure 3c*, one of the interesting features associated with needle-induced tracheid differentiation that, until recently, was unnoticed is the production of anomalous wall thickenings. These thickenings have been observed in circumstances where induction of tracheid differentiation is anticipated, that is, when an intact needle pair is present or the short-shoot apical meristem region of a needle-pair stump is wounded. The anomalous wall thickenings are localized to the trace-cambium junction and are most commonly seen in ray cells of the cambial zone (*Figure 3c*). The thickenings occasionally also are found in mature phloem and xylem rays. In addition, they occur associated with the primary walls of fusiform cambial cells.

A feature often exhibited by the secondary walls of TEs induced to differentiate at the trace-cambium junction by a single-needle pair is high opacity, when observed as unstained sections with interference contrast or brightfield optics. The opacity or light absorbing ability of the secondary walls often contrasts remarkably with the bright, light-reflecting secondary walls of nearby late wood tracheids (*Figures 1b, 1c, 2a and 2b*). The anomalous wall thickenings described above also tend toward opacity; thus, it may be that they are of similar chemical composition and produced through a pathway related to that generating the secondary walls during differentiation into TES of needle-induced cambial cells.

Variation in the qualitative differentiation response at the trace-cambium junction has been a problem necessitating substantial replication in experimental design, as noted above. Although TEs may or may not be present, the anomalous wall thickenings have been found at the trace-cambium junction in both cuttings producing and cuttings not producing new tracheids whenever needles are present. Thus, it appears that the wall thickenings, like the tracheids, are induced to develop at the junction by a factor originating in the needles. One possible interpretation is that the wall thickenings are indicative of gene expression gone amuck, giving rise to a novel phenotype, rather than to a tracheid.

2.5 *A chromophore associated with induced differentiation*

When a single needle pair is left intact on a dormant pine cutting, the wall colour of the existing tracheids in the unstained latewood, and also that of tracheids in the vascular trace, changes to become orange or rust-coloured in appearance. The colour is reminiscent of that of coniferyl aldehyde enriched lignin; however, application of coniferyl aldehyde to trace-cambium junctions using the method shown in *Figure 4b* has not been found to induce the colour change. In the vascular trace, the colour change occurs not only in the region where the tracheids pass from the stem axis outward into the needles, but also in the diagonal zone (in cuttings less than 1 year of age) or horizontal zone (in cuttings having two or more annual rings) connecting the vascular trace to the primary xylem adjoining the pith in the stem. In black and white images (*Figure 1f*), the colour change is seen as a darkening of the last few late wood tracheids.

The colour change occurs at the trace-cambium junction in response to intact needles whether or not new tracheids are produced. On the other hand, rare instances have been noted when the differentiation of cambial cells into tracheids at the trace-cambium junction has not been accompanied by a colour change.

3. More on IAA versus the tracheid-differentiation factor from mature needles

For higher plants in general, theory holds that indol-3-ylacetic acid (IAA, auxin) is produced in green leaves and extending shoots, is exported from these sites of biosynthesis, and is transported basipetally through the cambium (Savidge and Wareing, 1981b; Savidge, 1990, 1991). Considerable evidence has indicated that IAA tends to become concentrated in the region of xylem development (Savidge and Wareing, 1981b; Savidge *et al.*, 1982). However, Uggla *et al.* (1996) concluded that IAA is most concentrated in the cambial zone of *Pinus sylvestris*. Their conclusion was based on the analysis of IAA content in seven 30 μm sections (i.e., a 210 μm wide zone). The photomicrograph accompanying their data indicated that the cambial zone that was sectioned was much narrower, less than 60 μm wide; thus, for seven sections to contain cambial zone tissue the plane of sectioning could not have been parallel to the long axis of the cambium. Thus, it is probable that several tissues in addition to the cambial zone were present in each section. In this light, the conclusion of Uggla *et al.* (1996), which appear to contradict the findings of a number of previous investigations, deserves re-examination.

As noted above, bordered-pitted secondary-xylem tracheids are induced immediately at the trace-cambium junction, but with increasing distance above and below that junction an axial gradation in the extent of secondary-wall deposition occurs, varying from scalariform wall thickenings progressively to simple annular ribs and, beyond that, into fusiform cambial cells exhibiting no evidence for tracheid differentiation (Savidge and Wareing, 1981a). This increasing complexity in TE wall sculpturing mimics the developmental sequence occurring radially, from cell to cell, during primary-xylem development in extending shoots and roots, except that the response to intact needles involves an axial rather than a radial sequence. The amount of IAA exported from rapidly extending shoots is supposedly high, with basipetal transport of IAA being canalized to the cambium (Sachs, 1981). If this in fact is the case, and if IAA also is the factor initiating differentiation of cells into TEs, it is difficult to explain why the procambium and more of its adjoining cells do not become TEs. On the supposition that IAA is a major controlling factor promoting elongation of young stem parts, it is also not easy to reconcile the occurrence of strong cell extension (indicating high IAA content) in the sub-apical regions with the fact that no secondary xylem tracheids rather only primary xylem tracheids with scant secondary-wall thickening are produced in those regions. Gibberellins unquestionably also influence cell elongation in the sub-apical region; thus, it is not inconceivable that the auxin to gibberellin ratio may be a factor modulating the extent of differentiation occurring in primary xylem TEs (Kalev and Aloni, 1998). Gibberellic acid was noted to inhibit differentiation of cells into TEs (Savidge, 1983), but no evidence in support of an auxin-gibberellin combination modulating the extent of secondary-wall formation has yet been obtained in relation to needle-induced differentiation (Savidge, 1994).

The role of auxin in cambial growth can also be considered in relation to the overall pattern of reactivation occurring in the tree. Cambial reactivation in conifers more than a few metres tall characteristically resumes at about the time of budbreak in the foliated stem and branch leaders. This at first impression may seem to be in agreement with the concept of auxin export from leaves and buds leading to cambial reactivation (Savidge and Wareing, 1981b, 1984). However, reactivation at the bases of buds generally occurs several weeks later in springtime than it does in the lower bole (Savidge and Wareing, 1984; see also Chapter 1, *Figure 3*). In the cambia within the younger foliated regions of the tree, springtime initiation of cambial cell-division activity is clearly delayed by some as-yet-unexplained mechanism, relative to that in lower cambia. Nevertheless, once activity has commenced in the young stem regions there is minimal delay in the onset of differentiation of enlarged cambial derivatives into tracheids. In contrast, at lower (older) stem positions, although cambial reactivation and production of enlarged primary-walled derivatives occur considerably earlier in the season than in the tree top, initiation of differentiation of those enlarged derivatives into tracheids is delayed considerably beyond when it commences in the younger regions. Thus, when the stem of the conifer is considered as a whole, xylogenesis *per se* is initiated first in the foliated regions. When differentiation is initiated in foliated stem regions, the endogenous IAA content in their cambia is lower than that in older cambia (Savidge and Wareing, 1981b, 1984). Moreover, IAA is present in the overwintering conifer cambium, regardless of cambial age, although the levels are low (Savidge and Wareing, 1981b, 1982, 1984).

In the absence of a continuing supply of IAA (e.g., from intact buds or exogenous IAA applied to apical ends of disbudded segments), cambial fusiform cells typically expand and divide in the transverse plane, converting into thin-walled axial parenchyma (Savidge and Wareing, 1981b; Savidge, 1983). Conversely, normal fusiform-cell morphology is maintained when auxin is available (Savidge and Wareing, 1981b; Savidge, 1983). With both young and old cuttings, exogenous IAA is effective in preventing fusiform cambial cells from de-differentiating into axial parenchyma (see Chapter 1; see also Savidge and Wareing, 1981a, 1981b; Savidge, 1983). In nature, fusiform-cell morphology is maintained as such throughout the cambial zone over the tree, regardless of cambial age. This consideration in conjunction with the experimental findings indicates that although exogenous IAA is available to and moving basipetally through cambial cells of old as well as young stem segments at all times during the annual cycle, IAA by itself has little or no effect on growth in cambia more than three years old.

Exogenous IAA does not promote wood formation in pine stem cuttings older than two years (Savidge 1983). In contrast, three-year old and older needles when still green are effective in inducing differentiation at their trace-cambium junctions. Investigations into mature needles revealed that their endogenous IAA content decreased with needle age and that the highest values were considerably less than the content in young elongating needles (Savidge and Wareing, 1981b, 1984). Young elongating needles have not been found to induce TE differentiation in the cambial zone at the trace-cambium junction (Savidge and Wareing, 1981a).

De-budding of dormant evergreen conifer cuttings prevents cambial activity and wood formation immediately basal to the removed bud, but not at the cutting base (Savidge 1983). Similarly, debudding of whole trees in late winter reduces the cambial region's endogenous content of IAA, and that treatment also prevents cambial growth in the young stem regions in the coming summer. Nevertheless, wider than normal annual rings are produced in the basal root collar regions of those same debudded stems, despite the reduced IAA content (Savidge and Wareing, 1981b). Defoliated dormant cuttings with intact buds exhibit cambial activity and radially enlarged derivatives after several weeks of incubation in water at warm temperatures, but little or no wood formation occurs in such cuttings (Savidge and Wareing, 1981a, 1981b; Savidge, 1983). The same response is seen in whole trees that have been defoliated in late winter; the cambium reactivates but little or no new xylem is produced despite bud break and shoot extension occurring. IAA levels are reduced but still readily detected in cambia of such trees (Savidge and Wareing, 1981b). In the year following production of new needles by defoliated trees, wood formation resumes.

The quantities of IAA in cambium bordering tissues generally are many fold greater during growth than during dormancy (Savidge and Wareing, 1981b, 1984; Savidge, 1990). According to Sachs (1981), the flow of IAA in tissues becomes canalized, setting up auxin-flow channels and increasing IAA-transport capability. However, the endogenous IAA content of the cambium in conifers does not correlate well with initiation of xylogenesis during seasonal cambial growth (Savidge *et al.* 1982; Savidge and Wareing, 1984; Savidge 1991). Endogenous IAA content of the cambium in springtime is higher at the base of the live crown, where CZ reactivation and RE zone development first occur, than in young cambia; however, xylogenesis begins first in the foliated regions of the crown.

How much IAA is actually available to the cambium at any position at any time of the year cannot be answered simply by making point-in-time determinations of endogenous IAA content. Although no evidence has yet been obtained for the cambium having IAA biosynthetic competence, various IAA conjugates are known to occur in plants and conceivably one or more could be converted to IAA in stem tissues and stored as such, or exported as an IAA conjugate from needles. IAA-aspartate has been noted to promote differentiation in *Pinus* stem segments, but it has not been found to simulate the effect of intact needles (Savidge, 1994).

The ability of phytohormones to mobilize other substances toward sites where they are most concentrated also could be a factor influencing differentiation of cells into TEs (Savidge, 1983). Partitioning of assimilates between and within shoot and root systems in support of growth is well established to be under hormonal regulation (Thomas, 1985). Little *et al.* (1990) reported that IAA induced acropetal movement of photosynthate by promoting growth locally, beneath the IAA-application site, rather than by affecting radial growth or phloem-transport processes at a distance below the

IAA source. It follows that when exogenous phytohormones induce differentiation of TEs (e.g., in young stem material), the phytohormones may be acting only indirectly by facilitating the uptake of endogenous differentiation regulators (Savidge and Wareing, 1981b; Savidge, 1983, 1985).

Cambial cells basal to the trace-cambium junction most predictably respond to exogenous IAA by dividing and enlarging without commencing TE differentiation over a four-week or longer period (Savidge and Wareing, 1981a). Auxin-induced TE differentiation occurs in stem segments harvested in late winter or early spring but not in those harvested in late autumn, when the cambium is considered to be in the resting stage of dormancy. In contrast, needle-induced TE differentiation at the trace-cambium junction occurs regardless of the dormancy stage. Auxin-induced TEs resembled those of secondary xylem rather than those characteristic of the tdf response. Transmission electron microscopy research indicated that dormant cambial cells supplied with IAA became vacuolated (i.e., they transformed from the 'gel' to the 'sol' state) and underwent normal periclinal cell division followed by primary-wall radial expansion. Control cells also became vacuolated, enlarged and divided; however, division occurred in the transverse rather than the periclinal plane. TE differentiation did not occur in either IAA-treated or control cells. Protoplasmic fine structures of auxin-treated and control cambial cells were similar and both were very unlike the protoplasm seen within cells induced to differentiate into TEs. The presence of electron-dense exocytotic blebs at the plasma membrane together with a coarsely granular polyribosome-enriched protoplasm were diagnostic of needle-induced cytodifferentiation; however, the protoplasm of auxin-treated cells did not show those blebs and in most other respects was very dissimilar (Savidge and Barnett, 1993).

In many conifer species, competence for TE differentiation *in vitro* decreases and eventually disappears upon repeated sub-culturing (Savidge, 1993). In callus generated by cambial explants taken from mature stems, it is exceptionally rare for any TE differentiation to occur despite the callus being sub-culturable (Savidge, 1989, 1993). On the other hand, when *Larix laricina* stem 'chips' containing vascular cambium intact between mature phloem and xylem were cultured, wood formation was induced on a fully defined medium containing auxin (1-naphthalene acetic acid, NAA) as the sole growth promoter (Savidge, 1993). However, when the chips were sub-cultured on the same medium, cambial growth ceased and no further xylogenesis occurred (Savidge, 1993). Plausibly, *Larix* chips as primary explants contain one or more endogenous factors which, when depleted, limits growth and differentiation.

The possibility that IP_3 induction of tracheid differentiation is linked to auxin has been discussed in depth elsewhere (Savidge, 1994). Working with *Zinnia* mesophyll, Roberts *et al.* (1997) suggested that an oligosaccharide might be a factor controlling the nature of tracheid differentiation.

4. Conclusions

Mature needles of conifers house a regulatory mechanism for inducing secondary-wall formation characteristic of tracheary elements. The precise inductive mechanism appears to be distinct from auxin regulation and to involve D-*myo*-inositol 1,4,5-*tris*phosphate and also one or more additional factors, inducible by wounding, at non-green needle bases.

Induction of TE differentiation using pine needle pairs is an attractive experimental approach for understanding regulation of cellular differentiation for a number of reasons:

(i) A tissue that normally never differentiates, viz. the vascular cambium, can be induced to differentiate into TEs.
(ii) The location of TE differentiation can be accurately predicted, and the TE-differentiation response can be compared with that of adjoining, non-differentiating cambial cells of identical genotype.
(iii) *In vitro* (i.e., test-tube incubation of stem segments) research is possible, yet extensive preparation and complex media are unnecessary for the tdf response to develop.
(iv) The differentiating tissue is heterotrophic; hence, interpretations about regulation are not confounded by having to consider autotrophic metabolism.
(v) TE differentiation occurs in 'dormant' cells in the absence of preceding vacuolation and cell division; therefore, cellular differentiation *per se* can be investigated independently of processes which normally accompany but are not prerequisites for differentiation.
(vi) Polyribosomes, endoplasmic reticulum, dictyosomes, exocytotic blebs, and other organelles clearly fulfil roles related to TE differentiation in this system, and all appear worthy of further investigation in order to understand the biochemistry of this type of cellular differentiation.
(vii) The system is well suited for performance of *in situ* hybridization or immunolocalization research.
(viii) The tdf response can be simulated by wounding.
(ix) Interactions between the tdf and other regulatory molecules, or other chemical or physical factors, can readily be investigated.
(x) At the cellular level, developmental plasticity remains a relatively unexplored phenomenon.

The tdf response is an all-or-none event yet TE differentiation expresses itself in a variety of related phenotypes. With refinement, the tdf response has the potential to serve as a model system for future research in this area.

Acknowledgements

Early research into the tdf was supported by NERC, UK. Electron-microscopy research in 1991–92 was supported by both NSERC Canada and the Royal Society of England.

References

Budavari, S., ed. (1989) *The Merck Index*, 11th Ed., Merck & Co., Rahway, N.J.

Gersani, M. (1987) The induction of differentiation of organized vessels in a storage organ. *Ann. Bot.* **59:** 31–34.

Kalev, N. and Aloni, R. (1998) Role of auxin and gibberellin in regenerative differentiation of tracheids in *Pinus pinea* seedlings. *New Phytol.* **138:** 461–468.

Little, C.H.A. and Savidge, R.A. (1987) The role of plant growth regulators in forest tree cambial growth. *Plant Growth Regul.* **6**: 137–169.

Little, C.H.A. Sundberg, B. and Ericsson, A (1990) Induction of acropetal ^{14}C-photosynthate transport and radial growth by indole-3-acetic acid in *Pinus sylvestris* shoots. *Tree Physiol* **6**: 177–189.

Roberts, A.W., Donovan, S.G. and Haigler, C.H. (1997) A secreted factor induces cell expansion and formation of metaxylem-like tracheary elements in xylogenic suspension cultures of *Zinnia. Plant Physiology* **115:** 683–692.

Sachs, T. (1981) The control of the patterned differentiation of vascular tissues. In: *Advances in Botanical Research* (ed. H.W. Woolhouse), Academic Press, London), Vol 9, pp 152–262.

Savidge, R.A. (1983) The role of plant hormones in higher plant cellular differentiation. II. Experiments with the vascular cambium, and sclereid and tracheid differentiation in the pine, *Pinus contorta. Histochem J.* **15**: 447–466.

Savidge, R.A. (1985) Prospects for manipulating vascular cambium productivity and xylem cell differentiation. In: *Attributes of Trees as Crop Plants* (eds. M.G.R. Cannell and J.E. Jackson), NERC Press, Abbots Ripton, England, pp. 208–227.

Savidge, R.A. (1988) Auxin and ethylene regulation of diameter growth in trees. *Tree Physiology* **4**: 401–414.

Savidge, R.A. (1990) Phytohormonal regulation of cambial growth in trees. In: *Fast Growing Trees and Nitrogen Fixing Trees* (eds. D. Werner and P. Müller), Gustav Fischer, Jena, pp. 142–151.

Savidge, R.A. (1991) Seasonal cambial activity in *Larix laricina* saplings in relation to endogenous indol-3-ylacetic acid, sucrose, and coniferin. *Forest Sci.* **37:** 953–958.

Savidge, R.A. (1993) *In vitro* wood formation in 'chips' from merchantable stem regions of *Larix laricina. IAWA J.* **14**: 3–11.

Savidge, R.A. (1994) The tracheid-differentiation factor of conifer needles. *Int. J. Plant Sci.* **155**: 272–290.

Savidge, R.A. and Barnett, J.R. (1993) Protoplasmic changes in cambial cells induced by a tracheid-differentiation factor from pine needles. *J. Exp. Bot.* **44**: 395–405.

Savidge, R.A. and Farrar, J.L. (1984) Cellular adjustments in the vascular cambium leading to spiral grain formation in conifers. *Can. J. Bot.* **62**: 2872–2879.

Savidge, R.A., Heald, J.K., and Wareing, P.F. (1982) Non-uniform distribution and seasonal variation of endogenous indol-3-ylacetic acid in the cambial region of *Pinus contorta* Dougl. *Planta* **155**: 89–92.

Savidge, R.A. and Wareing, P.F. (1981a) A tracheid-differentiation factor from pine needles. *Planta* **153**: 395–404.

Savidge, R.A. and Wareing, P.F. (1981b) Plant growth regulators and the differentiation of vascular elements. In: *Xylem Cell Development* (ed. J.R. Barnett), Castle House Publ, Tunbridge Wells, England, pp. 192–235.

Savidge, R.A. and Wareing, P.F. (1982) Apparent auxin production and transport during winter in the non-growing pine tree. *Can. J. Bot.* **60**: 681–691.

Savidge, R.A. and Wareing, P.F. (1984) Seasonal cambial activity and xylem development in *Pinus contorta* in relation to endogenous indol-3-ylacetic and (*S*)-abscisic acid levels. *Can. J. For. Res.* **14**: 676–682.

Thomas, T.H. (1985) Hormonal control of assimilate movement and compartmentation. In: *Plant Growth Substances 1985* (ed. M. Bopp), Springer-Verlag, Berlin, pp. 350–359.

Uggla, C., Moritz, T., Sandberg, G. and Sundberg, B. (1996) Auxin as a positional signal in pattern formation in plants. *Proc. Natl. Acad. Sci. USA* **93**: 9282–9286.

19

The role of cytoskeleton in secondary xylem differentiation in conifers

R. Funada, O. Furusawa, M. Shibagaki, H. Miura, T. Miura, H. Abe and J. Ohtani

1. Introduction

Periclinal divisions of cells of the vascular cambium (cambium) produce the secondary xylem and phloem, and increase the diameter of the stem of the tree (see Catesson, 1990, 1994; Savidge, 1993; Larson, 1994). The amount of secondary xylem (wood) produced is generally much greater than the amount of secondary phloem.

As soon as cambial cells lose the ability to divide, they start to differentiate into xylem cells via the expansion and elongation of cells, cell wall thickening and lignification and, finally, cell death (cell autolysis). In secondary xylem cells, such as tracheids, fibres, and vessel elements, the orientation of newly deposited cellulose microfibrils changes as differentiation proceeds. Consequently, the secondary xylem cells of woody plants have cell walls with a highly organized, layered structure (see Harada and Côtè, 1985). The cellulose microfibrils reinforce the cell wall and resist expansion in response to turgor pressure. Thus, the orientation of newly deposited cellulose microfibrils on the inner surface of the primary wall determines the direction of cell expansion, determining, in turn, the final shape and size of each xylem cell. In addition, several of the mechanical properties of wood, such as its modulus of elasticity, are related to the orientation of microfibrils in the secondary wall.

Observations in a wide variety of plant cells have revealed that cortical microtubules play an important role in the orientation of newly deposited cellulose microfibrils (see Ledbetter and Porter, 1963; Giddings and Staehelin, 1991; Shibaoka 1994). Co-alignment of cortical microtubules and cellulose microfibrils has frequently been observed in the cells of both lower and higher plants. In addition, microtubule-depolymerizing agents often disrupt the orientation of cellulose microfibrils. These

Cell & Molecular Biology of Wood Formation, edited by R. Savidge, J. Barnett and R. Napier.

observations support the hypothesis that cortical microtubules control the orientation of newly deposited cellulose microfibrils.

Recent observations by immunofluorescence microscopy have revealed dynamic changes in the arrangement of cortical microtubules in the differentiating secondary xylem cells of woody plants (Uehara and Hogetsu, 1993; Abe *et al.*, 1995a, b; Prodhan *et al.*, 1995a; Chaffey *et al.*, 1997a, b, c; Funada *et al.*, 1997; Furusawa *et al.*, 1998). The cortical microtubules in fusiform cambial cells are arranged at random. The predominant orientation of cortical microtubules in differentiating tracheids of conifers changes from longitudinal to transverse during the radial expansion of cells (Abe *et al.*, 1995a, b; Funada *et al.*, 1997). Finally, ordered and transversely oriented cortical microtubules are observed when radial expansion ceases. During formation of the secondary walls of tracheids and wood fibres, after cell expansion has ceased, cortical microtubules are aligned in well-ordered arrays and their orientation changes progressively (Abe *et al.*, 1995a; Prodhan *et al.*, 1995a; Furusawa *et al.*, 1998). These progressive changes in orientation of cortical microtubules resemble the changes in the orientation of newly deposited cellulose microfibrils in tracheids and wood fibres during the formation of the primary wall and the secondary wall, providing evidence that cortical microtubules might control the orientation of cellulose microfibrils in tracheids and wood fibres (Abe *et al.*, 1995a, b; Prodhan *et al.*, 1995a). In addition, the localized appearance of cortical microtubules has been observed in tracheids and vessel elements during the formation of spiral thickenings and pit borders (Uehara and Hogetsu, 1993; Chaffey *et al.*, 1997b, c; Funada *et al.*, 1997; Furusawa *et al.*, 1998). These results reflect another function of cortical microtubules, namely, the control by localized cortical microtubules of the localized deposition of cellulose microfibrils in secondary xylem cells.

Our goal in the present study was to obtain further detailed information on the roles of cortical microtubules in the differentiation of secondary xylem cells in conifers. We examined arrays of cortical microtubules by confocal laser scanning microscopy after immunofluorescence staining in an attempt to visualize dynamic changes in the orientation and localization of cortical microtubules. We also examined the orientation and localization of actin filaments to investigate the relationship between cortical microtubules and actin filaments since it has been suggested that actin filaments are involved in the arrangement of cortical microtubules *in vitro* systems, for example, the differentiating tracheary elements of cultured mesophyll cells of *Zinnia* (see Fukuda and Kobayashi, 1989; Fukuda, 1996). However, significant differences in the behaviour of the cytoskeleton between cells *in vitro* and in natural systems have been recognized (Chaffey *et al.*, 1997a, b).

2. Material and methods

Samples from specimens of conifers, namely, *Abies sachalinensis* Maters, *Pinus densiflora* Sieb. et Zucc., and *Taxus cuspidata* Sieb. et Zucc., growing in the nursery of the Experimental Forest of Hokkaido University in Sapporo, were used. Sample blocks, containing mature and differentiating phloem, cambial zone cells and mature and differentiating xylem, were taken from stems at breast height during seasons of active and dormant cambial growth.

A procedure similar to those described previously (Abe *et al.*, 1995a, b; Prodhan *et al.*, 1995a; Funada *et al.*, 1997; Furusawa *et al.*, 1998) was used for immunofluorescence staining of the cytoskeleton. For examinations of actin filaments, small blocks of tissue were prefixed at room temperature in a solution of 0.2 mM *m*-maleimidobenzoic acid *N*-hydroxysuccinimide ester (MBS) in 2% DMSO and 0.05% Nonidet P-40 in distilled water (Sonobe and Shibaoka, 1989). Then, for examinations of cortical microtubules and actin filaments, the blocks were fixed at room temperature in a mixture of 3.6% paraformaldehyde plus 0.2% glutaraldehyde that contained 10% DMSO and 0.1% Nonidet P-40 in 50 mM PIPES buffer (pH 7.0), supplemented with 5 mM EGTA and 5 mM $MgSO_4$. All the small blocks were sectioned radially at a thickness of approximately 50 μm with a sliding microtome on a freezing stage (MA-101; Komatsu Electronics Inc., Tokyo, Japan). Each section was incubated with mouse monoclonal antibodies against chick brain α-tubulin or against chicken gizzard actin that had been diluted in PBSB (PBS containing 0.1% NaN_3 and 1 mg ml^{-1} bovine serum albumin). Then sections were incubated with fluorescein isothiocyanate (FITC)-conjugated sheep antibodies against mouse IgG that had been diluted in PBSB. All antibodies used in these experiments were purchased from Amersham Japan Co. (Tokyo, Japan). The sections were observed with a confocal laser scanning microscope with excitation light from a 488-nm argon ion laser (LSM-310 or LSM-410; Carl Zeiss Co., Oberkochen, Germany). Each series of confocal images was processed for construction of a single projection image.

3. Cortical microtubules in conifers

The orientation and localization of cortical microtubules and actin filaments in differentiating tracheids and ray parenchyma cells in conifers were observed by confocal laser scanning microscopy after immunofluorescence staining. This method is a powerful tool for visualizing the cytoskeleton in large areas within a single radial section.

The cortical microtubules in expanding tracheids were not well ordered (*Figure 1a*). The predominant orientation of cortical microtubules on the radial wall changed progressively from longitudinal to transverse as the tracheids expanded. In earlier reports, we demonstrated that the orientation of cortical microtubules changes during the formation of the primary wall in differentiating tracheids (Abe *et al.*, 1995a, b; Funada *et al.*, 1997). A similar rearrangement of cortical microtubules from random to helical has been observed when fusiform cambial cells differentiate to wood fibres in a hardwood (Chaffey *et al.*, 1997a, c).

During formation of the secondary wall, the orientation of cortical microtubules in differentiating tracheids changed rotatively from a flat S-helix to a steep Z-helix with clockwise rotation when viewed from the lumen side (*Figure 1b*). This shift in the orientation of cortical microtubules was completed within three or four tracheids in a radial file. Then, ten to fifteen tracheids, with cortical microtubules oriented in a steep Z-helix at almost the same angle, were aligned in a radial file. After further differentiation, the orientation of cortical microtubules changed from a steep Z-helix to a flat S-helix in the tracheids (*Figure 1c*). This shift was completed within one or two tracheids in a radial file. Our observations revealed that, during differentiation, the orientation of cortical microtubules in tracheids changed dynamically. Similar progressive changes in the orientation of cortical microtubules during the formation of the secondary wall in tracheids were reported by Abe *et al.* (1995a). By contrast, cortical microtubules in differentiating compression wood tracheids, which are formed on the lower side of

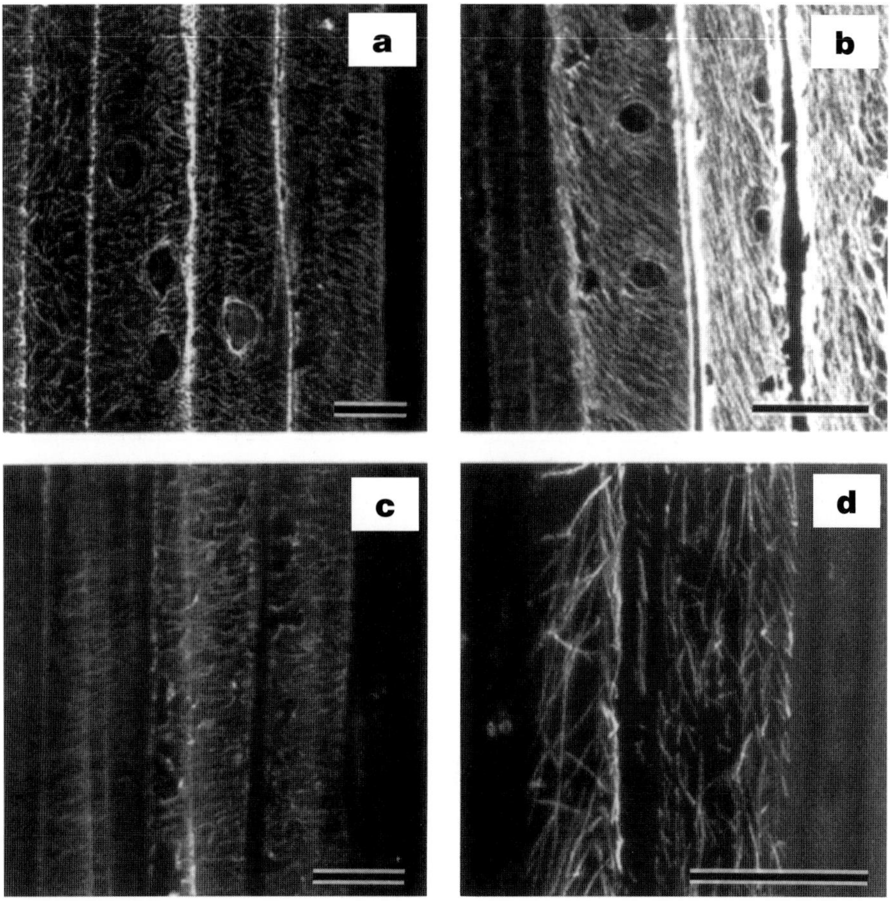

***Figure 1**. Immunofluorescence images, obtained by confocal laser scanning microscopy, showing changes in the orientation of cortical microtubules (viewed from the lumen side) in tracheids of* Abies sachalinensis. *(a) Cortical microtubules during formation of the primary wall. (b) and (c) Changes in the orientation of cortical microtubules from a flat S-helix to a steep Z-helix to a flat S-helix during formation of the secondary wall. (d) Cortical microtubules during period of cambial dormancy. The left side of this and other micrographs corresponds to the outer side of the tree. Bars = 25 μm.*

inclined stems of conifers, were oriented obliquely, at an angle of about 45° to the tracheid axis, during formation of the secondary wall (Furusawa *et al*. 1998). In such tracheids, the same orientation was retained until differentiation was completed, with little change in the oblique orientation at the final stage of xylem differentiation. These observations indicate that stimulation by inclination alters the pattern of alignment of cortical microtubules during the formation of secondary walls in tracheids. A similar effect of inclination on changes in the alignment of cortical microtubules in the secondary wall was observed in differentiating tension wood fibres in a hardwood (Prodhan *et al*., 1995a). The patterns of orientation of cortical microtubules during the formation of secondary walls might be altered by a variety of stimuli, such as gravity and plant hormones.

Numerous observations by high-resolution field emission scanning electron microscopy have revealed that the orientation of the newly deposited cellulose microfibrils on the innermost surfaces in tracheids and wood fibres changes during xylem differentiation (Abe *et al.*, 1991, 1992, 1994, 1995a, b, 1997; Prodhan *et al.*, 1995a, b). In brief, the cellulose microfibrils in the radial primary walls of tracheids are not well ordered and their predominant orientation is longitudinal with respect to the cell axis at the early stage of formation of the primary wall (Abe *et al.*, 1995b). These longitudinally oriented cellulose microfibrils, with their considerable tensile strength, restrain the longitudinal expansion of cells and, thus, they facilitate the lateral expansion of fusiform cambial derivatives. As tracheids expand, the orientation of cellulose microfibrils changes from longitudinal to transverse (Abe *et al.*, 1995b, 1997) and, finally, when cell expansion ceases, the cellulose microfibrils are well ordered and aligned parallel to one another (Abe *et al.*, 1991, 1992, 1994, 1995a, b, 1997; Prodhan *et al.*, 1995a, b). These cellulose microfibrils are considered to be the cellulose microfibrils of the secondary wall. The cellulose microfibrils change their orientation rotatively from a flat S-helix to a steep Z-helix during formation of the secondary wall. The successive changes in the orientation of cellulose microfibrils can be considered to reflect the semi-helicoidal structure of the cell wall (Roland *et al.*, 1987). With cessation of rotation, the thick middle layer develops because of repeated deposition of cellulose microfibrils with a consistent texture. Then the orientation of the cellulose microfibrils changes from a steep Z-helix to a flat S-helix.

The present results indicated that the dynamics of the changes in orientation of cortical microtubules were closely related to the successive changes in the orientation of newly deposited cellulose microfibrils in the differentiating tracheids. Our observations support the hypothesis that cortical microtubules control the orientation of cellulose microfibrils. The orientation of cortical microtubules might control the orientation of newly deposited cellulose microfibrils on the inner surface of the primary wall, thereby determining the direction of expansion of fusiform cambial cells. In addition, the behaviour of cortical microtubules, with respect, for example, to their angle to the cell axis and the speed of rotation during formation of the secondary wall, might control the orientation of cellulose microfibrils, in particular in the thick middle layer (S_2 layer) of tracheids. Such orientation is closely related to the mechanical properties of the wood. Thus, it seems very likely that, if we could manipulate the cortical microtubules, we would be able to control the orientation of microfibrils, with consequent improvements in the quality of wood and its products.

During formation of the primary wall, the cortical microtubules in differentiating tracheids disappeared locally from areas of prospective intertracheal bordered pits (*Figure 1a*). This localized disappearance of cortical microtubules might determine the site of a future pit region (Funada *et al.*, 1997). At a later stage, circular bands of cortical microtubules were observed around the edges of developing bordered pits (*Figure 1b*). Similar circular bands of cortical microtubules around the pit apertures of bordered pits have been observed in tracheids (Hirakawa and Ishida, 1981; Uehara and Hogetsu, 1993; Abe *et al.*, 1995b; Funada *et al.*, 1997) and in vessel elements (Robards and Humpherson, 1967; Chaffey *et al.*, 1997b, c). Concentrically oriented cellulose microfibrils on the inner surface of pit borders have been observed in tracheids (see Harada and Côtè, 1985). Therefore, circular bands of cortical microtubules might be involved in the deposition of concentrically oriented cellulose microfibrils at pit borders. The involvement of cortical microtubules in the formation of bordered pits suggests the importance of cortical microtubules for the localized deposition of cell wall materials.

Microtubules depolymerize in response to low temperatures and, thus, it is generally necessary to avoid low temperatures during the fixation of samples (see Materials and Methods). However, we found cortical microtubules in dormant fusiform cambial cells in mid-winter, when the temperature was below zero, namely, –5°C to –10°C, in Sapporo (*Figure 1d*). Similar observations of cortical microtubules have been reported in the dormant cambium of conifers by transmission electron microscopy (Itoh, 1971; Tsuda, 1975). Chaffey *et al.* (1998) reported that cortical microtubules were present in parallel helical arrays in dormant cambium within the taproots of a hardwood in frozen soil. These observations suggest the stability of cortical microtubules in dormant cambium at low temperatures. Such stability might be related to endogenous factors in cambial cells since the cold stability of microtubules can be modified by plant hormones, such as gibberellin and abscisic acid (Shibaoka, 1994). In addition, it is likely that microtubules in active and dormant cambial cells consist of different isotypes of tubulin whose polymerization exhibits different degrees of cold stability. More detailed investigations are needed to clarify the mechanisms of the cold stability of cortical microtubules in plant tissues.

The cortical microtubules in differentiating ray parenchyma cells were originally oriented transversely to the cell axis (*Figure 2a*). As cells elongated radially, the orientation of cortical microtubules changed from transverse to oblique or longitudinal to the cell axis. The densities of cortical microtubules in ray parenchyma cells after completion of cell elongation were lower than those in elongating cells. At a distance from elongated ray parenchyma cells, we observed ray parenchyma cells with ordered cortical microtubules at high densities (*Figure 2b*). These cortical microtubules were oriented transversely to the cell axis. Observations by light microscopy revealed that these cells were forming thick secondary walls (data not shown), an indication that, in contrast to events in tracheids and wood fibres, no subsequent formation of the secondary wall occurred soon after completion of cell elongation in the ray parenchyma cells of conifers, such as *Abies sachalinensis*. In further differentiated ray parenchyma cells, there were fewer cortical microtubules and many of them were fragmented. The cortical microtubules in ray parenchyma cells were present in the annual rings at least several years distant from the cambium.

The number of microtubules increases in association with the formation of the secondary wall in an experimental system consisting of tracheary elements differentiating from cultured mesophyll cells of *Zinnia* (Fukuda, 1996), as well as in differentiating cotton fibres *in vitro* (Seagull, 1992). In addition, cortical microtubules at high densities were found during formation of the S_2 layer in tracheids (Abe *et al.*, 1994) and in the gelatinous layer in gelatinous fibres of tension wood (Prodhan *et al.*, 1995a), in which closely packed cellulose microfibrils were deposited. Thus, the increases in the densities of distribution of cortical microtubules in differentiating ray parenchyma cells in conifers might be synchronized with the start of formation of the secondary wall, in which cellulose microfibrils are close to one another and oriented parallel to one another.

Our observations have revealed that the orientation of cortical microtubules in tracheids and ray parenchyma cells changes dynamically during differentiation of the xylem and the changes are closely related to this differentiation. The molecular

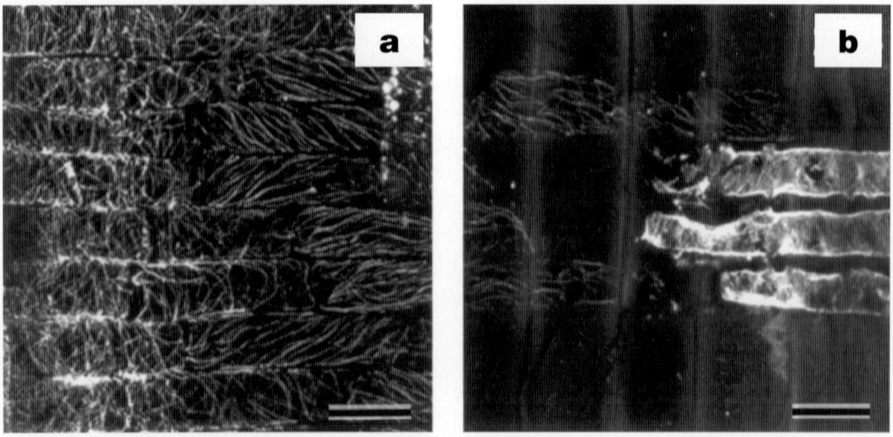

Figure 2. Immunofluorescence images, obtained by confocal laser scanning microscopy, of cortical microtubules (viewed from the lumen side) in differentiating ray parenchyma cells of Abies sachalinensis. *(a) During radial elongation of cells, the orientation of cortical microtubules changed from transverse to oblique or longitudinal with respect to the cell axis. (b) Ordered cortical microtubules were found at high density during formation of the secondary wall. Bars = 25 μm.*

mechanisms responsible for the changes in orientation of cortical microtubules are largely unknown. However, it seems likely that changes in the orientation of cortical microtubules during xylem differentiation are mediated by endogenous plant hormones, such as auxin, because the effects of plant hormones on the reorientation of cortical microtubules have been well documented (Shibaoka, 1994). A steep radial gradient in the level of indole-3-acetic acid (IAA), an endogenous auxin, has been shown to exist across the cambial region, which includes cambial cells, expanding cells and cells with thickening secondary walls (Uggla *et al.*, 1996, 1998; Tuominen *et al.*, 1997). Thus, it is possible that the radial distribution of auxin and other plant hormones across the cambial region might determine the orientation of cortical microtubules.

4. Actin filaments in conifers

Bundles of actin filaments were oriented axially in fusiform cambial cells, and their orientation in differentiating tracheids did not change during the formation of primary walls (*Figure 3a*) and secondary walls. In some tracheids, transversely or obliquely oriented actin filaments were observed at the final stage of xylem differentiation (*Figure 3b*), while in other tracheids the axial orientation was retained. These results indicate the absence of any clear relationship, in terms of orientation, between cortical microtubules and actin filaments in differentiating tracheids.

Bundles of microfilaments (actin filaments) have been observed in the cambial cells of some woody plants (Srivastava and O'Brien, 1966; Tsuda, 1975; Goosen-de Roo *et al.*, 1983; Chaffey *et al.*, 1997a, d), generally oriented parallel to the cell axis. Chaffey *et al.* (1997a, d) recently observed bundles of axially oriented microfilaments in fusiform cambial cells in a hardwood and noted that their axial orientation was retained in cambial derivatives during early stages of xylem differentiation, even though the orientation of the cortical microtubules did change. Our results, together with those

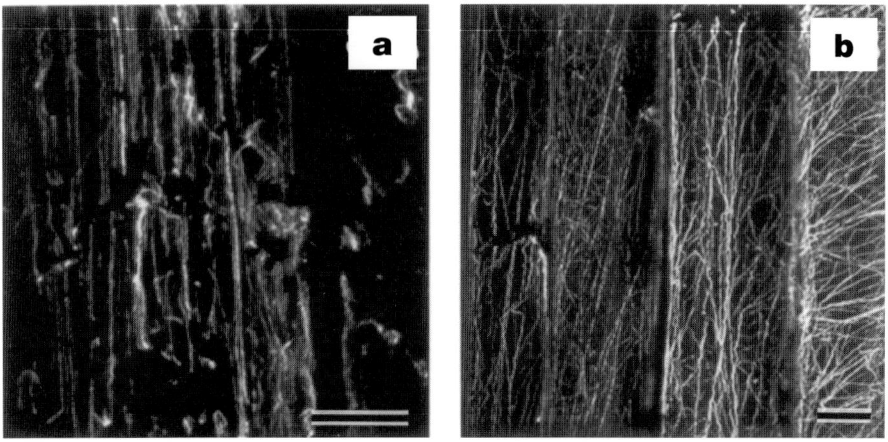

Figure 3. Immunofluorescence images, obtained by confocal laser scanning microscopy, of actin filaments (viewed from the lumen side) in differentiating tracheids of Pinus densiflora.
(a) Bundles of actin filaments were oriented axially during formation of the primary wall.
(b) The axial orientation of actin filaments was retained during the formation of secondary walls. Obliquely oriented actin filaments were observed at the final stage of xylem differentiation in some tracheids. Bars = 25 μm.

observations, suggest that actin filaments might not play a major role in the changes in orientation of cortical microtubules in differentiating cells derived from cambial cells. Much work clearly remains to be done in order to elucidate the role of actin filaments in secondary xylem differentiation.

Acknowledgements

This work was supported by Grants-in-Aid for Scientific Research from the Ministry of Education, Science and Culture, Japan (nos. 06404013, 08760157, and 09760152) and the Japan Society for the Promotion of Science (no. JSPS-RFTF 96L00605).

References

Abe, H., Funada, R., Imaizumi, H., Ohtani, J. and Fukazawa, K. (1995a) Dynamic changes in the arrangement of cortical microtubules in conifer tracheids during differentiation. *Planta* **197**: 418–421.

Abe, H., Funada, R., Ohtani, J. and Fukazawa, K. (1995b) Changes in the arrangement of microtubules and microfibrils in differentiating conifer tracheids during the expansion of cells. *Ann. Bot.* **75**: 305–310.

Abe, H., Funada, R., Ohtani, J. and Fukazawa, K. (1997) Changes in the arrangement of cellulose microfibrils associated with the cessation of cell expansion in tracheids. *Trees* **11**: 328–332.

Abe, H., Ohtani, J. and Fukazawa, K. (1991) FE-SEM observation on the microfibrillar orientation in the secondary wall of tracheids. *IAWA Bull. n.s.* **12**: 431–438.

Abe, H., Ohtani, J. and Fukazawa, K. (1992) Microfibrillar orientation of the innermost surface of conifer tracheid walls. *IAWA Bull. n.s.* **13**: 411–417.

Abe, H., Ohtani, J. and Fukazawa, K. (1994) A scanning electron microscopic study of changes in microtubule distributions during secondary wall formation in tracheids. *IAWA J.* **15**: 185–189.

Catesson, A.M. (1990) Cambial cytology and biochemistry. In: *The Vascular Cambium* (ed. Iqbal, M.), Research Studies Press, Taunton, pp. 63–112.

Catesson, A.M. (1994) Cambial ultrastructure and biochemistry: changes in relation to vascular tissue differentiation and the seasonal cycle. *Int. J. Plant. Sci.* **155**: 251–261.

Chaffey, N.J., Barlow, P.W. and Barnett, J.R. (1997a) Cortical microtubules rearrange during differentiation of vascular cambial derivatives, microfilaments do not. *Trees* **11**: 333–341.

Chaffey, N.J., Barnett, J.R. and Barlow, P.W. (1997b) Cortical microtubule involvement in bordered pit formation in secondary xylem vessel elements of *Aesculus hippocastanum* L. (Hippocastanaceae): a correlative study using electron microscopy and indirect immuno-fluorescence microscopy. *Protoplasma* **197**: 64–75.

Chaffey, N.J., Barnett, J.R. and Barlow, P.W. (1997c) Visualization of the cytoskeleton within the secondary vascular system of hardwood species. *J. Microsc.* **187**: 77–84.

Chaffey, N., Barnett, J. and Barlow, P. (1997d) Endomembranes, cytoskeleton, and cell wall: aspects of the ultrastructure of the vascular cambium of taproots of *Aesculus hippocastanum* L. (Hippocastanaceae). *Int. J. Plant Sci.* **158**: 97–109.

Chaffey, N.J., Barlow, P.W. and Barnett, J.R. (1998) A seasonal cycle of cell wall structure is accompanied by a cyclical rearrangement of cortical microtubules in fusiform cambial cells within taproots of *Aesculus hippocastanum* L. (Hippocastanaceae). *New Phytol.* **139**: 623–635.

Fukuda, H. (1996) Xylogenesis: initiation, progression, and cell death. *Annu. Rev. Plant Physiol. Plant Mol. Biol.* **47**: 299–325.

Fukuda, H., Kobayashi, H. (1989) Dynamic organization of the cytoskeleton during tracheary-element differentiation. *Dev. Growth Differ.* **31**: 9–16.

Funada, R., Abe, H., Furusawa, O., Imaizumi, H., Fukazawa, K. and Ohtani, J. (1997) The orientation and localization of cortical microtubules in differentiating conifer tracheids during cell expansion. *Plant Cell Physiol.* **38**: 210–212.

Furusawa, O., Funada, R., Murakami, Y. and Ohtani, J. (1998) Arrangement of cortical micro-tubules in compression wood tracheids of *Taxus cuspidata* visualized by confocal laser microscopy. *J. Wood Sci.* **44**: 230–233.

Giddings, T.H. Jr., Staehelin, L.A. (1991) Microtubule-mediated control of microfibril deposition: a re-examination of the hypothesis. In: *The Cytoskeletal Basis of Plant Growth and Form* (ed. Lloyd, C.W.), Academic Press, London, pp. 85–99.

Goosen-de Roo, L., Burggraaf, P.D. and Libbenga, K.R. (1983) Microfilament bundles associated with tubular endoplasmic reticulum in fusiform cells in the active cambial zone of *Fraxinus excelsior* L. *Protoplasma* **116**: 204–208.

Harada, H., Côtè, W.A. Jr. (1985) Structure of wood. In: *Biosynthesis and Biodegradation of Wood Components* (ed. Higuchi, T.), Academic Press, Orlando, pp. 1–42.

Hirakawa, Y., Ishida, S. (1981) A scanning and transmission electron microscopic study of layered structure of wall in pit border region between earlywood tracheids in conifer. *Res. Bull. College Exp. For. Hokkaido Univ.* **38**: 249–264.

Itoh, T. (1971) On the ultrastructure of dormant and active cambium of conifers. *Wood Res.* **51**: 33–45.

Larson, P.R. (1994) *The Vascular Cambium: Development and Structure.* Springer-Verlag, Heidelberg.

Ledbetter, M.C., Porter, K.R. (1963) A 'microtubule' in plant cell fine structure. *J. Cell Biol.* **19**: 239–250.

Prodhan, A.K.M.A., Funada, R., Ohtani, J., Abe, H. and Fukazawa, K. (1995a) Orientation of microfibrils and microtubules in developing tension-wood fibers of Japanese ash (*Fraxinus mandshurica* var. *japonica*). *Planta* **196**: 577–585.

Prodhan, A.K.M.A., Ohtani, J., Funada, R., Abe, H. and Fukazawa, K. (1995b) Ultrastructural investigation of tension wood fibre in *Fraxinus mandshurica* Rupr. var. *japonica* Maxim. *Ann. Bot.* **75**: 311–317.

Robards, A.W., Humpherson, P.G. (1967) Microtubules and angiosperm bordered pit formation. *Planta* 77: 233–238.

Roland, J.C., Reis, D., Vian, B., Satiat-Jeunemaitre, B. and Mosiniak, M. (1987) Morphogenesis of plant cell walls at the supermolecular level: internal geometry and versatility of helicoidal expression. *Protoplasma* **140**: 75–91.

Savidge, R.A. (1993) Formation of annual rings in trees. In: *Oscillations and Morphogenesis* (ed. Rensing, L.), Marcel Dekker Inc., New York, Basel, Hong Kong, pp. 343–363.

Seagull, R.W. (1992) A quantitative electron microscopic study of changes in microtubule arrays and wall microfibril orientation during *in vitro* cotton fiber development. *J. Cell Sci.* **101**: 561–577.

Shibaoka, H. (1994) Plant hormone-induced changes in the orientation of cortical microtubules: alterations in the cross-linking between microtubules and the plasma membrane. *Annu. Rev. Plant Physiol. Plant Mol. Biol.* **45**: 527–544.

Sonobe, S., Shibaoka, H. (1989) Cortical fine actin filaments in higher plant cells visualized by rhodamine-phalloidin after pretreatment with *m*-maleimidobenzoyl *N*-hydroxysuccinimide ester. *Protoplasma* **148**: 80–86.

Srivastava, L.M., O'Brien, T.P. (1966) On the ultrastructure of the cambium and its vascular derivatives 1. Cambium of *Pinus strobus* L. *Protoplasma* **61**: 257–276.

Tsuda, M. (1975) The ultrastructure of the vascular cambium and its derivatives in coniferous species. 1. Cambial cells. *Bull. Tokyo Univ. For.* **67**: 158–226.

Tuominen, H., Puech, L., Fink, S. and Sundberg, B. (1997) A radial concentration gradient of indole-3-acetic acid is related to secondary xylem development in hybrid aspen. *Plant Physiol.* **115**: 577–585.

Uehara, K., Hogetsu, T. (1993) Arrangement of cortical microtubules during formation of bordered pit in the tracheids of *Taxus*. *Protoplasma* **172**: 145–153.

Uggla, C., Moritz, T., Sandberg, G. and Sundberg, B. (1996) Auxin as a positional signal in pattern formation in plants. *Proc. Natl. Acad. Sci. USA* **93**: 9282–9286.

Uggla, C., Mellerowicz, E.J. and Sundberg, B. (1998) Indole-3-acetic acid controls cambial growth in Scots pine by positional signaling. *Plant Physiol.* **117**: 113–121.

20

Development of fibre and parenchyma cells in culms of bamboo (*Phyllostachys viridi-glaucescens* (Carr.) Riv. & Riv.) with emphasis on the cytoskeleton during cell wall deposition

E. Crow, R.J. Murphy and D.F. Cutler

1. Introduction

Bamboo represents a group of plants of high economic interest, particularly with respect to its structural properties. Mechanical strength is suggested as being optimized by an alternation of microfibril angle in consecutive wall lamellae of both fibre and parenchyma cells, providing balanced reinforcement to the culm (Parameswaran and Liese, 1980). Surprisingly, there has been no research into the process of cellulose microfibril orientation in fibre and parenchyma cells of bamboo. This report is the first to describe the arrangement of cytoskeletal arrays in relation to the orientations of newly deposited microfibrils in primary and early secondary walls in bamboo cells. The objective has been to elucidate a role for microtubules and microfilaments during wall deposition in bamboo.

Literature pertaining to cytoskeletal elements during wall formation in bamboo is limited to reports by Parameswaran and Liese (1980) that microtubules are common close to the plasma membrane in developing fibre cells seen under the transmission electron microscope.

Cell & Molecular Biology of Wood Formation, edited by R. Savidge, J. Barnett and R. Napier.

Investigations into the pattern of wall development have mostly been concerned with the polylamellate secondary wall, particularly that of fibre cells. Fibre wall development has been expressed in terms of its increasing thickness (Chengzhi and Gouen, 1985; Alvin and Murphy, 1988 (fibre and parenchyma cells); Murphy and Alvin, 1997). Another line of research has involved the monitoring of additional wall lamellae in fibre cells (Fujii, 1985; Kawase *et al.*, 1986; Liese and Weiner, 1996).

Studies of mature, fully elongated fibre and parenchyma cells have provided an insight into microfibril angles of primary and early secondary walls. The model for the polylamellate wall structure of the fibre cell put forward by Parameswaran and Liese (1980) is based upon observations made using transmission electron microscopy. It depicts the primary wall as having 'the usual network-like texture.' The microfibrils then become oriented 50° to the long axis of the cell in a lamella regarded by the authors as being transitional. This lamella, which is not present in all fibre cells, precedes the first 'true' secondary wall lamella in which microfibrils are oriented at an angle of 2–5°. In parenchyma cells, microfibrils of the primary wall are described as forming a flat helix, thereafter exhibiting ascending angles of 30–40° (Liese, 1998). In contrast, Preston and Singh (1950) using polarized light microscopy describe the outermost wall lamella (the primary wall) of fully elongated fibre cells as consisting of chains of cellulose in a flat helix, at about 35° to the long axis. Spirals of the adjacent lamella are much steeper, forming angles of 5–10°. Observations by Murphy *et al.* (1997) of soft rot decay in fibre and parenchyma cell walls led the authors to conclude that the microfibrils of parenchyma walls exhibit a typical helicoidal orientation. This was not the case in fibre cells, where broad lamellae of parallel microfibrils predominated but with some suggestion of a helicoidal orientation between lamellae. Direct visualization of ordered microfibrils of the primary and early secondary wall have been made in replicas of fibre cells in the course of lignification studies made by Suzuki and Itoh (in press).

2. Materials and methods

2.1 *Materials*

Elongating culms, around 1 m in length, were harvested in April/May from a stand of *Phyllostachys viridi-glaucescens* (accession number 1995–406) at the Royal Botanic Gardens, Kew. The stand was 4 years of age. Specimens were taken from 8 culms; 3 culms for light microscopy studies, 3 culms for both immunofluorescence of tubulin and field emission scanning electron microscopy work and 2 culms for immunofluorescence of actin.

2.2 *Light microscopy*

Material was fixed in the field in FAA (formalin acetic acid alcohol) (Johansen, 1940) for 2 days and stored in 70% ethanol. Internode samples were taken through an alcohol series before being taken through an alcohol:Histoclear series to 100% Histoclear. Material was embedded in Paraplast Regular and sectioned to a thickness of 8 μm on a Reichert rotary microtome. Sections were mounted on microscope slides and stained with 1% safranin in 50% alcohol and 1% aqueous alcian blue for 2 min each. They were then rinsed in distilled water, dehydrated in an alcohol series, cleared in Histoclear and mounted in DPX.

2.3 *Indirect immunofluorescence of tubulin and actin*

Unembedded material (tubulin). The method followed that of Abe *et al.* (1995). Material was cut into blocks and fixed in 3.6% paraformaldehyde, 0.2% glutaraldehyde, 10% dimethylsulfoxide (DMSO), 0.1% Nonidet P-40 in 50 mM piperazine-N,N'-bis-(2-ethylsuphonic acid) (PIPES) buffer at pH 7.2 and put under a vacuum of 500 millibars for 30 min. They were then removed from the vacuum and left overnight at room temperature. Each block was sectioned on a freezing stage at –23°C to a thickness of 30–50μm with a sliding microtome. Sections were immersed in phosphate-buffered saline (PBS) (1 l containing 8 g NaCl, 0.2 g KCl, 1.15 g Na_2HPO_4, 0.2 g KH_2PO_4, 0.2 g NaN_3, pH 7.6) application of the primary antibody mouse monoclonal antibody against chick brain α-tubulin (Amersham International plc, Bucks, UK), diluted 1:100 in PBS. Sections were washed in PBS before application of the secondary antibody, fluorescein isothiocyanate FITC-conjugated goat-antimouse IgG (Sigma Chemical Company, Dorset, UK), diluted 1:50 in PBS containing 1% (w/v) BSA, 0.1% (v/v) fish-skin gelatin and 0.2 g L^{-1} NaN_3. After washes with PBS, sections were stained with 0.01% (w/v) toluidine blue in PBS for 1 min and mounted in Vectashield. Exposure times followed those of Chaffey *et al.* (1997). Microtubule angles in cells at early stage 3 were measured from photographs. Observations were made using a Leitz Diaplan photomicroscope using filter 1 under epifluorescence and recorded on a Kodak Ektachrome 1600 ASA colour film. These observations were supported by transmission electron microscopy and scanning electron microscopy.

Resin-embedded material (actin). Material for immunofluorescence of actin was embedded in butyl-methyl methacrylate resin following a protocol employed by Chaffey *et al.* (1997). Briefly, longitudinal radial slivers of the culm wall at mid-internode positions, 3 mm in length, were cut under prefixative consisting of a 84 μM solution of *m*-maleimidobenzoic acid *N*-hydroxysuccinimide ester (MBS) (Sonobe and Shibaoka, 1989) in 25 mM PIPES buffer at pH 6.9, containing 0.05% DMSO. Specimens were then fixed in 3.7% paraformaldehyde with 0.2% glutaraldehyde in 25 mM PIPES buffer at pH 6.9 for 3.5 h. They were then rinsed in 25 mM PIPES buffer at pH 6.9 and dehydrated in an ethanol series to 100% with 30 min at each stage. Specimens were embedded in butyl-methyl methacrylate resin according to Gubler (1989).

Three longitudinal sections from 2 specimens per culm were cut using glass knives on a Reichert-Jung ultramicrotome to a thickness of 8 μm and mounted on slides coated with 10% (v/v) poly-L-lysine. Resin was removed from sections by immersion in acetone before rehydration with PBS. Immunolabelling was as for the previous technique but replacing the primary antibody with mouse monoclonal anti-actin from IgM (Amersham International plc, UK). Observations and recordings were made as for tubulin.

2.4 *Transmission electron microscopy*

Material was fixed in 2% paraformaldehyde and 2.5% glutaraldehyde in 0.1 M sodium cacodylate buffer at pH 8 overnight and postfixed in 1% osmium tetroxide at pH 6.8 for 2 h. The procedure for resin embedding in Spurr resin was followed (Spurr, 1969). Ultrathin sections were stained with 2% uranyl acetate/2% lead citrate and observed on a JEOL JEM 1210 electron microscope operating at 80 kV.

2.5 *Field emission scanning electron microscopy*

Material was fixed for 2 days in FAA and longitudinal sections were made to a thickness of 150 μm using a sliding microtome. Material at stages 1 and 2 of development was exposed to 7% sodium hypochlorite solution for 15 min whilst material at stage 3 was treated with a 40% aqueous methylamine solution for 5 min. Sections were dried in a critical point dryer and sputter coated using a gold palladium target in a Cressington sputter coater 208HR. Specimens were examined under the field emission microscope (FESEM) (Philips XL30 FE-SEM operating at 10 kV). The microfibrils of the inner walls of usually 3 to 6 fibre cells and 3 to 6 parenchyma cells at stages 1, 2 and 3 were observed. Fibre cells were selected at random, apart from those at stage 3 where the initial stages of secondary wall deposition were observed in 6 fibre cells adjacent to vascular elements. Six more fibre cells were observed elsewhere in the fibre strand. Parenchyma cells of similar lengths were observed. Wall areas were selected in central cell regions.

All figures are oriented with the long axis vertically unless otherwise stated.

3. Results

3.1 *Light microscopy*

Fibre and parenchyma cell differentiation occurs in a centrifugal direction across the culm radius. The development of fibre cells and surrounding parenchyma cells may be loosely divided into 3 stages.

Stage 1. Cell division occurring in both cell types; those in parenchyma cells mostly periclinal; in fibre cells mostly anticlinal. Fibre initials were elongated in the axial plane and had narrower diameters than the surrounding parenchyma cells. Fibre initials were storied, an arrangement which was further accentuated by their centrally positioned nuclei. Surrounding parenchyma cells of the ground tissue were usually rectangular in the radial plane and became arranged in axial rows due to numerous cell divisions occurring in a predominantly periclinal plane *(Figure 1)*.

Stage 2. Cell division continuing in parenchyma cells only; fibre cells elongating. End walls of elongating fibre cells became tapered and their bulging, slightly disfigured tips were possibly an indication that early intrusive growth was occurring in these cells. Cell contents were concentrated at the tips and the nucleus became pressed against the cell wall due to the presence of a large cell vacuole. Adjacent parenchyma cells continued to divide in a predominantly periclinal plane, accommodating an increase in length of adjacent fibre cells, or they occasionally showed initial signs of elongation *(Figure 2)*.

Stage 3. Cell division more or less complete; elongation occurring in all parenchyma and most fibre cells; initial stages of secondary wall deposition may be occurring in fibre cells adjacent to vascular elements. Fibre cells were long and needle-like in shape. Those adjacent to vascular elements usually showed initial stages of secondary wall deposition. However, elongation possibly continued in fibre cells situated further from the vascular elements. The majority of parenchyma cells were rectangular in the axial plane and were often at their longest adjacent to fibre cells where their end walls were occasionally sloped. Interspersed amongst the elongated parenchyma cells were isodiametric cells which did not elongate greatly *(Figure 3)*.

3.2 *Indirect immunofluorescence microscopy*

Microtubules

Stage 1. Cortical microtubule arrays in fibre and parenchyma cells at interphase exhibited an overall orientation transverse to the vertical axis *(Figure 4)*. Microtubules associated with end-walls of parenchyma cells, i.e. cell plate derived walls, as seen in transverse sections were orientated in a more or less radial direction with respect to the culm wall, or up to 30° from the radius.

Stage 2. A transverse orientation of microtubules remained in both cell types *(Figure 5)*. The apparently elongating tips of fibre cells exhibited transverse arrays with similar densities of microtubules as those at central regions of the cell. Microtubule organization at the very tips of fibre cells could not be assessed using this technique.

Stage 3. Microtubules in material at stage 3 were rarely seen due to poor infiltration of reagents and also to an increase in wall autofluorescence which was further intensified by the thickness of sections. However, isolated observations indicated that after a fairly lengthy transverse phase, microtubules became sloped in both cell types. Fibre cells *(Figure 6)* exhibited a slope steeper than that of the parenchyma cells — around 30° from the long axis of the cell as opposed to around 70°–80° in parenchyma cells (isodiametric cells were not observed in the study). It was not possible to assess whether microtubules were forming left or right hand helices.

Microfilaments

Stage 1. Material at this stage was not prepared for microfilament observations.

Stage 2. Microfilaments bundles in both cell types were axially aligned or very steeply sloped. They were more common in fibre cells where 1–3 bundles were seen, sometimes spanning considerable lengths of the cell. The arrangement of microfilaments at cell tips could not be assessed. Bundles were occasionally seen to branch into two, demonstrating a multi-filament composition *(Figure 7)*. Only a single bundle was seen in parenchyma cells and they were absent from cells which had obviously divided very recently.

Stage 3. Limited observations suggest that microfilament bundles in both cell types maintained similar orientations and incidence as those observed at stage 2.

3.3 *Transmission electron microscopy*

Stage 1 and 2. Glancing sections of material at stage 1 and 2 (not shown) revealed microtubules in LS, around 25 nm in diameter, in both cell types. They were quite often only 50 nm apart, occasionally running in pairs with a fairly constant distance of around 35 nm between them. In other areas of the wall they were as far apart as 150 nm. Microtubules were usually only several nm from the plasma membrane and occasionally cross-bridges between the two were visible. Populations of microtubules up to 150 nm from the plasma membrane were also observed. The majority were in parallel, transverse alignment, although an occasional microtubule adopted an entirely independent angle. Sometimes, a microfibril in the cell wall appeared to continue from a microtubule close to the plasma membrane, in precise linear co-alignment.

Stage 3. Transverse sections revealed an abundance of microtubules, in axial alignment, in fibre cells undergoing secondary wall deposition. Unlike the microtubules seen at stages 1 and 2, at stage 3, they were usually no further than several nm from the plasma

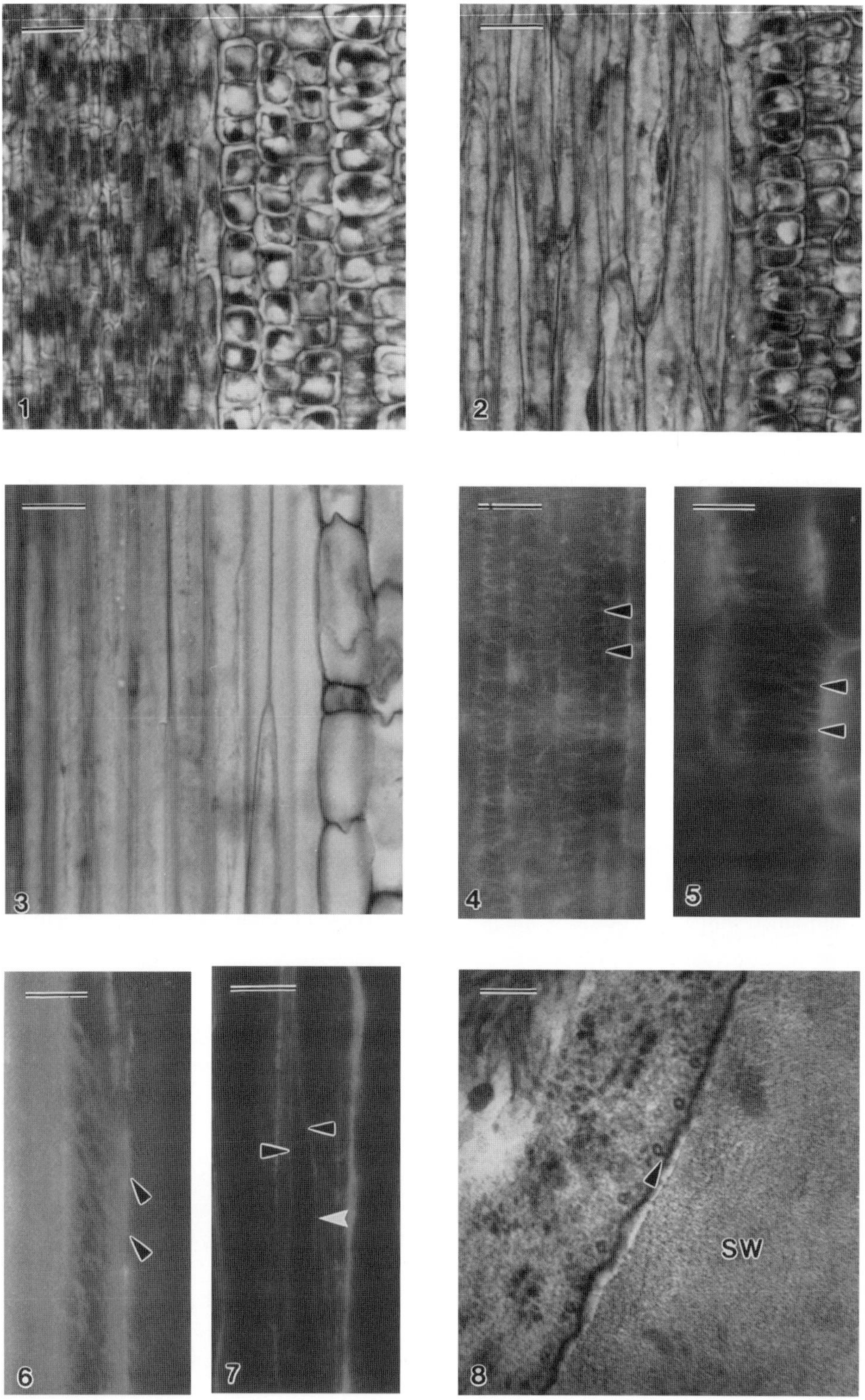
1
2
3
4
5
6
7
8
SW

membrane. Occasionally, cross-bridges were visible *(Figure 8)*. They were quite often only 50 nm apart, though in other areas of the wall they were as far apart as 150 nm. Microfilaments were not seen in TEM preparations.

3.4 *Field emission scanning electron microscopy*

Stage 1. The newly deposited cellulose microfibrils observed on the inner wall of both cell types usually demonstrated an overall transverse alignment. However, there was a slight degree of variation within this trend. *Figure 9* shows parallel, newly deposited microfibrils (black arrows), on the inner wall of a parenchyma cell, exhibiting an angle of around 60° from the long axis. In some instances, individual microfibrils adopted an entirely independent angle (white arrow). Of five parenchyma cells 10 mm in length, all exhibited wall texture similar to, or with a higher degree of microfibril parallelism as shown in *Figure 9*. However, in a sixth parenchyma cell of the same length, the orientation of microfibrils varied between successive lamellae. The most recently deposited microfibrils were transverse in orientation, adopting an angle of almost 90° to the previously deposited microfibrils beneath, which were more or less axially aligned (*Figure 10*). The end-wall of a parenchyma cell showed well-ordered microfibrils at an angle of 70° from the radius of the internode. The cross-fibrillar texture of the single parenchyma cell described above was not seen in the fibre cells.
Stage 2. Microfibril orientation remained in overall transverse orientation in both cell types. In areas around 5 mm from the tips of elongating fibre cells, microfibrils exhibited similar orientation and texture to those of the main body of the cell.
Stage 3. The microfibrils in a fibre cell adjacent to the metaxylem undergoing initial stages of secondary wall deposition were axially aligned *(Figure 12)*. In an adjacent fibre cell, next in line to undergo secondary wall deposition, the microfibrils were in transverse orientation. Microfibrils were sloped in other fibre cells of the same fibre

---

Figure 1. *Stage 1 of development. Dividing fibre and parenchyma cells. Scale bar = 25 μm.*

Figure 2. *Stage 2 of development. Fibre cells show possible signs of intrusive growth (slightly buckled cell tips) whilst parenchyma cells are either still dividing or have commenced upon elongation. Scale bar = 25 μm.*

Figure 3. *Stage 3 of development. Parenchyma cells and probably most fibre cells are undergoing elongation. Scale bar = 25 μm.*

Figure 4. *Elongating fibre cells at stage 2. Microtubules are in transverse orientation (arrows). Scale bar = 25 μm.*

Figure 5. *Parenchyma cells undergoing early stages of elongation at stage 2. Microtubules are in transverse orientation (arrows). Scale bar = 15 μm.*

Figure 6. *Fibre cell at early stage 3 showing steeply sloped microtubules (arrows). Scale bar = 15 μm.*

Figure 7. *Fibre cell at stage 2 showing axially aligned microfilament bundles (black arrows). White arrow shows the point at which a microfilament bundle is branching into two. Scale bar = 15 μm.*

Figure 8. *TEM micrograph showing a transverse section through a fibre cell at stage 3 undergoing secondary wall deposition. Microtubules are seen in abundance adjacent to the plasma membrane, occasionally with cross-bridges (arrow) forming connections with it. sw = secondary wall. Scale bar = 100 nm.*

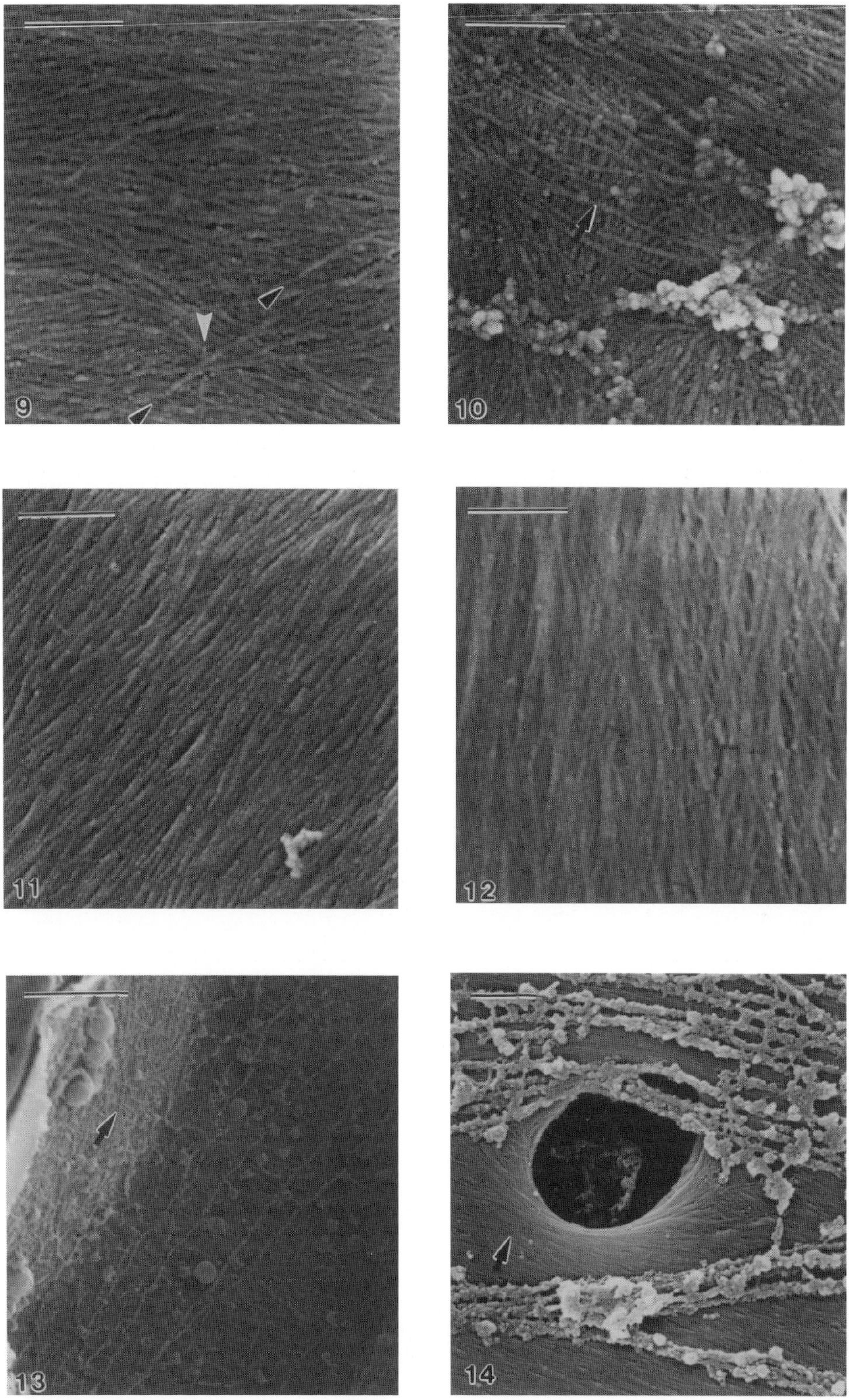
9
10
11
12
13
14

sheath *(Figure 11)*, with angles as little as 25° in some cells. Mostly left and some right hand helices were observed in the fibre cells. Parenchyma cells consistently exhibited a sloped microfibril orientation. Microfibrils formed an angle of between 35° and 45°, with the long axis of the cell with only right hand helices seen in these cells. The microfibrils demonstrated a consistently high degree of parallelism, not always present in the primary wall.

In the course of scanning electron microscopy work, occasional observations were made of a system of fine filaments on the inner walls of fibre and parenchyma cells at stages 1 and 2. A more or less transverse array of straight and sometimes curved filaments, around 100 nm in diameter was seen. In rare cases, an overlying longitudinal system of considerably fewer, thicker filaments was also observed *(Figure 13)*. The finer, transverse filaments were often co-aligned with microfibrils beneath *(Figure 14)*.

4. Conclusions

Cell divisions amongst initial cells and subsequent early differentiation of parenchyma and fibre cells in bamboo occur more or less at random. In order to follow systematically the involvement of the cytoskeleton during the development of the cell wall, it has therefore been necessary to define three stages of fibre and parenchyma cell development. Each stage merely provides an easily identifiable reference point for further studies and does not represent the developmental continuum which inevitably exists.

The comparison of microtubule and microfibril orientations using immunofluorescence techniques and scanning electron microscopy respectively (an approach following Abe *et al.*, 1995) clearly demonstrates co-alignment. There can be little doubt that microtubules play a key role in the guidance of synthetic complexes at the plasma membrane. Throughout most primary wall synthesis, a transverse arrangement of both microtubules and microfibrils predominates. This appears from the literature to be the usual situation in elongating cells of higher plants (McCann *et al.*, 1993; Holdaway *et al.*, 1995; McClinton and Sung, 1997) with elongation seemingly encouraged by the lateral reinforcement offered by transverse microfibrils (Green, 1980).

The transverse microfibrils deposited over axially aligned microfibrils seen in a single parenchyma cell at stage 1 was an isolated observation. This suggests that it

---

Figure 9. *Inner wall of a parenchyma cell at stage 1 showing a more or less transverse orientation of recently deposited microfibrils. Parallel microfibrils around 30° from the horizontal (black arrows). Microfibril apparently adopting an entirely independent angle (white arrow). Scale bar = 250 nm.*

Figure 10. *Inner wall of parenchyma cell at stage 1 showing differences in orientation between microfibrils of newly and previously deposited lamellae. The concretions are probably remnants of cell contents. The long axis is denoted by the arrow. Scale bar = 250 nm.*

Figure 11. *Inner wall of fibre cell at early stage 3 showing sloped microfibrils in a right hand helix. Scale bar = 250 nm.*

Figure 12. *FESEM micrograph showing inner wall of fibre cell at stage 3 during initial stages of secondary wall deposition. Scale bar = 250nm.*

Figure 13. *Inner wall of an elongating fibre cell at stage 2. A roughly transverse system of fine filaments is visible along with several, thicker, more or less axially aligned filaments. Arrow denotes the long axis of the cell. Scale bar = 5 μm*

Figure 14. *Parenchyma cell at stage 2 of development showing a pit on the inner wall. The fine, roughly transverse filaments are often co-aligned with microfibrils beneath (top right). Arrow denotes the long axis. Scale bar = 500 nm.*

represents a brief phase of microfibril deposition, possibly when cells have just completed or are about to embark upon a mitotic division. Axially aligned microtubules were not seen in immunofluorescence observations and so it is possible that axially aligned microfibrils were deposited in the absence of microtubule guidance. Alternatively, microtubules in axially alignment exist, but because of the brevity of this phase and variability in fixation quality, the probability of their observation was extremely low.

Towards the end of primary wall synthesis, microtubules appear to instigate a highly controlled sloping of microfibrils in both cell types. Gradients of microtubules and microfibrils are probably steeper in fibre cells than they are in the parenchyma cells. The fibre cells are first to undergo secondary wall deposition and their more steeply sloped microtubules and microfibrils may be the result of a coarser, more rapid transition from a transverse to an axial alignment occurring in these cells. The implications are that the orientations of microfibrils in the late primary walls of parenchyma cells differ from those of the fibre cells.

A system of filaments seen on the inner wall bears remarkable similarity in orientation and form to the cytoskeleton seen in immunolabelled material. Transverse filaments have a diameter of around 100 nm, considerably larger than the diameter of microtubules as seen under the TEM. However, it is possible that microtubule associated proteins may be partly responsible for the increase. The slight curving of filaments seen in *Figure 14* is reminiscent of those mapped by Seagull (1983) from serial TEM sections of a range of plant species, including corn and sorghum.

It seems surprising that the cytoskeleton could possibly endure a 15 min exposure to 7% sodium hypochlorite. However, there have been several reports, most recently by Scherfeld *et al.* in 1998 of an insoluble cytoskeleton capable of enduring quite vigorous washings with detergents; washings which were sufficient to delipidate the plasma membrane. If the transversely oriented filaments are in fact microtubules, that they remain firmly appressed to the cell wall whilst the plasma membrane has dissolved suggests a strong physical connection between the microtubules and the cell wall. Akashi and Shibaoka (1991) put forward the hypothesis that the cell wall stabilizes the microtubules via a transmembrane protein bridge. However, there may also be the possibility of productive cellulose synthetic complexes situated at the microtubules forming a link — with the wall on one side via newly polymerized microfibrils and with the microtubules on the other, perhaps via microtubule associated proteins or cross-bridges.

Observations of microfibrils in precise co-alignment with microtubules in TEM glancing sections (papers which include micrographs illustrating similar observations in other plants include those of Seagull and Heath (1982) (*Azolla* sp.), Seagull (1983) (*Zea mays*) and Eleftheriou (1990) (*Triticum aestivum*)) demonstrate that productive synthetic complexes are, at least on occasion, directly associated with the microtubules. Whether microtubules oriented at angles independent from the net orientation seen at stages 1 and 2 may be responsible for similarly arranged microfibrils seen on the inner wall remains unclear.

As well as the relatively thick axial microfilament bundles seen in immunolabelled material of the present study, some authors have described a possible array of finer microfilaments situated close to the plasma membrane and possibly associated with the microtubules. Seagull and Heath (1980), in radish root hair cells, concluded that these microfilaments may play a vital role in the mobility of synthetic complexes through the plasma membrane. More recently, Seagull (1993) suggests a role controlling microtubule orientation. In the present study, the absence of microfilaments in the vicinity of the microtubules does not support these claims, however it is feasible that an array of finer microfilaments may have escaped detection using immunofluorescence techniques.

In summary, a very close co-alignment of cortical, plasma membrane-associated microtubules with newly deposited cell wall microfibrils was found in these studies of primary and very early secondary wall formation in the bamboo *P. viridi-glaucescens*. The consistency of these observations at different times in wall development strongly suggests a functional role for microtubules in the positive orientation of microfibril synthetic complexes.

Acknowledgements

The authors wish to thank Dr Nigel Chaffey of Long Ashton Research Station, University of Bristol and Mr Ian Morris of Imperial College, London for their technical advice and assistance. Thanks are due to Dr Chris Jones and Dr Alex Ball at the Natural History Museum where FESEM work was conducted. The authors gratefully acknowledge support by the Biotechnology and Biological Sciences Research Council for this work in the form of a CASE award for E. Crow's PhD studies with the Royal Botanic Gardens, Kew as collaborating body.

References

Abe H., Funada R., Imaizumi H., Ohtani J. and Fukazawa, K. (1995) Dynamic changes in the arrangement of cortical microtubules in conifer tracheids during differentiation. *Planta* **197**: 418–421.

Akashi, T. and Shibaoka, H. (1991) Involvement of transmembrane proteins in the association of cortical microtubules with the plasma membrane in tobacco BY-2 cells. *J. Cell Sci.* **98**: 169–174.

Alvin, K.L. and Murphy, R.J. (1988) Variation in fibre and parenchyma wall thickness in culms of the bamboo *Sinobambusa tootsik. IAWA bulletin* **9**: 353–361.

Chaffey, N.J., Barnett J.R. and Barlow, P.W. (1997) Visualization of the cytoskeleton within the secondary vascular system of hardwood species. *J. Microscopy* **187**: 77–84.

Chengzhi S. and Guoen X. (1985). Fibre morphology and crystallinity of *Phyllostachys pubescens* with reference to age. In: *Recent Research on Bamboo. Proceedings of the International Workshop, Hangzou, People's Republic of China*, pp. 247–249 (eds. Rao, A.N., Dhanarajan, G. and Sastry, C.B.)

Eleftheriou, E.P. (1990) Microtubule and sieve plate development in differentiating protophloem sieve elements of *Triticum aestivum* L. *J. Exp. Bot.* **41**: 1507–1515.

Fujii, T. (1985) Cell-wall structure of the culm of *Azumanezasa* (*Pleiblastus chino* Max.). *J. Jap. Wood Res. Soc.* **31**: 865–872.

Green, P.B. (1980) Organogenesis — a biophysical view. *Annual Rev. Pl. Physiol.* **31**: 51–82.

Gubler, F. (1989) Immunofluorescence localisation of microtubules in plant root tips embedded in butyl-methyl methacrylate. *Cell Biol. Int. Rep.* **13**: 137–145.

Holdaway, N.J., White R.G. and Overall, R.L. (1995) Is the recovery of microtubule orientation in pea roots dependent on the cell wall? *Cell Biol. Int. Rep.* **19**: 916–918.

Johansen, D.A. (1940) *Plant Microtechnique.* 1st Edn. McGraw-Hill, New York.

Kawase, K., Sato, K., Imagawa, H. and Ujiie, H. (1986) Studies on utilisation of Sasa-bamboos as forest resources. 4. Pulping of young culms and histological changes of cell structure of the culms in growing process. *Res. Bull. Coll. Exp. Forests Hokkaido Imp. Univ.* **43**: 73–97.

Liese, W. (1998) *The Anatomy of Bamboo Culms.* INBAR, Beijing, Eindhoven, New Delhi.

Liese, W. and Weiner, G. (1996) Ageing of bamboo culms. A review. *Wood Sci. Techn.* **30**: 77–89.

McCann, M.C., Stacey, N.J., Wilson, R. and Roberts, K. (1993) Orientation of macromolecules in the walls of elongating carrot cells. *J. Cell Sci.* **106**: 1347–1356.

McClinton, R.S. and Sung, Z.R. (1997) Organisation of cortical microtubules at the plasma membrane in *Arabidopsis*. *Planta* **201**: 252–260.

Murphy, R.J. and Alvin, K.L. (1997) Fibre maturation in bamboos. In: *The Bamboos.* Linnean Society Symposium (ed. Chapman, G.P.), Academic Press, London, pp. 293–303.

Murphy, R.J., Sulaiman, O. and Alvin, K.L. (1997) Ultrastructural aspects of cell wall organisation in bamboos. In: *The Bamboos.* Linnean Society Symposium (ed. Chapman, G.P.), Academic Press, London, pp. 305–312.

Parameswaran, N. and Liese, W. (1980) Ultrastructural aspects of bamboo cells. *Cellulose Chem. Technol.* **14**: 587–609.

Preston, R.D. and Singh, K. (1950) The fine structure of bamboo fibres. I. Optical properties and X-ray data. *J. Exp. Bot.* **1**: 214–227.

Scherfeld, D., Schneider, G., Guttman, P. and Osborn, M. (1998) Visualization of the cytoskeletal elements in the transmission X-ray microscope. *J. Struct. Biol.* **123**: 72–82.

Seagull, R.W. (1983) The role of the cytoskeleton during oriented microfibril deposition. I. Elucidation of the possible interaction between microtubules and cellulose synthetic complexes. *J. Ultrastruct. Res.* **83**: 168–175.

Seagull, R.W. (1993) Cytoskeletal involvement in cotton fibre growth and development. *Micron* **24**: 643–660.

Seagull, R.W. and Heath, I.B. (1980) The organization of cortical microtubule arrays in the radish root hair. *Protoplasma* **103**: 205–229.

Seagull, R.W. and Heath, I.B. (1982) Oriented cellulose microfibrils and the cytoskeleton: a critical comparison of models. In: *The Cytoskeleton in Plant Growth and Development* (ed. Lloyd, C.W.), Academic Press, London, pp. 163–182.

Sonobe, S. and Shibaoka, H. (1989) Cortical fine actin filaments in higher plant cells visualized by rhodamine-phalloidin after pretreatment with *m*-maleimidobenzoyl *N*-hydroxysuccinimide ester. *Protoplasma* **148**: 80–86.

Spurr, A.R. (1969) A low viscosity epoxy resin embedding medium for electron microscopy. *J. Ultrastruct. Res.* **26**: 31–43.

Suzuki, K. and Itoh, T. (2000) The change of cell wall architecture during lignification of bamboo. '*Cell Walls, 98.*' *8th International Cell Wall Meeting*, John Innes Centre, 1st–5th September, 1998. In press.

21

Proteases in relation to the earliest stage of xylogenesis in *Pinus banksiana* Lamb. stems

I. Iliev and R. Savidge

1. Introduction

Proteases (E.C. class 3.4) are hydrolytic enzymes possessing the unique, physiologically important activity of catalysing peptide bond cleavage. They include diverse endopeptidases (proteinases) and exopeptidases and cleavage mechanisms. They occur both as discrete enzymes and as multi-catalytic complexes known as proteasomes (Neurath, 1986; Avila, 1997). Protease activity serves to modify peptide-chain length and the three dimensional structure of all types of proteins (e.g. enzymes, storage proteins, DNA-binding proteins and peptides). In catalysing such modifications, proteases can modify (enhance, retard or block) the catalytic activity of enzymes (including other proteases), and combinations of proteases also serve to remove damaged and foreign proteins from the protoplasm. The activity of proteases has emerged as one of the most important mechanisms controlling when and where growth and tissue differentiation occur in all biological systems that have been well investigated (Neurath, 1986; Avila, 1997).

Only limited information is available on protease activity in relation to wood formation (Sudachkova *et al.*, 1981; Zakrzewski and Rakowski, 1987; Rakowski and Wodzicki, 1994, 1995; Ye and Varner, 1996; Fukuda, 1997; Iliev and Savidge, 1999). Seasonal cambial growth in trees comprises a number of developmental stages and leads to the production of anatomically highly complex tissues (Savidge, 1996; Savidge *et al.*, 1998). By volume, the majority of cambial derivatives terminate their differentiation as dead, water-conducting and/or supporting tissues. In terms of cell numbers, on the other hand, sap wood in most woody species contains many times more living cells than dead elements; thus, only a small fraction of the cambial

Cell & Molecular Biology of Wood Formation, edited by R. Savidge, J. Barnett and R. Napier.

derivative population actually dies during xylogenesis (Savidge, 1996). Nevertheless, taking into consideration the volume of wood produced on earth each year, protoplasmic autolysis of otherwise fully differentiated xylem cells must be nature's most common form of programmed cell death. Proteolytic activity is undoubtedly important for autolysis to occur.

In our work, we observed that the phenological stages of seasonal cambial growth in *Pinus banksiana* involved an extensive assortment of diverse kinds and sizes of proteases, from resumption of cell division activity in springtime to the onset of complete destruction of the protoplast following differentiation of cambial derivatives into mature xylem elements (Iliev and Savidge, 1999). Unexpectedly, the strongest protease activity during the season of wood formation was associated with the earliest stage of xylogenesis, viz. radial expansion of primary-walled cambial derivatives, rather than with protoplasmic autolysis. We provide additional information on the protease activity of that early developmental stage here.

2. Historical perspective

Considerable research into both proteolytic enzymes and endogenous inhibitors of protease activity has been reported in seeds, leaves and other parts of perennial plants. However, only rudimentary research concerned with gross proteolytic activity of the cambial region has been reported, and minimal attention has been paid to the phenological stage of seasonal cambial growth (Kuprevich, 1949; Sudachkova *et al.*, 1981; Zakrzewski and Rakowski, 1987; Rakowski and Wodzicki, 1994, 1995).

Very little is known about the mechanisms controlling cell behaviour of primary-walled radially expanded cambial derivatives prior to their commencing to differentiate into early wood tracheids. These cells are able to expand radially but are prevented from entering into another cycle of cell division. Nevertheless, they can also experience a substantial delay in their entry into the pathway of tracheid differentiation (Savidge and Wareing, 1981). The cell fate of cambial derivatives at that early stage can readily be modified, such that the cell-death pathway is relinquished and living cells are incorporated into the xylem (Savidge, 1996). Such observations indicate that the derivatives are uncommitted to any particular cellular differentiation pathway. The regulatory mechanisms controlling the differentiation pathway pursued are very likely to involve proteases.

Research to characterize proteases specifically related to xylogenesis in model systems is also limited. Minami and Fukuda (1995) provided evidence that a cysteine endopeptidase was associated with protoplasmic autolysis in the *Zinnia elegans* mesophyll model system for *in vitro* tracheary element differentiation. Ye and Varner (1996) also reported a cysteine protease (22.7 kDa) and a serine protease (60 kDa) to be associated with tracheary element differentiation in *Zinnia* stems, as well as in cultured mesophyll cells. Both of those reports were concerned with late stages of xylogenesis, when tracheary element differentiation was well underway. To our knowledge, no research other than our own has focused on the earliest xylogenic stage, when the cell fate of cambial derivatives remains undetermined. The proteases in the vicinity of the cambial zone associated with this earliest stage of xylogenesis are described here further.

3. Materials and methods

3.1 *Trees*

Pinus banksiana trees growing in the University of New Brunswick Woodlot, Fredericton, NB, Canada, were harvested during May, when the cambium was dividing actively and producing primary-walled, radially enlarging and enlarged cambial derivatives. Few if any of the cambial derivatives had initiated bordered-pit development and none had initiated secondary-wall formation or lignification. After tree felling and branch removal, the main stems were transported to the laboratory. The bark was peeled, the exposed bark and wood surfaces scraped, the acquired scrapings combined and frozen at –196°C and the frozen tissue ground in liquid nitrogen to a fine powder using mortar and pestle.

3.2 *Samples 1 and 2*

Two protein fractions harvested from *P. banksiana* on May 20th were investigated. These fractions are referred to as 'sample 1' and 'sample 2'. Frozen, powdered samples were weighed and extracted with 0.05 M Tris-HCl buffer (pH 8.0) containing 5% glycerol for 1 h at 4°C under continuous stirring. The insoluble part was separated by centrifugation (10 000 × *g*, 2°C, 20 min). The supernatant was centrifuged again (24 000 × *g*, 2 °C, 20 min) and further concentrated using a Centriprep 10 (10 kDa cut-off membrane, Amicon) at 2500 × *g* and 2°C to yield 'Sample 1'. Centriprep 10 preparations were desalted using Sephadex G-25 columns (PD-10, Pharmacia, Sweden) and lyophilized. The re-dissolved sediment of the second centrifugation provided 'Sample 2'. When preparative IEF was done, the supernatant was decanted and treated with polyethylenimine (PEI, section 3.6).

Thus, sample 1 comprised macromolecules (> 10 kDa) that remained buffer-soluble, or -miscible, following centrifugation at both 10 000 × *g* and 24 000 × *g*. Sample 2 comprised soluble and/or miscible substances that remained in or at the surface of the supernatant following centrifugation at 10 000 × *g*, that pelleted when subjected to centrifugation at 24 000 × *g*, and that yielded a clear solution when reconstituted in buffer.

3.3 *Proteolytic activity*

The activity of proteases was defined as micrograms of tyrosine liberated per minute during hydrolysis of 1.2% haemoglobin (bovine) containing 0.1 M NaCl at 37°C. An aliquot of sample was added to 2.5 ml substrate solution (haemoglobin) and incubated at 37°C. The reaction was stopped by adding 2.5 ml 5% trichloroacetic acid (TCA). The sample was filtered through filter paper (Whatman 1), and the absorbance of the filtrate read at 280 nm against the absorbance of a blank sample. For blank samples, TCA was added immediately after addition of the sample. Two parallel samples were analysed and the results are averaged.

3.4 *Protease inhibitors*

In order to determine the kinds of proteases according to their catalytically active groups, different combinations of protease inhibitors specific for the four main kind of endopeptidases were used, viz. aprotinin for serine proteases, E-64 for cysteine proteases, pepstatin for aspartate proteases, and ethylenediaminetetraacetic acid (EDTA) for metallo proteases. The amino peptidase inhibitor bestatin was also used to investigate exopeptidase activity.

The inhibitors were used in the following concentrations: Aprotinin and Pepstatin (Sigma) — 1μg ml^{-1}; E-64 (Sigma) — 5.1 μg ml^{-1}; Bestatin (Sigma) — 40 μg ml^{-1}; EDTA (BDH) — 0.55 mg ml^{-1}. Leupeptin (Sigma) was also used in some experiments — 5 μg ml^{-1}.

3.5 *Ion exchange chromatography*

Ion exchange chromatography was performed on a Bakerbond Wide-Pore CBx, 5 μm column (7.75 × 100, Baker) equilibrated with 0.01 M sodium phosphate buffer (pH 7.5). Following initial isocratic elution with the same buffer, a linear NaCl gradient from 0 to 1.6 M was applied, eluting at 1 ml min^{-1}. The absorbance of the eluate at 280 nm was measured continuously and fractions were collected.

3.6 *Polyethylenimine removal of nucleic acids and phospholipids*

A 10% (w/v) PEI solution (Sigma) was prepared by diluting 10 ml of concentrate to 20% with distilled water and dialysing overnight in a half-filled dialysis bag (Spectra/Por, 6 000–8 000 MW cut-off, Spectrum) against distilled water. The pH was adjusted to pH 8.0 with HCl, diluted to 50 ml with distilled water, and filtered. This was added to the solubilized proteins to give a final concentration of 0.9% (w/v), with stirring at 4°C. Finally, the PEI-treated preparation was centrifuged for 15 min at 12 000 × *g*. The clear supernatant was collected.

3.7 *Ammonium sulfate precipitation*

Solid ammonium sulfate (Sigma) was added in increments with constant stirring at 4 °C until the supernatant obtained after PEI-treatment was brought to 85% saturation. The pH was followed and kept constant. After equilibration of the solvent it was centrifuged for 20 min at 22 000 × *g*. The sediment was re-dissolved in distilled water, desalted using a PD-10 column (Pharmacia) to 0.02 M potassium phosphate buffer (pH 7.5) and concentrated using a Centriprep 10 kDa cut-off membrane (Amicon).

3.8 *Preparative isoelectric focusing*

For preparative isoelectric focusing (IEF), Ultrodex granulated gel (4 g; Pharmacia) was added to 100 ml of 2% (w/v) ampholine solution (Pharmacia, pH 3.5–9.5), poured into an 11 × 24 cm gel tray and evaporated to 35% moisture. The sample was applied behind the cathode and IEF was performed on a 2117 Multiphor II unit (Pharmacia) for 18 h at 8 W. The paper print was stained with Coomassie Brilliant Blue R-250 (Sigma).

3.9 *SDS PAGE*

For sodium dodecyl sulfate polyacrylamide gel electrophoresis (SDS PAGE) 12.5% homogenous gels were run on a PhastSystem (Pharmacia). Silver staining and a LMW calibration kit were from Pharmacia. Samples were mixed with an equal volume of 10 mM Tris/HCl buffer (pH 8.0) containing 4% SDS and 0.01% bromophenol blue and incubated for 10 min at room temperature.

3.10 *Protease zymograms*

Agar gels (2% w/v) containing 1% (w/v) casein were made using the 'flap technique' (Höfelmann *et al.*, 1983) in 0.3 M Tris/HCl buffer (pH 7.5).

The SDS PAGE gel was immersed in 0.3 M Tris/HCl buffer (pH 7.5) for 10 min. Then the agar-casein gel was applied to the surface of the polyacrylamide, avoiding the formation of air bubbles. The sandwich was incubated for 30 min at 37°C. Clear bands (interpreted as proteolytic activity) on the opalescent gel were visualized after immersing the agar gel in 10% trichloroacetic acid.

4. Results and discussion

Protease inhibitors altered the proteolytic activities in both sample 1 and sample 2, at pH 4.0 and pH 7.5 (*Figure 1*), pointing to the presence of serine, cysteine, aspartate and metallo proteases in extracts of the cambial region. However, with all four endopeptidase inhibitors in combination, there was still detectable proteolytic activity in both samples at both pH values (*Figure 1*). When the endopeptidase cocktail was supplemented with bestatin, somewhat reduced proteolytic activity was observed in both samples, indicating that a small amount of amino peptidase activity was also present.

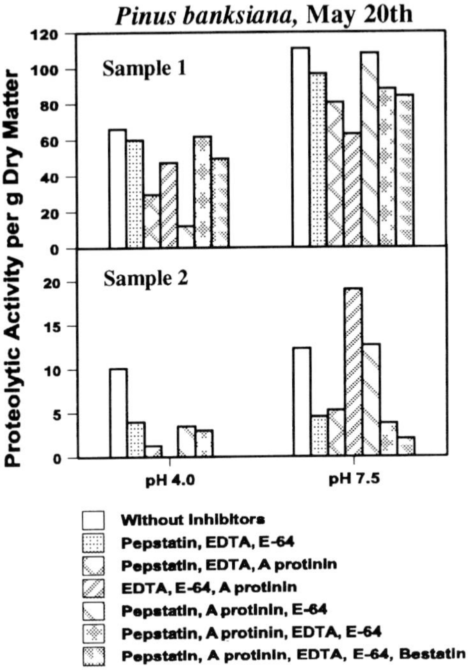

Figure 1. The effects of a number of distinct types of protease inhibitors on proteolytic activity in extracts from cells of the cambial region. Measurements were made at pH 4.0 and pH 7.5 in sample 1 and sample 2.

Samples 1 and 2 were subjected to ion exchange chromatography (*Figures 2* and *3*, respectively). Those substances eluting with starting buffer (*Figures 2a* and *3a*) were investigated separately from those eluting with a NaCl gradient (*Figures 2b* and *3b*). The total proteolytic activity eluting with starting buffer was higher than that retained on the column for both samples.

Proteases active at pH 7.5 and eluting with starting buffer did not affect each other to any great extent. Sample 1 activity in chromatography fraction 1 was due to the presence of serine/cysteine and metallo proteases (*Figure 2a*). Total activity at pH 4.0 was contributed by all the different kinds of proteases and, in contrast to pH 7.5, they affected each other specifically, such that activity could be observed only after inhibition of some kinds of proteases (*Figure 2a*). The specific interactions at pH 4.0 were also observed in fraction 2 (*Figure 2a*). Inhibition of serine/cysteine proteases led to loss of proteolytic activity (*Figure 2a*).

During chromatography of both samples, elution of bound proteolytic activity started at NaCl concentrations between 0.5–0.6 M (*Figures 2b* and *3b*). In sample 1, data on proteolytic activity measured at pH 7.5 indicated strong protease interactions, with total proteolytic activity in the inhibitor-treated fractions being much higher than that of the controls (*Figure 2b*).

Interactions of proteases were even more apparent at pH 4.0, particularly in the fractions of Sample 2 (*Figures 3a* and *3b*). Proteolytic activity measured at pH 4.0 was not detectable after inhibition of serine/cysteine proteases in fractions 1, 3, 5 (*Figure 3a*) and 8 (*Figure 3b*), nor was it detected in fraction 9 after inhibition of metallo proteases (*Figure 3b*). It was the opposite in fractions 2 and 4 (*Figure 3a*). Proteolytic activity at pH 4 in fraction 2 was observed only after EDTA inhibition of metallo proteases, and in fraction 4 only after inhibition of serine/cysteine proteases (*Figure 3a*).

The presence of a large number and variety of proteases in primary-walled enlarging cambial derivatives of *Pinus banksiana* and their interactions were confirmed by preparative IEF (*Figure 4*). For this data, larger masses of tissue were collected from four trees in early May and extracted in the same way as samples 1 and 2. In order to increase the concentration of the proteases and to eliminate the possible influences of non-protein compounds on proteolytic activity, the samples were subjected to additional purification. After removing nucleic acids and phospholipids by treatment with PEI and other compounds using ammonium sulfate precipitation, relatively pure protein fractions were obtained. These proteins were separated by preparative IEF in immobilized pH gradients and the presence of proteases was determined directly by assaying for proteolytic activity. The results show that proteases having p*I*s over a wide range, from 3.5 up to 9.5, were present in the cambial region of *Pinus banksiana* (*Figure 4*).

Comparison of the proteolytic activity before and after IEF indicated that significant interactions were occurring among the proteases (*Figure 4*). The activity of the starting sample at pH 4.0 dramatically decreased after IEF separation (*Figure 4*, pH 4.0, right). In contrast, at pH 7.5 the low activity of the starting sample (*Figure 4*, pH 7.5, right) was markedly increased after isoelectric focusing. Thus, the strong activity at pH 4 before IEF became the lowest following separation, whereas the almost negligible activity at pH 7.5 before IEF became the highest after separation. Thus, it is again apparent that considerable interactions, both positive and negative, occur between proteases of the cambial region. At pH 9.5, the activity of the starting sample could be accounted for by the summed activities present in the several

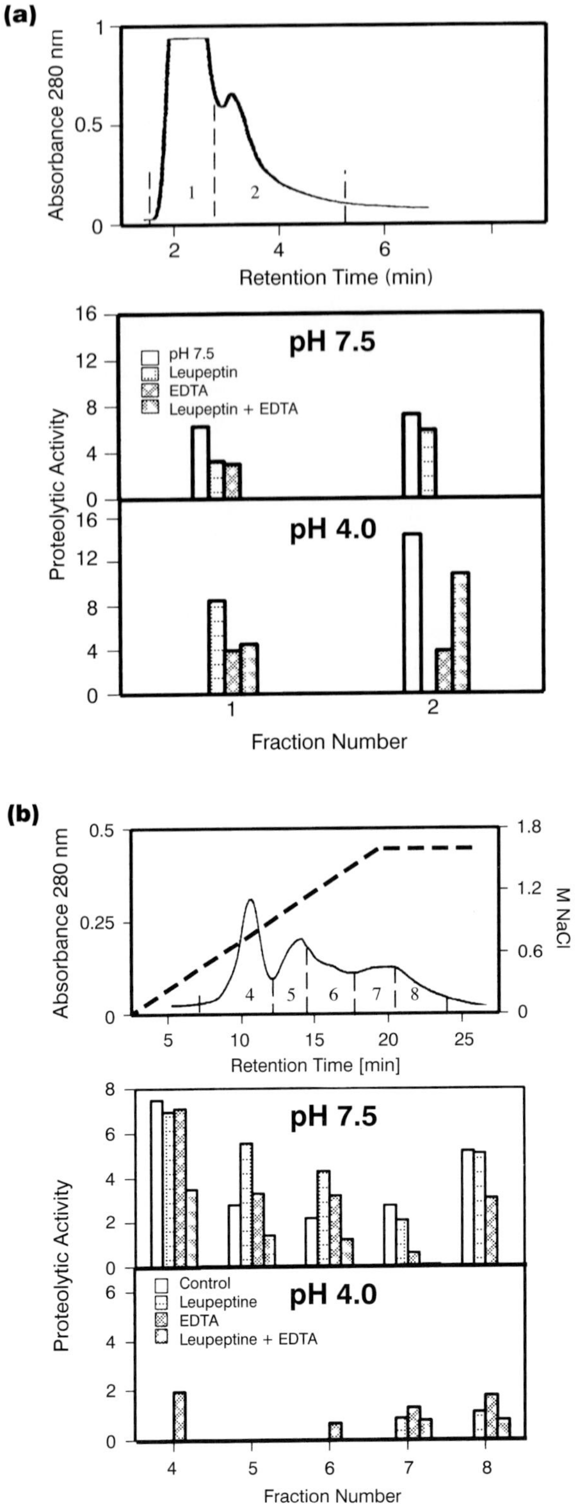

Figure 2. *Cation exchange chromatography of sample 1 showing variation in proteolytic activity as affected by leupeptin and EDTA. (a) Eluted with the starting buffer; (b) eluted by NaCl gradient.*

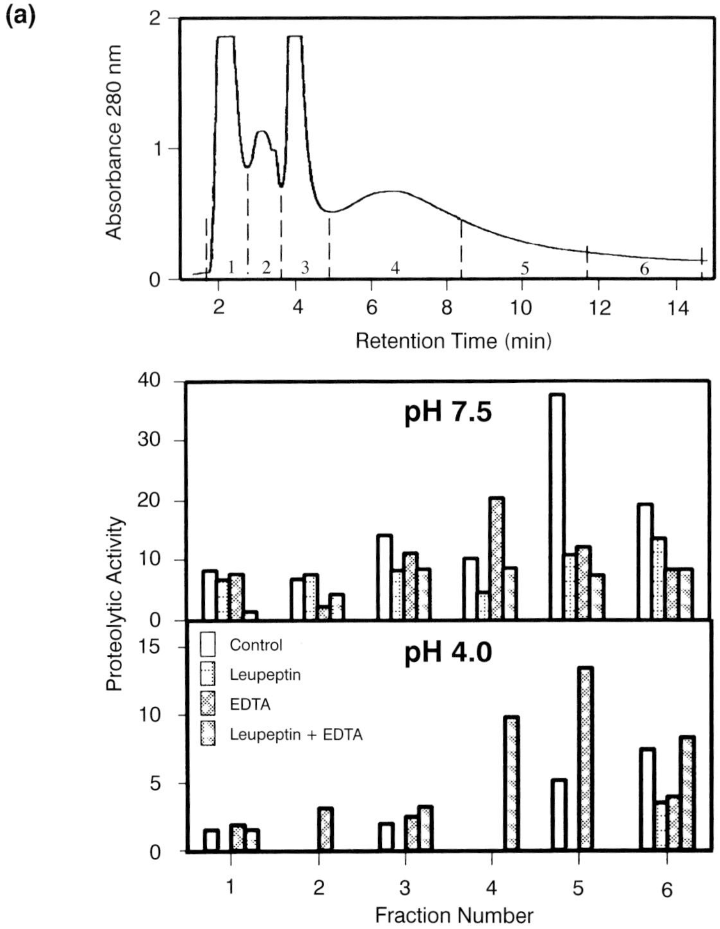

Figure 3. *Cation exchange chromatography of sample 2 showing variation in proteolytic activity as affected by leupeptin and EDTA. (a) Eluted with the starting buffer; (b) eluted by NaCl gradient.*

pH zones measured, indicating that the proteases associated with this activity were the most stable and least affected either by other proteases or by themselves.

The cation-exchange HPLC data (*Figures 2* and *3*) indicate that some protease combinations cancel one another, activity being observed only in the presence of protease inhibitors, thus corroborating the data of *Figure 4*. The inhibitor data indicated that inactivation was strongest for proteases associated with the activity at pH 7.5; not surprisingly, therefore, this activity was weakest before IEF separation (*Figure 4*), in agreement with the HPLC data. It can be suggested that at least some of the proteases active at pH 4.0 and 7.5 must be compartmentalized within or between cells. The protease activity at pH 9.5, previously noted to be present in both active and dormant cambium (Iliev and Savidge, 1999), appears to occur generally and stably throughout the seasons in both the cambial zone and its derivative cells.

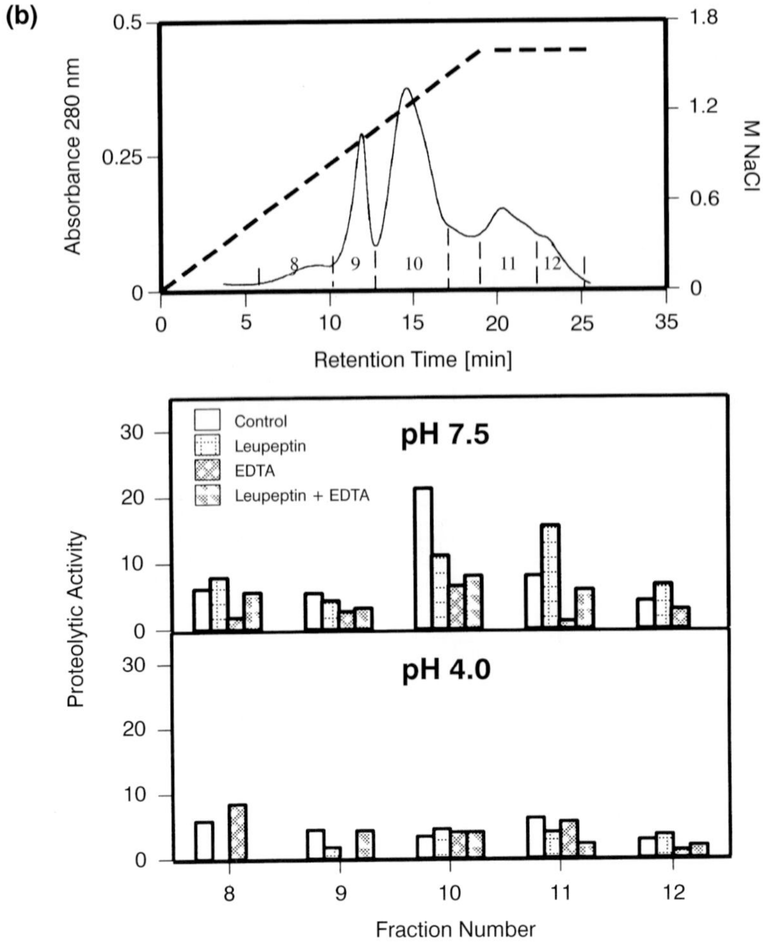

The proteases present in the IEF zones indicated at the bottom of *Figure 4* were eluted and further analysed by SDS PAGE with activity visualized in agar gels containing casein applied to the surface (*Figure 5*). Although the samples were not boiled in SDS prior to SDS PAGE, the method probably does not reveal all protease activities because of the denaturing effect of sodium dodecyl sulfate on many enzymes (Saitoh *et al.*, 1989). Nevertheless, more than one active protease was observed in each fraction, the molecular weights of detected proteases varying from 14 to more than 94 kDa (*Figure 5*).

In conclusion, the cambial derivatives of *Pinus banksiana* at the earliest stage of xylogenesis contain a substantial number of diverse and interacting proteases. Knowledge of cambial proteases remains rudimentary, and our findings indicate that the complexity of this type of catalytic activity in the cambium is such that a major research effort may well be needed before its physiological significance can be comprehended. In addition to helping to elucidate how secondary growth is regulated, protease research promises to yield insight into other important biological phenomena. Proteases evidently provide the tree with a defence mechanism thus

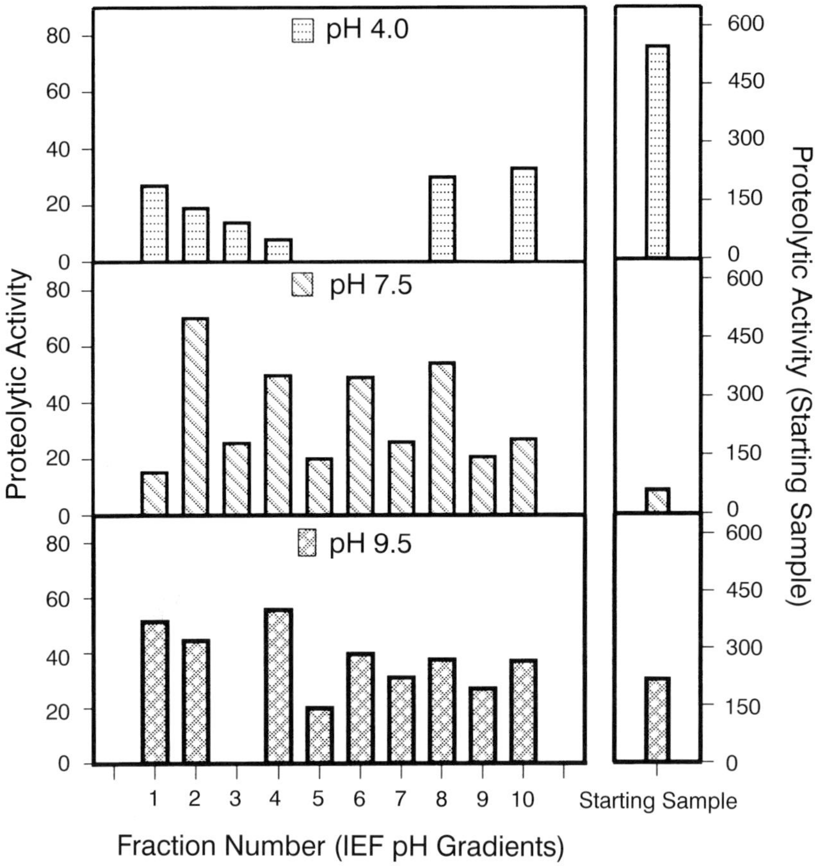

Approximate pH Range of IEF Fractions 1–10										
No.:	1	2	3	4	5	6	7	8	9	10
pH	9.5–7.8	7.8–7.5	7.5–7.3	7.3–7.1	7.1–6.7	6.7–6.1	6.1–5.7	5.7–5.3	5.3–4.8	4.8–3.5

Figure 4. *Proteolytic activity of partially purified protein from the cambial region of* P. banksiana. *Activity was measured at three pH values (pH 4.0, 7.5 and 9.5) before (right side) and after preparative IEF.*

augmenting tree health and survival (Hollick and Gordon, 1993; Lieutier and Berryman, 1988; Lyr, 1962; Saarikoski *et al.*, 1996). Moreover, invasive species employ their own proteases in attempts to overcome the defences of the tree (Breuil *et al.*, 1994; Gharibian *et al.*, 1996; Yoshitake *et al.*, 1993), and host-produced protease inhibitors likely have important roles here (Seldal *et al.*, 1994).

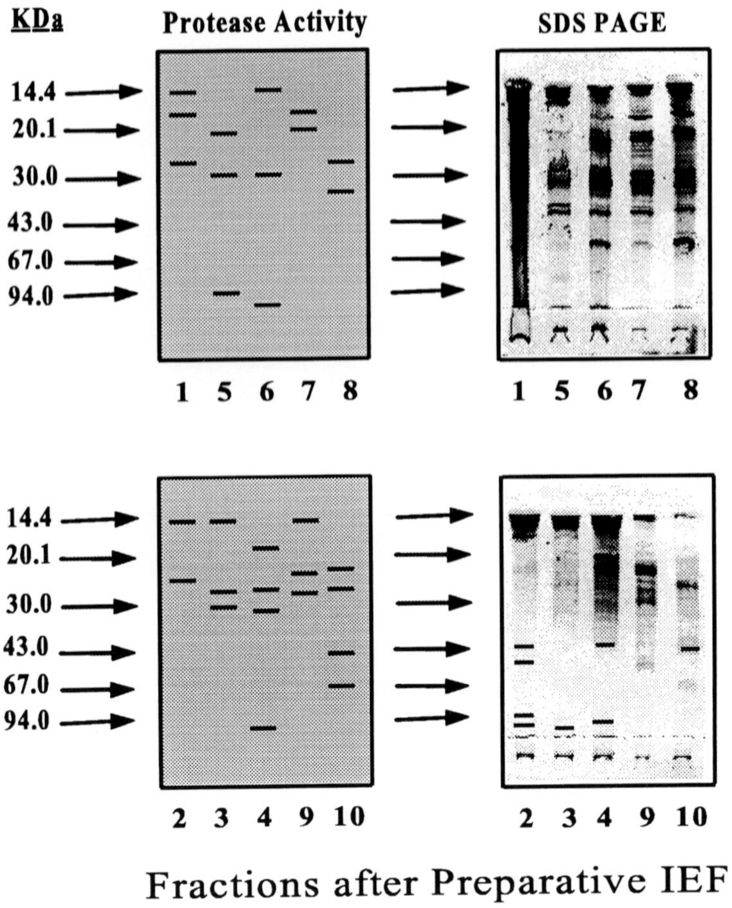

Figure 5. Agar/casein prints (left) of SDS PAGE gels (right) of the ten preparative IEF fractions indicated in Figure 4.

Acknowledgement

Iliyan Iliev acknowledges a North Atlantic Treaty Organization Science Fellowship enabling performance of this research at UNB.

References

Avila, J.L. (1997) Proteasomes: multicatalytic proteinase complexes. *Interciencia* **22**: 51–59.

Breuil, C., Yagodnik, C. and Abraham, L. (1994) Staining fungi growing in softwood produces proteinases and aminopeptidases. *Material und Organismen* **29**: 15–25.

Fukuda, H. (1997) Programmed cell death during vascular system formation. *Cell Death & Differentiation* **4**: 684–688.

Gharibian, S., Hoffert, C., Abraham, L.D. and Breuil, C. (1996) Localizing an *Ophiostoma piceae* proteinase in sapwood of four tree species using polyclonal antibodies. *New Phytol.* **133**: 673–679.

Höfelmann, M., Kittsteiner-Eberle, R. and Schreier, P. (1983) Ultrathin-layer agar gels: a novel print technique for ultrathin-layer isoelectric focusing of enzymes. *Anal. Bioch.* **128**: 217–222.

Hollick, J.B. and Gordon, M.P. (1993) A poplar tree proteinase inhibitor-like gene promoter is responsive to wounding in transgenic tobacco. *Plant Molec. Biol.* **4**: 561–572.

Iliev, I. and Savidge, R. (1999) Proteolytic activity in relation to seasonal cambial growth and xylogenesis in *Pinus banksiana*. *Phytochemistry* **50**: 953–960.

Kuprevich, V.F. (1949) 'Extracellular' enzymes of the conductive tissues of tree species. *Bot. Zh. S.S.S.R.* **34**: 613–617. (From *Forestry Abstracts* 1950–51 012–01752).

Lieutier, F. and Berryman, A.A. (1988) Preliminary histological investigations of the defense reactions of three pines to *Ceratocystis clavigera* and two chemical elicitors. *Can. J. Forest Res.* **18**: 1243–1247.

Lyr, H. (1962) Enzymatic inactivation of heartwood toxins. *Flora, Jena* **152**: 570–579.

Minami, A. and Fukuda, H. (1995) Transient and specific expression of a cysteine endopeptidase associated with autolysis during differentiation of *Zinnia* mesophyll cells into tracheary elements. *Plant Cell Physiol.* **36**: 1599–1606.

Neurath, H. (1986) The versatility of proteolytic enzymes. *J. Cellular Biochem.* **32**: 35–49

Rakowski, K.J. and Wodzicki, T.J. (1994) Proteolytic activity in the stem cambial region of *Pinus sylvestris* L. — A contribution to the specific differentiation of secondary xylem and phloem. *Acta Soc. Botan. Polon.* **63**: 247–253.

Rakowski, K.J. and Wodzicki, T.J. (1995) Effect of auxin (IAA) upon the proteolytic system in differentiating secondary xylem of pine (*Pinus sylvestris* L.). *Acta Soc. Botan. Polon.* **64**: 25–28.

Saarikoski, P., Clapham, D., von Arnold, S. and von Arnold, S. (1996) A wound inducible gene from *Salix viminalis* coding for a trypsin inhibitor. *Plant Molec. Biol.* **31**: 465–478.

Saitoh, Y., Yokosawa, H. and Ishii, S. (1989) Sodium dodecyl sulfate-induced conformational and enzymatic changes of multicatalytic proteinase. *Biochem. Biophys. Res. Commun.* **162**: 334–339.

Savidge, R.A. (1996) Xylogenesis, genetic and environmental regulation — a review. *IAWA J.* **17**: 269–310.

Savidge, R.A., Udagama-Randeniya, P.V., Xu, Y., Leinhos, V. and Förster, H. (1998) Coniferyl alcohol oxidase: a new enzyme spatio-temporally associated with lignifying tissues. In: *Lignin and Lignan Biosynthesis* (eds. N.G. Lewis and S. Sarkanen). *ACS Symposium Series* **697**: 109–130.

Savidge, R.A. and Wareing, P.F. (1981) Plant-growth regulators and the differentiation of vascular elements, In: *Xylem Cell Development* (ed. J.R. Barnett), Castle House Publ., Tunbridge Wells, England. pp. 192–235.

Seldal, T., Dybwad, E., Andersen, K.J. and Hogstedt, G. (1994) Wound induced proteinase inhibitors in grey alder (*Alnus incana*): a defence mechanism against attacking insects. *Oikos* **71**: 239–245.

Sudachkova, N.E., Kozhevnikova, N.N. and Lyubarskaya, T.G. (1981) Changes in activity and localization of intracellular enzymes during differentiation of the precambial zone in pine. *Soviet Plant Physiology* **28**: 586–593 (Engl. transl. from *Fiziologiya Rastenii* **28**: 802–810).

Ye, Z.H. and Varner, J. (1996) Induction of cysteine and serine proteases during xylogenesis in *Zinnia elegans*. *Plant Mol. Bio.* **30**: 1233–1246.

Yoshitake, A., Katayama, Y., Nakamura, M., Iimura, Y., Kawai, S. and Morohoski, N. (1993) *N*-linked carbohydrate chains protect laccase III from proteolysis in *Coriolus versicolor*. *J. Gen. Microbiol.* **139**: 179–185.

Zakrzewski, J. and Rakowski, K. (1987) The effect of cambial zone isolation upon the autolytic system in maturing tracheids of pine (*Pinus sylvestris* L.). *Acta Soc. Botan. Polon.* **56**: 399–405.

22

Pectin remodelling during cambial derivative differentiation

M.L. Follet-Gueye, F.F. Ermel, B. Vian, A.M. Catesson and R. Goldberg

1. Introduction

The cytoskeleton-cell wall continuum is thought to play a role in the perception and transduction of positional information (Wyatt and Carpita, 1993) whose importance in the maintenance of primary meristems has been recently underlined (Laux and Schoof, 1997; Schiefelbein *et al.*, 1997). It is more than likely that the identity of cambial cells and the fate of their derivatives are similarly regulated. The possible role of the cytoskeleton in these events has prompted studies such as those reported by Chaffey and by Funada *et al.* in this volume. The cell wall end of the continuum has been examined in parallel. Significant advances have been made towards understanding secondary wall deposition, lignification and their control (several chapters of this volume), but the early events involved in primary wall deposition in the cambial zone have not attracted much interest. Investigations have been made in attempts to fill this gap. At first, the study of changes in wall structure and composition in the cambial zone was mostly focused on pectins since these components are known to be abundant in young cell walls and to play a role in the regulation of cell wall extensibility (Goldberg *et al.*, 1996; McCann and Roberts, 1996). The immunolocalization results reported here, supplemented by biochemical analyses, present evidence of early pectin modifications occurring in cambial daughter cells entering either the phloem or the xylem pathway and of further remodelling during the course of primary wall maturation in cambial derivatives.

Cell & Molecular Biology of Wood Formation, edited by R. Savidge, J. Barnett and R. Napier.

2. Material and methods

2.1 *Material*

Samples were collected from hybrid aspen (*Populus tremula* L. × *P. tremuloides* Michx) grown at the Swedish University of Agricultural Sciences, in Umea (Nilsson *et al.*, 1997).

2.2 *Immunolocalization*

Samples were fixed, embedded, sectioned and treated for immunolocalization as described in a previous paper (Guglielmino *et al.*, 1997). Four monoclonal antibodies were used: (i) JIM 5 which binds to homogalacturonan chains with a low degree of methylation (DM<50%, Knox *et al.*, 1990); (ii) JIM 7 which recognizes homogalacturonan with a high degree of methylation (35%<DM<90%, Knox *et al.*, 1990); (iii) LM5 and (iv) LM6 antibodies raised respectively against the galactan and arabinan side chains of rhamogalacturonan 1 (Jones *et al.*, 1997; Willats *et al.*, 1998). After incubation in the presence of the antibody (or in the presence of pre-immune serum in the case of controls) semi-thin sections were treated with a goat anti-rat IgG coupled to fluorescein isothiocyanate (FITC) and observed in fluorescence. Other sections were treated for immunogold-silver intensification in order to prevent auto fluorescence artefacts.

2.3 *Pectin analysis*

Three fractions were scraped from shoots with an active cambium. (i) a_1 containing cortical parenchyma and mature phloem; (ii) a_2 comprising recently produced phloem and the cambial zone; (iii) a_3 containing enlarging future xylem cells. Two fractions, r_1 containing cortical parenchyma and the older phloem cells and r_2 the overwintering phloem and the cambium, were isolated from resting shoots. Cell walls were isolated from each fraction (Goldberg *et al.*, 1986), pectins extracted with boiling water (2× 1h and 1× 2h) and then lyophilized. Powders were analysed according to Schaumann *et al.* (1993).

3. Results and discussion

3.1 *Growth period*

Galacturonan labelling with JIM 5 and JIM 7, during this period, is shown in *Figures 1a* and *1b*. JIM 7 staining was strongest in the phloem and also strong in the cambial zone. It was also present in the thin walls of enlarging xylem derivatives but decreased with the onset of secondary wall deposition. JIM 5 cell wall labelling appeared fainter in the cambial zone than in the mature phloem and still fainter in the enlarging xylem cells. It disappeared almost completely with the deposition of secondary walls and the beginning of lignification. The distribution of JIM 5 and JIM 7 labelling agrees with the results reported previously for horsechestnut (Chaffey *et al.*, 1997) and for another species of poplar (Guglielmino *et al.*, 1997). A progressive masking of the reactive epitopes by hemicelluloses, cellulose and lignins would explain the disappearance of JIM 5 labelling. The pectic fraction, extracted with boiling water, accounted for 30% of cell wall dry weight in the cambial zone, but represented only 10% and 16% in the cortex and enlarging xylem respectively. The value obtained for the cambial zone is in accordance with the high pectin content previously reported for the cambium of *Populus tremuloides* (Simson and Timell, 1978). Galacturonic acid with a relatively low degree of methylation (around 25%), arabinose and galactose were the main sugars obtained after hydrolysis of the polymers solubilized with boiling water (*Figure 2*).

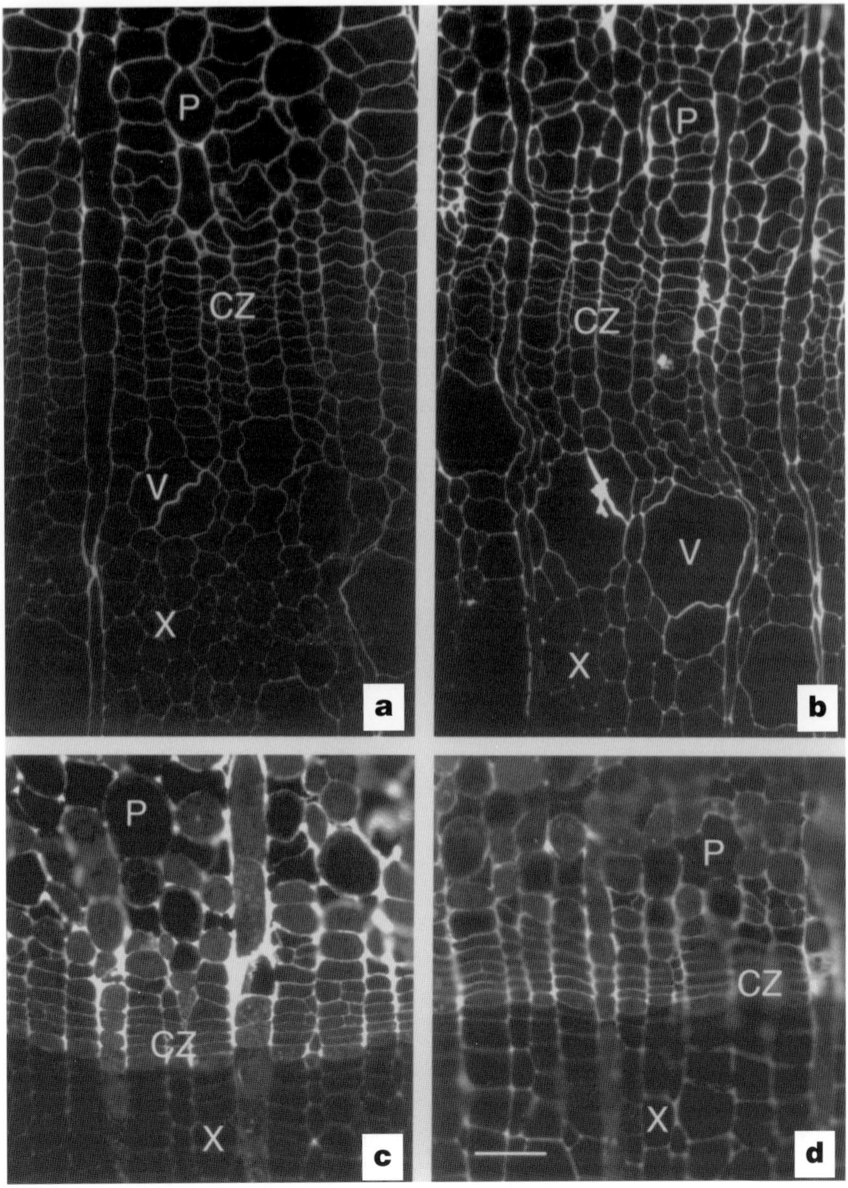

Figure 1. *Immunolocalization of galacturonan, using fluorescent labelled antibodies on transverse sections of active (a and b) and resting (c and d) hybrid aspen. (a) and (c) localization of acidic galacturonan with JIM 5. (b) and (d) localization of highly methylated galacturonan with JIM 7. P: phloem; V: enlarging vessel; X: xylem; CZ: cambial zone. Bar = 50 μm.*

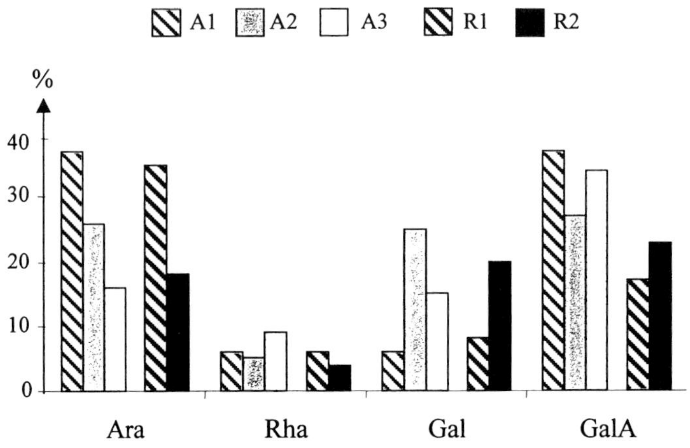

Figure 2. *Main sugars present in water soluble pectins. Data as percent of the sugars released after methanolysis of the pectins solubilized by boiling water. The tissue fractions were sampled from active (A_1, A_2, A_3) and resting (R_1, R_2) hybrid aspen shoots. A_1 and R_1: cortical parenchyma and older phloem; A_2 and R_2: recent phloem and cambial zone; A_3: enlarging xylem. Ara: arabinose; Rha: rhamnose; Gal: galactose; Gal A: galacturonic acid.*

The relative amount of galactan was higher in the tissue fractions containing either the cambial zone (A_2) or the enlarging xylem cells (A_3), that is, the fractions containing both dividing and expanding cells (*Figure 2*). Preliminary experiments with LM5 antiserum showed that galactan epitopes were mainly localized in the cambial zone and enlarging xylem. Labelling decreased rapidly on the phloem side, disappearing firstly in the maturing sieve tube-companion cell complexes and, more progressively, in cells differentiating into parenchyma or fibres. A gradual decrease in the relative amounts of arabinose residues (*Figure 2*) and in labelling intensity of arabinan epitopes was observed from the cortex to the xylem. Cell commitment to differentiate into phloem then means an increase in arabinan synthesis, possibly due to arabinan synthase activation. In contrast, a decrease in arabinan synthesis characterizes the cells committed to mature into xylem.

3.2 *Resting period*

Cessation of cambial activity was characterized by a thickening of the cambial walls, especially the radial walls, as already described in other species (see Larson, 1994). The thickening was accompanied, in hybrid aspen, by a marked decrease of the relative amount of pectins which did not exceed 5% of cell wall dry matter either in the cortical tissues or in the cambial zone and overwintering phloem. It was accompanied by a reduction of the degree of methylation culminating now at 10%. The decrease of the relative pectin content was probably linked to an increase in both hemicelluloses and cellulose such as previously described in another poplar species (Vietor *et al.*, 1995). This could explain, at least in part, the very weak reaction obtained both with LM5 and LM6, at a time when the high amount of non-pectic cell wall components could prevent the labelling of the reactive epitopes of galactan and arabinan. In contrast, homogalacturonan staining of non-lignified tissues remained strong (*Figures 1c* and

1d), especially with JIM 5 antibody, known to bind to pectin chains with a low degree of methylation (Knox *et al.*, 1990), which is precisely the case in resting cambial and phloem tissues.

4. Conclusion

The results presented here indicate that a continuous remodelling of pectin chains takes place during the maturation of the vascular tissues from the cambium. Such changes provide a means to differentiate, at a very early stage, between the cells of the cambial zone already committed either to the phloem or to the xylem pathway. Later modifications can also give some clues to the fate of the various cells already engaged in a given pathway. Thus, pectin modifications or, better, the changes in activity, location and gene expression of the enzymes involved in pectin biosynthesis might be considered as early markers of the commitment of cambial derivatives. As a consequence, they might provide the means to distinguish the 'elusive' (adjective used by Schmid, 1976) cambial initials from their daughter cells besides giving new insights into the regulation of cambial meristematic activity and the making of wood.

Acknowledgements

The present study has been carried out with financial support from the commission of the European Communities, Agriculture and Fisheries (FAIR) specific RTD programme, CT 98–3972, 'Wood formation processes: the key to improvement of the raw material'. It does not necessarily reflect its views and in no way anticipates the Commission's future policy in this area. MLFG and FFE gratefully acknowledge the fellowships obtained through this programme. The authors wish to thank Dr Björn Sundberg of the Swedish University of Agricultural Sciences in Umea, Sweden, for cultivating and providing the plant material and Dr J. Paul Knox from the University of Leeds, UK, for his generous gift of antibodies.

References

Chaffey, N., Barnett, J. and Barlow, P. (1997) Endomembranes, cytoskeleton, and cell walls: aspects of the ultrastructure of the vascular cambium of taproots of *Aesculus hippocastanum* L. (Hippocastanaceae). *Int. J. Plant Sci.* **158**: 97–109.

Goldberg, R., Morvan, C. and Roland, J.C. (1986) Composition, properties and localization of pectins in young and mature cells of the mung bean hypocotyl. *Plant Cell Physiol.* **27**: 419–427.

Goldberg, R., Morvan, C., Jauneau, A. and Jarvis, M.C. (1996) Methyl-esterification, de-esterification and gelation of pectins in the primary cell wall. In: *Pectins and Pectinases* (eds J.Visser and A.G.J. Voragen). Elsevier, Amsterdam, pp. 151–172.

Guglielmino, N., Liberman, M., Jauneau, A., Vian, B., Catesson, A.M. and Goldberg, R. (1997) Pectin immunolocalization and calcium visualization in differentiating derivatives from poplar cambium. *Protoplasma* **199**: 151–160.

Jones, L., Seymour, G.B. and Knox, J.P. (1997) Localization of pectic galactan in tomato cell walls using a monoclonal antibody specific to (1–4)-β-D-galactan. *Plant Physiol.* **113**: 1405–1412.

Knox, J.P., Linstead, P.J., King, J., Cooper, C. and Roberts, K. (1990) Pectin esterification is spatially regulated both within cell walls and between developing tissues of root apices. *Planta* **181**: 512–521.

Larson, P.R. (1994) *The Vascular Cambium: Development and Structure*. Springer Verlag, Berlin

Laux, T. and Schoof, H. (1997) Maintaining the shoot meristem. The role of *CLAVATA 1*. *Trends Plant Sci.* **2:** 325–327.

McCann, M.C. and Roberts, K. (1994) Plant cell wall architecture: the role of pectins. In: *Pectins and Pectinases* (eds J.Visser and A.G.J. Voragen). Elsevier, Amsterdam, pp. 91–107.

Nilsson, O., Tuominen, H., Sundberg, B. and Olsson, O. (1997) The *Agrobacterium rhizogenes rolB* and *rolC* promoters are expressed in pericycle cells competent to serve as root initials in transgenic hybrid aspen. *Physiol. Plant.* **100**: 456–462.

Schaumann, A., Bruyant-Vannier, M.P., Goubet, F. and Morvan, C. (1993) Pectic metabolism in suspension-cultured cells of flax, *Linum usitatissimum*. *Plant Cell Physiol.* **34**: 891–897.

Schiefelbein, J.W., Masucci, J.D. and Wang, H. (1997) Building a root: the control of patterning and morphogenesis during root development. *Plant Cell* **9**: 1089–1098.

Schmid, R. (1976) The elusive cambium — another terminological contribution. *IAWA Bull.* **1976/4**: 51–59.

Simson, B.W. and Timell, T.E. (1978) Polysaccharides in cambial tissues of *Populus tremuloides* and *Tilia americana.* I. Isolation, fractionation, and chemical composition of the cambial tissues. *Cellulose Chem. Technol.* **12**: 39–50

Vietor, R.J., Renard, C.M.G.C., Goldberg, R. and Catesson, A.M. (1995) Cell-wall polysaccharides in growing poplar bark tissue. *Int. J. Biol. Macromol.* **17**: 341–344.

Willats, W.G.T., Marcus, S.E. and Knox, J.P. (1998) Generation of a monoclonal antibody specific to (1–5)-α-L-arabinan. *Carbohydrate Res.* **308**: 149–152.

Wyatt, S.E. and Carpita, N.C. (1993) The plant cytoskeleton-cell wall continuum. *Trends Cell Biol.* **3**: 413–417.

23

Pectin methylesterases: possible markers for cambial derivative differentiation?

F. Micheli, M. Bordenave and L. Richard

1. Introduction

Plant cell wall architecture is not irreversibly fixed once cell division is completed but continues to undergo changes during growth and differentiation of the cell. Investigating the building up of the polysaccharide network around carrot protoplasts, Shea et al. (1989) observed that methylesterified pectins were laid down initially and later partially demethylesterified. During subsequent elongation, the architecture of the wall components is modified permitting cells to reach their definitive size and shape. One of the most important changes affects the degree of methylesterification of the polygalacturonan backbone, a process under the control of cell wall pectin methylesterases (PME; E.C. 3.1.1.11). Young elongating cells are characterized by having a high amount of methylesterified pectins and a low amount of acidic polygalacturonan within their cell wall. In contrast, the walls of stiffened older cells contained mainly acidic galacturonan and fewer methylesterified galacturonic acids (Goldberg et al., 1986; Li et al., 1995; McCann et al., 1993). In elongating cells, the described changes occur concomitantly with an increase of PME activity (Bordenave and Goldberg, 1994; Goldberg et al., 1992).

Most of the studies dealing with changes in pectin structure and metabolism concern the primary wall of cells originating from primary meristems, and only a few address the walls of cells originating from secondary meristems such as the cambium. Those few investigations report changes occurring at late stages of tracheid differentiation, that is, during secondary cell wall thickening, lignification and protoplasmic autolysis (Boudet, 1998; Chasan, 1994; Jones and Dangl, 1996). The early changes in cell wall features that precede lignin deposition are still poorly documented and remain unclear. We have focused our attention on changes affecting pectins and PMEs at

Cell & Molecular Biology of Wood Formation, edited by R. Savidge, J. Barnett and R. Napier.

earlier stages of vascular differentiation, when daughter cambial cells become committed to becoming xylem-type cells. Previous studies on poplar stems indicated that PMEs may play a role in the differentiation of cambial derivatives into xylem- and phloem-type cells during the seasonal cycle. In this paper, we present the overall data on this subject and propose some novel methodological approaches in an attempt to define better the biological role of the PMEs in cambial derivative differentiation.

2. First evidence for the involvement of PMEs in cambial derivative differentiation: biochemical and ultrastructural changes

Early biochemical studies (Northcote, 1963; Simson and Timell, 1978) established the presence of the pectin fraction in the primary walls of cells belonging to the cambial zone (7–20% of total cell wall polysaccharides). They underlined the occurrence of qualitative shifts between the pectin and hemicellulose biosynthetic pathways during cambial derivative differentiation (Bolwell and Northcote, 1981; Dalessandro and Northcote, 1977).

Some 10 years ago, it was shown that a cell committed to becoming a vessel element or sieve tube could be distinguished at an early stage by its primary cell wall features. These features can be analysed by the use of several polysaccharide extracting agents. For instance, the cell wall of a future vessel is sensitive to hot water extraction. In this case, the swollen middle lamella of radial walls is mostly solubilized whereas the middle lamellae of the tangential walls show resistance to such treatment (Catesson, 1989). Since hot water was known to extract mainly rhamnogalacturonan I and highly methylated homogalacturonan, these results raise the possibility of pectins and related enzymes having roles in cambial derivative differentiation.

In 1994, Baïer *et al.* used a biochemical approach to investigate the role of pectin methylesterases in cambial derivative differentiation in *Populus* × *euramerica*. Through the annual cycle, the composition of the pectin fraction extracted from the cambial zone and young phloem changed significantly, while the total amount of pectins per gram of cell wall was rather stable. The amount of esterified uronic acids increased during the transition from the active period to the rest stage of dormancy concomitantly with a decrease in unesterified uronic acids. These changes were correlated with PME activity, highest at the active stage. Substantial differences in PME isoform pattern also occurred during the seasonal cycle. The cell walls contained essentially neutral PMEs during the active period whereas the prevalent isoforms of the resting stage were basic (Guglielmino *et al.*, 1997a). Although these two studies indicated involvement of PMEs in the cambium's activity, they were limited to the cambial zone and recent phloem and did not take into consideration potential changes affecting the developing xylem. Therefore, although PMEs can be considered as seasonal markers in poplar, their role as early cellular markers of cambium derivative differentiation into xylem or phloem elements remains to be demonstrated.

Since acidic pectins are thought to arise from demethylesterification of highly methylesterified pectins by PMEs (Zhang and Staehelin, 1992), the use of antibodies targeted against highly methylesterified (JIM7) and low methylesterified homogalacturonan (JIM5) provide indirect information on PME activity (Knox, 1997). In *Populus* × *euramerica*, this immunolocalization approach revealed that low methylesterified pectins were restricted to cell junctions and middle lamellae of differentiating xylem derivatives while they were present in both primary walls and middle lamellae of differentiating and mature phloem cells. In this analysis, JIM5 labelling was also visible within the Golgi apparatus suggesting the existence of a secretion pathway for acidic pectins in parallel with the main secretory pathway of highly methyl-esterified pectins (Guglielmino *et al.*, 1997b). However, this interpretation could be compromised because the JIM5 antibody was capable of binding pectin with up to 50% methyl esterification and the JIM7 pectin with 35% methyl esterification and over (Knox, 1997). Overlapping in pectin recognition by the two JIMs therefore does not demonstrate unequivocally the occurrence of an alternative pathway for secretion of acidic pectins. The use of de-esterified homogalacturon block-specific antibodies, such as Pam1 (Willats *et al.*, 1999) could contribute toward testing this hypothesis. A co-localization of both the acidic polygalacturonan and the PMEs would strengthen the hypothesis of PME-catalysed demethylesterification. Obtaining specific antibodies raised against PME isoforms is, therefore, a prerequisite to spatio-temporal localization of PMEs in cell walls. These antibodies should also be useful for the isolation and biochemical characterization of PMEs, essential research for understanding their biological role in cambial derivative differentiation.

3. Novel approaches for studying the PMEs in cambial derivative differentiation

3.1 *Biochemical approaches*

We modified two methods of analysis in order to study the role of the PMEs in cambial derivative differentiation in *Populus tremula* × *tremuloides*. The first method was based on the use of tissue fractions scraped from active or resting stems, as described by Guglielmino *et al.* (1997a). Three tissue fractions were obtained from active stems: the first comprised cortical parenchyma and older phloem cells (named cortex fraction). The second fraction (named phloem fraction) included phloem derivatives with cambial cells and a small amount of immature xylary tissue. The third fraction (named xylem fraction) was esentially made of expanding unlignified xylem cells. At the dormant stage, two fractions were recovered. One contained cortical parenchyma and older phloem cells (named cortex fraction). The other contained cambial cells and phloem derivatives (named phloem fraction).

Cell walls were isolated according to the procedure described by Goldberg et al. (1986). Wall bound PME isoforms were solubilized with NaCl (1 M)–succinate (0.05 M) pH 5.0 buffer, and fractionated by isoelectrofocusing (IEF) on ultrathin polyacrylamide slab gels containing 10% ampholines pH range 3–10. Samples were calibrated to the same PME activity per volume before gel loading. A high level of PME activity was chosen in order to visualize a greater number of isoforms.

Active PMEs were revealed on gels under the conditions described by Bertheau *et al.* (1984) but with the agar-pectin sandwich replaced by an acrylamide-pectin sandwich. Apparent isoelectric points (p*I*) were determined by reference to some p*I* markers. PME activity of each fraction was measured titrimetrically by following the increase of free carboxyl groups. The carboxyl groups released by PMEs from 0.25% citrus pectin were titrated with 10 mM NaOH under nitrogen, the pH being maintained at 5.6 or 7.6 with an automatic titrator.

As shown in *Figure 1*, 10 PME isoforms (I1 to I10) were distinguished according to their apparent p*I*s. Most of the isoforms were basic (I7, I8, I9 and I10) or neutral (I3,

I4, I5, I6), and only two were distinguishable by their acidic p*I*s (I1 and I2). The isoforms I5, I6, I7, I8 and I10 were active in all fractions throughout the seasonal cycle; while I3, I4 and I9 were active in all fractions except in the active cortex. Isoform I1 was present in all fractions of the active stage and was almost undetectable during dormancy. Interestingly, isoform I2 was restricted to the active fractions, viz., 'phloem' and 'xylem'. Those fractions contained in addition to cells of the cambial zone the cambial derivatives entering into phloem and xylem differentiation, respectively, indicating that it may be more related to cambial derivative differentiation than any of the other isoforms detected.

Because the scraping method can provide high amount of tissues, it is suitable for preparative protein purification. However, this method of tissue sampling is less convenient for analytical studies when tissues are roughly isolated. In order to gain a more accurate definition of the distribution of PME activity within tissues, we resorted to the use of longitudinal cryosections of the stems, as described by Uggla *et al.* (1996). Blocks (3 × 10 mm) consisting of extraxylary tissues and a few annual rings were cut up during active and dormant periods from *Populus tremula* × *tremuloides* stems, and immediately frozen in liquid nitrogen. For PME activity assays, specific tissues from the outside to the centre were isolated by tangential centripetal cryosectioning at –20°C. The radial position of each section (25 μm) was determined by the help of transverse sections cut with a razor blade from both ends after every three sections. The two transverse sections were inspected under microscope and developmental stage of cells determined by their anatomical appearance. The percentage of each tissue was established with the support of an image processor. Extraction of cell wall proteins from three pooled consecutive tangential sections was done using a method we adapted for small amounts of plant material. After grinding the three pooled sections in Triton X-100 (0.1%), successive washing steps were done to remove any trace of detergent from the cell walls. Cell walls were then treated with NaCl (1 M) to extract proteins. PME activity was measured spectrophotometrically at 525 nm

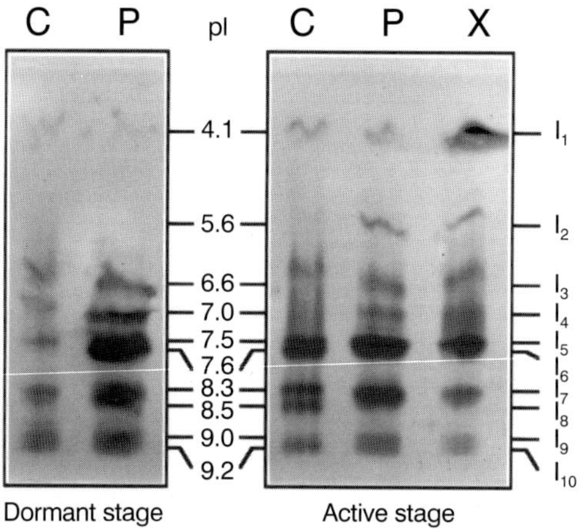

Figure 1*. Isoelectrofocusing of protein extracts solubilized from cell walls of* Populus tremula × tremuloides. *Cell walls proteins were extracted from cortex (C), phloem (P) and xylem (X) fractions during cambial rest and active periods. The apparent p*I *corresponding to the isoforms are indicated (centre). Designations of the isoforms are indicated on the right.*

using pectin methyl-red, pH 6.1, as pH indicator, as described by Richard *et al.* (1994). In parallel, cell wall masses were quantified by turbidimetry at 600 nm, using a standard reference curve where 1 OD was equivalent to 1 mg of cell wall.

The PME activity measured for each tissue throughout the active stem is shown in *Figure 2a*. The distribution of total PME activity reveals two peaks, one major peak in the cambial zone and another minor peak in the cortical parenchyma. Activity of the major peak may reflect the involvement of PME in xylem derivative differentiation. In cortical parenchyma, the PMEs may be related to the enlargement of the cells in response to physical constraint imposed by cambial activity. The slight activity observed in the young phloem may be correlated with cell wall rearrangement occurring during phloem differentiation. The PME activity is weak in old phloem and is undetectable in mature xylem. The weak PME activity in the old phloem agrees with the probable non functional state of this tissue which likely does not require enzymatic activity for undergoing passive mechanical deformation. In the mature xylem, lack of PME activity may be relevant to the death of the cells following lignification and loss of cellular content. In dormant stems, PME activity was weak in all tissues. In the cortical parenchyma as well as in the young phloem, the PMEs were two times less active during the rest period than during the active period. The low activities observed in the old phloem and mature xylem were similar across the seasons (data not shown).

After dialysis, cell wall protein extracted from each active fraction was separated by isoelectrofocusing as described above. The samples, calibrated to equivalent masses of cell walls, were loaded on an IEF gel. The IEF separation of cryosection proteins revealed eight PME isoforms differing by their apparent p*Is* (*Figure 2b*). The visible bands corresponded to isoforms I2, I3, I5, I6, I7, I8, I9 and I10 of *Figure 1*. Isoforms I3, I9 and I10 were only slightly visible in the cortical parenchyma whereas I1 and I4 were undetectable. The differences between *Figure 1* and *Figure 2b* may be attributed to the use of pH 5.0 succinate in the previous experiment (i.e., *Figure 1*), which improves elution of proteins from cell walls. The cortical parenchyma and the cambial zone peaks mainly contained the isoforms I5, I6, I7 and I8. In addition to these four isoforms, the cambial peak contained the isoform I2, as already detected in *Figure 1*. This isoform I2 is observed in fractions 9, 10 and 11 comprising the cambial cells and their closely related derivatives developing into phloem and xylem. The relationship of I2 with cellular differentiation is reinforced by the disappearance of that isoform in dormant samples (data not shown). Moreover, the intensity of I2 activity appeared greater in the fractions containing mainly derivatives in expanding xylem, indicating that I2 may be preferentially involved in xylem differentiation.

Cryosectioning combined with a new method of PME extraction adapted for small amounts of material allows refined detection of the PME isoforms at the tissue level. This approach brings interesting insights about involvement of some PMEs in cambial derivative differentiation. During the active stage, PME activity in the cambial zone is more important than during the rest period. This evidence confirms the PME feature as a seasonal marker, as suggested by Baïer *et al.* (1994) and Guglielmino *et al.* (1997a). Moreover, accurate analysis of the PME isoform pattern along tissues allowed us to detect a singular PME isoform (I2) which may be involved in xylem differentiation. Thus, it seems that PMEs may play a role in cambial derivative differentiation both at the quantitative and qualitative levels. These experiments will be completed by the analysis of PMEs at the level of a single instead of three 25 μm sections in order to refine the distribution of the isoform I2 in the vicinity of the cambial cells.

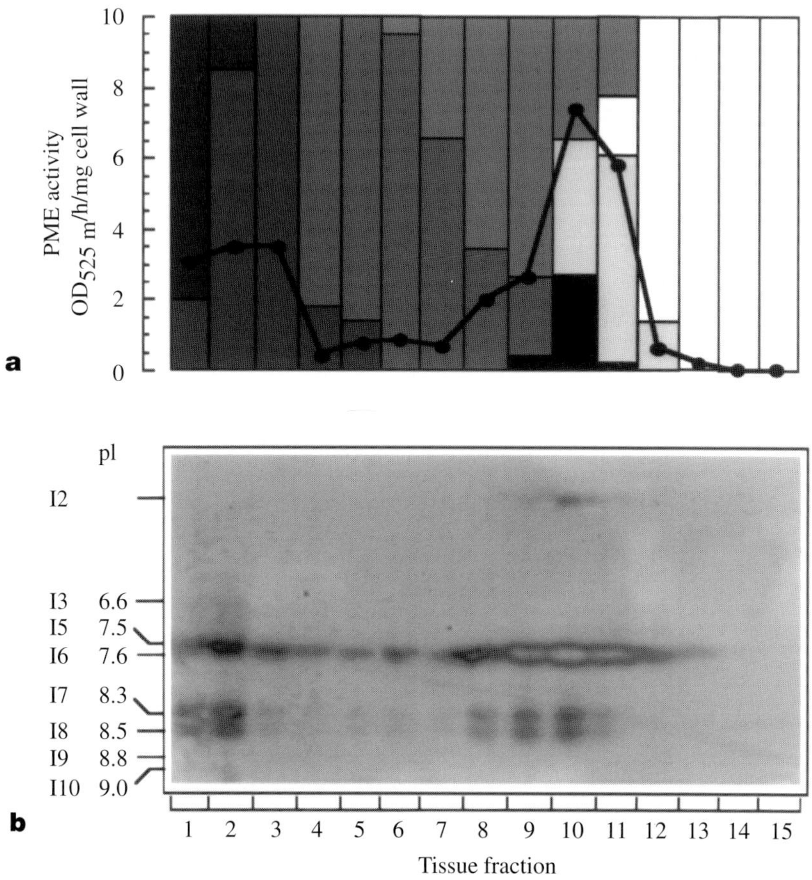

Figure 2. *Kinetic analyses of the PME isoforms along tissues of* Populus tremula × tremuloides *stem at the active stage. (a) Radial distribution of the PME activity. Each column represents a fraction made by pooling three tangential sections of 25 μm thickness. Shading gives the composition of the fraction in cell types from different developmental zones. Cambium is shown in black. The gradation from grey to white represents, from left to right, the stem tissues from outside to centre, that is, suberised bark tissue, cortical parenchyma, old phloem, young phloem, expanding xylem and mature xylem. (b) Isoelectrofocusing of protein samples extracted from cryosections. Each fraction corresponds to three tangential sections of 25 μm each. The apparent pIs and the designations of the isoforms are indicated on the left.*

3.2 *Towards molecular cloning of the* PME *genes*

During the five past years, particular attention devoted to the molecular studies of the PMEs led to the characterization of several *PME* genes in various higher plant species. Some of these genes were shown to be ubiquitously expressed (Gaffe *et al.*, 1997) whereas others denote a more specific expression towards some developmental processes such as fruit ripening (Harriman *et al.*, 1991), microsporogenesis (Albani *et al.*, 1991; Mu *et al.*, 1994) or organ elongation (Bordenave *et al.*, 1996). These data suggest that PMEs are encoded by a family of genes that are differentially regulated in cell types in response to specific developmental or environmental cues. In *Arabidopsis thaliana*, it has been shown that the *PME* gene family contained at least twelve genes.

Whereas most of these genes appeared to be more or less ubiquitously expressed throughout the plant, several genes appeared closely related to the development of the function of specific organs, such as flowers or roots (Micheli *et al.*, 1998; Richard *et al.*, 1994; Richard *et al.*, 1996). In *Populus tremula* × *tremuloides*, we focused our attention on the isolation and characterization of *PME* genes that may be involved in cambial derivative differentiation. Two complementary approaches are considered for the isolation of such genes.

The first approach relies on the use of a biochemical method of purification for microsequencing the proteins of interest and further cloning of the corresponding cDNAs, and it is being used to clone the cDNA related to isoform I2. Cloning of such cDNA should also bring interesting data about acidic PME isoforms, which remain to be well characterized in higher plant species.

The second approach is based on cloning of *PME*-related sequences from a *Populus tremula* × *tremuloides* cDNA library (Sterky *et al.*, 1998) made from active cambial-region (developing xylem, meristematic cambial zone, developing and mature phloem). We took advantage of the existence of consensus regions shared by all known PMEs to prepare a homologous probe for screening the library as described by Micheli *et al.* (1998). Two of these consensus regions encode the amino acids motifs GTIDFIFG and YLGRPWK, respectively, which have been recognized as signatures for PME products. Degenerate oligonucleotide primers targeted against nucleotide sequences encoding the motifs GTIDFIFG and YLGRPWK were designed in order to amplify by PCR a number of *PME*-related sequences from *Populus* genomic DNA. Nine distinct *PME*-related sequences of the expected size of 200 bp have been identified among the PCR product. These sequences were further used as homologous probes for screening the cDNA library and allowed isolation of five *PME* cDNAs, named *PttPME1* to *PttPME5*. The sequences of *PttPME1*, *PttPME2* and *PttPME3* are full length whereas those of *PttPME4* and *PttPME5* are slightly truncated in their 5′-terminal region. The corresponding genes should encode putative prepro-PME precursors structurally similar to other higher plant PME precursors. Those precursors carry a signal peptide and probably undergo post-translational proteolytic cleavage releasing the carboxy-terminal region of the precursor which corresponds to the mature enzyme. The five putative PME isoforms identified in *Populus* have a predicted molecular mass of about 34 kDa and a basic p*I* of 9.0 to 10.1. Although not yet ascertained experimentally, some of these putative gene products may correspond to the basic PME isoforms detected by IEF (*Figure 1*). These isoforms are present at both the active and dormant stages but show significantly higher activity in the phloem fraction than in other regions of the stem. Consequently, it should be interesting to follow the expression of the *PttPME* genes in order to assess their possible role in cambial derivative differentiation. Expression analysis of such genes will be made on cryosections with the use of RT-PCR.

3.3 *Immunological approach*

Up to now, immunological methods were mainly applied for the histological and cytological analyses of the pectins while a limited information concerns the distribution of the PME isoforms within tissues. It is therefore of interest to have antibodies raised against PMEs, in order to gain more insight into PME localization in relation to development.

The glycosylated state of PMEs likely explains the difficulty in obtaining reliably specific anti-PME antibodies that do not cross-react with unrelated proteins of the cell wall. To bypass this problem, we prepared antibodies raised against a synthetic oligopeptide whose sequence includes several consensus motifs known as signatures of the PMEs. We expect that such anti-oligopeptide antibodies will cross-react with a variety of PME isoforms. This feature should make it useful for covering the analysis of the whole PME isoform family.

We intend to use such poly-PME specific antibodies for mapping PMEs within cambial tissues, as well as for biochemical purposes. In this respect, the antibodies will support purification of some PMEs by affinity chromatography. Moreover, we expect that these antibodies will allow detection of several weakly active PME isoforms that cannot be easily revealed using the sandwich technique (Bertheau *et al.*, 1984). The specificity of such new isoforms towards cambial activity will then be assessed by Western blotting.

4. PMEs: early markers of cambial derivative differentiation?

The complexity of the PME family requires the convergence of different and complementary approaches which could bring new data for elucidating the relationship between PMEs and cambial derivative differentiation. With refinements, the preliminary analysis described here using cryosections will reveal the distribution pattern of the PME isoforms at the level of specific tissues. Using this approach it has been possible to detect an interesting acidic isoform whose occurrence is restricted to the expanding xylem. This particular isoform might be a potential marker of cambial derivative differentiation exciting interest for future biochemical and molecular studies.

As early markers of cambial derivative differentiation, the PMEs could play a role in the determination of some cambial daughter cells into xylem-cell types, or a role in some growth processes such as the enlargement of xylem-cell types vessel elements. To answer these intriguing and crucial questions, it will be necessary to take into consideration all future analyses concerning hormone levels and gene regulations within the cambium.

Acknowledgements

We are grateful to Dr B. Sundberg for supplying the plant material and the cambial region cDNA library of *Populus.* We thank Dr J. Barnett for his invitation to the SEB meeting and Pr A.-M. Catesson for helpful discussion and advice on the manuscript. This research was supported by the European Community.

References

Albani, D., Altosaar, I., Arnison, P.G. and Fabijanski, S.F. (1991) A gene showing sequence similarity to pectin esterase is specifically expressed in developing pollen of *Brassica napus*. Sequence in its 5′ flanking region are conserved in other pollen-specific promoters. *Plant Mol. Biol.* **16**: 501–513.

Baïer, M., Goldberg, R., Catesson, A.M., Liberman, M., Bouchemal, N., Michon, V. and Herv, du Penhoat, C. (1994) Pectin changes in samples containing poplar cambium and inner bark in relation to the seasonal cycle. *Planta* **193**: 446–454.

Bertheau, Y., Madgidi-Hervan, E., Kotoujanski, A., Nguyen The, C. and Coleno, A. (1984) Detection of depolymerase isoenzymes after electrophoresis. *Anal. Biochem.* **139**: 383–389.

Bolwell, G.P. and Northcote, D.H. (1981) Control of hemicellulose and pectin synthesis during differentiation of vascular tissue in bean (*Phaseolus vulgaris*) callus and in bean hypocotyl. *Planta* **152**: 225–233.

Bordenave, M. and Goldberg, R. (1994) Immobilized and free apoplastic pectinmethylesterases in mung bean hypocotyl. *Plant Physiol.* **106**: 1151–1156.

Bordenave, M., Breton, C., Goldberg, R., Huet, J.-C., Perez, S. and Pernollet, J.-C. (1996) Pectinmethylesterase isoforms from *Vigna radiata* hypocotyl cell walls: kinetic properties and molecular cloning of a cDNA encoding the most alkaline isoform. *Plant Mol. Biol.* **31**: 1039–1049.

Boudet, A.M. (1998) A new view of lignification. *TIPS* **3**: 67–71.

Catesson, A.M. (1989) Specific characters of vessel primary walls during the early stages of wood differentiation. *Biology of the Cell* **67**: 221–226.

Chasan, R. (1994) Tracing tracheary element development. *Plant Cell* **6**: 917–919.

Dalessandro, G. and Northcote, D.H. (1977) Changes in enzymic activities of nucleoside diphosphate sugar interconversions during differentiation of cambium to xylem in sycamore and poplar. *Biochem. J.* **162**: 267–279.

Gaffe, J., Tiznado, M.E. and Handa, A.K. (1997) Pectin methylesterase isoforms in tomato (*Lycopersicon esculentum*) tissues. Effects of expression of a pectin methylesterase antisense gene. *Plant Physiol.* **114**: 1547–1556.

Goldberg, R., Morvan, C. and Roland, J.C. (1986) Composition, properties and localization of pectins in young and mature cells of the mung bean hypocotyl. *Plant Cell Physiol.* **27**: 417–429.

Goldberg, R., Pierron, M., Durand L. and Mutaftshiev, S. (1992) *In vitro* and *in situ* properties of cell wall pectinmethylesterases from mung bean hypocotyls. *J. Exp. Bot.* **246**: 41–46.

Guglielmino, N., Liberman, M., Catesson, A.M., Marek, A., Prat, R., Mutaftschiev, S. and Goldberg, R. (1997a) Pectin methylesterases from poplar cambium and inner bark: localization, properties and seasonal changes. *Planta* **202**: 70–75.

Guglielmino, N., Liberman, M., Jauneau, A., Vian, B., Catesson, A.M., and Goldberg, R. (1997b) Pectin immunolocalization and calcium visualization in differentiating derivatives from poplar cambium. *Protoplasma* **199**: 151–160.

Harriman, R.W., Tieman, D.M. and Handa, A.K. (1991) Molecular cloning of tomato pectin methylesterase gene and its expression in rutgers, ripening inhibitor, nonripening, and never ripe tomato fruits. *Plant Physiol.* **97**: 80–87.

Jones, A.M. and Dangl, J.L. (1996) Logjam at the Styx: programmed cell death in plants. *TIPS* **1**: 114–119.

Knox, J.P. (1997) The use of antibodies to study the architecture and developmental regulation of plant cell walls. *Int. Rev. Cytology* **171**: 79–120.

Li, Y.Q., Fang, C., Faleri, C., Ciampoloni, F., Linskens, H.F. and Cresti, M. (1995) Presumed phylogenetic basis of the correlation of pectin deposition pattern in pollen tube walls and the stylar structure of angiosperms. *Proc. Kon. Ned. Akad. v. Wetensh.* **98**: 39–44.

McCann, M.C., Stacey, N.J., Wilson, R. and Roberts, K. (1993) Orientation of macromolecules in the walls of elongating carrot cells. *J. Cell Sci.* **106**: 1347–1356.

Micheli, F., Holliger, C., Goldberg, R. and Richard, L. (1998) Characterization of the pectin methylesterase-like gene *AtPME3*: a new member of a gene family comprising at least 12 genes in *Arabidopsis thaliana*. *Gene* **220**: 13–20.

Mu, J.-H., Stains, J.P. and Kao, T.-H. (1994) Characterization of a pollen-expressed gene encoding a putative pectin esterase of *Petunia inflata*. *Plant Mol. Biol.* **25**: 539–544.

Northcote, D.H. (1963) Changes in the cell walls of plants during differentiation. *Sym. Soc. Exp. Biol.* **17**: 157–174.

Richard, L., Qin, L.-X. and Goldberg, R. (1996) Clustered genes within the genome of *Arabidopsis thaliana* encoding pectin methylesterase-like enzymes. *Gene* **170**: 207–211.

Richard, L., Qin, L.-X., Gadal, P. and Goldberg, R. (1994) Molecular characterization of a putative pectin methylesterase cDNA and its expression in *Arabidopsis thaliana* (L.). *FEBS Lett.* **355**: 135–139.

Shea, E.M., Gibeaut, D.M. and Carpita, N.C. (1989) Structural analysis of the cell walls regenerated by carrot protoplasts. *Planta* **179**: 293–308.

Simson, B.W. and Timell, T.E. (1978) Polysaccharides in cambial tissues of *Populus tremuloides* and *Tilia americana.* I. Isolation, fractionation, and chemical composition of the cambial tissues. *Cellulose Chem. and Technol.* **12**: 39–50.

Sterky, F., Regan, S., Karlsson, J., Hertzberg, M., Rohde, A., Holmberg, A., Amini, B., Bhalerao, R., Larsson, M., Villarroel, R., Van Montagu, M., Sandberg, G, Olsson, O., Teeri, T.T., Boerjan, W., Gustafsson, P., Uhlen, M., Sundberg, B. and Lundeberg, J. (1998) Gene discovery in the wood-forming tissues of poplar: analysis of 5,692 Expressed Sequence Tags. *PNAS* **95**: 13 330–13 335.

Uggla, C., Moritz, T., Sandberg G. and Sundberg, B. (1996) Auxin as a positional signal in pattern formation in plants. *PNAS* **93**: 9282–9286.

Willats, W.G.T., Gilmartin, P.M., Mikkelsen, J.D. and Knox, J.P. (1999) Cell wall antibodies without immunization: generation and use of de-esterified homogalacturon block-specific antibodies from a native phage display library. *Plant J.* **18**: 57–65.

Zhang, G.F. and Staehelin, L.A. (1992) Functional compartmentation of the Golgi apparatus of plant cells. *Plant Physiol.* **99**: 1070–1083.

24

The structure of pectin methyl esterase: the relevance of parallel β helix proteins to xylem differentiation

P.W. Goodenough and C.E. Carter

1. Introduction

In the first 30 years of investigations into protein structures only four protein domains have been described. These are the all helical domains, alternating α helical/parallel β domains, antiparallel β domains and small domains stabilized by disulphide bonds or metal ion co-ordination. Thus the discovery that a unique domain, the parallel β helix, exists and binds pectin/pectate suggests the domain may control aspects of pectin/pectate biochemistry. This paper describes the unique features of the parallel β helix protein and discusses the hypothesis that pectin methyl esterase (PME) is a parallel β helix protein, in contradiction to recent published predictions. Finally it is postulated that this structure is an important factor controlling pectin biochemistry during xylem differentiation.

2. The discovery of the parallel β helix

Pectin/pectate is a carbohydrate polymer made up of C_6 monomers; aldonic acids formed by the oxidation of the corresponding UDP-sugar, arranged in the pyranose ring structure. The conventional method of display, the Haworth formulae, depicts the ring carbons as lying in one plane. This is not, however, the true case and the pyranose ring takes up the strainless C_1 structure. Two of the atoms of the ring lie in a different plane to the rest of those of the molecule. Pectic acid is a series of unbranched

Cell & Molecular Biology of Wood Formation, edited by R. Savidge, J. Barnett and R. Napier.

chains of galacturonic acid linked α-1,4. There is little free rotation about this linkage (known as a glycosidic linkage). When the C_6 is esterified with methanol the resultant molecule is known as pectin (*Figure 1*). The degree of methylation affects the stability and solubility of the polymer and is clearly linked to differentiation of xylem as will be described later.

Proteins that bind pectate and pectin fall into a small number of families. Pectate and pectin lyase enzymes, originally not thought to be found in higher plants, have recently been implicated in xylem development (Domingo *et al.*, 1999) and are important factors in plant necrosis being produced by several plant pathogens. These enzymes cleave the glycosidic bond by a β-elimination mechanism (*Figure 1*). The resultant products include a pyranose structure with an unsaturated C_4–C_5 bond at the non-reducing end of the galacturonyl chain. Calcium is essential for activity in pectate lyase enzymes but not in pectin lyase.

The bacterial pectate lyase enzymes from *Erwinia chrysanthemi* and *Bacillus subtillus* have been overexpressed and crystallized (Yoder *et al.*, 1990; Jenkins *et al.*, 1992). Crystals were grown from ammonium sulphate (*Erwinia*) and polyethylene glycol or 2-methyl-2,4-pentane diol (*Bacillus*). The *Erwinia* enzyme has 353 amino acids and a molecular mass of 37 676 whereas the *Bacillus* enzyme has 399 amino acids and a molecular mass of 43 505. This difference proved to be important in solving the final structure. The polypeptide chain of both enzymes folds into a single domain (shown for *Bacillus* pectate lyase in *Figure 2*).

The predominant secondary structural elements are parallel β sheets (flat yellow arrows in *colour Plate 3*) that are coiled into a large right-handed cylinder that was named a parallel β helix (Yoder *et al.*, 1993). The *Erwinia* enzyme has seven complete circuits in the molecule and it is 34 Å in length and 17–27 Å in width. Each single circuit is made up of three β sheets (PB 1–3) which are separated by turns (T1–3). In this way the complete parallel structure comprises three sets of seven β sheets with the

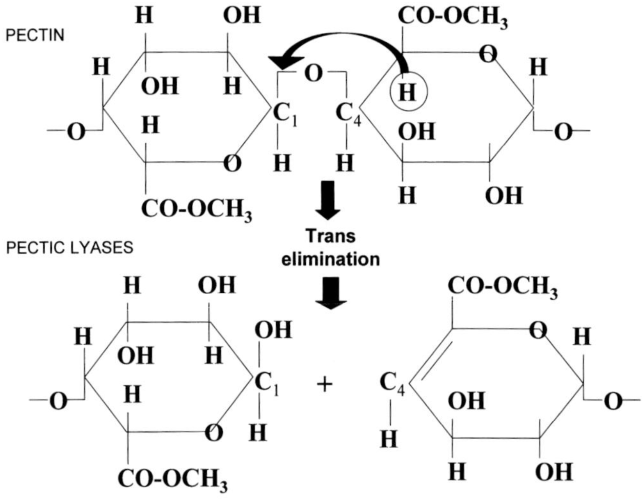

Figure 1. The action of pectin/pectate methyl esterase on pectin/pectate. The reaction is a trans elimination producing a double bond and cleaving the chain.

turns between the sheets facilitating the overall helical structure. In three cases the turns loop out from the structure forming one side of a groove with the exterior of the β helix forming the other side. The narrow groove is lined with surface charges and is approximately the shape and length of a dodecameric galacturonate substrate.

This domain structure was unique when discovered in the *Erwinia* enzyme (Yoder *et al.*, 1993). Most enzymes which have part of their structure made up of parallel β sheets are less stable than those with other types of structural domains. However the *Erwinia* pectate lyase structure is stable for long periods at room temperature. The reason for this is that the inner core of the parallel β helix is filled with hydrophobic and polar amino acids that are arranged in a particular way. At certain positions, particularly at junctions between sheets, the same amino acid residue occurs in each turn. A good example is the stacking of six asparagines (N) in consecutive turns of the β helix. The side chains of the asparagines are orientated so that the number of hydrogen bonds is maximized and the overall stacking resembles a ladder. One hydrogen of each asparagine NH_2 group forms a hydrogen bond with an adjacent asparagine side chain carbonyl oxygen, the second hydrogen bonds to a neighbouring main chain carbonyl oxygen. In total there are six examples of amino acid stacking including alanine, valine, isoleucine or leucine. There are also stacks of amino acids with aromatic ring side chains. The aromatic rings are tilted 50° relative to the polypeptide backbone and maintain an interplanar ring distance of 3.6 Å. A group of serine molecules also forms a stacking motif. In all cases the side chains form a linear arrangement parallel to the axis of the β helix.

In the crystal structure of the *Bacillus* enzyme the position of the essential calcium ion was identified using a molecule crystallized with bound calcium (Pickersgill *et al.*, 1994). The additional 46 amino acids of the *Bacillus* enzyme occur mainly in the first loop, which is therefore much longer than in the *Erwinia* enzyme. The sequence identity between the two enzymes is 30% and there are seven circuits in both molecules forming the β helix. In the *Bacillus* enzyme (1BN8, in *colour Plate 3*) it was found that the residues in the α_L confirmation occurred not only in the T_2 turns but also at the ends of β strands. This element of structure serves to rotate the strand through 180° and has not been identified in any other protein in the protein data bank. A pronounced active site cleft can be seen between the second structural region, formed by the three long loops and the PB1 sheet region of the parallel β helix. The calcium was found to be bound to three aspartates (D) in the active site cleft, one at the carboxy-terminal end of the second long T3 loop (D 184), one at the carboxy terminal end of the third long T3 loop (D 223) and one on the fourth strand of PB1 (D 227) on the surface of the parallel β helix. In total there are seven ligands for the calcium, one carboxy oxygen from each of D 223 and D 227, both carboxy oxygens of D 184 and the oxygens of three clearly defined water molecules. This clearly delineated site and the long loop, which includes some helical secondary structural elements (red cylinders in *colour Plate 3*), are completely different to the structure found in the *Erwinia* enzyme.

3. Structures of other enzymes binding pectate/pectin

The structure of pectin lyase from *Aspergillus niger* has been solved (Mayans *et al.*, 1997) and found to have a parallel β helix structure (1IDJ in *colour Plate 3*). The molecule was cloned into *E. coli* and overexpressed. After purification the protein was crystallized from 28% PEG 4000, 0.1 M Na acetate which was buffered with 1 M Tris/HCl at pH 8.5.

The molecule has 359 amino acids and the structure contains eight complete and one incomplete circuit. Three loops are also found in pectin lyase and one of them (T3) was found to be a long loop with two sections of β strand. A distinct N ladder stack was found with five residues and, as found in the *Erwinia* and *Bacillus* pectate lyase enzymes, these residues appear in the inner position of the T2 turn in the core of the helix.

Significantly, there is an inner aromatic stack composed of three residues (two phenylalanines (F) and one tryptophane (W) in the enzyme from one strain of the fungus and three F residues in another strain). As will be seen, this aromatic stack is found in other pectin binding enzymes. In comparison with the pectate lyase active site cleft, which has a number of charged amino acids (excluding those which bind calcium), the pectin lyase active site cleft is dominated by aromatic amino acids. These include four tryptophanes and three tyrosines. These are not stacked as found in the core of the enzyme but form three pairs of residues arranged in edge to face interactions. They may form part of the substrate binding complex. It has been well documented that part of the carbohydrate–protein interaction is ring stacking between aromatic amino acids and the pyranose ring. The three aspartate ligands necessary for calcium binding in pectate lyase are changed to arginine and valine but there remains one conserved aspartate. The only other residue conserved across the active site cleft is N 236.

Although both of these enzymes are closely related it was found that other hydrolases which bind pectin/pectic acid also had a similar structure. Polygalacturonase is classified as a family 28 glycosyl hydrolase (Henrissat and Bairoch, 1993). When the polygalacturonase from *Erwinia* was produced in *Bacillus* the 26 amino acid proregion was removed leaving a protein of 376 amino acids. Single crystals could be grown from polyethylene glycol and gave a space group of C_2. The structure has 10 complete circuits of right handed parallel β helix (1BHE in *colour Plate 3*; Pickersgill *et al.*, 1998). In addition to having more complete circuits than pectate and pectin lyase the molecule has more sections of parallel sheet in each coil. The three sections of parallel sheet making up each circuit are present, but this enzyme has an additional parallel β sheet, known as PB1a. As in the pectate and pectin lyase enzymes a pronounced substrate binding area (perhaps rather better described as a tunnel than a cleft) is formed by two long loops that precede strands two and three of PB1 and three loops that follow strands seven, eight and nine. The tunnel is lined with arginines and lysines which give a clear positive electrostatic potential. This is similar to pectate lyase but different to pectin lyase where the cleft has a negative electrostatic potential (made up of aspartates and glutamates (D and E)). The former residues in the active site are conserved, particularly D 202 and 223, whilst D/E 224 is semi conserved. The centre of the molecule has some amino acid stacks, but not N ladders or aromatic stacks.

Cell walls also contain rhamnogalacturonan, which is a polymer of both galacturonic acid and rhamnose. An enzyme which hydrolyses this material has been isolated from *Aspergillus aculeatus* (rhamnogalacturonase or RGA) and been shown to be a parallel β helix protein. The protein shares 23–29% homology with polygalacturonase, has 422 amino acids and so is larger overall than the lyase enzymes (1RMG in *colour Plate 3*). It has 13 complete circuits of parallel β helix, which has a structure of β strands similar to polygalacturonase with four parallel sheets, PB1, 1a, 2 and 3. Most of these strands are separated by a single residue in the α_L configuration, most of which are E or N. These form perfect hydrogen bonds to the residues above and below them forming an almost perfect turn (Peterson *et al.*, 1997). It is known that a perfectly circular parallel β helix structure would have 22 residues per turn (Cohen, 1993). The interior of the RGA molecule is packed with hydrophobic residues with at least one large hydrophobic cavity. As in polygalacturonase the number of identical internal amino acid residues with the same conformation as found in their neighbouring β strands is small. An aromatic stack, F 104, H 141, Y 164 and H 189, is found on the surface while a hydrophobic stack is found in the central cavity, V 140, 163, 188 and 211. Asparagine stacks do not occur within the central cavity although valine, isoleucine and leucine residues are aligned within the core. The original description of the protein does not mention the loops (Petersen *et al.*, 1997) and they are not at all prominent. The active site E residues are conserved and can be identified in the structure.

4. Predictions of an idealized parallel β sheet coil and its relevance to pectin methyl esterase

Attempts have been made to predict an idealized single helical coil of the parallel β sheet superfamily and to use this as a search algorithm to identify all such molecules in the protein databanks (Heffron *et al.*, 1998). This search algorithm relied heavily on recognition of the amino acids found in the turns between strands PB1–PB3 (T2). The loops that in these enzymes occur at turns, in particular from T1 or T3, never occur in T2. An idealized amino acid sequence was predicted for a perfect single coil. This was a coil of 23 amino acids which only contained a small bulge and no protruding loops. Loops are unpredictable in terms of size and shape and so an 'idealized perfect' single coil would not include any loops. A refined coil was 22 amino acids long, the same as described by Cohen (1993). It had three short regions of β strand, as found in the lyase enzymes, whereas polygalacturonase and rhamnogalacturonase have four. Strands B1 and B2 had four amino acids and were connected by a narrow T1 turn with four amino acids. B2 was connected to B3 by a turn with two amino acids and the B3 strand had three amino acids. Finally the T3 turn connected B1 and B3 by five amino acids. It was decided that asparagine at position 14 (T2) epitomized the idealized coil structure of this domain. The coil can be characterized by using 'a' for residues with their side chains pointing inwards to the coil centre and 'b' for those pointing outwards. The layout would then be ababbabbababbababbabba. When the amino acids found in each of these positions were calculated for all known members of the pectate lyase family it was shown that hydrophobic amino acids predominated in internal positions 1, 3, 9, 11, 19 and 22, aromatic amino acids were in position 16, N in position 14 and small amino acids in position 6 and 22. This arrangement was used as a search pattern to identify any similar protein motifs in the protein data banks; those which were identified would be predicted to have a parallel β helix structure. Although some proteins that had not had their tertiary structure determined were found to have similar sequences to the idealized coil (in particular internalin), several proteins whose tertiary structure had been identified as being the parallel β helix were not identified. These included polygalacturonase and rhamnogalacturonase, a virus tail spike protein and a proteinase. However none of these bind pectin. If a protein, such as PME, is known to bind pectin, it would seem highly likely that the algorithm would predict that it had a parallel β helix structure unless the similarity was very low. Pectin methyl esterases were not identified by the algorithm. This is of crucial importance as although PME binds pectin it does not modify C_1 but instead modifies C_6. Therefore it would be perfectly reasonable to expect PME to have a structure different to that of the pectin lyases.

5. Can pectin methyl esterase be modelled onto pectin/pectate lyase?

In order to predict the structure of PME we used a cDNA sequence of a plant PME from *Populus tremula* × *tremuloides* (Micheli *pers. comm.* 1999).

This enzyme sequence was aligned with 30 other plant PME enzymes taken from the SWISSPROT, PIR and SPTREMBL databases and conserved regions were identified. *Figure 2* shows a consensus sequence for the mature region (i.e., the proregion is absent) of a plant PME protein.

By use of the programme HOMOLOGY (MSI) it was possible to align the *Populus* PME sequence with pectin lyase A from *Aspergillus niger* (AC: Q01172; PDB: IDJ) There was low homology (ca. 20%). *Figure 3* shows the alignment that was used to model the PME structure as a parallel β helix (*see also colour Plate 6*).

The active site residues from pectin lyase, D 154 and R 236 were aligned with D 60 and R 140 of PME. R 236 is absolutely conserved across the pectin/pectate lyase enzymes and, in pectate lyase, D 154 is involved in binding calcium. Although these residues would not be expected to be active site residues in PME, it is noticeable that G 59 is fully conserved in all PME sequences known and 60 is always either D or E, indicating a possible catalytic role. T 79 is also highly (~97%) conserved in all PME sequences and the alignment we have used places it against T 171 of pectin lyase. In a similar way the following residues in PMEs are strictly conserved and aligned with homologous residues in pectin lyase (PME numbering) N 101, A 112, T 133, Y 135, R 140, A 183, G 185, Y (or F), L, G, R, 218–221. Of the last conserved sequence block 218–221 is also completely conserved in all pectin lyases. This conserved block of residues appears, in the model of the enzyme, to be on the surface of the cleft and is most likely involved in binding the pyranose ring of the pectin. In the active site of pectin lyase there are 4 tryptophanes (66, 81, 151 and 212) that are presumably involved as part of the mechanism of binding of the pyranose rings of polygalacturonate. As mentioned previously, this is a well known mechanism of ring–ring stacking between amino acids and carbohydrates. It is clear that in the model of PME there are also four highly conserved trytophane residues; these are on the loop regions that are not part of the parallel β helix strands, and are W 223, 248, 277 and 303 and are shown in *colour Plate 6 a–c*. Another W (246) is not found in all PMEs, in one case this position is occupied by a tyrosine. Pectin lyase has three tyrosines in the active site which have been identified as part of the binding mechanism (Y 85, 211 and 215), however Y 305, part of the conserved YLGR motif found in PME, is not identified as important in binding substrate in lyase enzymes. This latter fact is surprising as, amongst the four tyrosines in the PME model, Y 218 and Y 259 appear to be able to take part in substrate binding. Y 135 and 42 are also not part of the parallel β helix strands and are proposed to be part of the binding site (see *colour Plate 6 a–c*). There are six other tyrosines in PME (Y 15, 33, 130, 226, 262 and 280) which are not strictly conserved and are replaced by phenyalanine in PME enzymes from other plants (see *Figure 2*). Up to four other tyrosines are found in some sequences but these are quite variable.

Thus in summary we predict that PME has a parallel β helix, in contrast to Heffron *et al.* (1998) and that the active site of PME contains the D and N active site residues found in pectin lyase. A second N is absent. We also have indicated that at least four tryptophanes and four tyrosines are involved in ring to ring stacking with the substrate pyranose rings as found in pectin lyase. These residues are conserved across all PME proteins so far sequenced (*Figure 2*). In the model it is not possible to predict exactly how these residues are positioned to make a compact active site, but it is clear that they are not present in the β strands.

```
               *         *          *         *          *
        6  1 6VA DG3G 54T6 EA6A AP      R56656K G Y4EN6E6 K
1       I--N-VVA-DGSG-YKTL-EAVA-AP-----RYVIYVK-G-YKENVEV-K
     V  D  I     T  FR V   L           FIVFI      R  I I
                       I

               *         *          *         *          *
        4K N6666GDG   T663G  NV6DG33T5 3AT6A VG GF6A 16 F2
51      KK-NLMIVGDG---TIITG--NVVDGSTTF-SATVA-VG-GFIA-DI-FQ
     R    IVLI       VVS      I  TS Y T  L        L  NV  E
          V MM                M          I        M   L
            VL                                    V

               *         *          *         *          *
        NTAGP KHQA6ALR6G D S66 RC 6DA5QDTLY H3 RQF5R1  63G
101     NTAGP-KHQAVALRVG-D-SVI-RC-IDAYQDTLY-HS-RQFYRD--ITG
                L   I      IL    M  F        T    F N  VS
                                 V

               *         *          *         *          *
        T6DF6FG1AAVVFQ C 66A4KP   QKNM6TAQG4 DPN2 3G S6Q C
151     TVDFIFGNAAVVFQ-C-IVARKP---QKNMVTAQGR-DPNQ-TG-SIQ-C
      I  V   D         LL K          I    K     E S   L
                       VM            L
                        I

               *         *          *         *          *
         6 S31L  PV  3  35LGRPWK 5S4T66623 6  66 P G5 EW G
201     -I—SSDL--PV--S--TYLGRPWK-YSRTVVMES-I--LI-P-GW-EW-G
      V   TN       T  SF        F K IIIQT L  VV     Y
      L                               L   V  IL

               *         *          *         *          *
          AL 365YGE56N GPGAGT3 R6KW G5 66 3 A2A  5TV   62G
251     --AL-TLYYGEYMN-GPGAGTS-RVKW-GY-VI-S-AEA--FTV---IQG
           SIF  FL          T  I    F IV T  Q   Y      LE
                                      ML

               *      *
          W6  3GV 5  G6
301     --WL--TGV-Y--GL--
         I  S   F   M
```

Figure 2. *Consensus sequence for pectin methyl esterase mature polypeptide, derived from an alignment of 30 plant PME sequences. The upper consensus sequence, in italics, represents residues which are over 60% conserved. The consensus sequence also shows conservative substitutions. Where a substitution group rather than a single amino-acid remains conserved, a number code is given. The substitution groups are as follows: 1-DN, 2-EQ, 3-ST, 4-FYW, 6-LIVM. The bold sequence below shows the % conservation relating to each residue or substitution group at each position and, in the case of the latter, which residues in the substitution group are most common in the group of sequences analysed (the most common being uppermost). Bold groups are as follows: X 100% conserved, X 80% conserved, X 60% conserved, – less than 60% conserved.*

6. Parallel β strand structure of PME

We have modelled a protein with seven turns of β strands making up the parallel helix and there is no evidence for the stacking arrangement of asparagines found in the lyase proteins. There is only one conserved asparagine, N 101, in PMEs (shown in *Figure 2*). There are internal packages (stacks) of threonines found in other parallel β helix proteins and several threonines (four) are strictly conserved in the PME subfamily.

```
PMEPOP:
IDJ   : VGVSGSAEGFAEGVTGGGDATPVYPDTIDELVSYLGDDEARVIVLTKTFD

PMEPOP:
IDJ   : FTDSEGTTTGTGCAPWGTASACQVAIDQDDWCENYEPDAPSVSVEYYNAG

PMEPOP:       VTPNVVVAADGS-GDYKTVSEA-VAAAPKKSSKRYIIQIKAGVYR
IDJ   : VLGITVTSNKSLIGEGSSGAIKGKGLRIVSGAENI-----IIQNIA-VTD

PMEPOP: ENVEVPKDKHNIMFLGDGRKTTIITASRNVVDGSTTFKSATVAAVGQGFL
IDJ   : IN---PKYVWG----GDAI--TLDDCDLVWIDHVTTARIGRQHYVL-GTS

PMEPOP: ARG-VTFENTAGPSKHQAVALRVGSDLSAFYECDMLAYQDTL-----YVH
IDJ   : ADNRVSLTNNYIDGVSDYSATCDGYHYWGIYL-DGDADLVTMKGNYIYHT

PMEPOP: SNRQFFINCFVAGTVDFIFGNAAAVFQDCDIHARRPDSGQKNMVTAQGRT
IDJ   : SGRSPK----VQDNTLLHCVNNY--FYDISGHA--FEIGEGGYVLAEGNV

PMEPOP: DPNQNTGIVIQKSRIGA-TSDLLPVQSSFPTYLGRPWKEYSRTVIMQS--
IDJ   : -F-QNVDTVLETYEGAAFTVPSTTAGEVCSTYLGRDC------VINGFGC

PMEPOP: SITDVIQPAGWHEWSGSFALSTLFYAEYQNSGAGAGTS--SRVKWE-GYK
IDJ   : SGT----FS---EDSTSF-LSD-FEGKNIAS-ASAYTSVASRVVANAGQG

PMEPOP: VITSATEAQAFAPGNFIAGSSWLGSTSFPFSLGL
IDJ   : NL
```

Figure 3. *The amino acid sequence of* Populus *PME aligned with the amino acid sequence of pectin lyase from* Aspergillus. *The alignment was made using the software HOMOLOGY. The structure of this pectin lyase was used as the basis for the model* (colour Plate 6. *Co-ordinates of the pectin lyase were taken from the protein data bank, designation IDJ (Mayans* et al., *1997).*

Similarly two phenylalanines (F 142 and 154) are strictly conserved. F 156 is also highly conserved but is N in one sequence. F 143 is a tyrosine in some PME proteins. These form a clear arrangement within the turns of the β strands which must serve to stabilize the molecule. The conserved threonines are not shown on the model but would be situated mainly in the interior of the helix. Thus the mechanistic roles of many of the conserved amino acids in PME can be predicted from the model.

7. The significance to xylem development

The significance of PME as a marker of cambial development is discussed in another paper in this publication (see Chapter 23 by Micheli *et al.*). It is important to restate that pectic substances are a major component of the primary cell wall and middle lamella of dicotyledons. The primary cell-wall is a matrix of cellulose and xylans embedded in a pectic matrix. PME is ubiquitous in plants and is clearly involved in cell wall growth, extensibility and regeneration (Gaffe *et al.*, 1997). Correlations between meristematic activity and the degree of methylation of galacturonic acid have shown that the degree of esterification rises from 25% in spring to 35% in winter (Baier *et al.*, 1994). The increase in acidic carbohydrate in the growing season allows more interaction with cations such as calcium.

It would be expected that a reduction of esterified galacturonic acid would be mirrored by a change in PME. This has proved to be the case with PME isoenzymes from *Populus* × *euamericana* which showed a distinct difference between growth seasons. These enzymes have p*I*s either characterized as neutral or alkaline. The neutral p*I* enzyme has a much greater activity than the alkaline p*I* enzyme and is thought to be found in the resting period. The end result would be a more colloidal pectic acid–calcium matrix in the resting period which would not encourage active elongation (Guglielmino *et al.*, 1997).

The neutral isoform would appear to have a molecular mass of 55kDa whereas the alkaline isoform appears to have a molecular mass of approximately a 35 kDa. This latter figure is more typical of the known PME proteins. However, preliminary evidence (Micheli *et al.*, Chapter 23) does not indicate a difference in genes coding approximately for PME except, possibly, in the leader sequences.

8. Conclusion

We predict that the structure of PME is that of a parallel β helix. We have shown how an alignment of PME with pectin lyase can be used to produce a modelled structure with a conserved aspartic acid and asparagine in the active site as well as several conserved tryptophanes and tyrosines. These latter amino acids are known to be involved in generalized ring to ring stacking with the pyranose rings of carbohydrates. The structure does not contain an asparagine ladder. As this system of asparagines has been postulated to be a diagnostic feature of the parallel β helix it is possible that sequence diversity of proteins forming parallel β helices might be greater than expected. However there has to be some arrangement in these proteins to stabilize the structure and a group of phenylalanines within the helix seems to be a candidate for stabilizing PME from *Populus.*

9. Future research

Overexpressing clones of PME will be used to produce protein for purification and the structure reported here will be tested by protein crystallography. Glycosylation sites will be sought and the topography of the surface charge investigated. Genes altered to modify activity will be used in transgenic plants to provide the ultimate test of the veracity of the theory that rapid growth is correlated with higher concentrations of methylated pectin. The effect of PME on cell-wall growth will be tested by down-regulation of PME in transgenic plants.

References

Baier, M., Goldberg, R., Catesson, A-M., Liberman, M., Bouchemal, N., Michon, V. and du Penhoat, C.H. (1994) Pectin changes in samples containing poplar cambium and inner bark in relation to the seasonal cycle. *Planta* 193: 446–454.

Cohen, F.E. (1993) The parallel β helix of pectate lyase C: something to sneeze at. *Science* **260:** 1444–1445.

Domingo, C., Roberts, K., Stacey, N.J., Connerton, I., RuizTeran, F. and McCann, M.C. (1998) A pectate lyase from *Zinnia elegans* is auxin inducible. *Plant J.*, **13**, 17–28.

Gaffe, J., Tiznado, M.E. and Handa, A. K. (1997) Characterisation and functional expression of a ubiquitously expressed tomato pectin methylesterase. *Plant Physiol.* **114**: 1547–1556.

Guglielmino, N., Liberman, M., Catesson, A.M., Mareck, A., Prat, R., Mutaftschiev, S. and Goldberg, R. (1997) Pectin methylesterases from poplar cambium and inner bark: localization, properties and seasonal changes. *Planta* **202**: 70–75.

Heffron , S., Moe, G.R., Sieber,V., Mengaud, J., Cossart, P., Vitali, J. and Jurnak, F. (1998) Sequence profile of the parallel β helix in the pectate lyase superfamily. *J. Struct. Biol.* **122:** 223–235.

Henrissat, B. and Bairoch, A. (1993) New families in the classification of glycosyl hydrolases based on amino acid sequence similarities. *Biochem. J.* **293**: 781–788.

Jenkins, J.A., Nasser, W., Scott, M., Pickersgill, R., Vignon, J-C. and Robert-Baudouy, J. (1992) Crystallization and preliminary X-ray studies of the pectate lyase from *Bacillus subtilis. J. Mol. Biol.* **228**: 1255–1258.

Mayans, O., Scott, M., Connerton, I., Gravesen, T., Benen, J., Visser, J., Pickersgill, R. and Jenkins, J. (1997) Two crystal structures of pectin lyase A from *Aspergillus* reveal a pH driven conformational change and striking divergence in the substrate-binding clefts of pectin and pectate lyases. *Structure* 5: 677–689.

Peterson, T.H., Kauppinen, S. and Larsen, S. (1997) The crystal structure of rhamnogalacturonase A from *Aspergillus aculeatus:* a right-handed parallel β helix. *Structure* **5**: 533–544.

Pickersgill, R., Smith, D., Worboys, K. and Jenkins, J. (1998) Crystal structure of polygalacturonase from *Erwinia carotovora* ssp *carotovora. J. Biol. Chem.* **273**: 24 660–24 664.

Pickersgill, R., Jenkins, J., Harris, G., Nasser, W. and Robert-Baudouy, J. (1994) The structure of pectate lyase in complex with calcium. *Nature Struct. Biol.* **1**: 717–723.

Yoder, M.D., Keen, N.T. and Jurnak, F. (1993) New domain motif; the structure of pectate lyase C, a secreted plant virulence factor. *Science* **260**: 1503–1507.

25

Cellular patterns in dicotyledonous woods: their regulation*

S. Lev-Yadun

*Contribution from the Agricultural Research Organization, Institute of Field and Garden Crops, Bet Dagan, Israel, No. 113/99.

1. Introduction

Secondary xylem of woody dicotyledons comprises several types of cells: water-conducting tracheary elements (vessel members and tracheids), mechanically supporting elements (fibre-tracheids, libriform fibres and sclereids), parenchyma, secretory components (gum ducts and laticifers), cork cells, various idioblasts, and sometimes phloem. In each woody species only a particular subset of each of these types is found. The values for vessel volume in North American dicotyledonous trees range from 6 to 55%, for fibres from 29 to 76%, for axial parenchyma from 0 to 23%, and for rays from 6 to 31% (Gartner, 1995).

Two parenchyma systems are found in secondary xylem, rays and axial parenchyma. These can be viewed as three major patterns in the secondary xylem: (1) The axial component as seen in a cross section. (2) The axial component as seen in a tangential longitudinal section. (3) The structure and spacing of the radial component – the ray system. For each species, and sometimes for certain genera, there are characteristic sets of structural combinations that enable experts to identify taxa using wood anatomy. The fact that wood anatomy is taxon-specific is good evidence that the development of such patterns is under tight genetic and physiological control.

The regulation of differentiation of xylem axial parenchyma is largely unknown, and has not been studied for its own sake, but rather as a side effect of studying the axial water-conducting and supporting cells, or the rays. The regulation of ray differentiation was described in detail elsewhere (Lev-Yadun and Aloni, 1995), and therefore is described here only briefly.

Cell & Molecular Biology of Wood Formation, edited by R. Savidge, J. Barnett and R. Napier.

2. Wood anatomy

2.1 *Patterns of cell types in the axial system*

Wood anatomists distinguish between apotracheal parenchyma (axial parenchyma of the secondary xylem, usually not spatially associated with vessels, although occasionally contacting them) and paratracheal parenchyma (axial parenchyma of the secondary xylem, associated with vessels or vascular tracheids). Apotracheal parenchyma are divided into the following types, according to their distribution as seen in cross section: (1) Banded or metatracheal. Concentric uni- or multi-seriate bands, arcs or entire rings; (2) Diffuse. Single cells distributed irregularly among fibres; (3) Initial. Bands of parenchyma produced at the beginning of a growth ring; (4) Terminal. Bands of parenchyma produced at the end of a growth ring. Paratracheal parenchyma are also characterized in the same way: (1) Aliform. Expanded tangentially in the form of wings; (2) Confluent. Groups of aliform parenchyma that become continuous so as to form irregular tangential or diagonal bands; (3) Scanty. An incomplete sheath or a few parenchyma cells present around the vessels; (4) Vasicentric. Forming an entire sheath of variable width around individual vessels or groups of vessels (Fahn, 1990). These various patterns are illustrated in Carlquist (1988) and Wheeler *et al.* (1989).

A similar terminology for patterns of fibres in the secondary xylem does not exist, but the terminology used for parenchyma might be applied to fibres.

2.2 *Axial patterns of cell dimensions in the secondary xylem*

Most of the experimental studies on pattern formation in secondary xylem were conducted using vessels as their subject. The phenomenon of activation of the cambium in spring in the temperate zone (where most universities are located) led researchers to look at cambial activation and earlywood formation in detail. A basic aspect of woody stem ontogeny is the normal gradual increase in xylem cell length and width with age, with distance from the pith, and with distance from the young leaves. This was shown for fibres (Dinwoodie, 1961), tracheids (Bailey, 1958; Baas *et al.*, 1986), vessel members (Zimmermann and Potter, 1982; Aloni, 1987) and rays (Lev-Yadun and Aloni, 1995). Nevertheless, in many very old stems, after the length-on-age trend plateaus, a decrease in the size of cambial derivatives was observed (Dinwoodie, 1961). In woods having storeyed structure, fibre length does not increase regularly outwards from the pith (Metcalfe and Chalk, 1985).

2.3 *Patterns of vessel members as seen in cross section — the common view*

In a cross section of secondary xylem, three major types of vessel member distribution are distinguished: (1) Ring porous. Vessels in the earlywood are distinctly larger than those in the latewood of the previous or the same growth-ring and form a well-defined zone or ring; (2) Diffuse-porous. Vessels have more or less the same diameter throughout the growth-ring; (3) Semi-ring porous. Vessels in the earlywood are distinctly larger than those in the latewood of the previous growth-ring, but there is a gradual change to narrower vessels in the intermediate and latewood of the same growth-ring (Wheeler *et al.*, 1989). In addition to these gross definitions, vessels and other cell types of the secondary xylem have additional patterns of arrangement (see Carlquist, 1988; Wheeler *et al.*, 1989; Schweingruber, 1990).

The common view is that the first-formed earlywood vessels of ring-, semi-ring, and diffuse-porous woods are of the largest diameter; however, the actual situation is more variable. Data on sizes of the first-formed earlywood vessels were compiled using the plates in Schweingruber (1990). Of 77 ring-porous species, in seven species (16 wood samples out of 163) vessel members having the largest diameters were formed only after the first earlywood vessels had been produced (for example, *Eleagnus angustifolia* and *Hippophae rhamnoides*). Of 148 semi-ring-porous species, 19 (60 wood samples out of 351) also had their largest diameter vessel members placed later in the earlywood zone (for example, *Phlomis purpurea*). Of 215 diffuse-porous species, 42 (135 wood samples out of 552) produced their largest diameter vessel members in the earlywood later in the growing season (for example, *Alnus incana*). Altogether, in about 15% of the species (20% of the wood samples) described in Schweingruber (1990), the first formed earlywood vessels were not the widest within the earlywood.

2.4 *The ray system*

The ray system is the radial component of the secondary plant body. Rays have two origins: (1) Some are formed near the pith from interfascicular parenchyma and then become part of and are continued by the cambium; these are termed medullary rays, and they connect the pith with the cortex. (2) All other rays originate from the cambium and connect the secondary xylem and phloem (Larson, 1994). Two types of cells occur in the cambium: fusiform cells, which divide and differentiate to form the axial system, and ray cells, from which the radial system (the rays) differentiates (Philipson *et al.*, 1971; Larson, 1994). Fusiform cambial cells can reduce through cell division into ray cells, and ray cambial cells can elongate into fusiform cells (Philipson *et al.*, 1971). The ratio of fusiform to ray cells in the cambium varies greatly between species, and within the same plant at different locations, and following wounding. These ratios range from 100% fusiform cells in rayless plants, to only 25% fusiform initials in *Dillenia indica.*

Usually, rays are radially oriented multi-cellular sheets of parenchyma cells, but other elements such as ray tracheids (Larson, 1994), vessels (Carlquist, 1988), resin ducts (Fahn, 1990), sieve tubes (Lev-Yadun and Aloni, 1991a), sclereids (Carlquist, 1988), and fibres (Lev-Yadun, 1994a) can also be found in rays of various species. The parenchyma cells of rays that are axially longer than they are wide in the radial dimension are termed upright. Cells that are radially longer are termed procumbent. Cells that are about as tall as they are wide are termed square (Carlquist, 1988).

3 Regulation

3.1 *Hormonal regulation of differentiation in the secondary xylem*

The commonly accepted paradigm is that phytohormones, especially auxins, gibberellins, cytokinins and ethylene, regulate differentiation in the secondary xylem (Thimann, 1977; Wareing and Phillips, 1981; Sachs, 1981; Savidge and Wareing, 1981; Roberts *et al.*, 1988; Savidge, 1988; Lev-Yadun and Aloni, 1995; Little and Pharis, 1995; Zhong and Savidge, 1995; Jacobs, 1998). Two major types of data led to this current paradigm. (1) Identification of the hormones, and (2) experimental manipulation of their concentration or flow. The hormonal status of the cambium and its differentiating derivatives is not really known in relation to the regulation of pattern formation. Most studies were conducted to answer specific questions, such as the regulation of cambial activation or dormancy, the regulation of differentiation of a certain cell type, or the regulation of certain features in its secondary cell wall. Today, we stand at approximately the same point in knowledge about pattern formation where we were some 30 years ago. The power of classical developmental/physiological studies has not yet been exploited for the needed progress in understanding pattern formation in the axial components of the secondary xylem (and phloem). Moreover, in many experiments the externally applied hormone levels were very high compared to physiological levels, and it is probable that poisoning effects were interpreted as normal ones. Wound effects, seasonal influences, age effects and their combinations were not taken into consideration in almost all studies and, therefore, the relevant controls of all these aspects were never examined in a single coherent set of experiments.

The common techniques for manipulation of levels of hormones are: (1) surgical elimination of organs believed to produce them, (2) external phytohormone or hormonal antagonist application, (3) use of mutants and (4) use of transgenic plants. However, in spite of almost a century of studies of the ways in which phytohormones are involved in the regulation of differentiation, the process is not yet well understood and controversy remains concerning the mode of phytohormone action. For example, do they act following changes in concentration? (see Aloni, 1987), as a result of their cell-to-cell movement? (Sachs, 1981, 1991), or following changes in cellular sensitivity to their existence? (Trewavas, 1983). Experiments have indicated that polar auxin transport, rather than auxin concentration, is the activator of vascular differentiation (Sachs, 1981, 1991; Mattsson *et al.*, 1999), but its concentration might be related to xylem amount (Mattsson *et al.*, 1999). The distinction between auxin levels and auxin transport seems to be extremely important, but difficult to assess.

Accumulating experimental evidence indicates that the differentiation of certain cell types depends on specific hormones. Vessel members and sieve elements differentiate following an auxin signal (Wareing *et al.*, 1964; Digby and Wareing, 1966; Sachs, 1981); fibres differentiate following the action of a combination of auxin and GA (Digby and Wareing, 1966; Zhong and Savidge, 1985; Aloni, 1987); ray and axial parenchyma and cork cell components of the secondary xylem following the action of ethylene (Lev-Yadun, 1994b; Lev-Yadun and Aloni, 1990, 1995).

3.2 *The use of partial girdling to study the regulation of differentiation*

Narrow bridges of bark with cambium formed by partial girdling of trees can be used to study the hormonal regulation of xylem differentiation (Carmi *et al.*, 1972; Aloni and Zimmermann, 1984; Savidge and Farrar, 1984; Zagórska-Marek and Little, 1986; Lev-Yadun and Aloni, 1991b, 1992, 1995; Lev-Yadun, 1994b; Aloni *et al.*, 1997). In such bridges, wound effects clearly influence differentiation (Gregory and Romberger, 1975; Savidge and Farrar, 1984; Zagórska-Marek and Little, 1986; Harris, 1989; Lev-Yadun and Aloni, 1992, 1995; Lev-Yadun, 1994b). Vessels become narrower near wounds (Lev-Yadun and Aloni, 1992; Larson, 1994; Lev-Yadun, 1994b) or following exposure to high levels of ethylene (Savidge, 1988; Aloni *et al.*, 1998). The formation of resin ducts in Gymnosperms and of gum ducts in Dicotyledons is a well-known wound response, and reflects an ethylene effect (Fahn, 1988). Therefore, the formation in narrow bridges of numerous resin ducts in *Pinus halepensis*, and of traumatic gum ducts (gummosis) in *Ailanthus altissima* was interpreted as evidence of enhanced wound effects (Lev-Yadun and Aloni, 1992).

In the narrow cambial bridge formed by the partial girdling, there appears to be a delicate balance of two contrasting signal systems. All types of wounding presumably disturb the axial polar auxin flow in the cambium (Sachs, 1991; Lev-Yadun and Aloni, 1995). Wounding also induces the production of ethylene (Abeles *et al.*, 1992), which further disrupts the axial polar auxin flow (Mattoo and Aharoni, 1988). Therefore, it is undoubtedly simplistic to consider vessel members differentiating in narrow bridges as solely induced and influenced by high auxin levels (Aloni and Zimmermann, 1984; Aloni, 1987, 1991). Ethylene, and possibly other regulatory factors, need to be considered.

3.3 *Regulation of ray differentiation*

The following general hypothesis for the regulation of ray differentiation was recently proposed (Lev-Yadun and Aloni, 1995). Radial flow of developmental signals occurs through the ray initials of the vascular cambium. This radial signal flow is independent of the axial flow (Lev-Yadun, 1994b). The radial signal flow might occur bi-directionally: for example, ethylene originating in the xylem may flow outward and control the initiation of new rays and participate in controlling the enlargement of existing rays, and radial auxin may flow from the phloem inward and be involved in the induction of radial tracheids, vessels, sieve tubes and fibres. The radial flow of ethylene would disturb radial auxin transport, and thereby limit the differentiation of vascular elements and fibres in the rays, which are generally parenchymatous. In addition to the radial signal flow, interaction with the axial signal flow could also influence ray shape, direction, and size. The orientation of cells within rays might be controlled by the relationships between the axial and radial signal flows. Commonly, numerous cells in small rays and the marginal ray cells of large rays have an axial orientation (upright), while those in the centre of large rays are more likely to show radial orientation (procumbent). Temporary raylessness can be explained as the result of a strong dominant axial signal flow, which becomes less effective with the increasing distance from the young leaves.

The above hypotheses explain many known ontogenetic and experimentally, or wound-induced, changes in ray structure at the tissue level. They also explain the evolution of the ray system, which seems to reflect the changes in the relationships between the axial and radial developmental signal flows.

3.4 *Regulation of vessel differentiation and patterning*

The differentiation of vessel members involves several processes: (1) The signal (probably auxin) for differentiation of this cell type, (2) determination of final cell size, (3) regulation of patterns of vessel arrangement among other xylem cell types, (4) determination of secondary cell-wall sculpturing, and (5) regulation of programmed cell death. Thus, the current paradigm (induction by auxin) on differentiation of xylem seems to be an over-simplification of the situation. A 'black-box' of signal transduction events, and a chain of biochemical reactions, occurs between auxin application and vessel member differentiation. We are very far from even a basic understanding of this process.

Measurement of auxin as related to cambial activity is an important aspect of the regulation of differentiation. Relatively little has been done using precise new techniques. A major difficulty is to distinguish between various stages of differentiating xylem, cambial initials, and differentiating phloem (see Sundberg *et al.*, 1991). The small number of precision measurements (nevertheless limited to the thickness of sections) of auxin in the cambial region (e.g. Uggla *et al.*, 1996, 1998; Tuominen *et al.*, 1997) should stimulate others to study this extremely important issue and use similar techniques for other hormones.

3.5 *Vessel diameter*

The diameter of a vessel member can be determined by two major processes: (1) Cell wall loosening and expansion, which result in enlargement of the diameter, and (2) Deposition of secondary cell walls to arrest diameter enlargement. In each of these processes, dozens if not hundreds of genes are involved, and there are many options for their regulation (e.g. Cosgrove, 1996). The final diameter of a vessel member is the outcome of a balance between the operation of several mechanisms, some of which have contrasting effects. Variable physiological states that are obvious in a perennial woody species should shift the above balance according to their randomness and introduce unexpected anatomical variability. From what we know about regulation of gene expression there is no reason to assume that in all woody dicotyledons, only one mechanism operates. Similarly, we have no reason to assume that ring-porous and diffuse-porous species share the identical regulatory mechanism. Moreover, there is no reason to assume that within the same stem there will not be considerable variations in vessel grouping and diameters. The repeated evolution of patterns of porosity (e.g. Woodcock, 1989) since the emergence of woody dicotyledons some 150×10^6 years ago has allowed ample opportunity for different mutations to affect the regulatory mechanisms.

The concept that cambial reactivation and earlywood formation in the temperate regions in the spring is associated with increasing auxin levels was proposed long ago (see review in Savidge and Wareing, 1981). Using quantitative mass spectroscopy, Savidge and Wareing (1981) observed that endogenous auxin was relatively low during cambial reactivation and first earlywood development in springtime, reaching its maximum only later. The fact that in many woody species the largest diameter vessel members appear only after a certain amount of xylem is formed, and not in the beginning of the growth-ring (see section 2.3), and the fact that in species such as *Ulmus glabra*, *Hedera helix*, *Lonicera periclymenum*, *Sambucus nigra*, *Cornus sanguinea*, *Eleagnus angustifolia*, *Fagus sylvatica* (Schweingruber, 1990), and many other species, there are radial files containing successive large diameter vessel members formed over a longer duration than that of spring cambial reactivation indicates that cambial reactivation and enlargement of developing vessel members are not necessarily under the same form of regulation.

The early explanation (Wareing's hypothesis) for formation of large diameter vessels at the beginning of the growth season or later was that they resulted from high auxin levels in the beginning of the growth season (Wareing *et al.*, 1964; Digby and Wareing, 1966). If this hypotheis is correct, then the cases where vessel members attain their largest diameter only later in the growing season can be looked upon as an indication of a rise of auxin in the cambial region during the period of cambial reactivation, or else a gradual rise in sensitivity to auxin. However, the anatomical observations have also been interpreted in the opposite way (Aloni-Zimmermann hypothesis), i.e., the smaller diameter vessels in the beginning of the growth ring indicate a very high auxin level in the cambial region that gradually declines (Aloni and Zimmermann, 1983, 1984; Aloni, 1987, 1991). The few seasonal investigations into endogenous auxin content that have been done appear neither to support nor dismiss either Wareing's or the Aloni-Zimmermann hypothesis. In any case, the developmental data do not point for a single mechanism because earlywood production in ring-porous and some diffuse-porous species can start before budbreak or the expansion of new leaves (e.g. Atkinson and Denne, 1988).

The general pattern of a gradual increase in cell length and width with age, with distance from the pith, and with distance from the young leaves in vascular elements and fibres (see section 2.2) was attributed to decreasing auxin concentration (not flow) from leaves to roots (Aloni and Zimmermann, 1983; Aloni, 1987, 1991). Later, this axial apical-basal decrease in auxin was also proposed to be involved in the determination of ray characteristics (Lev-Yadun and Aloni, 1995). Precise auxin measurements along the tree axis (including the seasonal pattern) in the cambium itself and in the differentiating xylem of several woody species are almost non-existent. The studies by Savidge and Wareing (1981) indicated that auxin content was much lower in the young twigs than in larger diameter regions during cambial growth and wood formation in *Pinus contorta*. Until we have more such measurements in at least several species of woody plants, the proposal by Aloni and Zimmermann (1983) should be regarded as a working hypothesis and not considered as a clearly proven theory.

A special aspect of the Aloni-Zimmermann hypothesis is the distinction between ring-porous and diffuse-porous species (Aloni, 1991). It was proposed that in deciduous ring-porous species the cambium has become highly sensitive owing to decreased levels of hormonal stimulation. Conversely, in deciduous diffuse-porous species, it was proposed that the cambium remains less sensitive and therefore requires high levels of stimulation for its operation (Aloni, 1991). A problem with this reasoning is that in certain woody species, wood porosity changes from ring to diffuse. In woody flora of Israel, for example, five species *Populus euphratica* (deciduous), *Pistacia saporta* (evergreen), *Olea europaea* (evergreen), *Pistacia lentiscus* (evergreen) and *Quercus ithaburensis* (deciduous) change from ring- to diffuse-porous according to growth conditions, sometimes even within the same individual (Liphschitz and Waisel, 1970; Grundwag and Werker, 1976; Liphschitz, 1995). Therefore, the Aloni-Zimmermann hypothesis requires changing sensitivity to auxin within the cambium of the same tree as its porosity changes.

The difficulty in determining which of the two hypotheses is correct (or whether each is correct for a different group of species) is the outcome of the combination of both problems related to experimental design and to the clear lack of precise measurements of the hormonal status in the cambium and its differentiating derivatives. Wound effects influenced all experiments conducted to construct these two hypotheses. Wareing *et al.* (1964) and Digby and Wareing (1966) used decapitated stems and examined the newly formed tissues within the zone in which wound effects still existed. Similarly, Aloni and Zimmermann (1983, 1984) used partial girdling, in which wound effects dominate, and Aloni (1991) used decapitated stems and examined the tissues within the zone in which wound effects (including wound ethylene) still existed. Moreover, the high auxin levels applied must have resulted in auxin-induced ethylene production. High levels of ethylene are known to decrease vessel member diameter (Aloni *et al.*, 1998).

To conclude this section, earlywood does not always begin with the largest diameter vessel members being produced. Experiments conducted so far have been insufficient to enable the formulation of a hypothesis that can pass an objective review, and there is no reason to assume that the same mechanism determines vessel member diameter in all woody plants.

Acknowledgements

I thank Moshe Fleishman for his comments.

References

Abeles, F.B., Morgan, P.W. and Saltveit, M.E. Jr. (1992) *Ethylene in Plant Biology*. 2nd ed. Academic Press, San Diego, CA.

Aloni, R. (1987) Differentiation of vascular tissues. *Ann. Rev. Plant Physiol*, **38**: 179–204.

Aloni, R. (1991) Wood formation in deciduous hardwood trees. In: *Physiology of Trees* (ed A.S. Raghavendra). John Wiley & Sons, New York, pp. 175–197.

Aloni, R. and Zimmermann, M.H. (1983) The control of vessel size and density along the plant axis — a new hypothesis. *Differentiation* **24**: 203–208.

Aloni, R. and Zimmermann, M.H. (1984) Length, width and pattern of regenerative vessels along strips of vascular tissue. *Bot. Gaz.* **145**: 50–54.

Aloni, R., Alexander, J.D. and Tyree, M.T. (1997) Natural and experimentally altered hydraulic architecture of branch junctions in *Acer saccharum* Marsh. and *Quercus velutina* Lam. trees. *Trees* **11**: 255–264.

Aloni, R., Wolf, A., Feigenbaum, P., Avni, A. and Klee, H.J. (1998) The *Never ripe* mutant provides evidence that tumor-induced ethylene controls the morphogenesis of *Agrobacterium tumefaciens*-induced crown galls on tomato stems. *Plant Physiol.* **117**: 841–849.

Atkinson, C.J. and Denne, M.P. (1988) Reactivation of vessel production in ash (*Fraxinus excelsior* L.) trees. *Ann. Bot. n.s.* **61**: 679–688.

Baas, P., Schmid, R. and van Heuven, B.J. (1986) Wood anatomy of *Pinus longaeva* (bristlecone pine) and the sustained length-on-age increase of its tracheids. *IAWA Bull. n.s.* **7**: 221–228.

Bailey, I.W. (1958) The structure of tracheids in relation to the movement of liquids, suspensions and undissolved gases. In: *The Physiology of Forest Trees* (ed K.V. Thimann), The Ronald Press, New York, pp. 71–82.

Carlquist, S. (1988) *Comparative Wood Anatomy*. Springer, Berlin.

Carmi, A., Sachs, T. and Fahn, A. (1972) The relation of ray spacing to cambial growth. *New Phytol.* **71**: 349–353.

Cosgrove, D.J. (1996) Plant cell enlargement and the action of expansins. *BioEssays* **18**: 533–540.

Digby, J. and Wareing, P.F. (1966) The effect of applied growth hormones on cambial division and the differentiation of cambial derivatives. *Ann. Bot. n.s.* **30**: 539–548.

Dinwoodie, J.M. (1961) Tracheid and fibre length in timber: a review of literature. *Forestry* **34**: 125–144.

Fahn, A. (1988) Secretory tissues and factors influencing their development. *Phyton (Austria)* **28**: 13–26.

Fahn, A. (1990) *Plant Anatomy*. 4th edn. Pergamon, Oxford.

Gartner, B.L. (1995) Patterns of xylem variation within a tree and their hydraulic and mechanical consequences. In: *Plant Stems: Physiology and Functional Morphology* (ed B.L. Gartner), Academic Press, San Diego, CA, pp. 125–149.

Gregory, R.A. and Romberger, J.A. (1975) Cambial activity and height of uniseriate vascular rays in conifers. *Bot. Gaz.* **136**: 246–253.

Grundwag, M. and Werker, E. (1976) Comparative wood anatomy as an aid to identification of *Pistacia* L. species. *Isr. J. Bot.* **25**: 152–167.

Harris, J.M. (1989) *Spiral Grain and Wave Phenomena in Wood Formation*. Springer, Berlin.

Jacobs, W.P. (1998) Hormone transport and action in the green shoot: long-term studies of a clonal stock of *Coleus blumei* (Labiatae). *Am. J. Bot.* **85**: 1353–1363.

Larson, P.R. (1994) *The Vascular Cambium. Development and Structure*. Springer, Berlin.

Lev-Yadun, S. (1994a) Radial fibres in aggregate rays of *Quercus calliprinos* Webb. – evidence for radial signal flow. *New Phytol.* **128**: 45–48.

Lev-Yadun, S. (1994b) Experimental evidence for the autonomy of ray differentiation in *Ficus sycomorus* L. *New Phytol.* **126**: 499–504.

Lev-Yadun, S. and Aloni, R. (1990) Polar patterns of periderm ontogeny, their relationship to leaves and buds, and the control of cork formation. *IAWA Bull. n.s.* **11**: 289–300.

Lev-Yadun, S. and Aloni, R. (1991a) Polycentric vascular rays in *Suaeda monoica* and the control of ray initiation and spacing. *Trees* **5**: 22–25.

Lev-Yadun, S. and Aloni, R. (1991b) Natural and experimentally induced dispersion of aggregate rays in shoots of *Quercus ithaburensis* Decne. and *Q. calliprinos* Webb. *Ann. Bot. n.s.* **68**: 85–91.

Lev-Yadun, S. and Aloni, R. (1992) The role of wounding in the differentiation of vascular rays. *Int. J. Plant Sci.* **153**: 348–357.

Lev-Yadun, S. and Aloni, R. (1995) Differentiation of the ray system in woody plants. *Bot. Rev.* **61**: 45–88.

Liphschitz, N. (1995) Ecological wood anatomy: changes in xylem structure in Israeli trees. In: *Wood Anatomy Research 1995. Proceedings of the International Symposium on Tree Anatomy and Wood Formation*, (ed. Wu Shuming), Tianjin, China International Academic Publishers, Beijing, pp. 12–15.

Liphschitz, N. and Waisel, Y. (1970) Effects of environment on relations between extension and cambial growth of *Populus euphratica* Oliv. *New Phytol.* **69**: 1059–1064.

Little, C.H.A. and Pharis, R.P. (1995) Hormonal control of radial and longitudinal growth in the tree stem. In: *Plant Stems: Physiology and Functional Morphology* (ed. B.L. Gartner), Academic Press, San Diego, CA, pp. 281–319.

Mattsson, J., Sung, Z.R. and Berleth, T. (1999) Responses of plant vascular systems to auxin transport inhibition. *Development* **126**: 2979–2991.

Mattoo, A.K. and Aharoni, N. (1988) Ethylene and plant senescence. In: *Senescence and Aging in Plants* (eds L.D. Nooden and A.C. Leopold), Academic Press, San Diego, CA, pp. 241–280.

Metcalfe, C.R. and Chalk, L. (1985) *Anatomy of the Dicotyledons*, 2nd edn. Vol. II. The Clarendon Press, Oxford.

Philipson, W.R., Ward, J.M. and Butterfield, B.G. (1971) *The Vascular Cambium*. Chapman & Hall, London.

Roberts, L.W., Gahan, P.B. and Aloni, R. (1988) *Vascular Differentiation and Plant Growth Regulators*. Springer, Berlin.

Sachs, T. (1981) The control of patterned differentiation of vascular tissues. *Adv. Bot. Res.* **9**: 151–262.

Sachs, T. (1991) *Pattern Formation in Plant Tissues*. Cambridge University Press, Cambridge.

Savidge, R.A. (1988) Auxin and ethylene regulation of diameter growth in trees. *Tree Physiol.* **4**: 401–414.

Savidge, R.A. and Farrar, J.L. (1984) Cellular adjustments in the vascular cambium leading to spiral grain formation in conifers. *Can. J. Bot.* **62**: 2872–2879.

Savidge, R.A. and Wareing, P.F. (1981) Plant-growth regulators and the differentiation of vascular elements. In: *Xylem Cell Development* (ed J.R. Barnett), Castle House Publications, Tunbridge Wells, pp. 192–235.

Schweingruber, F.H. (1990) *Anatomy of European Woods*. Verlag Paul Haupt, Bern.

Sundberg, B., Little, C.H.A., Cui, K. and Sandberg, G. (1991) Levels of endogenous indol-3-acetic acid in the stem of *Pinus sylvestris* in relation to the seasonal variation of cambial activity. *Plant, Cell & Environ*. **14**: 241–246.

Thimann, K.V. (1977) *Hormone Action in the Whole Life of Plants*. The University of Massachusetts Press, Amherst, MA.

Trewavas, A.J. (1983) Is plant development regulated by changes in the concentration of growth substances or by changes in the sensitivity to growth substances? *TIBS* **8**: 354–357.

Tuominen, H., Puech, L., Fink, S. and Sundberg, B. (1997) A radial concentration gradient of indole-3-acetic acid is related to secondary xylem development in hybrid aspen. *Plant Physiol*. **115**: 577–585.

Uggla, C., Moritz, T., Sandberg, G. and Sundberg, B. (1996) Auxin as a positional signal in pattern formation in plants. *Proc. Natl. Acad. Sci. USA* **93**: 9282–9286.

Uggla, C., Mellerowicz, E.J. and Sundberg, B. (1998) Indole-3-acetic acid controls cambial growth in Scots pine by positional signaling. *Plant Physiol*. **117**: 113–121.

Wareing, P.F., Hanney, C.E.A. and Digby, J. (1964) The role of endogenous hormones in cambial activity and xylem differentiation. In: *The Formation of Wood in Forest Trees* (ed M.H. Zimmermann), Academic Press, New York, pp. 323–344.

Wareing, P.F. and Phillips, I.D.J. (1981) *Growth & Differentiation in Plants*. 3rd edn. Pergamon, Oxford.

Wheeler, E.A., Baas, P. and Gasson, P.E. (eds.) (1989) IAWA list of microscopic features for hardwood identification. *IAWA Bull. n.s.* **10**: 219–332.

Woodcock, D.W. (1989) Distribution of vessel diameter in ring-porous trees. *Aliso* **12**: 287–293.

Zagórska-Marek, B. and Little, C.H.A. (1986) Control of fusiform initial orientation in the vascular cambium of *Abies balsamea* stems by indol-3-ylacetic acid. *Can. J. Bot.* **64**: 1120–1128.

Zimmermann, M.H. and Potter, D. (1982) Vessel-length distribution in branches, stems and roots of *Acer rubrum* L. *IAWA Bull. n.s.* **3**: 103–109.

Zhong, Y. and Savidge, R.A. (1995) Effects of IAA and GA_3 on *in vitro* wood formation in merchantable stems of white ash (*Fraxinus americana* L.). *Proc. Plant Growth Regulator Soc. Am*. **22**: 231–236.

26

The commercial cambium: understanding the origin of wood property variation

G.M. Downes, R. Evans, L.R. Schimleck and H.C. Fritts

1. Introduction

The importance of wood products to the global economy cannot be overstated. Perlin (1991) argued that the rise and fall of civilization was largely related to the availability of the forest resource. At the present time, economic and environmental pressures are combining to increase the reliance of the forest industry on plantation resources. This provides a number of challenges, not the least of which is that of reducing variability in the wood resource, and production of wood with specific or improved properties suited to particular end products. Considerable research effort is being directed in our laboratories to defining wood properties required to optimize the properties of wood products.

The majority of genetic and silvicultural work on improving wood properties is directed ultimately at optimising conditions for cambial activity. In the past the primary goal has been to maximise merchantable stem volume, thus focusing only on the average rate of cambial activity. Wood quality has commonly been considered only in terms of large, straight stems, free of defect. Now, however, genetic and silvicultural trials are being evaluated in terms of fundamental wood and fibre properties (Nyakuengama *et al.*, 1997; Evans *et al.*, 1997; Matheson *et al.*, 1997).

Merchantable volume increment is the major factor that drives the economics of forest growth for pulp and paper, followed by density and pulp yield (Greaves *et al.*, 1997). This emphasis will not change until premiums are paid for stems with known and preferred wood properties, assuming that, (a) wood properties can be rapidly measured and included in resource assessment and (b) the effects of wood properties on product performance and processing costs are clearly defined.

The resource assessment for wood properties will require the extrapolation of wood properties determined at discrete points within the stem to predict whole tree values.

Cell & Molecular Biology of Wood Formation, edited by R. Savidge, J. Barnett and R. Napier.

This requires that the pattern of variation within the tree is known. Wood properties vary within trees from pith to bark and from base to apex. Much work has been directed at describing variation in both softwoods (e.g., Zobel and Van Buijtenen, 1989; Evans *et al.*, 1995; Tian *et al.*, 1995) and hardwoods (Downes *et al.*, 1997; Hudson *et al.*, 1997; Raymond *et al.*, 1998a, b). At any one height the radial variation in wood properties is the result of a complex interaction between genotype and climate, affecting the physiology of the whole tree, which in turn regulates cambial activity (Denne, 1981). The relationship between climate and cambial activity has been the focus of much study (Fritts, 1976; Fritts *et al.*, 1991; Fritts and Shashkin, 1995). Thus the cambial surface encompassing the woody stem produces a range of wood properties at any one point in time, the control of which is still poorly understood. More work is needed to elucidate the regulation of cambial activity within trees as a function of genotype, site, silviculture and climatic variation.

2. Within-tree variation in density, pulp yield and dollars

Genetically identical trees grown on different sites can produce wood with very different properties and a change from native forest conditions to intensive plantation silviculture can markedly change wood properties. For example, North Forest Products, in Tasmania, Australia obtains a 10% premium for wood chips from plantation eucalypts compared with those from native forests because of the higher Kraft pulp yield and greater consistency obtained from the former (R. Coffey, Director — Marketing, North Forest Products, Australia, personal communication). Higher Kraft pulp yield reflects higher cellulose and lower hemi-cellulose and lignin contents. The patterns of distribution in wood properties vary among trees and the relationships between wood properties and growth rate can often be conflicting and confusing (Downes *et al.*, 1997).

Within-tree variation in both average annual pulp yield (Schimleck and Michell, 1998) and density is considerable. Contour maps showing the radial and longitudinal variation in Kraft pulp yield, wood density, and dollar value are shown in *colour Plate 5* for a single *Eucalyptus nitens*. The value of the tree is a function of the amount of pulp it produces; a combination of volume, density, and pulp yield (Greaves *et al.*, 1997). The third map combines these into the mass of pulp produced per year, expressed as a dollar value. The latter is a means of visualizing the net commercial benefits of cambial activity. The components (volume, density and pulp yield) differ markedly in terms of within-tree variation. A major consideration in these studies is the relationship between a rapidly growing cambium and the properties of the wood it produces.

The problem is further complicated when one considers property variation at a sub-annual resolution. Cellulose content (a major determinant of pulp yield) varies across an annual ring, increasing from early season to late season wood (*Figure 1*). These data were obtained from five different trees representing the range of diameter classes in an irrigated *E. nitens* plantation at Lewisham in Tasmania, Australia (42° 49′S, 147° 36′E). Each annual ring contains a widely varying range of cellulose contents from which annual averages were determined. Variability in sub-annual growth patterns from year-to-year would be expected to contribute to a lot of the variability in the maps shown in *colour plate 5*.

A difficulty we face in trying to understand wood properties in terms of whole tree averages is relating the values we can measure non-destructively near breast height to

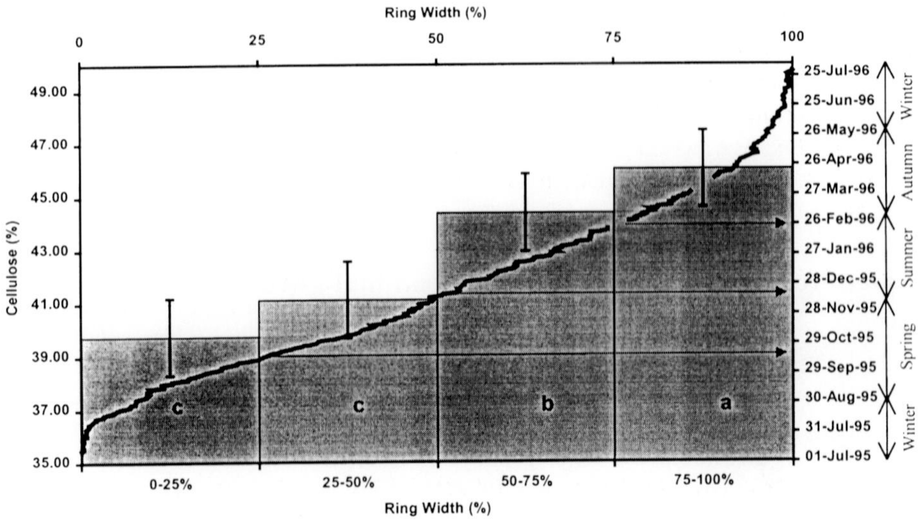

Figure 1. *The variation in cellulose content across the 1995–1996 annual ring from five irrigated* Eucalyptus nitens *at breast height. Wood produced later in the season had significantly higher cellulose content than that produced early in the season. Also shown (line) is the pattern of stem growth determined from the average growth pattern of three neighbouring trees. (Note: The axes describing the dendrometer data have been rotated 90° such that the amount of growth (% ring width) has been plotted on the upper x-axis to compare with within-ring cellulose data on the lower x-axis). Relating the sample width across the ring to the growth pattern allows the cellulose content to be related to the time period over which it was produced. (Bars represent standard errors. Columns with common letters are not significantly different at the 95% level of confidence.)*

those of the whole tree. To do this the patterns of wood property distribution need to vary in a measurably consistent and systematic manner. In plantation eucalypts, however, there is poor or inconsistent correlation between values of wood properties for the whole tree and those from increment cores. An increment core taken at breast height represents the radial pattern of wood property distribution. However, as we move from pith to bark we are also looking at wood that has been produced at successively lower proportions of tree height. The point of sampling moves down relative to the live crown, from juvenile to mature wood, that is wood produced at successively greater distances from growth regulators and photosynthate source. Improving our understanding of cambial activity should allow us to interpret these relationships better.

3. Automated assessment of wood properties

Extensive assessment of within-tree variation in wood properties requires methods for the rapid analysis of increment cores. The development of SilviScan for the rapid, automated analysis of wood properties has been described elsewhere (Evans *et al.*, 1995, 1996). This technology provides high resolution, radial profiles of wood properties (density, radial and tangential cell diameter, wall thickness and fibre coarseness and microfibril angle) from increment core samples. SilviScan is being increasingly used in the assessment of breeding trials (e.g. Nyakuengama *et al.*, 1997), fertilizer and thinning trials (Downes, unpublished data), resource assessment and studies relating wood properties to pulp and paper properties (Evans *et al.*, 1997). Examination of relationships between climate and properties of wood formed is in progress (Downes and Evans, 1994; Evans *et al.*, 1996) and will continue.

Effectively, radial profiles of wood properties reflect the history of cambial response to the environment. These profiles can be reduced to ring averages to generate within-tree contour maps of wood property variation as described above (see also Evans *et al.*, 1995). These, and other studies, have shown that wood properties, and therefore cambial activity, vary markedly with height and time.

4. Dendrometer measurements and cambial growth

SilviScan measurements of wood properties are spatial, but the events that regulate cambial activity are temporal. In order to relate the two data types, a template is required showing the spatial pattern or phenology of stem growth over time. In order to capture adequate radial resolution, measurements at sub-daily resolution are required, since measurements of cell production rates have shown they can reach six or more cells per day (Ridoutt and Sands, 1994; Ford and Robards, 1976).

Automated point dendrometers were installed on 24 eucalypt trees combining *Eucalyptus globulus* and *E. nitens* grown under four different conditions of water availability (Downes *et al.*, 1999a) and radial growth was measured hourly (*Figure 2*). Irrigated trees grew uniformly over the 3.5 years of monitoring. The other three treatments generated periods of rapid cambial activity interspersed with periods of low activity controlled by irrigation events or rainfall.

Relationships between weather variation and stem growth are being examined and statistical relationships determined to relate daily growth increments to weather variables (Downes *et al.*, 1999b). The differences between the treatments are marked, with rainfed trees experiencing periods of severe water stress, resulting in stem shrinkage. Pre-dawn water potentials down to –3.3 Mpa were measured and relationships between leaf water potential, stem shrinkage and soil water content are now being examined.

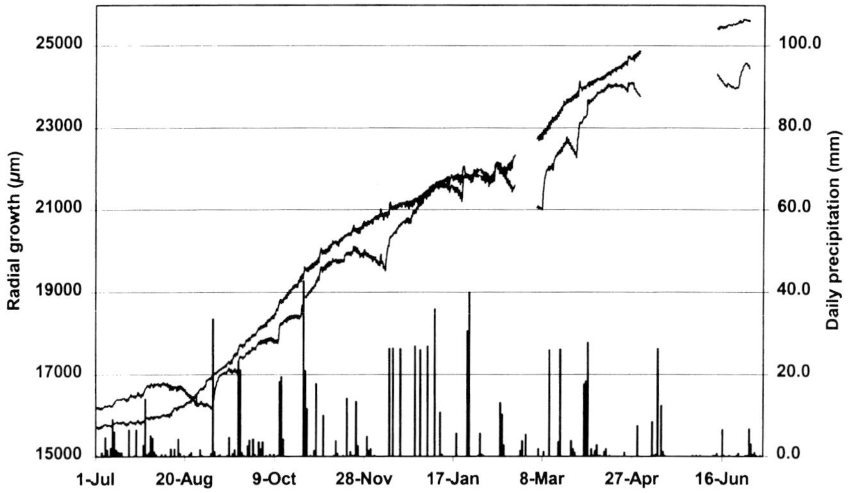

Figure 2. Stem growth patterns in response to water availability comparing an irrigated (smooth line) and rainfed (irregular line) tree over the 1996/1997 growing season. Daily precipitation is shown as columns.

The daily average rates of stem expansion varied seasonally. It is evident from this study, and others, that the daily radial stem increment reflects the daily average rate of cambial activity. Over a diurnal cycle (*Figure 3a*) the tree (cambium) experiences a continually changing set of limiting conditions. Therefore the rates of cell division and expansion/elongation would also be varying continually, predominantly as a function of water availability. These varying conditions, over both short and long term periods, make it difficult to measure accurately the rate of wood production.

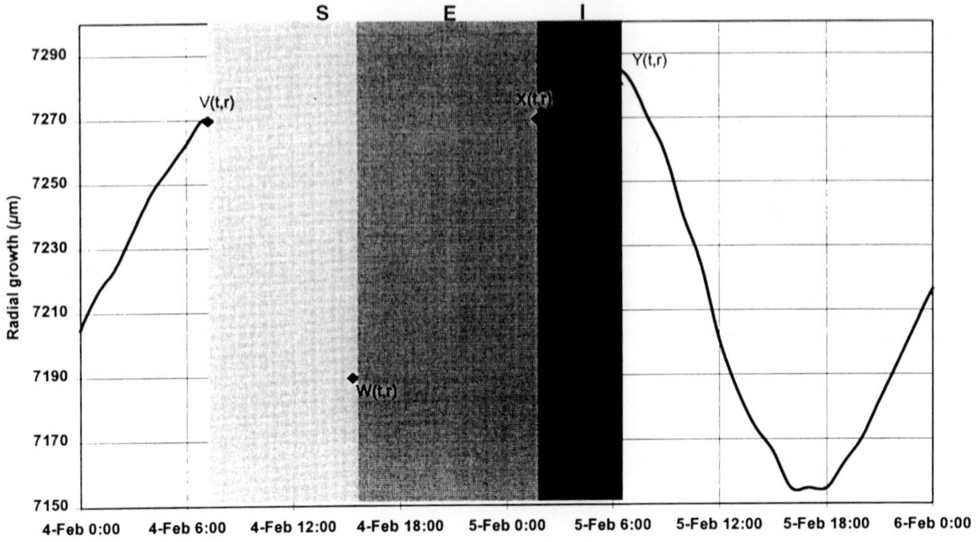

***Figure 3a**. Diurnal cycle showing defined phases of diurnal variation. Shrinkage (S), expansion (E) and increment (I) (Downes et al., 1999b).*

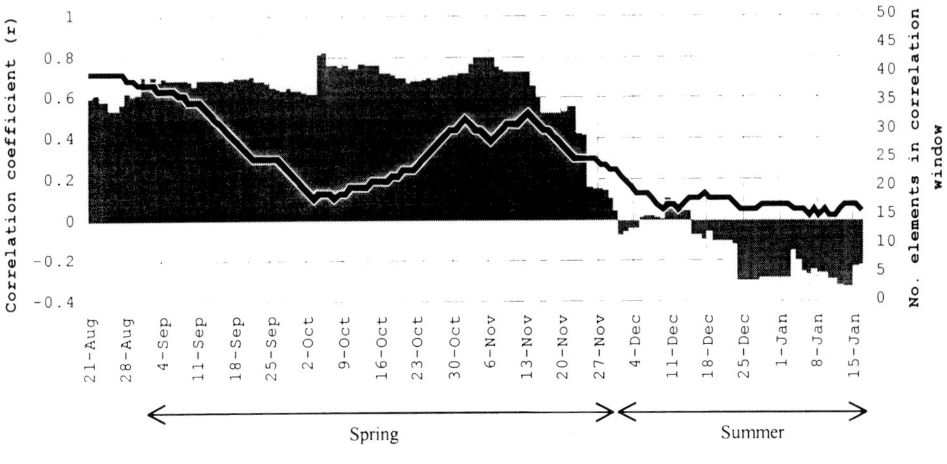

***Figure 3b**. The changing correlation between the rate of radius change and the average air temperature during the increment phase over the spring–summer growth period. The dates correspond to the centre of the 40-day correlation window.*

Herzog *et al.* (1995) examined similar diurnal patterns in relation to changing patterns of sapflow, and identified five distinct phases in the diurnal cycle signified by changing patterns of water flow into or out of the cambial region. One might expect cell expansion in particular to be confined to those periods when water is flowing into, rather than out of, the cambium. We have described the diurnal cycles in the eucalypt dendrometer data in terms of three distinct geometric phases (Downes *et al.*, 1999b; *Figure 3a*) labelled shrinkage, expansion and increment. Software was written to analyse the diurnal cycle in terms of this simplified three-phase model.

By definition, the daily increment is the magnitude of radius change in the increment phase (Y_r–X_r). We assume that the majority of net cell expansion occurs in this phase. While cell division cannot be assumed to predominate in this phase, cell division must be closely related to cell expansion, as net expansion (primary wall production) must generally occur prior to cell division. Each of these phases can be resolved into a rate and duration. A given daily increment can be generated by a stem radius expanding at a fast rate over a short duration, or at a slower rate over a longer duration. This has been shown to vary over a year in these trees (Downes *et al.*, 1999b) with conditions in early spring favouring a slower expansion rate over a longer duration, and the opposite in late spring and summer.

There has been a lot of confusion in the literature over the effect of growth rate on wood properties (see review in Downes *et al.*, 1997; Zobel and van Biujtenen, 1989), arising predominantly from this tendency to average growth rates over time. By extracting from the dendrometer data the increment phase as defined above, and resolving this into an expansion rate and duration, we have a more accurate estimate of cambial activity. It is probable that the rate of stem expansion (cambial activity) has a major effect on wood properties, and therefore the expansion rate defined above is of potentially greater interest than the average daily increment. Stem shrinkage, occurring over a period of two to many days, was a common feature of this data set, and is not yet adequately dealt with in the software.

The changing relationship between weather variables and cambial activity over time is illustrated in *Figure 3b* from a single *E. nitens*. The changing strength of the correlation between the rate of radius change and average air temperature during the increment phase was examined. The correlations were generated using only 40 days of data per correlation. This 40-day 'window' was incremented in daily steps over the first 150 days of the growing season. In *Figure 3b* the resultant series of correlations is shown in which the magnitude of the correlation was consistently positive in spring but became negative in the warmer, longer days of summer. Also shown is the number of elements within each 40-day correlation window. Often there was no increment phase from which to derive a rate, and these were treated as missing data. This pattern is consistent with that observed in 6 trees which had been continuously irrigated. The rate of expansion was shown to correlate with temperature more strongly than with the daily increment, and the strength of the correlation also varied over the season in a similar manner (Downes *et al.*, 1999b).

5. Relating growth patterns to wood properties

The wood density radial profiles were measured by X-ray densitometry using 12 mm increment cores obtained 150–300 mm below the point at which dendrometer measurements were taken. *Figure 4* (Downes, unpublished data) shows the association between the dendrometer profiles and the wood density data in two *E. nitens* representing typical responses to rainfed and irrigated treatments.

The work required to define these relationships more clearly is currently in progress. Radial profiles of microfibril angle, vessel frequency, vessel diameter, vessel free density, fibre diameter, wall thickness and fibre population (cells mm^{-1}) will be available from recent developments in the SilviScan technology.

6. Models as an interface

Popper, as cited in ReMine (1993), stated that scientific theories must be falsifiable. The process of science is to observe the natural world, develop hypotheses about how nature functions and make predictions, which can then be tested. Often biologists spend much time in observation and not enough in integrating knowledge into testable predictions. This is true of cambial research. We must now use our knowledge of cambial function to predict how it will behave under prescribed conditions. The success or failure of these predictions will serve to refine our understanding. Process-based models provide one means of doing this.

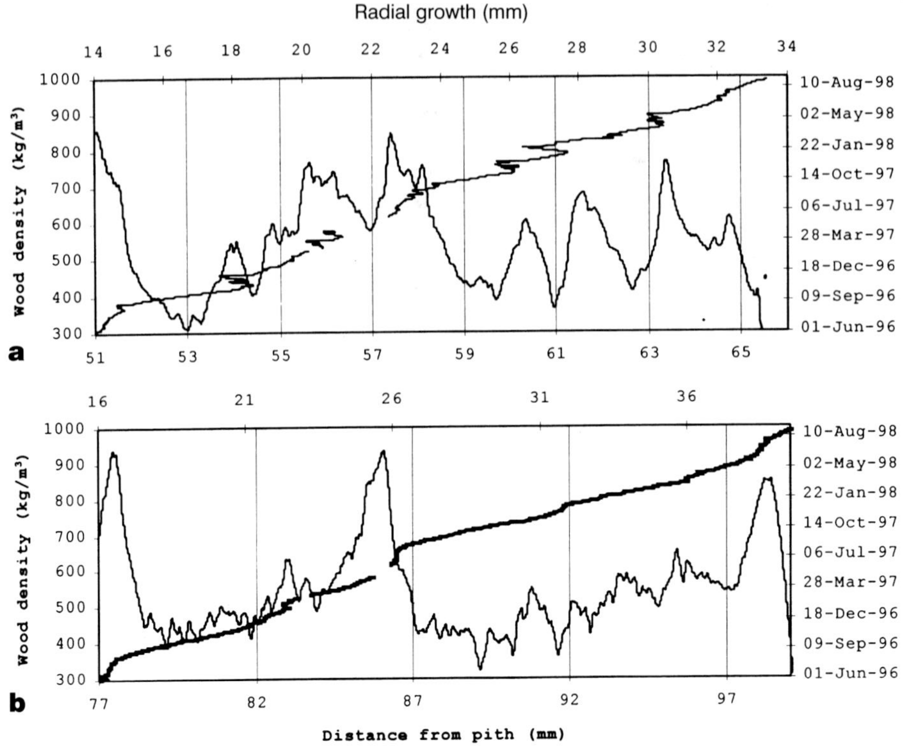

Figure 4. Comparison of dendrometer (thick line) and densitometer (thin line) profiles for (a) periodically water-stressed tree and (b) an irrigated tree. The axes describing the dendrometer data have been rotated 90° so that the radial growth (upper x-axis) may be compared with the position of the wood density measurement (lower x-axis). The correspondence between the periods of slower growth (dendrometer line more vertical) and increased wood density is apparent. In (a) increases in density (and slower growth) commonly relate to water stress events. In (b) the main density peaks correspond with the annual ring boundaries associated with slower winter growth.

Mayberry (1995) stated '*Models are how we think. They are how we understand the world works. As we go through life we build these very complex pictures in our minds of how the world works, and we are constantly referring back to them—matching incoming data against our models. That is how we make sense of things. One of the most important uses for models is in sorting incoming information to decide if it is important or not.*' This applies to cambial research as much as to any other field of activity.

The cambium is a unique meristem in that it leaves a recording of its growth activity in the form of wood. This makes it particularly amenable to modelling. One of the most convenient forms of expressing our knowledge is process-based modelling. Most modelling in forestry has been based around empirical models driven by statistical relationships between environmental and growth variables (see review by Battaglia and Sands, 1997). While they can often be very good, they are commonly either very general, or specific only to a narrow range of conditions. Process-based models facilitate understanding of the physiological interactions within the tree and between the tree and its environment. Obviously this is a complex and difficult task. However the advances in our knowledge that can be made, together with the identification of gaps in that knowledge, make this a potentially fruitful area of research. Another important advantage of models is that they encapsulate information in a more commercially useful form than do research papers alone.

6.1 *TREERING*

TREERING is a process-based model of cambial activity that has been under development for several years. In order to model cambial activity one requires a 'shell' in which the physiological interdependencies of the various tissues types with each other and with the environment can be modelled, to define their inputs to the cambium. This model has been described in detail in Fritts *et al.* (1999). At present TREERING is restricted to softwood species, and designed around slower-grown species. Some attention has been given to adapting the model to fast-grown radiata pine plantations (Downes pers. comm.), but much is still to be done here. As yet no efforts have been made to develop cambial modules for hardwoods.

6.2 *Towards a hardwood cambial model*

The hardwood cambium is responsible for the production of a much more heterogeneous tissue than is the softwood cambium, giving rise to vessels and parenchyma as well as rays and fibres. Models describing hardwood cambial activity will need to be quite different from those currently used in TREERING. The work described above using dendrometers to relate cambial growth phenology to wood properties is intended to provide a database for the development of the relationships required for such a model. There are a few principal differences that are worth noting here.

Difference in wood anatomy. Because of the greater diversity of tissue types, the hardwood model will need to be two-dimensional rather than one-dimensional (*Figure 5*). TREERING uses a rectangular model for the tracheid cross section (*Figure 6a*). This is less suitable for eucalypts as cells tend to a more hexagonal form. The factors which decide the frequency with which cambial initials develop into rays, fibres, vessels and parenchyma will need to be elucidated in some form and require a two dimensional approach (*Figure 6b*). Ridoutt and Sands (1993) also noted that fibres around vessels

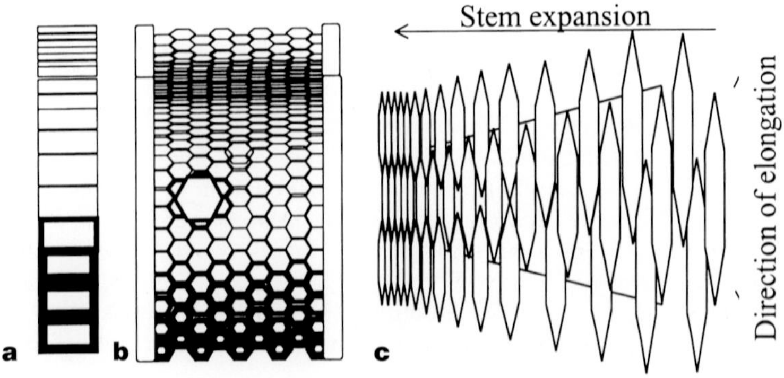

Figure 5. Diagrammatic comparison of hardwood and softwood cambial models. The (a) softwood model is one dimensional at present. Because of the heterogeneous nature of hardwood anatomy, a hardwood cambial model must be at least (b) two dimensional and (c) probably require a three dimensional component to explain radial growth attributable to fibre elongation.

tended to develop at a faster rate than other fibres which reflects a different pattern of expansion and secondary wall development to those further from vessels. Growth in eucalypts appears to be much less seasonal than in softwoods, often leading to a lack of distinct annual rings.

Fibre length effect. In gymnosperms the ratio of mature fibre length to cambial initial length has been shown to be commonly around 1.1 (Larson, 1991). In eucalypts the ratio is around 2 (Ridoutt and Sands, 1993). This means there must be more intrusive growth in the latter between fibres as they elongate. Therefore a much greater proportion of the radial growth measured by dendrometers is a result of fibre elongation rather than cell division and radial expansion (Figure 5c).

7. Conclusion

Increasingly, as technology becomes available to rapidly and non-destructively measure wood properties in standing trees, forestry research in Australia is addressing variation in wood properties. This allows wood properties to be included more readily in breeding and silvicultural programmes. As wood properties are the end result of cambial activity more effort is needed to understand the physiology of this meristem, to better interpret changing wood properties. In addition, models for predicting wood property variation under specific conditions will allow our knowledge to be continually tested and refined, and knowledge gaps efficiently identified. The commercial benefit of research will be more rapidly realized by the application of process-based models that can accurately predict wood property variation over time across a range of sites and genotypes. We believe these goals are now realizable given sufficient research effort.

Acknowledgements

The research activities described were conducted within the Cooperative Research Centre for Hardwood Fibre and Paper Science, Victoria, Australia.

References

Battaglia, M. and Sands, P.J. (1997) Process based forest productivity models and their application in forest management. *For. Ecol. Manag.* **102**: 13–32.

Denne, M.P. and Dodd, R.S. (1981) The environmental control of xylem differentiation. In: *Xylem Cell Development* (ed. Barnett, J.R.), Castle House Publ., London, pp. 236–255.

Downes, G.M. and Evans, R. (1994) Effects of environment on tracheid dimensions. *Proceedings of the Workshop 'Modelling of Tree-ring development — Cell Structure and Environment'* Frieburg 5–9th Sept. 1994 (eds. H. Spiecker and P Kahle). Institut für Waldwachstum, University of Freiburg, Germany.

Downes, G.M., Beadle, C., Gensler, W., Mummery, D. and Worledge, D. (1999a) Diurnal Variation and Radial Growth of Stems in Young Plantation Eucalypts: a Preliminary Study. In Press.

Downes, G.M., Beadle, C., and Worledge, D. (1999b) Stem growth — climate relationships in six irrigated plantation eucalypts. *Trees.* **14**: 102–111.

Downes, G.M., Hudson, I.L., Raymond, C.A., Dean, G.H., Michell, A.J., Schimleck, L.R., Evans, R. and Muneri A. (1997) *Sampling Plantation Eucalypts for Wood and Fibre Properties.* CSIRO Publishing, Melbourne, pp. 132.

Evans, R., Downes, G.M. and Murphy, J.O. (1996) Application of new wood characterization technology to dendrochronology. *In 'Tree Rings, Environment and Humanity'.* (Eds J.S. Dean, D.M. Meko and T.W. Swetnam.), *Radiocarbon* 1996. (Department of Geosciences: The University of Arizona, Tuscon), pp. 743–749.

Evans, R., Downes, G.M., Menz, D. and Stringer, S. (1995) Rapid measurement of variation in tracheid transverse dimensions in a radiata pine tree. *Appita J.* **48**: 134–138.

Evans, R., Kibblewhite, R.P. and Stringer, S. (1997) Kraft pulp fibre property prediction from wood properties in eleven radiata pine clones. *Appita J.* **50**: 25–33.

Ford, E.D. and Robards, A.W. (1976) Short-term variation in tracheid development in the earlywood of *Picea sitchensis*. *Lieden Botanical series* **No. 3**: 212–221.

Fritts, H.C. (1976) *Tree Rings and Climate*. Academic Press, New York.

Fritts, H.C., Shashkin, A.V. and Downes, G.M. (1999) TreeRing: A simulation model of conifer ring growth and cell structure. Accepted for publication in IUFRO Section 5 funded project (ed. Rupert Wimmer).

Fritts, H.C. and Shashkin, A.V. (1995) Modeling tree-ring structure as related to temperature, precipitation, and day length. In: *Tree Rings as Indicators of Ecosystem Health* (Lewis, T.E. ed.), CRC Press, Boca Raton. Chapter 2, pp. 17–57.

Fritts, H.C., Vaganov, E.A., Sviderskaya, I.V. and Shashkin, A.V. (1991) Climatic variation and tree-ring structure in conifers: Empirical and mechanistic models of tree-ring width, number of cells, cell size, cell-wall thickness and wood density. *Climate Res.* **1**: 97–116.

Greaves, B.L., Borralho, N.M.G. and Raymond, C.A. (1997) Breeding objective for plantation eucalypts grown for production of Kraft pulp. *For. Sci.* **43(4)**: 465–472.

Herzog, K.M., Hasler, R. and Thum, R. (1995) Diurnal changes in the radius of a subalpine Norway spruce stem: their relation to the sap flow and their use to estimate transpiration. *Trees* **10**: 94–101.

Hudson, I., Wilson, L. and van Beveren, K. (1997). Vessel and fibre property variation in *Eucalyptus globulus* and *Eucalyptus nitens*: some preliminary results. *IAWA J.* **19(2)**: 111–130.

Larson, P.R. (1994) *The Vascular Cambium*. Springer-Verlag, Berlin.

Matheson, A.C., Spencer, D.J., Nyakuengama, J.G., Yang, J.L., Cox, R.E. and Evans, R. (1997) Breeding for wood properties in radiata pine. In: *Genetics of radiata pine* (eds. R.D. Burdon and J.M. Moore), NZFRI Bulletin 203, pp. 169–79.

Mayberry, R. (1995) *Are you Liberal? Conservative or Confused?* Bluestocking Press, Placerville, CA.

Nyakuengama, J.G., Matheson, A.C., Evans, R., Spencer, D.J. and Vinden, P. (1997) Time trends in the genetic control of wood microstructure traits in *Pinus radiata*. *Appita J.* **50**: 486–494.

Perlin, J. (1991) Forest Journey: The role of Wood in the Development of Civilisation. Harvard University Press, Cambridge, USA

Raymond, C.A., Banham, P. and MacDonald, A.C. (1998a). Within-tree variation and genetic control of basic density, fibre length and coarseness in *Eucalyptus regnans* in Tasmania. *Appita J.* **51**: 299–305.

Raymond, C.A., Muneri, A. and MacDonald, A.C. (1998b). Non-destructive sampling for basic density in *Eucalyptus globulus* and *E. nitens*. *Appita J.* **51**: 224–228.

ReMine, W.J. (1993) *The Biotic Message. Evolution versus Message Theory*. St Paul Science, Saint Paul, Minnesota.

Ridoutt, B.G. and Sands, R. (1994) Quantification of the process of secondary xylem fibre development in *Eucalyptus globulus* at two height levels. *IAWA J.* **15(4)**: 417–424.

Ridoutt, B.G. and Sands, R. (1993) Within-tree variation in cambial anatomy and xylem cell differentiation in *Eucalyptus globulus*. *Trees* **8**: 18–22.

Schimleck, L.R. and Michell, A.J. (1998) Determination of within-tree variation of pulp yield. *Tappi*. **81(5)**: 229–236.

Tian, X., Cown, D.J. and McConchie, D.L. (1995) Modelling wood properties in radiata pine. Part 2: Basic density. *New Zealand J. For. Sci.* **25**: 214–230.

Zobel, B.J. and van Buijtenen, J.P. (1989) *Wood Variation: Its Causes and Control* (Springer-Verlag: Berlin).

27

Variation of the properties of tracheids in Norway spruce (*Picea abies* [L.] Karst.)

P. Saranpää, E. Pesonen, M. Sarén, S. Andersson, S. Siiriä, R. Serimaa and T. Paakkari

1. Introduction

Developmental dynamics of the cambium are recorded in the structure of secondary xylem. Differences in the cell length, diameter, cell wall thickness and cell arrangement reflect the changes occurring in the cambium and the environmental factors affecting it. In conifers, differentiation of cambial derivatives into xylem cells mainly involves limited radial growth, secondary wall-deposition, lignification,and autolysis of protoplasts Intrusive growth of the cambial derivatives is so limited that cellular alignment and arrangement in the mature tissues quite accurately reflects that prevailing in the cambium when those derivatives were formed.

The dimensions of tracheids have been a subject of investigation for more than 120 years. The pioneer work dates from Sanio (1872). Tracheid length increases rapidly and non-linearly during the first years of radial growth and then more gradually in the mature wood. There are numerous studies about variation of tracheid length and diameter within a growth ring or within a tree, mainly in pines (e.g., reviews by Spurr and Hyvärinen, 1954; Dinwoodie, 1961). Olesen (1982) states that the vascular cambium is subject to two types of maturation processes, namely (i) those transmitted to the cambium at its formation from the apical meristem (cyclophysis) and (ii) those which cambium undergoes after its formation. According to his investigations of tracheid width in Norway spruce morphological properties change in both radial direction, and with height in tree (Olesen, 1982).

Tracheid length of Norway spruce increases rapidly from the pith towards bark (Helander, 1933). There is no significant difference in tracheid length in *Picea abies* between the ten first shoots at different heights in the stem (Saranpää, 1994). Recent investigations have shown that environmental factors (light, water, nutrients) effect

Cell & Molecular Biology of Wood Formation, edited by R. Savidge, J. Barnett and R. Napier.

wood structure in Norway spruce (Lindström, 1997; Herman *et al.*, 1998; Dutilleul *et al.*, 1998). A slight decrease was observed in tracheid length of fast-grown trees after thinning (Herman *et al.*, 1998). Suppressed trees with a slow growth rate showed a smaller tracheid diameter than dominant trees (Lindström, 1997). However, some results are contradictory. The reason may be that several investigators have not compared secondary xylem of the same physiological age, that is, samples that are in the same position in the stem. Cambial growth is a complex phenomenon that is affected both by genetic and by environmental factors.

Our aim was to study in detail the changes in tracheid length, cell form, lumen diameter, cell wall thickness and microfibril angle from the pith to the bark and at various heights in the stems of Norway spruce. Modern image analysis techniques and X-ray diffraction make it possible to analyse properties of large numbers of samples in a short time. The effect of fertilization on wood structure of Norway spruce is also under investigation. Knowledge of the environmental factors that affect fibre characteristics is of great importance for the pulp and paper industry.

2. Material and methods

2.1 *Cell dimensions*

The tracheid length and diameter of a 34-year-old fertilized tree (irrigation water with complete nutrient solution for the last 12 years) were compared with a control tree without fertilization. Prof. Sune Linder (SLU, Uppsala) is gratefully acknowledged for providing the material from the experiment in Flakaliden (see Linder, 1995).

We determined tracheid length and diameter by image analysis (Image-Pro Plus, Media Cybernetics, USA). For length measurements, earlywood zone of every third annual ring from the pith was split into match stick-sized samples and macerated with glacial acetic acid/hydrogen peroxide (1:1, v/v) at 60°C overnight (Franklin, 1945). Suspensions of washed fibres were deposited on microscope slides. The lengths of 100 unbroken earlywood tracheids were measured. Parallel samples were analysed using Kajaani FS-200 and Kajaani FiberLab (Kajaani Elektroniikka Ltd., Valmet, Finland). Manual measurement of 100 unbroken tracheids gave almost identical values to the measurement of 30 000 fibres by Kajaani FS-200 or FiberLab. However, neither FS-200 or FiberLab were able to distinguish broken tracheids and ray parenchyma cells from whole tracheids. Thus, a weighted average tracheid length was used instead of the arithmetic mean. If the fines (<0.5 mm) were rejected, tracheid length frequencies formed almost a Gaussian distribution. Herman *et al.* (1998) reported that a Gaussian non-linear model could fit adequately the intra-ring variation of tracheid length. We decided to use manual measurement of 100 unbroken tracheids in order to study differences in tracheid length.

Tracheid diameter was analysed with Kajaani FiberLab simultaneously with length measurement of every 10th tracheid. Thus, 3000 measurements were made of each sample. In addition, cross-sections, 20 μm thickness were made of parallel wood samples using a cryo-microtome (–16°C). Average cell lumen diameter was determined from the measurement of 200 earlywood tracheids using image analysis. Both tracheid diameter measured by FiberLab and lumen diameter measured by image analysis showed a similar trend from the pith to the bark.

2.2 *Fourier analysis*

The cross-sectional shapes of the cells and the wall thickness were analysed with two-dimensional fast Fourier transformation (FFT, *Figure 1*). By Fourier transforming the optically scanned image data from real space to the wave number space, periodical structures such as thickness of wall and size of the cells could be isolated. These periodical structures were seen in the Fourier transformed image as emphasized frequency distributions (as darkened areas in *Figure 1b* and *Figure 1d*). By fitting the Gaussian distributions to the wave number distributions, the wall thicknesses of the original image can be measured. The distribution peak position and the standard deviation characterize the Gaussian distribution. When these values were transformed to the real space, the peak represented the wall thickness and the standard deviations represented the wall thickness distributions.

The average cell shapes were calculated from the Fourier transformations of the original scanned data. *Figure 1* shows two different images of wood samples both in real and wave number space. Because of symmetry only the upper halves of the Fourier transformed images are shown. The kernel of the transformation is seen at the bottom of the image (*Figure 1b* and *Figure 1d*), just in the middle of the horizontal axis. The vertical line dividing the images in two halves is formed by the wave numbers purely due to the vertical periodical structures in *Figure 1a* and *Figure 1c*.

The images show that the earlywood samples had cells with square cross-section, and producing highly oriented frequency distributions, whereas compression wood

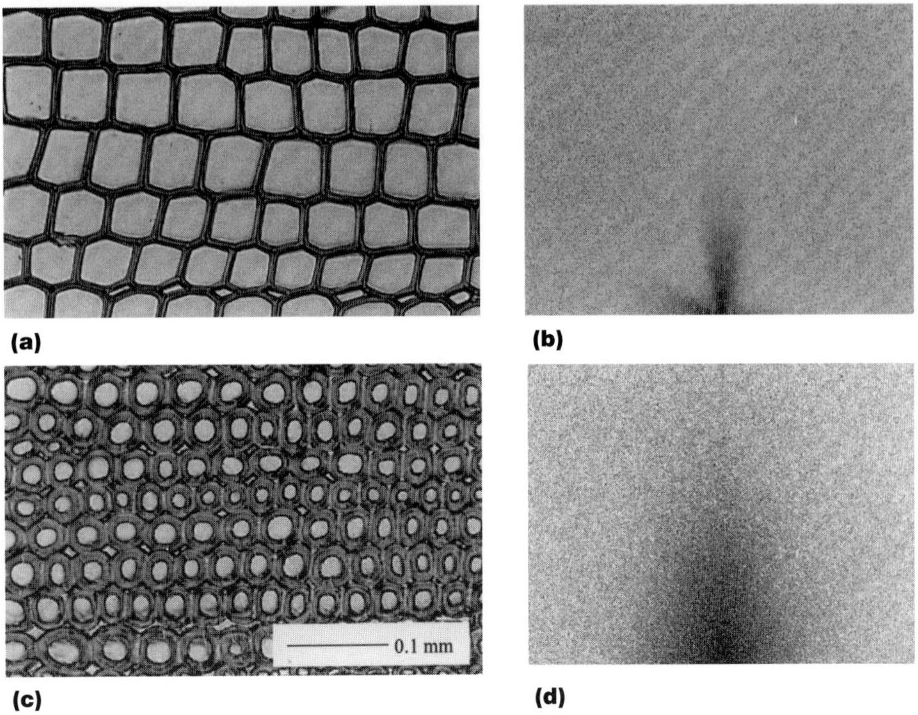

Figure 1a–d. The scanned images and the corresponding Fast Fourier transformations of mature earlywood tracheids (above) and compression wood (below) of Picea abies. *The grayscale of the FFT images is the logarithm of the absolute value of transformation.*

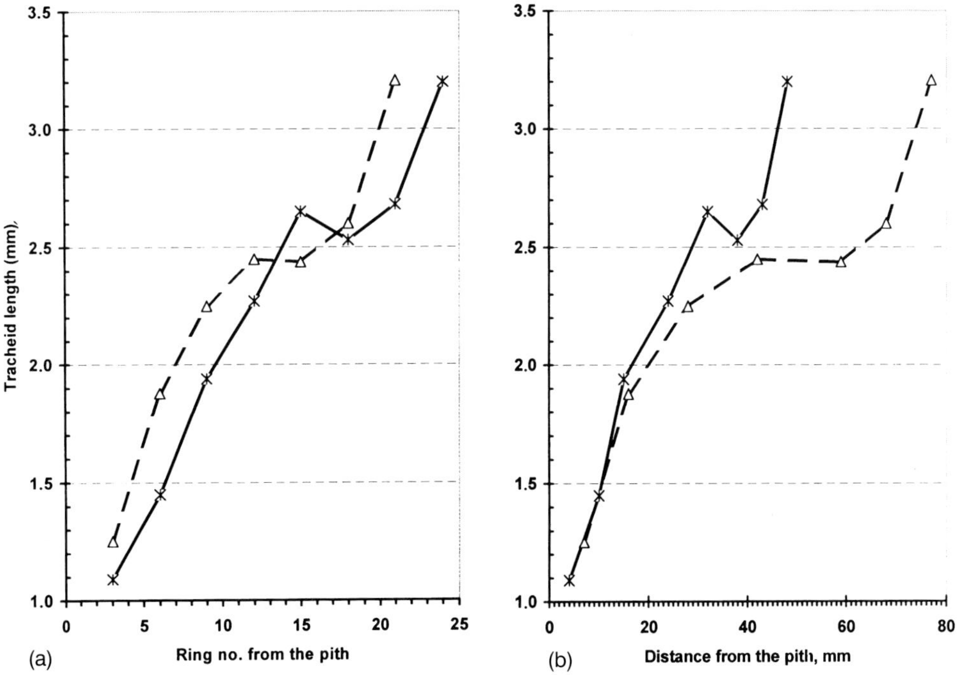

Figure 2. Tracheid length (mm) of a 34-year-old tree that was fertilized with irrigation water during the last 12 years (△ –) and of a control tree without irrigation or fertilization (*). Unbroken earlywood tracheids (n = 100) measured from every 3rd ring from the pith and presented as a function of both the (a) ring no. and the (b) distance from the pith.

had a more isotropic wave number distribution (*Figure 1d*). The average shape of the cross-section of the cells could be calculated by analysing the kernel part of the frequency distribution (Sarén, unpublished).

The demanded resolution in the original image sets the requirements for the resolution of the scanned image. The minimum resolution for the scanned image could be calculated using the sampling theorem. For example, if the required resolution was 1 μm, then the resolution of the source image had to be at least such that one pixel represented a length of 0.5 μm. The other side of the scanning problem is the size. For statistically valid measurement one must sample enough cells. The number of cells analysed should be more than 100 to reduce variance.

2.3 *Microfibril angle*

Sample disks were taken at breast height from three slow-grown and three fast-grown Norway spruce stems. Slow-grown trees were selected from a natural stand and had an average ring width of 2 mm. Fast-grown trees were from 20-year-old grafts that were planted in a former agricultural land (fertile site). The average ring width of the sampled fast-grown trees was 7 mm. Samples of 1 mm in thickness were further cut from every second or third ring at breast height.

Microfibril angles were determined by X-ray diffraction. In our work we used both reflections 002 and 040 and compared the obtained intensity curves. In this way we obtained information on both MFA and the cell cross section (Cave, 1997).

The diffractometer was used in symmetrical transmission mode. The measurements were performed with CuKα_1 radiation (λ = 1.541 Å) selected by a quartz monochromator. The divergence slit was circular (diameter 2 mm) and the width of the receiving slit was 0.2° in 2θ-scale. The scattered photons were detected by NaI (Tl) scintillation counter (Paakkari and Serimaa, 1984).

First the reflection was found by making a θ–2θ-scan. The detector was fixed to the reflection position. Then the sample was rotated around its normal and intensity was measured as a function of rotation angle ϕ.

In part of the samples the intensity curve of reflection 040 consisted of two well separated peaks. That of reflection 002 was broad and had its maximum at zero angle. This indicated that the shape of the cell cross section was circular (Cave, 1997). We concluded that the peaks arose from MFA distribution of S2 layer and fitted the 040 data with a pair of Gaussian functions.

In the other samples the reflections 002 and 040 gave identical intensity curves at angles smaller than 40 degrees, where the contribution of the other reflections was negligible. According to Cave (1966) this result is obtained for cells with rectangular cross section. For these samples the MFA distribution of each layer was assumed to be Gaussian and we fitted pairs of Gaussians to both 002 and 040 data. In this case the use of 002 reflection was preferable because it allowed the determination of MFAs larger than 40 degrees reliably (Saranpää *et al.*, 1998). The sum of the

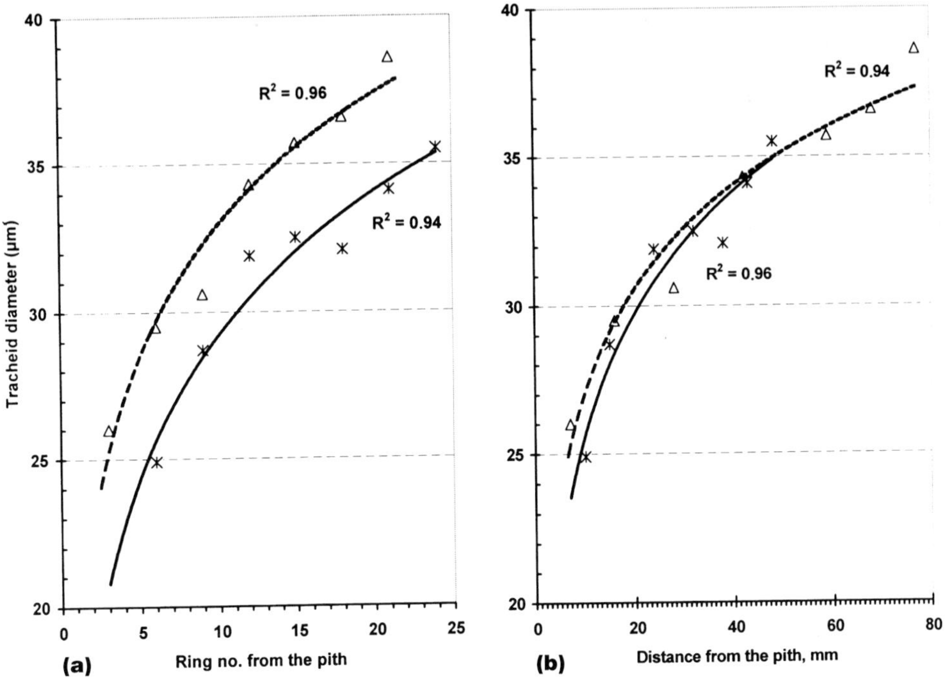

Figure 3. *Tracheid diameter (µm) of a 34-year-old tree that was fertilized with irrigation water during the last 12 years (△) and of a control tree without irrigation or fertilization (∗). Logarithmic regression lines and the R-squared values (R^2) are shown. Earlywood tracheids (n=3000) were measured from every 3rd ring from the pith and presented as a function of both (a) the ring no. and (b) the distance from the pith.*

Gaussians gave the MFA distribution of all cell wall layers. Only the contribution of S_3 layers could be identified reliably. It was omitted when calculating the average MFA in order to compare the results with those for cells with circular cross section. The average MFAs presented in this work thus include contributions of both S_2 and S_1 layers.

3. Results and discussion

3.1 *Tracheid length*

Tracheid length increased rapidly over the first 21–24 growth rings in Norway spruce (*Figure 2*) and showed a similar trend in both the control and in the fertilized trees. However, fertilization seemed to suppress the increase of tracheid length when correlated to the distance from the pith (*Figure 2b*). At 50 mm from the pith the difference in tracheid length was more than 0.5 mm. Closer to the bark the tracheid length increased again (*Figure 2*). Helander (1933) found shorter tracheids in Norway spruce trees with wide rings than in trees with narrow rings. Bannan (1950, 1960) studied trees of *Thuja occidentalis* which showed marked fluctuations in their growth rates. According to his results an increase in ring width is accompanied by a decrease in cell length at the time of anticlinal division.

Dutilleul *et al.* (1998) also found a negative correlation between ring width and fibre length in Norway spruce. They analysed 20 stems from both fast- and slow-

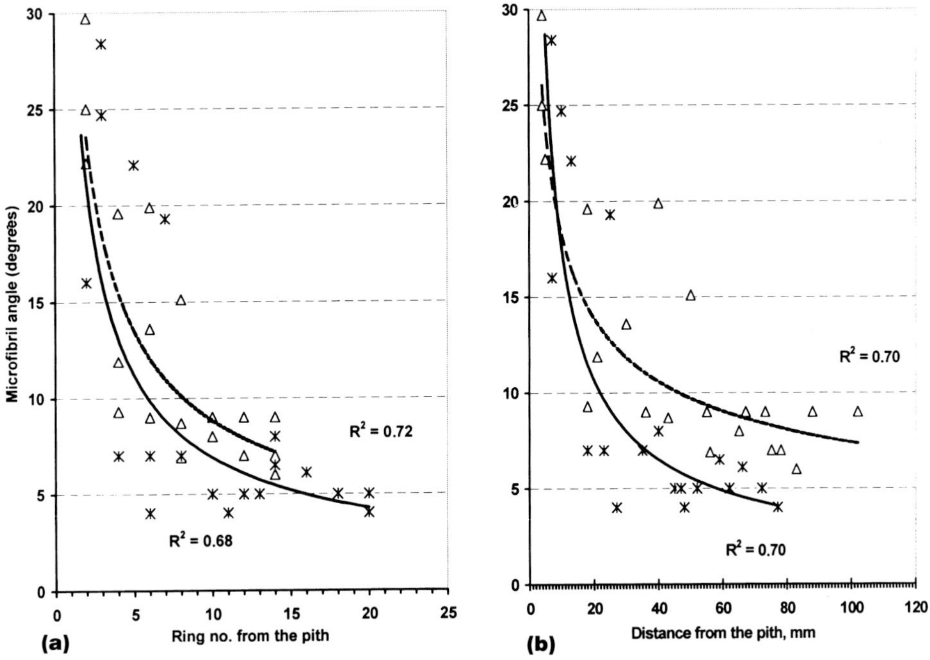

Figure 4. *Microfibril angle (degrees) of three slow-grown trees (average growth ring width 2 mm,) and of three fast-grown trees (average growth ring width 7 mm, △). Exponential regression lines and the R-squared values (R^2) are shown. Determined by X-ray diffraction of clear earlywood samples.*

grown categories and studied also the effect of first thinning. The correlation was stronger before the first thinning than after the first thinning, that is, stronger in the rings close to the pith than in rings of the outer wood. They also included a time-dependent correlation function in their analyses and demonstrated that the correlation between ring width and fibre length may depend on the year and growth category of the trees. In other words, ring no. from the pith, distance from the pith (number of cambial divisions from the pith) and climatic conditions during the year the ring was formed should be included in the statistical analysis. In our future work we will analyse more material and also include the effect of temperature and rainfall.

3.2 *Tracheid diameter*

Tracheid diameter increased rapidly over the first 21–24 growth rings (*Figure 3*). The fertilized tree with wide growth rings seemed to have much wider tracheids than the control tree with narrow rings. However, the relationship between tracheid diameter and distance from the pith was very similar both in the control tree and in the fertilized tree (*Figure 3*). Olesen (1978, 1982) reported that tracheid diameter was correlated more strongly with distance from the pith than with ring number from the pith in Norway spruce. He also showed a logarithmic relationship between cell diameter and distance from the pith, which was also true in our case.

Denne (1973) showed that mean tracheid diameter was correlated with ring width and rate of shoot elongation in *Picea abies*. However, only material from the second ring of shoots taken from branches was studied, and samples at the same distance from the pith were not compared. The difference may have been related to the fast increase in tracheid diameter close to the pith.

3.3 *Microfibril angle*

According to our results the MFA decreases rapidly from the pith towards the bark and is only 5–10 degrees by ring 6–7 from the pith in Norway spruce (*Figure 4a*). Sahlberg *et al.* (1997) and Lindström *et al.* (1998) have reported a similar trend and correlation.

However, growth rings in individual stems of our study are of different width and thus MFA is also shown in relation to the distance from the pith (*Figure 4b*). MFA reaches the value of 5–10 degrees at the distance of 20–40 mm from the pith. Longitudinal shrinkage of wood decreases to a value of 0.2% at the same distance in Norway spruce (Saranpää *et al.*, 1998). Thus, it seems there is a clear relationship between MFA and longitudinal shrinkage as Harris and Meylan (1965) had shown earlier.

The decrease of the MFA of fast grown trees showed almost a similar trend in the slow-grown stems when correlated to the ring no. from the pith (*Figure 4a*). However, MFA seemed to be larger at 40–80 mm distance from the pith in the fast grown trees than in the slow-grown trees. Growth conditions have been reported to have a slight influence on MFA (Lindström *et al.*, 1998). According to our results of X-ray diffraction the cross-sections of the cells of fast grown trees were circular up to about ring 10 from the pith. After that the cells have a rectangular cross section. In slow-grown wood nearly all the cells were rectangular. Only near the pith was the shape of the cross section of the cell circular (rings 2–3). MFA decreased further in rings 21–40 in a slow grown stem (data not shown).

4. Conclusions

Tracheid length and diameter increase rapidly over the first 20 growth rings in Norway spruce and there is a logarithmic relationship between cell diameter and distance from the pith. Microfibril angle (MFA) decreases rapidly from the pith towards the bark and is only 5–10 degrees by ring 6–7 from the pith. Tracheid length is shorter and MFA is larger in the fast grown trees with wide rings than in the slow grown trees with narrow rings when correlated to the distance from the pith.

Acknowledgements

The financial support from the Academy of Finland is gratefully acknowledged. We wish to thank Mrs Satu Järvinen, Mr Tapio Järvinen, Mrs Irmeli Luovula, Mr Carl Räihä, Ms Tanja Suni and Mrs. Päivi Supponen for skillful technical assistance Mr Antero Varhimo (The Finnish Pulp and Paper Research Institute, Espoo, Finland) is thanked for providing fibre analysis with Kajaani FS-200 and FiberLab. Prof. Sune Linder (SLU, Uppsala, Sweden) and Mr Sakari Vainikainen (Foundation for Forest Tree Breeding, Haapastensyrjä, Finland) are gratefully acknowledged for providing the experimental material.

References

Bannan, M.W. (1950) The frequency of anticlinal divisions in fusiform cambial cells of *Chamaecyparis*. *Amer. J. Bot.* **37**: 511–519.

Bannan, M.W. (1960) Cambial behavior with reference to cell length and ring width in *Thuja occidentalis* L. *Can. J. Bot.* **38**: 177–183.

Cave, I.D. (1997) Theory of X-ray measurement of microfibril angle in wood. *Wood Sci. Tech.* **31**: 225–234.

Denne, M.P. (1973) Tracheid dimensions in relation to shoot vigor in *Picea*. *Forestry* **46**: 117–124.

Dinwoodie, J.M. (1961) Tracheid and fibre length in timber. A review of literature. *Forestry* **34**: 125–144.

Dutilleul, P., Herman, M. and Avella-Shaw, T. (1998) Growth rate effects among ring width, wood density, and mean tracheid length in Norway spruce (*Picea abies*). *Can. J. For. Res.* **28**: 56–68.

Franklin, G. L. (1945) Preparation of thin sections of synthetic resins and wood resin composites and a new macerating method for wood. *Nature* **155**: 51.

Harris, J. M. and Meylan, B.A. (1965) The influence of microfibril angle on longitudinal and tangential shrinkage in *Pinus radiata*. *Holzforschung* **19**: 144–153.

Helander, A.B. (1933) Variations in tracheid length of pine and spruce. *Foundation For. Prod. Res. in Finland* **14**: 1–75.

Herman, M., Dutilleul, P. and Avella-Shaw, T. (1998) Growth rate effects temporal trajectories of ring width, wood density and mean tracheid length in Norway spruce (*Picea abies* (L.) Karst.). *Wood and Fiber Sci.* **30**: 6–17.

Linder, S. (1995) Foliar analysis for detecting and correcting nutrient imbalances in Norway spruce. *Ecol. Bull.* **44**: 178–190.

Lindström, H. (1997) Fiber length, tracheid diameter and latewood percentage in Norway spruce: development from pith outwards. *Wood and Fiber Sci.* **29**: 21–34.

Lindström, H., Evans, J.W. and Verrill, S.P. (1998) Influence of cambial age and growth conditions on microfibril angle in young Norway spruce (*Picea abies* (L.) Karst.). *Holzforschung* **52**: 573–581.

Olesen, P.O. (1982) The effect of cyclophysis on tracheid width and basic density in Norway spruce. *Forest Tree Improvement, Arboretet Horsholm* **15**: 1–80.

Paakkari, T. and Serimaa, R. (1984) A study of the structure of wood cells by X-ray diffraction. *Wood Sci. Tech.* **18**: 79–85.

Sanio, K. (1872) Ueber die Grösse der Holzzellen bei der gemeinen Kiefer (*Pinus sylvestris*). *Jahrbuch der Wissenschaftlichen Botanik* **8**: 401–420.

Saranpää, P. (1994) Basic density, longitudinal shrinkage and tracheid length of juvenile wood of *Picea abies* (L.) Karst. *Scand. J. For. Res.* **9**: 68–74.

Saranpää, P., Serimaa, R., Andersson, S., Pesonen, E., Suni, T. and Paakkari, T. (1998) Variation of microfibril angle of Norway spruce (*Picea abies* (L.) Karst.) and Scots pine (*Pinus sylvestris* L.) — comparing X-ray diffraction and optical methods. In: *Microfibril Angle in Wood. Proceedings of the IUFRO/IAWA International Workshop on the Influence of Microfibril Angle to Wood Quality* (ed. Butterfield, B.G.), University of Canterbury Printers, Canterbury, New Zealand, pp. 240–252.

Spurr, S.H. and Hyvärinen, M.J. (1954) Wood fiber length as related to position in tree and growth. *Bot. Rev.* **20**: 561–575.

28

Distribution, fluctuation and metabolism of food reserves in the wood of trees

W. Höll

1. Introduction

Wood forms the major supporting tissue of all the spermatophytes and pteridophytes. It is a secondary plant tissue produced by the vascular cambium. The formation of wood is under the control of numerous external and internal factors (see Roberts *et al.*, 1988; Aloni, 1991). Wood fulfills three major functions: support of plant through mechanical strength, conduction of water and minerals, and storage. The storage tissue in wood is represented by ray and vertical parenchyma cells of the sapwood. Storage tissue must be continuous with the conducting phloem in order to survive. Vertical xylem parenchyma connect with horizontal ray parenchyma to form a three-dimensional storage system. Storage materials are predominantly soluble sugars, fructosans, hydrolysable cell wall polysaccharides, fat (triacylglycerols), and specific proteins.

According to the predominant storage material, trees were classified by Fischer (1891) as 'starch trees' or 'fat trees'. Most ring-porous angiosperms, and also some gymnosperms (e.g., *Abies* and *Picea* spp.) belong to the 'starch trees', while most diffuse-porous angiosperms (and the genus *Pinus*) are regarded as 'fat trees'. Sinnott (1918) surveyed many tree species in this respect, but it does not seem possible to find a clear-cut classification; there is no well defined grouping along taxonomic or other lines. One can say that some tree species store variable amounts of triacylglycerols in addition to carbohydrates and protein.

Reserves in the wood of trees are accumulated during times of surplus production and are utilized as the demand arises. Storage materials in parenchyma cells of wood are important in providing both energy and the molecular building blocks for leaf growth, formation of new roots and twigs, and diameter increase of trunk and roots in early spring and during summer. The demand for mobilization of storage materials also arises in cases of emergency, for example when a tree is defoliated by insects.

Cell & Molecular Biology of Wood Formation, edited by R. Savidge, J. Barnett and R. Napier.

In the present article, results of investigations about the distribution of sugars, starch and fats across trunk and root cross sections as well as their seasonal fluctuations and their interconversion in both angiosperms and gymnosperms are reported. The investigations were performed during the last 20 years. For further information and details the original publications given as references should be consulted.

2 Distribution and seasonal fluctuations in carbohydrates

2.1 *Nature and radial distribution of non-structural carbohydrates*

In all trees investigated whether gymnosperms or deciduous trees, heartwood forming or non-heartwood forming, the dominant 'wood sugars' are stachyose, raffinose, sucrose, glucose and fructose (Kozlowski and Keller, 1966; Ashworth *et al.*, 1993; Terziev *et al.*, 1997). In some species, e.g., *Acer pseudoplatanus* L. (Höll, 1981) and *Populus × canadensis 'robusta'* (Sauter and Kloth, 1987; Sauter and van Cleve, 1991) maltose contributes significantly to the overall sugar pool. The occurrence of maltose seems to be temperature dependent (see below). In almost all cases, wood tissue nearest to the cambium contains highest amounts of soluble sugars (exception below). With increasing depth in the sapwood, the amounts of these carbohydrates decrease.

The heartwood proper is almost free of sugars. In *Pinus sylvestris* L. both arabinose and galactose were found to be present in the heartwood (Saranpää and Höll, 1989; Fischer and Höll, 1992). In contrast, arabinose/galactose are almost absent in the sapwood. The amount of both sugars increases markedly at the sapwood-heartwood boundary. Their level is considerably larger in inner (mean of 3.6 μmol g^{-1} dry weight) than in outer (1.5 μmol) heartwood. These sugars may have been hydrolysed from cell wall hemicelluloses during heartwood formation. Interestingly, galactose stimulates ethylene production in ripening tomato fruits (Kim *et al.*, 1987), and ethylene is thought to be closely associated with heartwood formation (Hillis, 1987) and formation of compression wood (Little and Eklund, 1999), respectively. The amount of starch across the sapwood varies and decreases towards the heartwood boundary, where it is absent (Kozlowski and Keller, 1966). On a few occasions small groups of cells containing starch have been observed in the heartwood of some tropical trees, for example *Shorea* species and in *Fagus sylvatica* (Dietrichs, 1964). The physiological meaning, if any, of starch in the dead heartwood is unclear.

The distribution of starch across stems of *Robinia pseudoacacia* L., *Pinus sylvestris* L., and *Tilia cordata* Mill. is depicted in *Figure 1.* Clearly within the two heartwood forming species (*Robinia* and *Pinus*) the concentrations of starch in individual annual rings of the sapwood follow a different pattern. In *Robinia* there is a steady decrease of starch towards the heartwood boundary whereas in *Pinus* there is a more or less even distribution within the sapwood, but a sharp drop when crossing the heartwood boundary. In both species, the heartwood contains negligible amounts of starch. On the other hand, xylem of the non-heartwood forming *Tilia* shows no pronounced starch gradient across the trunk. Note that the amounts of starch in *Pinus* and *Tilia*, both accumulating predominantly triacylglycerols (Fischer and Höll, 1992; Höll and Poschenrieder, 1975), are low.

Starch extracted from wood tissues of different age differs in its degree of branching. In all trees there is a shift (*Figure 2*) towards higher amounts of α (1→4) glycosyl chains (amylose) at the expense of α (1→6) glycosyl chains (amylopectin). In

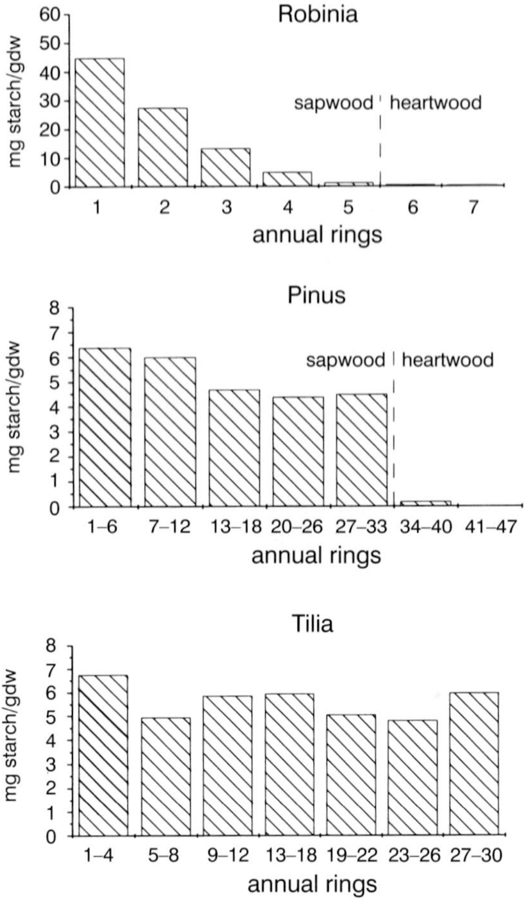

Figure 1. *Radial distribution of starch (determined with the iodine-method) across trunk cross sections of three tree species. The arabic numbers indicate annual ring tissues of increasing age. The heartwood (*Robinia *and* Pinus*) is almost free of starch.*

Robinia, the proportion of amylose is more or less constant in the older sapwood zones. It increases steadily towards the heartwood in *Pinus* whereas in *Tilia* only the innermost wood zone shows enhanced amounts of amylose. The physiological meaning of this shift in the starch fraction is unknown.

2.2 *Seasonal fluctuation*

The seasonal variation in quantity of stored material in the wood of trees was recognized long ago (older literature summarized by Kozlowski and Keller, 1966). In general, soluble sugars are high in winter and low in summer. In deciduous trees, starch levels decrease in spring, rise again after leaf growth and reach a maximum at the time of leaf shedding. During cold adaptation in winter, starch levels are comparatively low (see Sauter and Witt, 1997).

Figure 3 shows a thin layer chromatographic demonstration of the composition of soluble sugars in the outermost sapwood of *Acer pseudoplatanus* L. during an annual

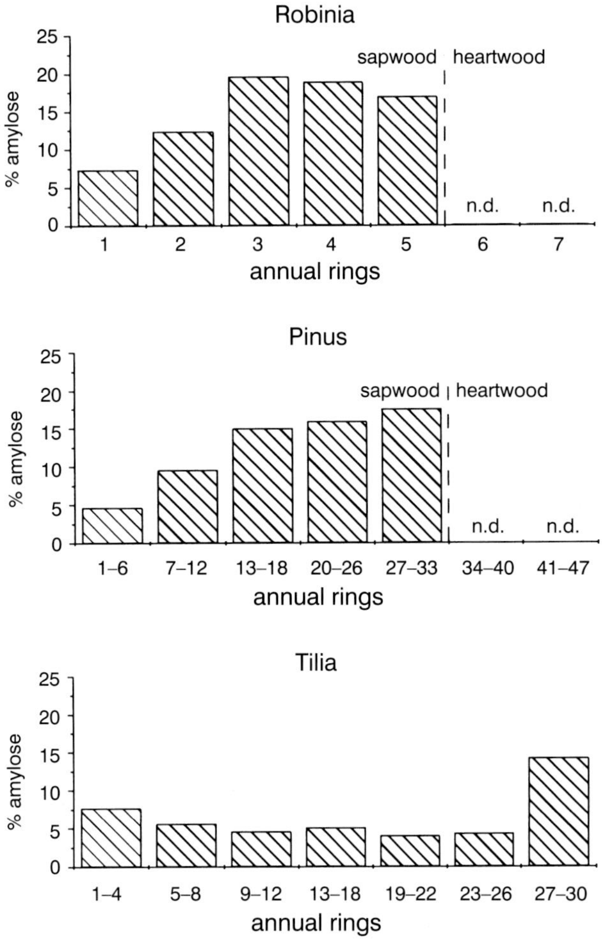

Figure 2. *Amylose content (as percent of total starch) extracted from different ages of trunk tissues of three tree species (see* Figure 1*).*

cycle (Höll, 1981). Stachyose, raffinose, maltose, sucrose, glucose and fructose are present. The intensity of the spot reflects the amounts of substance present. Sucrose is the dominant sugar throughout the year (up to 60 μmol g^{-1} dry weight). Clearly in the cold period of the year (November to December) soluble sugars accumulate. With the onset of bud burst and new growth ring formation (end of March to mid April) sugar concentrations decline. Raffinose and stachyose disappear almost completely in the summer months. The disaccharide maltose deserves special mention. Due to its rapid turnover, this sugar is normally present only in trace amounts in plant tissue. From *Figure 3* it becomes clear that maltose is accumulated at the beginning of September and again in mid November. According to the local weather station, temperatures decreased significantly a few days before sampling the wood. Thus, it may be that a cold stimulus induces or activates hydrolysing enzymes, for example β-amylase, releasing maltose from the starch molecule. This is in accordance with the fact that in poplar stem sections a very large pool of maltose was found when temperatures were lowered to 5 °C or 10 °C (Sauter, 1988).

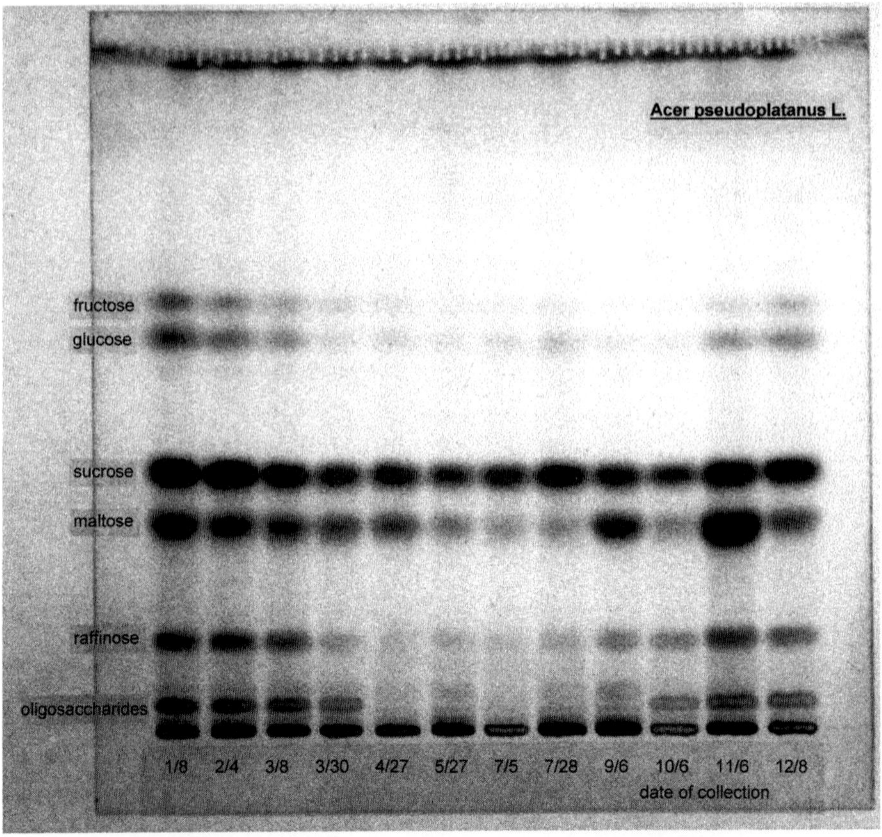

Figure 3. *Thin layer chromatographic demonstration of the annual fluctuation of soluble sugars in the trunkwood of one* Acer pseudoplatanus *L. tree. The outermost 5–10 annual layers of xylem were analysed.*

In wood cylinders obtained from trunks of 11 *Picea abies* (L.) Karst. trees, the fluctuation of reserve materials was followed quantitatively for one year (*Figure 4*). Regarding the soluble sugars (*Figure 4a*), wood contains higher concentrations in the cold period (from January until the beginning of March, also mainly December) than in spring and summer months. At the end of March sugars show a drastic decrease in concentration. Fructose, glucose, and raffinose concentrations are low during the warmer period of the year. At the beginning of April, raffinose is absent. During the warm period, sucrose is the dominant sugar. At the beginning of December, fructose, glucose, and sucrose show pronounced increases in concentration. One fact deserves special mention: specimens collected July 12th showed higher contents of fructose, glucose and sucrose than those collected three weeks earlier or three weeks later. Meteorological data indicated that about 10 days before July 12th a cold period accompanied by continuous rainfall occurred. One can conclude that lowering the temperatures in summer triggers metabolic changes in the trunkwood of *Picea abies*.

The higher molecular storage materials starch (='starch glucose') and fat ('glyceride glycerol') show, compared to soluble sugars and free glycerol, an opposite behaviour (*Figure 4b*). At the time when drastically reduced sugar and free glycerol

(a)

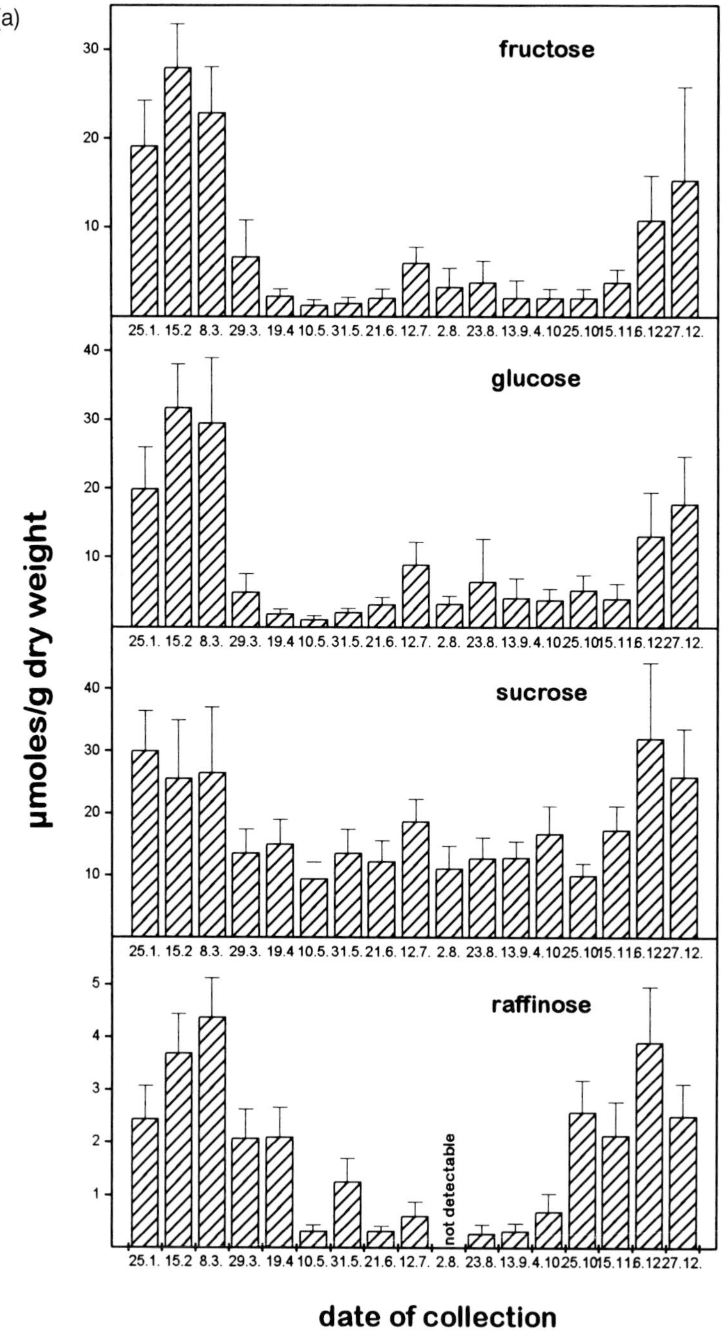

*Figure 4. Annual fluctuation of soluble sugars (a) and starch (= 'starch glucose'; expressed as glucose bound in starch), free glycerol and triacylglycerols ('glyceride glycerol'; expressed as glycerol bound in triacylglycerols; (b) in the outermost 5–7 rings of spruce sapwood (*Picea abies *(L.) Karst.). The bars represent mean values of 11 trees. Vertical lines indicate standard deviation.*

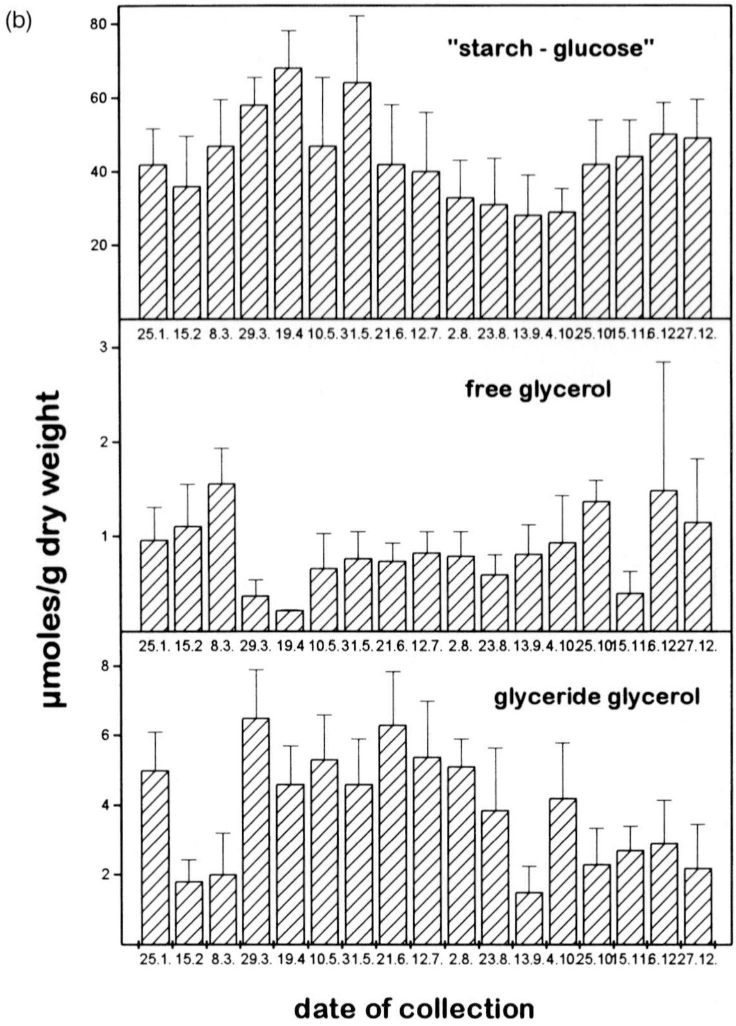

concentrations occur (end of March), the contents of starch and fat increase. Starch concentrations are more or less constant until the end of May and subsequently decrease until the beginning of October. From end of October until end of December, there is a slight increase in starch concentration. During the warm period of the year (end of March until end of August or beginning of October) the trunkwood of spruce trees contains more fat than in the cold period (see also Sauter, 1988; Sauter and van Cleve, 1994).

3. Distribution and seasonal fluctuations in fat reserves

3.1 *Nature and radial distribution*

In addition to carbohydrates, fats (triacylglycerols) play an important role as storage products. They contain more energy per unit weight than do carbohydrates or proteins. The heat of combustion of 1 g of fat is 9.3 kcal while that of carbohydrates is only 5.1 k-cal.

In comparison to carbohydrate reserves, investigations into the composition and annual fluctuations of fats in wood are scarce and contradictory. Since most of the early workers regarded materials which were stainable with fat reagents or extractable with organic solvents as 'true' fats, data expressing fat content in percent must be interpreted cautiously (older literature has been summarized by Kramer and Kozlowski, 1960). Extracts from wood tissue include such compounds as waxes, sterols and phospholipids in addition to true fats (triacylglycerols).

Using thin layer and gas chromatographic methods it becomes obvious that in both soft- and hardwood trees concentrations of triacylglycerols decline steadily across the trunkwood (Höll and Poschenrieder, 1975; Fischer and Höll, 1992; Hillinger *et al.*, 1996). Triacylglycerols are present in trace amounts in the heartwood. The fatty acid moieties of triacylglycerols vary in their chain length from 14 to 24 carbon atoms and have up to three double bonds.

An increase in free fatty acids towards the centre of the trunk seems to be specific within the genus *Pinus*. Free fatty acids are most abundant in the heartwood of *Pinus sylvestris* (Fischer and Höll, 1992), *P. muricata* and *P. contorta* (Lloyd, 1975; Anderson *et al.*, 1969).

The composition of fatty acids in the free fatty acid fraction differs from that of triacylglycerols across the trunkwood of *Pinus sylvestris*. Oleic acid (18:1) is the dominant free fatty acid and increases towards the centre of the trunk, while oleic and linoleic acid (18: 2) are dominant in the triacylglycerols but decrease in a centripetal direction (Fischer and Höll, 1992). Saranpää and Nyberg (1987) suggested that the free fatty acids in the heartwood of Scots pine may be the result of triacylglycerol degradation. If this is so, fatty acids are not only hydrolysed from triacylglycerols but also modified regarding their degree of unsaturation.

It has long been accepted that — even in 'fat trees' — almost no fat is produced as storage lipids in roots of trees (Kramer and Kozlowski, 1979). According to Ziegler (1964) only severe frost (many months below 0°C) induces fat accumulation in roots.

Figure 5 shows a thin layer chromatographic separation of lipid classes extracted from different aged zones of a root of *Pinus sylvestris* L. collected in the summer months. It is apparent that triacylglycerols (triglycerides) are present in the sapwood (annual rings 1 to 24 from the cambium). When crossing the heartwood boundary, however, amounts of triacylglycerols drop. Note that free fatty acids accumulate in the heartwood (*Figure 5*). Thus, the former statement that roots contain no (or only trace amounts) of triacylglycerols is no longer tenable.

3.2 *Seasonal fluctuation*

So far little information has been obtained by modern analytical techniques on the changes in fat content during the course of a growing season. Fischer and Höll (1992) followed the fluctuations of the neutral lipid composition of *Pinus sylvestris* L. sapwood (tissue of the outermost 10 annual rings) during an annual cycle. Steryl esters, triacylglycerols, free fatty acids, diacylglycerols and free sterols were present throughout the year. Amounts of triacylglycerols were about 24 μmoles g^{-1} dry weight (expressed as glycerol bound in the triacylglycerols). Seasonal changes in total amounts could not be detected in the diacylglycerol and free fatty acid fractions. Triacylglycerol levels were 10 times higher than diacylglycerol levels in all months investigated. Sapwood collected in February contained slightly higher amounts of fat

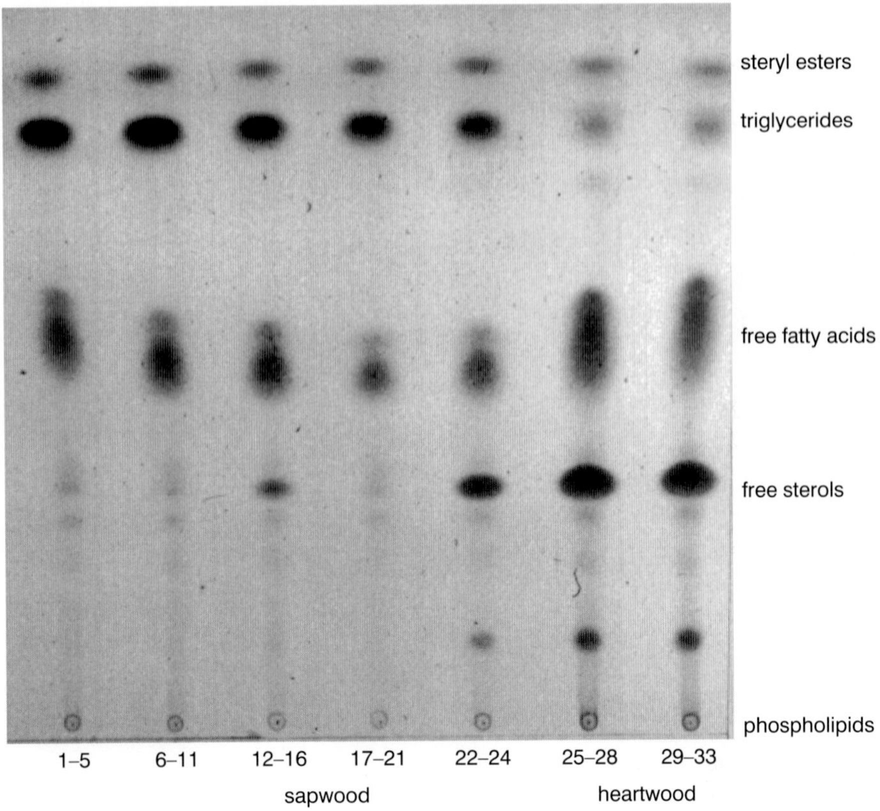

Figure 5. *Thin layer chromatographic demonstration of the radial distribution of neutral lipids across a cross section of one of the main roots of* Pinus sylvestris *L. The arabic numbers indicate annual ring tissues of increasing age. When crossing the heartwood boundary (annual ring 25 inward from the cambium), amounts of triacylglycerols (triglycerides) drop and those of free fatty acids increase.*

than sapwood collected from June to October. Differences between all other months were not significant. Oleic acid and linoleic acid represented 90% of all triacylglycerol fatty acids.

Figure 6 shows changes of storage lipids from the beginning of December until beginning of May in the trunk- and rootwood of *Tilia cordata* Mill. (Höll and Priebe, 1985). The thin layer chromatograms show that triacylglycerols (triglycerides) are abundant in the trunkwood at the beginning of December (upper left in *Figure 6*). Fatty acids are not visible at that time. The wood of one of the main roots also contains triacylglycerols (lower left in *Figure 6*). A gradient in the distribution of triacylglycerols is visible: the outermost wood tissue (zone I) contains considerably less triacylglycerols than the innermost tissue (zone III).

In addition to triacylglycerols, free fatty acids are present in the rootwood. In March, there is a great change in free fatty acid and triacylglycerol concentrations in the trunk (upper middle in *Figure 6*). Triacylglycerols seem to disappear almost completely while free fatty acids become visible. Compared to the trunk, rootwood

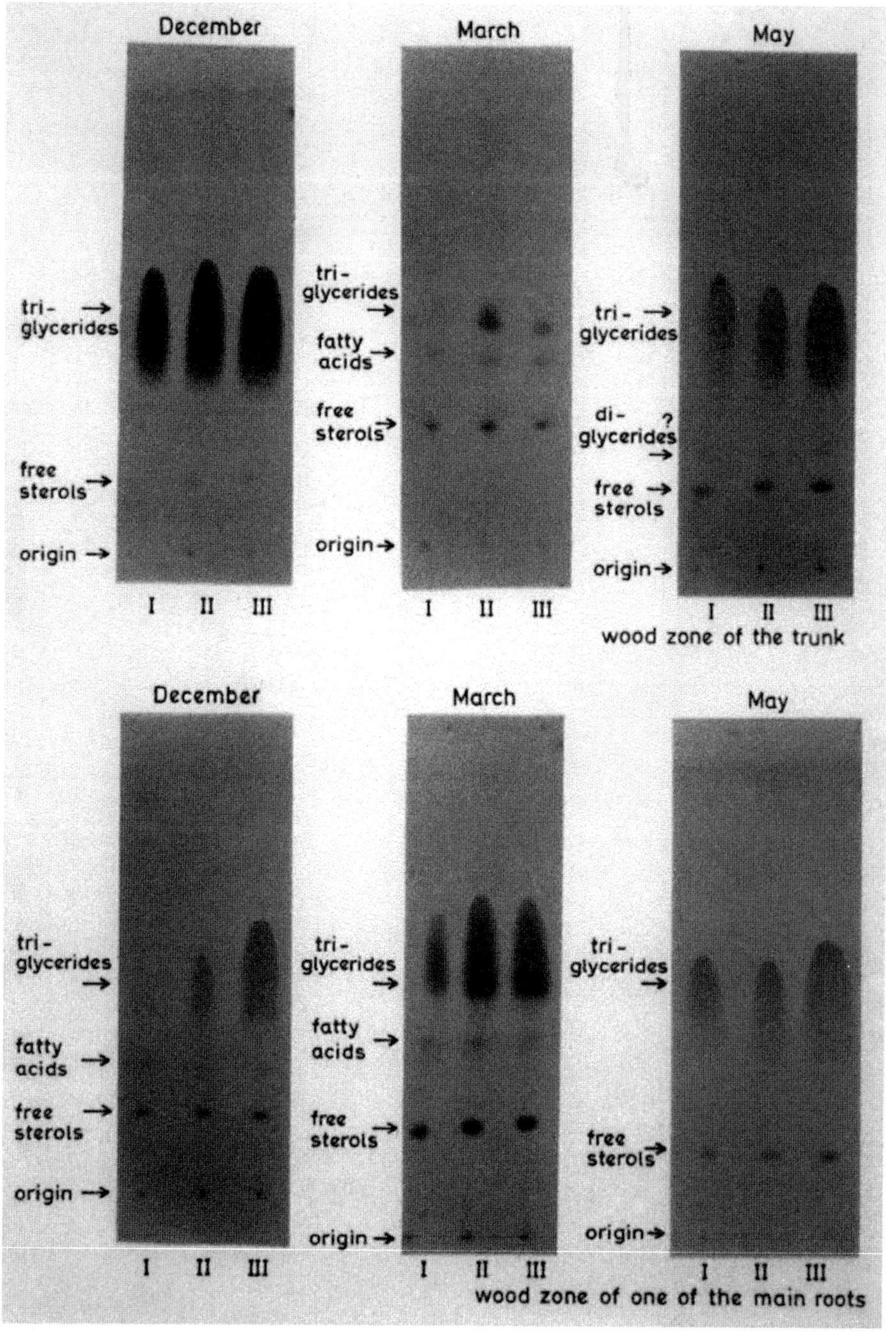

Figure 6. *Thin layer chromatographic demonstration of free sterols, fatty acids and triacylglycerols (triglycerides) in different-aged wood zones (Roman numerals) of the trunk and one of the main roots collected in December, March and May from* Tilia cordata *Mill. Zone I represents annual rings 1–12 of increasing age; II: 13–20; III: 21–29. Root zone I: 1–6; II: 7–16; III: 17–29 years.*

contains considerable amounts of triacylglycerols at that time of the year. Free fatty acids are also present in the root (lower middle in *Figure 6*). The trunkwood contains more triacylglycerols at the beginning of May than in March (upper right in *Figure 6*). Free fatty acids are almost absent, but diacylglycerols might be present. Root lipids contain no free fatty acids in May (lower right in *Figure 6*).

The intensity of the spots on the plates (*Figure 6*), however, do not reflect the true amounts of substances present. Since the spray reagent used detects only unsaturated compounds, this might lead to underestimation. To eliminate these inadequacies, carbon content was measured in chromatogram scrapings corresponding to the triacylglycerol and fatty acids (data not shown). This quantification approach revealed that triacylglycerols in the trunk were high in December, low in March and again present at the beginning of May. In contrast to the trunk, the root triacylglycerol content was much greater in March. In May, storage cells of the trunk were refilled with triacylglycerols whereas those of the root showed only a slight increase (*Figure 6*). From December until May, the root exhibited a pronounced triacylglycerol gradient from the younger (zone I) to the older wood (zone III). In March both trunk and root contained much higher amounts of free fatty acids than wood tissue collected at the beginning of December. In May the root yielded approximately 10 times less free fatty acids than in March. In the March specimen, there was a discrepancy between the intensity of the spots on the chromatographic plate (upper middle in *Figure 6*) and the carbon values measured in the corresponding chromatogram scrapings. On the stained chromatogram, it seems that the outermost zone of the trunkwood (zone I) contained almost no triacylglycerols while the inner zones (zone II and III) did contain this lipid class. The carbon values (data not shown), however, indicated the opposite situation: chromatogram scrapings corresponding to triacylglycerols from the outermost trunkwood (March, zone I) contained more carbon than those from the inner wood zones. This result indicates that triacylglycerols extracted from the peripheral trunkwood are made up of saturated fatty acids (which are not detectable with the spray reagent used, see above) while triacylglycerols from the inner parts of the trunk contain higher amounts of unsaturated compounds. From *Figure 6* it becomes clear that long before budbreak and unfolding of the leaves, storage lipids of the linden tree undergo conspicuous changes. It seems that most of the triacylglycerols are hydrolysed in the trunkwood in early spring. Again (see above) it is demonstrated that rootwood contains fat.

3.3 *Fat-sugar conversion at the cambial zone*

It is widely acknowledged that lipids in the storage cells are formed from either transport sugars or from starch. Observations at the cellular level indicate a starch-to-fat conversion and vice versa (Sauter and Neumann, 1994). In wood tissue these transformations are not fully understood biochemically.

Figure 7 shows the activities of lipase (an enzyme which hydrolyses triacylglycerols) and the concentrations of sucrose, glucose, and fructose in the outermost zone of the trunks of linden trees (*Tilia cordata* Mill.). Specimens were collected in December, April, and June. The wood zone investigated was the outermost 0.5 mm of the trunk. With the onset of cambial activity (May/June) lipase activity increased dramatically. The increasing lipolytic activity from the dormant (December) to the growing period in June was accompanied by a peak in sugar (sucrose, glucose, and fructose) concentrations (*Figure 7*). High lipase activity and the accumulation of sugars in the cambial region might be an indication that fats are converted into sugars.

According to the results obtained with fat-accumulating seeds, acetyl coenzyme A, arising from the oxidative breakdown of the fatty acids (which are liberated from triacylglycerols), is transformed into oxaloacetate via the glyoxylic acid cycle. Two specific enzymes, isocitrate lyase and malate synthase, are involved in these processes. These enzymes are located in glyoxysomes where storage fats are broken and converted into carbohydrates. Preparations from the cambial region (December, April, June) of *Tilia cordata* Mill. trees (see *Figure 7*) were incubated with ^{14}C glyoxylic acid. *Figure 8b* shows that upon incubation with glyoxylic acid and acetyl-CoA at least three labelled compounds are formed. The compound with the R_f value of 0.31 is malate. Neither boiled preparations (*Figure 8a*) prior to incubation nor those obtained in December or April catalysed the conversion of glyoxylic acid into malate. Thus, the results show that during cambial activity and wood formation, storage lipids seem to be converted into sugars via the glyoxylate pathway.

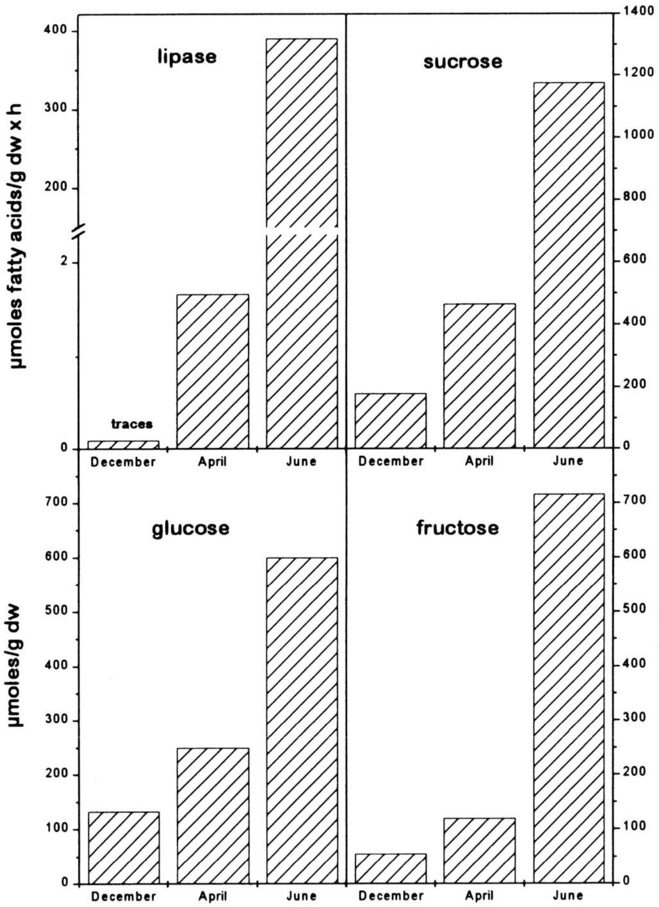

Figure 7. Activities of lipase and concentrations of soluble sugars in the cambial zone (layer between phloem and xylem, approx. the outermost 0.5 mm of the xylem) in trunks of Tilia cordata *Mill. from December to June. In the June specimen, this zone is a sweet tasting, soft layer.*

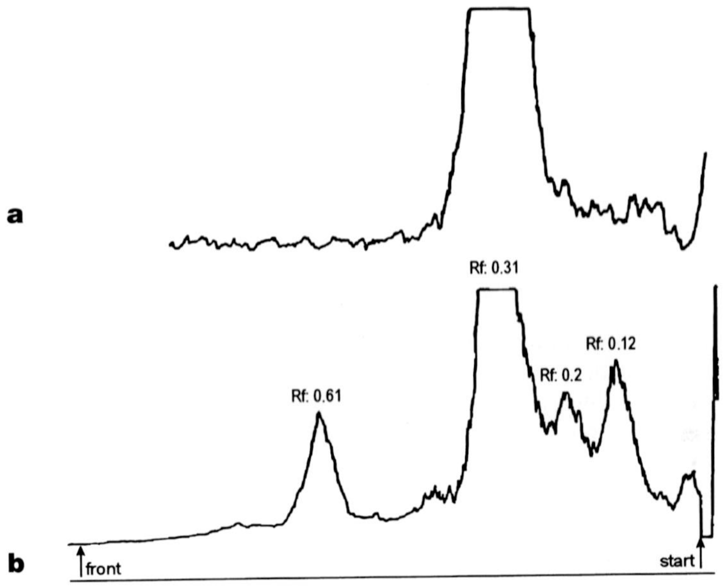

Figure 8. *Transformation of* ^{14}*C-glyoxylic acid by a preparation from the cambial zone of* Tilia cordata *Mill. collected in June (see* Figure 7*). (a) Preparation boiled prior to incubation. (b) Active preparation. Compounds with* R_f *0.31 and* R_f *0.61 are glyoxylic acid and malic acid (malate), respectively. The other compounds (*R_f *0.12 and* R_f *0.2) have not been identified.*

The nature, distribution and annual fluctuations of reserves in the wood of trees have been studied for a long time. However, little is still known about the regulation, details of initiation, genetics and the biochemical and molecular basis of the accumulation events.

4 Experimental procedures

4.1 *Wood material*

Increment borings (0.5 mm in diameter) were collected from trunks and roots of *Robinia pseudoacacia* L., *Pinus sylvestris* L., *Picea abies* Karst., *Acer pseudoplatanus* L., and *Tilia cordata* Mill. in a forest stand near Munich. The deciduous trees were approximately 40 years old, the conifers approximately 80 years old. Immediately after collection, samples were frozen in dry ice pellets and lyophilized. After removal of the bark, the wood zones of interest (see respective Figures) were scraped off and homogenized to a fine powder. Prior to extraction of carbohydrates and lipids, the wood powder was heated for 10 min at 100 °C to inactivate hydrolytic enzymes.

4.2 *Extraction and determination of soluble sugars and starch*

For the data of *Figures 1, 2, 4a, 4b*, and *7*, 20 mg of each sample were extracted with 2 mL bi-distilled water for 30 min. For the determination of fructose, glucose, sucrose and raffinose, 100 μL of the supernatant was assayed. The sugars were determined using specific enzyme assays. Starch was determined either by the iodine method (Krisman, 1962, *Figure 1*) or by enzymatic assay of glucose after hydrolysing the starch with amyloglucosidase (*Figure 4b*). Free glycerol and glyceride glycerol (liberated from triacylglycerols by 0.5 N aqueous KOH for 30 min at 70°C followed by neutralization with perchloric acid) were determined enzymatically (*Figure 4b*). The amylose/amylopectin ratios (*Figure 2*) were determined colorimetrically according to Magel (1991).

4.3 *TLC to evaluate annual fluctuation of sugars*

Twenty milligrams of wood powder (*Acer pseudoplatanus* L., *Figure 3*) were extracted with 300 μL 75% methanol for 1 h. After centrifugation, 50 μL of the clear supernatant were applied to TLC plates (E. Merck, 0.20 mm Silica gel 60 on aluminium). Development was four times in chloroform/acetic acid/water (6/7/1, v:v:v). Spots were visualized by spraying with diphenylamine/aniline/ortho-phosphoric acid/acetone (1/1/0.1/100, w/v/v/v) and heating at 130°C (De Stefanis and Ponte, 1968).

4.4 *Extraction and separation of neutral lipids from* **Pinus sylvestris** *L. (root) and* **Tilia cordata** *Mill. (trunk and root)*

Twenty milligrams of the *Pinus sylvestris* (*Figure 5*) wood powder from the respective wood zone were extracted with 1 mL chloroform/methanol (2: 1, v/v) for 30 min and washed twice with 0.5 mL chloroform/methanol (see above). The combined extracts were dried under a stream of nitrogen. The extracts were redissolved in 50 μL chloroform/methanol (see above) and applied to TLC plates (see above). Plates were developed three times in petroleum ether (boiling range 40–60°C)/diethyl ether/acetic acid (90:10:1, v/v/v). Plates were visualized by spraying with sulphuric acid/acetic acid (1:1, v/v) and heating (40 s at 130 °C).

The wood powder (350 mg) of the respective trunk and root zones of *Tilia cordata* Mill. (*Figure 6*) was extracted with acetone. The extracts were taken to dryness, redissolved in *n*-hexane and applied to TLC plates (see above). Triacylglycerols and free fatty acids were quantified by determination of carbon. Silica gel containing triacylglycerols and free fatty acids, respectively, was packed into tin caps and carbon content was determined with a modified Nitrogen Analyzer (Carlo Erba, Ort). Acetanilide served as standard.

4.5 *Lipase activity*

Enzyme activity (*Figure 7*) was determined as described by Schmidt *et al.* (1970). Olive oil (emulsified with gum arabic) served as standard. Reaction vessels contained 600 mg of the respective wood powder. Mixtures boiled prior to incubation served as blanks. Incubation was for 1 h and 40 °C. The liberated fatty acids were detected as their copper salts with diethyldithiocarbaminate.

4.6 *Malate formation*

Wood powder from differentiated zones (collected in December and April) and thawed cambial tissue (collected in June) were homogenized (Potter Elvejhem) with

K-phosphate buffer (pH 7.4) and 50 mM $MgCl_2$. Clear supernatants were used. The incubation mixture contained, in μmoles: 100 K-phosphate, 50 Mg^{2+}, 1 acetyl-SCoA, 1 [U-^{14}C] glyoxylic acid (1 μCi). Blanks contained boiled wood preparations. Incubation 1 h, 37°C. Aliquots were chromatographed on Whatman No. 1 paper in the organic phase of *tert*-amyl alcohol/formic acid/water (3:1:3, v/v/v). Radioactivity (*Figure 8*) was located by scanning the paperstrips (Berthold LB 2723). Malate was identified by co-chromatography with authentic [U-^{14}C] malate and by enzymatic test with malate dehydrogenase and NAD.

References

Aloni, R. (1991) Wood formation in deciduous hardwood trees. In: *Physiology of Trees* (ed. Raghavendra A.S.), John Wiley & Sons, New York, pp. 175–197.

Anderson, A.B., Riffer, R. and Wong, A. (1969) Monoterpenes, fatty and resin acids of *Pinus contorta* and *Pinus attenuata*. *Phytochemistry* **8**: 2401–2403.

Ashworth, E.N., Stirm, V.E. and Volence, J.J. (1993) Seasonal variations in soluble sugars and starch within woody stems of *Cornus sericea* L. *Tree Physiol.* **13**: 379–388.

De Stefanis, V.A. and Ponte, J.G., Jr. (1968) Separation of sugars by thin layer chromatography. *J. Chromatogr.* **34**: 116–120.

Dietrichs, H.H. (1964) *Chemisch-physiologische Untersuchungen über die Splint-Kern-Umwandlung der Rotbuche* (*Fagus sylvatica*). Mitt. Bundesforschungsanstalt Forst-Holzwirtsch. **Nr. 58**, pp. 149.

Fischer, A. (1891) Beiträge zur Physiologie der Holzgewächse. *Jahr. wiss. Bot.* **22**: 73–160

Fischer, C. and Höll, W. (1992) Food reserves of Scots pine (*Pinus sylvestris* L.). II. Seasonal changes and radial distribution of carbohydrate and fat reserves in pine wood. *Trees* **6**: 147–155.

Hillinger, C., Höll, W. and Ziegler, H. (1996) Lipids and lipolytic enzymes in the trunkwood of *Robinia pseudoacacia* L. during heartwood formation. I. Radial distribution of lipid classes. *Trees* **10**: 366–375.

Hillis, W.E. (1987) *Heartwood and Tree Exudates*. Springer Series in Wood Science (ed. Timell, T.E.). Springer Verlag, Berlin/Heidelberg/New York.

Höll, W. (1981) Eine dünnschichtchromatographische Darstellung des Jahresgangs löslicher Zucker im Stammholz von drei Angiospermen und einer Gymnosperme. *Holzforschung* **35**: 173–175.

Höll, W. and Poschenrieder, G. (1975) Radial distribution and partial characterization of lipids in the trunk of three hardwoods. *Holzforschung* **29**: 119–123.

Höll, W. and Priebe, S. (1985) Storage lipids in the trunkwood of *Tilia cordata* Mill. from the dormant to the growing period. *Holzforschung* **39**: 7–10.

Kim, J., Gross, K.C. and Solomos, T. (1987) Characterization and stimulation of ethylene production by galactose in tomato (*Lycopersicon esculentum* Mill.) fruit. *Plant Physiol.* **85**: 804–807.

Kozlowski, T.T. and Keller, T. (1966) Food relations of woody plants. *Bot. Rev.* **32**: 293–382.

Kramer, P.J. and Kozlowski, T.T. (1960) *Physiology of Trees.* Vol. 1. McGraw Hill, New York.

Kramer, P.J. and Kozlowski, T.T. (1979) *Physiology of Woody Plants.* Academic Press, New York, San Francisco, London.

Krisman, C.R. (1962) A method for the colorimetric estimation of glycogen with iodine. *Anal. Biochem.* **4**: 17–23.

Little, C.H.A. and Eklund, L. (1999) Ethylene in relation to compression wood formation in *Abies balsamea* shoots. *Trees* **13**: 173–177.

Lloyd, J.A. (1975) Fatty acids, resin acids and phenols in *Pinus muricata*. *Phytochemistry* **14**: 483–485.

Magel, E. (1991) Qualitative and quantitative determination of starch by a colorimetric method. *Starch/Stärke* **43**: 384–387.

Roberts, L.W., Gahan, P.B. and Aloni, R. (1988) *Vascular Differentiation and Plant Growth Regulators*. Springer Series in Wood Science (ed. Timell, T.E.). Springer Verlag, Berlin/Heidelberg/New York.

Sauter, J.J. (1988) Temperature induced changes in starch and sugars in the stem of *Populus* × *canadensis* Moench 'robusta'. *Planta* **132**: 608–612.

Sauter, J.J. and van Cleve, B. (1991) Biochemical and ultrastructural results during starch-sugar-conversion in ray parenchyma cells of *Populus* during cold adaption. *J. Plant Physiol.* **139**: 19–26.

Sauter, J.J. and Kloth, S. (1987) Changes in carbohydrates and ultrastructure in xylem ray cells of *Populus* in response to chilling. *Protoplasma* **137**: 45–55.

Sauter, J.J. and Neumann, U. (1994) The accumulation of storage materials in ray cells of poplar wood (*Populus* × *canadensis* 'robusta'): Effect of ringing and defoliation. *J. Plant Physiol.* **143**: 21–26.

Sauter, J. J. and Witt, W. (1997) Structure and function of rays: storage mobilization, transport. In: *Trees – Contributions to Modern Tree Physiology* (eds. Rennenberg, H., Eschrich W. and Ziegler), H. Backhuys Publishers, Leiden, The Netherlands, pp. 177–195.

Saranpää, P. and Höll, W. (1989) Soluble carbohydrates of *Pinus sylvestris* L. sapwood and heartwood. *Trees* **3**: 138–143.

Saranpää, P. and Nyberg, H. (1987) Lipids and sterols of *Pinus sylvestris* L. sapwood and heartwood. *Trees* **1**: 82–87.

Schmidt, F.H., Stork, H. and van Dahl, K. (1970) Lipase, photometrische Messung. In: *Methoden der enzymatischen Analyse* (ed. Bergmeyer, H.U.) Verlag Chemie, Weinheim/ Bergstr, pp. 786–791.

Sinnott, E.W. (1918) Factors determining character and distribution of food reserve in woody plants. *Bot. Gaz.* **66**: 162–175.

Terziev, N., Boutelje, J. and Larson, K. (1997) Seasonal fluctuation of low molecular weight sugars, starch and nitrogen in sapwood of *Pinus sylvestris* L. *Scandinavian J. Forest Res.* **12**: 216–224.

Ziegler, H. (1964) Storage, mobilization and distribution of reserve materials in trees. In: *The Formation of Wood in Forest Trees* (ed. Zimmermann, M.H.), Academic Press, New York, pp. 303–320.

29

Biochemistry and physiology of heartwood formation

E.A. Magel

1. Introduction

Heartwood formation, a kind of programmed cell death in the inner parts of woody axes, is the final step in the life cycle of living xylem cells. Due to its ecological and economical roles, heartwood formation is one of the more important secondary differentiation processes of trees. In the transition zone between living sapwood and dead heartwood, parenchyma cells die genetically determined and extractives are accumulated. Despite the importance of this process, only little is known, up to now, about the origin of these extractives.

During recent years we intensively studied the biochemical processes of heartwood formation. Based on very recent findings, at least two different types of heartwood formation can be distinguished. Type 1, or the *Robinia*-Type of heartwood formation is characterized by the accumulation of phenolic extractives starting in the tissues transient between sapwood and heartwood (Roux and Paulus, 1962; Magel *et al.*, 1994). In the second type, Type 2 or *Juglans*-Type of heartwood formation, phenolic precursors of the heartwood extractives gradually accumulate in the ageing sapwood tissues. In the tissues transient between sapwood and heartwood, characteristic heartwood phenolics are formed by transformation (oxidation, hydrolysis) of the phenolic precursors, and by *in situ* synthesis (Dellus *et al.*, 1997; Burtin *et al.*, 1998).

This chapter summarizes our knowledge of the biochemistry and physiology of the Type 1 or *Robinia*-Type of heartwood formation and gives proof for the *in situ* biosynthesis of the extractives. The first section addresses the distribution of phenolics across a radial stem transect and their role in 'pseudo lignification'. In the second part, the potential of the transition zone to synthesize the heartwood extractives is demonstrated by showing both the catalytic activity and enzyme-specific protein of key enzymes for the biosynthesis of heartwood extractives — phenylalanine ammonia lyase (PAL) and chalcone synthase (CHS). The third section provides information about the source of the carbon-skeletons which are sequestered as phenols within the heartwood. A close correlation between phenol-yielding secondary metabolism and

Cell & Molecular Biology of Wood Formation, edited by R. Savidge, J. Barnett and R. Napier.

carbon-providing primary metabolism was found in the transition zone. It became obvious that sucrose and sucrose cleaving enzymes (invertases and sucrose synthase, SuSy) play a key role. Our studies reveal that SuSy can be taken as a measure for sink strength in tissues with extensive synthesis of phenolic components, such as the sapwood-heartwood transition zone. During heartwood formation, at least for this enzyme, regulation at the transcription level and thus by gene expression was shown for the first time. Thus, the gene coding for SuSy was evoked as a marker gene for heartwood formation.

2. Distribution of heartwood phenolics and their relation to 'pseudo lignification'

2.1 *Heartwood formation and extractives*

In cross-sections of freshly felled trunks of many mature trees several zones can usually be observed. Neglecting the rhythidome, inner bark and cambium, a pale coloured outer zone, the sapwood, surrounds a dark-coloured inner core, the heartwood. Located between these two major zones, there is a third, the transition zone. Depending on the species, this tissue can be recognized by varying colours or by its transitory location between sapwood and heartwood.

In a living tree, sapwood, 'the living, physiological active wood of pale colour' which 'contains living cells and reserve material' (Anonymous, 1957), has important functions such as conduction of sap, provision of a strong and rigid stem, and storage of reserve substances. The heartwood, in contrast, is formed by the 'inner layers of the wood, which in the growing tree, have ceased to contain living cells, and in which the reserve materials (e.g., starch) have been removed or converted into heartwood substances' (Anonymous, 1957). Moreover, heartwood formation is accompanied by pit aspiration in conifers, and by the formation of tyloses in many hardwood species. The colour of heartwood covers almost the entire spectrum, such as the jet-black of ebony and violet-brown or blue of *Mansonia* species. Heartwood formation results from physiological death due to internal factors or normal phenomena associated with the annual cycle of the life of the tree. This type of regularly formed heartwood will be addressed in this chapter. Similarly altered wood, but with different chemical and biological properties which results from external factors (e.g. pathological death or injury) will not be considered.

An excellent review on what is known about heartwood and similarly altered wood was given by Hillis (1987). This book includes a detailed description of the chemical and biological features of heartwood, and summarizes what was known about the function, formation and control of heartwood and heartwood extractives up to the late 1980s. In the following I will concentrate on new information obtained during the last years with respect to the accumulation of heartwood extractives.

The most prominent feature of heartwood is the presence of extractives which in regularly formed heartwood infiltrate the cell wall, reduce its shrinkage and swelling capacity, and account for its durability. These substances render heartwood valuable for different purposes, such as building material for construction, furniture, or boats. Despite the importance of the extractives, only little has been known about their origin. In the late 1950s, import of these substances from sites of their biosynthesis, leaves and cambium, was a favoured idea (Hillis, 1958; Hergert and Goldschmid,

1958). Some years later, *in situ* synthesis at the sapwood–heartwood boundary was postulated (Hillis *et al.*, 1962; Higuchi and Fukazawa, 1966; Ziegler, 1968), but the direct proof, the detection of enzyme activities involved, failed.

For wood, the successful application of biochemical and molecular methods is hampered by the very low percentage of living cells together with the rigidity and high content of phenolics of this material. Overcoming these difficulties, we succeeded in monitoring enzyme activities linked inevitably with the biosynthesis of phenolic components (Magel *et al.*, 1991, 1994) in the sapwood–heartwood transition zone of a deciduous tree, *Robinia pseudoacacia* L.

2.2 *Heartwood phenolics and 'pseudo lignification'*

Heartwood phenolics. The high durability of the yellow-brown heartwood of black locust (*Robinia pseudoacacia* L.) is accounted for by the presence of flavonoids in free and not in conjugated forms as, for example, glycosides (Roux and Paulus, 1962). Two main compounds are dominant: the flavanonol (i.e. dihydroflavonol) dihydrorobinetin (3,7,3′,4′,5′ pentahydroxy-dihydroflavonol) and the corresponding flavonol robinetin (3,7,3′,4′,5′ pentahydroxyflavonol) (Roux and Paulus, 1962). HPLC analyses of heartwood extracts carried out with 80% acetone revealed a third prominent component which was tentatively identified as a hydroxycinnamic acid derivative (Magel *et al.*, 1994). The radial distribution of these compounds showed that they were located almost exclusively in the inner parts of the trunk (*Figure 1a* and *b*). In the sapwood, dihydrorobinetin was the only detectable phenolic component. It may act here as repellent for insects and fungal attack. Even very low amounts of this compound are sufficient for active protection. The heartwood extractives, dihydrorobinetin and robinetin, accumulated in high amounts in the sapwood–heartwood transition zone. As robinetin was present in significant amounts in the tissue which showed discolouration, one can conclude that the precursor dihydrorobinetin was converted into robinetin right in the inner part of the transition zone. This might have been due to enzyme catalysis or to spontaneous oxidation. Details of this flavanonol-flavonol oxidation are still unknown (Hahlbrock, 1981). Highest levels of the hydroxycinnamic acid derivative in the tissues of the transition zone and the outermost heartwood, with decreasing contents within the heartwood towards the pith, might indicate this component to be an intermediate of robinetin synthesis.

'Pseudo lignification.' In relation to heartwood formation, the chemical reactivity of heartwood extractives and preexisting cell wall components is poorly understood. The term 'secondary lignification' was coined in 1977 by Hergert in consequence of a comprehensive comparison of total lignin content in heartwood versus sapwood of conifer trees. Indeed, total Klason lignin contents were reported to be higher in heartwood than in sapwood tissues without, however, addressing the biological basis of this phenomenon. Only very few investigations have been published relating to changes in the lignification patterns and differences between sapwood and corresponding heartwood quality, in spite of their possible technical significance (Monties, 1991).

In the heartwood of black locust, the Klason lignin content is significantly increased compared to sapwood and transition zone (*Table 1*). However, both the lignin

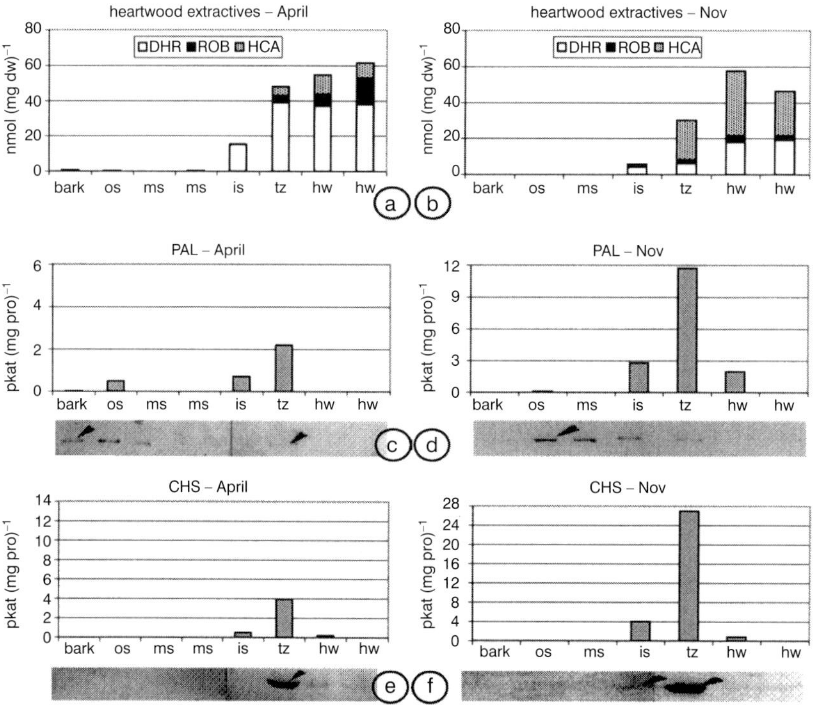

Figure 1. *Heartwood formation in* Robinia pseudoacacia *L. Amount of heartwood extractives (a and b), catalytic activities and enzyme specific protein of phenylalanine ammonia-lyase (PAL; c and d) and chalcone synthase (CHS; e and f) in stem tissues of trunks collected in April and November, respectively. (os=outer sapwood, ms=middle sapwood, is=inner sapwood, tz=sapwood-heartwood transition zone, hw=heartwood).*

monomeric composition, expressed as the syringyl (S)/guaiacyl (G) ratio, and depolymerization yields in non-condensed monomeric units, expressed by the total S+G yield, were not significantly changed in total *in situ* lignin from extractive-free samples of sapwood, heartwood and transition zone (Magel *et al.*, 1995).

Based on results of preliminary experiments, it can be speculated that during heartwood formation the accumulation of phenolic compounds might involve enzymatically initiated but chemically driven copolymerization between phenolic derivatives and the preexisting cell-wall macromolecular components, including lignins and lignin–polysaccharide complexes, as suggested by Monties (1987, 1989). Such a mechanism would explain the occurrence of a relative increase in lignin content of heartwood dependent on the biological conditions. Similar results as given here for *Robinia* were also reported for *Quercus* (Jouin *et al.*, 1988) and *Quebracho* (Fengel, 1991). Corroborating the concept of 'secondary lignification' (Hergert, 1977), they failed to indicate any quantitative and qualitative changes in the lignification pattern of the investigated woods. These data indicate the occurrence of associations between phenolic extractives, cell-wall and the lignin network in it. Thus, at least in angiosperm trees, *de novo* synthesis of lignin during heartwood

Table 1. Total lignin content (% by weight of extractive-free, lyophilized xylem) and monomeric composition (in nmoles [mg extractive-free xylem powder]$^{-1}$, and as the ratio of syringyl/guaiacyl units) of in situ *lignin from the sapwood, transition zone, and heartwood of a 12-year-old* R. pseudoacacia *L. tree (mean values n=3; ± SD)*

	Sapwood	Transition zone	Heartwood
Lignin content (%)	15.6 ± 0.5	16.8 ± 0.2	19.2 ± 0.2
Syringyl (nmoles [mg dw]$^{-1}$)	278 ± 4	317 ± 23	311 ± 17
Guaiacyl (nmoles [mg dw]$^{-1}$)	110 ± 3	128 ± 8	133 ± 10
Syringyl/guaiacyl ratio	2.5	2.4	2.4

formation obviously can to be excluded. The term 'secondary lignification' should therefore be changed into '*pseudo lignification*'.

3. *In situ* synthesis of heartwood extractives at the sapwood–heartwood transition zone

The biosynthesis of flavonoids, prominent heartwood extractives of *Robinia* and other tree species, branches from the core reactions of general phenylpropanoid metabolism. Key enzymes are phenylalanine ammonia lyase (PAL) and chalcone synthase (CHS). PAL (EC 4.3.1.5) catalyses the first irreversible step in the general phenylpropanoid metabolism, the deamination of L-phenylalanine to *t*-cinnamic acid (Hanson and Havir, 1981). The first committed step in the biosynthesis of flavonoids is the condensation of 4-coumaroyl-CoA and malonyl-CoA, catalysed by the key enzyme of the various flavonoid pathways CHS (EC 2.3.1.74; Hahlbrock and Scheel, 1989).

Some years ago we succeeded in monitoring the catalytic activities of both key enzymes in flavonoid biosynthesis, PAL and CHS, in the trunkwood of *Robinia* (Magel *et al.*, 1991). Catalytic activity of PAL was present at two sites of the trunk: the youngest part of the sapwood and the sapwood–heartwood transition zone (*Figure 1c* and *d*). Experiments performed with wood material collected over a season, supported the hypothesis that PAL in the trunk of *Robinia* was involved in two events, differing qualitatively from each other. In *Figure 1* (*c* and *d*) this is shown for a spring and an autumn harvested sample. During early spring through late summer, highest PAL activities in the outermost sapwood provide lignin precursors for vigorous lignin formation in newly formed wood. At the second site of PAL activity, the sapwood–heartwood boundary, no lignin formation occurred (see above). Here, PAL was related to the formation of heartwood extractives. In contrast to PAL, CHS, the second key enzyme of flavonoid biosynthesis, was active exclusively in the transition zone (*Figures 1e* and *f*). Thus, as in herbaceous plants, CHS activity peaks in time and space with the accumulation of high amounts of flavonoids. Highest catalytic activities of both enzymes were measured in trunks felled in autumn, just at the time when heartwood extractives were formed (*Figures 1a* and *b*; and references in Hillis, 1987). Heartwood formation is regarded to be a senescence process, which as in herbaceous plants seems to be characterized by ethylene production (Shain and Hillis, 1973). Although the precise site of ethylene emanation is unknown, there are indications that ethylene production, PAL (and perhaps CHS) activities and polyphenol formation are interlinked during heartwood formation.

In standing trees, the amounts of protein decrease with increasing age of the trunkwood tissue (Magel *et al.*, 1991). Heartwood itself contains no protein. Therefore, one should assume a dramatic increase in the specific activity for enzymes which are active in the transition zone, or a disturbance of the equilibrium in different enzyme systems, but not a *de novo* formation of enzymes (Hillis and Inoue, 1966). However, micro-gel electrophoretic separation of total protein extracts revealed pronounced alterations in the protein banding pattern of ageing and dying wood tissues, including newly emerging polypeptides within the sapwood–heartwood transition zone. Our results provide evidence that the two enzymes inevitably linked with the accumulation of heartwood extractives, differ in their mode of regulation. Western blots showed that, in wood, specific amounts of PAL protein are very similar, independent of tissue-specific activity or season (*Figures 1c* and *d*; r^2(coefficient of correlation between amount of enzyme protein and catalytic activity) <0.142; Magel and Hübner, 1997). Posttranslational modification of the enzyme protein and changes in tissue pool sizes of competent inhibitors of PAL activity, such as *t*-cinnamic acid, are possible means for modulation of catalytic activity of PAL during heartwood formation. Thus, the regulation of PAL with regard to flavonoid accumulation during genetically determined heartwood formation seems to be different from its regulation upon pathogen attack, which is characterized so far for herbaceous plants by a transient increase in gene expression and translation (Habereder *et al.*, 1989). In contrast, enzyme protein levels and catalytic activity of CHS are restricted to main sites and times of heartwood formation, the sapwood–heartwood transition zone during fall and the beginning of dormancy (*Figures 1e* and *f*; r^2 >0.981; Magel and Hübner, 1997). CHS activity is thus regulated by the amount of enzyme protein present in the dying wood tissues, thus by *de novo* protein synthesis.

Our results indicate that contrary to literature data, phenylalanine and its ammonia lyase are obligatory parts of the biosynthetic pathway of phenolics during heartwood formation, and it seems unlikely that there are other pathways than the accepted ones which lead to the synthesis of phenolics.

4. Source of carbon-skeletons for the biosynthesis of heartwood extractives

4.1 *Role of reserve substances (e.g. lipids, starch and other nonstructural carbohydrates)*

Light or electron microscopic studies revealed, that during sapwood heartwood transformation storage material, mainly starch, is consumed or withdrawn (e.g. Kumar and Datta, 1989; Nair, 1988; Nobuchi *et al.*, 1987a, b). The composition of the heartwood substances formed is under genetic control and the amount accumulated is significantly influenced by the physiological conditions impinging on the parenchyma at the time of formation. Therefore, it is attractive to speculate that the building blocks of these secondary substances derive from storage material and/or photoassimilates. At least in *Robinia* wood, lipid reserves constitute less than 0.1% of the total storage pool. Qualitatively, however, there are significant changes in the lipid fractions with respect to wood ageing and dying, revealing the importance of lipids in the process of heartwood formation (for detailed information refer to Hillinger *et al.*, 1996a,b). Carbohydrates are, by far, the dominant reserve substances (close to 100% of the total storage pool). Therefore, they are exclusively addressed in the discussion of the role of reserve substances for the biosynthesis of heartwood constituents in *Robinia pseudoacacia* L.

Investigations on the nature and radial distribution of reserve substances in woody axes showed that, in general, the youngest sapwood contains greatest pools of soluble sugars (sucrose, glucose and fructose are dominant) and starch, while the heartwood is almost free of storage material (*Figures 2a* and *b*). This is valid for both softwood and hardwood species (Dietrichs, 1964; Höll, 1972; Magel et al., 1997). Small amounts of sugars like mannose and arabinose in the heartwood of Scots pine might originate from the breakdown of hemicelluloses during heartwood formation (Saranpää and Höll, 1989).

The distribution of storage carbohydrates and flavonoids across trunk cross-sections is complementary with sites of flavonoid accumulation containing only trace amounts of carbohydrates (*Figures 1* and 2). A tremendous imbalance between carbon-skeletons available in the transition zone and carbon-skeletons fixed as phenolics within the heartwood was manifested (*Figures 2a* and *b*). A first indication that this gap might be filled by the steady import of carbohydrates from the outer sapwood (and

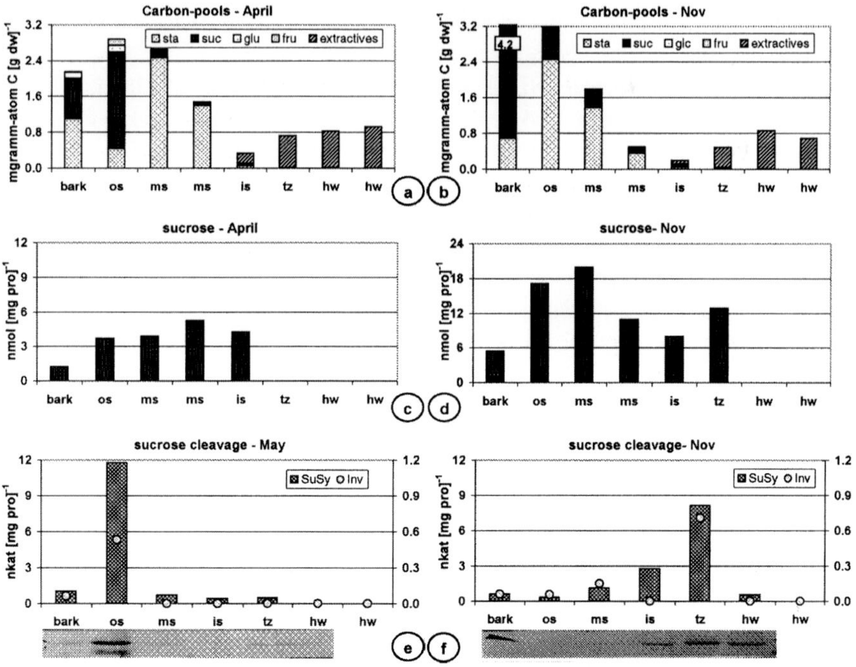

Figure 2. *Heartwood formation in* Robinia pseudoacacia *L. Carbon-pools fixed as nonstructural carbohydrates and extractives (a and b), amount of sucrose (expressed as nmol sucrose [mg protein]$^{-1}$; c and d), catalytic activities of sucrose synthase (SuSy) and neutral invertase (Inv) and enzyme specific protein of SuSy (e and f) in stem tissues of trunks collected in April or May, and November, respectively. (Abbreviations, see Figure1, and sta=starch, suc=sucrose, glc=glucose, fru=fructose.)*

phloem) to the sapwood–heartwood transition zone was found in the literature. After application of ^{14}C-glucose to the phloem of *Eucalyptus sieberi* labelled phenolic compounds were detected at the heartwood boundary (Hillis and Hasegawa, 1963). Further proof was given by the result that, in *Robinia*, sucrose was detectable in the transition zone predominantly during times of heartwood formation (*Figures 2c* and *d*; and Magel *et al.*, 1995). Since the amounts of sucrose and thus reserve substances depend on climatic factors, it was not surprising that the deposition of heartwood compounds varied from one year to the other. After the cessation of heartwood formation no further extractives could be produced (Hillis *et al.*, 1962). This explained why the amounts of phenolics were not evenly distributed within the heartwood.

4.2 *Enzymes involved in the breakdown of starch and sucrose*

Starch cleaving enzymes. In the trunkwood of most tree species, starch constitutes the main storage product. In *Robinia*, the decline of starch is due to both reduced capacity for starch synthesis and activated starch breakdown. Activities of ADP-glucose pyrophosphorylase (ADPGPPase), the key enzyme of starch synthesis, vanished within the sapwood, and were absent in the transition zone (Höll, 1972). Whether the decline of ADPGPPase has to be correlated with the accumulation of an (unidentified) inhibitor of enzyme activity, or to the loss of potassium, a cofactor of enzyme activity, has to be elucidated. The loss in starch synthesizing capacity, however, was coupled with intensified starch degradation (Höll, 1972; Magel *et al.*, 1997). Our results indicated that both phosphorolytic and amylolytic activities occur (Magel *et al.*, 1997). It seems likely that the phosphorolytic breakdown dominates in the younger wood tissues, while amylolytic activities seem to be more abundant in older tissues. From investigations on herbaceous plants it is known that endoamylases are induced or stimulated by factors such as water stress, sucrose starvation, calcium accumulation, or ethylene production. Interestingly, all of these events were observed during heartwood formation in woody axes (Hillis, 1987). Exoamylases, yielding maltose as the only reaction product, do not seem to play any role during heartwood formation. The role of a debranching enzyme as well as the appearance of highly unbranched starch in older sapwood tissues and the transition zone of *Robinia* is still unknown.

The active hydrolysis of starch, concomitant with reduced amounts of both starch and soluble sugars in the sapwood–heartwood transition zone, indicate that these products function as precursors for the formation of extractives which are deposited during the process of heartwood formation (Magel *et al.*, 1991, 1994). Only a minor part of the vigorous synthesis of heartwood extractives can be driven by this *in situ* carbon-pool. The synthesis of the bulk of the phenolics, however, was dependent on imported carbon. As sucrose is the major transport form of carbon, we proposed that sucrose (see above) and sucrose metabolism might play a crucial role not only during cambial differentiation (Sung *et al.*, 1993; Higuchi, 1997), but also during processes of secondary differentiation of wood tissues, such as heartwood formation.

Sucrose metabolizing enzymes. In woody axes of *Robinia* and other tree genera such as *Populus*, *Juglans*, *Pinus*, and *Picea*, sucrose turnover is accomplished by the activity of the sucrose-metabolizing enzymes, sucrose-phosphate synthase, sucrose synthase, and invertases. In *Robinia* trunkwood, as in other well characterized plant tissues, sucrose is predominantly synthesized by sucrose-phosphate synthase (SPS, Ec 2.4.1.14). However, sucrose is cleaved mainly by sucrose synthase (SuSy; Ec 2.4.1.13) with limited contribution of a neutral invertase (NI; Ec 3.2.1.26). In radial direction, the distribution of sucrose-synthesizing activities is opposite to that of sucrose-cleaving activities: SPS is preferably located in the storage part of the sapwood, the mature middle and inner wood, whereas the sucrose-cleaving activities dominate in the outermost and the innermost living wood (Hauch and Magel, 1998). The activities and amounts of enzyme protein of the sucrose metabolizing enzymes changed with the season. These changes were related to physiological events such as cambial growth, starch accumulation and heartwood formation. At the sapwood–heartwood transition zone, the capacity to cleave sucrose was prominent. Most obvious for autumn samples, the activities of SuSy and NI increased dramatically in this tissue (*Figures 2e* and *f*). Probing with heterologous antibodies revealed that the catalytic activity of SuSy was linearly correlated with the amount of SuSy specific protein both across a radial stem-transect and in seasonal course ($r^2 > 0.93$; *Figures 2e, f* and *3*). Semi-quantitative RT-PCR showed that in both, the cambial growth zone and the sapwood–heartwood transition zone, SuSy transcripts are correlated with the amounts of enzyme specific protein and its catalytic activity, both in time and type of tissue (*Figure 3*).

The enhanced enzymatic degradation of sucrose coincided in time and location with increased activities of key enzymes of the pathways involved in the formation of phenolic compounds, PAL and CHS and the accumulation of phenolic heartwood extractives (see above and *Figure 3*). We take this as evidence that in these samples sucrose and secondary metabolism are linked to each other. Our data imply that during heartwood formation a concerted action of sucrose cleavage and metabolism occurs via the sucrose synthase pathway (Xu *et al.*, 1989) and NI is needed to meet the high demand for energy and substrates for the *in situ* biosynthesis of heartwood extractives. Thus we propose that the sucrose synthase pathway is not only characteristic for metabolic sinks associated with polysaccharide biosynthesis (e.g., Ap Rees, 1974) but is also connected to the biosynthesis of phenolic heartwood extractives during heartwood formation (and lignin during cambial xylem differentiation). At least for heartwood formation in the *Robinia*-Type, the gene coding for sucrose synthase has to be considered as a marker gene of heartwood formation. As autumn samples exhibit highest rates of SPS in the innermost, and elevated SPS/SuSy ratios (data not shown) in the outer sapwood, we suggest that these tissues are the locations where sucrose is formed, which is then needed in the transition zone.

Taken together these data indicate a close involvement of sucrose metabolism during heartwood formation. Moreover, we demonstrated that SuSy can be taken as a measure for sink strength in tissues with extensive synthesis of phenolic components. At least for this enzyme, regulation is at the transcript level.

5. Conclusion

Summarizing our data and those given in the literature, heartwood formation of Type 1 or the *Robinia*-Type can be described in biochemical terms as follows (see *Figure 4*). Heartwood formation in the sapwood–heartwood transition zone is characterized by transient activation and shift of metabolic activities. It seems likely that cell death starts with the depletion of storage substances due to an increase of the activity of

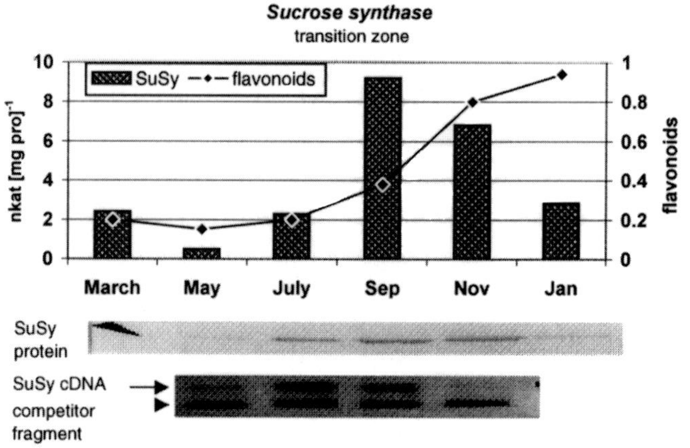

Figure 3. *Catalytic activity (closed bars), content of enzyme specific protein and specific cDNA of sucrose synthase and the accumulation of flavonoids (given in relative units) within the sapwood–heartwood transition zone in stems of* Robinia pseudoacacia *L. from March until January of the following year.*

hydrolysing enzymes. Additionally, sucrose is imported by this sink tissue and is primarily cleaved by sucrose synthase. Due to the correlation of enhanced catalytic activities, enzyme protein levels and SuSy-specific mRNA pools, regulation of this enzyme is by gene expression. In the beginning, products of sucrose breakdown i.e., fructose and UDP-glucose, are predominantly used for glycolytic and respiratory energy production (Höll and Lendzian, 1973; Magel and Höll, 1993), and to a minor extent, for the synthesis of extractives. Flavonoids, probably accumulating within the cytosol, might be strong inhibitors of mitochondrial electron flow. Thus mitochondria are regarded to be the first organelles ceasing activity (Ziegler, 1968). Carbon-skeletons accumulating due to non-functional mitochondria together with products of the accelerated oxidative pentose-phosphate pathway might be increasingly used for the synthesis of phenolics via general phenyl propanoid metabolism (Hillis, 1987, and references therein; Magel *et al.*, 1992, 1993). Concomitantly, a transient expression and activation of key enzymes of the biosynthesis of secondary substances, PAL and CHS, is correlated with the accumulation of heartwood extractives. From these results we conclude that the type of programmed cell death that occurs during heartwood formation shares some features with, for example, the type of programmed cell death that mature leaves undergo during senescence. Both processes require activation of hydrolytic enzymes, gene expression and *de novo* protein synthesis.

Only few data are available for the second type of heartwood formation, the *Juglans*-Type. At least from the scarce literature data it seems likely that this type of heartwood occurs in genera such as *Juglans* (Burtin *et al.*, 1998), *Pinus*, *Pseudotsuga* (Dellus *et al.*, 1997) and *Platycarya* (Tanaka *et al.*, 1998). Whether parts of the above given scheme (*Figure 4*) can also be applied to this type of heartwood formation will be shown in the near future.

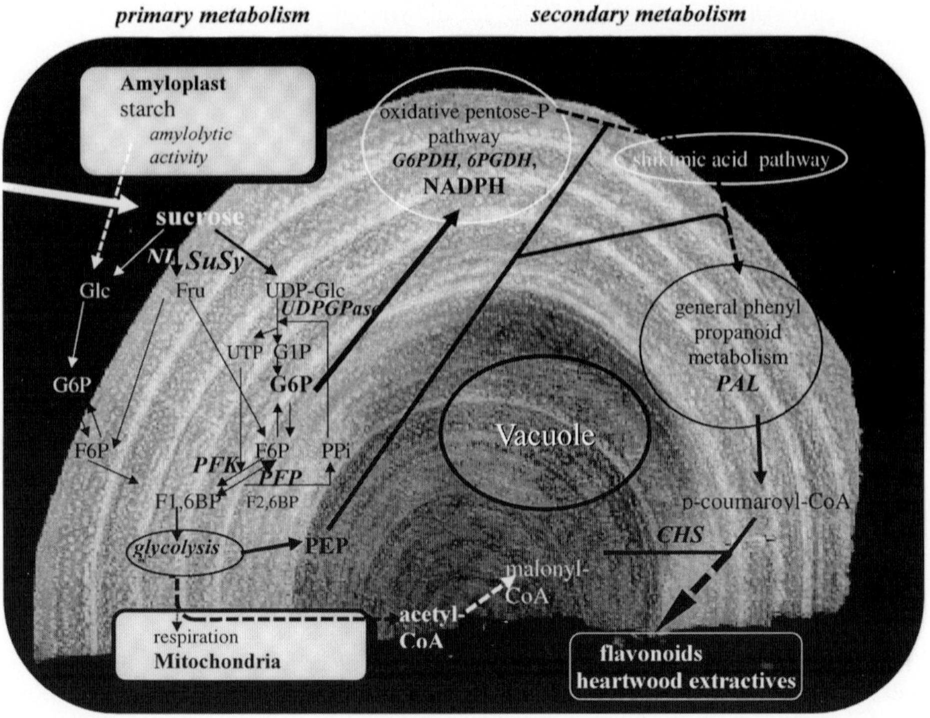

Figure 4. Biochemical processes leading to the formation of heartwood in Robinia pseudoacacia *L. are summarized (modified from Ziegler, 1968). The scheme represents metabolic processes in dying ray and axial parenchyma cells of the sapwood–heartwood transition zone. NI = neutral invertase, PFK = ATP-dependent, PFP = PPi-dependent phosphofructokinase, SuSy = sucrose synthase, UDPGPase = UDP-glucose pyrophosphorylase, G6PDH = glucose-6-phosphate dehydrogenase, 6PGDH = 6 phosphogluconate dehydrogenase, PAL = phenylalanine ammonia-lyase, CHS = chalcone synthase.*

Acknowledgements

I thank S. Hauch, B. Hübner, and C. Klier for contributing to this work, and Prof. Dr R. Hampp for critical reading of the manuscript. Part of this work was done in collaboration with Prof. Dr B. Monties, INRA-Grignon, Dr C. Jay-Allemand, INRA-Orleans, and Prof. Dr A. Drouet, University of Orleans, France. This work was supported by grants from the Deutsche Forschungsgemeinschaft, EUROSILVA and the EC. This chapter is dedicated to Prof. Dr Drs h.c. Hubert Ziegler on the occasion of his 75th birthday.

References

Anonymous (1957) International glossary of terms used in wood anatomy: prepared by the Int. Assoc. of Wood Anatomists. *Trop. Woods* **107**: 1–36.

Ap Rees, T. (1974) Pathways of carbohydrate breakdown in higher plants. In: *International Review of Biochemistry* (ed. Northcote, D.H.), Buttersworth, London, pp. 90–127.

Burtin, P., Jay-Allemand, C., Charpentier, J.P. and Janin, G. (1998) Natural wood colouring process in *Juglans* sp. (*J. nigra, J. regia* and hybrid *J. nigra 23* × *J. regia*) depends on native phenolic compounds accumulated in the transition zone between sapwood and heartwood. *Trees* **12**: 258–264.

Dellus, V., Mila, I., Scalbert, A., Menard, C., Michon, V. and Herve du Penhoat, C. (1997) Douglas-fir polyphenols and heartwood formation. *Phytochem.* **45**: 1573–1578.

Dietrichs, H.H. (1964) Das Verhalten von Kohlenhydraten bei der Holzverkernung. *Holzforschung* **18:** 14–24.

Fengel, D. (1991) Chemical studies on the wood of *Quebracho colorado* (part 1). *Holzforschung* **45**: 259–263.

Habereder, H., Schröder, G. and Ebel, J. (1989) Rapid induction of phenylalanine ammonia-lyase and chalcone synthase mRNAs during fungus infection of soybean (*Glycine max* L.) roots or elicitor treatment of soybean cell cultures at the onset of phytoalexin synthesis. *Planta* **177**: 58–65.

Hahlbrock, K. (1981) Flavonoids. In: *The Biochemistry of Plants: A Comprehensive Treatise, Vol. 7, Secondary Plant Products* (eds. Stumpf, P.K. and Conn, E.E.), Academic Press, New York, London, Toronto, pp. 425–456.

Hahlbrock, K. and Scheel, D. (1989) Physiology and molecular biology of phenylpropanoid metabolism. *Ann. Rev. Plant Physiol. Mol. Biol.* **40**: 347–369.

Hanson, K.R. and Havir, E.A. (1981) Phenylalanine ammonia-lyase. In: *The Biochemistry of Plants: A Comprehensive Treatise, Vol. 7, Secondary Plant Products* (eds. Stumpf, P.K. and Conn, E.E.), Academic Press, New York, London, Toronto, pp. 577–625.

Hauch, S. and Magel, E.A. (1998) Extractable activities and protein content of sucrose phosphate synthase, sucrose synthase and neutral invertase in trunk tissues of *Robinia pseudoacacia* L. are related to cambial wood production and heartwood formation. *Planta* **207:** 266–274.

Hergert, H.L. (1977) Secondary lignification in conifer trees. *Amer. Soc. Symp. Series*, Vol. **48**, 'Cellulose Chemistry and Technology': 227–243.

Hergert, H.L. and Goldschmid, O. (1958) Biogenesis of heartwood and bark constituents. I. Taxifolin glucoside. *J. Org. Chem.* **23**: 700–704.

Higuchi, T. (1997) *Biochemistry and Molecular Biology of Wood.* Springer Series Wood Science, Springer, Berlin.

Higuchi, T. and Fukazawa, K. (1966) Study on the mechanism of heartwood formation. III. On the role of phenylalanine deaminase. *Mokuzai Gakkaishi* **12**: 135–139.

Hillinger, C., Höll, W. and Ziegler, H. (1996a) Lipids and lipolytic enzymes in the trunkwood of *Robinia pseudoacacia* L. during heartwood formation. I. Radial distribution of lipid classes. *Trees* **10**: 366–375.

Hillinger, C., Höll, W. and Ziegler, H. (1996b) Lipids and lipolytic enzymes in the trunkwood of *Robinia pseudoacacia* L. during heartwood formation. I. Radial distribution of lipases and phospholipases. *Trees* **10**: 376–381.

Hillis, W.E. (1958) Formation of condensed tannins in plants. *Nature* **182:** 1371.

Hillis, W.E. (1987) *Heartwood and Tree Exudates*. Springer, Berlin, München.

Hillis, W.E. and Hasegawa, M. (1963) The formation of polyphenols in trees I. Administration of ^{14}C-glucose and subsequent distribution of radioactivity. *Phytochemistry* **2:** 195–199.

Hillis, W.E. and Inoue, T. (1966) The formation of polyphenols in trees — III. The effect of enzyme inhibitors. *Phytochemistry* **5**: 483–490.

Hillis, W.E., Humphreys, F.R., Bamber, R.K. and Carle, A. (1962) Factors influencing the formation of phloem and heartwood polyphenols. Part II. The availability of stored and translocated carbohydrates. *Holzforschung* **16:** 114–121.

Höll, W. (1972) Stärke und Stärkeenzyme im Holz von *Robinia pseudoacacia* L. *Holzforschung* **26:** 41–45.

Höll, W. and Lendzian, K. (1973) Respiration in the sapwood and heartwood of *Robinia pseudoacacia* L. *Phytochemistry* **12:** 975–977.

Jouin, D., Tollier, M.-T. and Monties, B. (1988) Lignification of oak wood. I. Lignin determinations in sapwood and heartwood. *Cellulose Chem. Technol.* **22:** 231–243.

Kumar, A. and Datta, S.K. (1989) Histochemical and histoenzymological changes during heartwood formation in a timber tree *Shorea robusta*. *Proc. Indian Sci. (Plant Sci)* **99:** 21–27.

Magel, E.A. and Höll, W. (1993) Storage carbohydrates and adenine nucleotides in trunks of *Fagus sylvatica* in relation to discolored wood. *Holzforschung* **47:** 19–24.

Magel, E. and Hübner, B. (1997) Distribution of phenylalanine ammonia-lyase (PAL) and chalcone synthase (CHS) within trunks of *Robinia pseudoacacia* L. *Botanica Acta* **110:** 314–322.

Magel, E.A., Claudot, A.C., Drouet, A. and Ziegler, H. (1991) Formation of heartwood substances in the stem of *Robinia pseudoacacia* L. I. Distribution of phenylalanine-ammonialyase and chalcone synthase across the trunk. *Trees* **5:** 203–207.

Magel, E.A., Höll, W. and Ziegler, H. (1992) Radial distribution of pyridine nucleotides in the trunk of hardwoods and its relation to heartwood formation. *Plant Physiol. Suppl.* **148:** No. 891.

Magel, E.A., Hillinger, C. and Höll, W. (1993) Activities of key enzymes of the oxidative pentose-phosphate pathway in stems of *Robinia pseudoacacia* L. in relation to heartwood formation. *Plant Physiol. Suppl.* **179:** No. 1033.

Magel, E.A., Jay-Allemand, C. and Ziegler, H. (1994) Formation of heartwood substances in the stem of *Robinia pseudoacacia* L. II. Distribution of nonstructural carbohydrates and wood extractives across the trunk. *Trees* **8:** 165–171.

Magel, E.A., Monties, B., Drouet, A., Jay-Allemand, C. and Ziegler, H. (1995) Heartwood formation: biosynthesis of heartwood extractives and 'secondary' lignification. In: *EURO-SILVA — Contribution to Forest Tree Physiology* (eds. Sandermann, H., Bonnet-Masimbert, M.), INRA edition, Paris, pp. 35–56.

Magel, E.A., Hillinger, C., Höll, W. and Ziegler, H. (1997) Biochemistry and physiology of heartwood formation: Role of reserve substances. In: *Trees — Contribution to Modern Tree Physiology* (eds. Rennenberg, H., Eschrich, W., Ziegler, H.), SFB Academic Publisher, The Hague, pp. 477–506.

Monties, B. (1987) Composition chimique des bois de chenes: Relations avec la qualite des eaux de vie. *Connaissance Vigne Vin* **21:** 169–190.

Monties, B. (1989) Lignins. In: *Methods in Plant Biochemistry. Vol.1, Plant Phenolics* (ed. Harborne, J.B), Academic Press, pp. 119–155.

Monties, B. (1991) Plant cell wall as lignocellulosic composite: relations with lignin structures and functions. *Anim. Feed Sci. Technol.* **32:** 159–175.

Nair, M.N.B. (1988) Wood anatomy and heartwood formation in Neem (*Azadirachta indica* A. Juss.). *Bot. J. Lin. Soc.* **97:** 79–90.

Nobuchi, T., Takai, K. and Harada, H. (1987a) Distribution of heartwood phenols in the trunk of Sugi *(Cryptomeria japonica* D. Don) and partial characterisation of heartwood formation. *Mokuzai Gakkaishi* **33:** 88–96.

Nobuchi, T., Tokuchi, N. and Harada, H. (1987b) Variability of heartwood formation and cytological features in broad-leaved trees. *Mokuzai Gakkaishi* **33:** 596–604.

Roux, D.G. and Paulus, E. (1962) Condensed tannins 13. Interrelationships of flavonoid components from the heartwood of *Robinia pseudoacacia*. *Biochem. J.* **82:** 324–330.

Saranpää, P. and Höll, W. (1989) Soluble carbohydrates of *Pinus sylvestris* L. sapwood and heartwood. *Trees* **3:** 138–143.

Shain, L. and Hillis, W.E. (1973) Ethylene production in xylem of *Pinus radiata* in relation to heartwood formation. *Can. J. Bot.* **51:** 1331–1335.

Sung, S.S., Kormanik, P.P. and Black, C.C. (1993) Vascular cambial sucrose metabolism and growth in loblolly pine (*Pinus taeda* L.) in relation to transplanting stress. *Tree Physiol.* **12:** 243–258

Tanaka, T., Jiang, Z.H. and Kouno, I (1998) Distribution of ellagic acid derivatives and a diarylheptanoid in wood of *Platycarya strobilacea. Phytochem.* **47:** 851–854.

Xu, D.P., Sung, S.J., Loboda, T., Kormanik, P.P. and Black, C.C. (1989) Characterisation of sucrolysis via the uridine diphosphate and pyrophosphate-dependent sucrose synthase pathway. *Plant Physiol.* **90:** 635–642.

Ziegler, H. (1968) Biologische Aspekte der Kernholzbildung. *Holz als Roh- und Werkstoff* **26:** 61–68.

Valuable wood as a result of abnormal cambial activity — the case of *Betula pendula* var. *carelica*

P. Velling, A. Viherä-Aarnio, R. Hagqvist and J. Lehto

'The Devil once asked a peasant to give him some wood. The man agreed, but added that the tree must be neither straight nor crooked, but something in between. All day long the Devil searched for such a tree in the forest, but without success. By the evening he was so angry that he grabbed a birch by its trunk, squeezing the surface so hard that it became gnarled and the inside curly. Thus the Devil got his tree, neither straight nor crooked'. According to a Finnish folk story this is how the curly birch was born (Pätiälä, 1980).

1. Introduction

Having the name 'the Devil's' tree, the external appearance of curly birch (*Figures 1 and 2*), and that of its wood (*Figure 3*), is often rather ornamental. This 'wooden marble' has for long attracted forest scientists, as well as practical foresters and artisans who work with wood. Curly birch, like birch trees in general, is popular in the Finnish cultural landscape, parks and gardens. From about 1910 curly birch wood became popular and sought-after in northern and central Europe (Vesterinen, 1957). It was soon in such demand that thefts in the forest became common, and the punishments correspondingly harsh: one stolen log could result in a prison sentence of several months. The price of 1 kg of wood was close to the price of 1 kg of sugar. Curly birch wood, with a high green density (around 950 kg m^{-3}, according to Salmi, 1987) and sold by weight, is still today the most highly-priced native forest tree species in Finland.

The formation of curly-grained wood is most common in silver birch (*Betula pendula* Roth), but it has also been reported in downy birch (*Betula pubescens* Ehrh.),

Cell & Molecular Biology of Wood Formation, edited by R. Savidge, J. Barnett and R. Napier.

Figure 1. *Mature trees of curly birch.*

black and grey alder (*Alnus glutinosa* (L.) Gaertner and *Alnus incana* (L.) Moench), rowan (*Sorbus aucuparia* L.), oak (*Quercus*) and goat-willow (*Salix caprea* L.) (e.g., Heikinheimo, 1951; Saarnijoki, 1961). According to Saarnijoki, curly grain formation can occur in coniferous species like Scots pine and Norway spruce. Obviously the grain pattern of different species, like curly wood in birch and 'birdseye' in maple (*Acer*), even if they are visually similar, have different background and development mechanisms.

The primary cause of abnormal cambial activity in curly birch is not clear, even though there has been considerable speculation about the phenomenon (see e.g. Hintikka, 1922; Mühle-Larsen, 1940; Ruden, 1954; Atanasoff, 1967; Jevdokimov, 1984; Novitskaya, 1998). Hintikka (1922), for example, emphasized nutritional and climatic influences responsible for abnormal development and considered 'the Wisa disease' a kind of latent gummosis. Later, a number of researchers attributed the

Figure 2. *The stem of a young curly birch tree showing the irregular form of the stem.*

phenomenon to pathogens or viruses (e.g., Atanasoff, 1967). Now, however, the formation of curly-grained wood in birch is generally considered a special, hereditary feature rather than a consequence of disease.

Over the years researchers have given curly birch several Latin names, for example *Betula verrucosa* f. *carelica* Sok., *B. alba* L. var. *carelica* Mercklin, *B. verrucosa* Ehrh. f. *gibbosa* Lindqvist, *B. verrucosa* Ehrh. *maserica* Ruden, *B. verrucosa* Ehrh. *callosa* Svoboda (see e.g. Etholén and Huuri, 1982; Pagan and Paganová, 1994). A number of generic names have also been used (curly or curled birch, Masur or Mazer birch, speckled birch, Carelian birch, see e.g., Ruden, 1954; Václav *et al.*, 1969). The current 'official' name in Finland at least is *Betula pendula* var. *carelica* (Mercklin) Hämet-Ahti (Hämet-Ahti, 1987; Hämet-Ahti *et al.*, 1992).

In this article 'curly-birch' is primarily considered from the point of view of inheritance, wood quality and cultivation practices.

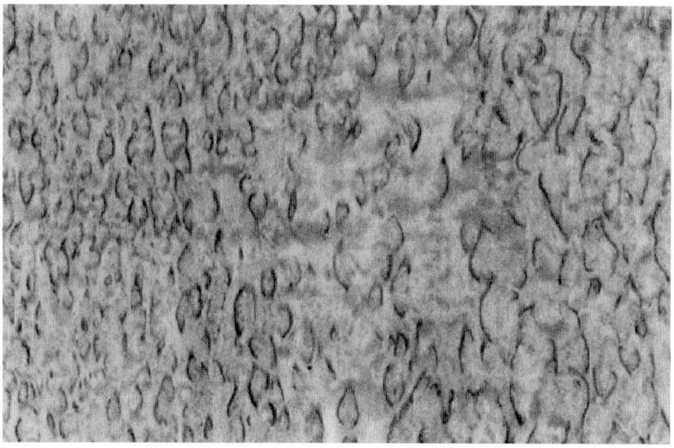

Figure 3. Photograph of a piece of veneer produced from curly birch.

2. Geographical distribution of curly birch

The natural distribution of curly birch covers southern Finland and Sweden, south-eastern Norway, the Baltic countries and the western parts of Russia, especially Carelia, Belorussia and Ukraine (e.g., Scholz, 1963; Václav, 1963; Jevdokimov, 1984; Pagan and Paganová, 1994). It also occurs sporadically in Poland, Germany and Slovakia. In Finland most of the natural curly birch stands have been grown, like birch in general, in areas where shifting agriculture was widely practised up until the early 1900s. Being a pioneer species, birch occupied the growing space that became available after burning and the rotation of crop plants (turnip, barley, rye or flax).

Due to its wide distribution, a lot of research on curly birch was carried out in the former Soviet Union (see e.g., Etholén and Huuri, 1982; Jevdokimov, 1984; Novitskaya, 1998). Unfortunately, however, these studies are poorly known abroad. The same is perhaps true as regards the curly birch literature published in Finland and in the former Czechoslovakia.

3. Formation of curly-grained wood

The development of curly-grained wood is ascribed to disturbances in the functioning of the cambium. Groups of cambium cells cease to divide, the surrounding cells divide rapidly and wound callus and bark parenchyma become incrusted in the wood (Ruden, 1954; Johnsson, 1974). The annual rings become wavy (*Figure 4*), the tissue exceptionally oriented, and the pith rays dilated. V-shaped figures, which can form a pronounced and closed 'curly flower', appear in cross sections (*Figure 5*). A longitudinal cut shows a lens-like configuration. When the bark is peeled off, the surface of the wood has a granulated appearance with small oblong swellings and depressions which, according to Hintikka and Soinne (1937), look like the horns of a wild deer or the relief map of a mountain.

Artificially induced disturbances of cambial activity in both normal silver birch and Carelian (curly) birch have recently been studied in Russian Carelia (Novitskaya, 1998). Such purpose-designed disturbances were obtained by removing the bark and

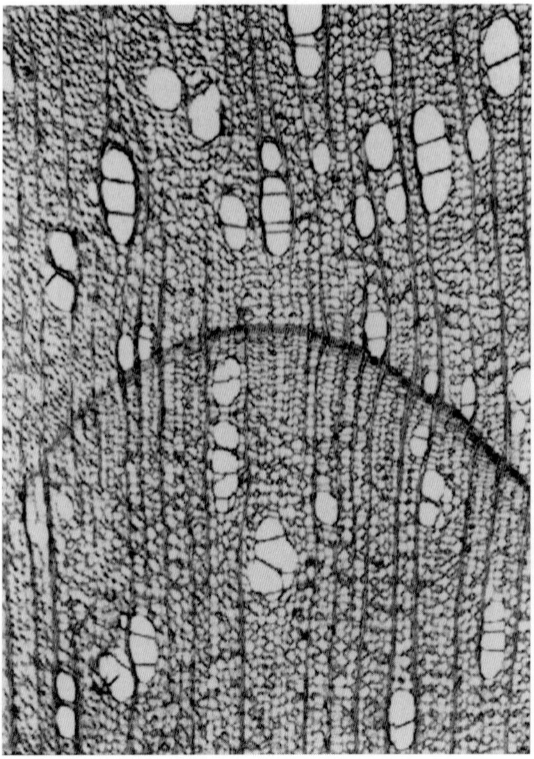

Figure 4. *Transverse section through curly birch wood showing the wavy nature of the annual ring.*

cambial layer by stripping some bark from wood during the period of active radial growth. Many structural aberrations recorded in the wounded wood of normal silver birch were also characteristic of the figured wood of curly birch. The formation of abnormal wood in both cases was discussed in relation to disturbances in the transport of assimilates.

The external appearance of curly birch trees differs from that of the normal silver birch. The bark is often dark or almost black, thick and cracked (*Figure 1*). The trunk is shorter than that of normal silver birch, usually forked and leaning. In extreme cases the whole tree is rather bushy.

Saarnio (1976) divided the curly birch stems into four types on the basis of their visible external characteristics:

(i) with protuberances (P, 'paukura'); stem surface full of small 'peas';
(ii) with necks (K, 'kaula'); strongly suberized swollen 'muffs', alternating with thinner 'necks';
(iii) with stripes (J, 'juomu'); longitudinal 'ridges' on the stem surface;
(iv) with rings (R, 'rengas'); transversal, smooth rings along the stem.

Most of the trees, however, are mixtures of these types. The protuberance type (P), either pure or mixed, is the most common (and most valuable), and that with rings (R) the rarest (Saarnio, 1976; Oskarsson and Nikkanen, 1998). P- and K-type trees have pronounced brown figuring in the wood, and those of the J-type clearly less. The R-type lacks the brown colour. However, this is only 'a general rule': it is often difficult

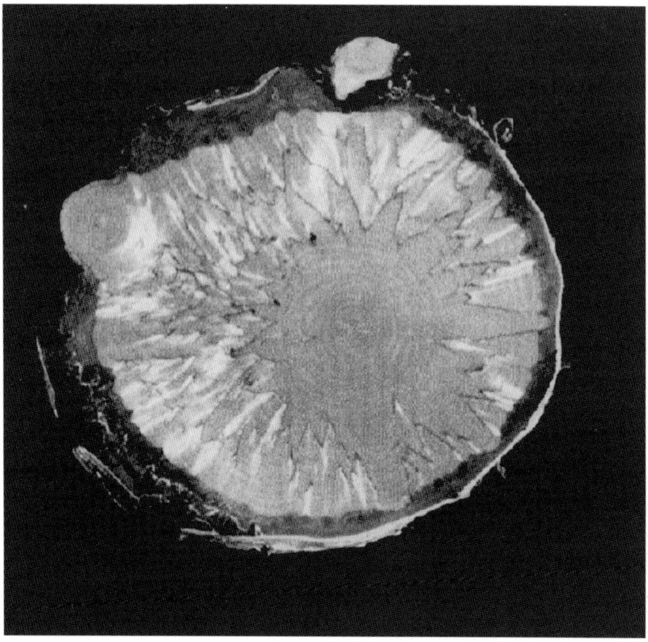

Figure 5. *Transverse surface of a curly birch stem showing the characteristic 'flower' pattern.*

to recognize whether a growing birch tree really has curly-grained wood or not. In spite of the external signs of curly grainedness, curly grain formation is not always found in the cross-section of the wood when the tree is cut down (Saarnio, 1976; Jevdokimov, 1984). Similarly, there may be no signs on the stem surface, but the cross-section has a 'curly flower'. It is also questionable whether the processes involved in the formation of the four types are related to each other. Sometimes the brown-coloured wood associated with curly-grained wood formation is confused with the larval tunnels made by the birch cambium fly, *Phytobia (Dendromyza) betulae*, which is common in both silver and downy birch (see e.g., Kangas, 1935; Bhat, 1980; Ylioja *et al.*, 1998).

Studies available on the biochemistry of curly-grained wood formation are few. When comparing the spring sap of curly and normal silver birch, Ahokas (1985) found higher cytokinin content in curly birch, which may explain the poorer apical dominance of curly birch trees and possibly even the curly-grained wood formation itself. Pätiälä *et al.* (1978) reported on differences in the carbohydrate composition of the sap: sorbitol was identified only in normal trees. The accumulation of sucrose in the phloem of curly birch has been discussed by Novitskaya (1998).

According to the external appearance of the stem, the formation of curly-grained wood commences at an average age of 5–6 years (Johnsson, 1951; Ruden, 1954; Saarnio, 1976). However, there is great variation with respect to the age when the curly trait appears (Jevdokimov, 1984). For example, in a progeny trial series consisting of selected, curly birch plus trees, the authors found that 32% and 48% of the individuals expressed external signs of curly-grainedness at the age of 7 years and 15 years, respectively. The larger the spacing between the trees and the more fertile the site, the younger is the age at which the curly-grain formation commences (Pätiälä, 1980). The range can be as large as 2–30 years. This is the reason why thinnings, which are of great importance when growing curly birch trees, should be light and repeated at short intervals.

4. Inheritance, breeding and production of cultivation material

In Finland the first cultivation experiments and studies on curly birch were carried out in the 1920s and 1930s, mainly by Prof. Olli Heikinheimo. The Curly Birch Society — 'Visaseura' — was founded in 1956 in order to promote the cultivation and use of curly birch, and to co-ordinate the activities of curly birch cultivators, the forest industry and research (Huuri, 1978). Since the 1940s some 200 curly birch plus trees have been selected, most of them during the 1970s and 1980s. Today there is also a curly birch seed orchard (in a plastic greenhouse like all birch seed orchards) and some progeny and clone tests. Besides genetically improved seed, *in vitro* micropropagation can be used to produce high-quality cultivation material (Ryynänen and Ryynänen, 1986). There is thus a rather good basis for studying the inheritance, cultivation and wood quality of this valuable curiosity.

The inheritance of the curly-grained pattern has been studied using different types of breeding material (e.g., Mühle-Larsen, 1940; Johnsson, 1951 and 1974; Ruden, 1954; Václav, 1965; Jakuszewski, 1973; Jevdokimov, 1984; Hagqvist, unpublished; Velling and Viherä-Aarnio, unpublished). When the seed is collected from solitary curly birch trees growing in the forest in Finland, the percentage of individuals expressing curly-grainedness in the progeny varies considerably. Sometimes it is 10–20%, and sometimes 50%, depending on the mother tree and how much of the pollen is derived from curly birches. Seed orchard seed produces 60–70% trees with curly-grained wood, and vegetative micropropagation 100%.

Ruden (1954) explained the genetical background of curly-grain formation on the basis of a dominant 'masur gene', that is lethal in homozygotic conditions. In such a case the seed, after open pollination, should in most cases represent the cross masur (curly) × normal and result in progeny with 50% masur individuals. The cross masur × masur should give 2/3 masur individuals (heterozygotes), and the homozygotes should die. Johnsson (1951) earlier proposed that curly grain formation might depend on a few recessive main genes, and that other genes, which *per se* do not induce curly-grain formation, could have modificatory influences. Later on Johnsson (1974) suggested that it might be a question of a few dominant-semidominant genes, a series of multiple alleles in one locus, or genes in different loci. Using modern biotechnology it may be possible to shed some more light on, if not solve, the inheritance pattern of curly-grained wood.

5. Cultivation

The establishment procedure for a curly birch stand is similar to that for normal silver birch. Some 1600–2000 seedlings are planted ha^{-1}. Using micropropagated plantlets as low as 800–1000 plantlets ha^{-1} can be justified by considerably lowered costs. However, if there is a risk of pest damage, then it is recommended to plant up to 2500 seedlings ha^{-1} and/or fence the plantation. The most harmful browsing animals are, as for birch in general, moose (*Alces alces*), hare (*Lepus* spp.) and vole (*Microtus* spp.).

Sites suitable for curly birch are fertile, light soils. Heavy clay or peaty soil should be avoided. In addition to high-quality cultivation material and a suitable growing site, well-timed silvicultural measures are needed to achieve good results, that is, the maximum amount of turnable wood. Cleaning is needed to remove other species, especially broadleaved sprouts. Normal silver birch trees without any visible signs of curly-grained formation are removed at the age of about 10–13 years, that is, when the dominant height is 7–9 m (Heikinheimo, 1951; Sarvas, 1966). This measure is important, even if often contradictory, because the largest (non-curly) trees must be removed. Repeated thinnings from below of the remaining curly birch individuals are needed to ensure enough light (more than for normal birch).

Artificial pruning of the stems must be started at the same time as the first thinning at the latest, but preferably much earlier, that is, 2–3 years after establishment and then continued in stages (Heikinheimo, 1951; Sarvas, 1966; Raulo *et al.*, 1978; Hagqvist, 1997). The best time to carry out pruning in Finland is in late July to early August. When growing curly birch, profitability 10 times greater than that for normal forest trees is possible if the silvicultural measures are well-timed and carried out in the correct way and damage caused by pests avoided (Raulo and Sirén, 1978: Hagqvist, 1996).

References

Ahokas, H. (1985) Cytokinins in the spring sap of curly birch (*Betula pendula* f. *carelica*) and the non-curly form. *J. Plant Physiol.* **118**: 33–39.

Atanasoff, D. (1967) Virus stem pitting of birch (Wisa and Maser Disease). *Zeitschrift für Pflanzenkrankheiten und Pflanzenschutz* **74**: 205–208.

Bhat, K.M. (1980) Pith flecks and ray abnormalities in birch wood. *Silva Fennica* **14(3)**: 277–285.

Etholén, K. and Huuri, L. (1982) Visakoivua käsittelevä kirjallisuus. Bibliography on curly birch *Betula pendula* var. *carelica* (Mercklin). *Folia Forestalia* **502**: 1–24.

Hagqvist, R. (1996) Visakoivun kasvatuksen kannattavuus. [Profitability of growing curly birch.] Lecture at the Forest Week's seminar. Suomen Metsäyhdistysry — Finnish Forestry Association. Mimeographed, pp. 1–4.

Hagqvist, R. (1997) Visakoivun latvuksen hoito: karsinta ja oksien typistäminen. [Crown management in curly birch: Pruning and shortening of branches.] Lecture at the Finnish Curly Birch Society. Mimeographed, pp. 1–2.

Hämet-Ahti, L. (1987) Suomen koivujen taksonomisia ongelmia. Abstract: Taxonomical problems of the Finnish birches. *Sorbifolia* **18(3)**: 115–117.

Hämet-Ahti, L., Palmén, A., Alanko, P. and Tigerstedt, P.M.A. (1992) Suomen puu- ja pensaskasvio. [Woody Flora of Finland.] Dendrologian Seura — Dendrologiska Sällskapet r.y. Helsinki, 1992, p. 373.

Heikinheimo, O. (1951) Kokemuksia visakoivun kasvatuksesta. Summary: Experiences in the Growing of Curly Birch. *Communicationes Instituti Forestalis Fenniae* **39.5**, 1–26.

Hintikka, T.J. (1922) Die 'Wisa'-Krankheit der Birken in Finnland. *Zeitschrift für Pflanzenkrankheiten und Gallenkunde* **32(5/6)**: 193–210.

Hintikka, T.J. and Soinne, U.T. (1937) Curly and flamy birch. *Wood*, January 1937: 27–29.

Huuri, O. (1978) Visaseura. Summary: Curly Birch Society. *Silva Fennica* **12(4)**: 241–244.

Jakuszewski, T. (1973) Badanie zmienosci i dziedziczności populacji brzozy karelskiej z Gorców. Summary: Studies on the variation and inheritance in a Gorce population of the Karelian birch (*Betula pendula* Roth var. *carelica* (Mercklin) Hejtmánek). *Arboretum Kórnickie* 18: 73–79.

Jevdokimov, A.P. (1984) Visakoivun kasvatus Neuvostoliiton luoteisosassa. Summary: Experiences of curly birch growing in northwestern Russia. *Silva Fennica* **18(3)**: 303–311.

Johnsson, H. (1951) Avkommor av masurbjörk. [Progenies of curly birch.] *Svensk Skogsvårdsföreningens Tidskrift* **49(1)**: 34–45.

Johnsson, H. (1974) Genetic characteristics of *Betula verrucosa* Ehrh. and *Betula pubescens* Ehrh. *Annales Forestales. Anali za Šumarstvo* **6(4)**: 1–149.

Kangas, E. (1935) Die Braunfleckigkeit des Birkenholzes und ihr Urheber *Dendromyza (Dizygomyza) betulae* n. sp. Vorläufige Mitteilung. *Communicationes Instituti Forestalis Fenniae* **22(1)**: 1–31.

Mühle-Larsen, C. (1940) Masurbirk. [Masur birch] *Dansk Skovforenings Tidskrift* **25**: 33–72.

Novitskaya, L. (1998) Regeneration of bark and formation of abnormal birch wood. *Trees* (1998) **13**: 74–79.

Oskarsson, O. and Nikkanen, T. (1998) Metsäpuiden erikoismuotoja kultakuusesta luutakoivuun. [Special forms of forest trees from golden spruce to broom birch]. *Metsäntutkimuslaitoksen Tiedonantoja* **670**: 1–54.

Pagan, J. and Paganová, V. (1994) Breza biela svalcovitá (*Betula alba* L. var. *carelica* Merk.) na Slovensku. Summary: Curly birch in Slovakia. *Vedecké a pedagogické aktuality* **10/1994**: 1–75. Vydala Technická univerzita vo Zvolene v Edicnom stredisku.

Pätiälä, R-V. (1980) Pirun puristama visakoivu. [Birch tree squeezed by the Devil]. *Suomen Luonto* **1/80**: 1–5.

Pätiälä, R-V., Blomberg, K., Piepponen, S. and Paakkanen, J. (1978) Havaintoja raudus- ja visakoivun mahlan sokeripitoisuudesta. Summary: On carbohydrates of the sap of silver birch and its form curly birch. *Silva Fennica* **12(4)**: 253–256.

Raulo, J., Saarnio, R. and Ylitalo, T. (1978) Visakoivun karsittujen oksien kyljestyminen ja värivian leviäminen niistä runkoon. Summary: Sealing-off of pruned branch stumps of curly birch and subsequent spread of discolouration into the stemwood. *Silva Fennica* **12(4)**: 257–263.

Raulo, J. and Sirén, G. (1978) Neljän visakoivikon päätehakkuun tuotos ja tuotto. Summary: Yield and return of final cutting in four curly birch stands. *Silva Fennica* **12(4)**: 245–251.

Ruden, T. (1954) Om valbjørk og endel andre unormale veddannelser hos bjørk. Summary: On speckled birch ('mazer-birch') and some other forms of curled birch. *Meddelelser fra Det Norske Skogforsøksvesen* **43**, pp. 455–505.

Ryynänen, L. and Ryynänen, M. (1986) Propagation of adult curly-birch succeeds with tissue culture. *Silva Fennica* **20(2)**: 139–147.

Saarnijoki, S. (1961) On muutakin visaa kuin koivun visa! Summary (p. 276): Curly wood is not met in birch alone. *Metsätaloudellinen Aikakauslehti* **78**: 257–259.

Saarnio, R. (1976) Viljeltyjen visakoivikoiden laatu ja kehitys Etelä-Suomessa. Summary: The quality and development of cultivated curly-birch (*Betula verrucosa* f. *carelica* Sok.) stands in southern Finland. *Folia Forestalia* **263**, pp. 1–28.

Salmi, J. (1987) Koivun puuaineen ominaisuudet ja käyttö. Abstract: Wood properties and uses of birch, *Betula* spp. *Sorbifolia* **18(3)**: 123–132.

Sarvas, R. (1966) Visakoivikon perustaminen ja hoito [Establishment and silviculture of a curly birch stand.] *Metsätaloudellinen Aikakauslehti* **83(8)**: 331–333.

Scholz, E. (1963) Das Verbreitungsgebiet der Maserbirke. Summary: The natural range of brown-grain Birch. *Archiv für Forstwesen* **12(12)**: 1243–1253.

Václav, E. (1963) Rozšíření, stanovištní podmínky a rust svalcovité (karelské) břízy v Evrope. Zusammenfassung: Verbreitung, Standortsansprüche und Wachstum der Braunmaserbirke in Europa. *Sborník Lesnické Faculty Vysoke Školy Zemedelské v Praze* **6 1963**: 217–237.

Václav, E. (1965) Dedicné vlastnosti a šlechtení svalcovité (karelské) břízy. Summary: Hereditary characteristics and improvement selection of the Karelian (curly) birch. *Sborník Vedeckého Lesnického Ústavu Vysoké Školy Zemedelské v Praze* **8 1965**: 269–290.

Václav, E., Kučera, B. and Řežábková, J. (1969) Anatomické, fyzikální a mechanické znaky a vlastnosti dřeva svalcovité, očkové a plamenne břízy. Summary: Anatomical, physical and mechanical properties of curly birch, maser birch and flamy birch wood. *Sborník Vedeckého Lesnického Ústavu Vysoké Školy Zemedelské v Praze* **12 1969**: 111–127.

Vesterinen, E. (1957) Hat die Maserbirke Zukunft in Europa? *Schweizerische Zeitschrift für Forstwesen Journal Forestier Suisse* **108(2)**: 110–113.

Ylioja, T., Saranpää, P., Roininen, H. and Rousi, M. (1998) Larval tunnels of *Phytobia betulae (Diptera: Agromyzidae)* in birch wood. *Journal of Economic Entomology* **91(1)**: 175–181.

31

Heterogeneous, branched, zigzag and circular vessels: unexpected but frequent forms of tracheary element files: description — localization — formation

J.P. André

1. Introduction

The xylem stores the histogenetic memory of the main events in the growth of the plant, of the effects of the environmental factors and also of the countless disturbances in its cell files. Because the conducting function confers a structuring role to the vascular system within the whole plant, the vessels gather a great part of that memory both at the scale of the plant and of each vessel member. In order to have access to this rich xylogenetic information, we have to observe the vascular system on its entire length. On this point, Zimmermann wrote: 'It is rarely possible to see vessels throughout their entire length, because they consist of small cells which need a microscope for observation, but at the same time they are so long that a microscope is far too myopic to grasp their extent. They can never be seen on single sections or even on short series of transverse or longitudinal sections.' He and his collaborators recorded on film the enlargements of long series of transverse serial sections of wood (Zimmermann, 1983). With the same aim in view, we use an intracellular microcasting technique, which probably offers an easier solution to the problem. The method has been fully described elsewhere (André, 1993, 1998, 1999). In short, the vessel lumina of a wood sample are materialized under the form of silicon elastomer filaments with a resolution better than one micrometer. The lumina casts, here named vessel casts, keep

Cell & Molecular Biology of Wood Formation, edited by R. Savidge, J. Barnett and R. Napier.

the original form and position of each element of the vascular system in the three dimensions.

The 'classical' vessel pattern is a linear chain of biperforated elements having the same secondary wall ornamentation, with uniperforation at each end. Observation of thousands of vessel casts in some 50 plants species of about 15 'classical' angiosperm orders has revealed a more complex picture. We have found vessel shapes which are rarely or never mentioned in the literature and which are notably different from the typical pattern. These have been identified, localized and studied by means of microcasting (*Figure 1*). Some of them seem to characterize particular plant families (ramified shapes in Poaceae) or positions in the plant (zigzag, circular and heterogeneous shapes, in the connecting zones between organs). In this chapter, I describe these non-classical vessel types, paying attention to their localization and to their abundance. In addition we have investigated the previously described circular-shaped vessels and propose hypotheses on their formation.

2. Circular vessels

The existence of circular vessels has been known for about a century (Larson, 1994) but hypotheses on their formation are no more than 15 years old. These anatomical structures seem relatively poorly known to wood anatomists, except to a few students of functional anatomy and morphogenesis, Aloni and Kalev (1995), Aloni and Wolf (1984), Hejnowicz and Kurczynska (1987), Kurczynska and Hejnowicz (1991), Lev-Yadun and Aloni (1990) and Lev-Yadun (1996).

Circular vessels are described as closed endless ring-shaped vessels. They abound along the junction between the wood of two living axes, for example a stem and a branch, and especially in the wood formed during healing around a branch stump (*Figure 2*). They have also been found in crown galls induced by *Agrobacterium tumefaciens* (Aloni *et al.*, 1995; Schurr *et al.*, 1996).

We have particularly studied a piece of healing wood taken from a eucalyptus tree (*Figure 2* right). The circular vessels often formed concentric series always inserted in open concentric vessel loops (*Figure 8a* and *b*). The wood sample was cut with a saw into thin slices — 1 mm thick — parallel to the surface, before casting. In this way we

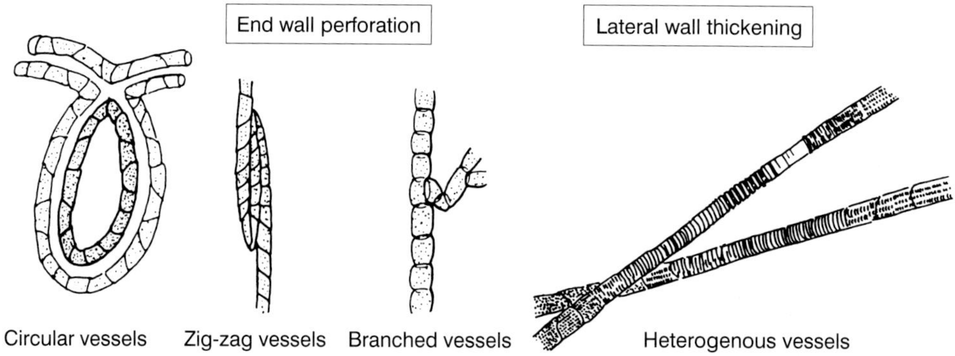

Figure 1. *Schematic aspect of non classical vessel shapes.*

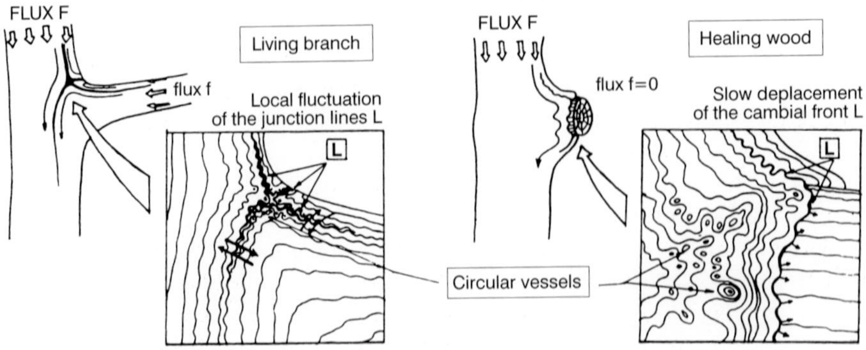

Figure 2. *Localization of the circular vessels.*

could observe a slow shift of the wood grain in the successive slices, that is, a slow deformation and displacement of circular vessels and loops in the course of time. Different characteristics of the closed and open vessel types were accurately compared, as for example, size and shape of the vessel members, intervascular distances and arrangement of rays. No particular histological difference was found. Moreover, overlapping vessel ends were similarly found in each type, that is, concentric circular vessels either were really endless with biperforated elements only or were like short coiled up vessels (*Figure 8c*). That important morphological detail allows us to hypothesize an almost common formation process for circular and open vessel types. We consider the wood junction lines or wood grain limits between a stem and a branch as the last position of the contact lines between their respective cambial edges, and therefore as equilibrium lines between tangential pressure forces exerted on each other by two opposite cambial fronts. As open loops and circular vessels are formed on each side of these wavy lines, we assume that the cambial divisions are more frequent in the convex than in the concave borders of each front, increasing the local curvatures. The waves would progressively change in loops, until the sides of the most internal loops come into contact. The closure of the first circular vessel would occur during the vessel differentiation process, through a slight modification in the location of one end wall perforation only (*Figure 3*). Larger concentric circular vessels would form later in further xylem layers.

The pruning of the branch breaks the balance and causes the slow displacement of the surviving cambial front toward the stump and an increase of the radii of the waves, loops and circular vessels.

This hypothesis, as yet unproven, refutes the assumption made by the authors mentioned above, that a ‘circular flow’ of auxin must take place in the cambium before the differentiation of circular vessels can occur.

3. Zigzag vessels

In young woody stems, the secondary vessels commonly present short segments, with more or less complicated ‘zigzag’ shapes, which are located in the nodes, on each side of the bulging insertion zone of petioles and axillary branches (*Figures 4* and *8d*). They

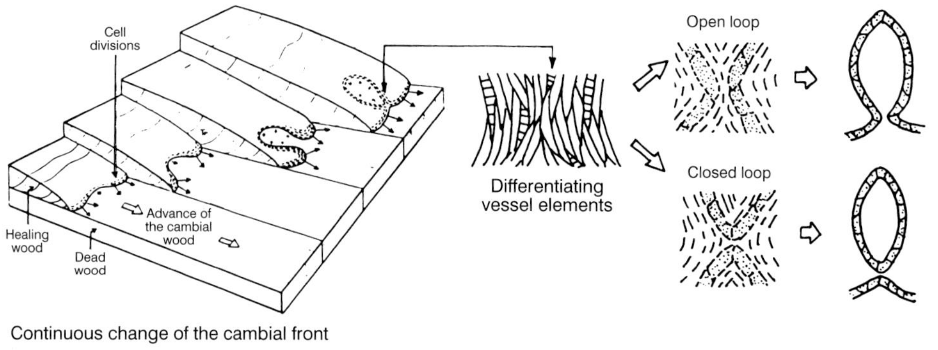

Figure 3. *Hypotheses on the circular vessel formation.*

have been found in numerous dicot species from different families, for example, *Lycopersicum esculentum* M. (Solanaceae), and *Quercus robur* L. (Fagaceae). Such vessels can often be seen in tangential sections of normal wood (B. Sundberg, personal communication 1999).

Careful stretching of vessel casts with a pair of sharp tweezers clearly revealed the arrangement of the vessel elements in the so-called zigzag segment and the continuity of the biperforated element files. When released, the elastic cast recovers its original shape of a compact cluster. Such a vessel seems to consist of two linear parallel, but non-adjacent vessels, A and B, joined by a set, C, of short adjacent member files (*Figure 4*). The particular position of the zigzag segments allows us to suppose that the growth of adjacent organs (petioles, branches) exerts tangential mechanical strains, which upset the linear differentiation process of the vessels produced in their vicinity. The presence of numerous zigzag vascular segments in the nodes may locally increase the hydraulic resistance and the available water volume, but probably has a small effect on the ascent of sap.

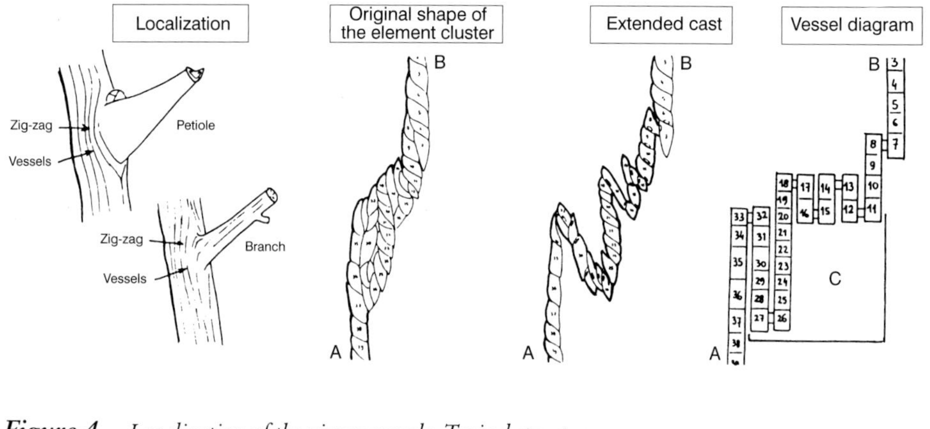

Figure 4. *Localization of the zigzag vessels. Typical structure.*

4. Branched vessels

Triperforated vessel elements were observed in xylem macerations (Van der Schoot and Van Bel, 1989; Yulong and Liese, 1997), but in general little attention has been given to the vessel branching, which necessarily originates in each triperforated element. Normal histological techniques are inappropriate for finding and observing branched vessels, whereas the microcasting method is ideally suited to this work. Using this technique, branched vessels were found in the secondary xylem of dicots and in the protoxylem and metaxylem of Poaceae species (*Figure 5*). The branching was single or multiple according to the abundance of triperforated elements.

Such elements were scarce in the wood of dicotyledons — less than one per ten thousand elements — but present in almost every species studied. Localization and orientation of the vessel forks seemed to be random, although the two branches of the fork were always parallel and in contact along their whole length, suggesting they consisted of two sister-cell files. On the other hand, countless triperforated elements were found in the xylem of the Poaceae species examined, which included 25 bamboos from 11 genera and 6 common cereals. Their conducting bundles, which extended through several internodes, had their lower and upper ends in nodes. There, the protoxylem and later the metaxylem vessels ramified in complex structures looking like a plexus made of hundreds of perforated and pitted tracheary elements (André, 1998; Yulong and Liese, 1997). Casting gave a striking picture of their three-dimensional shape. The sap ascending in the culm and the branches presumably passed from one bundle to the next through the large pitted contact surface of their end ramifications. The xylem of the culm and of its roots was similarly connected through pitted contact surfaces. The culm and the rhizome, however, were directly connected through branched vessels made of perforated elements (*Figures 5* and *8e*).

In the growing culm, the size of the casts of the metaxylem ramifications was larger in basal than in distal nodes. During the maturation of the nodes, the metaxylem differentiation probably increased, thus increasing the number of conducting elements in each ramification. During this time, the protoxylem ramifications which were already formed showed similar structures to those in metaxylem ones, but with reticulated secondary wall thickening.

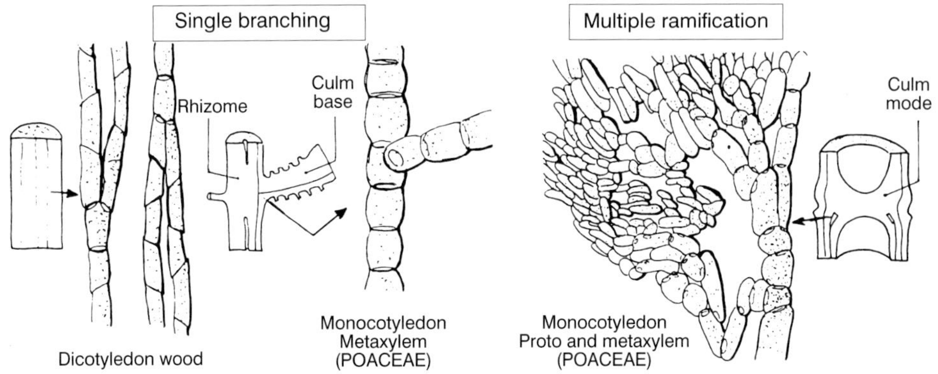

Figure 5. *Localization of the branched vessels.*

5. Heterogeneous vessels

The term 'heterogeneous' was proposed in order to denote a vessel type in which a short sequence of about 5–20 members with different ornamentation was inserted between two files of pitted members. This vessel type was recently discovered both in monocotyledon (Poaceae family) and dicotyledon species, using vessel casting (André, 1998; André *et al.*, 1999). Complete and symmetrical sequences of ornamentation, that is, scalariform–reticulate–helical–annular–helical–reticulate–scalariform, were found in the xylem of the branches of Bambuseae species (*Figure 8f*). Elsewhere, the sequences were generally incomplete.

Heterogeneous vessels were always present passing through the connecting or abscission zone between two organs or two organ parts. The pitted parts of their length were distributed on each side of this zone. They were found in the foliar traces of numerous dicotyledons (e.g. maple, oak, plane tree and clematis), in the rachis of compound leaves (walnut) and in foliar and fruit pedicels (tomato, pimento) and in each case, the heterogeneous sequence was located in the abscission zone (*Figure 6 left*). Their position between protoxylem and secondary xylem suggests they belonged to metaxylem. In Poaceae, every metaxylem vessel possessed a heterogeneous sequence located at the base of each internode and at the junctions between leaf sheaths and lamina, which were probably relics of intercalary meristems (*Figures 6 right* and *8f*). Such sequences were lacking in the rhizomes, and often also in the first basal internodes of the culms. They became more and more complete in the distal internodes and branches. Generic differences were also observed (André, 1998).

A priori, two morphogenetic processes might be proposed in order to explain the formation of a heterogeneous vessel (*Figure 7*).

The first could be related to the existence of the procambium–metacambium–cambium continuum and to the slowing down of the rate of cell maturation along that continuum (Larson, 1983). As a consequence of the delayed transition from procambium to cambium, the cambium proper may be interrupted either definitely as in walnut petioles or only during the early stages of fruit development as in tomato pedicels, giving rise to a longitudinal sequence of cambial–procambial–cambial cells, and later to a heterogeneous sequence (André *et al.*, 1999).

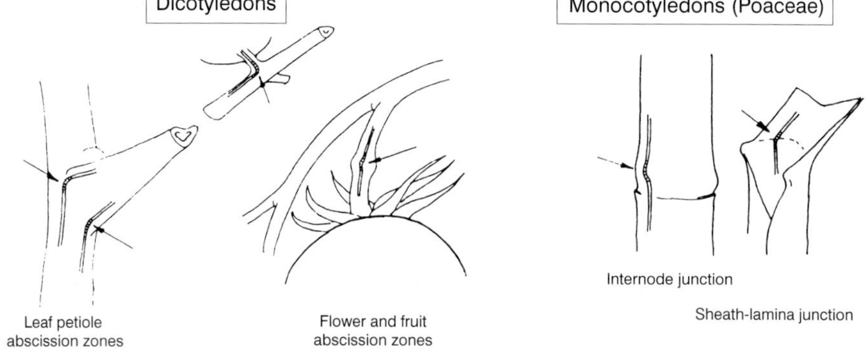

Figure 6. *Localization of the heterogeneous vessels.*

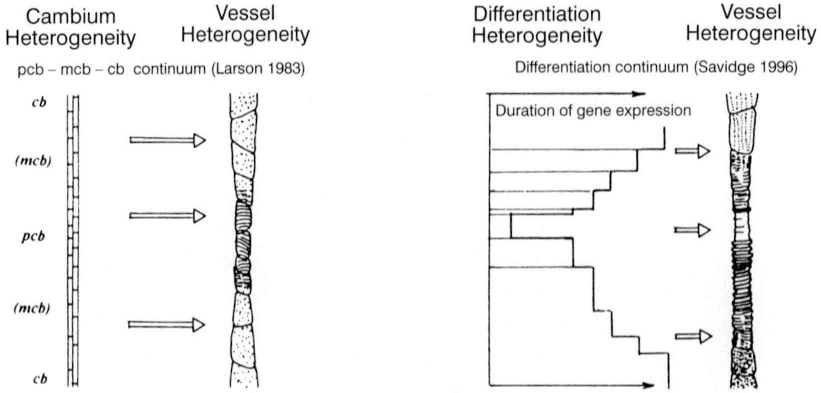

Figure 7. *Hypotheses on the heterogeneous vessel formation.*

The second cause could be found in the continuum differentiation hypothesis proposed by Savidge (1996). It offers an interesting and valuable explanation for the formation of the heterogeneous vessels deriving from an intercalary meristem in Poaceae. The central element of the symmetric heterogeneous sequence would be the last cell to differentiate (*Figure 8f*).

6. Discussion and conclusion

An accurate overview of vascular systems may contribute to present research in xylogenesis and systematic anatomy. We can briefly try to link the above observations to these vast topics.

6.1 *Lateral wall thickening*

In the growing leafy sprouts of dicot species, the limit between procambium and cambium continuously shifts from the base to the apex. During the transitional period, it is foreseeable and has been observed that some vessels of the successive foliar traces present morphological heterogeneities. Is it a rule that the heterogeneous sequences characterize any metaxylem vessel? Does the hypothesis square with that of the continuum of differentiation?

6.2 *End wall perforations*

The factors which control the differentiation of the uni-, bi- and triperforated elements consequently determine the length and the shape of every known — and still unknown — vessel type. Access to the exact features of immature and mature vascular systems at various scales could help achieve a realistic model of the vascular organization for a better understanding of its physiological functions.

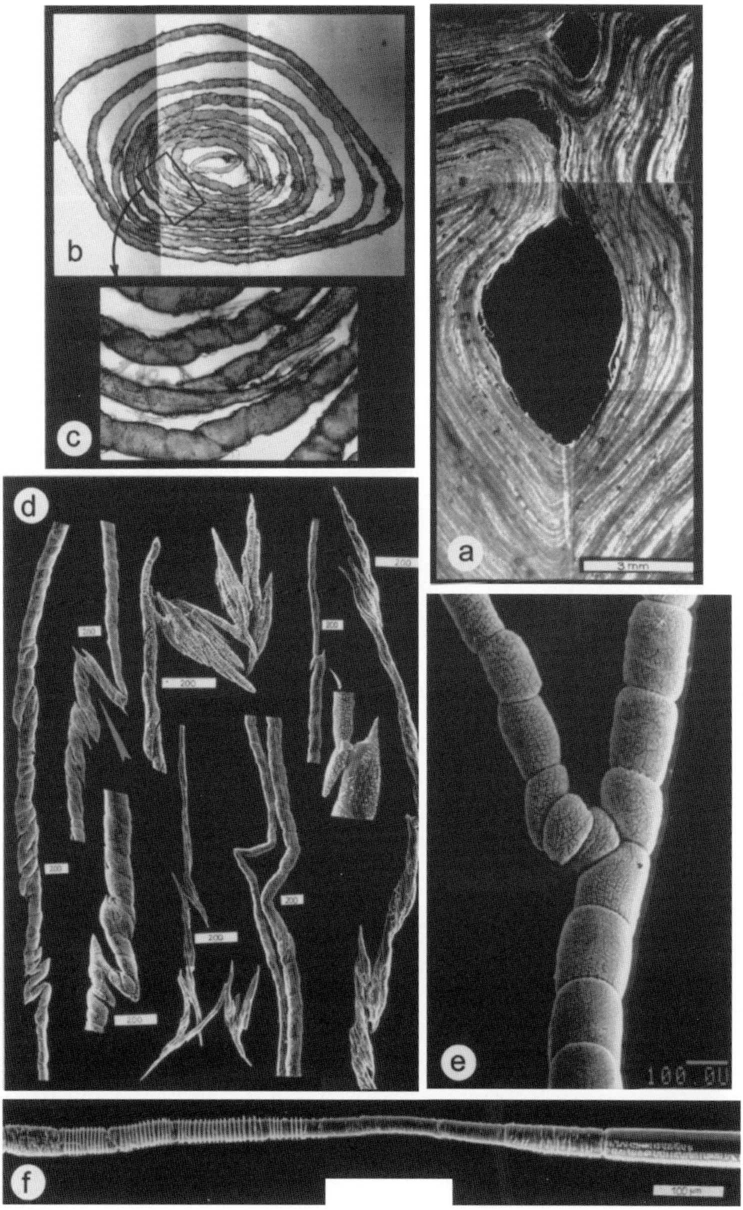

Figure 8 a–f. *SEM micrographs of the casts of different vessel types is visible (a) Open vessel loops in a eucalyptus healing wood sample. About 20 concentric circular vessels were taken off. At the bottom, a vertical mark of the tiny scalpel cut through which the elastomer has penetrated the vessel lumina is visible. (b) Some concentric circular vessels at a twofold higher scale. (c) Partial enlargement showing the overlapping ends of an 'untrue' circular vessel. (d) Zigzag segments in oak vessels, some of them being partially stretched (bottom), others not (left). The second from left (arrow) is sketched in* Figure 4 *(bars in μm). (e) Single vessel ramification in* Phyllostachys viridis *(Bambuseae), connecting the metaxylem of the rhizome and the culm (lateral branch). Note the triangle-shaped triperforated element. (f) Heterogeneous segment located at the base of a branch internode. Metaxylem vessel in* Bambusa tuldoïdes *(basal end to the left).*

6.3 *Systematic anatomy*

Numerous anatomical details such as, for example, the size and shape of vessel end ramifications, seem to be specific or generic characters for the Paoles order. By means of microcasting, interesting observations were made of vascularization in *Quercus* and *Lycopodium* species. Between these, the field of applications seems quite wide!

References

Aloni, R. and Wolf, A. (1984) Suppressed buds embedded in the bark across the bole and the occurrence of their circular vessels in *Ficus religiosa. Amer. J. Bot.* **71**: 1060–1066.

Aloni, R. and Kalev, N. (1995) Role of auxin, gibberelin and ethylene in the size, pattern and orientation of regenerative tracheids in *Pinus pinea L. Congress report,* IUFRO XX World Congress. Tampere, Finland.

Aloni, R., Pradel, K.S. and Ullrich, C.I. (1995) The three-dimensional structure of vascular tissues in *Agrobacterium tumefaciens*-induced crown galls and in the host stems of *Ricinus communis* L. *Planta.* **196**: 597–605.

André, J.P. (1993) Micromoulages des espaces vides intra et intercellulaires dans les tissus végétaux. *C.R. Acad. Sci. Paris.* **316**: 1336–1341.

André, J.P. (1998) A study of the vascular organisation of bamboos (Poaceae — Bambuseae) using a microcasting method. *IAWA J.* **19**: 265–278.

André, J.P., Catesson A.M. and Liberman M. (1999) Characters and origin of vessels with heterogeneous structure in leaf and flower abscission zones *Can. J. Bot.* **77**: 253–261.

Hejnowicz, Z. and Kurczynska, E.V. (1987) Occurrence of circular vessels above axillary buds in stems of woody plants. *Acta Soc., Bot. Pol.* **56**: 415–419.

Kurczynska, E.U. and Hejnowicz, Z. (1991) Differentiation of circular vessels in isolated segments of *Fraxinus excelsior. Physiologia plantarum* **83**: 275–280.

Larson, P.R. (1982) The concept of cambium. In: *New Perspectives in Wood Anatomy* (ed. Baas, P.) Martinus Nijhoff Publ., The Hague, Boston, London, pp 85–121.

Larson, P.R. (1994) *The Vascular Cambium. Development and Structure*. Springer Verlag, Heidelberg, Berlin, New York

Lev-Yadun, S. and Aloni, R. (1990) Vascular differentiation in branch junction of trees: circular patterns and functional significance. *Trees* **4**: 49–54.

Lev-Yadun, S. (1996) Circular vessels in the secondary xylem of *Arabidopsis thaliana* (Brassicaceae). *IAWA* J. **17**: 31–35.

Savidge, R.A. (1996) Xylogenesis, genetic and environmental regulation. A review. *IAWA J.* **17**: 269–310.

Schurr, V., Schuberth, B., Aloni, R., Pradel, K., Schmundt, D., Jähne, B. and Ullrich, C. (1996) Structural and functional evidence for xylem-mediated water transport and high transpiration in *Agrobacterium tumefaciens* induced tumors of *Ricinus communis. Bot. Acta* **109:** 405–411.

Van der Schoot, C. and Van Bel, A.J.E. (1989) Morphogram: a novel diagram to organise the transitive secondary xylem elements of basal tomato internodes. *Am. J. Bot.* **76**: 475–486.

Yulong, D. and Liese, W. (1997) Anatomical investigations on the nodes of bamboos. In: *The Bamboos* (ed. Chapman, G.P.), Academic Press, London, pp. 269–283.

Zimmerman, M.H. (1983) *Xylem Structure and the Ascent of Sap*. Springer Verlag, Berlin, Heidelberg, N.Y., Tokyo.

32

Plant laccases: are they involved in lignification?

P. Ranocha, D. Goffner and A.M. Boudet

1. Introduction

Because of their involvement in mechanical resistance and hydrophobicity of the cell wall, lignins probably played a crucial role in plant evolution, especially in the acquisition of an efficient vascular system. For this reason, it is generally thought that lignins were a key factor in the colonization of terrestrial habitats by plants 450 million years ago. Despite their overall importance in plant biology, many aspects of lignin biosynthesis are still partially or completely unsolved. Lignins are derived from the oxidation and polymerization of three phenylpropanoid units, referred to as 'monolignols'. Whereas the monolignol biosynthetic pathway is fairly well established, despite increasing evidence for alternative routes by which they may be synthesized, the nature of the enzymes catalysing the oxidative polymerization of monolignols, and consequently lignins biosynthesis *stricto sensu* is still a matter of controversy. Until recently, peroxidases have been considered the sole class of enzyme involved in the final step of lignification. However, in recent years, convincing evidence has been reported that obliges the scientific community to consider other oxidases and particularly laccases as enzymes potentially involved in lignification.

For those readers with a significant background in the field of lignification, the title of this chapter may seem ambitious, since it suggests that we can provide a clear answer. In fact, it is deliberately provocative. Indeed, our main objective is not to unambiguously solve the difficult problem of understanding the role of laccases in lignification, but, more reasonably, to present recent results obtained in our laboratory that may provide insight as to laccase function in plants.

2. Laccases, ubiquitous enzymes with multiple putative functions

Laccases (*p*-diphenol: O_2 oxidoreductase, EC 1.10.3.2) were first mentioned by Yoshida (1883) in studies on sap of *Rhus* sp.(lacquer tree), and this work is certainly at the origin of the enzyme's name. Laccases, ascorbate oxidases and ceruloplasmins

Cell & Molecular Biology of Wood Formation, edited by R. Savidge, J. Barnett and R. Napier.

(mammalian plasmatic proteins) constitute a large and diverse family of copper-containing metalloproteins called 'blue copper oxidases' (Reinhammar and Malström, 1981). These enzymes are unique, along with cytochrome c oxidase and a cyanide-resistant oxidase of an alternate pathway of plant mitochondria (Siedow *et al.*, 1995) because they are able to catalyse the tetraelectronic reduction of oxygen to water (*Figure 1*). This reduction is coupled to the monoelectronic oxidation of a substrate, generally a phenolic compound, leading to the formation of an unstable free radical (Reinhammar and Malström, 1981).

The active site of laccases contains four copper atoms bound to apoprotein via histidine and cysteine residues that are strictly conserved amongst copper oxidases (Messerschmidt and Huber, 1990). These copper atoms are essential to enzyme function. Indeed, for example, Bligny *et al.* (1986) demonstrated that *Acer* cells cultured in a copper-depleted medium produce an inactive apoproteic laccase.

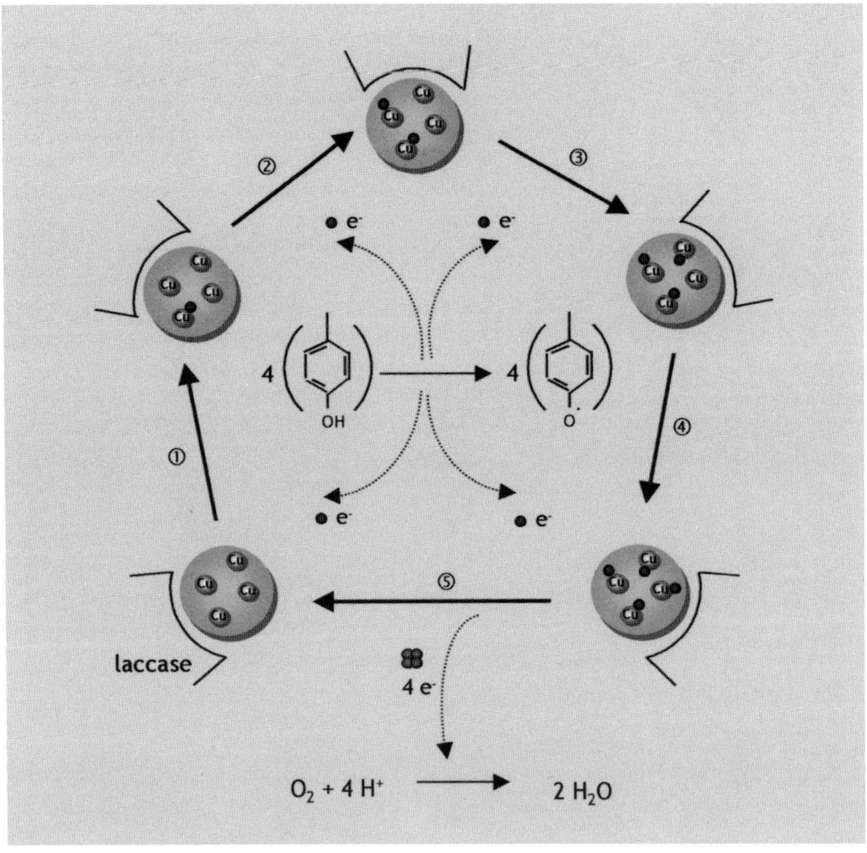

Figure 1. *The oxidation of a phenolic compound and O_2 reduction by laccases. Schematically, laccase is analogous to a battery, successively accumulating an electron produced by each of four monoelectronic oxidation reactions of a phenolic molecule (reactions 1, 2, 3 and 4). Then it discharges four electrons (reaction 5), allowing the tetraelectronic reduction of O_2 to water (Call and Mücke, 1997).*

Laccases are wide-spread enzymes in living organisms. Indeed, they can be found in plants and fungi (Mayer, 1987), but also in a bacterium (Givaudan *et al.*, 1993) and some insects (Sidijanski *et al.*, 1997; Sugumaran *et al.*, 1992; Thomas *et al.*, 1989). However, despite this large distribution, relatively little is known about the biological functions of these enzymes. Laccases are presumably involved in numerous biological processes (*Figure 2*), but, as we are going to see, what is known of this enzyme family is highly variable from one type of organism to the other.

2.1 *Bacterial, insect and fungal laccases*

To our knowledge, laccase-like phenol oxidase activity was detected only in one bacterial species, *Azospirillum lipoferum* (Givaudan *et al.*, 1993). The authors suggested that this enzyme could be involved in mechanisms of bacterial protection by encystment. However, neither structural nor functional data was reported to confirm this hypothesis.

Laccase activity has also been detected in haemolymph and the cuticle of numerous insects including the silk worm and drosophila (Binnington and Barret, 1988; Sugumaran *et al.*, 1992), and in the digestive epithelium of a tropical mosquito, *Anopheles stephensi* (Sidijanski *et al.*, 1997). These enzymes may be involved in carapace melanization (oxidative polymerization of L-dopamine) in response to wounding (Binnington and Barret, 1988), and in isolation and rejection mechanisms of parasites based on their encapsulation in melanin (Christensen and Tracy, 1989).

By far, laccases of fungal origin are the most extensively studied. The large number of fungi producing laccases and the diversity of laccase isoforms they produce suggest that these enzymes play an important role in fungal biology (for review, see Thurston (1994). The implication of laccases in fungal morphogenesis is probably the most commonly accepted of the numerous putative functions of these enzymes. Indeed, they are believed to be involved in *Aspergillus nidulans* pigment biosynthesis (Clutterbuck, 1972; Hermann *et al.*, 1983), in *Armillaria mellea* rhizomorphs formation (Worral *et al.*, 1986) and more precisely in the polymerization of the polyphenolic cement that bind hyphae to each other.

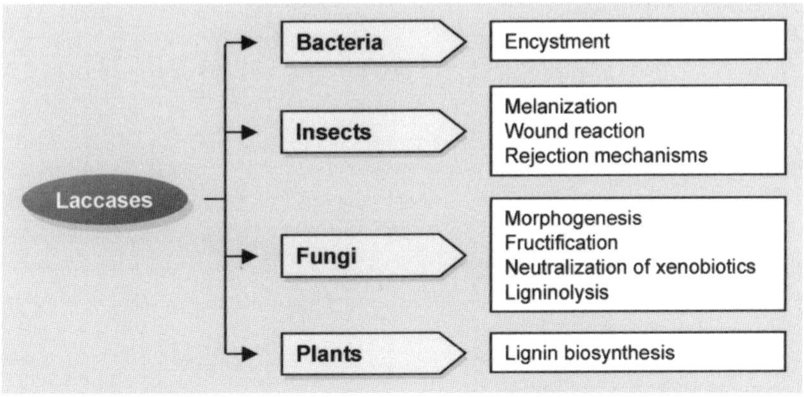

Figure 2. *Distribution of laccases in living organisms and possible functions of these enzymes. (References corresponding to each laccase function are given in the text.)*

Several reports indicate that laccases may also be involved in virulence mechanisms of numerous phytopathogenic fungi (Geiger *et al.*, 1986; Marbach *et al.*, 1984; Rigling and Van Alfen, 1991; Thurston, 1994; Viterbo *et al.*, 1993). For example, laccases are likely to be involved in the virulence of *Cryphonectria parasitica*, the fungus responsible for Chestnut blight (MacDonald and Fullbright, 1991), although their precise role has not yet been established (Rigling *et al.*, 1989; Kim *et al.* 1995).

Fungal laccases are also thought to be involved in lignin degradation. Most ligninolytic fungal strains, including *Phlebia radiata, Phanerochaete chrysosporium* and *Rigidiporus lignosus* possess both peroxidases (Datta *et al.*, 1991; Karhunen *et al.*, 1990a) and laccases (Dittmer *et al.*, 1997; Karhunen *et al.*, 1990b; Niku-Paavola *et al.*, 1988; Saloheimo *et al.*, 1991). The presence of both classes of enzymes is an obstacle in elucidating their precise roles in lignin biodegradation (Addleman *et al.*, 1995). In this regard, *Pycnoporus cinnabarinus* is an interesting model species since it produces laccases but does not possess any of the peroxidases (LiP and MnP) commonly associated with lignin degradation. Interestingly, this fungus is perfectly capable of hydrolysing lignin polymers (Eggert *et al.*, 1996). The same authors showed that a laccase-deficient *P. cinnabarinus* mutant was practically unable to hydrolyse monolignol polymers (Eggert *et al.*, 1997). Together, these results suggest that laccases are essential for the ligninolytic function of this fungus.

3. Plant laccases and lignification

Despite the fact that laccases were first discovered in plants over 100 years ago (Yoshida, 1883), relatively little is known about the role of these enzymes *in vivo*. Plant laccases are cell wall-associated enzymes that are relatively difficult to purify, and consequently, biochemical and molecular data are extremely rare. Due to their capacity to oxidize phenolic compounds, laccases were postulated to be involved in the final step of lignification. However, the implication of laccases in lignin biosynthesis is still a matter of controversy. Since plant laccase research has recently been reviewed in O'Malley *et al.* (1993), Dean and Eriksson (1994), Dean *et al.* (1998) and Lewis *et al.* (1999), we will just highlight here some of the most significant findings.

In 1958, Freudenberg and coworkers were first to demonstrate that fungal (*Agaricus campestris*) and plant (*Araucaria excelsa*) laccases were capable of oxidizing cinnamyl alcohols to phenoxy radicals *in vitro*, leading to the hypothesis of a possible involvement of these enzymes in lignins biosynthesis. However, this idea was rapidly discounted when Nakamura (1967) reported that *Rhus vernicifera* laccase could not oxidize coniferyl alcohol and when Harkin and Obst (1973) presented histochemical evidence of the exclusive participation of peroxidases in lignification in various tree species. Today, it is known that *Rhus* laccase is capable of polymerizing monolignols (Okusa *et al.*, 1996) and that the sensitivity of laccases to freeze/thaw cycles (Dean and Eriksson, 1994) is probably responsible for the negative results of Harkin and Obst. However, these results led to the dogmatic view of the exclusive participation of peroxidases in the lignification process.

In recent years, convincing evidence has been reported that obliged us to re-evaluate this hypothesis and to reconsider laccases and other H_2O_2-independent oxidases as potentially involved in lignification. Several early studies indicated that laccases and laccase-like activities were closely correlated with lignin deposition in developing xylem (Davin *et al.*, 1992; Savidge and Udagama-Randeniya, 1992). Since then, it has been demonstrated that a laccase originally isolated from *Acer* cell suspension cultures (Bligny and Douce, 1983) was not only immunolocalized in lignifying tissues of *Acer* (Driouich *et al.*, 1992), but was also capable of oxidizing and polymerizing monolignols *in vitro* (Sterjiades *et al.*, 1992). In 1993, a laccase was purified for the first time from a lignifying tissue, *Pinus taeda* xylem (Bao *et al.*, 1993). In conifers, a catechol oxidase, coniferyl alcohol oxidase (CAO) specifically associated with lignin deposition during wood formation has also been purified and characterized (for review, see Savidge *et al.*, 1998). Although a full length cDNA sequence has not yet been reported, CAO appears to be distinct from laccases with respect to its physical and biochemical properties.

3.1 *Difficulties associated with studying plant laccases*

The study of plant laccases is complicated by the fact that crude extracts contain a large number of H_2O_2-independent phenoloxidases with overlapping substrate specificities. Among them are polyphenol oxidases (Mayer, 1987), catalysing the oxidation of monophenols such as tyrosine and *o*-diphenols such as L-dopamine, leading to quinones (Steffens *et al.*, 1996). These oxidases are believed to play a role in the plant defence *vis-à-vis* herbivores and insects (Felton *et al.*, 1989; Kowalski *et al.*, 1992), in the regulation of chloroplastic O_2 levels (Lax and Vaughn, 1991; Tolbert, 1973) and non phenolic pigment synthesis (betalain, betacyanin, betaxanthin) in fruits and flowers (Joy *et al.*, 1995). Polyphenol oxidases are also responsible for browning that rapidly occurs when a plant tissue is cut or ground (Walker, 1975). Thus, the large number of broad spectra oxidases in crude extracts and difficulties associated with the solubilization of cell wall laccases (McDougall *et al.*, 1994) are two main reasons why relatively few data related to laccase function in plants are available.

Recently, we undertook the study of plant laccases in poplar, a woody species, by an integrated biochemical and molecular approach in order to gain a better understanding of plant laccases and their potential role in lignin deposition. According to Lewis and Yamamoto's definition (1990), if a laccase plays a role in lignification, it must meet the following criteria:

(i) It must be temporally and spatially correlated with lignification.
(ii) Purified laccase must oxidize monolignols *in vitro* into their free-radical derivatives.
(iii) Plants with modified levels of laccase should have quantitatively or qualitatively modified lignins.

Taking into consideration the aforementioned criteria, we have attempted to provide evidence either for or against the implication of laccases in the final step of lignification.

3.2 *Identification of two poplar xylem laccases capable of polymerizing coniferyl alcohol*

We originally reported that only one protein of approximately 110 kDa localized in the ionically bound cell wall fraction of poplar young xylem was able to polymerize coniferyl alcohol in the absence of H_2O_2, using an in-gel detection method (Sterjiades *et al.*, 1996). However, by increasing the incubation time of the gels with substrate, we were also able to detect H_2O_2-independent oxidase activity in the form of a doublet of approximately 90 kDa. Therefore, this method enabled us to identify, *à priori*, two phenol oxidases (110 kDa and 90 kDa) that may be potentially involved in lignification.

The 90 kDa phenol oxidase was purified by a relatively simple procedure including hydrophobic and hydroxylapatite chromatographic separation (Ranocha *et al.*, 1999). The near identical HPLC profiles of tryptic digestion of both members of the doublet seemed to indicate a very close structural relationship between the two proteins. The 110 kDa phenol oxidase was still detectable after hydrophobic interaction chromatography, but further attempts to purify this enzyme were unsuccessful. Although the biochemical characterization of this oxidase was not possible, we were able to obtain peptide sequence data to enable cDNA cloning by excision of the corresponding protein band from an electrophoretic gel at a partial purification stage (Ranocha *et al.*, 1999).

Several structural and functional properties of the 90 kDa coniferyl alcohol-oxidizing enzyme were determined. This enzyme is a glycoprotein with high-mannose-type N-glycan moieties, α(1,3)-fucose and β(1,2)-xylose residues as indicated by affinodetection with concanavalin A and immunodetection respectively. Moreover, the 90 kDa enzyme cross-reacted with Lewis a-specific antibodies. The purified 90 kDa enzyme (whose mass and p*I* were estimated at 81 kDa and 9.2 respectively), can oxidize a wide range of aromatic compounds (Ranocha *et al.*, 1999). As a general rule, laccase-type, *p*-diphenol oxidases oxidize both *o*- and *p*-dihydroxyphenolic compounds, whereas *o*-diphenol oxidases oxidize *o*-dihydroxyphenolic compounds only (Ferrar and Walker, 1996). Thus, based on its ability to efficiently oxidize hydroquinone, a *p*-diphenol, the 90 kDa enzyme appeared to be more like a *p*-diphenol oxidase. All three monolignols and their corresponding aldehydes were readily oxidized by the 90 kDa enzyme. However, as judged by the ratio of V/K_m, coniferyl alcohol seems to be the preferred substrate (*Table 1*). It is interesting to note that the K_m values are lower than those obtained for the equivalent *Pinus taeda* enzyme (Bao *et al.*, 1993).

Peptide sequences were obtained for both the 90 kDa and 110 kDa proteins. Putative full length cDNA clones (*lac90* and *lac110*) were isolated from a poplar xylem cDNA library by a PCR-based approach (Ranocha *et al.*, 1999). Sequence comparison of the two cDNA clones confirmed that the two coniferyl alcohol oxidizing proteins isolated from poplar xylem are genetically distinct laccases. Both cDNAs encode proteins of about 60 kDa, suggesting that the difference in migration of the corresponding proteins on SDS-PAGE is likely due to the extent and/or the type of carbohydrate moieties. The putative full-length cDNA clones of the 90 kDa and 110 kDa laccases exhibited significant homology (50% identity + 12% similarity) and contained the characteristic features of laccases including a hydrophobic signal peptide, four putative copper binding sites and several putative N-glycosylation sites.

3.3 *Poplar laccases, a divergent multigene family*

In parallel, by using an *Acer* laccase cDNA to screen a poplar xylem cDNA library, three additional laccase clones, *lac1*, *lac2* and *lac3*, were isolated (Ranocha *et al.*, 1999). Therefore, at least five different laccase genes are present in poplar xylem. In general, they exhibit a relatively low degree of homology amongst themselves. Four different

Table 1. *Substrate specificity of the 90 kDa poplar laccase (Ranocha et al., 1999)*

Substrate	K_m (μM)	V (nkat mg^{-1})	V/K_m	V/K_m (%)
Coniferyl alcohol	1.5	90.5	60.3	100
Sinapaldehyde	2.7	81	30	49.8
Hydroquinone	0.8	13.2	16.5	27.4
Sinapyl alcohol	17	100	5.9	9.8
Coniferaldehyde	25	69	2.8	4.6
p-Coumaryl alcohol	16	31	1.9	3.2
4-Methyl catechol	1560	1528	1.0	1.7
p-Anisidine	10	7.2	0.7	1.2
ABTSa	26	13	0.5	0.8
Guaiacol[b]	√	√		
Syringaldazine[b]	√	√		
p-Coumaraldehyde[b]	√	√		

[a] ABTS = 2,2′-azinobis(3-ethylbenzothiazoline-6-sulfonate).
[b] (√) indicates that activity toward these compounds was detected but was too low to be spectrophotometrically quantified.

laccase cDNAs have been also characterized in *Liriodendron tulipifera*, yellow poplar, by LaFayette *et al.* (1999). Southern experiments suggest that six to eight closely related laccase genes exist in the yellow poplar genome.

A phylogenetic tree was constructed with several multicopper oxidases including plant and fungal laccases and plant ascorbate oxidases (*Figure 3*). Although the five poplar laccases are most closely related to other plant laccases (as opposed to fungal laccases or plant ascorbate oxidases), they do not form a species-related cluster as yellow poplar laccases, and are found dispersed amongst other plant laccases.

Laccase gene expression in poplar was analysed by Northern blot analysis using the divergent 3′ untranslated region of each gene as probe (*Figure 4*). In these experiments all of the laccases were exclusively and abundantly expressed in stem tissue as compared to leaves and roots. RT PCR experiments confirmed high levels of laccase expression in stem tissue, however, transcripts of each laccase gene were also detected in other organs (data not shown).

3.4 *Antisense laccase poplars do not exhibit changes in lignin profiles but an increase in soluble phenolic content*

With the objective of elucidating the potential role of laccases in lignification, we have transformed poplar with four different laccase genes (*lac1*, *lac3*, *lac90* or *lac110*) in the antisense orientation, under the control of a strong and constitutive promoter, 35S CaMV (unpublished data). The four populations of antisense laccase transformants were screened by Northern blot analysis of stem RNA using specific 3′UTR as probe. Transformants with low levels of laccase mRNA were selected for further analysis.

No significant differences in growth or development were observed between antisense laccase and control poplars (i.e., height, stem diameter, phyllotaxy). Moreover, tissue sections stained for lignin (phloroglucinol-HCl) and oxidase activity were identical to those of controls.

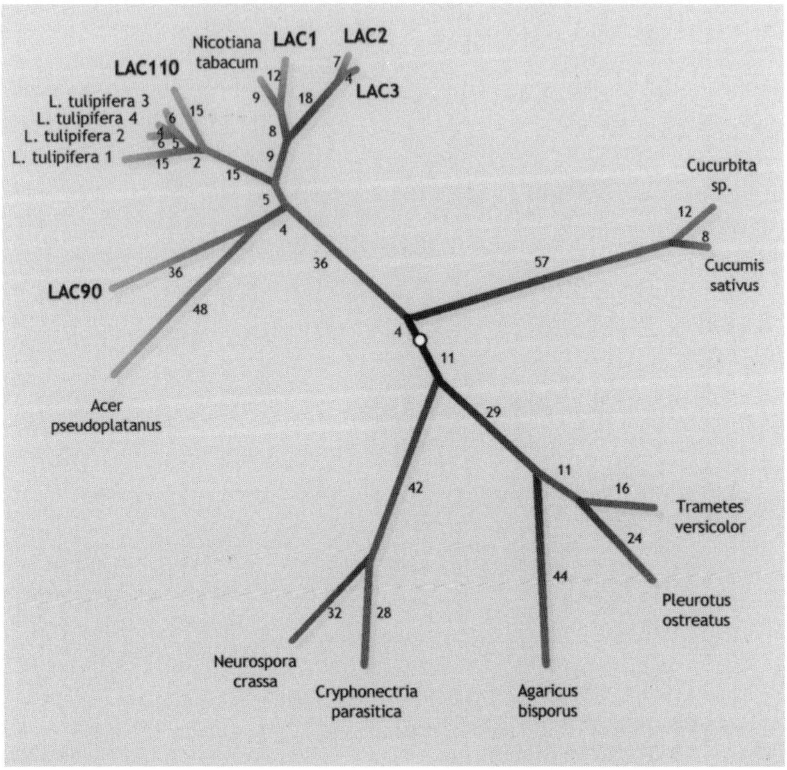

Figure 3. *Phylogenetic relationships between laccases and other blue copper oxidases. This phylogenetic tree was constructed using the PhyloTree program available on the Swiss Federal Institute of Technology Zurich web site (http: //cbrg.inf.ethz.ch). Data base accession numbers are as follows. For plants:* Acer pseudoplatanus *laccase (U12757),* Liriodendron tulipifera *laccases 1, 2, 3 and 4 (U73103-U73106),* Nicotiana tabacum *laccase 3 (U43542),* Populus trichocarpa *laccases 1, 2, 3, 90 and 110 (Y13769-Y13773),* Cucumis sativus *ascorbate oxidase (J04494),* Cucurbita *sp. ascorbate oxidase (X55779). For fungi:* Agaricus bisporus *laccase (L10664),* Cryphonectria parasitica *laccase (S38903),* Neurospora crassa *laccase (M18333),* Pleurotus ostreatus *laccase (Z22591),* Trametes versicolor *laccase (L78077).*

Lignin content was determined by two different methods: micro-Klason and acetyl bromide. None of the transformants had a significatively lower lignin content as compared to control poplars (*Table* 2). In all cases, lignins represented about 20% to 25% of extracted xylem residues. These values are similar to those obtained by Baucher *et al.* (1996) for the same species.

Lignin monomeric composition was determined by thioacidolysis. This method enables the quantification of non-condensed β-*O*-4 bonded subunits. By this method, we demonstrated that the amount of syringyl and guaiacyl residues is not significantly different in antisense plants as compared to controls (*Table* 2). Together, these results agree with previous data showing that lignin content and composition of *Liriodendron tulipifera* laccase antisense lines were unchanged as compared to the wild type (Dean *et al.*, 1998). An in-depth analysis of transgenic lignins should be carried out in order to determine if subtle changes in the structure of lignins have occurred as a result of manipulating laccase levels in poplars.

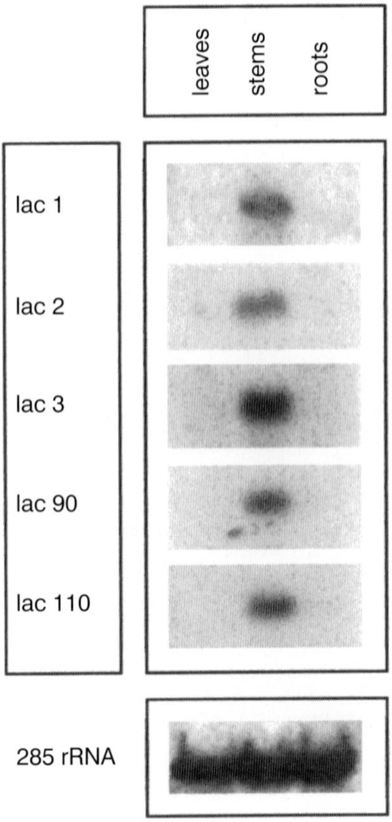

Figure 4. *Northern analysis of laccase gene spatial expression (Ranocha* et al., *1999). Total RNA from poplar leaves, stems and roots was electrophoresed on denaturing gels. Hybridization was performed with specific probes of* lac1, lac2, lac3, lac90 *and* lac110. *Ribosomal RNA (28S) was visualized with UV light in order to control that equal amounts of RNA were loaded on each lane.*

One potential difficulty in interpreting the results of our antisense experiments is the fact that transformants were screened and selected based on low levels of laccase mRNA. In some cases, it has been clearly demonstrated that there is not always a tight correlation between mRNA and protein amount. Therefore, the candidates that were selected based on reduced laccase mRNA may actually have relatively normal amounts of laccase. This could contribute to the fact that no phenotypic differences were observed in antisense plants.

Another critical point is that an RNA antisense approach may not be the most judicious approach for determining the physiological role of individual members of multigene families. It is conceivable that even if a given target has been successfully knocked-out, as appears to be the case in our experiments, either another laccase sufficiently distant in terms of sequence homology so that it is not affected by antisense transgene, or another genetically distinct phenoloxidase, may compensate to take over the suppressed enzyme's function. This may be especially problematical in the case of poplar laccases in light of their relatively low level of homology. If compensation occurs, the expected phenotype (in this study, effect on lignin content or composition) may not be observed. One solution to overcome this obstacle would be to use an antisense construct containing several transgenes in tandem in order to knock out a wide range of laccases in a single transformation experiment.

For these two main reasons, it is difficult for the moment to draw any clear-cut conclusions as to the role of laccases in lignification. Indeed, even if no effect on either

Table 2. *Lignin content, lignin composition and soluble phenolic compounds content of antisense laccase poplars.*

Individual[a]	Micro-Klason lignins[b] (% EXR)	Acetyl bromide lignins[c] (% EXR)	S+G units[d] (μmoles. g^{-1} lignins)	S/G units[d]	Soluble phenolics[e] (mg/gfw)
C1	20.7 ± 0.6	23.9 ± 1.1	1309 ± 175	2.07 ± 0.04	2.0 ± 0.1
C2	20.8 ± 0.7	24.3 ± 1.5	1273 ± 155	1.90 ± 0.12	2.1 ± 0.2
lac3.3as	20.5 ± 0.6	23.6 ± 1.1	1461 ± 98	2.14 ± 0.21	2.6 ± 0.1
lac3.4as	20.1 ± 0.7	23.4 ± 0.6	1187 ± 52	2.41 ± 0.04	5.6 ± 0.2
lac90.13as	20.5 ± 0.7	25.3 ± 0.8	1272 ± 78	1.76 ± 0.08	2.8 ± 0.2
lac110.4as	21.4 ± 0.8	25.4 ± 0.7	1358 ± 95	1.98 ± 0.05	2.0 ± 0.04
lac110.6as	19.3 ± 0.4	25.2 ± 0.4	1272 ± 89	1.95 ± 0.04	1.8 ± 0.05

[a] C1 and C2 are both controls: they correspond to untransformed poplars and to poplars transformed with empty vector respectively. Other individuals are designated by the name of the underexpressed laccase gene (*lac3*, *lac90* or *lac110*) followed by an arbitrarily attributed number.
[b] Lignin content was determined by the micro-Klason method (modified from Whiting *et al.*, 1981) and expressed as a percentage of the Extracted xylem residue (EXR).
[c] Lignin content was determined by the acetyl bromide method (Iiyama and Wallis, 1988; Iiyama and Wallis, 1990) and are expressed as a percentage of the extracted xylem residue (EXR).
[d] Monomeric composition (syringyl and guaiacyl yield and ratio) of the non-condensed lignin fraction was determined by thioacidolysis according to Lapierre *et al.*, 1986.
[e] Soluble phenolic compounds may be defined according to three criteria: solubility in methanol, reactivity with Folin's reagent and adsorption on PVPP.

lignin content or composition was observed as a result of individual laccase down-regulation, it is possible that laccases are not involved in lignification or that the knocked-out laccase is not the sole enzyme responsible for the final step of lignin biosynthesis.

Laccase antisense poplars were also analysed for total soluble phenolic content. In one line of antisense laccase transformants (poplars underexpressing the *lac3* gene), a two- to three-fold increase in soluble phenolic compounds was observed. Interestingly, Dean *et al.* (1998) showed that in yellow poplar transgenic lines overexpressing the *Acer pseudoplatanus* laccase gene, there was a general increase in the fluorescence of all cell walls, suggesting an increased deposition of phenolic material or a change in the cell wall phenolic components. The overexpressed laccase isoform could conceivably play a role in the polymerization/cross-linking of simple phenolic units. The increase in soluble phenolics observed in our antisense poplars would therefore be in agreement with this putative role for laccases in the cell wall. Several studies have demonstrated the importance of cell wall phenolic compounds in response to pathogen attack (Ingham, 1973; Nicholson and Hammerschmidt, 1992). Indeed, it would be of interest to determine the nature of the modified phenolic(s) and evaluate if these antisense laccase transformants are affected in their response to pathogens.

4. Conclusion

Several indirect lines of evidence including a tight spatio-temporal correlation between oxidase activities (or corresponding cofactor availability) and lignin formation have been provided (see Lewis *et al.*, 1999 for an extensive, recent review), suggesting the involvement of peroxidases and/or laccases in lignification. However, no unequivocal proof has ever been obtained for the role of a particular laccase or peroxidase isoenzyme in lignification through loss-of-function experiments in transgenic plants. Although many purified enzymes are capable of oxidizing monolignols *in vitro*, the physiological significance of these findings *in planta* remains to be elucidated.

There is an increasing body of evidence demonstrating that lignin deposition is a carefully orchestrated process. This is indicated by a highly specific monomeric composition or/and interunit linkages as a function of different tissues, cells and even cell wall layers. It can be postulated that a wide range of oxidase isoenzymes may be necessary in close association with regulatory proteins such as the recently characterized dirigent protein (Lewis *et al.*, 1999) to provide the appropriate type of lignin in a given tissue at a given time. This is one reason why global approaches based on enzymatic oxidase patterns or lignin analysis of plant organs and tissues are most likely insufficient in providing a clear answer as to the role of individual oxidases in lignification.

In contrast to peroxidases, laccases are specifically localized in the cell wall. Therefore they may be involved either in lignification or in the oxidation of other wall-associated, phenolic substrates. Phenolic acids that are cross-linked at the cell wall have a major impact on cell wall properties including mechanical resistance, cell adhesion, texture and wall stiffness. Further investigations of plant laccase function will be necessary in order to determine their precise role in cell wall biochemistry. These results will enable us to assess the suitability of laccases as targets in biotechnological strategies aimed at plant biomass improvement.

Acknowledgements

The authors wish to acknowledge Gisèle Borderies for excellent technical assistance in laccase purification, and Matthieu Chabannes and Julien Fabregue for help in lignin and soluble phenolic compounds content determination. We also thank J. Dean (University of Georgia, USA) and L. Faye (Université de Rouen, France) for *Acer* laccase cDNAs, and W. Boerjan (Universiteit Gent, Belgium) for the gift of the poplar xylem cDNA library. Ph.R. was supported by a fellowship from the Ministère de l'Enseignement Supérieur et de la Recherche. This research has been supported by the European Community (AIR Program: AIR2-CT93–1661), the Conseil Régional Midi-Pyrénées, the Centre National de la Recherche Scientifique and the Université Paul Sabatier, Toulouse.

References

Addleman, K., Dumonceaux, T., Paice, M.G., Bourbonnais, R. and Archibald, F.S. (1995) *Appl. Environ. Microbiol.* **61**: 3687–3694.

Bao, W., O'Malley, D.M., Whetten, R., Sederoff, R.R. (1993) A laccase associated with lignification in loblolly pine xylem. *Science* **260**: 672–674.

Baucher, M., Chabbert, B., Pilate, G., Van Doorsselaere, J., Tollier, M.-T., Petit-Conil, M., Cornu, D., Monties, B., Van Montagu, M., Inzé, D., Jouanin, L. and Boerjan, W. (1996) Red xylem and higher lignin extractability by down-regulating a cinnamyl alcohol dehydrogenase in poplar (*Populus tremula* × *P. alba*). *Plant Physiol.* **112**: 1479–1490.

Binnington, K.C. and Barrett, F.M. (1988) Ultrastructural localization of phenoloxidases in cuticle and haemopoietic tissue of the blowfly *Lucilia cuprina. Tissue Cell* **20**: 405–419.

Blanchette, R.A. (1991) Delignification by wood-decay fungi. *Ann. Rev. Phytopath.* **29**: 381–398.

Bligny, R. and Douce, R. (1983) Excretion of laccase by sycamore (*Acer pseudoplatanus* L.) cells. Purification and properties of the enzyme. *Biochem. J.* **209**: 489–496.

Bligny, R., Gaillard, J. and Douce, R. (1986) Excretion of laccase by sycamore (*Acer pseudoplatanus*) cells. Effects of a copper deficiency. *Biochem. J.* **237**: 583–588.

Call, H.P and Mücke, I. (1997) History, overview and applications of mediated lignolytic systems, especially laccase-mediator-systems. *J. Biotechnol.* **53**: 163–202.

Christensen, B.M. and Tracy, J.W. (1989) Arthropod-transmitted parasites: mechanisms of immune interaction. *Am. Zool.* **29**: 387–398.

Clutterbuck, A.J. (1972) Absence of laccase from yellow spored mutants of *Aspergillus nidulans. J. Gen. Microbiol.* **70**: 423–435.

Datta, A., Betterman, A. and Kirk, K.T. (1991) Identification of a specific manganese peroxidase among lignolytic enzymes secreted by *Phanerochaete chrysoporium* during wood decay. *Appl. Environ. Microbiol.* **57**: 1453–1460.

Davin, L.B., Bedgar, D.L., Katayama, T. and Lewis, N.G. (1992) On the stereoselective synthesis of (+)-pinoresinol in *Forsythia suspensa* from its achiral precursor, coniferyl alcohol. *Phytochem.* **31**: 3869–3874.

Dean, J.F.D. and Eriksson, K.-E.L. (1994) Laccase and the deposition of lignin in vascular plants. *Holzforschung* **48**: 21–33.

Dean, J.F.D., LaFayette, P.R., Rugh, C., Tristram, A.H., Hoopes, J.T., Eriksson, K.-E.L. and Merckle, S.A. (1998) Laccases associated with lignifying vascular tissues. In: *Lignin and Lignan Biosynthesis* (ed. Lewis, N.G. and Sarkanen, S.), American Chemical Society, pp. 96–108.

Dittmer, J.K., Patel, N.J., Dhawale, S.W. and Dhawale, S.S. (1997) Production of multiple laccase isoforms by *Phanerochaete chrysosporium* grown under nutrient sufficiency. *FEMS Microbiol. Lett.* **149**: 65–70.

Driouich, A., Laine, A.C., Vian, B. and Faye, L. (1992) Characterization and localization of laccase forms in stem and cell cultures of sycamore. *Plant J.* **2**: 13–24.

Eggert, C., Temp, U. and Eriksson, K.-E.L. (1996) The ligninolytic system of the white-rot fungus *Pycnoporus cinnabarinus*: purification and characterization of the laccase. *Appl. Environ. Microbiol.* **62**: 1151–1158.

Eggert, C., Temp, U. and Eriksson, K.-E.L. (1997) Laccase is essential for lignin degradation by the white-rot fungus *Pycnoporus cinnabarinus. FEBS Lett.* **407**: 89–92.

Eriksson, K.-E., Blanchette, R.A. and Ander, P. (1990) Biodegradation of lignin. In: *Microbial and Enzymatic Degradation of Wood and Wood Components*. Springer-Verlag, Berlin.

Felton, G., Donato, K., Del Vecchio, R. and Duffey, S. (1989) Activation of plant foliar oxidases by insect feeding reduces nutritive quality of foliage for noctuid herbivores. *J. Chem. Ecol.* **15**: 2667–2694.

Ferrar, P.H. and Walker, J.R.L. (1996) Inhibition of diphenol oxidases: a comparative study. *J. Food Biochem.* **20**: 15–30.

Freudenberg, K., Harkin, J.M., Reichert, M. and Fukuzumi, T. (1958) Die ander Verholzung beteiligten Enzym. Die Dehydrierung des Sinapilalkohols. *Die Chem. Berlin* **91**: 581–590.

Geiger, J.P., Nicole, M., Nandris, D. and Rio, B. (1986) Root rot disease of *Hevea brasiliensis* I. Physiological and biochemical aspects of aggression. *Eur. J. Pathol.* **16**: 22–37.

Givaudan, A., Effosse, A., Faure, D., Potier, P., Bouillant, M. and Bally, R. (1993) Polyphenol oxidase from *Azospirillum lipoferum* isolated from the rhizosphere: evidence for a laccase in non-motile strains of *Azospirillum lipoferum*. *FEMS Microbiol. Lett.* **108**: 205–210.

Harkin, J.M. and Obst, T.R. (1973) Lignification in trees: indication of exclusive peroxidase participation. *Science* **180**: 296–297.

Hermann, T.E., Kurtz, M.B. and Champe, S.P. (1983) Laccase localized in Hülle cells and cleistothecial primordia of *Aspergillus nidulans*. *J. Bacteriol.* **154**: 955–964.

Iiyama, K. and Wallis, F.A. (1988) An improved acetyl bromide procedure for determining lignin in woods and wood pulps. *Wood Sci. Technol.* **22**: 271–280.

Iiyama, K. and Wallis, F.A. (1990) Determination of lignin in herbaceous plants by an improved acetyl bromide procedure. *J. Sci. Food Agric.* **51**: 145–161.

Ingham, J.L. (1973) Disease resistance in higher plants. The concept of preinfectional and postinfectional resistance. *Phytopath.* **78**: 314–315.

Joy, R.W., Sugiyama, M., Fukuda, H. and Komamine, A. (1995) Cloning and characterization of polyphenol oxidase cDNAs of *Phytolacca americana. Plant Physiol.* **107**: 1083–1089.

Karhunen, E., Kantelinen, A. and Niku-Paavola, M.L. (1990a) Mn-dependent peroxidase from the lignin-degrading white rot fungus *Phlebia radiata. Arch. Biochem. Biophys.* **279**: 25–31.

Karhunen, E., Niku-Paavola, M.L., Viikari, L., Haltia, T., Van Der Meer, R.A. and Duine, J.A. (1990b) A novel combination of prosthetic groups in a fungal laccase; PQQ and two copper atoms. *FEBS Lett.* **267**: 6–8.

Kim, D.H., Rigling, D., Zhang, L. and Van Alfen, N.K. (1995) A new extracellular laccase of *Cryphonectria parasitica* is revealed by deletion of *Lac1*. *Mol. Plant Microbe Interact.* **8**: 259–266.

Kowalski, S.P., Eannetta, N.T., Hirzel, A.T. and Steffens, J.C. (1992) Purification and characterization of polyphenol oxidase from glandular trichomes of *Solanum berthaultii. Plant Physiol.* **100**: 677–684.

Lapierre, C, Rolando, C. and Monties, B. (1986) Thioacidolysis of poplar lignins: identification of monomeric syringyl products and characterization of guaiacyl syringyl fractions. *Hotzforschung* **40**: 113–118.

Lafayette, P.R., Eriksson, K.E.L. and Dean, J.F.D. (1999) Characterization and heterologous expression of laccase cDNAs from xylem tissues of yellow-poplar. *Plant Mol. Biol.* **40**: 23–35.

Lax, A.R. and Vaughn, K.C. (1991) Colocalization of polyphenol oxidase and photosystem II proteins. *Plant Physiol.* **96**: 26–31.

Lewis, N.G., Davin, L.B. and Sarkanen, S. (1999) The nature and function of lignins. In: *Comprehensive Natural Products Chemistry* (ed. Sir D.H.R. Barton, Nakaski, K. and Meth-Cohn, O.), Pergamon Press, Vol. 3, pp. 617–745.

Lewis, N.G. and Yamamoto, E. (1990) Lignin: occurrence, biogenesis and biodegradation. *Ann. Rev. Plant Mol. Biol.* **41**: 455–496.

MacDonald, W.L. and Fullbright, D.W. (1991) *Plant Disease* **75**: 656–661.

McDougall, G.J., Stewart, D. and Morisson, I.M. (1994) Cell wall-bound oxidases from tobacco (*Nicotiana tabacum*) xylem participate in lignin formation. *Planta* **194**: 9–14.

Marbach, I., Harel, E. and Mayer, A.M. (1984) Molecular properties of extracellular *Botrytis cinerea* laccase. *Phytochem.* **23**: 2713–2717.

Mayer, A.M. (1987) Polyphenol oxidases in plants — recent progress. *Phytochem.* **26**: 11–20.

Messerschmidt, A. and Huber, R. (1990) The blue oxidases, ascorbate oxidase, laccase and ceruloplasmin. *Eur. J. Biochem.* **187**: 341–352.

Nakamura, W. (1967) Studies on the biosynthesis of lignins. I. Disproof against the catalytic activity of laccase in the oxidation of coniferyl alcohol. *J. Biochem.* **62**: 54–61.

Nicholson, R.L. and Hammerschmidt, R. (1992) Phenolic compounds and their role in disease resistance. *Ann. Rev. Phytopathol.* **30**: 369–389.

Niku-Paavola, M.L., Karhunen, E., Salola, P. and Raunio, V. (1988) Ligninolytic enzymes of the white-rot fungus *Phlebia radiata. Biochem J.* **254**: 877–884.

O'Malley, D.M., Whetten, R., Bao, W., Chen, C.-L. and Sederoff, R.R. (1993) The role of laccase in lignification. *Plant J.* **4(5)**: 751–757.

Okusa, K., Miyakoshi, T. and Chen, C.-L. (1996) Comparative studies on dehydrogenative polymerization of coniferyl alcohol by laccases and peroxidases. Part 1. Preliminary results. *Holzforschung* **50**: 11–16.

Ranocha, Ph., McDougall, G., Hawkins, S., Sterjiades, R., Borderies, G., Stewart, D., Cabanes-Macheteau, M., Boudet, A.-M. and Goffner, D. (1999) Biochemical characterization, molecular cloning and expression of laccases — a divergent gene family — in poplar. *Eur. J. Biochem.* **259**: 485–495.

Reinhammar B. and Malström B.G. (1981) 'Blue' copper-containing oxidases. In *Copper Proteins — Metal Ions in Biology.* John Willey & Sons, New-York, pp. 109–149.

Rigling, D., Heiniger, H and Hohl, H.R. (1989) Reduction of laccase activity in dsRNA-containing hypovirulent strains of *Cryphonectria (Endothia) parasitica. Phytopath.* **79**: 219–223.

Rigling, D. and Van Alfen, N.K. (1991) Regulation of laccase biosynthesis in the plant-pathogenic fungus *Cryphonectria parasitica* by double-stranded RNA. *J. Bacteriol.* **24**: 8000–8003.

Saloheimo, M., Niku-Paavola, M.L. and Knowles, J.K. (1991) Isolation and structural analysis of the laccase gene from the lignin-degrading fungus *Phlebia radiata. J. Gen. Microbiol.* **137**: 1537–1544.

Savidge, R.A. and Udagama-Randeniya, P.V. (1992) Cell wall-bound coniferyl alcohol oxidase associated with lignification in conifers. *Phytochem.* **31**: 2959–2966.

Savidge, R.A., Udagama-Randeniya, P.V., Xu, Y., Leinhos, V. and Förster, H. (1998) Coniferyl alcohol oxidase: a new enzyme spatio-temporally associated with lignifying tissues. In: *Lignin and Lignan Biosynthesis* (ed. Lewis, N.G., Sarkanen, S.), American Chemical Society, pp. 109–130.

Sidijanski, S., Mathews, G.V. and Vanderberg, J.P. (1997) Electrophoretic separation and identification of phenoloxidases in hemolymph and midgut of adult *Anopheles stephensi* mosquitoes. *J. Parasitol.* **83**: 686–691.

Siedow, J.N., Umbach, A.L. and Moore, A.L. (1995) The active site of the cyanide-resistant oxidase from plant mitochondria contains a binuclear iron center. *FEBS Lett* **362**: 10–14.

Steffens, J.C., Harel, E., Hunt, M.D. and Thipyapong, P. (1996) Polyphenol oxydase. Dans *Polyphenols 96* (Vercauteren, J, Cheze, C, Triaud, J, éditeurs), INRA Editions, Versailles, France, pp. 223–250.

Sterjiades, R., Dean, J.F.D. and Eriksson, K.-E.L. (1992) Laccase from sycamore maple (*Acer pseudoplatanus*) polymerizes monolignols. *Plant Physiol.* **99**: 1162–1168.

Sterjiades, R., Ranocha, Ph., Boudet, A.-M. and Goffner, D. (1996) Identification of specific laccase isoforms capable of polymerizing monolignols by an 'in-gel' procedure. *Analytical Biochem.* **242**: 158–161.

Sugumaran, M., Giglio, L., Kundzicz, H., Saul, S. and Semensi, V. (1992) Studies of the enzymes involved in puparial cuticle sclerotization in *Drosophilia melanogaster. Arch. Insect Biochem. Physiol.* **19**: 271–283.

Thomas, B.R., Yonekura, M., Morgan, T.D., Czapla, T.H., Hopkins, T.L. and Kramer, K.J. (1989) A trypsin-solubilized laccase from pharate pupal integument of the tobacco hornworm, *Manduca sexta. Insect Biochem.* **19**: 611–622.

Thurston, C.F. (1994) The structure and function of fungal laccases. *Microbiol.* **140**: 19–26.

Tolbert, N.E. (1973) Activation of polyphenol oxidase of chloroplasts. *Plant Physiol.* **51**: 234–244.

Viterbo, A., Yagen, B. and Mayer, A.M. (1993) Cucurbitacins, 'attack' enzymes and laccase in *Botrytis cinerea. Phytochem.* **32**: 61–65.

Walker, J.R.L. (1975) Enzymic browning in foods; a review. *Enz. Tech. Digest* **4**: 80–100.

Whiting, P., Favis, B.D., St-Germain, F.G.T and, Goring, D.A.I. (1981) Fractional separation of middle lamella and secondary wall tissue from spruce wood. *J. Wood Chem. Tech.* **1**: 29–42.

Worrall, J.J., Chet, I. and Hüttermann, A. (1986) Association of rhizomorph formation with laccase activity in *Armillaria* spp. *J. Gen. Microbiol.* **132**: 2527–2533.

Yoshida, H. (1883) Chemistry of lacquer (Urishi) Part I. *J. Chem. Soc.* **43**: 472–486.

33

Cell-wall-associated proteins and wall biogenesis in developing compression and non-compression woods

G.J. McDougall

1. Introduction

1.1 *Developing xylem and the biochemical study of wall biogenesis*

The formation of new wood or xylem probably represents the most complex and sustained programme of cell wall synthesis, assembly and modification in plant cell development. In conifers, new xylem cells (> 90% tracheids by volume) are produced by the division of the lateral meristems and initially have plastic, thin cell walls that allow elongation and expansion. However, after expansion, tracheids deposit multi-layered secondary walls that are subsequently encrusted with lignin (Brown, 1971; Wardrop, 1981). During maturation and apoptosis, the walls undergo selected modification before the formation of the functional xylem elements. The cambial meristem continues to divide and a single radial file, from cambial zone to the mature wood, may contain cells at all stages of development from division and enlargement through secondary cell wall deposition and lignification to maturation and apoptosis. As large amounts of tissue can be easily collected, the developing xylem of conifers represents an ideal system to study the biochemical events involved in wall biogenesis.

1.2 *Cell wall enzymes, wall biogenesis and xylogenesis*

Cell-wall-associated enzymes influence wall structure in two ways. Firstly, although many components are synthesized intracellularly and extruded into the wall (McDougall *et al.*, 1993), certain macromolecular components (e.g., lignin) are polymerized in the wall by cell-wall-associated enzymes (Boudet *et al.*, 1995). Secondly, it has become clear that cell-wall-associated enzymes alter the structural properties or

Cell & Molecular Biology of Wood Formation, edited by R. Savidge, J. Barnett and R. Napier.

architecture of the wall by modifying pre-existing components, for example the depolymerization of pectins during ripening (Steele *et al.*, 1997) or the cross-linking of components to limit growth (Fry, 1986).

I have used two different approaches to study the role of cell-wall-associated proteins in xylogenesis. The first approach was a biochemical study that targeted cell-wall-associated oxidases in the developing xylem of Sitka spruce (*Picea sitchensis*). Over the last decade, evidence has accumulated that oxidase activity capable of oxidizing monolignols is specifically associated with the lignifying tissues of a range of woody plants (Sterjiades *et al.*, 1992; Savidge and Udagama-Randenyia, 1992; Bao *et al.*, 1993; Liu *et al.*, 1994; McDougall and Morrison, 1996; Richardson and McDougall, 1997; Richardson *et al.*, 1997; Richardson *et al.*, 1998; Deighton and McDougall, 1998). This challenged the long-held assumption that peroxidases exclusively catalysed this reaction in lignifying tissues (see reviews by O'Malley *et al.*, 1993 and Dean and Eriksson, 1994). Genes encoding laccase-type polyphenol oxidases (EC 1.3.10.2) have been identified in *Acer pseudoplatanus* (LaFayette *et al.*, 1995), tobacco (Kiefer-Mayer *et al.*, 1996) and laccase genes have been found to be specifically expressed in the developing xylem of *Liriodendron tulipifera* (LaFayette and Dean, 1997) and poplar (Ranocha *et al.*, 1999). Surprisingly, considering that much of the early descriptive work concerned oxidase activity associated with the lignifying xylem of conifer species (Savidge and Udagama-Randenyia, 1992; Bao *et al.*, 1993), the nature of the oxidase activity in conifers has not been confirmed. I report the culmination of the targeted approach; the purification and amino-terminal sequencing of an oxidase, which proved to be the first sequence of a laccase-like polyphenol oxidase from a conifer.

The second approach compared the expression of cell-wall-associated proteins in the developing xylem of compression and non-compression wood of branches of Sitka spruce. Compression wood is mainly formed in trunks or stems of conifers as a corrective reaction to a bending from the perpendicular and is also universally formed on the underside of branches to maintain their branch angle (Timell, 1973, 1986). Compression wood differs from non-compression wood in the ultrastructure of its tracheids and in the structure and composition of their secondary cell walls (see Timell, 1982, 1983, 1986; Parham and Gray, 1984). Compression wood tracheids are generally shorter, more rounded and have characteristic intercellular spaces. Their secondary cell walls comprise only two layers (S1 and S2) compared to the three layers of non-compression wood tracheids. The S1 layer is composed of concentric lamellae of cellulose microfibrils in near-transverse orientation to the cell axis which renders this layer birefringent in polarized light (Timell, 1983). The S2 layer is also birefringent, but to a lesser extent as the cellulose microfibrils are deposited at 30–50° c.f. 10–30° of non-compression wood.

The cell walls of compression wood tracheids also differ from non-compression wood tracheids in their chemical composition. They contain more lignin, less cellulose and altered levels of matrix hemicelluloses, including unique polysaccharides (Timell, 1983; Parham and Gray, 1984). Approximately 90% of the lignin is present in the secondary cell wall layers, particularly in the lumen side of the S2 layer (S2(L)), with correspondingly less lignin in the middle lamella and primary cell wall. Compression wood lignin also has an elevated content of *p*-hydroxyphenylpropane units (Monties, 1989) and is more condensed than non-compression wood (Nimz *et al.*, 1981).

The ultrastructure of compression wood is indicative of accelerated development to overcome and alter the direction of growth of the trunk or branch. The tracheids are shorter and have bent, forked and deformed tips characteristic of rapid, intrusive growth (Münch, 1940). The S1 and S2 layers are laid down as loose convoluted lamellae that become compacted, forming the characteristic helical cavities (Timell, 1973). Indeed, the high lignin content of the S2(L) layer may be a consequence of the lignin deposition between the (initially) loosely packed cellulose microfibrils. As a consequence of these differences in structure and composition, compression wood has altered physical properties that impinge on its value as timber or pulp. The wood is generally denser but more brittle and suffers from enhanced longitudinal shrinkage. Groundwood and sulphite pulps are inferior to those from non-compression wood (Timell, 1973).

Although the chemical structure and physical properties of compression wood have been largely mapped out, the biochemical processes that control its formation are less well understood. The altered lignin structure outlined above is brought about by the enhanced supply of *p*-coumaryl alcohol via *p*-coumaryl glucoside (Fukushima *et al.*, 1997) and is influenced by elevated levels of the lignin biosynthetic enzymes, 4-coumarate:CoA ligase and cinnamoyl CoA oxidoreductase, in lignifying compression wood (Kutsuki and Higuchi, 1981; Zhang and Chiang, 1997).

Non-compression and compression wood could be said to represent the culmination of two regimes of development and illustrate the plasticity inherent in wood cell development. As compression and non-compression wood tracheids differentiate from cambial initials with the same genotype, the great phenotypic differences in ultrastructure, cell wall structure and composition must result from the differential expression of genes. However, little work has been done to identify the differentially expressed genes. One report (Allona *et al.*, 1996) intimated a comparison of cDNAs from normal and compression wood of *Pinus taeda* but no genes were identified. I describe a novel means to identify genes involved in wall biogenesis during xylogenesis through the comparison of proteins, especially cell-wall-associated proteins, that are differentially expressed in the developing xylem on the compression and non-compression sides of branches of Sitka spruce.

2. Materials and methods

2.1 *Plant material*

Branches of Sitka spruce (*Picea sitchensis*) were obtained from the Forest Research clonebank at Ledmore, Perth and Kinross, UK in mid May 1998. In the laboratory, the bark was removed and the developing xylem scraped from the old wood. For the comparative study, the extent of formation of compression wood was estimated by the dark colour on the underside of branches and the compression and non-compression xylem tissue carefully removed.

2.2 *Extraction procedure*

Xylem tissue was extracted by the method outlined previously (Deighton and McDougall, 1998). The comparative extraction of the compression and non-compression xylem was carried out in an identical fashion. In brief, the tissue was placed in a five-fold excess of ice-cold homogenization buffer (25 mM MOPS pH 7.0 containing 0.5% v/v insoluble polyvinylpyrrolidine (PVP) and 0.5 mM phenylmethyl sulphonyl fluoride (PMSF)) and homogenized in a Waring blender, then in an Ultra-Turrax disintegrator. The soluble extract was clarified by filtration through muslin, then glass fibre filters. The insoluble residue was re-homogenized in fresh ice-cold buffer that lacked both PVP and PMSF. These filtrates were discarded and the entire procedure was repeated three times. The insoluble crude cell wall residue was then extracted with 25 mM MOPS pH 7.0 containing 200 mM $CaCl_2$ (ConA buffer) for 1 h on ice with stirring. The cell wall extract was clarified by filtration.

2.3 *Purification of oxidase*

Portions of the cell wall extracts were applied to individual small Concanavalin-A Sepharose columns and the eluate discarded. After a wash with a five-fold column volume of ConA buffer, the ConA-bound extract, which is enriched in cell-wall-associated glycoproteins (McDougall, 1997), was eluted using equal volumes of ConA buffer containing 100 mM α-methyl mannoside. The use of immobilized metal affinity chromatography and gel permeation chromatography to purify oxidase activity is detailed in McDougall (1998).

2.4 *Assays for oxidase, monolignol oxidase activity and protein content*

Oxidase activity was determined by measuring the oxidation of 2,2′-azino-bis(3-ethylbenzothiazoline-6-sulfonic acid) (ABTS) over a fixed period (Sterjiades *et al.*, 1992). Monolignol oxidase activity was measured by the decline in A_{MAX} of the three monolignols (coniferyl alcohol at 262 nm, sinapyl alcohol at 270 nm and *p*-coumaryl alcohol at 260 nm) over a fixed period (Deighton and McDougall, 1998). Protein content was measured by the dye-binding method (Bradford, 1976).

2.5 *Analytical SDS-PAGE*

Proteins in extracts were routinely precipitated by the addition of a 2.5-fold (v/v) excess of cold (–20°C) acetone, mixed well then stored overnight at –20°C and collected by centrifugation (× 10 000 **g**, 5 min). The precipitated proteins were re-dissolved in SDS-PAGE sample buffer, boiled for 5 min then applied to SDS-PAGE (Laemmli, 1970). A Mini-Protean II slab gel system was used according to the makers' instructions (Bio-Rad Ltd., Hemel Hempstead, UK) with 10% (w/v) polyacrylamide gels. Pre-stained markers and silver stain protein markers were used for molecular weight estimation and gels were stained using the Bio-Rad silver stain method.

2.6 *Non-denaturing SDS-PAGE (ND SDS-PAGE)*

This procedure was carried out as described before (Richardson *et al.*, 1997). After electrophoresis, oxidase-like activity was detected using 0.2 mM diaminofluorene (DAF) in 100 mM sodium acetate buffer pH 5.0. Staining was normally complete within 20 min then, after oxidase-like bands had been noted, peroxidase activity was visualized by the addition of 2.5 mM hydrogen peroxide.

2.7 *Amino-terminal protein sequencing*

Protein samples (100–200 μg) were precipitated using acetone then prepared in Nu-PAGE sample buffer according to the manufacturer's instructions (NOVEX Ltd, San Diego, USA). The samples were electrophoresed in pre-cast gradient (4–12% acrylamide) gels with suitable marker proteins using the Nu-PAGE MES running buffer. The gel was then blotted onto PVDF membrane using the Nu-PAGE blotting system and stained with Amido black.

For the comparative study, the abundance of target bands was checked against initial SDS-PAGE results. The amino-terminal sequence of excised bands was determined using a PROCISE 492 protein sequencer using a modified Edman degradation method optimized for samples from PVDF blots (Applied Biosystems Ltd, Warrington, UK).

Sequence similarities were determined using the 'BLITZ' method against the non-redundant protein and DNA databases maintained by the European Bioinformatics Institute (at http://www2.ebi.ac.uk/bic_sw). Similarities to non-redundant EST sequences were determined using the BLASTN series of programmes using the database of expressed sequences tags maintained by the National Centre for Biotechnology Information (see http://www2.ncbi.nih.gov/dbEST). Where required, the positions of amino-terminal residues of DNA sequences were predicted using the SignalP programme (see http://www.cbs.dtu.dk/services/SignalP).

3. Results

3.1 *Purification of oxidase activity from developing xylem of Sitka spruce*

As noted previously (Richardson *et al.*, 1997), oxidase activity is specifically expressed in the lignifying, developing xylem of Sitka spruce (*Figure 1*). This oxidase activity was enriched in the ConA-bound extract which was obtained by a procedure known to select cell-wall-associated glycoproteins (McDougall, 1997). Oxidase activity was then purified by immobilized zinc metal affinity chromatography followed by gel permeation chromatography (McDougall, 1998). The zinc-bound oxidase fraction represented only a 1.4-fold increase in specific activity but, more importantly, it contained few enriched protein bands (*Figure 2*, lane c). Gel permeation resulted in a further 2.4-fold increase in specific activity and further enriched the band at ~85 kDa (*Figure 2*, lane d). The co-purification of this band with the oxidase activity strongly suggests that it was responsible for the oxidase activity. The ~85 kDa band gave an amino-terminal sequence that had homology with plant laccases (results not shown, manuscript in preparation). This confirms that the oxidase activity in the lignifying xylem was, at least in part, due to the expression of laccases.

3.2 *Comparison of cell wall proteins expressed by developing xylem from compression and non-compression wood*

Proteins in the soluble, cell-wall and ConA-bound extracts from developing compression and non-compression xylem were separated by SDS-PAGE and stained with silver (*Figure 3*). Bands with M_rs of 66, 62, 58 and 47 kDa were more abundant in the soluble extracts of the compression tissue and the bands at 66 kDa and 62 kDa may be unique to compression tissue. The bands at 60, 54, 46, 44 and a doublet at 42 kDa were more abundant in the soluble extracts of the non-compression tissue (*Figure 3*, compare lanes a and b).

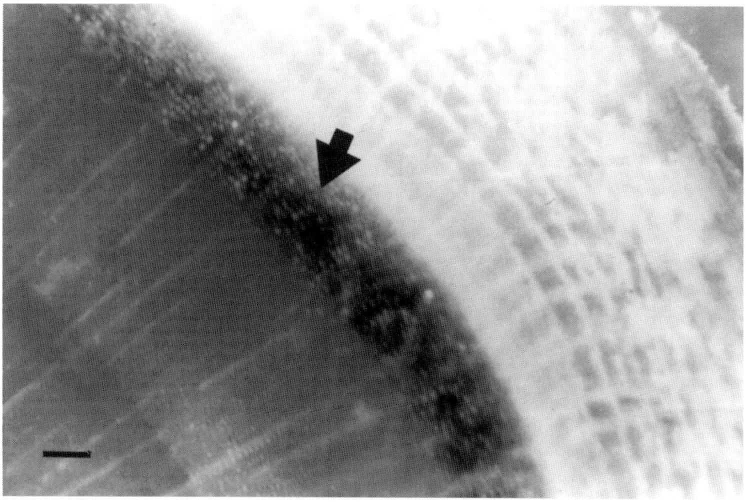

Figure 1. *Oxidase activity in lignifying, developing xylem of Sitka spruce. Cross sections of branches of Sitka spruce from mid May were stained with 1 mM o-dianisidine in 100 mM acetate buffer pH 5.0 containing 50 μg ml⁻¹ catalase. The dark colouration in the developing new xylem (see arrow) is due to the presence of the oxidized product of o-dianisidine. Bar = 100 μm.*

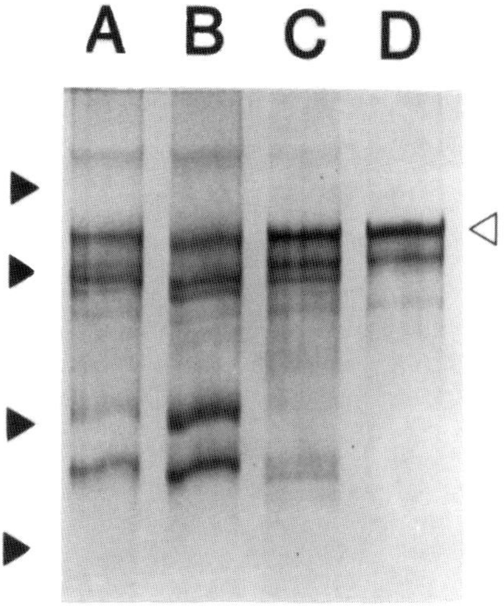

Figure 2. *SDS-PAGE of samples from purification of oxidase. Each lane contains 2.5 μg of protein. A = cell wall extract, B = ConA-bound fraction, C = Zn-bound IMAC fraction and D = oxidase-containing fraction after gel permeation on Superose-6. Specific oxidase activities were A = 233 ± 15, B = 881 ± 16, C = 1195 ± 55 and D = 2930 ± 85 nmol min^{-1} mg^{-1} protein. The migration of molecular mass marker proteins (97.4, 66.2, 45 and 30 kDa) is denoted by the closed arrows and the putative oxidase band by the open arrow.*

Bands with M_rs of 120, 78, 59, 54 and 48 kDa were more abundant in the cell wall extract from compression tissue but no band was consistently more abundant in the equivalent non-compression extract (*Figure 3*, compare lanes c and d). There were notable differences in the pattern of bands in the cell wall extracts and their ConA-bound counterparts (compare lanes c and e, non-compression; lanes d and f, compression) which indicated that not all of the proteins in the cell-wall-associated extracts bind to this lectin. For example, the band at 78 kDa, which was abundant in the cell wall extract from the compression tissue (*Figure 3*, lane d second arrow from top), was not present in the corresponding ConA-bound extract (*Figure 3*, lane f). A

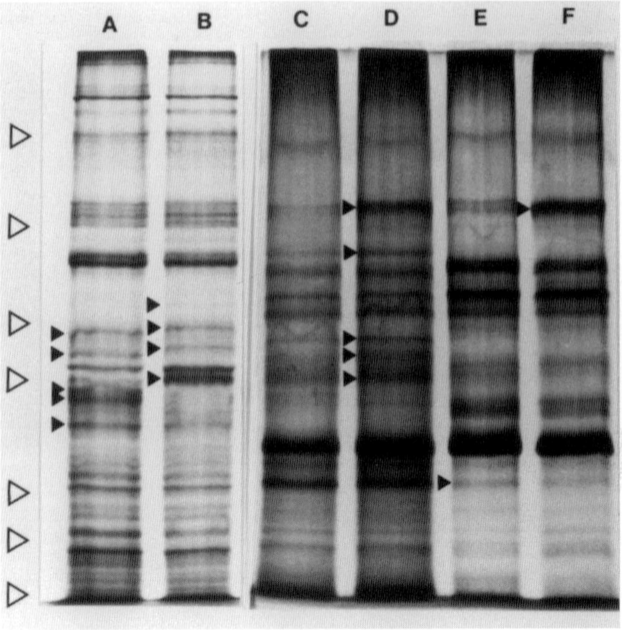

Figure 3. *Comparison of proteins in extracts from compression and non-compression xylem by SDS-PAGE. Each lane contains 2.5 μg of protein. Lane A = soluble extract from non-compression xylem, lane B = soluble extract from compression xylem, lane C = cell wall extract from non-compression xylem, lane D = cell wall extract from compression xylem, lane E = ConA-bound fraction from non-compression xylem, lane F = ConA-bound fraction from compression xylem. The closed arrows denote bands that are more abundant in either the compression or non-compression extracts. The open arrows denote the migration of molecular mass marker proteins (175, 83, 62, 47.5, 32.5, 25 and 16 kDa).*

band with an M_r of 120 kDa was much more abundant in the ConA-bound extract of the compression tissue, whereas a band at 34 kDa was more abundant in the ConA-bound extract of the non-compression tissue (*Figure 3*, compare lanes e and f).

The compression-specific 120 kDa band gave an amino-terminal protein sequence that was homologous with a number of plant laccase genes (*Table 1*). The greatest homology was with one of the laccase genes from the developing xylem of *Liriodendron tulipifera*; LAC2–4 (65% identity and 85% similarity), whereas the other *Liriodendron* genes had lower homology. The Sitka sequence had high homology (71% identity and 85% similarity) with a putative diphenol oxidase (F16M14.1) from *Arabidopsis*, but this was based on only 14 residues. Laccases from *Arabidopsis* (T914.21), tobacco (Kiefer-Meyer *et al.*, 1996) and poplar xylem (LAC-110) (Ranocha *et al.*, 1999) also showed considerable homology. Moreover, the amino-terminal sequence of the Sitka protein can be aligned exactly with the amino-terminal residues quoted or predicted (Nielsen *et al.*, 1997) for these proteins. The blank cycle at residue 17 (X) coincides with an asparagine residue in a potential N-glycosylation site (consensus sequon = Asn-Xaa-Ser/Thr where Xaa can be any amino acid except Pro; Hubbard and Ivatt (1981)) in nearly all the other sequences and is therefore likely to be due to a glycosylated asparagine residue.

Table 1. *Comparison of the amino-terminal protein sequence of the 120 kDa band with database sequences*

	1				5					10					15					20			
SITKA	K	H	W	Q	T	T	**R**	N	**Y**	K	**F**	N	V	R	L	K	X	V	T	**R**	ID	S	N
Liriodendron LAC2–4	K	H	A	G	I	T	**R**	H	**Y**	K	**F**	D	I	R	L	Q	*N*	V	T	**R**	65	85	20
Liriodendron LAC2–3	K	H	A	R	I	T	**R**	H	**Y**	K	**F**	D	I	Q	M	R	*N*	V	T	**R**	50	85	20
Liriodendron LAC2–1	T	H	T	G	S	T	W	H	**Y**	K	**F**	D	V	Q	L	K	*S*	V	K	**R**	50	70	20
Liriodendron LAC2–2	N	Q	E	G	I	T	**R**	H	**Y**	K	**F**	D	I	R	L	Q	*N*	V	T	**R**	50	60	20
Arabidopsis T914.21	A	S	A	G	I	T	**R**	H	**Y**	Q	**F**	D	I	Q	L	K	*N*	I	T	**R**	53	93	20
Arabidopsis F16M14.1	–	–	–	–	–	–	**R**	H	**Y**	K	**F**	N	V	V	M	K	*N*	V	T	**R**	71	85	14
Tobacco U43542	–	–	–	–	R	I	**R**	H	**Y**	K	**F**	N	V	V	M	K	*N*	T	T	**R**	64	86	14
Poplar LAC-110	–	–	–	–	–	T	**R**	Q	**Y**	T	**F**	N	I	T	H	K	*N*	F	T	**R**	53	73	15

Near-invariant residues are in bold, asparagine residues in potential N-glycosylation sequons are in italic. ID denotes identity, S denotes similarity and N denotes number of residues compared.

The total and specific oxidase activities of the cell-wall-associated extracts from the developing compression and non-compression xylem were similar (*Table 2*). This suggests that the non-compression extracts contained a different set of oxidases that compensate for the greater abundance of the 120 kDa laccase in the compression extracts. However, the ConA-bound extracts from both the compression and non-compression tissue yielded the same three oxidase-like bands on ND SDS-PAGE with M_rs of 128, 85 and 50–60 kDa (*Figure 4*, lanes a and b). The bands at 128 and 85 kDa did not darken (results not shown) upon addition of hydrogen peroxide. This indicates that these oxidase-like bands were not caused by peroxidase activity. The 85 kDa oxidase-like band was more abundant in the non-compression tissues (*Figure 4*; lane a) and the 128 kDa oxidase-like band was more abundant in the compression sample (*Figure 4*, lane b).

Although the ConA-bound extracts from the compression and non-compression tissues had similar oxidase activity against ABTS, they had different abilities to oxidize monolignols (*Table 3*). Each extract oxidized the monolignols in the order of preference sinapyl alcohol > coniferyl alcohol > *p*-coumaryl alcohol, but the compression extract was comparatively less able to oxidize coniferyl alcohol than the non-compression extract. This is demonstrated by the ratios in *Table 3*.

4. Discussion

Two laccases (M_r = ~85 kDa and ~120 kDa) were discovered in the cell-wall-associated extracts of lignifying, developing xylem of Sitka spruce by different approaches. These are the first laccase sequences from a conifer. The 85 kDa laccase was identified using a classical biochemical approach and was purified using its affinity for immobilized zinc ions. The nature of this oxidase from Sitka was only confirmed by the homology of its amino-terminal protein sequence with plant laccase genes. The purified 85 kDa oxidase could not be classified by its substrate and inhibitor profiles (Richardson *et al.*, 1997) as these were intermediate between those ascribed to laccase-like (EC. 1.10.3.2) or catechol oxidase-like (EC. 1.10.3.1) polyphenol oxidases (Mayer, 1987). However, substrate and inhibitor profiles are insufficient to unambiguously assign the nature of the oxidase activity and supporting information is required. For example, Bao *et al.* (1993) confirmed that the oxidase from the lignifying xylem of

Table 2. *Oxidase activity in extracts from compression and non-compression xylem*

	Oxidase activity			
	Non-compression		Compression	
	Specific activity (nmols ABTS h^{-1} mg^{-1})	Total activity (nmols ABTS h^{-1})	Specific activity (nmols ABTS h^{-1} mg^{-1})	Total activity (nmols ABTS h^{-1})
Soluble extract	250 ± 13	39 500 ± 100	200 ± 17	56 520 ± 240
Cell wall extract	3630 ± 90	44 060 ± 260	4140 ± 120	33 285 ± 210
ConA-bound extract	12 590 ± 220	28 550 ± 350	15 090 ± 330	29 030 ± 410
Triton extract	190 ± 14	16 600 ± 420	230 ± 16	17 685 ± 585

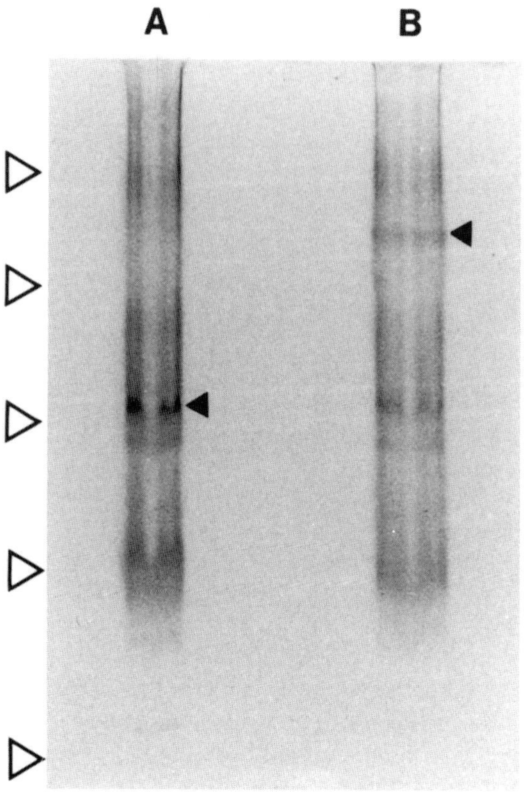

Figure 4. *Analysis of oxidase activities in compression and non-compression xylem by non-denaturing SDS-PAGE. Lane A = ConA-bound fraction from non-compression xylem, lane B = ConA-bound fraction from compression xylem. Each lane contains an equal amount of oxidase activity against ABTS. The gel was stained using diaminofluorene. The closed arrows denote the position of the 85 kDa (lane A) and the 120 kDa (lane B) oxidase bands respectively. The open arrows denote the migration of molecular mass marker proteins (206, 108, 82, 52 and 31 kDa).*

Pinus taeda was a laccase by its substrate and inhibitor profiles and the blue colour and electron paramagnetic resonance (EPR) spectra of the purified enzyme, which are characteristic of type I copper. However, protein or gene sequences for this enzyme have not been published. On the other hand, the coniferyl alcohol oxidase (CAO) activity associated with the lignifying xylem of *Pinus strobus* (Savidge and Udagama-Randeniya, 1992) had substrate and inhibitor profiles more akin to a catechol oxidase-type (EC 1.10.3.1) than a laccase-type (EC 1.10.3.2) polyphenol oxidase (Udagama-Randeniya and Savidge, 1994, 1995). In fact, the general biochemical properties of CAO were not characteristic of laccases and the authors have speculated that this activity may be due to a catechol oxidase-type enzyme.

By chance, the 120 kDa protein identified, amongst others, by a comparative survey of proteins expressed in the developing xylem of compression and non-compression wood, was also found to be a laccase. The 120 kDa laccase was more abundant in the cell wall extracts of the developing xylem from the compression-side of branches of Sitka spruce and was only barely expressed in similar extracts from the developing xylem of the non-compression side. In comparison, the major oxidase-like activity in the non-compression extracts had an M_r of 85 kDa (*Figure 4*) and was probably the same enzyme as the 85 kDa laccase noted above.

Different sets of laccase isoforms have previously been identified in extracts of the lignifying xylem of the upper and basal portions of tobacco stems (Richardson and McDougall, 1997). In addition, Lafayette and Dean (1997) have reported that four laccase genes were differentially expressed in the lignifying xylem of *Liriodendron*

Table 3. Monolignol oxidation by ConA-bound extracts from compression and non-compression xylem

Substrate	Oxidase activity (μmol h^{-1} mg^{-1} protein)			
	Non-compression		Compression	
ABTS	13.0 ± 0.1	(1.00)	14.8 ± 0.3	(1.00)[a]
Coniferyl alcohol	16.8 ± 0.5	(1.29)	10.9 ± 0.5	(0.74)
Sinapyl alcohol	41.8 ± 1.5	(3.21)	44.7 ± 2.1	(3.02)
p-coumaryl alcohol	9.9 ± 0.9	(0.76)	10.0 ± 1.0	(0.68)

[a] Figures in parentheses represent the ratio of specific activity against each monolignol over the specific activity against ABTS.

tulipifera and it is likely that other families of laccase genes in poplar (Ranocha *et al.*, 1999) and tobacco (Kiefer-Meyer *et al.*, 1996) are also expressed in a site- or developmentally-specific manner.

The differential expression of laccases in the developing compression and non-compression xylem must fulfil some function. As well as an elevated lignin content, compression wood tracheids are known to have a higher content of *p*-hydroxyphenylpropane units, derived from *p*-coumaryl alcohol, and lower content of guaiacyl units, derived from coniferyl alcohol, than non-compression wood (Timell, 1983; Monties, 1989). Therefore, it is intriguing that the compression extracts, which are enriched in the 120 kDa laccase, have a lower ability to oxidise coniferyl alcohol and a higher comparative ability to oxidize *p*-coumaryl alcohol than the non-compression extracts, which are enriched in the 85 kDa laccase. This suggests that these laccases have different abilities to oxidize monolignols.

Other groups have studied differentially expressed genes in the developing xylem of reaction versus normal wood. Allona *et al.* (1996) compared cDNA libraries from developing xylem of compression and non-compression wood of *Pinus taeda.* These mRNA-based methods and our quasi-proteomic (Humphery-Smith and Blackstock, 1997) approach share some basic features. They require a comparative method to screen out genes/proteins involved in mundane 'house-keeping' functions and they rely on homology searches against protein, EST and DNA databases to identify differentially expressed genes/proteins.

The main limitations of the quasi-proteomic approach are sensitivity and resolution. A finite amount of protein is required for sequencing and, as no amplification step (such as in cDNA-AFLP methods) can be used, more tissue may be required. The quality of resolution can limit the detection of differentially expressed proteins. However, the resolving power of 2D-electrophoretic methods is steadily improving (Hochstrasser, 1998) and new tools for protein identification, such as matrix-assisted laser desorption-ionization time of flight mass spectrometry (MALDI-TOF MS) (see Anderson & Anderson (1998) and other articles in that volume), can reduce the amount of protein required for identification.

On the other hand, unlike methods based on cDNA/mRNA, the study of expressed proteins can give some idea of post-translational modifications. Indeed, in this study, binding to concanavalin A produced a subset of cell-wall-associated glycoproteins (McDougall, 1997). The use of other specific extraction methods could enrich proteins from other subcellular locales, such as membranes (Tani *et al.*, 1998). Such an approach not only yields information on the probable locale of the proteins but the subsets of proteins obtained may be more easily and completely resolved. For example, simple one-dimensional electrophoretic separation was sufficient to identify target proteins in the subsets of cell-wall-associated proteins, but more powerful, two dimensional separations using isoelectric focusing (IEF) and SDS-PAGE would be required to resolve the large number of proteins in the soluble extracts.

The comparative, quasi-proteomic approach identified a number of other cell-wall-associated proteins that were more abundantly expressed in the developing xylem of the compression or non-compression tissues of branches of Sitka spruce. These proteins may influence the development of the differences in wall structure characteristic of compression wood tracheids and they are obvious targets for further investigations. The quasi-proteomic approach offers a novel route to the discovery of proteins involved in important but poorly understood developmental processes such as xylogenesis. Many enzymes that have powerful effects on the properties of primary cell wall (e.g. xyloglucan endotransglycosylase (Fry, 1995) and expansins (Cosgrove 1997)) have proved to be cell-wall-associated proteins. In addition, a recent survey of cell-wall-associated proteins from suspension-cultured cells (Robertson *et al.*, 1997) found that the majority of proteins selected for amino-terminal protein sequencing had no homology with sequences in protein, DNA or expressed sequence tag (EST) databases. Even taking into account that the amino-terminal regions of homologous proteins are often the least conserved domains, it is clear that cell-wall-associated proteins are under-represented in such databases and these unknown proteins may represent a large reservoir of enzymes with potentially important roles in cell wall biogenesis.

Acknowledgements

I wish to thank Andrew Richardson and Julie Duncan for their hard work, Heather Ross for help with protein sequencing and Professor Howard Davies for his support.

I also thank Dr Steve Lee of Forest Research for access to the Sitka spruce clonebank. The Scottish Crop Research Institute receives grant-in-aid from the Scottish Office Agriculture Fisheries and Environment Department.

References

Allona, I., Quinn, M., Swope, K., Retzel, E., Whetten, R. and Sederoff, R.R. (1996) Xylem expressed cDNAs in *Pinus taeda* normal and compression wood. *Plant Physiol.* **111**: S217.

Anderson, N.L. and Anderson, N.G. (1998) Proteome and proteomics: New technologies, new concepts, and new words. *Electrophoresis* **19**: 1853–1861.

Bao, W., O'Malley, D.M., Whetten, R. and Sederoff, R.R. (1993) A laccase associated with lignification in Loblolly pine xylem. *Science* **260**: 672–674.

Boudet, A.M., Lapierre, C. and Grima-Pettenati, J. (1995) Biochemistry and molecular biology of lignification. *New Phytol.* **129**: 203–236.

Bradford, M.M. (1976) A rapid and sensitive method for the detection of microgram amounts of protein utilising the principle of dye-binding. *Analyt. Biochem.* **7**: 248–256.

Brown, C.L. (1971) Secondary growth. In: *Trees, Structure and Function* (ed. Zimmerman, M.H, Brown, C.L. and Tyree, M.T), Springer-Verlag, New York, pp 67–123.

Cosgrove, D.J. (1997) Assembly and enlargement of the primary cell wall in plants. *Ann. Rev. Cell Dev. Biol.* **13**: 171–201.

Dean, J.F.D. and K-E.L. Eriksson (1994) Laccase and the deposition of lignin in vascular plants. *Holzforschung* **48 suppl**: 21–33.

Deighton, N. and McDougall, G.J. (1998) Coniferyl alcohol oxidase operates via a free radical intermediate. *Phytochemistry* **48**: 601–606.

Fry, S.C. (1995) Polysaccharide-modifying enzymes in the plant-cell wall. *Ann. Rev. Plant Physiol. Plant Mol. Biol.* **46**: 497–513.

Fry, S.C. (1986) Cross-linking of matrix polymers in the growing cell-walls of angiosperms. *Ann. Rev. Plant Physiol. Plant Mol. Biol.* **37**: 165–186.

Fukushima, K., Taguchi, S., Matsui, N. and Yasuda, S. (1997) Distribution and seasonal changes of monolignol glucosides in *Pinus thunbergii*. *Mokuzai Gakkaishi* **43**: 254–259.

Hochstrasser, D.F. (1998) Proteome in perspective. *Clin. Chem. Lab. Med.* **36**: 825–836.

Hubbard, S.C. and Ivatt, R.J. (1981) Synthesis and processing of asparagine-linked oligosaccharides. *Ann. Rev. Biochem.* **50**: 555–583.

Humphery-Smith, I. and Blackstock, W. (1997) Proteome analysis: Genomics via the output rather the input code. *J. Protein Chem.* **16**: 537–544.

Kiefer-Meyer, M-C., Gomord, V., O'Connell, A., Halpin, C. and Faye, L. (1996) Cloning and sequence analysis of laccase-encoding cDNA clones from tobacco. *Gene* **178**: 205–207.

Kutsuki, H. and Higuchi, T. (1981) Activities of some enzymes of lignin formation in reaction wood of *Thuja orientalis*, *Metasequoia glyptostroboides* and *Robinia pseudoacacia*. *Planta* **152**: 365–368.

Laemmli, U.K. (1970) Cleavage of structural proteins during the assembly of bacteriophage T4. *Nature* **227**: 680–685.

Lafayette, P.R., Eriksson, K-E.L. and Dean, J.F.D. (1995) Nucleotide sequence of a cDNA clone encoding an acidic laccase from sycamore maple (*Acer pseudoplatanus* L.). *Plant Physiol.* **107**: 667–668.

Lafayette, P.R. and Dean, J.F.D. (1997) Characterisation of four laccase genes which are differentially expressed in cambium/lignifying tissue of yellow poplar (*Liriodendron tulipifera*). *Plant Physiol.* **114**: S132.

Liu, L., Dean, J.F.D., Friedman, W.E. and Eriksson, K-E.L. (1994) A laccase-like phenoloxidase is correlated with lignin biosynthesis in *Zinnia elgans* stem tissues. *The Plant Journal* **6**: 213–224.

Mayer, A.M. (1987) Polyphenol oxidases in plants — recent progress. *Phytochemistry* **26**: 111–120.

McDougall, G.J. (1997) Procedure for selection of cell-wall-associated glycoproteins. *Phytochemistry* **45**: 633–636.

McDougall, G.J. (1998) Purification of coniferyl alcohol oxidase from lignifying xylem of Sitka spruce using immobilised metal affinity chromatography. *J. Plant Physiol.* **153**: 539–544.

McDougall, G.J. and Morrison, I.M. (1996) Extraction and partial purification of a cell-wall-associated coniferyl alcohol oxidase from developing xylem of Sitka spruce. *Holzforschung* **50**: 549–553.

McDougall, G.J., Morrison, I.M., Stewart, D., Weyers, J.D.B. and Hillman, J.R. (1993) Plant fibres: Botany, chemistry and processing for industrial use. *J. Sci. Food Agric.* **62**: 1–20.

Monties, B. (1989) Lignins. In: *Methods in Plant Biochemistry*. Volume 1. (ed. Harborne J.B.), Academic Press, London, pp 113–157.

Münch, E. (1940) Weitre untersuchungen uber druckholz und zugholz. *Flora* **34**: 45–57.

Nielsen, H., Engelbrecht, J., Brunak, S. and von Heijne, G. (1997) Identification of prokaryotic and eukaryotic signal peptides and prediction of their cleavage sites. *Protein Engineering* **10**: 1–6.

Nimz, H.H., Robert, D., Faix, O. and Nemr, D.W. (1981) Carbon-13 NMR spectra of lignin VIII. Structural differences between the lignins of hardwoods, softwoods, grasses and compression wood. *Holzforschung* **35**: 16–26.

O'Malley, D.M., Whetten, R., Bao, W., Chen, C-L. and Sederoff, R.R. (1993) The role of laccase in lignification. *The Plant Journal* **4**: 751–757.

Parham, R.A. and Gray, R.L. (1984) Formation and structure of wood. In: *The Chemistry of Solid Wood* (ed. Rowell, R.), Advances in Chemistry series no. 207. American Chemical Society, Washington DC, USA, pp 3–56.

Ranocha, P., McDougall, G.J., Hawkins, S., Sterjiades, R., Borderies, G., Stewart, D., Cabanes-Macheteau, M., Boudet, A.M. and Goffner, D. (1999) Biochemical characterization, molecular cloning and expression of laccases — a divergent gene family in poplar. *Eur. J. Biochem.* **259**: 485–495.

Richardson, A. and McDougall, G.J. (1997) A laccase-like polyphenol oxidase from lignifying tobacco xylem. *Phytochemistry* **44**: 229–235.

Richardson, A. and McDougall, G.J. (1998) Identification and partial characterisation of an oxidase from lignifying xylem of Leyland cypress (× *Cupressocyparis leylandii*). *Trees* **12**: 146–152.

Richardson, A., Stewart, D. and McDougall, G.J. (1997) Identification and partial characterisation of a coniferyl alcohol oxidase from lignifying xylem of Sitka spruce (*Picea sitchensis*). *Planta* **203**: 35–43.

Robertson, D., Mitchell, G.P., Gilroy, J.S., Gerrish, C., Bolwell, G.P. and Slabas, A.R. (1997) Differential extraction and protein sequencing reveals major differences in patterns of primary cell wall proteins from plants. *J. Biol. Chem.* **272**: 15 841–15 848.

Savidge R.A. and Udagama-Randenyia, P.V. (1992) Cell wall-bound coniferyl alcohol oxidase associated with lignification in conifers. *Phytochemistry* **31**: 2959–2966.

Steele, N.M., McCann, M.C. and Roberts, K. (1997) Pectin modification in cell walls of ripening tomatoes occurs in distinct domains. *Plant Physiol.* **114**: 373–381.

Sterjiades, R., Dean, J.F.D. and Eriksson, K-E.L. (1992). Laccase from Sycamore maple (*Acer pseudoplatanus*) oxidises monolignols. *Plant Physiol.* **99**: 1162–1168.

Tani, H., Kamidate, T. and Watanabe, H. (1998) Aqueous micellar two-phase systems for protein separation. *Analytical Sciences* **14**: 875–888.

Timell, T. E. (1973) *Ultrastructure of the dormant and active cambial zones and the dormant phloem associated with formation of normal and compression woods in* Picea abies *(L.) Karst.* Technical Publication No. 96. College of Environmental Science and Forestry. State University of New York, Syracuse, New York.

Timell, T.E. (1982) Recent progress in the chemistry and topochemistry of compression wood. *Wood Science and Technology* **16**: 83–122.

Timell, T.E. (1983) Origin and evolution of compression wood. *Holzforschung* **37**: 1–10.

Timell, T.E. (1986) *Compression Wood in Gymnosperms*. Springer-Verlag, Heidelberg, Germany.

Udagama-Randenyia, P.V. and Savidge R.A. (1994) Electrophoeretic analysis of coniferyl alcohol oxidase and related laccases. *Electrophoresis* **15**: 1072–1077

Udagama-Randenyia, P.V. and Savidge R.A. (1995) Coniferyl alcohol oxidase — a catechol oxidase? *Trees* **10**: 102–107.

Wardrop A.B. (1981) Lignification and xylogenesis. In: *Xylem Cell Development* (ed. Barnett, J.R.), Castle House Publications, Kent, UK, pp 115–151.

Zhang, X.-H. and Chiang, V.L. (1997) Molecular cloning of 4-coumarate coenzyme A ligase in Loblolly pine and the roles of this enzyme in the biosynthesis of lignin in compression wood. *Plant Physiol.* **113**: 65–74.

34

Investigating lignin biosynthesis using transgenic mutant plants

C. Halpin, J. Abbott and A. Barakate

1. Introduction

Just seven years have passed since the first cloning of genes dedicated to monolignol synthesis, yet the pathway has already been genetically manipulated in a variety of species and field trials underpinning commercial exploitation are underway. 'Designer' fibres, manipulated to offer optimum raw material quality for processes as diverse as papermaking and forage production, are becoming a reality. Despite this headlong rush towards biotechnological applications, many basic aspects of the lignin biosynthetic pathway are still poorly understood. The spatial organization of the pathway, how it is regulated and even the exact sequence and identity of enzymes involved, are all areas of intensive current research. Detailed investigation of mutant plants with altered lignification is helping to elucidate these questions.

2. Native lignins

2.1 *Function, composition and structure*

Lignification, a basic developmental process of higher plants, is fundamental to the adaption of plants to life on land. Lignin waterproofs and strengthens many types of plant cell and is particularly important in fibres and xylem. It is a complex aromatic polymer, heterogeneous both in structure and composition, and huge natural variability exists in the amount and type of lignin produced by different plants or even different cell types within a plant.

Lignin is predominantly a polymer of cinnamyl alcohols or monolignols (*Figure 1*). The proportions of each monolignol incorporated into lignin varies in different types of plant. Gymnosperm lignin is made of coniferyl alcohol units, yielding a guaiacyl (G) lignin. Angiosperm lignin contains sinapyl alcohol, or syringyl (S) units, in addition to

Cell & Molecular Biology of Wood Formation, edited by R. Savidge, J. Barnett and R. Napier.

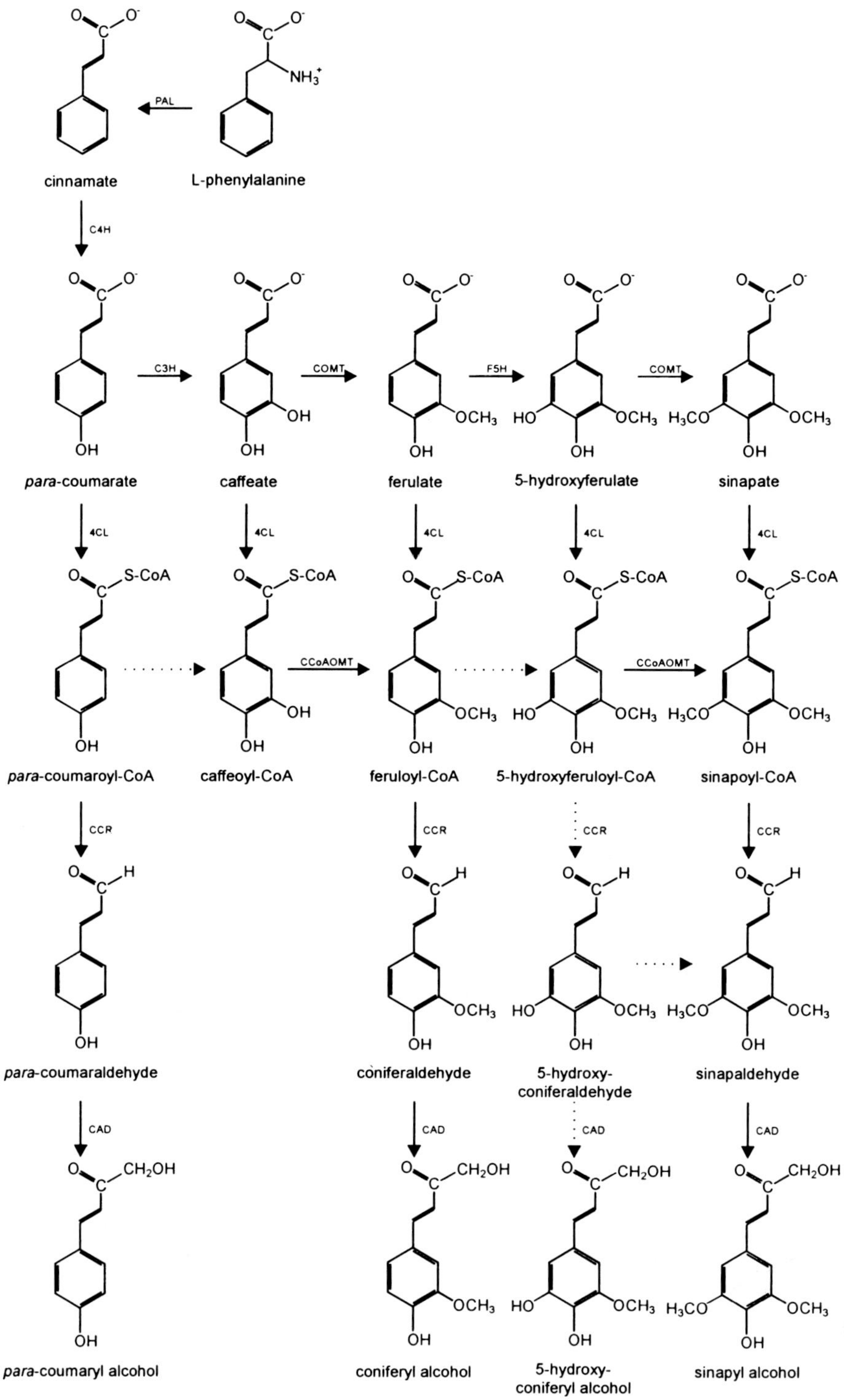

cinnamate
L-phenylalanine
PAL
C4H
C3H
COMT
F5H
COMT
para-coumarate
caffeate
ferulate
5-hydroxyferulate
sinapate
4CL
CCoAOMT
CCoAOMT
para-coumaroyl-CoA
caffeoyl-CoA
feruloyl-CoA
5-hydroxyferuloyl-CoA
sinapoyl-CoA
CCR
para-coumaraldehyde
coniferaldehyde
5-hydroxy-coniferaldehyde
sinapaldehyde
CAD
para-coumaryl alcohol
coniferyl alcohol
5-hydroxy-coniferyl alcohol
sinapyl alcohol

G units and is known as guaiacyl-syringyl lignin. Grass lignin additionally contains *p*-hydroxyphenyl (H) units from *p*-coumaryl alcohol.

It is not currently possible to fully elucidate lignin structure and composition in any species. The polymer is immensely complex and incorporates both non-condensed ether bonds and condensed C-C bonds as well as cross-links to other cell wall components. Most techniques for lignin characterization rely on initial solubilization of a lignin fraction under relatively harsh chemical conditions. It is important to realize that the resulting soluble fraction does not represent native lignin either quantitatively or qualitatively. In particular, condensed lignin fractions resistant to degradation are absent from such preparations. These problems mean that no single technique is completely reliable or unbiased. Lignin analysis techniques also vary considerably in sensitivity, and minor lignin components that can be detected by one technique may appear absent if using a different technique. A variety of techniques must therefore be used before any firm conclusions can be drawn about lignin content, composition or structure. This is particularly important when dealing with 'novel' lignins from genetically manipulated plants.

2.2 *The lignin biosynthetic pathway*

The pathway for lignin biosynthesis is outlined in *Figure 1*. Lignin is a product of the phenylpropanoid pathway which also has branches leading to other important metabolites including flavonoids, coumarins and phytoalexins. Phenylalanine ammonia-lyase (PAL) controls entry to the pathway and is highly regulated during plant development. It deaminates phenylalanine to cinnamate which is subsequently hydroxylated to *para*-coumarate by cinnamate 4-hydroxylase (C4H). After PAL and C4H, the pathway has traditionally been thought to progress through a series of hydroxylation and methylation reactions that produce ferulate and sinapate from *para*-coumarate. These reactions are catalysed by the enzymes cinnamate 3-hydroxylase (C3H), ferulate 5-hydroxylase (F5H) and bi-specific caffeate/5-hydroxyferulate O-methyltransferase (COMT). All three acids can be activated by 4-coumarate: CoA ligase (4CL) to yield the corresponding CoA thioesters. The thioesters undergo 2 successive reduction reactions catalysed by cinnamoyl-CoA reductase (CCR) and cinnamyl alcohol dehydrogenase (CAD) to give rise to the three monolignols, *p*-coumaryl, coniferyl and sinapyl alcohol. The monolignols are thought to be exported to the cell wall (as glucosylated derivatives) where they undergo peroxidase- or laccase-catalysed polymerization.

3. Genetic manipulation of lignin biosynthesis

It has long been appreciated that lignin can limit the efficiency with which plant materials are used in a variety of agricultural and industrial processes. Lignin affects the

---

Figure 1. *Pathway of lignin biosynthesis in higher plants. A hatched arrow indicates a step whose existence has been inferred but not conclusively proven. PAL, phenylalanine ammonia lyase; C4H, cinnamate 4-hydroxylase; C3H, p-coumarate 3-hydroxylase; COMT, caffeic acid O-methyltransferase; F5H, ferulate 5-hydroxylase; 4CL, 4-coumarate:CoA ligase; CCoA-3H, coumaryl-CoA 3-hydroxylase; CCoAOMT, caffeoyl-CoA O-methyltransferase; CCR, cinnamoyl-CoA reductase; CAD, cinnamyl alcohol dehydrogenase.*

digestibility of forage crops, for example, and interferes with cellulose extraction in pulp and paper production. The exact content, composition and structure of lignin and the frequency of connections to other cell components determine the extent to which lignin affects these processes. The natural variability in lignin suggests that it should be possible to manipulate its content, composition and structure in plants, with possible commercial benefits, without adversely affecting plant health and fitness. Over the past few years considerable research effort has been expended in this area. Most of the genes involved in central phenylpropanoid metabolism (PAL, C4H and 4CL) and lignin precursor methoxylation (COMT and F5H) have now been cloned, as have both genes dedicated to monolignol synthesis (CAD and CCR). This has enabled genetic approaches to lignin quality manipulation. The majority of subsequent work has concentrated on introducing antisense or partial sense genes for lignin biosynthetic enzymes into transgenic plants to impose specific blocks at critical points in the biochemical pathway. These experiments have had varying degrees of success for applications to plant improvement, but all have been invaluable in increasing our understanding of the workings of the lignin biosynthetic pathway and in defining the limits of that understanding. These experiments on manipulation of genes involved in monolignol synthesis are reviewed below.

3.1 *Manipulation of central phenylpropanoid metabolism*

PAL and C4H. Tobacco plants expressing a heterologous bean PAL gene were produced by Elkind *et al.* (1990) and shown to have reduced PAL activity as a result of co-suppression. The transgenic plants exhibited a number of unusual and aberrant phenotypes including stunted growth, altered leaf shape and texture, reduced pollen viability and altered flower morphology and pigmentation. Subsequent characterization of the lignin in these plants showed that Klason lignin was reduced to half that of wild-type plants when PAL was reduced by 85% (Sewalt *et al.*, 1997). This change in lignin content was accompanied by an increase in the S/G ratio, measured by pyrolysis GC-MS. Antisense or sense suppression of a heterologous C4H in tobacco also caused reductions in lignin content and unexpectedly had the opposite effect to PAL (the preceding enzyme on the pathway) on S/G ratio (Sewalt *et al.*, 1997). The observed decrease in S/G was, however, only evident by pyrolysis GC-MS and could not be demonstrated by wet chemical methods. Over-expression of PAL or C4H had no effect on lignin content or composition.

4CL. The expression of 4CL has recently been suppressed in both tobacco and *Arabidopsis* by sense and/or antisense expression of the corresponding cDNA (Kajita *et al.*, 1996; Lee *et al.*, 1997). In both cases, reduction of 4CL activity to below 20% resulted in a modest decrease in lignin content. In *Arabidopsis* this was accompanied by significant decreases in G units while S units were unchanged. The G/S ratio was therefore greatly decreased (i.e., S/G was increased). The inference is that, in *Arabidopsis* at least, 4CL is not essential for S lignin synthesis. There may be alternative routes for S monomer synthesis that can by-pass 4CL. Alternatively, there could be distinct and divergent 4CL isoforms for G and S lignin synthesis with only the G-specific form being affected in these plants.

In tobacco, sense suppression of 4CL also led to a decrease in G units but, in this case, S units were also clearly reduced and H units increased (Kajita *et al.*, 1996; Kajita *et al.*, 1997a). An increase in S/V (i.e. syringyl/vanillin similar to S/G) was measured in most plants with reduced 4CL (Kajita *et al.*, 1996) although a further more detailed study of one transgenic line found a decrease in S/G (Kajita *et al.*, 1997a). Increased amounts of hydroxycinnamic acids, including sinapic acid, were found to be associated to lignin by ester and/or ether bonds. The changes in lignin in tobacco plants low in 4CL were concentrated on areas of wood that were brown in colour. Normal coloured wood had normal levels of 4CL expression (Kajita *et al.*, 1997b). Some of the tobacco plants with reduced 4CL expression were stunted.

3.2 *Manipulation of precursor methylation*

COMT. Atanassova *et al.* (1995) severely suppressed COMT activity in transgenic tobacco by introducing antisense or partial sense COMT transgenes. Plants with as little as 3% remaining COMT activity had a normal phenotype and a normal xylem colour and Klason lignin determination showed no significant change in total lignin content. Staining with Mäule reagent, which is specific for free syringyl units, suggested large changes in lignin composition. Stem sections from wild-type plants stained bright red but sections from transgenics with low COMT activity stained orange or yellow. This pointed to a marked decrease in free syringyl units in the lignin of low COMT plants. Lignin analysis by more sophisticated methods confirmed this. Thioacidolysis showed a marked decrease in syringyl moieties released from low-COMT plants while guaiacyl moieties were unchanged or even slightly increased. The S/G ratio was therefore greatly decreased. Most interestingly, unusual 5-OH guaiacyl units not normally seen in lignin were detected.

A similar strategy was used to decrease COMT activity in poplar with almost identical results (Van Doorsselaere *et al.*, 1995). Plants with 50% reduced activity showed only slight changes in lignin monomer composition by thioacidolysis, but plants where COMT was reduced by 95% showed a six-fold decrease in the S/G ratio. Syringyl units were reduced and guaiacyl units increased and novel 5-OH guaiacyl units were again detected in thioacidolysis products from very low COMT plants. Two independent procedures failed to detect any significant change in total lignin content.

Earlier reports describing the use of heterologous sequences to manipulate COMT activity in tobacco are in some respects at variance with the studies described above. Dwivedi *et al.* (1994) reported changes in both the composition and content of lignin in tobacco plants expressing an antisense aspen COMT gene, while Ni *et al.* (1994) described reduced lignin content but no change in composition in tobacco transformed with antisense alfalfa COMT. The reason for the discrepancies between reports is unclear, but may be something to do with the use of heterologous transgenes in some studies.

It is likely that plants engineered to reduce COMT will be useful for particular agricultural purposes. Low COMT tobacco plants have been shown to be more digestible (in terms of dry matter disappearance) than wild-type plants, offering an obvious application to forage crop improvement (Vailhe *et al.*, 1996). Low COMT poplars, on the other hand, have recently been shown to be less suitable for pulp production than normal counterparts (Lapierre *et al.*, 1999).

Role of CCoAOMT. COMT can methylate both caffeate and 5-hydroxyferulate (Bugos *et al.*, 1992), so the observation that guaiacyl lignin increases slightly in low

COMT plants is very surprising. It has been suggested that another enzyme, caffeoyl-CoA O-methyltransferase (CCoAOMT) might also be involved in the methylation reactions important to lignin biosynthesis (Ye *et al.*, 1994) but until very recently only circumstantial evidence supported this idea (Ye, 1997; Ye and Varner, 1995; Ye *et al.*, 1994). Genetic manipulation experiments have recently confirmed it. Zhong *et al.* (1998) generated transgenic tobacco plants with substantial reductions in CCoAOMT and plants with simultaneous reductions in CCoAOMT and COMT. Reduction of CCoAOMT alone resulted in decreased lignin content. This affected the amount of both G and S units but G units were preferentially reduced leading to an increase in the S/G ratio. Plants with suppression of both CCoAOMT and COMT showed a further reduction in lignin content confirming that both enzymes are indeed involved in methylation reactions in lignin biosynthesis (see *Figure 1*).

F5H. Ferulate-5-hydroxylase (F5H), the enzyme preceding COMT in the conversion of ferulate to sinapate, has proved to be one of the most difficult lignin biosynthetic enzymes to work with and only the *Arabidopsis* cDNA has been isolated (Chapple *et al.*, 1992). *Arabidopsis* plants defective in the F5H gene, were found not to make syringyl lignin (Meyer *et al.*, 1998) demonstrating an important role for F5H in determining lignin monomer composition. Over-expression of F5H (from the C4H promoter) in *Arabidopsis* led to an increase in syringyl lignin content and abolished the tissue specificity of syringyl deposition which is normally restricted to sclerified parenchyma (Meyer *et al.*, 1998). This suggests that F5H may catalyse a rate-limiting step in syringyl lignin biosynthesis.

3.3 *Manipulation of genes dedicated to monolignol synthesis*

CCR. CCR has long been thought to control a regulatory or rate-limiting step on the lignin biosynthetic pathway. This view is supported by recent transgenic studies that indicate that severe phenotypes result in tobacco plants with reduced CCR expression. Piquemal *et al.* (1998) produced tobacco plants with 30% of normal CCR activity that were stunted and had reduced leaf size, altered leaf shape and delayed flowering. Xylem in these plants had an orange-brown colour and the xylem vessels were beginning to collapse. Lignin content in CCR down-regulated plants was substantially reduced, down to 47% of wild-type levels. Thioacidolysis indicated there were also changes in lignin composition with decreased amounts of β-O-4 linked S units, and radically decreased (up to 70%) β-O-4 linked G units. Ferulic acid and sinapic acid were both incorporated into the cell wall of low CCR plants and NMR analysis indicated that tyramine ferulate was also incorporated and may represent a sink for the accumulating feruloyl-SCoA that should result from CCR down-regulation (Ralph *et al.*, 1998). These results have been confirmed by independent work on similar plants (O'Connell *et al.*, 1999) although, in this case, not all low CCR plants were stunted.

CAD. CAD, the final enzyme of lignin monomer synthesis, was one of the first lignin enzymes to be purified, cloned and manipulated in tobacco (Halpin *et al.*, 1992; Halpin *et al.*, 1994; Knight *et al.*, 1992). An antisense CAD gene was effective in reducing CAD activity by up to 95%. Despite this, antisense plants developed normally. Plants with less than 16–20% CAD activity had a novel red colour integrally associated with lignin in the xylem at certain developmental stages. Total lignin determination by two methods (Klason; acetyl bromide) demonstrated that antisense plants incorporated almost the same amount of lignin as wild-type plants. However treatment of antisense and wild-type lignin with sodium hydroxide or thioglycolic acid demonstrated that significantly more lignin could be extracted from antisense plant samples by both treatments. This suggests structural changes in the low CAD lignin that make it more extractable.

The composition of lignin from antisense plants was investigated by temperature-resolved in-source pyrolysis-mass spectrometry (PYMS). Wild-type xylem showed high intensity mass peaks for coniferyl and sinapyl alcohol, indicative of a normal angiosperm lignin, but xylem from antisense plants gave a different profile. Peaks corresponding to coniferyl and sinapyl alcohol decreased in amplitude while those for coniferyl and sinapyl aldehyde increased. This suggests that, when the lignin pathway is blocked at CAD, the aldehyde substrates of the enzyme are incorporated into lignin in place of the usual reaction products, the monolignol alcohols. The altered aldehyde/alcohol incorporation was most obvious for the syringyl lignin component.

These results have been confirmed by a number of similar but independent reports of CAD suppression in tobacco. Yahiaoui *et al.* (1998) describe low CAD plants that display a significant decrease in S/G ratio in thioacidolysis products, higher lignin extraction by alkali and a 10-fold increase in C-6–C-1 aldehydes, including coniferyl aldehyde. NMR characterization of the altered lignin confirmed the incorporation of fewer coniferyl and sinapyl alcohol-derived units and elevated incorporation of benzaldehydes and cinnamaldehydes (Ralph *et al.*, 1998). Hibino *et al.* (1995) used a heterologous CAD antisense gene in tobacco to produce plants with moderate degrees of CAD down-regulation (20–55% reduction). The transformant with lowest CAD activity had altered xylem colour and increased cinnamyl aldehydes in lignin with a slightly higher S/G ratio (Higuchi *et al.*, 1994).

CAD has also been suppressed in transgenic poplar using both antisense and sense genes (Baucher *et al.*, 1996). Resulting changes in lignin were broadly similar to those already seen in transgenic tobacco. Plants with over 70% reduction in CAD had a red colour in xylem tissue and structural changes in lignin. The lignin was more extractable with alkali and had an increased proportion of aldehydes although neither the lignin content nor the overall S/G ratio were substantially changed. Kraft pulping experiments showed more extensive delignification of low-CAD plants than wild-type plants, demonstrating the benefit of this particular genetic manipulation for pulp and paper-making (Lapierre *et al.*, 1999). Field trials of these novel trees are currently being performed in France and the UK.

Alfalfa transgenics with down-regulated CAD activity have recently been described (Baucher *et al.*, 1999). Like the tobacco and poplar transgenics, these plants showed a novel red colour in stems. Lignin content, determined by three different methods, was not altered but the amounts of G and S (particularly S) monomers in thioacidolysis products were reduced and the amount of lignin extracted by alkali increased. Most importantly, these plants like low CAD tobacco plants, were more digestible than normal counterparts, confirming the benefits of CAD down-regulation for forage improvement (Baucher *et al.*, 1999; Vailhe *et al.*, 1998).

Double and triple mutants. Our recent work has concentrated on the production and characterization of double and triple mutants of the lignin pathway where activity of two/three biosynthetic genes is reduced. Such plants may be useful for both basic research and for applied purposes by (i) revealing alternative pathways for monolignol synthesis, if they exist, and (ii) yielding new and useful lignins. Double mutants can be made in a number of different ways and we have been comparing different methods to determine which is most effective. In general, combining two antisense transgenes by crossing existing plants with single antisense genes is only partially successful, due to the prevalence of gene silencing in the progeny. Typically we find that the effect of one or both transgenes is either lost or considerably weakened in progeny plants, even when well characterized and stable homozygote plants are used as the parents. In our experiments, this homology-based transgene silencing has probably been caused by the repeated use of the CaMV 35S promoter in different transgenic constructs.

An alternative strategy that we have found to be superior to crossing for producing double mutants is the use of transgenes containing partial sense fragments of multiple genes under the control of a single promoter. This strategy was inadvertently discovered by Seymour *et al.* (1993) and has been further tested by the same group (Jones *et al.*, 1998) but has not been widely adopted by other researchers. Transgenes containing partial fragments of genes in a sense orientation act to down-regulate the activity of homologous endogenous genes, probably by a mechanism akin to co-suppression. We have tested this strategy with 4 different constructs designed to achieve co-ordinate down-regulation of two or three of the lignin biosynthetic genes COMT, CCR and CAD (see *Figure 2*). In all cases we have been able to identify plants with reduced activity of both, or all three, targeted genes. In most cases, these plants show greater degrees of down-regulation than corresponding plants produced by crossing parents with single antisense transgenes. Many of these plants look completely normal although some plants with all three genes down-regulated show severe stunting and altered leaf phenotypes. We are currently characterizing these plants in terms of their lignin and soluble phenolics content and composition in order to further elucidate all possible routes of lignin monomer biosynthesis.

4. Perspectives

The experiments outlined above demonstrate that it is possible to manipulate both lignin content and/or lignin composition and structure by transgenic approaches. The exact changes in lignin depend on the particular gene or combination of genes

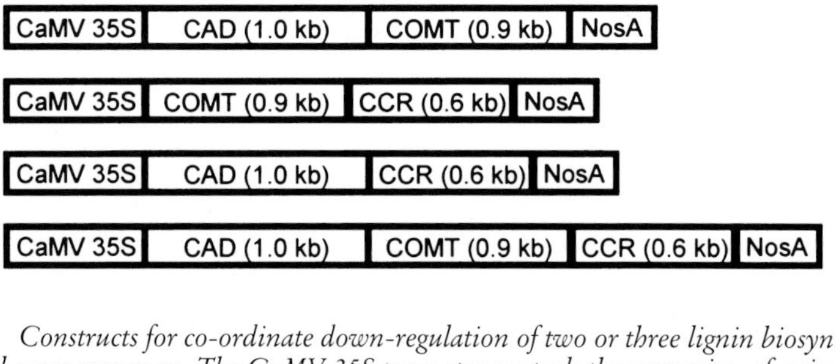

Figure 2. *Constructs for co-ordinate down-regulation of two or three lignin biosynthetic genes using partial sense sequences. The CaMV 35S promoter controls the expression of a single transcript with partial sense sequences of two or three different genes. The length of each gene is shown in kilobases (kb). The 3′ sequence of the NosA gene is used as a terminator. For explanation of abbreviations see Figure 1 or text.*

targeted. In many cases, lignin can be altered without affecting xylem ultrastructure or plant health and fitness. Some changes are already proving useful for plant raw material improvement. More genes have been manipulated in tobacco than in any other single species (see *Table 1*). Comparing the results obtained in tobacco by down-regulating different genes should allow us a clearer insight into the working and organization of the pathways and illuminate the role and importance of each gene product. Such a comparison, however, exposes a number of interesting anomalies in that the observed changes in lignin are not always consistent between different experimenters and are sometimes not predictable from the current pathway model (*Figure 1*). A number of particular points require explanation.

(i) How can blocks at supposedly sequential steps on the pathway have apparently completely opposite effects on lignin S/G ratios? This occurs both at the beginning and end of the pathway if results for down-regulation of PAL and C4H, or CCR and CAD, are compared.
(ii) Is 4CL activity necessary for S lignin synthesis in angiosperms other than *Arabidopsis*?
(iii) At what point do the methylation reactions occur?

Some of these apparent anomalies are likely to result from the differing technical approaches taken by different groups working in this field. For example, full length cDNA sequences have been used by some groups to trigger co-suppression and thereby reduce activity of a given enzyme. Such constructs are also capable of over-expression so it is quite possible that plants with an overall reduction in activity might contain some tissues/cells that were in fact overproducing that transcript. The resulting changes in lignin might be different to those expected and difficult to explain on the basis of down-regulation alone. Differing lignin analysis techniques may also contribute to apparently anomalous results when comparing one experiment with another. Lignin techniques vary both in their sensitivity and in the portion of the lignin

Table 1. *Summary of main results obtained on down-regulation of genes involved in monolignol synthesis in tobacco. Selected data taken from the papers cited has been used to derive a precentage increase or decrease compared to wild-type plants for lignin content, G, S, and S/G. + indicates an increase, – indicates a decrease. (a) values for line 106P3a, Sewalt* et al. *(1997); (b) values for plant 13B, Sewalt* et al. *(1997); (c) Zhong* et al. *(1998); (d) Atanassova* et al. *(1995); (e) average of values for plants S1–S4, Kajita* et al. *(1996); (f) Piquemal* et al. *(1998); (g) Halpin* et al. *(1994). For explanation of abbreviations see Figure 1 or text*

Enzyme	Lignin content	G units	S units	S/G	Unusual component
PAL[(a)]	–48%	Reduced	No change	+70%	
C4H[(b)]	–63%	Reduced	V. reduced	–91%	
CCoAOMT[(c)]	–47%	–53%	–43%	+22%	
COMT[(d)]	No change	+17%	–91%	–92%	
4CL[(e)]	–25%	–68%	–50%	+83%	5-OH G units
CCR[(f)]	–47%	–84%	–64%	+116%	Cinnamic acids Feruloyl tyramine
CAD[(g)]	No change	–27%	–70%	–45%	Aldehydes

polymer that is accessible to them and so comparing S/G ratios obtained by different methods, for example, can be misleading. It is also important to remember that the S/G ratio merely indicates the relative proportions of S and G units in whatever subfraction of lignin that is accessible to the particular technique being used. Clearly an increase in S/G, for instance, could be achieved by increasing the proportion of S units, by decreasing the proportion of G units, or by various combinations of both. The S/G ratio on its own is therefore not as informative as specific information on the amounts of S and G. A good illustration of this point is the data of Kajita *et al.* (1996) who generated a number of transgenic tobacco lines with reduced 4CL activity. The S/G ratio in these plants varied widely from 0.45–2.0 (wild-type 0.8), but consideration of the amounts of V/G and S showed a more consistent pattern with both monomer types considerably reduced in all plants. It is also important to consider that the lignin fraction inaccessible to various degradative techniques might be significant and have a different composition. The time of sampling is also crucial because lignin composition changes as lignification progresses. Not enough consideration is given to these factors in many publications, and factors such as the timing of sampling are sometimes not even mentioned. Finally we do not yet know how universal the details of the pathway are in different species. Results of genetic manipulation experiments in tobacco and poplar often seem generally comparable but not enough work has been done on other angiosperm species (e.g., *Arabidopsis*) to determine to what extent they behave similarly. Consequently a degree of caution is needed when attempting to compare the results obtained in different species and even the results of different research groups on the same species.

Despite these cautions it seems likely that at least some of the questions prompted by the experiments summarized in *Table 1* indicate inadequacies or inaccuracies in the current pathway model. Double and triple mutants, where lignin biosynthesis has been severely blocked at two or three distinct positions, may help illuminate these questions. The existence and route of potentially minor alternative pathways might be revealed if the flux of precursors through them is increased by cutting off the 'normal' major pathway. Double mutants will also help in determining the sequence of reactions on the pathway as the particular changes observed in lignin will indicate which of the two down-regulated genes causes the earliest block on the pathway. Finally, double mutants may produce lignins with novel composition and structure that might be useful for commercial applications. Characterization of such double and triple mutants is underway.

References

Atanassova, R., Favet, N., Martz, F., Chabbert, B., Tollier, M.T., Monties, B., Fritig, B. and Legrand, M. (1995) Altered lignin composition in transgenic tobacco expressing O-methyltransferase sequences in sense and antisense orientation. *Plant J.* **8**: 465–477.

Baucher, M., Chabbert, B., Pilate, G., VanDoorsselaere, J., Tollier, M.T., PetitConil, M., Cornu, D., Monties, B., VanMontagu, M., Inze, D., Jouanin, L. and Boerjan, W. (1996) Red xylem and higher lignin extractability by down-regulating a cinnamyl alcohol dehydrogenase in poplar. *Plant Physiol.* **112**: 1479–1490.

Baucher, M., B.-V., Chabbert, B., Besle, J.-M., Opsomer, C., Van Montagu, M. and Botterman, J. (1999) Down-regulation of cinnamyl alcohol dehydrogenase in transgenic alfalfa (*Medicago sativa* L.) and the effect on lignin composition and digestibility. *Plant Mol. Biol.* **39**: 437–447.

Bugos, R.C., Chiang, V.L.C. and Campbell, W.H. (1992) Characterization of bispecific caffeic acid/5-hydroxyferulic acid O-methyltransferase from aspen. *Phytochemistry* **31**: 1495–1498.

Chapple, C.C.S., Vogt, T., Ellis, B.E. and Somerville, C.R. (1992) An *Arabidopsis* mutant defective in the general phenylpropanoid pathway. *Plant Cell* **4**: 1413–1424.

Dwivedi, U.N., Campbell, W.H., Yu, J., Datla, R.S.S., Bugos, R.C., Chiang, V.L. and Podila, G.K. (1994) Modification of lignin biosynthesis in transgenic *Nicotiana* through expression of an antisense o-methyltransferase gene from *Populus. Plant Mol. Biol.* **26**: 61–71.

Elkind, Y., Edwards, R., Mavandad, M., Hedrick, S.A., Ribak, O., Dixon, R.A. and Lamb, C.J. (1990) Abnormal plant development and down-regulation of phenylpropanoid biosynthesis in transgenic tobacco containing a heterologous phenylalanine ammonia-lyase gene. *Proc. Natl Acad. Sci. USA* **87**: 9057–9061.

Halpin, C., Knight, M.E., Grima-Pettenati, J., Goeffner, D., Boudet, A.M. and Schuch, W. (1992) Purification and characterisation of cinnamyl alcohol dehydrogenase from tobacco stems. *Plant Physiol.* **98**: 12–16.

Halpin, C., Knight, M.E., Foxon, G.A., Campbell, M.M., Boudet, A.M., Boon, J.J., Chabbert, B., Tollier, M-T. and Schuch, W. (1994) Manipulation of lignin quality by downregulation of cinnamyl alcohol dehydrogenase. *Plant J.* **6**: 339–350.

Hibino, T., Takabe, K., Kawazu, T., Shibata, D. and Higuchi, T. (1995) Increase of cinnamaldehyde groups in lignin of transgenic tobacco plants carrying an antisense gene for cinnamyl alcohol-dehydrogenase. *Biosci. Biotechnol. Biochem.* **59**: 929–931.

Higuchi, T., Ito, T., Umezawa, T., Hibino, T. and Shibata, D. (1994) Red-brown color of lignified tissues of transgenic plants with antisense CAD gene — wine-red lignin from coniferyl aldehyde. *J. Biotechnol.* **37**: 151–158.

Jones, C.G., Scothern, G.P., Lycett, G. W. and Tucker, G.A. (1998) The effect of chimeric transgene architecture on co-ordinated gene silencing. *Planta* **204**: 499–505.

Kajita, S., Katayama, Y. and Omori, S. (1996) Alterations in the biosynthesis of lignin in transgenic plants with chimeric genes for 4-coumarate:Coenzyme A ligase. *Plant Cell Physiol.* **37**: 957–965.

Kajita, S., Hishiyama, S., Tomimura, Y., Katayama, Y. and Omori, S. (1997a) Structural characterization of modified lignin in transgenic tobacco plants in which the activity of 4-coumarate:Coenzyme A ligase is depressed. *Plant Physiol.* **114**: 871–879.

Kajita, S., Mashino, Y., Nishikubo, N., Katayama, Y. and Omori, S. (1997b) Immunological characterization of transgenic tobacco plants with a chimeric gene for 4-coumarate:CoA ligase that have altered lignin in their xylem tissue. *Plant Sci.* **128**: 109–118.

Knight, M.E., Halpin, C. and Schuch, W. (1992) Identification and characterisation of cDNA clones encoding cinnamyl alcohol dehydrogenase from tobacco. *Plant Mol. Biol.* **19**: 793–801.

Lapierre, C., Pollet, B., PetitConil, M., Toval, G., Romero, J., Pilate, G., Leple, J.C., Boerjan, W., Ferret, V., DeNadai, V. and Jouanin, L. (1999) Structural alterations of lignins in transgenic poplars with depressed cinnamyl alcohol dehydrogenase or caffeic acid O- methyltransferase activity have an opposite impact on the efficiency of industrial kraft pulping. *Plant Physiol.* **119**: 153–163.

Lee, D., Meyer, K., Chapple, C. and Douglas, C.J. (1997) Antisense suppression of 4-coumarate:Coenzyme A ligase activity in *Arabidopsis* leads to altered lignin subunit composition. *Plant Cell* **9**: 1985–1998.

Meyer, K., Shirley, A.M., Cusumano, J.C., Bell-Lelong, D.A. and Chapple, C. (1998) Lignin monomer composition is determined by the expression of a cytochrome P450-dependent monooxygenase in *Arabidopsis*. *Proc. Natl Acad. Sci. USA* **95**: 6619–6623.

Ni, W.T., Paiva, N.L. and Dixon, R.A. (1994) Reduced lignin in transgenic plants containing a caffeic acid O-methyltransferase antisense gene. *Transgenic Res.* **3**: 120–126.

O'Connell, A., Holt, K., Halpin, C., Piquemal, J., Grima-Pettenati, J., Boudet, A., Lapierre, C., Petit-Conil, M. and Schuch, W. (1999) Gene silencing of cinnamoyl CoA reductase in tobacco plants significantly modifies lignin content and structure and improves pulping characteristics. *Plant Mol. Biol.* In press.

Piquemal, J., Lapierre, C., Myton, K., O'Connell, A., Schuch, W., GrimaPettenati, J. and Boudet, A.M. (1998) Down-regulation of cinnamoyl-CoA reductase induces significant changes of lignin profiles in transgenic tobacco plants. *Plant J.* **13**: 71–83.

Ralph, J., Hatfield, R.D., Piquemal, J., Yahiaoui, N., Pean, M., Lapierre, C. and Boudet, A.M. (1998) NMR characterization of altered lignins extracted from tobacco plants down-regulated for lignification enzymes cinnamyl-alcohol dehydrogenase and cinnamoyl-CoA reductase. *Proc. Natl. Acad. Sci. USA* **95**: 12 803–12 808.

Sewalt, V.J.H., Ni, W.T., Blount, J.W., Jung, H.G., Masoud, S.A., Howles, P.A., Lamb, C. and Dixon, R.A. (1997) Reduced lignin content and altered lignin composition in transgenic tobacco down-regulated in expression of L-phenylalanine ammonia-lyase or cinnamate 4-hydroxylase. *Plant Physiol.* **115**: 41–50.

Seymour, G.B., Fray, R.G., Hill, P. and Tucker, G.A. (1993) Down-regulation of 2 nonhomologous endogenous tomato genes with a single chimeric sense gene construct. *Plant Mol. Biol.* **23**: 1–9.

Vailhe, M.A.B., Migne, C., Cornu, A., Maillot, M.P., Grenet, E., Besle, J.M., Atanassova, R., Martz, F. and Legrand, M. (1996) Effect of modification of the O-methyltransferase activity on cell wall composition, ultrastructure and degradability of transgenic tobacco. *J. Sci. Food Agric.* **72**: 385–391.

Vailhe, M.A.B., Besle, J.M., Maillot, M.P., Cornu, A., Halpin, C. and Knight, M. (1998) Effect of down-regulation of cinnamyl alcohol dehydrogenase on cell wall composition and on degradability of tobacco stems. *J. Sci. Food Agric.* **76**: 505–514.

VanDoorsselaere, J., Baucher, M., Chognot, E., Chabbert, B., Tollier, M.T., PetitConil, M., Leple, J.C., Pilate, G., Cornu, D., Monties, B., VanMontagu, M., Inze, D., Boerjan, W. and Jouanin, L. (1995) A novel lignin in poplar trees with a reduced caffeic acid 5-hydroxyferulic acid O-methyltransferase activity. *Plant J.* **8**: 855–864.

Yahiaoui, N., Marque, C., Myton, K.E., Negrel, J. and Boudet, A.M. (1998) Impact of different levels of cinnamyl alcohol dehydrogenase down-regulation on lignins of transgenic tobacco plants. *Planta* **204**: 8–15.

Ye, Z.H. (1997) Association of caffeoyl Coenzyme A 3-O-methyltransferase expression with lignifying tissues in several dicot plants. *Plant Physiol.* **115**: 1341–1350.

Ye, Z.-H. and Varner, J.E. (1995) Differential expression of 2 O-methyltransferases in lignin biosynthesis in *Zinnia elegans*. *Plant Physiol.* **108**: 459–467

Ye, Z.-H., Kneusel, R.E., Matern, U. and Varner, J.E. (1994) An alternative methylation pathway in lignin biosynthesis. *Plant Cell* **6**: 1427–1439

Zhong, R., Morrison, W.H., Negrel, J. and Ye, Z.-H. (1998) Dual methylation pathways in lignin biosynthesis. *Plant Cell* **10**: 2033–2045.

35

MYB proteins and xylem differentiation

L.J. Newman and M.M. Campbell

1. Introduction

Approximately 400 to 430 million years ago, plants acquired the ability to occupy a terrestrial environment (Kenrick and Crane, 1997). Appearance of these plants coincided with the evolution of a vascular system that could both transport water and solutes from the roots to the aerial tissues, and provide structural support for the plant body. The vascular tissue upon which this conductive system is based is known as xylem, and it has been postulated that the evolution of xylem was crucial for plants to occupy a terrestrial habit (Kenrick and Crane, 1991; Cook and Friedman, 1998).

Xylem composition varies between plant species. In trees such as loblolly pine (*Pinus taeda* L.), secondary xylem is a relatively simple tissue consisting mainly of tracheid cells (Esau, 1965). By comparison, herbaceous angiosperms, such as *Arabidopsis*, produce less xylem and form only limited amounts of secondary xylem. Angiosperm xylem is generally more complex than that of gymnosperms and contains not only tracheids but also vessel elements (Esau, 1965). Nevertheless, it is likely that the differentiation of these different conducting cells proceeds by a very similar pathway (Esau, 1965; Torrey *et al.*, 1971).

The formation of xylem, xylogenesis, is an example of terminal cellular differentiation. Xylogenesis consists of four overlapping ontogenetic events: cell expansion; cell wall thickening; cell wall waterproofing and reinforcement by lignification; and autolysis, including wall lysis (Esau, 1965; Torrey *et al.*, 1971; see also Chapter 1, Savidge). Despite the importance of xylogenesis in plant development and evolution, relatively little is known about the biochemical, molecular, and genetic mechanisms that regulate this process.

Developmental processes like xylogenesis depend upon the regulated expression of many thousands of genes (Goldberg *et al.*, 1978; Kamalay and Goldberg, 1980; Kamalay and Goldberg, 1984; Doebley and Lukens, 1998). Regulation occurs at the level of transcription and is mediated through the activity of transcription factors. The importance of transcription factors in the formation of plant organs, particularly

Cell & Molecular Biology of Wood Formation, edited by R. Savidge, J. Barnett and R. Napier.

flowers, is becoming increasingly well understood (Doebley and Lukens, 1998). Paradoxically, very little is known about transcriptional regulatory mechanisms important in the formation of specific tissues like xylem. This chapter describes some of the latest work in the area.

2. Lignification

In terms of understanding transcriptional regulation as it relates to xylogenesis, the most extensively characterized of all the processes is lignification. Lignification involves the deposition of the aromatic polymer, lignin, in the walls of specialized cells, including those of xylem, sclerenchyma, phloem fibres, and periderm tissues (Campbell and Sederoff, 1996). The primary function of lignin is to reinforce cell walls, rendering them rigid, hydrophobic and impermeable to water. Lignin is also known to play a role in plant defence by providing a physicochemical barrier against pathogens (Vance *et al.*, 1980). In trees, lignin represents 15–36% of the dry weight of secondary xylem (wood) depending upon the taxa (Campbell and Sederoff, 1996). The abundance of lignin in xylem makes lignin the second most abundant polymer on earth.

In addition to its importance in plant growth and development, lignin is also important in an agro-industrial context. For example, in the pulp and paper industry, lignin must be removed from the wood as it limits accessibility to cellulose. In addition, lignin that remains after the pulping process can be photochemically oxidized by sunlight, resulting in reduced paper quality. Unfortunately, removal and bleaching of lignin is an expensive process, both in terms of cost, and due to the release of toxic pollutants. On the other hand, high levels of lignin can be especially useful when wood is used as a fuel source. Since lignin has a high carbon content, and therefore holds more potential energy than any of the other major wood components, high lignin levels can enhance the energy content of the wood. For these reasons, there is great interest in the potential to modify lignin composition in plants.

There have been many attempts to genetically improve plant species by modifying their lignin composition using a molecular-genetics approach. For this reason, many of the genes that encode enzymes of the lignin biosynthetic pathway have been isolated and cloned. In order to modify lignin quality or quantity, numerous cDNAs that encode enzymes of the pathway have been overexpressed in both sense and antisense orientation in various species (Baucher *et al.*, 1998) and in various tissues (Meyer *et al.*, 1998). This approach, coupled with analysis of lignin mutants (MacKay *et al.*, 1997) has revealed an unsuspected level of complexity and plasticity in the regulation of the lignin biosynthetic pathway (Sederoff *et al.*, 1999).

3. Promoter motifs and the lignification pathway

Despite the apparent complexity in the lignin biosynthetic pathway, analysis of promoters from the genes encoding enzymes on the lignin biosynthetic pathway has revealed that a limited subset of motifs are important in the regulation of these genes. To date, all of these promoters contain conserved motifs (reviewed by Douglas, 1996). Amongst the motifs, one group shown to be significant in the context of xylem-localized gene expression is the AC elements, also known as the H-boxes or PAL motifs. They are referred to as AC elements or H-boxes due to their adenosine and cytosine rich nature and due to the fact that they do not contain guanine residues on the coding strand.

AC elements have been found in the promoters of the genes encoding enzymes involved in general phenylpropanoid metabolism, and, more specifically, monolignol biosynthesis. This includes the genes encoding: phenylalanine ammonia-lyase (PAL), cinnamate 4-hydroxylase (C4H), caffeate O-methyltransferase (COMT), hydroxycinnamate:CoA ligase (4CL), and coniferyl alcohol dehydrogenase (CAD) (reviewed by Douglas, 1996; Bell-Lelong *et al.*, 1997) see *Figure 1*, Halpin *et al.*, this volume. In experiments that involved the analysis of transgenic plants expressing promoter:: *beta*-glucuronidase (GUS) fusions, AC elements had demonstrable importance in the regulation of tissue-specific gene expression (Hatton *et al.*, 1995). By mutating individual AC elements in the bean PAL2 promoter, and then using these mutant promoters to drive GUS expression, AC elements were demonstrated to be important in xylem-localized gene expression (Hatton *et al.*, 1995).

4. MYB proteins may regulate transcription related to lignification

Proteins that bind to AC elements have been identified. Sablowski *et al.* (1994) demonstrated that members of the MYB family of transcription factors were able to bind to AC elements in a sequence specific fashion. Furthermore, MYB proteins have been shown to activate transcription from promoters containing AC elements in an AC element-dependent fashion (Hatton *et al.*, 1995). To date, however, the relationship between MYB proteins and AC elements has only been demonstrated in the context of flavonoid biosynthesis during floral development. MYB proteins are also good candidates for regulating AC element-mediated gene expression during xylem development.

The MYB family of transcription factors was first identified as a *v-myb* oncogene from the avian myeloblastosis virus, which is responsible for causing leukaemia in chickens (Lipsick, 1996). MYB transcription factors are bipartite proteins that can be structurally dissected into a highly conserved N-terminal DNA-binding domain and a C-terminal transcriptional activation domain.

In plant MYB proteins, the DNA-binding domain contains two imperfect repeats of a 50–53 amino acid helix-turn-helix motif. These repeats, R2 and R3, are believed to function head-to-tail in the major groove of the double helix. The recognition helices contact each other and bind in a co-operative manner (Lipsick, 1996; Paz-Ares and Martin, 1997). The C-terminal domain is highly variable among all MYB proteins. The only common motif of the carboxy terminus is an acidic, *alpha*-helical domain that has been implicated in transcriptional activation (Sablowski *et al.*, 1994; Solano *et al.*, 1995). In plants, it has been estimated that there may be over 100 different *MYB* loci in each species (Paz-Ares and Martin, 1997). This is supported by the finding that there are over 100 different genes encoding R2R3-MYB proteins in *Arabidopsis thaliana* (Romero *et al.*, 1998; Kranz *et al.*, 1998). It is hypothesized that this vast array of *MYB* genes have evolved in conjunction with the development of novel functions in plants.

Recently, the importance of MYB proteins in the regulation of lignification during xylogenesis was demonstrated serendipitously. Two MYB transcription factors (AmMYB308 and AmMYB330), isolated from *Antirrhinum* flowers, were shown to transactivate genes related to lignin biosynthesis, such as PAL, in tobacco and in *Antirrhinum* (Sablowski *et al.*, 1994, 1995; Moyano *et al.*, 1996). Work on AmMYB305 provided detailed evidence for the role of this transcription factor in the regulation of phenylpropanoid metabolism (Sablowski *et al.*, 1994; Sablowski *et al.*, 1995; Moyano *et al.*, 1996; Tamagnone *et al.*, 1998a and b).

Overexpression of the *Antirrhinum* flower-expressed MYB transcription factors in tobacco resulted in an unexpected reduction in lignin by at least 17% in the vascular tissue (Tamagnone *et al.*, 1998b). These results suggested that when the weak transactivators, AmMYB308 and AmMYB330, were overexpressed in tobacco, they titrated the endogenous, strong transcriptional activators related to lignin biosynthesis (Tamagnone *et al.*, 1998a and b). This effect can be viewed as a dominant-negative effect that resulted in a net decrease in the transcription of the genes encoding enzymes of the lignin biosynthetic pathway and a decrease in lignification. The interpretation suggests that there is an endogenous MYB that functions in xylem to regulate the genes encoding enzymes of the lignin biosynthetic pathway. It is not yet known which MYB functions in this capacity.

5. Xylem-abundant MYB proteins

In order to test the hypothesis that MYB transcription factors play a role in regulating gene expression during xylogenesis, a screen was initiated to identify xylem-expressed MYB genes from pine. Trees like pine prove to be particularly amenable to the study of the molecular control of xylogenesis. For example, kilogram quantities of differentiating xylem can be isolated from individual, genetically-characterized trees. Tissue from rapidly growing *P. taeda* can be collected throughout the summer in the Southeastern Coastal Plain of the USA. Trees between 10 and 30 years of age are felled, and the bark peeled from the de-limbed stems. When the bark is peeled off, it splits the vascular cambium such that the differentiating xylem remains attached to the wood while the phloem stays with the bark. Using vegetable peelers, the differentiating xylem is stripped from the wood and frozen in liquid nitrogen and stored until use. Differentiating xylem tissue is a rich source of proteins and gene transcripts involved in the formation of xylem.

Using the differentiating xylem of *P. taeda*, two full-length cDNAs that encoded *P*inus *t*aeda *x*ylem *MYB* orthologues (*PtxMYB*) were cloned. Reverse transcriptase PCR of mRNA from differentiating pine xylem was used to generate a probe that would hybridize to sequences encoding the MYB conserved amino-terminal DNA binding domain. This probe was used to screen an unamplified cDNA library constructed from the mRNA from differentiating pine xylem. In this manner, the two full length cDNAs, *PtxMYB126* and *PtxMYB413*, were obtained.

Northern blots prepared with polyadenylated RNA suggest that the MYB orthologues are preferentially expressed in differentiating xylem (Campbell, in preparation). This finding was substantiated by data from ribonuclease protection assays (Campbell, in preparation). We propose that the PtxMYB proteins regulate transcriptional events during xylogenesis. We are currently testing this hypothesis.

6. Xylem-abundant MYB proteins activate transcription from conserved promoter elements

The PtxMYB proteins are able to bind to and activate transcription from conserved elements found in the promoters that exhibit xylem-abundant expression. Using the well-characterized bean PAL2 promoter as the basis for our analyses, we were able to demonstrate that two elements, known as AC-I and AC-II, were targets for DNA binding and transactivation by the PtxMYB proteins (Campbell, Bevan and Sederoff, in preparation). In elegant and thorough studies, the AC-I and AC-II motifs were demonstrated to be essential for xylem localized expression of the minimal PAL2 promoter (Hatton *et al.*, 1995). Furthermore, elements displaying significant sequence conservation with the AC boxes have been found in the promoters of 4CL and CAD in species as diverse as *Arabidopsis*, parsley, and *Eucalyptus* (Ohl *et al.*, 1990; Leyva *et al.*, 1992; Wanner *et al.*, 1995; Logemann *et al.*, 1995; Hauffe *et al.*, 1991, 1993; Feuillet *et al.*, 1995). The abundant expression of these genes in differentiating xylem is presumably linked to the large amount of lignin biosynthesis in this tissue. It may be that the PtxMYB proteins co-ordinate the expression of lignin biosynthetic enzymes in this tissue. An understanding of how phenylpropanoid metabolism is regulated in xylem should point to genetic strategies through which lignin could be modulated and thereby enhance the agroindustrial utilization of this tissue.

7. *Arabidopsis* orthologues of the xylem-abundant *MYB* genes

Recently, cDNA and genomic clones encoding the putative *Arabidopsis thaliana* orthologues of both *PtxMYB* genes were cloned (Newman and Campbell, unpublished). The identification of orthologous transcription factors in pine and *Arabidopsis* provides a unique opportunity to address several important questions relating to the growth, development and evolution of land plants. Molecular evidence suggests that angiosperms and gymnosperms shared their last common ancestor some 275 to 290 million years ago (Savard *et al.*, 1994). Since that time, gymnosperms such as pines and angiosperms such as *Arabidopsis thaliana* have evolved dramatically different strategies for growth, development and survival. How has this occurred?

Evidence is mounting that mutations in a relatively few loci can account for very large changes in plant morphology (Doebley *et al.*, 1990; Doebley and Stec, 1993; Doebley *et al.*, 1995; Bradshaw *et al.*, 1995; Bradshaw *et al.*, 1998). Some authors have proposed that such loci are likely to encode proteins that regulate other genes or gene products; as such their effects would be pleiotropic (Doebley, 1993; Doebley and Lukens, 1998). Two types of mutations in such regulatory loci could result in dramatic effects on plant phenotype: (i) those affecting the function of the regulatory protein and (ii) those affecting the location or timing of expression of the regulatory protein. In fact, in a recent review of the role of homeotic transcription factors in animal evolution, Carrol (1995) found that morphological diversity was largely a consequence of changes in the expression pattern of genes encoding transcription factors and changes in the functions of the homeotic proteins themselves. Wang *et al.* (1999) have demonstrated that alterations in the upstream regulatory regions of a gene encoding a transcription factor resulted in misregulation. This in turn resulted in substantial change in plant development.

8. Perspectives: determining the role of xylem-abundant MYB proteins in diverse genera

Functional comparison of two gene products that have very similar sequences should address a fundamental question related to sequence comparisons: 'Are sequences that appear to be homologous truly functionally equivalent?' This is an important question in light of the recent efforts to sequence expressed genes in a large number of organisms, including plants. A basic tenet in these cDNA sequencing efforts is that gene product function will be revealed by what is known about a similar sequence in another organism (e.g., Allona *et al.*, 1998). For many gene products, especially those of a regulatory nature, this remains to be tested.

While it is known that both the *Arabidopsis* and pine sequences described in this paper encode MYB proteins and that both gene products are likely to bind to DNA and activate transcription, it is not known if both target the same binding site, activate transcription from the same promoters or interact with the same proteins. The answers to these questions will indicate the extent to which similar regulatory function is conserved for MYBs that display a high degree of sequence similarity. Additionally, these experiments should indicate the degree of conservation in promoter elements that serve as targets for transcriptional factors. These data should shed light on the conservation of gene regulatory mechanisms between highly diverged plants.

Our experiments will also reveal if there are regions of the xylem MYB proteins and regulatory sequences in their promoters that would be particularly susceptible to functional changes if mutated.

Current experiments are aimed at testing the hypothesis that differences between pine and *Arabidopsis* xylem architecture and composition may be related, in part, to differences in MYB function. Pine xylem and *Arabidopsis* xylem share some similar features but also exhibit some important differences. Most notably, pines make large quantities of secondary xylem whereas *Arabidopsis* does not. Differences in the regulation and function of the MYB homologues in pine *versus Arabidopsis* may point to ways that the xylem in these two plants became different. Alternatively, total conservation of function may point to the fact that these MYB proteins regulate gene expression that is common to xylem development in many plants. A comparison of the function, location and timing of expression of apparently orthologous regulatory proteins from diverged species would allow us to determine if these proteins have been recruited for different functions in the two species or whether basic regulatory mechanisms have been conserved. These data would then begin to reveal whether a particular regulatory protein has played a role in the evolution of differences in xylem differentiation within the plant kingdom.

References

Allona, I., Quinn, M., Shoop, E., Swope, K., St. Cyr, S., Carlis, J., Riedl, J., Retzel, E., Campbell, M.M., Sederoff, R. and Whetten, R.W. (1998) Analysis of xylem formation in pine by cDNA sequencing. *Proc. Natl. Acad. Sci. USA* **95**: 9693–9698.

Baucher, M., Monties, B., van Montagu, M. and Boerjan, W. (1998) Biosynthesis and genetic engineering of lignin. *Crit. Rev. Plant Sciences* **17**: 125–197.

Bradshaw, H.D., Wilbert, S.M., Otto, K.G. and Schemske, D.W. (1995) Genetic mapping of floral traits associated with reproductive isolation in monkeyflowers (*Mimulus*). *Nature* **376**: 762–765.

Bradshaw, H.D. Jr., Otto, K.G., Frewen, B.E., McKay, J.K. and Schemske, D.W. (1998) Quantitative trait loci affecting differences in floral morphology between two species of monkeyflower (*Mimulus*). *Genetics* **149**: 367–382.

Bell-Lelong, D.A., Cusumano, J.C., Meyer, K. and Chapple, C. (1997) Cinnamate-4-hydroxylase expression in *Arabidopsis*. Regulation in response to development and the environment. *Plant Physiol.* **113**: 729–738.

Campbell, M.M. and Sederoff, R. (1996) Variation in lignin content and composition. *Plant Physiol.* **110**: 3–13.

Carrol, S.B. (1995) Homeotic genes and the evolution of arthropods and chordates. *Nature* 376: 479–485.

Cook, M.E. and Friedman, W.E. (1998) Tracheid structure in a primitive extant plant provides an evolutionary link to earliest fossil tracheids. *Int. J. Plant Sci.* **159**: 881–890.

Doebley, J. (1993) Genetics, development and plant evolution. *Curr. Opin. Gene. Dev.* **3**: 865–872.

Doebley, J. and Lukens, L. (1998) Transcriptional regulators and the evolution of plant form. *Plant Cell* **10**: 1075–1082.

Doebley, J. and Stec, A. (1993) Inheritance of the morphological differences between maize and teosinte: comparison of results for two F_2 populations. *Genetics* **134**: 559–570.

Doebley, J., Stec, A. and Gustus, C. (1995) *teosinte branched1* and the origin of maize: Evidence for epistasis and the evolution of dominance. *Genetics* **141**: 333–346.

Douglas, C.J. (1996) Phenylpropanoid metabolism and lignin biosynthesis: from weeds to trees. *Trends Plant Sci.* **1**: 171–178.

Esau, K. (1965) *Vascular Differentiation in Plants*. Holt, Rinehart and Winston, New York

Feuillet, C., Lauvergeat, V., Deswarte, C., Pilate, G., Boudet, A. and Grima-Pettenati, J. (1995) Tissue- and cell-specific expression of a cinnamyl alcohol dehydrogenase promoter in transgenic poplar plants. *Plant Molec. Biol.* **27**: 651–667.

Goldberg, R.B., Hoschek, G., Kamalay, J.C. and Timberlake, W.E. (1978) Sequence complexity of nuclear and polysomal RNA in leaves of the tobacco plant. *Cell* **14**: 123–131.

Hatton, D., Sablowski, R., Yung, M.-H., Smith, C., Schuch, W. and Bevan, M. (1995) Two classes of *cis* sequences contribute to tissue-specific expression of a *PAL2* promoter in transgenic tobacco. *Plant J.* **7**: 859–876.

Hauffe, K.D., Lee, S.P., Subramaniam, R. and Douglas, C.J. (1993) Combinatorial interactions between positive and negative *cis*-acting elements control spatial patterns of *4CL-1* expression in transgenic tobacco. *Plant J.* **4**: 235–253.

Hauffe, K.D., Paszkowski, U., Schulze-Lefert, P., Hahlbrock, K., Dangl, J.L. and Douglas, C.J. (1991) A parsley 4CL-1 promoter fragment specifies complex expression pattterns in transgenic tobacco. *Plant Cell* **3**: 435–443.

Kamalay, J.C. and Goldberg, R.B. (1980) Regulation of structural gene expression in tobacco. *Cell* **19**: 935–946.

Kamalay, J.C. and Goldberg, R.B. (1984) Organ-specific nuclear RNAs in tobacco. *Proc Natl. Acad. Sci. USA* **81**: 2801–2805.

Kenrick, P. and Crane, P.R. (1991) Water conducting cells in early fossil land plants: implications for the early evolution of tracheophytes. *Bot. Gaz.* **152**: 335–356.

Kenrick, P. and Crane, P.R. (1997) The origin and early evolution of plants on land. *Nature* **389**: 33–39.

Kranz, H., Denekamp, M., Greco, R., Jin, H., Leyva, A., Meissner, R., Petroni, K., Urzainqui, A., Bevan, M., Martin, C., Smeekens, S., Tonelli, C., Paz-Ares, J. and Weisshaar, B. (1998) Towards functional characterisation of the members of the *R2R3-MYB* gene family from *Arabidopsis thaliana*. *Plant J.* **16**: 263–270.

Leyva, A., Liang, X., Pintor-Toro, J.A., Dixon, R.A. and Lamb, C.J. (1992) *cis*-Element combinations determine phenylalanine ammonia-lyase gene tissue-specific expression patterns. *Plant Cell* **4**: 263–271.

Lipsick, J.S. (1996) One billion years of Myb. *Oncogene* **13**: 223–235.

Logemann, E., Parniske, M. and Hahlbrock, K. (1995) Modes of expression and common structural features of the complete phenylalanine ammonia-lyase gene family in parsley. *Proc. Natl. Acad. Sci. USA* **92**: 5905–5909.

MacKay, J.J., O'Malley, D.M., Presnell, T., Booker, F.L., Campbell, M.M., Whetten, R.W. and Sederoff, R.R. (1997) Inheritance, gene expression and lignin characterization in a mutant pine deficient in cinnamyl alcohol dehydrogenase. *Proc. Natl. Acad. Sci. USA* **94**: 8255–8260.

Meyer, K., Shirley, A.M., Cusumano, J.C., Bell-Lelong, D.A. and Chapple, C. (1998) Lignin monomer composition is determined by the expression of a cytochrome P450-dependent monooxygenase in arabidopsis. *Proc. Natl. Acad. Sci. USA* **95**: 6619–6623.

Moyano, E., Martinez-Garcia, J.F. and Martin, C. (1996) Apparent redundancy in *myb* gene function provides gearing for the control of flavonoid biosynthesis in *Antirrhinum* flowers. *Plant Cell* **8**, 1519–1532.

Ohl, S., Hedrick, S.A., Chory, J. and Lamb, C.J. (1990) Functional properties of a phenylalanine ammonia-lyase promoter from *Arabidopsis*. *Plant Cell* **2**: 837–848.

Paz-Ares, J. and Martin, C. (1997) MYB transcription factors in plants. *Trends Genet.* **13**: 67–72.

Romero, I., Fuertes, A., Benito, M.J., Malpical., J.M., Leyva, A. and Paz-Ares, J. (1998) More than 80 *R2R3-MYB* regulatory genes in the genome of *Arabidopsis thaliana*. *Plant J.* **14**: 273–284.

Sablowski, R.W.M., Moyano, E., Culianez-Macia, F.A., Schuch, W., Martin, C. and Bevan, M. (1994) A flower-specific MYB protein activates transcription of phenylpropanoid biosynthetic genes. *EMBO J.* **13**: 128–137.

Sablowski, R.W.M., Baulcombe, D.C. and Bevan, M. (1995) Expression of a flower-specific MYB protein in leaf cells using a viral vector causes ectopic activation of a target promoter. *Proc. Natl. Acad. Sci. USA* **92**: 6901–6905.

Savard, L., Li, P., Strauss, S.H., Chase, M.W., Michaud, M. and Bousquet, J. (1994) Chloroplast and nuclear gene sequences indicate Late Pennsylvanian time for the last common ancestor of extant seed plants. *Proc. Natl. Acad. Sci. USA* **91**: 5163–5168.

Sederoff, R.R., MacKay, J.J., Ralph, J. and Hatfield, R.D. (1999) Unexpected variation in lignin. *Curr. Op. Plant Biol.* **2**: 145–152.

Solano, R., Nieto, C., Avila, J., Canas, L., Diaz, I. and Paz-Ares, J. (1995) Dual DNA binding specificity of a petal epidermis-specific MYB transcription factor (MYB.Ph3) from *Petunia hybrida*. *EMBO J.* **14**: 1773–1784.

Tamagnone, L., Merida, A., Parr, A., Mackay, S., Culianez-Macia, F.A., Roberts, K. and Martin, C. (1998a) The AmMYB308 and AmMYB330 transcription factors from *Antirrhinum* regulate phenylpropanoid and lignin biosynthesis in transgenic tobacco. *Plant Cell* **10**: 135–154.

Tamagnone, L., Merido, A., Stacey, Plaskitt, Parr, Chang, Lynn, D., Dow, Roberts, K. and Martin, C. (1998b) Inhibition of phenolic acid metabolism results in precocious cell death and altered cell morphology in leaves of transgenic tobacco plants. *Plant Cell* **10**: 1801–1816.

Torrey, J.G., Fosket, D.E. and Hepler, P.K. (1971) Xylem formation: a paradigm of cytodifferentiation in higher plants. *Am. Scientist* **59**: 338–352.

Vance, C.P., Kirk, T.K. and Sherwood, R.T. (1980) Lignification as a mechanism of disease resistance. *Ann. Rev. Phytopathol.* **18**: 259–288.

Wang, R.-L., Stec, A., Hey, J., Luken, L. and Doebley, J. (1999) The limits of selection during maize evolution. *Nature* **398**: 236–239.

Wanner, L.A., Li, G., Ware, D., Somssich, I.E. and Davis, K.R. (1995) The phenylalanine ammonia-lyase gene family in *Arabidopsis thaliana*. *Plant Molec. Biol.* **27**: 327–338.

36

Role of the *ATHB-8* gene in xylem formation

S. Baima, M. Tomassi, A. Matteucci, M.M. Altamura, I. Ruberti and G. Morelli

1. Introduction

Transcription is thought to be one of the major points at which biological processes are regulated. Among the various classes of transcription factors, the animal homeodomain proteins play key roles in developmental processes, including cell fate and pattern formation (Affolter *et al.*, 1990). A large number of homeobox genes have been identified in several plant species. These homeobox genes can be divided into two large groups encoding homeodomain and homeodomain-leucine zipper (HD-Zip) proteins, respectively (Morelli *et al.*, 1998). By analogy with animal homeobox genes, plant homeobox genes are expected to encode transcription factors involved in the regulation of key developmental processes.

This chapter focuses on the use of an *Arabidopsis* HD-Zip protein, known as ATHB-8, as a molecular tool to study vascular development. The first section reviews the expression pattern of ATHB-8 during vascular development and regeneration. The next section describes the changes in the development of vascular tissues induced by elevated levels of ATHB-8 in transgenic plants of tobacco and *Arabidopsis*, and presents some more recent data aimed at demonstrating a role of ATHB-8 in promoting xylem formation. The review ends by considering the implications of these studies for the comprehension of vascular development.

2. *ATHB-8*, an early marker of vascular development

2.1 ATHB-8 *gene expression during plant development*

The *Arabidopsis ATHB-8* gene encodes an HD-Zip protein of about 92 kDa (Baima *et al.*, 1995). A search for similarity revealed that ATHB-8 shares some homologies throughout its entire amino acid sequence with GLABRA2 (Sessa *et al.*, 1998), a HD-Zip protein required for regulation of trichome development, stomata and root hair

Cell & Molecular Biology of Wood Formation, edited by R. Savidge, J. Barnett and R. Napier.

formation (Berger *et al.*, 1998; Di Cristina *et al.*, 1996; Hung *et al.*, 1998; Masucci *et al.*, 1996; Rerie *et al.*, 1994).

In situ hybridization measurements of the localization of *ATHB-8* mRNAs revealed a specific signal clearly associated with the vascular regions of flowering stem and pedicel. In particular, the signal is restricted to the layer of cells located between the xylem and the phloem elements. This region, formed by procambial cells, is the meristem responsible for the formation of the vasculature in the primary body of the plant. Hybridization signals associated with the procambium were also observed in the embryo, flower and root. In the embryo at the linear cotyledon stage, a low but clear hybridization signal was detected in the core central region corresponding to the procambium, which extends into the cotyledons in this stage embryo. In the root, the hybridization signal was associated with the procambial cells in a region corresponding to the meristem and elongation zone. Furthermore, the *ATHB-8* mRNAs were localized in lateral root primordia. Lateral root formation begins by periclinal and anticlinal divisions in pericycle cells of the primary root (Dolan *et al.*, 1993). Upon formation of a dome-shaped structure (primordium) by pericycle derivatives, a new root meristem is derived from a subset of these cells. A significant level of *ATHB-8* expression was found associated with the centre of the primordium, just above the periphery of the vascular cylinder (Baima *et al.*, 1995).

The *in situ* analysis was compared with histochemical localization of the glucuronidase activity encoded by the *GUS* reporter gene (Jefferson *et al.*, 1987), driven by the *ATHB-8* promoter. In this chimaeric gene, the leader sequence of the *ATHB-8* gene, which contains two short open reading frames and one intron, was entirely included in the construct. In this way, the *GUS* gene should undergo the same transcriptional and translational regulation as *ATHB-8*. Transgenic *Arabidopsis* seedlings harbouring the chimaeric gene revealed a GUS expression pattern always associated with procambial cells. The only exception was a strong staining of the columella cells of the root (see *colour Plate 7*). Since the GUS activity is very high in both the procambial and columella cells, it is not clear whether the *ATHB-8* promoter is also active in the quiescent centre. Transverse sections of the root elongation zone revealed that GUS expression is restricted to a region of the stele from which the xylem originates (*colour Plate 7*) (Baima *et al.*, 1995; Morelli *et al.*, unpublished).

An expression pattern coincident with that of the *ATHB-8* gene in *Arabidopsis* was also detected in transgenic tobacco plants carrying the same *ATHB-8*::*GUS* chimaeric gene. In fact, the GUS staining was visible in the vascular system of the shoot, in the elongation zone of the primary root, and in lateral root primordia. GUS activity was also detectable in the elongated central cells of tobacco embryos starting at the heart stage (Baima *et al.*, 1995). Histological studies have revealed that the central cells of the embryo start to divide and elongate to form the provascular tissue during the transition from the globular to the heart stage (West and Harada, 1993). Therefore, a significant GUS activity driven by the *ATHB-8* promoter becomes evident almost coincident with the appearance of procambial cells. Unfortunately, the heart stage embryos have not been analysed by GUS staining in *Arabidopsis* because the developing seeds are difficult to manipulate compared to those of tobacco.

An important conclusion from the *in situ* and GUS analyses is that *ATHB-8* expression is not linked to mitotic activity *per se*, but is specifically associated with meristems giving rise to the vascular system. In fact, the expression of the gene has never been observed in vegetative nor floral meristems of *Arabidopsis* nor tobacco.

2.2 *ATHB-8 gene expression during vascular regeneration after damage*

One of the most widely used approaches to study determinants of vascular differentiation has been the analysis of revascularization after severing vascular bundles. When a vascular bundle is severed, nearby parenchyma cells resume cell division and redifferentiate into both xylem and phloem (Steeves and Sussex, 1989). Therefore, differentiation of tracheary elements after wounding has been widely used as an experimental model system to study vascular development without the superimposition of spatial and temporal factors such as in the apical meristem.

As tobacco is easy to manipulate in wounding experiments, transgenic tobacco plants harbouring the *ATHB-8::GUS* chimaeric gene were analysed. In particular, to determine whether the *ATHB-8* promoter is active during revascularization, young internodes wounded by a double cut were analysed. The two partially overlapping transverse cuts (about 3 mm apart) severed all the vascular bundles but maintained a narrow bridge of pith cells connecting the apical and basal portion of the stem (*Figure 1a*). After wounding, the apical portion withered and stopped growing, but it regained a healthy appearance in approximately seven days, suggesting that functional connections had been re-established. The expression of the *ATHB-8::GUS* gene in wounded stems was detected by histochemical staining at different time points after cutting. Although unwounded stems did not show any specific staining, stems stained one hour after the wound appeared diffusely blue coloured, indicating that the *ATHB-8* promoter is wound-inducible. Eight hours after wounding, GUS activity showed a polar localization with respect to the main axis of the plant, being mainly restricted to the portion of the stem above the cut region, whilst no staining was visible in the bridge (*Figure 1b*). About seven days after wounding, when functional connections had been re-established, the expression of GUS activity in recovered plants was still polarized, but mainly present in the bridge and in a narrow region immediately above the margin of the upper cut (*Figure 1c*). The analysis of transverse sections of the bridge region showed that the blue stain was associated with a wound cambium. Moreover, sections of stems at earlier phases of recovery showed that many single or grouped parenchyma cells, which were not distinguishable from the neighbouring cells for any other morphological aspect, positively stain for GUS activity. These cells were most likely parenchyma cells at very early steps of transdifferentiation, which can occur in these cells without the progression of cell division. In fact, occasionally, large tracheary cells which had both a pale blue colour and the typical pattern of secondary wall deposition were observed (Baima *et al.*, 1995).

The stimulus for vascular redifferentiation after wounding is thought to be the auxin released from the severed bundle itself (Steeves and Sussex, 1989). The local increase of auxin descending along the leaf traces and coming out from severed vascular bundles (Gersani, 1985; Sussex *et al.*, 1972) might explain the polar and asymmetrical GUS staining of wounded *ATHB-8*::GUS plants (*Figure 1*). This hypothesis is supported by the finding that the expression of *ATHB*-8 is modulated by auxin (Baima *et al.*, 1995).

3. *ATHB-8*, a positive regulator of vascular development

3.1 *Elevated levels of ATHB-8 in transgenic plants enhance xylem formation*

To study the function of *ATHB-8*, constructs of the gene in sense and antisense orientation driven by the Cauliflower Mosaic Virus (CaMV) 35S promoter were generated and used to transform *Arabidopsis* and tobacco. In both plants, overexpression of *ATHB-8* anticipates primary xylem differentiation, and overall increases xylem formation. Most of the contribution to xylem production is derived from an increased secondary growth. The data imply a role for this HD-Zip protein in the regulation of both procambial and cambial cell differentiation. Auxin is considered as a major factor controlling the differentiation of the vascular system in plants. It has been suggested that xylem differentiation is positively controlled by the amount of auxin both qualitatively and quantitatively (Aloni and Zimmermann, 1983). Therefore, it could be argued that the ATHB-8 protein may be a key component of the signalling pathway(s) mediated by auxin leading to vascular cell differentiation.

The transgenic *Arabidopsis* and tobacco plants also share other phenotypic traits, such as an upward curling of cotyledon and leaf margins, and a reduction in internode elongation (Baima *et al.*, submitted). A phenotypic trait specific for the tobacco plants is a defect in the formation of the first pair of leaves. Three tobacco lines expressing the transgene at high levels showed either a partial or a complete fusion of the first pair of leaves, or developed a small radially symmetric pin-like structure containing a provascular strand in continuity with the hypocotyl provascular tissue. The formation of fused leaves and the development of a pin-like structure have been reported in *Arabidopsis* mutants impaired in auxin polar transport, such as *pin-formed* and *pinoid* (Bennet *et al.*, 1995; Gälweiler *et al.*, 1998; Okada *et al.*, 1991), or in plants treated with auxin polar transport inhibitors (Hadfi *et al.*, 1998; Liu *et al.*, 1993; and reference therein). Indeed, fused leaves can be obtained *in vitro* by treating tobacco plants with either auxin polar transport inhibitors or 2,4-D, a synthetic auxin (Baima, unpublished). Therefore, the phenotypes observed in both *ATHB-8* transgenic *Arabidopsis* and tobacco plants might be arisen from an alteration of auxin signalling or transport, due to the activity of the transgene. Consistently with the 'canalization hypothesis' proposed by Sachs (1969, 1981), a precocious differentiation of the vascular system in the first stages of post-embryonic development in the transgenic tobacco plants could be responsible for draining out the auxin from the apical part of the seedling. This, in turn, might produce an alteration of the auxin balance within the shoot apical meristem which, altering the proper placement of leaf primordia, results in the formation of fused leaves or pin-like structures. In support of this hypothesis, an increase of auxin polar transport has been observed at early stages of inflorescence stem elongation of *35S::ATHB-8 Arabidopsis* plants (Baima *et al.*, submitted).

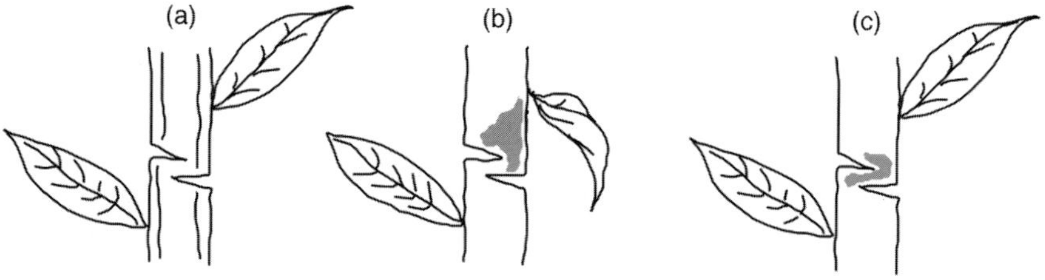

Figure 1. ATHB-8::GUS *gene expression during wound recovery. Schematic representation of young internodes from* ATHB-8::GUS *plants wounded by a double cut. At the time of cutting (a), at 8 hours (b) and 7 days (c) after cutting. The stems are oriented with the apical end upward. The shadowed area indicates the polar staining derived from* GUS *activity expressed by the* ATHB-8 *promoter.*

The fact that the overexpression of *ATHB-8* perturbs neither the formation of the cotyledons nor the following leaves might suggest the involvement of factor(s) whose concentration is critical only at specific stages. A similar phenotype was never observed in *ATHB-8* transgenic *Arabidopsis* plants, either because the expression level of the transgene is low compared to that of the tobacco plants, or because the position of the first two leaf primordia, being already established during embryogenesis (Medford *et al.*, 1994), cannot be affected during germination.

None of the transformed *Arabidopsis* or tobacco plants overexpressing *ATHB-8* in the antisense orientation showed any evident phenotypic alterations. The lack of phenotypes in these plants might be explained by the presence in the vasculature of proteins highly related to *ATHB-8* with redundant function(s) (Steindler *et al.*, 1998).

3.2 *Elevated levels of* ATHB-8 *enhance xylogenesis and shoot formation in tobacco thin cell layers*

To investigate whether auxin signalling is altered in *ATHB-8* transgenic plants, the effects were analysed of *ATHB-8* overexpression on morphogenesis in thin cell layer (TCL) tobacco explants. TCLs are small tissue explants, approximately six cell layered, consisting of fully differentiated epidermal, subepidermal, and parenchyma cells. Explants stripped from tobacco stems initiate organogenesis from the subepidermal cells when incubated on a suitable tissue culture medium (Tran Thanh Van *et al.*, 1974). Depending on the hormonal composition of the medium and on the physiological stage of the plant, TCLs can be directed to form shoots (S) from vegetative plants or roots (R) from reproductive plants (Altamura *et al.*, 1995; Torrigiani *et al.*, 1989). In general, the number and type of organs formed, and the overall growth, are dependent upon the initial concentrations of auxin and cytokinin (Mohnen *et al.*, 1990). In the absence of added hormones (hormone free medium, HF), the organogenic response of the wild-type tobacco was very poor (*Table 1*). When TCLs of transgenic plants were grown on HF medium, the production of shoots was significantly increased, at least when the explants were excised from vegetative plants (*Table 1*), when analysed as the number of responding explants as well as the number of shoots per explant (*Tables 1* and *2*). Most of the shoots had curled leaves similar to those produced *in planta*. Conversely, no alterations were observed in the controls.

At the end of the HF culture, histological analyses of the transgenic explants revealed both the presence of shoot apices with a morphology typical of a vegetative meristem (*Figure 2a*), and altered apices producing a pin-like structure (*Figure 2b*). Moreover, the transgenic explants excised from vegetative plants produced more vascular nodules and strands than the controls (*Table 2*). The nodule is a vascular bundle-like structure with a central core of vascular tissue surrounded by a zone of 'neocambial' cells (*Figure 2c*; Tran Thanh Van and Dien, 1975). The strand is composed of long files of 'neocambial cells' flanking *de novo* vascular elements (*Figure 2d*). Finally, the average numbers of meristemoids and domes produced in the TCL transgenic explants were similar to those of the control explants (*Table 2*).

TCLs were also cultured in media containing hormones in which explants preferentially produced either shoots (S, [auxin]/[cytokinin] equal to 0.1) or roots (R, [auxin]/[cytokinin] equal to 10). Under S-inducing medium no significant differences were observed between transgenic and wild-type explants. On R-inducing medium the number of roots produced by the transgenic TCLs was similar to that of

Table 1. *Organogenic response of tobacco thin cell layers (TCLs). Organogenic and callogenic responses of TCLs from wild type (SR1) and transgenic (*ATHB-8*) tobacco plants after 30 days of culture on hormone free (HF), shoot-inducing (S) or root-inducing (R) medium given as percent of cultured explants which responded* in vitro *(R.E.) and mean number (n) of organs per each responding explant (O/E) (±s.e.m.).*

		HF		S		R	
		SR1	*ATHB-8*	SR1	*ATHB-8*	SR1	*ATHB-8*
	Vegetative	2.7	20*	24.2	39.2	24	34**
R.E.	Veg.+ roots	2.7	0	0	0	33	38
(%)	Roots	–	–	–	–	13	12
	Callus	94.6	80	75.8	60.8	30	16
O/E	Shoots	2.0±0.6	6.4±0.9*	6.2±0.9	6.0±0.8	5.1±0.3	9.4±0.6**
(*n*)	Roots	–	–	–	–	4.1±0.3	4.7±0.3

* $P<0.05$ ** $P<0.01$

the controls, whilst the number of shoots was about 2-fold higher; moreover, the percentage of explants with shoots was higher (*Table 1*). In addition, the formation of vascular nodules showed a significant increase in transgenic TCLs compared to the controls (*Figure 3*). Altogether, the amount of xylem tissue produced by transgenic TCLs showed a very strong enhancement, considering that the average area of these vascular nodules was 2.5–3.3-fold higher than that of the controls (*Figure 3*). The data suggest that explants from transgenic tobacco plants expressing elevated levels of *ATHB-8* might have an increased sensitivity to auxin in the formation of vascular tissue.

Table 2. *Histological analysis of tobacco TCL explants on HF medium. Mean number of vascular nodules, meristemoids, domes and vegetative buds formed by wild-type and* ATHB-8 *tobacco TCL explants after 30 days of culture on hormone-free medium. Meristemoids, domes, vegetative buds, vascular nodules and vascular strands present in the sectioned TCLs were counted under a light microscope and expressed as mean number/explant ± s.e.m. In the HF treatment, the mean area of both control and* 35S::ATHB-8 *vascular nodules was about the same (9.15×10^4 mm^2); therefore, to quantify the total production of vascular structures, vascular strands were counted as n-fold the mean area of vascular nodules. The area of the vascular nodules in the R medium was measured at 100 μm intervals on the entire pool of 5 explants*

	Wild type	*ATHB-8*
Vascular nodules	2.7±1.3	10.2±0.6**
Meristemoids	2.3±0.3	1.7±1.2
Domes	1.0±0.6	0.7±0.3
Vegetative buds	1.3±0.3	7.3±0.3**

** $P<0.01$

4. Perspectives

The expression studies indicated that *ATHB-8* has to be considered a specific molecular marker of early vascular development. In fact, the *ATHB-8* gene is active in provascular cells of the embryo where these cells are not differentiating into more mature vascular cells and at early stages of vascular cell determination during revascularization. These findings strongly suggest that early steps in differentiation are similar in *de novo* and regenerating (wounded) vascular regions.

Several mutants that interfere with various aspects of vascular development have been isolated (reviewed by Nelson and Dengler, 1997). However, no conclusions on the role of the corresponding genes in vascular pattern formation could be drawn with the exception of *MONOPTEROS* whose action is required in the establishment of tissue continuity (axialization) within the vascular system (Hardtke and Berleth, 1998; Przemeck *et al.*, 1996). Furthermore, none of the mutants isolated so far specifically affects vascular cell differentiation. This might be due to the essential requirement of this process or to the action of regulatory factors with redundant functions.

To date, several genes encoding HD-Zip proteins have been isolated and the corresponding gene products have been grouped into four families of related proteins, named HD-ZIP I-IV (Morelli *et al.*, 1998). GLABRA2, a protein of the HD-ZIP IV family, is required for regulation of trichome development, stomata and root hair formation (Berger *et al.*, 1998; Di Cristina *et al.*, 1996; Hung *et al.*, 1998; Masucci *et al.*, 1996; Rerie *et al.*, 1994); ATML1 and O39, the other two members of the same family, share with GLABRA2 a common dermal-specific pattern of expression (Lu *et al.*, 1996). Altogether, these data suggest that HD-ZIP IV proteins might be involved in the regulation of epidermal cell fate and differentiation.

Conversely, at least 3 out of the 5 members of the HD-ZIP III family, such as ATHB-8, ATHB-9 and ATHB-14, are expressed in the vascular system (Baima *et al.*, 1995). It is noteworthy that ATHB-8, ATHB-9 and ATHB-14 share significant similarities with GLABRA2, ATML1 and O39, indicating that they could have evolved from a common ancestor protein (Sessa *et al.*, 1998). Therefore, it is tempting to speculate that proteins of the HD-ZIP III family may participate in the regulatory network of transcription factors whose interactions are required to determine the cell fate of procambial and cambial initials from which xylem and phloem derive. The finding that overexpression of *ATHB-8* in *Arabidopsis* and tobacco transgenic plants results in a higher production of primary and secondary xylem strongly supports this hypothesis.

New basic techniques for studying gene function, such as transposon and T-DNA insertion mutagenesis, are being developed and they will be employed to identify the function of all the genes in *Arabidopsis*. However, the majority of the insertions isolated so far does not result in a mutant phenotype, probably because of gene redundancy. The presence in the *Arabidopsis* genome of several genes highly related to *ATHB-8* as well as the vascular expression of at least two of them suggest that some HD-ZIP III factors might indeed have redundant functions. Therefore, it seems relevant to characterize the expression pattern of the other two HD-ZIP III protein encoding genes, *ATHB-15* and *ATHB-16*, and to identify mutants in all the HD-ZIP III genes specifically expressed in the vascular system. The subsequent production of double or multiple mutants and their analysis should significantly contribute to the understanding of the role of ATHB-8 and other HD-ZIP III proteins in the regulation of vascular development.

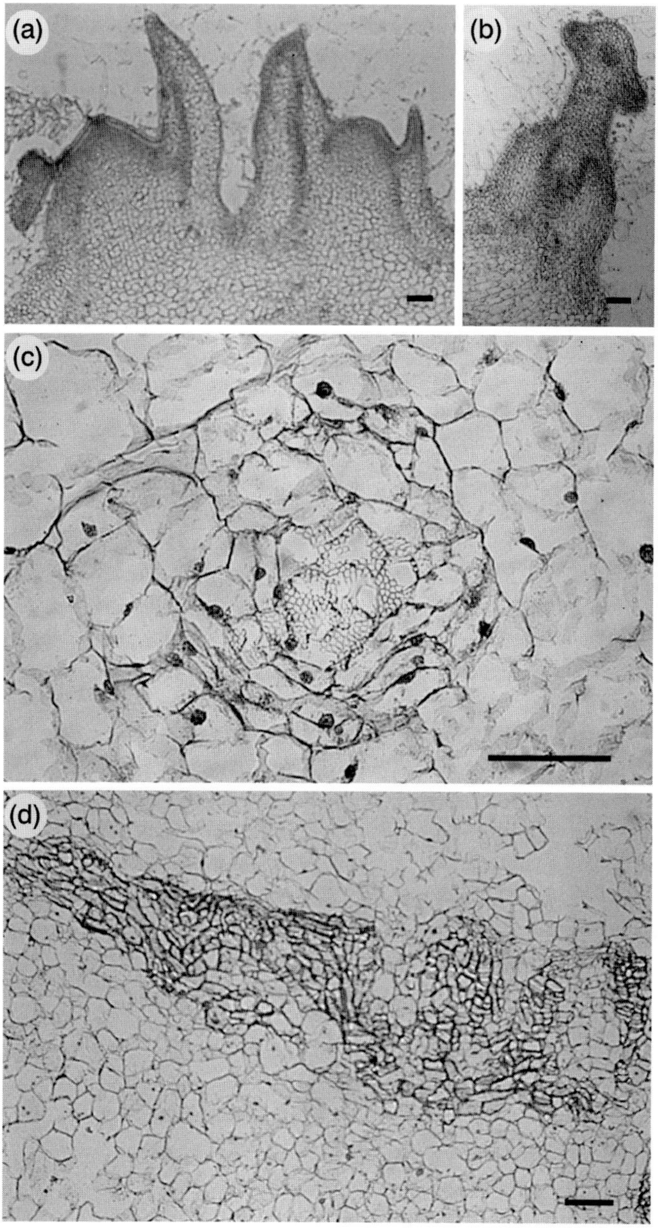

Figure 2. *Histological analysis of morphogenesis on tobacco TCLs from* 35S::ATHB-8 *transgenic plants cultured on HF medium. (a) Normally-shaped vegetative buds. (b) An anomalous vegetative bud resembling a pin-like structure. (c) Vascular nodule. (d) Vascular strand typically produced by* ATHB-8 *TCLs under HF conditions. Sections were stained with eosin and Carazzi's haemalum. Scale bar: 100 μm. (Altamura* et al., *unpublished).*

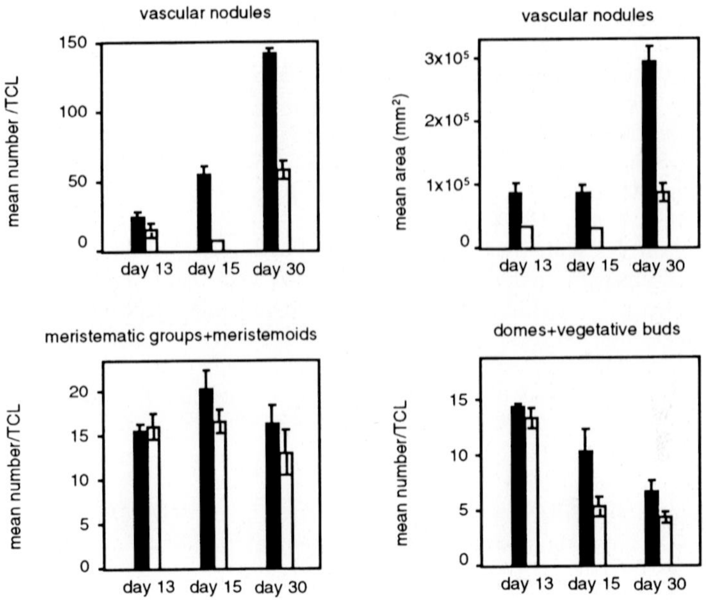

Figure 3. *Morphogenesis on tobacco TCLs cultured on root-inducing medium. The histological analysis was performed after 13, 15 and 30 days of culture of ATHB-8 (black columns) and wild type (white columns) explants. Meristematic groups and meristemoids are composed of 4–8 cells and more than 8 cells, respectively. The bar represents the standard error of the mean.*

Acknowledgements

We would like to acknowledge funding from the European Union Biotechnology Program (BIO4-CT960217) and Piano Nazionale 'Biotecnologie Vegetali', Ministero per le Politiche Agricole.

References

Affolter, M., Schier, A. and Gehring, W.J. (1990) Homeodomain proteins and the regulation of gene expression. *Curr. Opin. Cell Biol.* **2**: 485–495.

Altamura, M.M., Capitani, F., Falasca, G., Gallelli, A., Scaramagli, S., Bueno, M., Torrigiani, P. and Bagni, N. (1995) Morphogenesis in cultured thin layers and pith explants of tobacco. I. Effect of putrescine on cell size, xylogenesis and meristemoid organization. *J. Plant Physiol.* **147**: 101–106.

Altamura, M.M., D'Angeli, S. and Capitani, F. (1998) The protein of *RolB* gene enhances shoot formation in tobacco leaf explants and thin cell layers from plants in different physiological stages. *J. Exp. Bot.* **49**: 1139–1146.

Aloni, R. and Zimmermann, M.H. (1983) The control of vessel size and density along the plant axis — a new hypothesis. *Differentiation* **24**: 203–208.

Baima, S., Nobili, F., Sessa, G., Lucchetti, S., Ruberti, I. and Morelli, G. (1995) The expression of the *ATHB-8* homeobox gene is restricted to provascular cells in *Arabidopsis thaliana*. *Development* **121**: 4171–4182.

Baima, S., Matteucci, A., Possenti, P., Pane, P., Ruberti, I., Altamura, M.M. and Morelli, G. (2000) Overexpression of the *ATHB-8* homeobox gene increases xylem formation in *Arabidopsis* and tobacco transgenic plants. Submitted to *The Plant Cell.*

Bennet, S.R.M., Alvarez, J., Bossinger, G. and Smyth, D.R. (1995) Morphogenesis in *pinoid* mutants of *Arabidopsis*. *Plant J.* **8**: 505–520.

Berger, F., Linstead, P., Dolan, L. and Haseloff, J. (1998) Stomata patterning on the hypocotyl of *Arabidopsis thaliana* is controlled by genes involved in the control of root epidermis patterning. *Dev. Biol.* **94**: 226–234.

Di Cristina, M., Sessa, G., Dolan, L., Linstead, P., Baima, S., Ruberti, I. and Morelli, G. (1996) The *Arabidopsis* ATHB-10 (GLABRA2) is a HD-Zip protein required for regulation of root hair development. *Plant J.* **10**: 393–402.

Dolan, L., Janmaat, K., Willemsen, V., Linstead, P., Poethig, S., Roberts, K. and Scheres, B. (1993) Cellular organisation of the *Arabidopsis thaliana* root. *Development* **119**: 71–84.

Gälweiler, L., Changui, G., Müller, A., Wisman, E., Mendgen, K., Yephremov, A. and Palme, K. (1998) Regulation of polar auxin transport by AtPIN1 in *Arabidopsis* vascular tissue. *Science* 282: 2226–2230.

Gersani, M. (1985) Appearance of new transport capacity in wounded plants. *J. Exp. Bot.* **36**: 1809–1816.

Hadfi, K., Speth, V. and Neuhaus, G. (1998) Auxin-induced developmental patterns in *Brassica juncea* embryos. *Development* **125**: 879–887.

Hardtke, C.S. and Berleth, T. (1998) The *Arabidopsis* gene *MONOPTEROS* encodes a transcription factor mediating embryo axis formation and vascular development. *EMBO J.* **2**: 1405–1411.

Hung, C.Y., Lin, Y., Zhang, M., Pollock, S., Marks, M.D. and Schiefelbein, J. (1998) A common position-dependent mechanism controls cell-type patterning and *GLABRA2* regulation in the root and hypocotyl epidermis of *Arabidopsis*. *Plant Physiol.* **117**: 73–84.

Jefferson, R.A., Kavanagh, T.A. and Bevan M.W. (1987) *GUS* fusions: β-glucuronidase as a sensitive and versatile gene fusion marker in higher plants. *EMBO J.* **6**: 3901–3907.

Liu, C.-M., Xu, Z.-H. and Chua, N.-H. (1993) Auxin polar transport is essential for the establishment of bilateral symmetry during early plant embryogenesis. *Plant Cell* **5**: 621–630.

Lu, P., Porat, R., Nadeau, J.A. and O'Neil, S.D. (1996) Identification of a meristem L1 layer-specific gene in *Arabidopsis* that is expressed during embryonic pattern formation and defines a new class of homeobox genes. *Plant Cell* **8**: 2155–2168.

Masucci, J.D., Rerie, W.G., Foreman, D.R., Zhang, M., Galway, M.E., Marks, M.D. and Schiefelbein, J.W. (1996) The homeobox gene *GLABRA2* is required for position-dependent cell differentiation in the root epidermis of *Arabidopsis thaliana*. *Development* **122**: 1253–1260.

Medford, J.I., Callos, J.D., Behringer, F.J. and Link, B.M. (1994) Development of the vegetative shoot apical meristem. In *Arabidopsis* (eds. E.M. Meyerowitz and C.R. Sommerville), Cold Spring Harbor Laboratory Press, pp. 355–378.

Mohnen, D., Eberhard, S., Marfa, V., Doubrava, N., Toubart, P., Gollin, D.J., Gruber, T.A., Nuri, W., Albersheim, P. and Darvill, A. (1990) The control of root, vegetative shoot and flower morphogenesis in tobacco thin cell-layer explants (TCLs). *Development* **108**: 191–201.

Morelli, G., Baima, S., Carabelli, M., Di Cristina, M., Lucchetti, S., Sessa, G., Steindler, C. and Ruberti, I. (1998) Homeodomain-leucine zipper proteins in the control of plant growth and development. In *Cellular Integration of Signalling Pathways in Plant Development* (eds. Lo Schiavo, F., Last, R.L., Morelli, G. and Raikhel, N.V.), Berlin: Springer, pp 251–262.

Nelson, T. and Dengler, N. (1997) Leaf vascular pattern formation. *Plant Cell* **9**: 1121–1135.

Okada, K., Ueda, J., Komaki, M.K., Bell, C.J. and Shimura, Y. (1991) Requirement of the auxin polar transport system in early stages of *Arabidopsis* floral bud formation. *Plant Cell* **3**: 667–684.

Przemeck, G.K., Mattsson, J., Hardtke, C.S., Sung, Z.R. and Berleth, T. (1996) Studies on the role of the *Arabidopsis* gene *MONOPTEROS* in vascular development and plant cell axialization. *Planta* **200**: 229–237.

Rerie, W.G., Feldmann, K.A. and Marks, M.D. (1994) The *GLABRA2* gene encodes a homeodomain protein required for normal trichome development in *Arabidopsis*. *Genes Dev.* **8**: 1388–1399.

Sachs, T. (1969) Polarity and the induction of organized vascular tissues. *Annu. Bot.* **33**: 263–275.

Sachs, T. (1981) The control of the patterned differentiation of vascular tissues. *Adv. Bot. Res.* **9**: 151–262.

Sessa, G., Steindler, C., Morelli, G. and Ruberti, I. (1998) The *Arabidopsis ATHB-8*, *-9* and *-14* genes are members of a small gene family coding highly related HD-Zip proteins. *Plant Mol. Biol.* **38**: 609–622.

Steeves, T.A. and Sussex, I.M. (1989) *Patterns in Plant Development*, 2nd ed. Cambridge: Cambridge University Press.

Sussex, I.M., Clutter, M.E. and Goldsmith, H.M. (1972) Wound recovery by pith cell redifferentiation: structural changes. *Amer. J. Bot.* **59**: 797–804.

Torrigiani, P., Altamura, M.M., Capitani, F., Serafini-Fracassini, D. and Bagni, N. (1989) *De novo* root formation in thin cell layers of tobacco: changes in free and bound polyamines. *Physiol. Plantarum* **77**: 294–301.

Tran Thanh Van, K., Dien N.T. and Chlyah, A. (1974) Regulation of organogenesis in small explants of superficial tissue of *Nicotiana tabacum* L. *Planta* **119**: 149–159.

Tran Thanh Van, K. and Dien N.T. (1975) Etude au niveau cellulaire de la differéntiation *in vitro* et *de novo* bourgeons végétatifs, de racines, ou de call a partir de couches minces de cellules de type épidermique de *Nicotiana tabacum* Wisc. 38. *Can. J. Bot.* **53**: 553–559.

West, M.A.L. and Harada, J.J. (1993) Embryogenesis in higher plants: an overview. *Plant Cell* **5**: 1361–1369.

37

Tracheary element formation in an *in vitro* system

M.C. McCann, C. Domingo, N.J. Stacey, D. Milioni, and K. Roberts

1. Introduction

All living plant cells are spherical when undressed. It is largely by their clothes, or constraining cell walls, that forty or so different plant cell types are recognized. We classify plant cells by the shape, size and composition of the extracellular matrix within which each cell lives. The plant cell wall is a highly organized composite of many different polysaccharides, proteins, and aromatic substances that undergo dynamic changes during cell division, expansion and differentiation. Cell walls become highly specialized for cell function, and an extreme example of this is in the formation of tracheary elements (TEs), where the cell contents autolyse to form a hollow water-conducting tube as unit structures of the plant's xylem. In this chapter, we will review the use of an *in vitro* system, the *Zinnia* mesophyll cell system, in exploring the construction of a secondary wall and the signals that initiate this developmental pathway.

1.1 *The Zinnia mesophyll cell system*

In intact plants, TEs normally develop from proliferating cells of the procambium and the vascular cambium. However, in certain instances such as wounding, lateral root formation, host–graft interactions, or even, host–parasite interactions (as with mistletoes and oaks), cells that are already differentiated trans-differentiate to form vascular tissue and to establish continuity of water supply. TEs can be induced to form from various cell types including cells of the phloem parenchyma and the cortex in roots, the pith parenchyma in shoots, the tuber parenchyma, and the mesophyll and epidermis in leaves (Fukuda, 1992). Many populations of cells, either in liquid culture or calli, spontaneously produce a low percentage of TEs, but such systems are not amenable to biochemical analyses (Fukuda, 1996). A breakthrough came in 1975, when Kohlenbach and Schmidt discovered that mesophyll cells from the leaves of *Zinnia elegans* trans-differentiate into TEs when cultured in a medium containing auxin and

Cell & Molecular Biology of Wood Formation, edited by R. Savidge, J. Barnett and R. Napier.

cytokinin. The culture conditions were subsequently optimized by Fukuda and Komamine to turn a scientific curiosity into a model system with synchronous differentiation of up to 60% of the living cell population (Fukuda and Komamine, 1980).

TE formation is characterized by a series of cytological changes (Fukuda, 1996; McCann, 1997). First, microtubules rearrange into a transverse cortical banding pattern. This reflects the position of future secondary thickenings, formed by the subsequent deposition of cellulose and other secondary wall components including structural proteins and xylan. A process of lignification is initiated at a stage when the thickenings are clearly visible by light microscopy, and finally cell autolysis results in a mature TE.

In practice, the *Zinnia* system is sensitive to many factors including the age and condition of the plants, the cultivar, culture dish and specific culture conditions (Fukuda, 1992; Roberts *et al.*, 1992). However, its utility is a consequence of two features that, in combination, are unique to this system. First, unlike most other plants, cells can be isolated intact from the leaves. By grinding leaves gently in mannitol in a mortar and pestle, mesophyll and palisade cells are released as single intact cells, and the epidermis and vasculature remain as large intact sheets of cells that are easily separated (Fukuda and Komamine, 1980). The scanning electron micrograph of the *Zinnia* leaf in *Figure 1* shows that the cells have remarkably few points of contact with each other. Second, the semi-synchronous differentiation is an inducible event triggered by the addition of exogenous auxin and cytokinin. Thus, the hormones act as a developmental switch that turns on the developmental pathway of TE formation.

1.2 *Induction of TE formation*

Three factors, wounding, auxin and cytokinin, are involved in initiating TE formation in *Zinnia*. Isolated cells may be regarded as wounded cells, and wounding is known to induce TE formation in intact plants. When a vascular bundle in a stem is severed, a connection is re-established by the trans-differentiation of intervening pith parenchyma as vascular elements. Likewise, auxin is implicated in all known examples of TE formation, and the disruption of the polar transport of auxin by wounding may induce new pathways of auxin transport to form, along which trans-differentiation to TE fate is induced (Roberts *et al.*, 1988).

Cytokinin is also required to induce TE formation in *Zinnia,* but exogenously-added cytokinin is not required for TE formation in other plants. Cytokinin only promotes TE differentiation in the presence of auxin, and it has been speculated that cytokinins increase the sensitivity of cambial initials and their derivatives to auxin (Baum *et al.*, 1991). Plant condition and age of the leaf are critical parameters for developmental plasticity in the *in vitro* system, and this may reflect differences in the levels of endogenous hormones. In the absence of experiments to assay for intracellular hormone concentrations, the levels of exogenous hormones required to induce TE formation are set at somewhat arbitrary levels at present (McCann, 1997).

Other hormones have also been invoked as signalling molecules for TE differentiation including ethylene, gibberellin, abscisic acid and brassinosteroids. These cannot substitute for auxin and cytokinin, and thus may be involved downstream, or as subtle endogenous modulators (Fukuda, 1996). For example, brassinosteroids at very low concentrations counteract the inhibitory effect of uniconazole, a synthetic plant-growth retardant, that blocks TE formation but not cell division (Iwasaki and

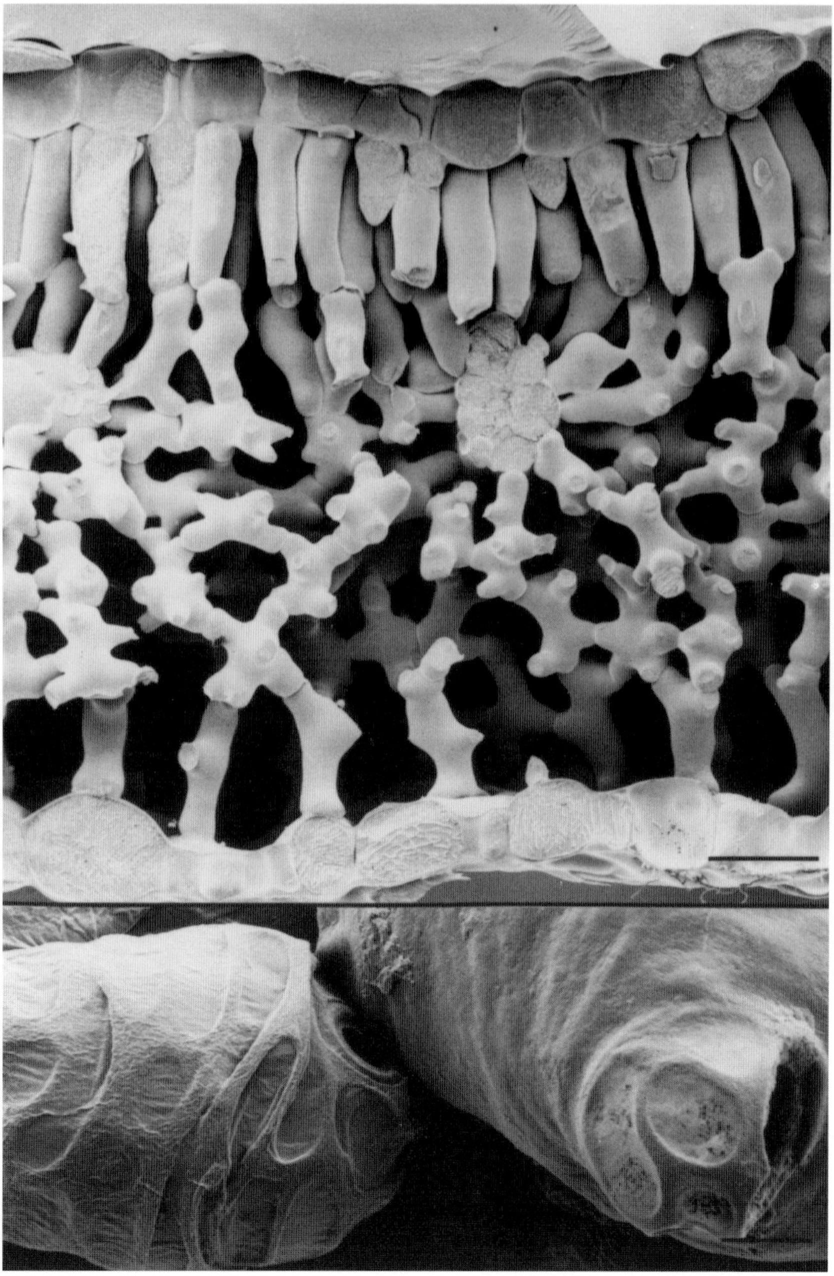

Figure 1. *(top) Scanning electron micrograph of the mesophyll and palisade cells in the* Zinnia *leaf showing few points of cell–cell contact, a feature that allows them to be isolated as single intact cells. Scale bar represents 50 μm. (bottom) Scanning electron micrograph of the TEs formed at 96 h of culture. Some of the cells have intact end-walls whereas others appear perforated. Scale bar represents 5 μm. Micrographs courtesy of K. Findlay.*

Shibaoka, 1991). Uniconazole blocks expression only of genes up-regulated at late times in TE formation, a peroxidase *ZPO-C*, and a cysteine protease *ZCP4*, together with genes involved in lignification (Yamamoto *et al.*, 1997).

Calcium has also been implicated in the regulation of secondary wall formation. Membrane-associated Ca^{2+} is elevated during differentiation, and differentiation can be inhibited using Ca^{2+} channel blockers or calmodulin antagonists (Fukuda, 1996). Methylxanthines, which act to reduce intracellular sequestration of Ca^{2+}, inhibit differentiation prior to secondary wall formation (Roberts and Haigler, 1992). Additionally, secondary wall thickening correlates with the expression of two calmodulin-binding proteins and an increase in calmodulin levels (Kobayashi and Fukuda, 1994).

2. Components of the secondary cell wall

The production of a thick secondary wall of carbohydrate, structural protein, and lignin is essential for TE function, as the wall must be reinforced to resist compressive forces from the other cells that arise as a consequence of the extreme negative pressures that may develop in actively transpiring plants. The most visible changes in TE differentiation occur at the cell surface, but changes in wall architecture and the influence of these changes on subsequent processes of differentiation have not been analysed in detail, with the exception of the late event of lignification.

2.1 *Cellulose*

Bands of cellulose microfibrils form one of the major elements of the thickening. An *Arabidopsis* mutant, *irx3*, has a collapsed xylem phenotype and is thought to result from a mutation in a xylem-specific cellulose synthase (Turner and Somerville, 1997). In the *Zinnia* system, secondary wall deposition is a hierarchical process in which the deposition of cellulose is patterned by cortical microtubules. Disassembly of microtubules by depolymerizing agents results in random deposition of secondary wall material over the entire surface of the inner wall. Plasma-membrane sucrose synthase, thought to channel the UDP-glucose substrate for cellulose synthase, has been immunolocalized to sites of secondary thickenings (Harrison *et al.*, 1997) indicating that the cellulose synthase complexes may be restricted to regions of the plasma membrane by the presence of cortical microtubules. Deposition of lignin in ordered patterns depends on the prior deposition of cellulose (Suzuki *et al.*, 1992), and inhibition of cellulose synthesis by the herbicide 2,6-dichlorobenzonitrile disrupts lignin patterning and causes the loss of xylans from the cellulose-depleted thickenings (Taylor *et al.*, 1992; Taylor and Haigler, 1993). In contrast, inhibitors of lignification do not affect the patterning of thickenings or subsequent cell autolysis (Ingold *et al.*, 1990).

2.2 *Matrix polymers*

In addition to increased cellulose and lignin synthesis, an increase in total carbohydrate and xylose in alkali-extractable fractions and a change in the relative proportion of pectic polysaccharides that are EDTA-extractable are also correlated with the time course of TE formation in the *Zinnia* system (Ingold *et al.*, 1988). Sugar and linkage analysis of culture media show that a relatively unbranched rhamnogalacturonan is enriched in inductive medium from 48 h and increases rapidly in concentration (Stacey *et al.*, 1995). Secretion of pectic polysaccharides and fucosylated xyloglucans into inductive culture medium has been detected with monoclonal antibodies (Stacey *et al.*, 1995) and the activity of a xylan synthase increases (Suzuki *et al.*, 1991).

2.3 *Structural proteins*

Many proteins are specifically associated with secondary thickenings. In bean, glycine-rich proteins are synthesized in the xylem parenchyma cells and exported into the walls of protoxylem vessels (Ryser and Keller, 1992). Arabinogalactan proteins (AGPs) (Schindler *et al.*, 1995), an extensin-like protein (Bao *et al.*, 1992) and a tyrosine and lysine-rich protein (Domingo *et al.*, 1994) have been found in maize, loblolly pine and tomato xylem, respectively. In the *Zinnia* system, twenty-four hours before secondary thickenings are visible, an AGP is present in the primary walls of a sub-population of cells, and is secreted into the culture medium. This molecule is present in the secondary thickenings of mature TEs but not in their surrounding primary walls (Stacey *et al.*, 1995). Many hydrolytic enzyme activities have also been detected including polygalacturonase, pectin methylesterase, fucosidase, and xyloglucan endo-transglycosylase (XET) (Stacey *et al.*, 1995). These enzymes, which appear to increase in activity at a late stage, may have a role in perforating the end-walls of the TEs (*Figure 1*).

3. A molecular approach to understanding TE formation

A clear description of the genes that encode the biosynthetic and hydrolytic enzymes involved in secondary wall formation and lignification, in addition to the regulatory genes that are responsible for inducing and coordinating TE formation, is required. The *Zinnia* mesophyll cell system is ideal for comparing transcripts expressed at different stages of development by subtractive hybridization or differential screening.

In the first 48 h of culture, there are genes encoding proteins involved in the wound response, proteinase inhibitors *ZePI1* and *2*, and in protein synthesis, the ribosomal proteins *ZeRPS3* and *3a*, *ZeRPPO*, and an elongation factor *ZeEF1b* (Fukuda, 1997). *ZeTubB1* and *3* encode β-tubulins (Yoshimura *et al.*, 1996). Several cDNAs have also been isolated for transcripts that are specifically up-regulated after 48 h in medium with auxin and cytokinin (Demura and Fukuda, 1993, 1994). These include *TED4* (a putative lipid transfer protein, also isolated as *p48h-10*; Ye and Varner, 1994), *TED2* with homology to alcohol dehydrogenases and a *zeta*-crystallin protein in guinea pig lens, and *TED3* which encodes a hydrophilic protein with no known homologies, but 15 repeats of Asn-Gly-Tyr (Demura and Fukuda, 1993). Expression of the *Zinnia TED3* promoter in developing TEs of transgenic *Arabidopsis* (Igarashi *et al.*, 1998) gave GUS expression in TEs, immature TEs, stipules and trichomes. At later times in the culture, genes relevant to lignification (cinnamyl alcohol dehydrogenase *ZCAD1*, phenylalanine ammonia lyases *ZePAL1*, *2*, and *3*, cinnamic acid-4-hydroxylase *ZC4H*, and peroxidase *ZPO-C* (Fukuda, 1997)) and cell death (an endonuclease *ZEN1* (Fukuda, 1997), a ribonuclease *ZRNAse1* (Ye and Droste, 1996), *ZCP4* (Fukuda, 1997) and *p48h-17* (Ye and Varner, 1996) are up-regulated.

Each of these genes provides a useful molecular marker for the developmental pathway. However, with the notable exception of genes involved in lignification, very few cell-wall related genes, biosynthetic enzymes or structural proteins or hydrolases have been found to date. One exception to this is a pectate lyase (Domingo *et al.*, 1998).

3.1 *Pectate lyase*

Pectate lyases (E.C. 4.2.2.2) have previously been described as microbial extracellular enzymes that assist pathogenesis by cleaving polygalacturonate blocks in the plant host cell wall (Davis *et al.*, 1984; Collmer and Keen, 1986). Enzymatic cleavage of α-1,4-linked galacturonosyl residues occurs at a pH optimum of 8 to 11 through a β elimination mechanism, resulting in an unsaturated C-4–C-5 bond in the galacturonosyl moiety at the non-reducing end of the polysaccharide produced at the cleavage site (Rombouts and Pilnik, 1980). This mechanism is common to another class of pectin-degrading enzymes, pectin lyases, but pectate lyase is distinguished by its preference for a glycosidic linkage next to a free carboxyl group rather than to an esterified carboxyl group and by its pH optimum (Pilnik, 1990). All pectate lyases show calcium dependence. Given the complex structure of pectic polysaccharides, their enzymic degradation by pathogens is usually accomplished by the synergistic action of enzymes such as pectin methylesterase, pectin lyase, endo- and exo-polygalacturonase and pectate lyase. Pectate lyase is particularly effective, and soft-rot symptoms in various plant species can be induced by applying recombinant pectate lyases from *E. coli* (Keen and Tamaki, 1986; Bartling *et al.*, 1995). In plants, pectate lyase genes have been described that are expressed primarily in pollen and in the transmitting tissue of plants, and the enzyme may facilitate pollen tube growth (Ori *et al.*,1990; McCormick, 1991).

In a search for genes involved in modifications to cell-wall architecture before any overt signs of cell differentiation, a differential hybridization of a 72 h cDNA library from the *Zinnia* system with probes from mRNA at time-points of 24 h and 72 h was done revealing a number of transcripts up-regulated between these times. One of these cDNAs showed homology to pectate lyase (Domingo *et al.*, 1998). The complete cDNA sequence (*ZePel*) corresponds to a translated protein of 44 kDa with an N-terminal signal peptide of about 2 kDa, and one potential N-glycosylation site. Northern analysis showed that the strong expression of this gene during TE induction occurs at a very early stage of the process and is due solely to the presence of auxin in the induction medium. Gene expression was correlated with increased enzyme activity in the *in vitro* cultures. *In situ* hybridization studies in young *Zinnia* stems show that *ZePel* expression is associated with vascular bundles and shoot primordia. Recombinant protein made in *E. coli* possessed calcium-dependent pectate lyase activity and was used to raise a polyclonal antisera that labels an epitope in cell walls of *Zinnia*.

Although expression of *ZePel* is dramatically up-regulated by auxin, it is not a primary auxin response as defined by the appearance of the gene product within minutes of the inductive signal (Abel and Theologis, 1996) and it is not related to the other auxin-regulated genes isolated from the *Zinnia* system (Ye and Varner, 1993, 1994; Demura and Fukuda, 1993), that appear later in the time course of determination and differentiation. With a pH optimum of 10, and only residual enzymatic activity below pH 7.5, the pectate lyase is likely to be sub-optimally active *in vivo*, and this may be a necessary regulatory control for an enzyme that can produce soft-rot symptoms in plants. The natural substrates of pectate lyases are not clearly defined. Three *Erwinia* PL isoforms obtained by expression of three *Pel* genes in *E. coli* act synergistically to extend the range of pectin substrates which the bacterium can degrade (Bartling *et al.*, 1995). It will be an important task to characterize the substrate specificity for *ZePel* and other plant pectate lyases.

In the *Zinnia* plant we have shown that *ZePel* expression is correlated with sites of vascular differentiation and with cells that are recent products of meristematic divisions. The localization of pectate lyase gene expression to new primordia on the flanks of the shoot meristem reveals a transient expression of the gene, as only some primordia are labelled. Auxin induces vascular differentiation as well as being necessary for final maturation of xylem (Jacobs, 1952). Sachs (1981) proposed that the pattern of vascular system results from initial local differences in the flow of auxin through cells, which leads to the establishment of preferred channels of auxin transport. These channels become progressively improved pathways of auxin movement and drain the surrounding regions at the same time that their cells are induced to undergo differentiation as vascular elements. In *Pinus* trees, there is a steep gradient of IAA across the vascular cambium and its derivatives, with a peak at the cambium (Uggla *et al.*, 1996). The pattern of gene expression, where the formation of new vascular strands are taking place, is consistent with auxin regulation of *ZePel*.

In the *Zinnia* system, *ZePel* expression is up-regulated by auxin prior to either cell elongation or cell differentiation. We have observed that the newly-synthesized methyl-esterified pectin deposited in the wall during growth becomes de-esterified when cells reach their mature size (Kim and Carpita, 1992; McCann and Roberts, 1994). This de-esterified and presumably Ca^{2+} cross-linked pectin may be a substrate for pectate lyase in cells which are responding to new growth signals after growth has normally ceased, or where the wall architecture is to become functionally adapted, as in TE formation. During TE formation, there is much evidence that a range of different polysaccharides and proteins become deposited in the walls at an early stage (Ingold *et al.*, 1988; Stacey *et al.*, 1995). *ZePEL* may assist in the removal and modification of an existing pectin matrix in order to allow the deposition of newly-synthesized wall polymers for a specialized function or to create an architecture that is extensible. A further possibility is that pectic fragments released by pectate lyase may act as oligosaccharins — polysaccharide fragments that act as cell-signalling molecules to elicit a range of cellular responses (Mohnen and Hahn, 1993). The enzymes from *Erwinia* liberate oligosaccharides from purified pectic substances and higher plant cell walls that function as elicitors of active plant defence reactions (Gardner and Kado, 1976; Davis *et al.*, 1984).

Different time courses for differentiation in the *Zinnia* system have been reported, perhaps as a consequence of the different cultivars and cell isolation procedures adopted by the various laboratories, and so it is important to define cellular events with respect to markers related to developmental state rather than simply time in culture (Stacey *et al.*, 1995). Also, although differentiation is semi-synchronous in the *Zinnia* system, other processes of wound response, division and elongation are occurring and timing of TE formation can vary considerably between different cell preparations. Given that the expectation is for a large number of cell-wall-related genes to be involved in the process of TE formation, we considered that one problem might be the synchrony of the system.

4. Improving the synchrony of the *Zinnia* system

Differentiation to TEs is a terminal fate and it is possible to define a time-point at which the cells become irreversibly committed (Church and Galston, 1988; Stacey *et al.*, 1995), by removing the stimulus for differentiation and observing if differentiation proceeds in the absence of the stimulus. We have previously defined a point at which the cells no longer require the presence of auxin and cytokinin: cells transferred into non-inductive medium after at least 48 h in inductive medium still differentiate with no delay in their usual time-course (Church and Galston, 1988; Stacey *et al.*, 1995). Although it appears, therefore, that cell commitment takes 48 hours, hormonal involvement in the early phase of culture is less than clear.

Church and Galston (1988) reported that auxin had to be present for 56 h while cytokinin was required for 24 h. Fukuda and Komamine (1985) reported that the hormones are not required in the first 12 h of culture. We investigated the precise times in the differentiation time-course at which the hormones are required in the culture medium (*Figure 2*). Cells were isolated from *Zinnia* leaves and cultured at 10^6 cells ml^{-1} in maintenance medium (medium without hormones) for 1, 21, 26 or 47 h before transfer into inductive medium (medium with 1.0 mgl^{-1} of both auxin and cytokinin). The cells then remained in inductive medium until the 48 h time-point when they were transferred into fresh maintenance medium. Before transfer, the cells were washed twice in 0.2 M mannitol to minimize carry-over of exogenous hormone. The density of cells was maintained at 10^6 ml^{-1} after transfer into fresh medium. Progressively increasing numbers of TEs were observed at 120 h of culture as the length of time of exposure to inductive medium decreased. As just 1 h of exposure to exogenous hormones at 48 h was sufficient to induce TE formation by 96 h, we continued to narrow this window, finding that just 5 to 10 min exposure to inductive medium at 48 h is sufficient to induce differentiation by 96 h. Further, the brief exposure to hormones at around 48 h resulted in an enhanced level of differentiation (about 80% of living cells), with the culture apparently more synchronous and with a higher final percentage of tracheids than if the cells are cultured continuously in inductive medium (about 60% of living cells). *Figure 3* shows a typical field of view in cultures where cells have been cultured in maintenance medium for 48 h before being transferred into inductive medium, and the early response of *ZePel* up-regulation just 9 h after hormone addition. The molecular marker *ZePel1* is expressed at 48 h in inductive medium (Domingo *et al.*, 1998), and the *TED2* and *TED3* genes are expressed at 36 and 60 h of culture, respectively (Demura and Fukuda, 1994). Transcribed from these genes were mRNAs isolated at various times during the time-course when cells were transferred from maintenance medium into inductive medium after 48 h. Both *ZePel1* and *TED3* are expressed at 57 h, just 9 h after transfer, whereas *TED2* is expressed later (*Figure 4*). These results demonstrate the improved synchrony of TE formation under these culture conditions.

Some early events clearly occur in the absence of both hormones. These include the rearrangement of actin (Kobayashi, 1996), repair-type DNA synthesis (Sugiyama *et al.*, 1995) and tubulin synthesis (Yoshimura *et al.*, 1996). The increased synchrony that we observe may be a consequence of such processes having been completed prior to the cells receiving hormonal signals. Our working hypothesis is that the first 48 h is a period of increasing competence of cells to respond to auxin and cytokinin. This is likely to be triggered as a wound response to cell isolation. The short duration that the

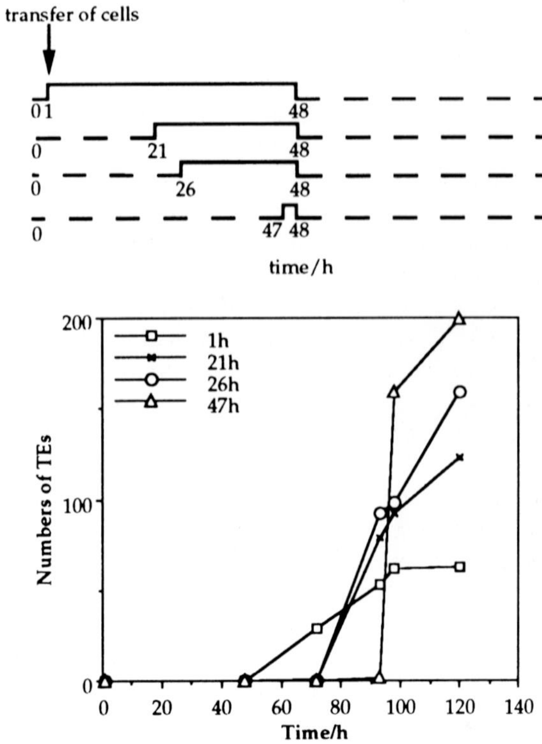

Figure 2. *Graph showing the number of TEs formed when cells are transferred (indicated by step) from maintenance medium (dashed line), after 1, 21, 26 or 47 h to inductive medium (solid line), and subsequently transferred to fresh maintenance medium at 48 h. The number of tracheids per square micron was averaged from three fields of view in each of four independent experiments for each time-point. Error limits for each point are less than 25 TEs.*

cells can be exposed to the hormones and still differentiate implies that the hormones function as a switch, increasingly effective near 48 h. However, further signals are required to complete TE formation. Inhibitors of brassinosteroid synthesis block TE formation until a very late stage of differentiation (Yamamoto *et al.*, 1997). The competence to differentiate is also highly dependent on cell concentration up to at least 72 h of culture, indicating that cell–cell signalling mechanisms are involved and that the process is not cell-autonomous. Negative and positive regulators of TE formation are likely to be present.

It is not clear whether the response to auxin and cytokinin alone is sufficient to commit the cell to TE fate or if determination occurs as a series of commitments to intermediate cell fates, with the final commitment to a TE fate potentially as late as cell autolysis. The idea of determination being a multistep process is based on the expression patterns of the three *TED* transcripts (Demura and Fukuda, 1994). These genes are expressed in a temporal order during *in vitro* culture that corresponds to a spatial distribution in the living plant associated with progressive restriction in cell fate. In intact plants, *TED2* (expressed at 24 h in cultures where secondary wall deposition has occurred by 72 h) is present in the procambial regions of root tips and in

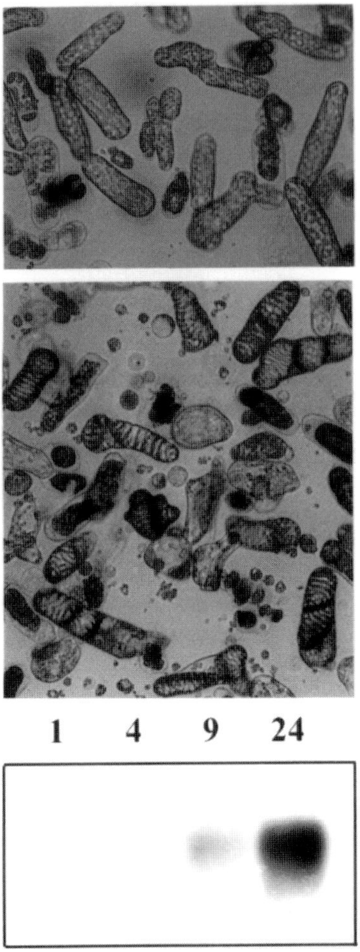

Figure 3. Zinnia *mesophyll cells freshly isolated from young leaves (top) are induced to form TEs* in vitro *at 96 h of culture, with a 48 h pre-incubation in non-inductive medium (middle). The expression of the pectate lyase gene begins at 9 h after hormone addition.*

both immature primary xylem and phloem; *TED4* (found at 36 h in culture) is in the immature primary xylem (parenchyma, fibres and TEs), and *TED3* (found at 36 h in culture) is present in cells that are destined to become TEs (Demura and Fukuda, 1994). Clearly, *TED2* and *TED4* are markers for stages in vascular development but not specifically for TE fate. There are two possibilities: either these intermediate stages are simply stages in TE formation that the cell, having been determined, must transit, or the cell is only committed to an intermediate fate, and must receive additional signals to differentiate further. If the latter is the case, then it should be possible to provide other signals that would divert the cell into a different developmental pathway — for example to make phloem fibres from cells expressing *TED2*. If the cell is already fully determined, then it can have no requirement for external stimuli, and should differentiate to a TE in a cell-autonomous manner. However, the dependence of TE formation on cell concentration and on brassinosteroid biosynthesis (Yamamoto *et al.*, 1997) until a late stage in the process, indicates that the cells are only committed to a precursor TE cell fate by the inductive auxin and cytokinin. Prior to this point, the cells are not committed. Relative to this marker, we have found that the *TED3* gene is

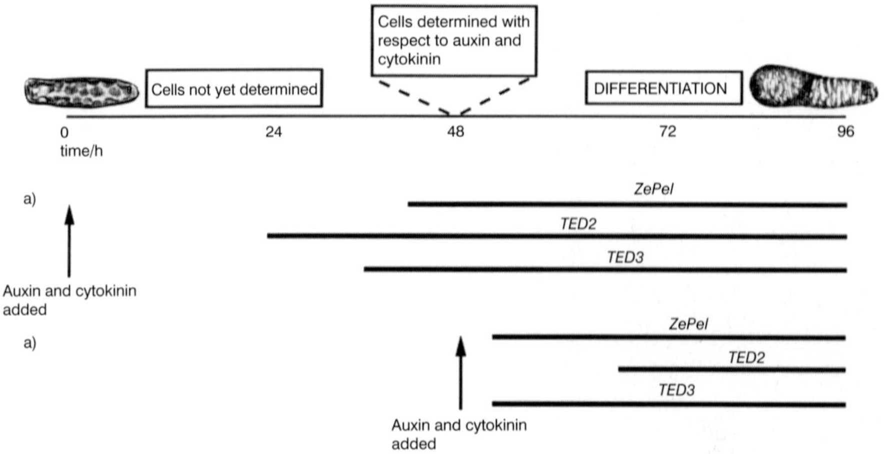

Figure 4. *Schematic of the expression timecourses of* ZePel, TED2 *and* TED3 *mapped with respect to: the early phase in which the cells are not yet determined; the point at which the cells are determined with respect to auxin and cytokinin; and the late phase of differentiation in: (a) a standard timecourse in which auxin and cytokinin are added at time 0; (b) with a pre-incubation in non-inductive medium in which auxin and cytokinin are added at 48 h.*

expressed at 9 h whilst the *TED2* gene is expressed after this but before 24 h; the improved synchrony of the time-course will allow us a finer-grained mapping of cytological, biochemical and molecular markers.

References

Abel, S. and Theologis, A. (1996) Early genes and auxin action. *Plant Physiol.* **111**: 9–17.

Bao, W., O'Malley, D.M. and Sederoff, R.R. (1992) Wood contains a cell-wall structural protein. *Proc. Natl. Acad. Sci. USA* **86**: 6604–6608.

Bartling, S., Wegener, C. and Olsen, O. (1995) Synergism between *Erwinia* pectate lyase isoenzymes that depolymerize both pectate and pectin. *Microbiology* **141**: 873–881.

Baum, S.F., Aloni, R. and Peterson, C.A. (1991) The role of cytokinin in vessel regeneration in wounded Coleus internodes. *Ann. Bot.* **67**: 543–548.

Church, D.L. and Galston, A.W. (1988) Kinetics of determination in the differentiation of isolated mesophyll cells of *Zinnia elegans* to tracheary elements. *Plant Physiol.* **88**: 92–96.

Collmer, A. and Keen, N.T. (1986) The role of pectic enzyme in plant pathogenesis. *Annu. Rev. Phytopathol.* **24**: 383–409.

Davis, K.R., Lyon, G.D., Darvill, A.G. and Albersheim, P. (1984) Host–pathogen interactions. XXV. Endopolygalacturonic acid lyase from *Erwinia carotovora* elicits phytoalexin accumulation by releasing plant cell wall fragments. *Plant Physiol.* **74**: 52–60.

Demura, T. and Fukuda, H. (1993) Molecular cloning and characterization of cDNAs associated with tracheary element differentiation in cultured *Zinnia* cells. *Plant Physiol.* **103**: 815–821.

Demura, T. and Fukuda, H. (1994) Novel vascular cell-specific genes whose expression is regulated temporally and spatially during vascular system development. *The Plant Cell* **6**: 967–981.

Domingo, C., Gomez, M.D., Canas, L., Hernandez-Yago, J., Conejero, V. and Vera, P. (1994) A novel extracellular matrix protein from tomato associated with lignified secondary cell walls. *Plant Cell* **6**: 1035–1047.

Domingo, C., Roberts, K., Stacey, N.J., Connerton, I., Ruíz-Teran, F. and McCann, M.C. (1998) A pectate lyase from *Zinnia elegans* is auxin inducible. *Plant J.* **13**: 17–28.

Fukuda, H. and Komamine, A. (1980) Establishment of an experimental system for the tracheary element differentiation from single cells isolated from the mesophyll of *Zinnia elegans*. *Plant Physiol.* **52**: 57–60.

Fukuda, H. and Komamine, A. (1985) Cytodifferentiation. In: *Cell Culture and Somatic Cell Genetics of Plants 2* (ed. Vasil, I.K.), Academic Press, New York, pp 149–212.

Fukuda, H. (1992) Tracheary element formation as a model system of cell differentiation. *International Review of Cytology* **136:** 289–332.

Fukuda, H. (1996) Xylogenesis: initiation, progression, and cell death. *Annu. Rev. Plant Physiol. Plant Mol. Biol.* **47**: 299–325.

Fukuda H. (1997) Tracheary element differentiation. *Plant Cell* **9:** 1147–1156.

Gardner, J.M. and Kado, C.I. (1976) Polygalacturonic acid trans-eliminase in the osmotic shock fluid of *Erwinia rubrifaciens*: characterisation of the purified enzyme and its effect on plant cells. *Journal of Bacteriology* **127**: 451–460.

Harrison, M.J., Delmer, D.P., Amor, Y., Grimson, M.J., Johnson, S. and Haigler, C.H. (1997) Localization of sucrose synthase in differentiating tracheary elements of *Zinnia elegans*. *Plant Physiology* **114**(**3**) *suppl*: 349.

Igarashi, M., Demura, T. and Fukuda, H. (1998) Expression of the *Zinnia TED3* promoter in developing tracheary elements of transgenic *Arabidopsis*. *Plant Molec. Biol.* **36**: 917–927.

Ingold, E., Sugiyama, M. and Komamine, A. (1988) Secondary cell wall formation: changes in cell wall constituents during the differentiation of isolated mesophyll cells of *Zinnia elegans* to tracheary elements. *Plant Cell Physiol.* **29**: 295–303.

Ingold, E., Sugiyama, M. and Komamine, A. (1990) L-α-aminooxy-β-phenylpropionic acid inhibits lignification but not the differentiation to tracheary elements of isolated mesophyll cells of *Zinnia elegans*. *Physiol. Plant.* **78**: 67–74.

Iwasaki, T. and Shibaoka, H. (1991) Brassinosteroids act as regulators of tracheary element differentiation in isolated *Zinnia* mesophyll cells. *Plant Cell Physiol.* **32**: 1007–1011.

Jacobs, W.P. (1952) The role of auxin in differentiation of xylem around a wound. *Am. J. Botany* **39**: 301–309.

Keen, N.T. and Tamaki, S. (1986) Structure of two pectate lyase genes *from Erwinia chrysanthemi* EC16 and their high-level expression in *Escherichia coli*. *Journal of Bacteriology* **168:** 595–606.

Kim, J.-B. and Carpita, N.C. (1992) Changes in esterification of the uronic acid groups of cell wall polysaccharides during elongation of maize coleoptiles. *Plant Physiol.* **98**: 646–653.

Kobayashi, H. and Fukuda, H. (1994) Involvement of calmodulin and calmodulin-binding proteins in the differentiation of tracheary elements in *Zinnia* cells. *Planta* **194**: 388–394.

Kobayashi, H. (1996) Changes in the relationship between actin filaments and the plasma membrane in cultured *Zinnia* cells during tracheary element differentiation investigated by using plasma membrane ghosts. *J. Plant Res.* **109**: 61–65.

Kohlenbach, H.W. and Schmidt, B. (1975) Cytodifferenzierung in Form einer direkten Umwandung isolierter Mesophyll-Zellen zu Tracheiden. *Z. Pflanzenphysiol.* **75**: 369–374.

McCann, M.C. and Roberts, K. (1994). Changes in cell wall architecture during cell elongation. *J. Exp. Bot.*, **45**: 1683–1691.

McCann, M.C. (1997) Tracheary element formation: building up to a dead end. *Trends in Plant Science* **2:** 333–338.

McCormick, S. (1991) Molecular analysis of male gametogenesis in plants. *Trends Genet.* **7:** 298–303.

Mohnen, D. and Hahn, M.G. (1993) Cell wall carbohydrates as signals in plants. In: *Seminars in Cell Biology 4*, Academic Press, pp 93–102.

Ori, N., Sessa, G., Lotan, T., Himmelhoch, S. and Fluhr, R. (1990) A major stylar matrix polypeptide (sp41) is a member of the pathogenesis-related proteins superclass. *EMBO J.* **9:** 3429–3436.

Pilnik, W. (1990) Pectin — a many splendoured thing. In: *Gums and Stabilizers in the Food Industry 5* (eds. Phillips, G.O., Wedlock, D.J. and Williams, P.A.), Oxford University Press, pp. 209–221

Roberts, A.W., Koonce, L.T. and Haigler, C.H. (1992) A simplified medium for *in vitro* tracheary element differentiation in mesophyll suspension cultures from *Zinnia elegans* L. *Plant Cell, Tissue and Organ Culture* **28**: 27–35.

Roberts, A.W. and Haigler, C.H. (1992) Methylxanthines reversibly inhibit tracheary element differentiation in suspension cultures of *Zinnia elegans* L. *Planta* **186**: 586–592.

Roberts, L.W., Gahan, P.B. and Aloni, R. (1988) *Vascular Differentiation and Plant Growth Regulators*. Springer, Berlin.

Rombouts, F.M. and Pilnik, W. (1980) Pectic enzymes. In: *Economic Microbiology*, Vol. 5: *Microbial Enzymes and Bioconversions* (eds Rose, A.H.), Academic Press, New York, pp, 222–282.

Ryser, U. and Keller, B. (1992) Ultrastructural localisation of a bean glycine-rich protein in unlignified primary walls of protoxylem cells. *Plant Cell* **4**: 773–783.

Sachs, T. (1981) The control of the patterned differentiation of vascular tissues. *Adv. Bot. Res.* **9**: 151–262.

Schindler, T., Bergfeld, R. and Schopfer, P. (1995) Arabinogalactan proteins in maize coleoptiles: Developmental relationship to cell death during xylem differentiation but not to extension growth. *Plant J.* **7**: 25–36.

Stacey, N.J., Roberts, K., Carpita, N.C., Wells, B. and McCann, M.C. (1995) Dynamic changes in cell surface molecules are very early events in the differentiation of mesophyll cells from *Zinnia elegans* into tracheary elements. *Plant J.* **8**: 891–906.

Sugiyama, M., Yeung, E.C., Shoji, Y. and Komamine, A. (1995) Possible involvement of DNA-repair events in the transdifferentiation of mesophyll cells of *Zinnia elegans* into tracheary elements. *J. Plant Res.* **108**: 351–361.

Suzuki, K., Ingold, E., Sugiyama, M. and Komamine, A. (1991) Xylan synthase activity in isolated mesophyll cells of *Zinnia elegans* during differentiation to tracheary elements. *Plant Cell Physiol.* **32**: 303–306.

Suzuki, K., Ingold, E., Sugiyama, M., Fukuda, H. and Komamine, A. (1992) Effects of 2,6-dichlorobenzonitrile on differentiation to tracheary elements of isolated mesophyll cells of *Zinnia elegans* and formation of secondary cell walls. *Physiol. Plant.* **86:** 43–48.

Taylor, J.G., Owen, T.P., Koonce, L. and Haigler, C.H. (1992) Dispersed lignin in tracheary elements treated with cellulose synthesis inhibitors provides evidence that molecules of the secondary cell wall mediate wall patterning. *Plant J.* **2**: 959–970.

Taylor, J.G. and Haigler, C.H. (1993) Patterned secondary cell-wall assembly in tracheary elements occurs in a self-perpetuating cascade. *Acta. Bot. Neer.* **42**: 153–163.

Turner, S.R. and Somerville, C. (1997) Collapsed xylem phenotype of *Arabidopsis* identifies mutants deficient in cellulose deposition in the secondary cell wall. *Plant Cell* **9**: 689–701.

Uggla, C., Moritz, T., Sandberg, G. and Sundberg, B. (1996) Auxin as a positional signal in pattern formation in plants. *Proc. Natl. Acad. Sci. USA* **93:** 9282–9286.

Yamamoto, R., Demura, T. and Fukuda, H. (1997) Brassinosteroids induce entry into the final stage of tracheary element differentiation in cultured *Zinnia* cells. *Plant Cell Physiol.* **38**: 980–983.

Ye, Z.-H, and Varner, J.E. (1993) Gene expression patterns associated with *in vitro* tracheary element formation in isolated single mesophyll cells of *Zinnia elegans. Plant Physiol.* **103**: 805–813.

Ye, Z.-H, and Varner, J.E. (1994) Expression of an auxin and cytokinin regulated gene in cambial region in *Zinnia*. *Proc. Natl. Acad. Sci. USA* **91**: 6539–6543.

Ye, Z.-H. and Droste, D.L. (1996) Isolation and characterization of cDNAs encoding xylogenesis — associated and wounding-induced ribonucleases in *Zinnia elegans*. *Plant Mol. Biol.* **30**: 697–709.

Ye, Z.-H. and Varner, J.E. (1996) Induction of cysteine and serine proteases during xylogenesis in *Zinnia elegans*. *Plant Molecular Biology* **30**: 1233–1246

Yoshimura, T., Demura, T., Igarashi, M. and Fukuda, H. (1996) Differential expression of three genes for different β-tubulin isotypes during the initial culture of *Zinnia* mesophyll cells that divide and differentiate into tracheary elements. *Plant and Cell Physiol.* **37**: 1167–1176.

38

Genetic analysis of xylem development in the primary root of *Arabidopsis thaliana*

A. Caño-Delgado, K. Metzlaff and M. Bevan

1. Introduction

The primary root of *Arabidopsis thaliana* has become a favourite tissue for genetic and developmental studies of cell differentiation. The discrete cell types and their relatively invariant number and position permit genetic analysis of cell differentiation, pattern formation and organogenesis (Dolan *et al.*, 1993; Schiefelbein *et al.*, 1997). Cell lineages can be traced to a small number of meristematic cells in the root tip, allowing genetic analyses of both cell interactions during development and the various stages of differentiation of cell types.

Cells are arranged in two general orientations in the primary root. Longitudinal patterns define the different stages of cell differentiation from their origins in the meristematic cells of the root tip, through a zone of cell elongation to a zone characterized by mature cell types and the formation of secondary meristematic cells giving rise to lateral roots and secondary thickening. The radial pattern consists of three concentric layers of cell types: the epidermis, cortex and vascular tissue. This architecture is established early in embryogenesis, when initial cells become arranged in a radial pattern in the torpedo stage (Scheres *et al.*, 1994). During secondary thickening, a vascular cambium is formed that establishes a second radial pattern of xylem and phloem cells.

This chapter first describes the present understanding of the development of the vascular system, emphasizing the insights gained from genetic analysis of cell interactions in the *Arabidopsis* primary root (reviewed by Weigel and Doerner, 1996; Wysocka-Diller and Benfey, 1997), the central role of auxins in axis formation and vascular development, and the molecular genetic analysis of gene expression profiles during xylem development. Recently described mutants altering terminal stages of xylem and fibre cell development are also assessed. The next section deals with a

Cell & Molecular Biology of Wood Formation, edited by R. Savidge, J. Barnett and R. Napier.

description of new screens to detect mutants altering both radial and longitudinal patterns in xylem development, and a novel screen for mutants with altered xylem transport properties. Finally, prospects for extending the scope of these studies following the isolation of key regulatory genes are discussed.

2. Development of the root vascular system

The vascular tissue of the *A. thaliana* root consists of a single layer of 12 pericycle cells surrounding central vascular cells. Two protoxylem strands and two opposing protophloem strands at 90° to the xylem axis establish a diarch pattern (*Figure 1a*). The protoxylem cells define the xylem axis and they are always next to a pericycle cell that faces two endodermal cells (*Figure 1c*), and the protophloem cells generally face a pericycle cell adjacent to only one endodermal cell. Later in longitudinal development, the three vascular cells in between the protoxylem strands differentiate into metaxylem (*Figure 1b*), (Dolan *et al.*, 1993). In the mature root, parenchyma cells between the xylem file and phloem poles form the vascular cambium that, subsequently, lays down secondary xylem on the inside (xylem side) and secondary phloem on the outside (phloem side). This elaboration of the vascular system of the primary root is relatively invariant (as shown in *Figure 1*), and is therefore amenable to genetic analysis (Dolan and Roberts, 1995).

A variety of mutants affecting many aspects of root development have provided important insights into pattern formation and cell differentiation. Mutants in vascular development can be broadly categorized into: (i) those affecting the radial patterning such as the appropriate numbers of cells and their spatial relationships in layers; (ii)

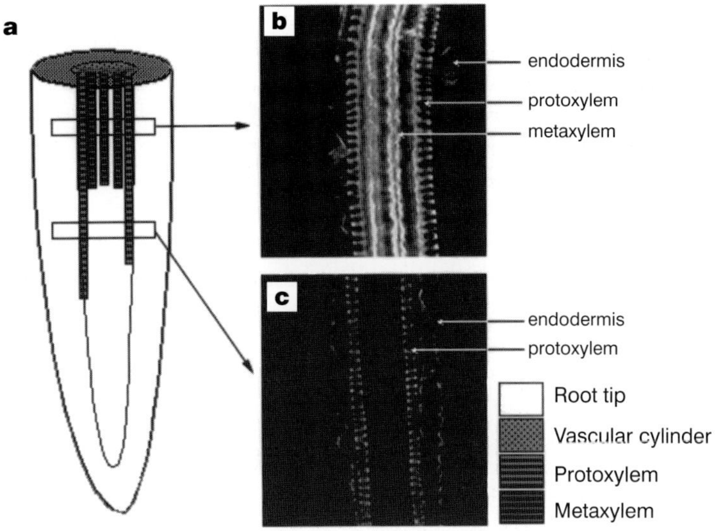

Figure 1. *Xylem organization in the* Arabidopsis thaliana *primary root.*
(a) Schematic representation of the xylem strands. (b) Confocal scanning laser microscopy (CSLM) image of protoxylem and metaxylem in the late differentiation zone; (c) CSLM image of protoxylem in the early differentiation zone.

those affecting longitudinal processes such as cell lineage, the rate of differentiation and the formation of vascular cambium; and (iii) those affecting terminal stages of xylem and phloem differentiation such as secondary thickening and other vascular cell-specific processes.

2.1 *Mutants affecting radial patterning*

The search for mutants altered in radial patterning uses screens for defects in root growth, based on the supposition that most alterations in cell type and number will generally decrease normal growth. Microscopical examination of plantlets with retarded root growth revealed several defects in radial patterning (Scheres *et al.*, 1995). One class of mutant is missing individual cell layers. *Short root* (*shr*) lacks an endodermal layer, and *scarecrow* (*scr*) and *pinocchio* (*pic*) lack the cortical layer (Benfey *et al.*, 1993). *Gollum* (*glm*) lacks a complete pericycle layer and has disorganized xylem and phloem elements. The cells of the vascular cylinder of *woodenleg* (*wol*) are fewer than normal and they all develop into xylem elements early in root growth (Scheres *et al.*, 1995). The region of the pericycle in *wol* contains fewer cells than wild type. The *fass* mutant (Torres-Ruiz and Jürgens, 1994) has additional cortical cells and an enlarged vascular cylinder in which more than two xylem poles may be present. The mutant *hydra1* is similar to *fass*, as it also exhibits an abnormal radial pattern with disrupted vascular strands (Topping *et al.*, 1997). These mutants suggest a key role for the pericycle in determining the number and radial pattern of xylem cells. In general all of these alterations in root architecture originate in the embryo and extend into the hypocotyl.

2.2 *Mutants affecting cell expansion*

Longitudinal and, to a lesser extent, radial cell expansion occur in a concerted manner to preserve radial patterning in the cell lineages derived from the root meristem as the frequency of cell divisions decreases into the elongation zone. This regular pattern of development was used in genetic screens for mutants with deregulated cell expansion leading to increased root diameter. Several mutant loci were characterized (Hauser *et al.*, 1995), all of which were conditional upon the rate of root growth. *Lion's tail* (*lit*) is not able to increase its cell volume and elongate and expand normally. This effect was seen most strongly in the stele. The mutant *cobra* shows a shift in the polarity of cell growth rather than increased cell volume while the *pom-pom* mutants show an increased expansion of the cortex and epidermis. In *sabre*, the only mutant phenotype not influenced by the rate of root growth, the cortex was the most affected.

2.3 *Mutants in longitudinal growth*

This aspect of root development is strongly influenced by the environment. Numerous mutants with altered root responses to gravitropism, obstacle avoidance, light, temperature and chemical environment have been characterized. Interestingly, some gravitropism and obstacle-avoidance mutants such as *shy2* (Tian and Reed, 1999) are allelic to auxin resistant mutants, demonstrating the central importance of auxin in defining cell polarity and axis formation in the vascular system. Auxin is synthesized in aerial regions of the plant and transported basipetally in the vascular system. Recently a basipetal-specific auxin transport protein encoded by the *AtPIN1* gene and expressed in xylem cells has been defined (Gälweiler *et al.*, 1998). Auxin transport inhibitors disrupt vascular differentiation, and redifferentiation of xylem strands after disruption occurs along the path of basipetal auxin transport. The *lop1* mutant (Carland and McHale, 1996) has disrupted leaf veins and root vasculature amongst other defects, and this is thought to be due to disrupted basipetal auxin transport. The *MONOPTEROS* (*MP*) gene encodes a transcription factor (Hardtke and Berleth, 1998) previously identified as IAA24 (Ulsamov *et al.*, 1997) that binds to response elements in promoters of auxin-regulated genes. *MP* is expressed in vascular tissue early in embryo formation, and less severe *mp* alleles result in disorganized vascular differentiation. Strong *mp* alleles fail to establish embryo polarity, resulting in loss of the basal region (Berleth and Jürgens, 1993). A recently isolated auxin insensitive mutant called *bodenlos* (*bdl*) results in a very early defect in embryogenesis, supporting the role of auxins in apical-basal patterning during embryogenesis (Hamann *et al.*, 1999).

Defining the molecular mechanisms of auxin transport and gene regulation continues to provide important new insights into development of the vascular system.

2.4 *Terminal differentiation of xylem elements*

Secondary thickening of the cell wall and loss of cellular contents are the two main characteristics of xylem cell development. The longitudinal continuity of differentiated xylem cell files forms a conduit for solute and water transport extending to all regions of the plant.

Identifying genes active during xylem development is possible in mechanically isolated mesophyll cells from *Zinnia* which, in appropriate conditions, differentiate synchronously into isolated tracheids. An increasingly detailed view of the identity and timing of gene expression during development is being obtained (reviewed by Fukuda, 1997). The cells traverse an irreversible stage after which they are committed to undergo xylogenesis. Several calmodulin binding proteins (CaM) are transiently induced during this transition, and CaM antagonists inhibit the transition to xylogenesis, implicating CaM in the process (Kobayashi and Fukuda, 1994). Among the set of genes expressed early in the committed phase are those encoding hydrolytic enzymes, lipid transfer proteins, lignin biosynthetic enzymes, a cysteine protease, and a peroxidase (Demura and Fukuda, 1993; Ye and Varner, 1993). The transcription factors orchestrating the timing and cell specificity of gene expression include members of the Myb family of transcription factors (Hatton *et al.*, 1995). Interestingly the brassinosteroid inhibitor uniconazole inhibits expression of genes during the committed stage of xylogenesis, and exogenous brassinolide promotes tracheid formation (Iwasaki and Shibaoka, 1991). Many of the proteins expressed are known to be involved in remodelling the endomembrane system in preparation for secondary thickening of the longitudinal walls, synthesis of lignin precursors, and lysis of cell contents. In xylem, the disruption of the tonoplast leads quickly to organelle disruption, DNA degradation and complete autolysis of cell contents (Fukuda, 1997). This terminal differentiation step has been likened to programmed cell death (PCD) in animal cells, although the specific mechanisms associated with PCD in animals have not yet been found in plant cells (reviewed by Greenberg, 1996).

Cell wall modifications specific to developing xylem cells have been studied extensively. Particular proline- and glycine-rich proteins are located specifically in lignifying cell walls, where they are deposited both during lignification and after terminal differentiation by adjacent xylem parenchyma cells (Ryser *et al.*, 1997). Arabinogalactan proteins (AGPs) are associated with the walls of xylem cells, and are deposited before cell death (Schindler *et al.*, 1995). A potential regulatory gene encoding a homeodomain-leucine zipper DNA binding protein domain called *Athb-8* is expressed in developing xylem and in wounded cells redifferentiating to xylem and is regulated by auxin, all indicating a potential role in regulating xylem development (Baima *et al.*, 1995 and Chapter 36 this book). Obtaining mutations with altered *Athb-8* function will provide important insights into the regulatory hierarchies governing xylem differentiation.

2.5 *Mutants disrupting terminal stages of xylem development*

Many mutations that alter different aspects of polysaccharide synthesis and cell wall structure have been identified in *A. thaliana*. Two specifically alter cells of the vascular system. Irregular xylem mutants (*irx*) are characterized by a reduction in cellulose in stem tissue, primarily associated with xylem cells (Turner and Somerville, 1997 and Chapter 39 this book). The cell walls of *irx* xylem are not resistant to normal compressive forces, resulting in mechanically weak stems. These mutants revealed the importance of cell wall architecture for normal lignification. An interfascicular fibre mutant (*ifl*) disrupted normal development of interfascicular fibres leading to flaccid and mechanically weak stems (Zhong *et al.*, 1997). The identification of further genes contributing to fibre development and normal cell wall formation will permit a more detailed analysis of the terminal stages of xylem and fibre cell development.

3. Screen for mutations in xylem development

The molecular mechanisms of xylem formation are poorly understood. One reason for this may be the inner location of the vascular cylinder, which prevents the direct visualization of the xylem strands and other components of the vascular cylinder. To overcome this problem, it is necessary to design specific screens to search for mutants with altered xylem development.

In this chapter, we present a straightforward method to screen for mutants showing abnormalities in the organization of xylem strands in the primary root of *A. thaliana*. Differentiation of the xylem cells involves the formation of a secondary thickened wall, composed mainly of cellulose and lignin. Lignins are phenolic compounds responsible for the strength and waterproofing of particular cell walls. Coniferaldehyde groups of lignin can be stained with a phloroglucinol-HCl solution (the Wiesner reaction), enabling us to visualize lignin in the differentiated xylem tracheids (*Figures 2b and 2c*).

Using three different mutagenesis systems, a total of seven thousand families were screened (*Table 1*). Four thousand families from an EMS-mutagenized population in a Columbia background, one thousand families from a Fast Neutron mutagenesis in a Landsberg erecta background and approximately two thousand families carrying the autonomous maize transposable element (En) from the AMAZE collection (Wisman *et al.*, 1998).

Seeds from individual M_2 families were germinated on vertical plates, and seven-day-old roots were squeezed in water, stained with a phloroglucinol:HCl solution and observed by light microscopy. Since the screening method is destructive, heterozygous siblings from segregating populations were kept in order to maintain the mutations.

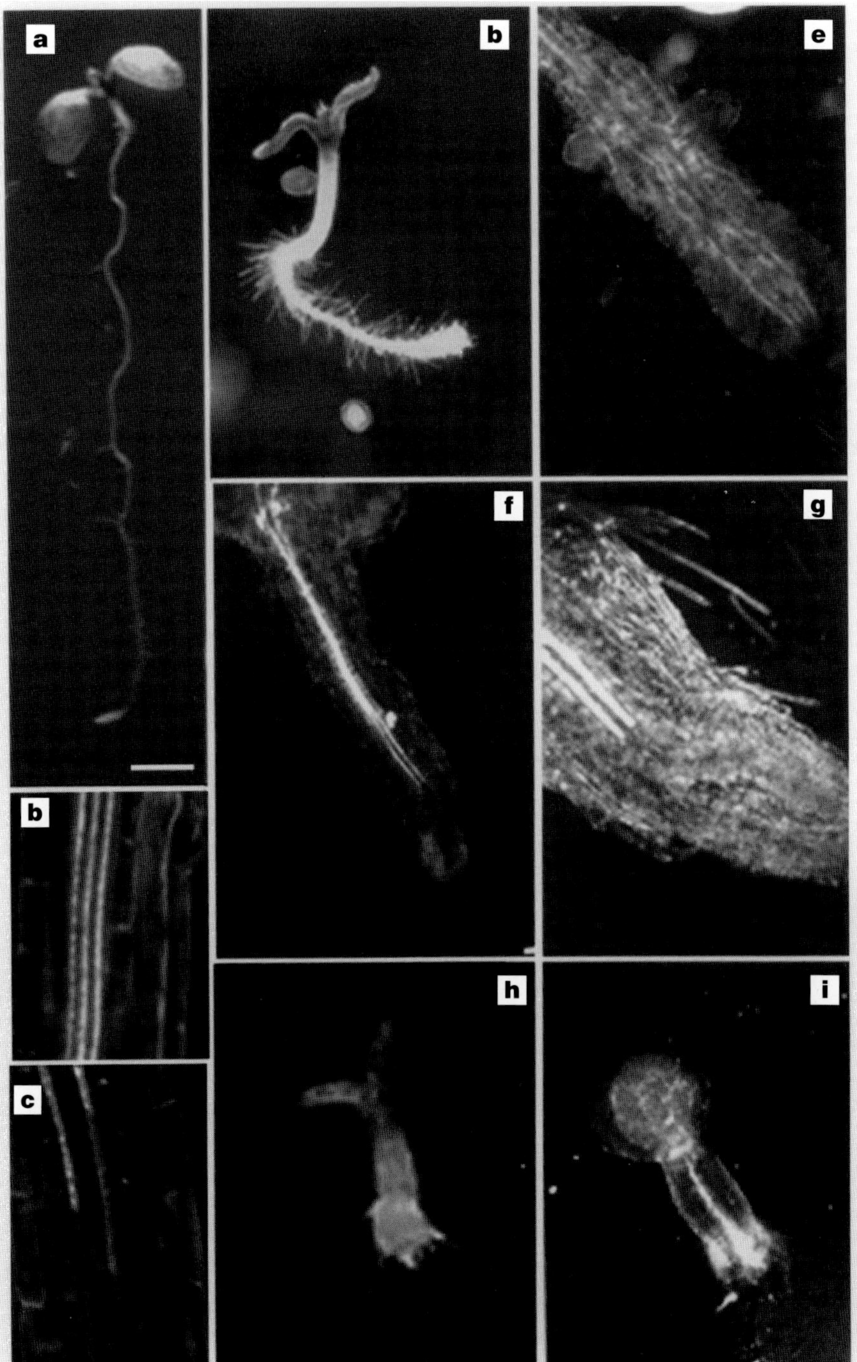

Figure 2. *Phenotypes of 7-day-old wild type and mutant seedlings: (a) Columbia wild type Bar 10 mm; (d) eli (×10); (f) tpx (×40); (h) mux (×40). Roots stained with phloroglucinol: (b) wild type protoxylem and metaxylem (×100); (c) wild type protoxylem (×100); (e) eli (×40); (g) tpx (×80); (i) mux (×40). Magnification in parenthesis.*

Table 1. Classification of mutants with altered xylem in the primary root

Class of mutant	Screening method	Description of phenotype	Number of mutants in M2[2]			
			0.5% EMS	En -Transposons	FN[1]	Total
eli	Phloroglucinol	Ectopic lignin	6 (4000)	1 (2100)	–	7
mux	Phloroglucinol	Multiple xylem strands	3 (4000)	–	1 (1000)	4
tpx	Phloroglucinol	Time of protoxylem development	11 (4000)	–	–	11
spt	Pyranine transport	Slower pyranine transport	9 (2000)	–	–	9
fpt	Pyranine transport	Faster pyranine transport	2 (2000)	–	–	2
Total						33 [1]

[1] Fast neutrons.
[2] In ( ) the number of families screened.

This screen permitted the isolation of a number of mutants that exhibit defective vascularization in the primary root. These have been grouped into three different classes according to the degree of cell lignification, the number of xylem elements and the timing of development of protoxylem cells. Establishing different complementation groups for all these mutants is part of our current investigation, and a preliminary characterization of the different mutant classes is summarized here.

3.1 eli *(ectopic lignification) mutants*

Seven mutants (six EMS-induced and one from the En-mutagenized lines) from independent families have been identified which show an increased amount of lignin based on phloroglucinol staining. Under light microscopy, the primary root of these mutants showed lignin in cell types that are not normally lignified in the wild type, as well as lack of lignin in some xylem cells (compare *Figures 2e* with *2b* and *2c*). This abnormal pattern of lignification is associated with reduced cell elongation that is reflected in shorter and thicker primary roots (*Figure 2d*) compared to the wild type (*Figure 2a*).

The *eli* seedlings continued to develop giving rise to adult plants with a reduced fertility. The homozygous mutant plants have a stunted phenotype with smaller siliques and flowers. In four of the *eli* mutants, the segregation in the progeny of the heterozygous plants showed a 3:1 ratio, indicating that these mutations are inherited as single, recessive, nuclear loci. At the time of writing the inheritance of the other three mutants has not been clarified.

3.2 mux *(multiple xylem strands) mutants*

These mutants show an increased number of xylem elements in the primary root. For practical purposes, two subgroups have been distinguished. One group, comprising three mutants, has multiple lignified cells in the protoxylem strands. They were isolated in an EMS mutagenesis screen, and share a seedling phenotype of shorter and thicker primary roots. The homozygous mutant plants show slow growth, but are still able to produce small flowers on a dwarf plant. These mutations are inherited as single recessive gene loci, although there is an inter-sibling variation among the homozygous plants, indicating that their phenotype may be influenced by environmental factors. Experiments to determine the gene penetrance of these mutations have yet to be carried out.

The other type of *mux* mutant shows abnormalities in the overall architecture of the seedling (*Figures 2h and 2i*). Phloroglucinol staining of the roots revealed an increased number of xylem strands. Furthermore, these xylem strands were disorganized, rather than following continuous cell files as occurs in the wild type, leading to an impaired vascular pattern. The mutations are recessive. The developmental lesions that caused this phenotype may occur very early in development, during embryogenesis.

None of the *mux* mutants are viable in the homozygous state, so heterozygous siblings are used to propagate the lines.

3.3 tpx *(timing of protoxylem development) mutants*

Eleven mutants have been isolated from the EMS-mutagenized lines. They differ from wild type by having a different timing of xylem development in the primary root. That is, the differentiation of the xylem cells is, to a varied extent, altered, although the pattern of vascularization seems to be correct compared to wild type. All the mutations are in the Columbia background.

In one line described in this chapter (*Figures 2f* and *2g*), the differentiation of the two protoxylem strands starts at the same time rather than in the sequential way commonly observed in the wild type (*Figure 2c*). Backcrosses to establish the inheritance patterns of these mutations are in progress.

4. Screen for mutations altering xylem function

The function of xylem is to transport and distribute water and solutes from the root to the aerial parts of the plant. We postulated that mutants altered in xylem organization and/or function will transport water and solutes in abnormal ways. Therefore, we have designed a novel screening method able to detect mutants with altered xylem transport in the primary root.

The membrane-impermeant, pH-sensitive fluorescence probe 8-hydroxy-1,3,6-pyrenetrisulfonate (pyranine) has been previously used for monitoring pH changes within anionic vesicles (Kano and Fendler, 1978). Pyranine is a hydrophilic, non-toxic, low molecular weight compound. These features make pyranine suitable to monitor apoplastic transport through the xylem in a non-destructive screening method.

In seedlings grown in conditions permitting full transpiration, pyranine was taken up into the roots and transported through the xylem vessels to the cotyledons. Using a low magnification fluorescence microscope with blue light, we could detect the presence of pyranine in the cotyledons by a green colouration in the vascular system (*Figure 3a*). After a longer exposure to pyranine, the cotyledons appeared totally green (as shown in *Figure 3b*). In the absence of pyranine, the cotyledons were red as a result of the chlorophyll autofluorescence (see *Figure 3c*).

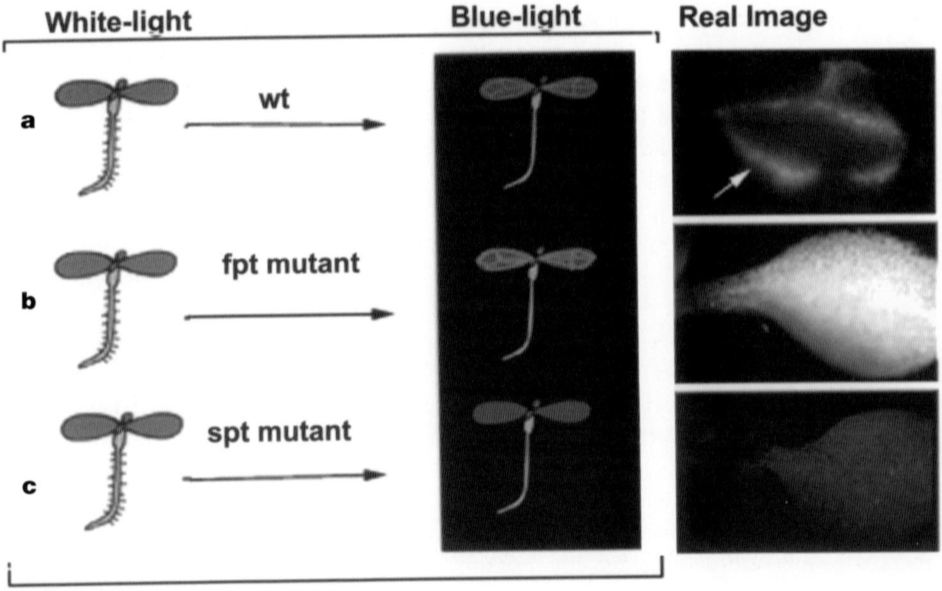

Figure 3. *Pyranine transport in* Arabidopsis *seedlings. Visualization of pyranine in the cotyledons in a representation in white-light (left), blue-light (centre), and microscopical image with blue-light (right). (a) Wild-type pyranine transport. The arrow indicates presence of pyranine in the vascular system. (b) Fast pyranine transport mutants. (c) Slower pyranine transport mutants.*

Different mutants with alterations in the vascular system already described have been assayed with the pyranine transport method. Some of them have shown differences in the time taken to transport the pyranine from the root to the cotyledons. For example *wol*, which has no phloem cell specification and fewer cells in the vascular cylinder of the primary root (Scheres *et al.*, 1995), has a shorter transport time than wild type seedlings. In other words, when all the wild type seedlings showed a green vasculature of the cotyledons (as shown in *Figure 3a*), the *wol* homozygous mutant seedlings showed higher levels of staining (similar to those observed in the wild type at later times).

4.1 *Pyranine transport mutants*

To date, two thousand individual families from an EMS mutagenized population have been screened for differences in timing of pyranine transport. According to this assay, two new classes of mutant have been isolated (see *Table 1*).

spt ***(slower pyranine transport) mutants.*** Nine families were found to show an absence of pyranine fluorescence when all wild type control plants showed pyranine in the vascular cells of the cotyledons (*Figure 3c*). Only the autofluorescence of the chlorophyll could be detected, indicating that the mutants did not transport pyranine at the normal rate.

fpt ***(faster pyranine transport) mutants.*** In this class we have included those mutants that showed a short timing for pyranine transport. Two mutants have been isolated that exhibit strong intensity of pyranine staining in the cotyledons. This phenomenon was observed in wild type plants when they were exposed to pyranine for a longer time (*Figure 3b*). This pyranine transport phenotype is similar to that of *wol.*

It may be possible that not all the mutants isolated with this method will show a defect in xylem organization. For example, some of them could be specifically affected in the capability to take up the stain. However, this method has advantages, such as high capacity, clear results, and it is a non-destructive screen.

The re-screen and characterization of these mutants is part of our current work.

5. Conclusions and prospects

The work described in this chapter provides a foundation for the genetic and molecular genetic analysis of the xylem. Mutants with altered xylem development have been isolated in whole-plant screens using phenotypes that might be predicted to lead to alterations in the mechanical properties of xylem (Turner and Somerville, 1997; Zhong *et al.*, 1997). The two screening methods used in the work presented here have added further phenotypes to this category. The first uses the invariant morphology of the *A. thaliana* primary root to screen for mutants with altered xylem cell shape, potential inductive effects from adjacent tissues, and altered patterns of secondary thickening. The second screen is for xylem function, in terms of solute transport and apoplastic flow, using a fluorescent dye.

The classes of mutant obtained from the former screen all have profound effects on cell morphology and development but their phenotypes are not specific to xylem or vascular cells. While the mutant screen is not close to saturation, our screen could detect such mutations if they occur. Therefore it seems that fundamental processes common to many cell types operate during xylem development, and it may be that only terminal differentiation processes such as secondary thickening and lignification may be xylem/fibre specific.

Future work aims to establish in molecular detail the regulatory processes that control xylem cell specification and differentiation. Candidate genes will be found from screens such as those described here and also from the *Zinnia* system. It is clear that xylem development, which has previously not been approached genetically, can now be analysed with a fuller complement of modern biological tools. The genes identified in these studies might also prove to be useful for modifying wood production in desirable ways.

Acknowledgements

The technical assistance of Xiao-He Wu and Sarah Gutteridge is gratefully acknowledged. We thank Fred Rook for valuable comments on the manuscript. This work is supported by a John Innes Foundation studentship to A.C.-D.

References

Baima, S., Nobili, F., Sessa, G., Lucchetti, S., Ruberti, I. and Morelli, G. (1995) The expression of the *Athb-8* homeobox gene is restricted to provascular cells in *Arabidopsis thaliana*. *Development* **121**: 4171–4182.

Benfey, P.N., Linstead, P.J., Roberts, K., Schiefelbein, J.W., Hauser, M.-T. and Aeschbacher, R.A. (1993) Root development in *Arabidopsis*: four mutants with dramatically altered root morphogenesis. *Development* **119**: 57–70.

Berleth, T. and Jürgens, G. (1993) The role of the *monopteros* gene in organizing the basal body region of the *Arabidopsis* embryo. *Development* **118**: 575–587.

Carland, F.M. and McHale, N.A. (1996) *LOP1*: a gene involved in auxin trasport and vascular patterning in *Arabidopsis*. *Development* **122**: 1811–1819.

Demura, T. and Fukuda, H. (1993) Molecular cloning and characterization of cDNAs associated with tracheary element differentiation in cultured *Zinnia* cells. *Plant Physiol.* **103**: 815–821.

Dolan, L., Janmaat, K., Willemsen, V., Linstead, P., Poethig, S., Roberts, K. and Scheres, B. (1993) Cellular organisation of the *Arabidopsis thaliana* root. *Development* **119**: 71–84.

Dolan, L. and Roberts, K. (1995) Secondary thickening in roots of *Arabidopsis thaliana*: anatomy and cell surface changes. *New Phytol.* **131**: 121–128.

Fukuda, H. (1997) Tracheary element differentiation. *Plant Cell* **9**: 1147–1156.

Gälweiler, L., Guan, C., Müller, A., Wisman, E., Mendgen, K., Yephremov, A. and Palme, K. (1998) Regulation of polar auxin transport by AtPIN1 in *Arabidopsis* vascular tissue. *Science* **282**: 2226–2230.

Greenberg, J.T. (1996) Programmed cell death: A way of life for plants. *Proc. Natl. Acad. Sci. USA* **93**: 12094–12097.

Hamann, T., Mayer, U. and Jürgens, G. (1999) The auxin-insensitive *bodenlos* mutation affects primary root formation and apical-basal patterning in the *Arabidopsis* embryo. *Development* **126**: 1387–1395.

Hardtke, C.S. and Berleth, T. (1998) The *Arabidopsis* gene *MONOPTEROS* encodes a transcription factor mediating embryo axis formation and vascular development. *EMBO J.* **17**: 1405–1411.

Hatton, D., Sablowski, R., Yung, M.H., Smith, C., Schuch, W. and Bevan, M. (1995) Two classes of cis sequences contribute to tissue-specific expression of a PAL2 promoter in transgenic tobacco. *Plant J.* **7**: 859–876.

Hauser, M.-T., Morikami, A. and Benfey, P.N. (1995) Conditional root expansion mutants of *Arabidopsis*. *Development* **121**: 1237–1252.

Iwasaki, T. and Shibaoka, H. (1991) Brassinosteroids act as regulators of tracheary-element differentiation in isolated *Zinnia* mesophil cells. *Plant Cell Physiol.* **32**: 1007–1014.

Kano, K. and Fendler, J.H. (1978) Pyranine as a sensitive pH probe for liposome interiors and surfaces. pH gradients across phospholipid vesicles. *Biochim. Biophys Acta* **509**: 289–299.

Kobayashi, H. and Fukuda, H. (1994) Involvement of calmodulin and calmodulin-binding proteins in the differentiation of tracheary elements in *Zinnia* cells. *Planta* **194**: 388–394.

Ryser, U., Schorderet, M., Zhao, G.F., Studer, D., Ruel, K., Hauf, G. and Keller, B. (1997) Structural cell-wall proteins in protoxylem development: evidence for a repair process mediated by a glycine-rich protein. *Plant J.*, **12**: 97–111.

Scheres, B., Di Laurenzio, L., Willemsen, V., Hauser, M.-T., Janmaat, K., Weisbeek, P. and Benfey, P.N. (1995) Mutations affecting the radial organisation of the *Arabidopsis* root display specific defects throughout the embryonic axis. *Development* **121**: 53–62.

Scheres, B., Wolkenfelt, H., Willemsen, V., Terlouw, M., Lawson, E., Dean, C. and Weisbeek, P. (1994) Embryonic origin of the *Arabidopsis* primary root and root meristem initials. *Development* **120**: 2475–2487.

Schiefelbein, J.W., Masucci, J.D. and Wang, H. (1997) Building a root: the control of patterning and morphogenesis during root development. *Plant Cell* **9**: 1089–1098.

Schindler, T., Bergfeld, R. and Schopfer, P. (1995) Arabinogalactan proteins in maize coleoptiles: developmental relationship to cell death during xylem differentiation but not to extension growth. *Plant J.* **7**: 25–36.

Tian, Q. and Reed, J.W. (1999) Control of auxin-regulated root development by the *Arabidopsis thaliana SHY2/IAA3* gene. *Development* **126**: 711–721.

Topping, J.F., May, V.J., Muskett, P.R. and Lindsey, K. (1997) Mutations in the *HYDRA1* gene of *Arabidopsis* perturb cell shape and disrupt embryonic and seedling morphogenesis. *Development* **124**: 4415–4424.

Torres-Ruiz, R. and Jürgens, G. (1994) Mutations in the *FASS* gene uncouple pattern formation and morphogenesis in *Arabidopsis* development. *Development* **120**: 2967–2978.

Turner, S.R. and Somerville, C.R. (1997) Collapsed xylem phenotype of *Arabidopsis* identifies mutants deficient in cellulose deposition in the secondary cell wall. *Plant Cell* **9**: 689–701.

Ulsamov, T., Hagen, G. and Guifoyle, T.J. (1997) ARF1, a transcription factor that binds to auxin response elements. *Science* **276**: 1865–1868.

Weigel, D. and Doerner, P. (1996) Cell–cell interactions: taking cues from the neighbors. *Curr. Biol.* **6**: 10–12.

Wisman, E., Cardon, G.H., Fransz, P. and Saedler, H. (1998) The behaviour of the autonomous maize transposable element En/Spm in *Arabidopsis thaliana* allows efficient mutagenesis. *Plant Mol. Biol.* **37**: 989–999.

Wysocka-Diller, J.W. and Benfey, P.N. (1997) Root development: signalling down and around. *BioEssays* **19**: 959–965.

Ye, Z.-H. and Varner, J.E. (1993) Gene expression patterns associated with *in vitro* tracheary element formation in isolated single mesophyll cells of *Zinnia elegans*. *Plant Physiol.* **103**: 805–813.

Zhong, R., Taylor, J.J. and Ye, Z.-H. (1997) Disruption of interfascicular fiber differentiation in an *Arabidopsis* mutant. *Plant Cell* **9**: 2159–2170.

39

The use of *Arabidopsis* mutants to study secondary cell wall formation

S.R. Turner, N. Taylor, M. Hall and L. Jones

1. Introduction

The deposition of the plant secondary cell wall occurs once the cell has stopped growing. This deposition of secondary wall material often results in a very thick wall and is responsible for many of the structural properties associated with plants. Consequently, the many industrial processes that utilize plant material, which are as diverse as papermaking and food processing, depend heavily on the properties of the secondary cell walls of these plants.

Despite lignin and cellulose being two of the most abundant biopolymers, their biosynthesis is relatively poorly understood. It has not been possible to isolate the higher plant cellulose synthase complex (reviewed in Delmer and Amor, 1995). Consequently, our understanding of how the complex is regulated and how cellulose deposition is orientated remains relatively poor. In addition, although the pathway for the biosynthesis of monolignols, the monomers required for lignin synthesis, is relatively well understood, many questions about the structure and formation of lignin remain (reviewed in Whetton and Sederoff, 1995; Lewis and Yamamoto, 1990). The role that many other components play in determining the properties and the assembly of the cell wall is unknown (McCann and Roberts, 1994). The biochemical complexity of the cell wall and the difficulty of working with some of the individual components has prevented more rapid progress in understanding its higher order structure and the contribution that different components make to its properties.

In common with many problems that have proved intractable by conventional biochemical approaches, the structure and assembly of the cell wall is a problem for which a molecular genetic approach is likely to be productive. Brittle culm mutants of barley have reduced cellulose in the cell wall and, as a consequence, the walls have large changes in their physical properties when compared to wild type (Kokubo *et al.*, 1991).

Cell & Molecular Biology of Wood Formation, edited by R. Savidge, J. Barnett and R. Napier.

More recently, mutants have been identified in the model plant species *Arabidopsis thaliana*. Reiter *et al.* (1993) used direct analysis of the chemical composition of the primary cell wall to isolate a number of mutants. Only seven complementation groups were identified, a relatively small number given the biochemical complexity of the wall (Reiter *et al.*, 1997). In the temperature-sensitive radial swelling (*rsw1*) mutants (Baskin *et al.*, 1992) *rsw1* plants die at an early stage at the restrictive temperature and have only half the cellulose content of the wild type (Arioli *et al.*, 1998). In addition a mutant with altered patterns of cellulose deposition in the trichomes has been reported (Potikha and Delmer, 1995) as well as a mutant with altered lignin monomer composition (Chapple *et al.*, 1992).

The differentiation of tracheary elements, the large water conducting cells of the xylem, has been extensively used as a model for plant cell differentiation and secondary wall deposition (reviewed in Fukada, 1992). The process of carrying water from the roots to the leaves causes a large compressive force to be placed on tracheary elements of the xylem. To prevent their collapse, the walls of these cells are strengthened by an elaborate pattern of secondary thickening, which includes the deposition of cellulose and lignin. Any process which prevents the formation of an integral cell wall has a dramatic effect, for example the addition of phenylalanine ammonia lyase (PAL) inhibitors which prevent lignin deposition (Amrhein *et al.*, 1983; Smart and Amrhein, 1985). The cells are no longer able to maintain their proper shape and collapse inward, forming irregular and contorted cells when viewed in cross section.

The characteristics exhibited by xylem cells unable to attain their proper strength has formed the basis of a genetic screen for novel cell wall mutants. *Arabidopsis* mutants in which the walls of the xylem elements were defective causing the cells to collapse inward were selected (Turner and Somerville, 1997). This screen had the advantage over previous genetic screens in that it selected directly for mutants with dramatic alterations in the properties of the cell wall. These mutants, together with mutants isolated in similar genetic screens, will not only identify genes involved in biosynthesis of major components such as lignin and cellulose, but also identify other minor components which are required for the structure or assembly of a functional secondary cell wall.

2. Irregular xylem (*irx*) mutants

2.1 *Identification of secondary cell wall mutants*

Development of the inflorescence stem in *Arabidopsis* plants is accompanied by a large increase in the deposition of secondary cell walls. This formation of heavily-lignified, thick-walled cells occurs both in the xylem and in the interfascicular region and is accompanied by increases in the deposition of cellulose, lignin and xylan, all of which accumulate to very high levels (Turner and Somerville, 1997).

As the xylem element grows and develops it loses its contents and is dead when the element reaches maturity (*Figure 1*). This is accompanied by the deposition of a secondary cell wall, which is localized only on the lateral sides of the cell. In addition, the primary wall is removed at the end of the cell to form a perforation plate and secondary cell wall material is excluded from these ends of the cell (*Figure 1*). The xylem elements of mutant plants are unable to withstand the negative pressure generated in the xylem and are readily identified by their collapsed appearance (*Figure 2a, b*).

This approach allowed the initial identification of mutants from 3 complementation groups with deficient secondary cell walls (Turner and Somerville, 1997), further *irx* complementation groups have been identified subsequently (N. Taylor and S. Turner, unpublished).

Sections of relatively young xylem elements from *irx* lines show that initially the cells appeared to grow and expand normally and attain their wild type shape (Turner and Somerville, 1997). The collapse of the tracheary elements occurs only late during differentiation, consistent with the collapse being due to a defect in the secondary cell wall, which would normally maintain the proper cell shape after the loss of cell contents (*Figure 1*). Although first identified in stems, the xylem in mature roots, hypocotyls and petioles also show the same phenotype.

2.2 *Cellulose deficient mutants*

Several lines of evidence indicate that *irx1, 2* and *3* show a specific decrease in secondary cell wall cellulose. The cellulose and non-cellulose carbohydrate composition of the plant cell wall can be distinguished readily due to the fact that cellulose is resistant to routine acid hydrolysis. The composition of the non-cellulose fraction of the wall has been examined by separating the alditol acetate derivatives by gas chromatography (Blackeney *et al.*, 1983). No significant differences between *irx1, 2* and *3* and the wild type were found in any of the seven sugars measured when wall material from stems of mature hypocotyl was analysed (Turner and Somerville, 1997). In contrast when cellulose was measured using the method of Updegraff (1969), dramatic differences were observed between *irx 1, 2* and *3* and the wild type (Turner and Somerville, 1997). Stems of *irx3* plants contain less than one fifth of the cellulose content of the wild type. Differences between wild type and *irx3* are much greater in tissues with a large proportion of secondary cell wall material such as mature stems or hypocotyls, whilst tissue with few cells with thick secondary thickening, such as leaves, show little difference. This correlation suggests strongly that the *irx* mutants are specifically defective in cellulose synthesis in the secondary cell wall. This hypothesis is supported by recent work on *rsw1* mutants. These mutants are defective in cellulose synthesis in the primary cell wall and at the restrictive temperature they fail

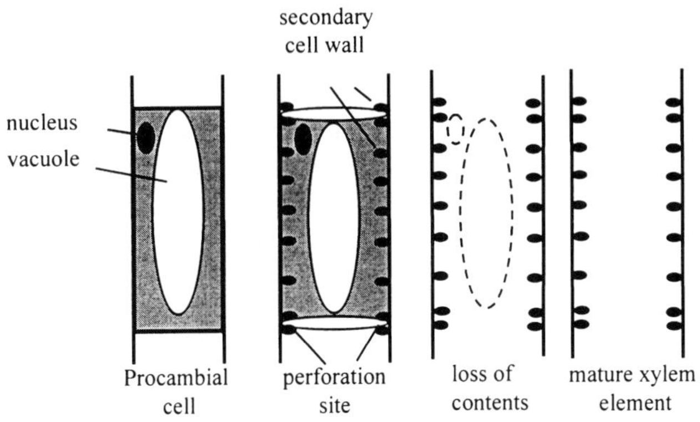

Figure 1. Diagram showing the differentiation of a xylem element from a procambial cell.

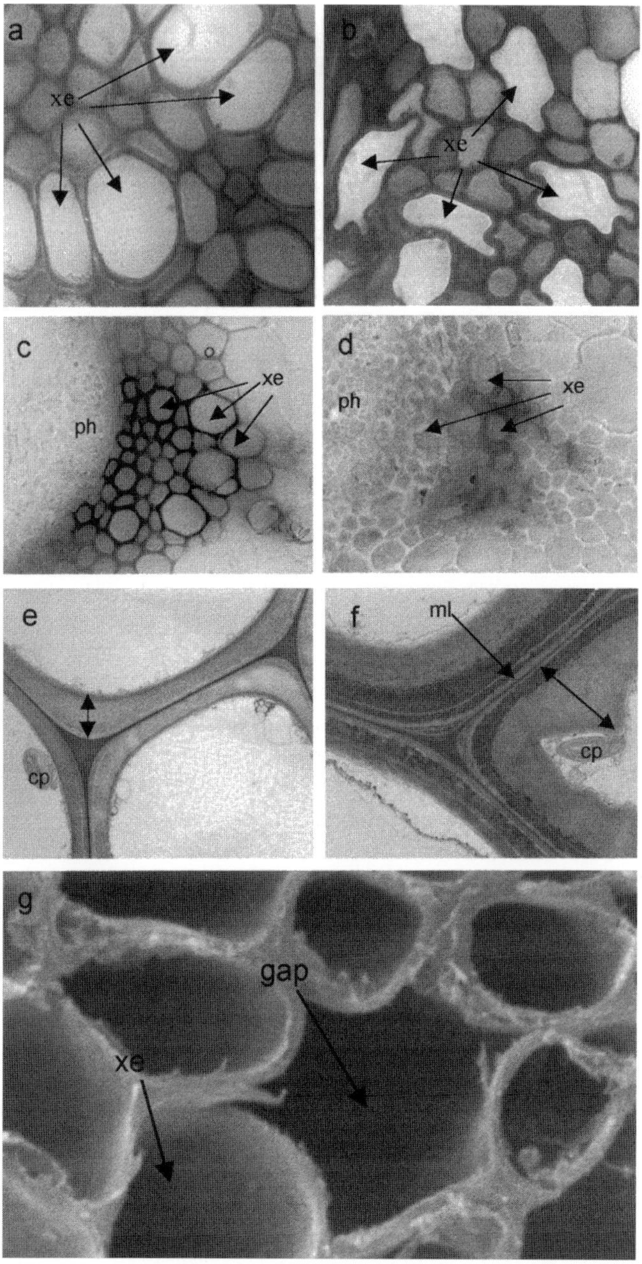

Figure 2. *Micrographs showing xylem cells from wild type (a, c, e) and mutant (b, d, f, g)* Arabidopsis *plants. Light micrographs of stem sections showing vascular bundles from wild type (a, c)* irx3 *(b) and* irx4 *(d) stained with toluidine blue (a, b) and phloroglucinol (d, e). Transmission electron micrographs showing secondary cell walls from xylem elements of wild type (e) and* irx4 *(f). Scanning electron micrograph showing part of the xylem in a stem vascular bundle from a* gpx1 *plant. Xylem elements (xe), phloem (ph), middle lamellae (ml) and chloroplasts (cp) are indicated. Double-headed arrow is used to emphasize the width of the cell wall.*

to develop past the seedling stage (Arioli *et al.*, 1998). In contrast, *irx* seedlings grow almost normally suggesting that the primary cell wall in these plants is unaffected (Turner and Somerville, 1997).

2.3 *Cloning the* irx3 *gene*

Cellulose synthesis is relatively well understood in two bacterial species: *Acetobacter xylinum* and *Agrobacterium tumefaciens*. Four and five genes, respectively, are essential for cellulose synthesis in these bacteria. Interestingly, only one gene is homologous in *Agrobacterium* and *Acetobacter*. This gene (*BcsA/celA*) is believed to be the catalytic subunit, which adds successive glucan units (reviewed in Delmor and Amor, 1995). In higher plants, however, progress in purifying and identifying the cellulose synthase complex has been slow, largely because of the fragile nature of the cellulose synthase complex. Prolonged synthesis of cellulose has still not been achieved *in vitro*. Evidence that cellulose is synthesized on the plasma membrane by large complexes known as rosettes has come primarily from electron microscopy studies, which suggest that each rosette has six subunits. Each of these subunits is itself a hexamer, such that the rosette is a 36mer with 36 copies of several different polypeptides (Delmor and Amor, 1995).

Expressed sequence tags (EST) and genomic sequences have been used to identify more than 27 members of a *celA*-like family from *Arabidopsis* (Cutler and Somerville, 1997). In order to test the hypothesis that one of these genes might correspond to *irx3*, polymerase chain reaction (PCR) based markers were used to map *irx3* to the top arm of chromosome 5 to a region spanned by only 4 yeast artificial chromosomes (YACs) (Taylor *et al.*, 1999). PCR primers were then designed to amplify each potential *celA* homologue and used in a PCR reaction with the 4 YAC clones. Only one set of primers, corresponding to EST 75G11, gave a product.

Several lines of evidence indicate that 75G11 does indeed correspond to the *IRX3* gene. There is a good correlation between 75G11 expression and secondary cell wall synthesis. In addition, the expression of mRNA corresponding to 75G11 is much lower in *irx3* mutant plants that in the wild type. Cosmids containing the gene corresponding to 75G11 can complement the *irx3* mutation. Comparison of the cDNA sequence of 75G11 between wild type and *irx3* indicates that *irx3* plants carry a mutation that results in the loss of 168 amino acids from the carboxy terminus of the *IRX3* protein. This data conclusively demonstrates that the gene identified by 75G11 is the gene defective in *irx3* plants (Taylor *et al.*, 1999).

2.4 *Comparison of* irx3 *with other cellulose synthases*

Until very recently no genes for any of the subunits of the higher plant cellulose synthase subunit had been cloned. This situation changed when Pear *et al.* (1996) described a clone from cotton (*celA*) which showed homology to the catalytic subunit of bacterial cellulose synthases and contained several conserved sequences indicative of a processive glucosyl transferase. Many similar sequences (described as *celA*-like) have been identified in the *Arabidopsis* EST and genomic databases (Cutler and Somerville, 1997). Conclusive proof that a member of this family of genes represents the higher plant cellulose synthase was provided by studies on a temperature sensitive, cellulose deficient mutant of *Arabidopsis* (*rsw1*). *Rsw1* has a mutation in a member of the *celA*-like family of genes (Arioli *et al.*, 1998).

The *IRX3* gene contains 12 exons and the cDNA encodes for a protein of 1025 amino acids with a predicted molecular weight of 116 kDa. *IRX3* contains the four conserved features identified in processive glucosyl transferases: three regions centred on an aspartate residue and the fourth consists of QXXRW, where x represents any amino acid. In addition, *IRX3* contains two predicted transmembrane domains at the amino terminus followed by a large cytoplasmic domain and a further six transmembrane domains close to the carboxy terminus. These structural features indicate that *IRX3* is a member of the *celA* family. The similarity is further confirmed by sequence comparison, *IRX3* shows between 70% and 80% identity with the *Arabidopsis* gene products *RSW1*, *Ath-A* and *CELA1* from cotton. Sequence comparison with the wider family of *Arabidopsis* genes showing homology to *CelA* indicates that there are 3 distinct subfamilies of genes. It is suggested that these families be designated *Csa, Csb* and *Csc* (Taylor *et al.*, 1999). *IRX3* falls into the Csa family (*Figure 3*) together with *RSW1* and *Ath-A* and *Ath-B*. This subfamily has already been suggested to represent the true cellulose synthases (Arioli *et al.*, 1998). The function of genes from other families (*Csb* and *Csc*) remains to be determined. Interestingly, *IRX3* appears to show more similarity to the cotton *CELA1* gene than it does to *RSW1*. Since both *IRX3* and cotton *CELA1* are involved in secondary cell wall formation it is possible that they contain structural features which would distinguish them from other cellulose synthase genes involved in primary wall biosynthesis.

The mutation in *irx3* plants causes the gene product to be truncated by 168 amino acids and it lacks the last four transmembrane domains. This mutation would almost certainly result in a non-functional protein. Interestingly, *irx3* plants show no increase in the acid soluble cell wall components (Turner and Somerville, 1997). This result is significant since *rsw1* plants are reported to show an increase in a non-crystalline β (1–4) linked glucose. Our data would indicate that no similar polymer is synthesized in *irx3* plants. The fact that *irx3* represents a null mutation should allow the structure of other members of the family to be analysed. Any gene may be placed in front of the *irx3* promoter and tested to see whether it can complement the *irx3* mutation.

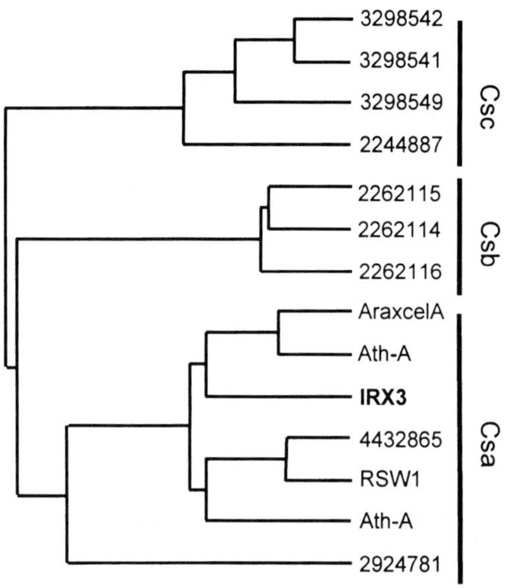

Figure 3. Relationship between Arabidopsis *sequences showing homology to known cellulose synthase genes. Numbers shown are Genbank accessions; the tree was generated using the GCG program pileup.*

2.5 Irx *mutants defective in lignin biosynthesis*

More recently, a fourth *irx* complementation group has been identified. *Irx4* plants show almost no difference in cellulose deposition (L. Jones and S. Turner, unpublished data). In contrast lignin deposition in *irx4* plants is clearly disrupted. The phenotype of *irx4* plants does not resemble any previously described plants with altered lignin. Our evidence suggests that *irx4* is a novel kind of lignin mutant.

Several different approaches have been used to analyse lignin composition in *irx4* plants. The results demonstrated that stem segments from *irx4* possess only ~50% of the total cell wall phenolics of the wild type (L. Jones and S. Turner, unpublished data). This data has been extended using solid state NMR which also shows that *irx4* plants have around 50% of the lignin content of the wild type (M. Jarvis, pers. comm.).

Phloroglucinol is considered a diagnostic stain for lignin. The walls of the xylem cells in the stems of wild-type *Arabidopsis* plants stain a bright red colour, indicative of the presence of lignin. In contrast to the wild type and *irx1, 2* or *3*, a comparable section from *irx4* plants shows a much weaker reaction (*Figure 2c, d*). In addition, the xylem elements of *irx4* plants have an irregular and collapsed appearance. Other stains which are considered diagnostic for lignin, such as Mäule's reagent, also give a much weaker reaction with sections from *irx4* plants compared to the wild type (L. Jones and S. Turner, unpublished data).

Under the light microscope the secondary cell walls of *irx4* plants do not appear as discrete regions in the way that they do in wild type. Walls from *irx4* plants are very diffuse, the middle-lamellae are prominent and surrounded by very poorly stained material which is almost translucent. The reason for this can be seen clearly under the electron microscope (*Figure 2e and f*). Wild type walls have an amorphous appearance with only the middle-lamellae prominent as a separate layer within the wall (*Figure 2e*). In *irx4* plants the cell wall is hugely expanded and is divided into several discretely stained regions (*Figure 2f*). There is a distinct outer domain that spreads towards the centre of the cell. It is assumed that the large expansion of the cell wall accounts for the translucent looking material that can be seen under the light microscope.

A large number of different genes such as cinnamyl alcohol dehydrogenase (CAD), phenylalanine ammonia lyase (PAL) and caffeic acid 3-O-methyltransferase (COMT) have been used to modify lignin in herbaceous plants and trees (reviewed in Campbell and Sederoff, 1996). In addition, a number of mutants with altered lignin have been described, such as the brown-midrib (bm) mutants of maize and sorghum. Plants which have levels of cinnamoyl-CoA reductase reduced to 25% of wild type show a distortion in the shape of the xylem elements (Piquemal *et al.*, 1998). None of the previously described modified plants or any of the characterized lignin mutants exhibit the severe collapse of the xylem and the huge expansion of the secondary cell wall seen in *irx4* plants.

Measurements of cell wall phenolics and solid state NMR both indicate that lignin is reduced to approximately 50% of the wild type, but not further it is surprising that the effects on physical properties (L. Jones and S. Turner, unpublished data) and ultra-structure (*Figure 2*) of the wall are so dramatic. One explanation for this is that the lignin is deposited early on in secondary wall formation, but not later in development. This would be consistent with histochemical studies where early forming protoxylem shows some staining, severely reduced or absent in later forming metaxylem.

Despite extensive biochemical analysis of irx4 the nature of the defect remains unclear. Future work will involve cloning the *IRX4* gene in order to gain a better understanding of the critical role of the *IRX4* gene in lignin deposition.

3. Other xylem mutants

During the course of stem differentiation, cells in both the xylem and in the highly thickened cells of the interfascicular region differentiate from cells surrounded by thin primary cell wall. Cross sections of vascular bundles in an *Arabidopsis* stem (*Figure 2a*) show that the xylem elements fill the bundle and all have similar degrees of secondary wall thickening. We have identified a gapped xylem (*gpx1*) mutant of *Arabidopsis* in which the vascular bundles contain both thick walled xylem elements and areas in which xylem elements appear absent. The shape of these spaces is made up only of the walls of adjacent cells (*Figure 2g*).

Another characteristic for *gpx* plants is the irregular pattern of thickening in the interfascicular region. In wild type plants, adjacent cells in the interfascicular region show a similar degree of secondary wall thickening (Turner and Somerville, 1997). In *gpx1* plants, interfascicular cells with very thick walls occur adjacent to cells which undergo little, or no, secondary thickening (M. Hall and S. Turner, unpublished data).

A possible explanation for this phenotype may be found in the different developmental pathways of xylem elements and interfascicular cells. Only xylem elements undergo a process of developmentally-regulated cell death. It appears that *gpx1* plants are defective in the pathway that leads to secondary cell wall deposition. The phenotype is not completely penetrant as some cells still undergo normal thickening. In the interfascicular region this presents no particular problem and plants containing both thick and thin walled cells remain relatively unaffected. In xylem elements, however, the cell contents are removed during cell death and this process includes the release of hydrolases required for removal of the perforation plate (*Figure 1*). It appears that in the absence of a secondary cell wall the remains of the primary cell walls are unprotected from cell wall degrading enzymes. This results in the cell almost completely digesting itself and leaves little evidence of its existence.

The *gpx1* mutant clearly demonstrates the essential role of the secondary cell wall in protecting the wall of xylem elements during cell death. It also shows that there is little interaction between the pathways leading to secondary cell wall deposition and that leading to cell death. Even in the absence of a secondary cell wall, xylem elements from *gpx1* plants still invoke the pathway of programmed cell death. Nevertheless, once invoked the process of secondary cell wall deposition occurs in a strictly developmentally regulated manner and clearly requires all the genes required for secondary cell wall synthesis to be co-ordinately expressed. How this process is regulated remains unclear, but a better understanding of it would clearly offer an enormous opportunity for manipulating secondary cell wall structure and wood deposition.

References

Amrhein N., Frank, G., Lemm, G. and Luhmann, H.B. (1983) Inhibition of lignin formation by *L*-alpha-aminooxy-beta-phenylpropionic acid, an inhibitor of phenylalanine ammonia-lyase Mungbean, *Vigna radiata. Eur. J. Cell Biol.* **29**: 139–144.

Arioli, T., Peng, L., Betzner, A.S., Burn, J., Wittke, W., Herth, W., Camilleri, C., Hofte, H., Plazinski, J., Birch, R., Cork, A., Glover, J., Redmond, J. and Williamson, R.E. (1998) Molecular analysis of cellulose biosynthesis in *Arabidopsis*. *Science* **279**: 717–720.

Baskin, T.I., Betzner, A.S., Hoggart, R., Cork, A. and Williamson, R.E. (1992) Root morphology mutants in *Arabidopsis thaliana*. *Aust. J. Plant Physiol.* **19**: 427–437.

Blackeney, N.B., Harris, P.J., Henry, R.J. and Stone, B.A. (1983) A simple and rapid preparation of alditol acetates for monosaccharide analysis. *Carbohydr. Res.* **182**: 291–299.

Campbell, M.M. and Sederoff, R.R. (1996) Variation in lignin content and composition — Mechanism of control and implications for the genetic improvement of plants. *Plant Physiol.* **110**: 3–13.

Chapple, C.C.S, Vogt, T., Ellis, B.E. and Somerville, C.R. (1992) *Arabidopsis* Mutant defective in the general phenylpropanoid pathway. *Plant Cell* **4**: 1413–1424.

Cutler, S. and Somerville, C.R. (1997) Cellulose synthesis: Cloning in silico. *Current Biology* **7**: 108–111.

Delmer, D.P. and Amor, Y. (1995) Cellulose biosynthesis. *Plant Cell* **7**: 987–1000.

Fukada, H. (1992) Tracheary element formation as a model system for cell differentiation. *Int. Rev. Cytol.* **136**: 289–332.

Kokubo, A., Sakurai, N., Kuraishi, S. and Takeda, K. (1991) Culm brittleness of barley (*Hordeum vulgare* L.) mutants is caused by smaller number of cellulose molecules in cell wall. *Plant Physiol.* **97**: 509–514.

Lewis, N.G. and Yamamoto, E. (1990) Lignin occurrence biogenesis and biodegradation. *Annu. Rev. Plant Physiol.* **41**: 455–496.

McCann, M.C. and Roberts, K. (1994) Changes in cell wall architecture during cell elongation. *J. Exp. Bot.* **45**: 1683–1691.

Pear, J.P., Kawagoe, Y., Schreckengost, W.E., Delmer, D.P. and Stalker, D.M. (1996) Higher plants contain homologs of the bacterial *celA* genes encoding the catalytic subunit of cellulose synthase. *Proc. Natl. Acad. Sci. USA* **93**: 12 637–12 642.

Piquemal, J., Lapierre, C., Myton, K., O'Connell, A., Schuch, W., GrimaPettenati, J. and Boudet, A.M. (1998) Down-regulation of cinnamoyl-CoA reductase induces significant changes of lignin profiles in transgenic tobacco plants. *Plant J.* **13**: 71–83.

Potikha, T. and Delmer, D.P. (1995) A mutant of *Arabidopsis thaliana* displaying altered patterns of cellulose deposition. *Plant J.* **7**: 453–460.

Reiter, W.D., Chapple, C. and Somerville, C.R. (1993) Altered growth and development in a fucose-deficient cell wall mutant of *Arabidopsis*. *Science* **261**: 1032–1035.

Reiter, W.D., Chapple, C. and Somerville, C.R. (1997) Mutants of *Arabidopsis* thaliana with altered cell wall polysaccharide composition. *Plant J.* **12**: 335–345.

Smart C.C. and Amrhein, N. (1985) The influence of lignification on the development of vascular tissue in *Vigna radiata* L. *Protoplasma* **124**: 87–95.

Taylor, N.G., Scheible, W., Cutler, S., Somerville, C.R. and Turner, S.R. (1999) The irregular *xylem 3* locus of *Arabidopsis* encodes a cellulose synthase gene required for secondary cell wall synthesis. *Plant Cell,* in press.

Turner, S.R. and Somerville, C.R. (1997) Collapsed xylem phenotype of *Arabidopsis* identifies mutants deficient in cellulose deposition in the secondary cell wall. *Plant Cell* **9**: 689–701.

Updegraff, D.M. (1969) Semi-micro determination of cellulose in biological materials. *Anal. Biochem.* **32**: 420–424.

Whetton, R. and Sederoff, R. (1995) Lignin biosynthesis. *Plant Cell* **7**: 1001–1013.

40

Cell, tissue and organ culture for the study of cambial activity and wood formation — a resurgence of interest in an old technique

M.A. Leitch and R.A. Savidge

1. Introduction

Numerous practical and scientific applications exist for *in vitro* cell, tissue and organ culture in forestry, arboriculture and horticulture (e.g., Bajaj, 1986, 1989, 1991; Grey *et al.*, 1987; Hanover and Keathley, 1987; Smith, 1986). It is now routinely used in plant propagation and is essential in genetic transformation of plants.

Wood has diverse anatomical, chemical and physical traits, requiring for its formation highly sophisticated sub-cellular control mechanisms in both the cambium and its derivative cells (Savidge, 1996). Identification and characterization of the extrinsic and intrinsic factors regulating formation of the various sub-cellular features, cellular differentiation pathways, and polarities of cell and tissue organization present major challenges. Reductionistic *in vitro* approaches hold forth promise for unravelling individual components of that formative complexity and for identifying underlying control mechanisms. When the physiology of secondary growth is truly well understood, *in vitro* simulation of cambial growth and development as it occurs in nature should be readily demonstrable.

1.1 *Advantages and disadvantages of the culture approach*

Advantages. Whether the development of interest be an aspect of cambial cell division, cell expansion, or differentiation, it can be investigated *in vitro* free from the complication that physiological processes occurring elsewhere within the whole

Cell & Molecular Biology of Wood Formation, edited by R. Savidge, J. Barnett and R. Napier.

organism might be influencing the response (Street and Henshaw, 1966). Culture approaches are particularly useful for enabling the full set of cellular/sub-cellular competencies to be, as well as for demonstrating that the physiological 'whole' is indeed the sum of its parts. *In vitro* responses not uncommonly include, in addition to normal developmental expressions, novel anatomical phenotypes. Past research has revealed that strong growth occurring in explanted stem tissues, with cell division and expansion activities occurring in the cambium, is not necessarily accompanied by attending xylogenesis (Savidge and Wareing, 1981; Savidge, 1983). Thus, the *in vitro* approach permits separation of distinct developmental processes. Once cellular phenotypes have been observed, subsequent experimental investigation to discover the factors controlling their development can most readily be undertaken using a culture approach.

The ability to investigate the physiological effects of nutrients, plant hormones and other chemical constituents in defined culture media under controlled conditions is one of the most important applications of tissue culture. Suspension or protoplast cultures permit singular cells to be manipulated as discrete entities. Culturing mass replicates of singular genotypes permits identification of genotypic responses to varied extrinsic conditions. Thus, culture methodologies enable more confident identification and interpretation of factors of the intrinsic microenvironment than is possible when investigating whole trees grown in the growth chamber, greenhouse or field.

Where resources permit, the culture approach also makes it possible for a variety of genotypes to be grown under constant environmental conditions and compared, rapidly, for inherent variation, expressed either as molecular or cellular phenotypic variants. The validation of empirical ideas on the physiological processes attending secondary growth is an important contribution that will be forthcoming from *in vitro* research. The method may also one day find practical application in an area such as 'vat' fibre production (Savidge, 1997).

In addition to the *in vitro* approach, there are of course other methods that can be used to investigate phenomena of secondary growth and how they are regulated (Savidge, 1996). However, as explained more fully below, *in vitro* methodology permits one to distinguish those essential metabolites synthesized in the cambial region from those which must be provided from sources such as root, shoots, xylem and phloem storage reserves, etc. Thus, *in vitro* findings have the potential to bridge from biochemistry/cell biology to the physiology of whole stem (or root) organogenesis, probably more effectively than any other scientific approach.

The aim of simulating cambial growth as it normally occurs *in arbor* by growing stem pieces *in vitro* has so far not been achieved, other than marginally, with any species (e.g., Leitch and Savidge, 1995; Leitch *et al.*, 1999; Savidge, 1983, 1993a; Zajączkowski, 1973). When such simulations can be fully achieved, and even surpassed, forestry management practices concerned with sustaining wood supplies will begin to have a solid scientific basis. Innovative means to manipulate both wood quality and quantity will undoubtedly emerge, possibly leading to industrial innovations (Savidge, 1985, 1997).

Disadvantages. The tissue culture literature is full of assumptions about how constituents in culture media serve — physiologically and biochemically — and, therefore, why it is important to use some ingredients and exclude others (e.g., Dell and Robinson, 1993; Gauch, 1972; Salisbury and Ross, 1991; Smith, 1992). This is ironic because neither the physiology nor the biochemistry of plant growth and development begins to approach a state of 'perfect' knowledge. Thus, in relation to the needs of the cambium, assumptions about how culture media should be formulated can, more correctly, be regarded merely as hypotheses awaiting investigation. In order to design physiologically 'correct' media and perform *in vitro* research into cambial growth and development intelligently, there is a definite need to know more about molarities and ratios of endogenous organic metabolites and inorganic ions within the several formative zones of the cambial region, and as they vary with the changing seasons.

The physical environment experienced by the cultured explant may be abnormal in relation to what the cambium in stem or root experiences and, plausibly, uses as signalling messages *in situ*. For example, although there is considerable information on the ascent of water through intact vascular plants (see Amin, 1982; Canny, 1995; Zimmermann *et al.*, 1995), it remains unclear if water-potential relations in cultured explants behave at all similarly (Savidge 1996). Phloem loading from mesophyll, water uptake by roots and transpiration from stomata are not part of the cultured system. Thus, it seems unlikely that even explants having cambium sandwiched between mature phloem and xylem (e.g., Sheriff, 1983; Savidge, 1983, 1993; Leitch and Savidge, 1995) when cultured could experience negative pressure potential in the wood, or strong osmotic forces in the phloem, as commonly occurs in the whole tree. If this is correct, then rapid physiological transmission of metabolites — such as axial translocation through phloem/xylem or radial translocation through rays as normally occurs *in arbor* — probably does not occur. Apoplastic/symplastic diffusion through the cultured tissues probably is the principal mechanism of distributing nutriment. To our knowledge, no definitive data has been collected in relation to these issues (Savidge 1993a, 1996).

Wounding during preparation of primary explants may elicit epigenetic or physiological changes in the cultured cells (Savidge, 1993a; Savidge and Leitch, 1995). Somaclonal variation is known to be common in plant cell cultures, but no information evidently exists on this in relation to the cambium. As noted in the section on plant growth regulators below, habituation can arise as cells are cultured. Induction of mutation, gene transpositions, and DNA methylation, have scarcely been investigated in relation to wood formation *in vitro*.

Many woody species have been observed to be intractable in culture (e.g. Gautheret, 1934; Brown, 1964; Jacquiot, 1964). In this light, more easily cultured genera such as *Populus* and *Salix* emerge as curiosities, rather than as model systems. Possibly, species easy to culture manifest only basic physiological control mechanisms, with more recalcitrant species exhibiting additional complexity which must be satisfied by having just the right balance of media constituents and culture conditions. Alternatively, it could be that 'easy' species are the more complex, being better able to adapt their metabolism rapidly to changes in substrate supply and other environmental conditions. The question of whether 'easy' species are physiologically the more simple or complex is important to answer if the knowledge base developing from *in vitro* experimentation is to have general applicability to the Tracheophyta. Comparative *in vitro* research among species is still very much needed to answer such questions, and the information will also have relevance to ecological biochemistry/physiology.

The 'disadvantages' attending culture investigations would be minor if there were not limitations in research funding, personnel and infrastructure. The advantages are major and immensely important for furthering knowledge on how trees regulate their secondary growth.

1.2 *Quantitative and qualitative challenges for* in vitro *research*

The vascular cambium was one of the first plant tissues to be cultured (Gautheret 1934), and many subsequent investigations dealt with one or more aspects of *in vitro* xylogenesis in explants of woody species (e.g., Ball, 1950; White, 1943, 1954; Jacquiot, 1966; Barnett 1978; Brown and Wodzicki, 1969; Dalessandro and Roberts 1971; Savidge 1983, 1993a). As referenced in Konar and Nagmani (1974), early cytodifferentiation investigations were reported by Vöchting in 1892, and the first attempts to culture isolated cells were made by Haberlandt (1902). Since then, tissue culture has been routinely used to study many aspects of cellular development (see Bajaj, 1986, 1989, 1991; de Fossard, 1976; Dodds, 1983; Evers *et al.*, 1988; Gautheret, 1934, 1959, 1983; Jacquiot, 1964, 1966; Reinert and White, 1956; Savidge, 1983, 1989; Street, 1977; White, 1954, 1967; Yeoman and Truman, 1982).

The problems remaining to be solved nevertheless are numerous. They can be considered quantitatively or qualitatively, in terms of how adjoining cells or tissues react to *in vitro* treatments. In addition, comparisons can of course be made between *in vitro* biochemistry, growth and developmental responses and those occurring *in vivo.*

On the qualitative side, the issues are no less numerous than the immense number of catalysts, metabolites, sub-cellular compartments and anatomical phenomena that occur in the cambial region. The fact that there are some 80 000 tree species on earth with each producing a more or less unique kind of wood presents an interesting challenge to differentiate between the general and the specific. For example, does the cambium of every species have the same, or even similar, nutriment requirements when grown *in vitro*? Identifying or devising media that will support growth of the cambium remains to be done for the vast majority of species.

When attempting to mimic *in arbor* cellular development under *in vitro* conditions, the intrinsic circumstances existing within tissues at the time of explanting vary with cambial age, position, and date of harvest, and the physiological status of the explant influences how its tissues respond when cultured (Leitch and Savidge, 1995; Savidge, 1993a, 1996). In the studies by McComb and Wroth (1986), an important discovery was the existence of temporal windows for growth during the year. Similar windows were also seen in culture experiments conducted by Leitch (1995), Leitch and Savidge (1995) and Savidge (1993a) using *Larix laricina*.

Some examples of intrinsic factors which are known to vary phenologically and, plausibly, epigenetically include sub-cellular organization (e.g., organelle number, extent of vacuolation, cell-wall chemistry), content and character of storage reserves and other intracellularly produced compounds, ratios and absolute molarities of endogenous metabolites (e.g., sugars, amino acids, vitamins), hormones and other transmissible molecules, pH and buffering capacity, as well as content of H_2O, O_2 and CO_2. Many of those factors vary qualitatively (i.e., present or not present), as well as quantitatively, depending on when and where an explant is taken from a tree, tissue age, genotype and phenology. Boulay (1989) suggested that obtaining chemical and biochemical information to determine the ontogenetic age of explants, subsequent to *in vitro* transfers, could lead to an understanding of processes linked with maturation. So far, concepts for distinguishing phenological stages and states in tissues appear to be relatively crude and lacking in clear definition; thus, it remains difficult to compare and contrast findings from one experiment to the next (Savidge 1993a, 1996; Leitch and Savidge, 1995).

In the few favoured woody species where there is a methodological foundation for doing *in vitro* research, there is a great deal of quantitative research still wanting. The priority challenge identified by us is to gain sufficient control and technological innovation in methodology to enable not merely achievement of similar magnitudes of growth and development as normally occur *in arbor* in mature stems, but to go beyond that to optimization of intrinsic conditions, such that the full potential of a genotype for growth and differentiation can be known. To our knowledge, no such optimization research has yet been attempted with the cambium.

Successful induction of wood formation is most readily achieved in cultures derived from cambial tissue younger than one year. Considering that the cambium in the stem of a healthy tree produces new wood and phloem for a hundred years, or more, the fact that cultured explants taken from stems more than a few years old grow little, if at all, on media supporting growth of younger tissue should be a clarion call for explaining the differences (Zajączkowski, 1973; Savidge and Wareing, 1981; Savidge, 1983). Moreover, the cambium explanted from branches generally is less vigorous than the same age of cambium taken from a tree's main axis (R.A. Savidge, unpublished data). When secondary growth in a cambial explant can be maintained, by sub-culturing or other means, such that it at least equals annual-ring formation as it would have occurred had the explant been left in the tree stem or branch, the claim that the metabolic requirements for vascular development are understood will have begun to be validated.

It is often desirable that natural variation be non-existent within explants of the source genotype(s) in order to minimize variation in phenotypic expression among replicates, and this can be readily accomplished through *in vitro* research. On the other hand, the imposition of genotypic uniformity can also inadvertently produce a qualitative bias. For example, particular genotypes may be favoured (and even labelled as 'model systems') after they are found to exhibit superior responsiveness *in vitro* in relation to whatever phenomenon happens to be under investigation. Selection for responsiveness obviously is convenient for progress, but the concomitant decision not to investigate genotypes which either respond weakly or that are intractable sidesteps the more difficult question of why some are less responsive than others. Gaining new knowledge about issues such as genetic diversity or changing epigenetic status, particularly in relation to variation in physiological control mechanisms, appears to be essential and promising in relation to achieving improvements in both biotechnology and forest management.

2. Methodological considerations

2.1 *Historic perspective*

Street and Henshaw (1966) provided a useful historical overview of the methods employed for plant tissue culture in general. A variety have been used to investigate the vascular cambium and xylogenesis (Gautheret, 1934, 1983; Jacquiot, 1964; Brown and Wodzicki, 1969; Zajączkowski, 1973; Sheriff, 1983; Savidge, 1983, 1993a; Leinhos and Savidge, 1993; Leitch *et al.*, 1999).

2.2 *The basal medium and extrinsic conditions*

Street and Henshaw (1966) considered that media devised by White (1943) and Heller (1953) were probably the most suitable for attempting the culture of new species or tissues, since both had already been found to support the growth of a wide range of tissues. There are many other formulations available (for example: Gautheret, 1934, 1959; Reinert and White, 1956; Murashige and Skoog, 1962; Gamborg *et al.*, 1968; Harvey and Grasham, 1969; Schenk and Hildebrant, 1972; Reilly and Washer, 1977; Lloyd and McCown, 1981; Allen *et al.*, 1988; Savidge, 1993a). Different media can yield contrasting growth or developmental responses for a given species, tissue composition, or age of tissue(s) explanted.

When a functional culture system (including selection of explant material, preparation and surface sterilization, explant dissection techniques and defined basal medium) is realized, then modifications in the chemistry of the basal medium enable detection of cellular responses dependent on particular constituents. For example, Savidge (1983) grew stem 'chips' (i.e., stem pieces containing cambium sandwiched between bark and wood) of *Pinus contorta* on the basal media of Murashige and Skoog (1962) and Reinert and White (1956). Chips grew strongly, in terms of callus production, on both media; however, they produced different types of callus (compact green and friable white, respectively). Within the chips, cambial cell-division activity and differentiation of cambial derivatives into tracheids occurred only on the Murashige and Skoog (1962) medium (see Savidge, 1983 for examples of the tracheids produced). Both media had auxin at the same concentration as the sole phytohormone additive; thus, the differences in calli and the distinctive cambial responses evidently were due to differences between other constituents in the two media. Because both media contain a substantial number of ingredients, deciphering which of those basal ingredients actually underlies the observed developmental differences is a major challenge for systematic investigation.

When the cambium is dissected from the stem as a thin, nearly transparent layer and explanted as a 'free' meristem, strong callus production occurs but little, usually no, new xylem is generated on media (such as Murashige and Skoog, 1962) supporting xylogenesis in stem chips (Savidge, 1983). Thus, the physical pressure experienced by the cambium between wood and bark could be a factor controlling the nature of vascular development (Savidge and Wareing, 1981; Savidge, 1985, 1996). Cambium has been grown under different osmotic conditions and physical constraints, and the evidence indicates that external pressure does have a role in determining formation and functioning of the cambium (Brown and Sax, 1962; Savidge and Wareing, 1981); however, much more research is needed to clarify the roles of external and internal pressure more precisely (Savidge, 1996). Some of the other external factors influencing cultured plant tissues include temperature, water, light, gas composition and partial pressures, gravitational and electromagnetic forces and acidity (Roberts, 1983). Obviously, the greater the control over all aspects of a culturing system, the greater will be the reliability of the data in aiding us to better understand cambial development and wood formation in trees.

In 1865 Knop (referenced in Liefert *et al.*, 1995) began to address the nutritional requirements of plants under controlled conditions. Street (1966a, 1966b) provided detailed historical perspective on the early work and its limitations. Pirt (1975) summarized the nutriment required for growth into four classes: (i) major elements C,

H, O and N; (ii) minor elements P, K, S and Mg; (iii) vitamins, hormones and other growth factors (e.g., amino acids) and (iv) trace elements.

In principal, when the objective is to use an *ex arbor* system to model *in arbor* growth, only those molecular constituents normally transmitted *in arbor* should have to be provided to the growing explant. Cellular metabolism should be able to produce all of the secondary metabolites and end products from the starting material. Street (1966a) provided an extensive review of nutrition and metabolism in plant cultures. The quality and quantity of inorganic nutriment needed for whole-tree growth has been well researched (Gauch, 1972; Ingestad, 1987; Dell and Robinson, 1993). Information has also been accumulated on organic molecules and ions moving in the axial phloem and xylem streams (Salisbury and Ross, 1991; Leifert *et al.*, 1995, and see chapters herein), but unloading to the cambium and how it is regulated continue to be speculated upon. Supplying nutrients to the explants in the correct molar ratios obviously is crucially important, but very little research to establish those ratios has yet been done with the cambium.

Williams (1993) described several aspects of *in vitro* mineral nutrition attempting to answer: (i) How much of the element is available to the explant?; (ii) is uptake dictated by a sink effect or by passive uptake? and (iii) how does availability of elements change over the culture period? Gelling agents were found to affect availability of media components to the explant in various ways: mineral content of the gelling agent itself (Williams, 1993), inhibitory effects of the gelling agent (MacRae and van Staden, 1990), buffering capacity of the gelling agent (Skirvin *et al.*, 1986; Williams *et al.*, 1990), binding of ions to the gelling agent, and restrictions on diffusive mobility of ions and molecules through the gelling agent (Williams, 1993).

Most plant cultures respond best on media having a pH between 5.0 and 7.0 (de Fossard, 1977; Islam *et al.*, 1980). It should be considered, however, that many tree species grow on acid forest soils below pH 4.0, while others do well on alkaline soils exceeding pH 8.0. Thus, although concentrations of inorganic ions tend to be reduced at both high and low acidity, intrinsic buffering capacity of the tissues of the cambial region could vary among species and be an important factor controlling culture responses.

De Fossard (1977) noted that pH influences the form and solubility of medium components (i.e., phosphate at pH 4.0 is $H_2PO_4^-$ while at pH 9.0 is HPO_4^{2-}), affects hydrogen and hydroxyl ions which in turn can affect membranes involved in ion transport and can be a problem in relation to the setting of gelling agents. A pH above 7.0 lowers availability of Fe and Mn; below pH 6.0 availability of P, Ca and Mg is lowered; and availability of most elements is limited below pH 5.0 (Minocha, 1987; Williams, 1993).

The processes of adding gelling agents and autoclaving (Murashige and Skoog, 1962; Williams *et al.*, 1990) and storing media (Williams, 1993) can change medium pH. Following explanting of biological material to the medium, pH changes until the explant reaches an equilibrium (Skirvin *et al.*, 1986). Williams *et al.* (1990) suggested that movement of ions out of an explant could occur over several days if the tissues were sub-cultured or the suspension-culture medium replaced frequently. Differential diffusion rates of ions through the medium and into the explant can also affect the balance and concentrations of ions in close proximity to the explant (Williams, 1993). Systematic testing of medium components, molarities and molar ratios is time-consuming but essential for progress in cambial tissue culture.

2.3 *Plant growth regulators (PGRs)*

Intrinsic regulation of cell-division activity in the cambium and of secondary phloem and xylem development has frequently been addressed in terms of phytohormone, or plant growth regulator (PGR) control (Savidge and Wareing, 1981; Little and Savidge, 1987; Little and Pharis, 1995). A broader perspective on plant growth regulation has also been advocated (Savidge, 1996). PGRs believed to regulate cambial growth include auxins (Leitch and Savidge, 1995; Savidge, 1993a), gibberellins (Digby and Wareing, 1966; Zhong and Savidge, 1995a,b), abscisic acid (Little *et al.*, 1978; Little and Wareing, 1981), ethylene (Barker, 1979; Lev-Yadun and Aloni, 1995; Telewski and Jaffe, 1986), cytokinins (Aloni *et al.*, 1990; Zakrzewski, 1983), brassinosteroids (Bellincampi and Morpurgo, 1988; Iwasaki and Shibaoka, 1991), jasmonates (Miyamoto *et al.*, 1997; Sembdner and Parthier, 1993) and oligosaccharins (Aldington and Fry, 1993; Côté and Hahn, 1994).

Auxins, gibberellins, abscisic acid and cytokinins have been identified in the cambial region of some conifers (Savidge and Wareing, 1981; Little and Savidge, 1987; Little and Pharis, 1995) while auxins, gibberellins and abscisic acid have been identified in the cambial region of some hardwood species (Savidge and Wareing, 1981; Savidge, 1990; Hasan *et al.*, 1994; Ridoutt *et al.*, 1996; Wang *et al.*, 1999). The ethylene precursor 1-aminocyclopropane-1-carboxylic acid (ACC) was identified in the cambial region of *Pinus contorta* in relation to compression wood formation (Savidge *et al.*, 1983), and it was observed that both auxin and ACC had to be provided to *in vitro* explants in order for ethylene to be generated by incubated cambia of *Abies balsamea* (Savidge, 1988).

Little and Pharis (1995) reported that characterizations of endogenous cytokinins, brassinosteroids and jasmonates in the cambial region of woody dicotyledons are still wanting. Brassinolide has been identified in cambial extracts of *Pinus sylvestris* (Kim *et al.*, 1990), and reviews by Sasse (1997) and Creelman and Mullet (1997) detailed current knowledge on this relatively new plant hormone.

PGRs have been described as necessary for cambial activity to occur (Smith, 1992). However, Minocha (1987) suggested that the effect of including PGRs in tissue culture medium is important only during the first few hours (or days) of incubation, and beyond this the need for them is minimal, and that they may also be inhibitory. Street (1966a) emphasized that it was not unusual for cultures to have no PGR requirement, and that such systems offered opportunity for better understanding of the synthesis and involvement in metabolism of the PGRs. An important discovery in angiosperm research was the need to transfer callus to a medium with and without PGRs for induction of shoot and root production (de Fossard, 1974). Similar findings have been reported in propagation studies (see Bajaj, 1986, 1989, 1991; de Fossard, 1974; de Fossard *et al.*, 1974; Pryor and Willing, 1963; Sommer and Wetzstein, 1984; Street, 1977; Winton, 1970, 1971).

In callus cultures derived from the conifer cambium, it has been repeatedly noted that even a small amount of cytokinin in the medium strongly inhibits cell division and parenchyma proliferation (R.A. Savidge, unpublished data). Auxin alone promotes growth and differentiation. On the other hand, addition of an auxin–cytokinin combination to the medium was essential for the callus to metabolize *E*-coniferin (Savidge, 1989).

2.4 *Culturing cambial cells*

It is currently possible to culture cells, tissues and organs of plants for prolonged periods; however, the cambium of higher plants is one tissue that has been notoriously difficult to culture due to complex requirements (Gamborg *et al.*, 1968; Williams, 1993). Certain species are more readily cultured than others (Brown, 1964).

Protoplast culture. Protoplasts have been isolated from cotyledons, roots, needles, callus and cell suspensions, pollen, leaves, shoot cultures and cotyledons of many woody plants (Ahuja, 1984; Burger and Hackett, 1981; David *et al.*, 1982; Duhoux, 1980; Faye and David, 1983; Hakman and Von Arnold, 1983; Hurwitz and Agrios, 1984; Kirby and David, 1988; Kirby and Cheng, 1979; Smith and McCown, 1982/3).

Leinhos and Savidge (1993) isolated protoplasts from cambial zone and developing xylem of *Pinus* spp. and provided data on *E*-coniferin content. Protoplasts of ray and fusiform cells were readily distinguishable on the bases of size and structure. Protoplast isolation and separation could enhance biochemistry and physiology understanding of cellular differentiation. The method could also serve as a means to investigate cambial commitment and competence for totipotent expression.

David (1987) suggested that protoplast systems may permit nuclear and cytoplasmic reprogramming facilitating cellular rejuvenation allowing totipotent expression. However, very few totipotency successes, in terms of regenerating whole trees from protoplasts, have yet been reported (Kobayashi *et al.*, 1983; Sticklen *et al.*, 1985; Vardi *et al.*, 1975).

Cell suspension culture. *Acer pseudoplatanus* suspension cultures, originating from an explant of 'cambium', have been used extensively for cell studies and shown to subculture repeatedly for long periods. Both free cells and cell aggregates of batch and continuous suspension cultures have been described (Davey and Street, 1971; King *et al.*, 1973; Roberts and Northcote, 1970; Sutton-Jones and Street, 1968; Wilson, 1980; Withers, 1976). From one culture passage to the next, similar patterns of growth and cell yield occur, allowing various tests and comparisons to be made (Street, 1977).

Liquid or suspension culture methods have been described by several authors (Steward *et al.*, 1958; Parham and Kaustinen, 1977; Ramsden and Northcote, 1987; Reynolds, 1987; Leitch *et al.*, 1999). The methods have been used not only to investigate xylogenesis but also to study development of specific features in wood, such as production of tyloses in the mature sapwood region adjacent to the vascular cambium of *Eucalyptus globulus* stem as shown in *Figure 1(h)*. Leitch *et al.* (1999) observed tyloses to form in 80% of vessel elements in sapwood when cultured on a liquid medium containing completely defined nutriment without hormones. NAA in the medium had a slight, but significant, inhibitory effect on tylosis formation. This response was not repeatable when *E. globulus* stem chips were cultured on a static gelled medium containing identical components.

Dougall (1980) described batch suspension cultures as advantageous because subcultures can occur at 7 to 21-day intervals and, following a number of subcultures, cells can be grown exponentially without any lag in the batch cultures. Dougall (1980) further described chemostat and turbidostat cultures, relating the former to growth rate limitations generated by the concentration of nutrients in the medium and the latter to growth rate limitations due to the intrinsic capabilities of the cell.

Tissue culture. Various approaches for culturing the cambium, or the cambium's derivative cells, as tissues were reviewed by Savidge (1983, 1993a). The method of bark peeling during active cambial growth, followed by dissection and explanting of nearly

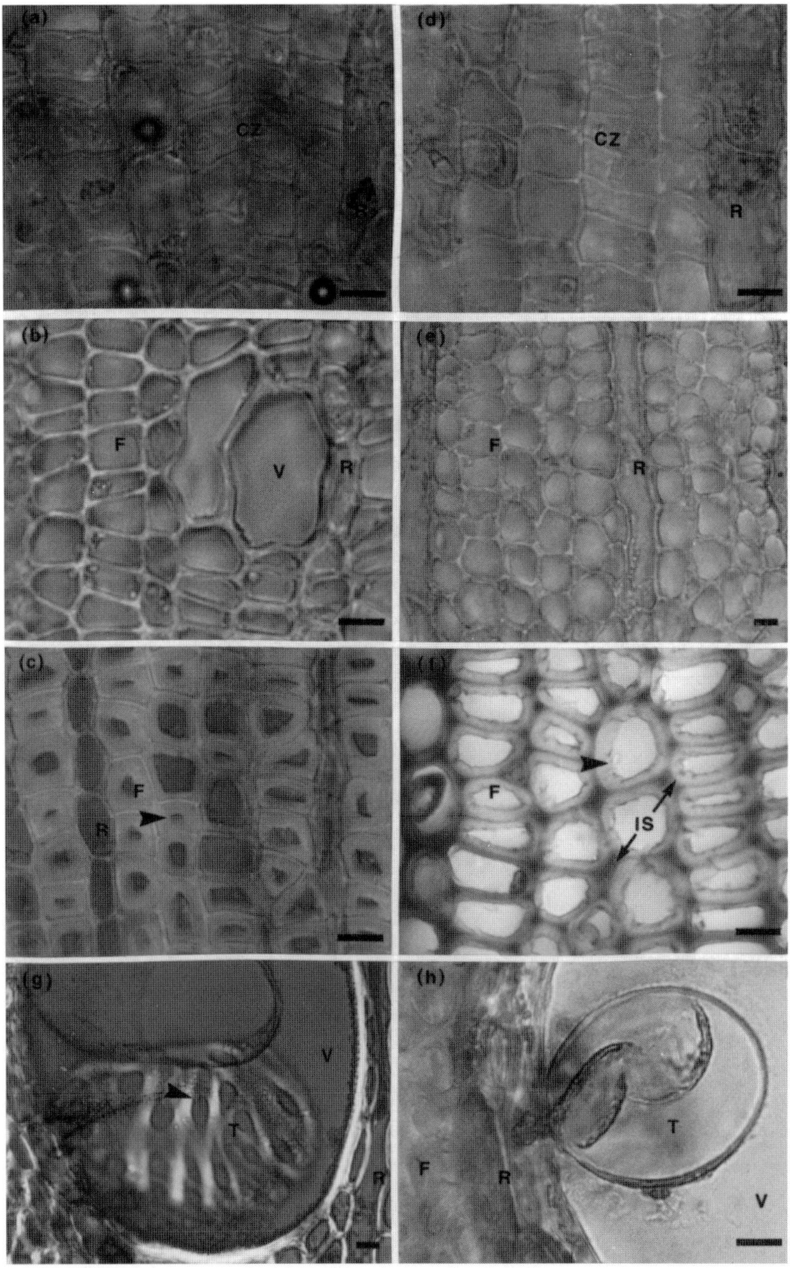

transparent thin layers or strips of cambial zone from the exposed inner surface of the bark, is conveniently used. The radial thickness of the cambial zone is rarely more than 50 μm; thus, fine needle-tipped forceps are best for this work. Living cambial derivatives from the surface of the xylem can also be readily lifted and explanted (Gautheret, 1934, 1959, 1983; White, 1943, 1954, 1967; Jacquiot, 1964; Brown, 1964). Static semi-solid media are useful for culturing strips of the vascular cambium or its derivatives (Savidge, 1983, 1988, 1989, 1993a; Leitch, 1995; Leitch and Savidge, 1995). Dipping an end of a strip of cambial cells into liquid media and allowing the medium to migrate upwards or downwards, as in paper chromatography, has also been tried but without much success in terms of inducing macro growth responses (R.A. Savidge, unpublished).

Organ culture. 'Organ' is used here in the sense that more than a single tissue is contained in the explant. Brown and Wodzicki (1969) developed a procedure for supplying stem segments apically with sterilized liquid media through tubing. That method was investigated extensively by Zajączkowski (1973) and found to yield a differential cambial response, depending on stem segment age. The differential age response was confirmed in other species using the same procedure (R.A. Savidge, unpublished data). The Brown and Wodzicki (1969) method is most readily used with young (i.e., small diameter) stem segments. It requires specialized glassware, considerable benchspace and ancillary equipment, and it can be troublesome in terms of maintaining the medium aseptic. On the other hand, the medium can be delivered under different pressures (i.e., gravitational heads) and, in addition to the cambium, both shoot and root growth and development can be investigated using this approach (R.A. Savidge, unpublished data). Sheriff (1983) described an alternative *in vitro* approach for culturing small diameter segments that evidently was quite successful.

It is difficult to investigate the cambium *in vitro* within complete cylinders of large diameter stems or roots. However, small pieces, or 'chips', of dormant cambium sandwiched between bark and wood can be explanted from stems. In chips, in contrast to explanted cambial 'strips', it was found that ray cells did not proliferate to overwhelm the cambium's internal structure with callus (Savidge, 1983, 1993a). Thus, the structure of the fusiform cambial cells was maintained, provided auxin was included in the medium (Savidge and Wareing, 1981). Subsequent research resulted in development of a medium supporting periclinal division and production of almost normal appearing xylem within stem chips of *P. contorta* (Savidge, 1983). Cambial cells underwent periclinal divisions to produce radial files of secondary-walled, lignified, bordered-pitted and autolysed earlywood tracheids (Savidge, 1983, 1993a). Remarkably, traumatic resin canals, which usually form in response to wounding of intact conifers stems, did not differentiate when cambium was induced to grow and produce new xylem within chips (Savidge, 1983, 1993a).

The chip organ-culture method offers hope for progress with cambia dissected from large diameter stems or roots. A limitation of the method is that it has not yet been found possible to sustain ongoing cambial growth within the chip beyond that occurring during the first culture period. The cambium evidently enters a type of

Figure 1. *Hand sections (unstained) displaying* in vitro *responses from young apical* (a–c) *and 3-year-old main stem* (d–h) *regions of* Eucalyptus globulus *trees cultured on a defined basal medium with PGRs.* (a), *Cell division activity on 0.5 mg l^{-1} NAA + 2.5 mg l^{-1} GA_3;* (b) *Induction of varied cell types displaying secondary walls on 1.0 mg l^{-1} IAA + 1.0 mg l^{-1} GA_4;* (c) *Induction of tension wood (arrowhead) formation on 1.0 mg l^{-1} IAA + 1.0 mg l^{-1} GA_4;* (d) *1.0 mg l^{-1} NAA + 0.1 mg l^{-1} L-methionine);* (e) *Induction of varied cell types displaying secondary walls on 2.5 mg l^{-1} NAA + 0.5 mg l^{-1} GA_3);* (f) *Induction of tension wood (arrowhead) formation on 0.5 mg l^{-1} NAA + 2.5 mg l^{-1} GA_3;* (g) *Crossed polarizing filters image of pitted and secondary walled tylosis formation (arrowhead) in static culture using main-stem material in static liquid culture (2.0 mg l^{-1} GA_3 + 1.0 mg l^{-1} 2,4-D + 1.0 mg l^{-1} l-methionine);* (h) *Liquid culture (1.0 mg L^{-1} NAA). CZ = cambial zone, F = fibre, R = ray, V = vessel element, T = tylosis, IS = intercellular spaces. Scale bars are 10.0 μm.*

dormancy within the first few weeks of its *in vitro* growth, possibly because of stored endogenous reserves becoming exhausted.

Formation of bordered-pits requires marked developmental changes in the primary walls of cambial derivatives. Leitch and Savidge (1995) used the chip method to study the regulation of bordered-pit formation in *Larix laricina*. Tangential wall pitting is not a feature of the normal earlywood of *L. laricina*, but bordered pits were produced on both radial and tangential walls *in vitro* (Leitch and Savidge, 1995; Savidge and Leitch, 1995). Tangential wall pitting, pit diameters and frequencies all varied depending on the auxin level added to the basal medium of Savidge (1993a).

Zhong and Savidge (1995a) used a chip-culture technique with *Fraxinus americana* cambial explants and produced xylogenic responses that varied according to the IAA and GA_3 content of the basal medium. A high fibre to vessel element ratio was favoured by a high GA_3 to IAA ratio, and a wood with enhanced content of vessel elements developed in response to a low GA_3 to IAA ratio.

Leitch (1999) developed techniques based on those of Savidge (1993a) for *E. globulus* tissues harvested from the apical (less than one-year-old) and lower main stem (several-year-old) regions (*Figure 1 a–g*). In agreement with what was observed by Leitch (1995), Leitch and Savidge (1995) and Savidge (1993a), cambial cells underwent periclinal and anticlinal divisions to produce organized radial files of secondary-walled, lignified, bordered-pitted and autolysed earlywood xylem cells, particularly in the younger *E. globulus* explants. Vessel elements (*Figure 1b*) and gelatinous tension wood fibres (*Figures 1c and 1f*) were also produced. The *E. globulus* results are in agreement with the concept of young cambium responding and having distinct requirements compared to older cambia (see Little and Savidge, 1987; Zhong and Savidge, 1995a). Bennet and McComb (1982), Creswell and Nitsch (1975), LeRoux and van Staden (1991), McComb and Wroth (1986), Oka *et al.* (1982), Paton *et al.* (1970) and Trindade *et al.* (1990) have provided additional information on culture of eucalypts.

Many '*in vitro*' experiments have involved applications of lanolin pastes containing hormones to the apical cut surface of stem segments cultured in non-sterilized water (e.g., Digby and Wareing, 1966; Kurczynska and Hejnowicz, 1991; Savidge 1983, 1994; Savidge and Wareing, 1981). Interpreting the precise regulatory roles of PGRs supplied in lanolin pastes is difficult because the method does not account for the probability that metabolites stored in the segments, at the time of collecting them from the tree, could vary in concentration and/or mobility with the changing seasons or with age. Non-sterilized lanolin enriched with sugar and other nutrients does not become overtly contaminated with either bacteria or fungi, despite weeks of exposure to air (R.A. Savidge, unpublished data). To circumvent the possibility that endogenous factors become limiting when stem segments are treated with PGRs, Zhong and Savidge (1995b) used lanolin containing the constituents of the medium of Savidge (1993a), but at concentrations one tenth those normally used in tissue culture, together with different ratios of auxin to gibberellic acid to manipulate xylogenesis in *F. americana* stem segments.

Callus: organized versus friable. Calli have been found to originate from the cambium (Jacquiot, 1966; White, 1967; Savidge, 1983; Savidge, 1993a), phloem (White, 1967), rays (Gautheret, 1959) and other parenchymatous cells (White, 1967). Early studies revealed that proliferation of callus on a nutrient medium devoid of any PGRs was possible (see Gautheret, 1983). Ball (1950) described callus growing from the apical ends of stem pieces to have originated from the vascular cambium, and that from the basal ends almost entirely from cortical cells. Barnett and Asante (chapter herein) comment further about the origins of callus.

The presence of visible callus on the surface of a stem segment or cultured explant is an indicator that cambial growth may be occurring. However, callus formation by itself does not permit the conclusion that xylogenesis is occurring (Savidge, 1983, 1993a; Leitch and Savidge, 1995).

Several types of calli growing from primary explants of the cambial region have been reported (Ball, 1950; Barnett, 1978; de Fossard *et al.*, 1974; Harvey and Grasham, 1969; Jacquiot, 1966; McCown *et al.*, 1988; Niedergang-Kamien and Skoog, 1956; Reinert and White, 1956; Sakai *et al.*, 1994; Savidge, 1993a; Tsutsumi *et al.*, 1994; Wetmore and Rier, 1963; White, 1967). Granular friable calli consisting mainly of parenchyma cells are reported frequently in the literature (Ball, 1950; Grant and Fuller, 1968; Durzan *et al.*, 1973; Barnett, 1978; Gautheret, 1983; Savidge, 1983, 1993a; Leitch, 1995) and have been described as containing normal cell organelles and thin-walls with plasmodesmata (Barnett, 1978). Tannins (Ball, 1950; Barnett, 1978) and other metabolites have been observed in such callus (Durzan *et al.*, 1973). Savidge (1983, 1989) noted that explanted strips of pine cambium grew well through proliferation of the ray cell component, but that they did not produce tracheids. *E*-Coniferin, normally associated with cambial activity within the tree (see chapters by Förster *et al.* and Steeves and Savidge herein), was not found in such callus, indicating that its biosynthesis was not occurring (Savidge 1989).

Compact striated calli were noted to be associated with, and predictably diagnostic of, xylogenesis occurring within chip cultures of *Larix laricina* (Savidge, 1993a; Leitch and Savidge, 1995). Tracheary elements were also produced within the compact striated callus produced around the margins of those chips (Savidge, 1983). Leitch (1995) described such calli as containing tracheid-like cells interior to a cambium-like layer that arose *de novo* within the callus. Centrifugal to that cambium were phloem and epidermal-like layers. Similar descriptions of callus differentiation have been noted by others (Ball, 1950; White, 1967; Barnett, 1978).

Callus-derived cambial cells (Geissbühler and Skoog, 1957; Brown and Sax, 1962; Aitchison, *et al.*, 1977) and tracheids displaying bordered and simple pits on all walls have also been described (Barnett, 1978; Savidge, 1983; Leitch, 1995). White (1967) considered that proliferation of parenchyma cells into cambium involves dedifferentiation followed by re-differentiation of the cells to produce a meristematic zone. Barnett (1978) suggested periclinal divisions occurred predominantly in callus because radial walls in the developing cambium were thicker than tangential walls.

Long-term cultures of calli have been described for *Populus tremuloides* (Mathes, 1964), *P. alba* (Tsutsumi *et al.*, 1994), *Sequoia sempervirens* (Ball, 1950), *Fraxinus americana*, *Syringa vulgaris*, *Ligustrum vulgare*, *Salix purpurea*, *Parthenocissus tricuspidata* and *Helianthus tuberosus* (Wetmore and Rier, 1963). Jacquiot (1966) described several angiosperms that had been sub-cultured for several years. Callus cells once produced appear to have greater competence for ongoing growth than cells contained internally within tissues of the cambial region.

Etienne *et al.* (1997) emphasized the importance of calcium in long term proliferation of callus. In *Hevea brasiliensis*, calcium has been demonstrated to control many physiological processes at the cellular level in callus cultures (Cheung, 1980).

3. Concluding remarks

Culture techniques enable the physiology, biochemistry and biophysics of secondary growth to be investigated *in vitro*. This avenue of research can facilitate increased understanding of cambial cell division and differentiation processes in relation to both phloem and wood formation. Modifying culture medium components and/or environmental conditions to elicit specific responses will increase our understanding of how both wood quality and quantity can be manipulated. Plant cell, tissue and organ cultures permit studies into diverse aspects of secondary growth, including cambium formation, division of fusiform cambial cells, differentiation of varied cell types, control of bordered-pit development, induction of reaction-wood formation, and tylosis development. Current transgenic experiments with woody plants require lengthy tests and field trials in order to decide whether a gene or biochemical process is specific to a particular wood trait. In future, culture of stem pieces *in vitro* will accelerate progress.

References

Ahuja, M.R. (1984) Protoplast research in woody plants. *Silvae Genet.* **33**: 32–37.

Aldington, S. and Fry, S. (1993) Oligosaccharins. *Adv. Bot. Res.* **19**: 1–101.

Allen, E.A., Blenis, P.V. and Hiratsuka, Y. (1988) Axenic culture of *Endocronartium harknessii*. *Mycologia* **80**: 120–123.

Aloni, R., Baum, S.F. and Peterson, C.A. (1990a) The role of cytokinin in sieve tube regeneration and callose production in wounded *Coleus* internodes. *Plant Physiol.* **93**: 982–989.

Amin, M. (1982) Ascent of sap in plants by means of electrical double layers. *J. Biol. Phys.* **10**: 103–109.

Bajaj, Y.P.S. (1986) *Trees I: Biotechnology in Agriculture and Forestry 1*, Vol. 1. Springer-Verlag, Berlin, Heidelberg.

Bajaj, Y.P.S. (1989) *Trees II: Biotechnology in Agriculture and Forestry*, Vol. 5. Springer-Verlag, Berlin, Heidelberg.

Bajaj, Y.P.S. (1991) *Trees III: Biotechnology in Agriculture and Forestry 1*, Vol. 16. Springer-Verlag, Berlin, Heidelberg.

Ball, E. (1950) Differentiation in a callus culture of *Sequoia sempervirens*. *Growth* **14**: 295–325.

Barker, J.E. (1979) Growth and wood properties of *Pinus radiata* in relation to applied ethylene. *NZ. J. For. Sci.* **9**: 15–19.

Barnett, J.R. (1978) Fine structure of parenchymatous and differentiated *Pinus radiata* callus. *Ann. Bot.* **42**: 367–373.

Bellincampi, D. and Morpurgo, G. (1988) Stimulation of growth in *Daucas carota* L. cell cultures by brassinosteroids. *Plant Sci.* **54**: 153–156.

Bennett, I.J. and McComb, J.A. (1982) Propagation of Jarrah (*Eucalyptus marginata*) by organ and tissue culture. *Aust. For. Res.* **12**: 121–127.

Boulay, M. (1989) Redwood (*Sequoia sempervirens*). In: *Trees II: Biotechnology in Agriculture and Forestry*, Vol. 5. (ed. Bajaj, Y.P.S.), Springer-Verlag, Berlin, Heidelberg, pp. 549–573.

Brown, C.L. (1964) The influence of external pressure on the differentiation of cells and tissues cultured *in vitro*. In: *The Formation of Wood in Forest Trees* (ed. Zimmermann, M.H.), Academic Press, New York, pp. 389–404.

Brown, C.L. and Sax, K. (1962) The influence of pressure on the differentiation of secondary tissues. *Am. J. Bot.* **49**: 683–691.

Brown, C.L. and Wodzicki, T.J. (1969) A simple technique for investigating cambial activity and the differentiation of cambial derivatives. *Forest Sci.* **15**: 27–29.

Burger, D.W. and Hackett, W.P. (1981) Protoplast culture of several citrus trees. *Hort. Science* **16**: 417.

Canny, M.J. (1995) A new theory for the ascent of sap — cohesion supported by tissue pressure. *Ann. Bot.* **75**: 343–357.

Cheung, W.Y. (1980) Calmodulin plays a pivotal role in cellular regulation. *Science* **207**: 19–27.

Côté, F. and Hahn, M. (1994) Oligosaccharins: structures and signal transduction. *Plant Mol. Biol.* **26**: 1379–1411.

Creelman, R.A. and Mullet, J.E. (1997) Oligosaccharins, brassinolides, and jasmonates: Non-traditional regulators of plant growth, development, and gene expression. *Plant Cell* **9**: 1211–1223.

Cresswell, R. and Nitsch, C. (1975) Organ culture of *Eucalyptus grandis* L. *Planta* **125**: 87–90.

Dalessandro, G. and Roberts, L.W. (1971) Induction of xylogenesis in pith parenchyma explants of *Lactuca*. *Am. J. Bot.* **58**: 378–385.

Davey, M.R. and Street, H.E. (1971) Studies on the growth in culture of plant cells. IX. Additional features of the fine structure of *Acer pseudoplatanus* L. cells cultured in suspension. *J. Exp. Bot.* **22**: 90–94.

David, A. (1987) Conifer protoplasts. In: *Cell and Tissue Culture in Forestry*, Vol. 2, *Specific Principles and Methods: Growth and Developments* (eds. Bonga, J.M. and Durzan, D.J.), Martinus Nijhoff Publ., Dordrecht, The Netherlands, pp. 2–15.

David, H., David, A. and Mateille, T. (1982) Evaluation of parameters affecting the yield, viability and cell division of *Pinus pinaster* protoplasts. *Physiol. Plant.* **56**: 108–113.

de Fossard, R.A. (1974) Tissue culture of Eucalypts. *Aust. For.* **37**: 43–54.

de Fossard, R.A. (1976) *Tissue Culture for Plant Propagators*. University of New England Printery, Armidale, NSW.

de Fossard, R.A. (1977) *The Horizons of Tissue Culture Propagation*. Proceedings of seminar by Dr. R.A. de Fossard for N.S.W. Assoc. Nurserymen Ltd., Sydney. pp. 1–132.

de Fossard, R.A., Nitsch, C., Cresswell, R.J. and Lee, E.L.M. (1974) Tissue and organ culture of eucalypts. *NZ J. For. Sci.* **4**: 267–278.

Dell, B. and Robinson, J.M. (1993) Symptoms of mineral nutrient deficiencies and the nutrient concentration ranges in seedlings of *Eucalyptus maculata* Hook. *Plant and Soil* **155/156**: 255–261.

Digby, J. and Wareing, P.F. (1966) The effect of applied growth hormones on cambial division and the differentiation of the cambial derivatives. *Ann. Bot.* **30**: 539–549.

Dodds, J.H. (1983) *Tissue Culture of Trees*. Croom Helm Ltd., Kent, London.

Dougall, D.K. (1980) The use of tissue cultures in studies of metabolism. In: *The Biochemistry of Plants* (ed. Davies, D.D.), Academic Press, New York, pp. 627–642.

Duhoux, E. (1980) Protoplast isolation of Gymnosperm pollen. *Z. Pflanzenphysiol.* **99**: 207–214.

Durzan, D.J., Chafe, S.C. and Lopushanski, S.M. (1973) Effects of environmental changes on sugars, tannins, and organized growth in cell suspension cultures of white spruce. *Planta* **113**: 241–249.

Etienne, H., Lartaud, M., Carron, M.P. and Michaux-Ferriere, N. (1997) Use of calcium to optimize long-term proliferation of friable embryogenic calluses and plant regeneration in *Havea brasiliensis* (Müll. Arg.). *J. Exp. Bot.* **48**: 129–137.

Evers, P.W., Dinkers, J., Prat, A. and Vermeer, E. (1988) *Micropropagation of Forest Trees Through Tissue Culture*. Centre for Agricultural Publishing and Documentation (Pudoc), Wageningen.

Faye, M. and David, A. (1983) Isolation and culture of Gymnosperm root protoplasts (*Pinus pinaster*). *Physiol. Plant.* **59**: 359–362.

Gamborg, O.L., Miller, R.A. and Ojima, K. (1968) Nutrient requirements of suspension cultures of soybean root cells. *Exp. Cell Res.* **50**: 151–158.

Gauch, H.G. (1972) *Inorganic Plant Nutrition*. Dowden, Hutchinson & Ross Inc., Stroudsburg, PA.

Gautheret, R.J. (1934) Culture du tissu cambial. *C. R. Ac. Sc.* **198**: 2195–2196.

Gautheret, R.J. (1959) *La Culture des Tissus Végétaux. Techniques et Réalisations*. Masson, Paris.

Gautheret, R.J. (1983) Plant tissue culture: A history. *Bot. Mag. Tokyo* **96**: 393–410.

Geissbühler, H. and Skoog, F. (1957) Comments on the application of plant tissue cultivation to propagation of forest trees. *TAPPI* **40**: 257–262.

Grant, M.E. and Fuller, K.W. (1968) Tissue culture of root cells of *Vicia faba*. *J. Exp. Bot.* **19**: 667–680.

Grey, D., Stepan-Sarkissian, G. and Fowler, M.W. (1987) Biochemistry of forest tree species in culture. In: *Cell and Tissue Culture in Forestry; Vol. 2. Specific Principles and Methods: Growth and Development* (eds. Bonga, J.M. and Durzan, D.J.), Martinus Nijhoff, London, pp. 31–60.

Haberlandt, G.F.J. (1902) *Physiologische Pflanzenanatomie*, translated from the 4th German (ed. M. Drummond), 1965 reprint edition, Today & Tomorrow's Book Agency, New Delhi.

Hakman, I.C. and Von Arnold, S. (1983) Isolation and growth of protoplasts from cell suspensions of *Pinus contorta* Doug. ex Loud. *Plant Cell Rep.* **2**: 92–94.

Hanover, J.W. and Keathley, D.E., editors (1987) *Genetic Manipulation of Woody Plants*. Basic Life Sciences, Volume 44, Plenum Publ., New York.

Harvey, A.E. and Grasham, J.L. (1969) Procedures and media for obtaining tissue cultures of 12 conifer species. *Can. J. Bot.* **47**: 547–549.

Hasan, O., Ridoutt, B.G., Ross, J.J., Davies, N.W. and Reid, J.B. (1994) Identification and quantification of endogenous gibberellins in apical buds and the cambial region of *Eucalyptus*. *Physiol. Plant.* **90**: 475–480.

Heller, R. (1953) Recherches sur la nutrition minéral des tissus végétaux cultivés *in vitro*. *Ann. Sci. Nat. Bot. Biol. Veg. Ser.* **11**: 1.

Hurwitz, C.D. and Agrios, G.N. (1984) Isolation and culture of protoplasts from apple callus and cell suspension cultures. *J. Am. Soc. Hort. Sci.* **109**: 348–350.

Ingestad, T. (1987) New concepts on soil fertility and plant nutrition as illustrated by forest trees and stands. *Geoderma* **40**: 237–252.

Islam, A.K.M.S., Edwards, D.G. and Asher, C.J. (1980) pH optima for crop growth – results of a flowering solution culture experiment with six species. *Plant and Soil* **54**: 339 – 357.

Iwasaki, T. and Shibaoka, H. (1991) Brassinosteroids act as regulators of tracheary-element differentiation in isolated *Zinnia* mesophyll cells. *Plant Cell Physiol.* **32**: 1007–1014.

Jacquiot, C. (1964) Application de la technique de culture des tissus végétaux l'étude de quelques problèmes de la physiologie du l'arbre. *Ann. Sci. For.* **21**: 317–473.

Jacquiot, C. (1966) Plant tissues and excised organs culture and their significance in forest research. *J. Inst. Wood Sci.* **16**: 22–34.

Kim, S.-K., Abe, H., Little, C.H.A. and Pharis, R.P. (1990) Identification of two brassinosteroids from the cambial region of Scots pine (*Pinus silvestris*) by gas chromatography-mass spectrometry, after detection using a dwarf rice lamina inclination bioassay. *Plant Physiol.* **94**: 1709–1713.

King, P.J., Mansfield, K.J. and Street, H.E. (1973) Control of growth and cell division in plant cell suspension cultures. *Can. J. Bot.* **51**: 1807–1823.

Kirby, E.G. and Cheng, T. (1979) Colony formation from protoplasts derived from Douglas fir cotyledons. *Plant Sci. Lett.* **14**: 145–154.

Kirby, E.G. and David, A. (1988) Use of protoplasts and cell cultures for physiological and genetic studies of conifers. In: *Genetic Manipulation of Woody Plants* (eds. Hanover, J.W. and Keathley, D.E.), Plenum Press, New York and London, pp. 185–197.

Kobayashi, S., Uchimiy, H. and Ikeda, I. (1983) Plant regeneration from *Trovita orange* protoplasts. *Jpn. J. Breed.* **33**: 119–122.

Konar, R.N. and Nagmani, R. (1974) Tissue culture as a method for vegetative propagation of forest trees. *NZ J. For. Sci.* **4**: 279–290.

Kurczynska, E.U. and Hejnowicz, Z. (1991) Differentiation of circular vessels in isolated segments of *Fraxinus excelsior*. *Physiol. Plant.* **83**: 275–280.

Leifert, C., Murphy, K.P. and Lumsden, P.J. (1995) Mineral and carbohydrate nutrition of plant cell and tissue cultures. *Crit. Rev. Plant Sci.* **14**: 83–109.

LeRoux, J.J. and van Staden, J. (1991) Micropropagation and tissue culture of Eucalypts – a review. *Tree Physiol.* **9**: 435–477.

Leinhos, V. and Savidge, R.A. (1993) Isolation of protoplasts from developing xylem of *Pinus banksiana* and *Pinus strobus*. *Can. J. For. Res.* **23**: 343–348.

Leitch, M.A. (1995) *Bordered-pit Development in* Larix laricina *(Du Roi) K. Koch*. M.Sc.F. Thesis, Univ. New Brunswick, New Brunswick, Canada.

Leitch, M.A. (1999) The development of tissue caltal techniques to study wood formation in *Eucalyptus globular* Labill. PhD. Thesis, University of Melbourne, Victoria, Australia.

Leitch, M.A. and Savidge, R.A. (1995) Evidence for auxin regulation of bordered-pit positioning during tracheid differentiation in *Larix laricina* (Du Roi) K. Koch. *IAWA J.* **16**: 289–297.

Leitch, M.A., Savidge, R.A., Downes, G.M. and Hudson, I.L. (1999) Induction of tyloses in *Eucalyptus globulus* 'chips'. *IAWA J.* **20**: 193–210.

Lev-Yadun, S. and Aloni, R. (1995) Differentiation of the ray system in woody plants. *Bot. Rev.* **61**: 46–84.

Little, C.H.A. and Pharis, R.P. (1995) Hormonal control of radial and longitudinal growth in the tree stem. In: *Plant Stems: Physiology and Functional Morphology* (ed. B.L. Gartner), Academic Press, San Diego, pp. 281–319.

Little, C.H.A. and Savidge, R.A. (1987) The role of plant growth regulators in forest tree cambial growth. *Plant Growth Reg.* **6**: 137–169.

Little, C.H.A. and Wareing, P.F. (1981) Control of cambial activity and dormancy in *Picea sitchensis* by indole-3-ylacetic and abscisic acids. *Can. J. Bot.* **59**: 1480–1493.

Lloyd, G. and McCown, B. (1981) Commercially-feasible micropropagation of mountain laurel, *Kalmia latifolia*, by use of shoot tip culture. *Int. Plant Prop. Soc. Proc.* **30**: 421–427.

MacRae, S. and van Staden, J. (1990) *In vitro* culture of *Eucalyptus grandis*: effect of gelling agents on propagation. *Plant Physiol.* **137**: 249–251.

Mathes, M.C. (1964) The *in vitro* formation of plantlets from isolated aspen tissues. *Phyton* **21**: 137–141.

McComb, J.A. and Wroth, M. (1986) Vegetative propagation of *Eucalyptus resinifera* and *Eucalyptus maculata* using coppice cuttings and micropropagation. *Aust. For. Res.* **16**: 231–242.

Minocha, S.C. (1987) Plant growth regulators and morphogenesis in cell and tissue culture of forest trees. In: *Cell and Tissue Culture in Forestry, Vol. 1: General Principles and Biotechnology* (eds. Bonga, J.M. and Durzan, D.J.), Martinus Nijhoff, London, pp. 50–66.

Miyamoto, K., Oka, M. and Ueda, J. (1997) Update on the possible mode of action of the jasmonates: Focus on the metabolism of cell wall polysaccharides in relation to growth and development. *Physiol. Plant.* **100**: 631–638.

Murashige, T. and Skoog, F. (1962) A revised medium for rapid growth and bioassay with tobacco tissue cultures. *Physiol. Plant.* **15**: 473–497.

Niedergang-Kamien, E. and Skoog, F. (1956) Studies on polarity and auxin transport in plants. I. Modification of polarity and auxin transport by triodobenzoic acid. *Physiol. Plant.* **9**: 60–73.

Oka, S., Yeung, E.C. and Thorpe, T.A. (1982) Shoot formation in *Eucalyptus globulus* hypocotyl explants. *NZ. J. For. Sci.* **12**: 501–509.

Parham, R.A. and Kaustinen, H.M. (1977) On the site of tannin synthesis in plant cells. *Bot. Gaz.* **138**: 465–467.

Paton, D.M., Willing, R.R., Nichols, W. and Pryor, L.D. (1970) Rooting of stem cuttings of *Eucalyptus*: a rooting inhibitor in adult tissue. *Aust. J. Bot.* **18**: 175–183.

Pirt, S.J. (1975) *Principles of Microbe and Cell Cultivation*. John Wiley & Sons, New York.

Pryor, L.D. and Willing, R.R. (1963) The vegetative propagation of *Eucalyptus* — an account of progress. *Aust. For.* **27**: 52–62.

Ramsden, L. and Northcote, D.H. (1987) Tracheid formation in cultures of pine (*Pinus sylvestris*). *J. Cell Sci.* **88**: 467–474.

Reilly, K.J. and Washer, J. (1977) Vegetative propagation of radiata pine by tissue culture: plantlet formation from embryonic tissue. *N. Z. J. For. Sci.* 7: 199–206.

Reinert, J. and White, P.R. (1956) The cultivation *in vitro* of tumor tissues and normal tissues of *Picea glauca*. *Physiol. Plant.* 9: 177–189.

Reynolds, T.L. (1987) Cytodifferentiation in cell suspension cultures of *Solanum carolinense*: effects of 2,4-D and kinetin. *Plant Physiol.* **130**: 373–383.

Ridoutt, B.G., Pharis, R.P. and Sands, R. (1996) Fibre length and gibberellins A1 and A20 are decreased in *Eucalyptus globulus* by acylcyclohexanedione injected into the stem. *Physiol. Plant.* **96**: 559–566.

Roberts, K. and Northcote, D.H. (1970) The structure of sycamore callus cells during division in partially synchronised suspension culture. *J. Cell Sci.* **6**: 299–321.

Roberts, L.W. (1983) The influence of physical factors on xylem differentiation *in vitro*. In: *Tissue Culture of Trees* (ed. J.H. Dodds), AVI Publ., Connecticut, pp. 88–102.

Sakai, K., Kusaba, K., Tsutsumi, Y. and Shiraishi, T. (1994) Secondary metabolites in cell culture of woody plants III. formation of β-thujaplicin in *Cupressus lusitanica* callus cultures treated with fungal elicitors. *Mokuzai Gakkaishi* **40**: 1–5.

Salisbury, F.B. and Ross, C. (1991) *Plant Physiology*. Wadsworth Publ. Co., Belmont, CA.

Sasse, J. (1997) Recent progress in brassinosteroid research. *Physiol. Plant.* **100**: 696–701.

Savidge, R.A. (1983) The role of plant hormones in higher plant cellular differentiation. II. Experiments with the vascular cambium, and sclereid and tracheid differentiation in the pine, *Pinus contorta*. *Histochem. J.* **15**: 447–466.

Savidge, R.A. (1985) Prospects for manipulating vascular cambium productivity and xylem cell differentiation. In: *Attributes of Trees as Crop Plants* (eds. Cannell, M.G.R. and Jackso, J.E.). Institute of Terrestrial Ecology, Natural Environment Research Council, Abbots Ripton, England, pp. 207–227.

Savidge, R. A. (1988) Auxin and ethylene regulation of cambial growth in forest trees. *Tree Physiology* **4**: 401–414.

Savidge, R.A. (1989) Coniferin, a biochemical indicator of commitment to tracheid differentiation in conifers. *Can. J. Bot.* **67**: 2663–2668.

Savidge, R.A. (1990) Characterization of indol-3-ylacetic acid in developing secondary xylem of 26 Canadian species by combined gas chromatography — mass spectrometry. *Can. J. Bot.* **68**: 521–523.

Savidge, R.A. (1993a) *In vitro* wood formation in chips from merchantable stem regions of *Larix laricina*. *IAWA J.* **14**: 3–11.

Savidge, R.A. (1993b) Formation of annual rings in trees. In: *Oscillations and Morphogenesis* (ed. Rensing, L.), Marcel Dekker, New York, pp. 343–363.

Savidge, R.A. (1994) The tracheid-differentiation factor of conifer needles. *Int. J. Plant Sci.* **155**: 272–290.

Savidge, R.A. (1996) Xylogenesis, genetic and environmental regulation. A review. *IAWA J.* **17**: 269–310.

Savidge, R.A. (1997) A healthy forest is a productive forest — a physiological viewpoint. *UNB Forestry Focus* **22**(2): 1–4.

Savidge, R.A. and Leitch, M.A. (1995) In *Proc. Plant Growth Regul. Soc. Am.*, 22nd annu. mtg,. (ed. Fritz, C.), P.O. Box 12014, Research Triangle Park, NC, pp. 160–165.

Savidge, R.A. and Wareing, P.F. (1981) Plant-growth regulators and the differentiation of vascular elements. In: *Xylem Cell Development* (ed. Barnett, J.R.), Castle House Publications, UK, pp. 192–235.

Savidge, R.A., Mutumba, G.M.C., Heald, J.K. and Wareing, P.F. (1983) Gas chromatography-mass spectroscopy identification of 1-aminocyclopropane-1-carboxylic acid in compression wood vascular cambium of *Pinus contorta* Dougl. *Plant Physiol.* **71**: 434–436.

Schenk, R.U. and Hildebrandt, A.C. (1972) Medium and techniques for induction and growth of monocotyledonous and dicotyledonous plant cell cultures. *Can. J. Bot.* **50**: 199–204.

Sembdner, G. and Parthier, B. (1993) The biochemistry and the physiological and molecular actions of jasmonates. *Annu. Rev. Plant Physiol. Plant Mol. Biol.* **44**: 569–589.

Sheriff, D.W. (1983) Control by indole-3-acetic acid of wood production in *Pinus radiata* D. Don segments in culture. *Aust. J. Plant Physiol.* **10**: 131–135.

Skirvin, R.M., Chu, M.C., Mann, M.L., Young, H., Sullivan, J. and Fermanian, T. (1986) Stability of tissue culture medium pH as a function of autoclaving, time and cultured plant material. *Plant Cell Reports* **5**: 282–294.

Smith, D.R. (1986) Radiata pine (*Pinus radiata* D. Don). In: *Trees I, Biotechnology in Agriculture and Forestry, Vol. 1* (ed. Bajaj, Y.P.S.), Springer-Verlag, Berlin, Heidelberg, pp. 274–291.

Smith, M.A.L. and McCown, B.H. (1982/3) A comparison of source tissue for protoplast isolation from three woody plant species. *Plant Sci. Lett.* **28**: 149–156.

Smith, R.H. (1992) *Plant Tissue Culture: Techniques and Experiments*. Academic Press, San Diego, CA.

Sommer, H.E. and Wetzstein, H.Y. (1984) Hardwoods. In: *Handbook of Plant Cell Culture: Vol. 3, Crop Species* (eds. Ammirato, P.V., Evans, D.A., Sharp, W.R. and Yamada, Y.), MacMillan Publ. Co., New York, pp. 511–540.

Steward, F.C., Mapes, M.O. and Mears, K. (1958) Growth and organized development of cultured cells. II. Organization in cultures grown from freely suspended cells. *Am. J. Bot.* **45**: 705–708.

Sticklen, M.B., Lineberger, R.D. and Domir, S.C. (1985) Isolation and culture of *Ulmus* × Homestead protoplasts. *Hortscience* **20**: 571.

Street, H.E. (1977) *Plant Tissue and Cell Culture*, Botanical Monographs Vol. 11, 2nd Edn. Blackwell Scientific Publications, Oxford.

Street, H.E. (1966a) The nutrition and metabolism of plant tissue and organ cultures. In: *Cells and Tissues in Culture, Methods, Biology and Physiology, Vol. 3* (ed. E.N. Willmer), Academic Press, London, New York, pp. 533–629.

Street, H.E. (1966b) Growth, differentiation and organogenesis in plant tissue and organ cultures. In: *Cells and Tissues in Culture, Methods, Biology and Physiology, Vol. 3* (ed. E.N. Willmer), Academic Press, London, New York, pp. 631–689.

Street, H.E. and Henshaw, G.G. (1966) Introduction and methods employed in plant tissue culture. In: *Cells and Tissues in Culture, Methods, Biology and Physiology, Vol. 3* (ed. Willmer, E.N.), Academic Press, London, New York, pp. 459–532.

Sutton-Jones, B. and Street, H.E. (1968) Studies on the growth in culture of plant cells. III. Changes in fine structure during the growth of *Acer pseudoplatanus* L. cells in suspension culture. *J. Exp. Bot.* **19**: 114–118.

Telewski, F.W. and Jaffe, M.J. (1986) Thigmomorphogenesis: the role of ethylene in the response of *Pinus taeda* and *Abies fraseri* to mechanical perturbation. *Physiol. Plant.* **66**: 227–233.

Trindade, H., Ferreira, J.G., Pais, M.S. and Aloni, R. (1990) The role of cytokinin and auxin in rapid multiplication of shoots of *Eucalyptus globulus* grown *in vitro*. *Aust. For.* **53**: 221–223.

Tsutsumi, Y., Nishida, T. and Sakai, K. (1994) Lignin biosynthesis in woody angiosperm tissues. III. Isolation of substrate-specific peroxidases related to the dehydrogenative polymerization of synapyl and coniferyl alcohols from *Populus* callus cultures. *Mokuzai Gakkaishi* **40**: 1348–1354.

Vardi, A., Speigel-Roy, P. and Galun, E. (1975) *Citrus* cell culture: Isolation of protoplasts, plating densities, effect of mutagens and regeneration of embryos. *Plant Cell Lett.* **4**: 231–236.

Wang, Q., Leitch, M.A. and Bossinger, G. (1999) The role of plant hormones in cambial activity of *Eucalyptus globulus*. *Appita J.* in press.

Watt, M.P., Gauntlett, B.A. and Blakeway, F.C. (1996) Effect of anti-fungal agents on *in vitro* cultures of *Eucalyptus grandis*. *S.A. For. J.* **175**: 23–27.

Wetmore, R.H. and Rier, P. (1963) Experimental induction of vascular tissues in callus of angiosperms. *Am. J. Bot.* **50**: 418–430.

White, P.R. (1943) *A Handbook of Plant Tissue Culture*. J. Cattell, Lancaster, PA, USA.

White, P.R. (1954) *The Cultivation of Animal and Plant Cells*. The Ronald Press Company, New York.

White, P.R. (1967) Sites of callus production in primary explants of spruce tissue. *Am. J. Bot.* **54**: 1055–1059.

Williams, R.R. (1993) Mineral nutrition *in vitro* — a mechanistic approach. *Aust. J. Bot.* **41**: 237–251.

Williams, R.R., Taji, A.M. and Winney, K.A. (1990) The effect of *Ptilotus* plant tissue on pH of *in vitro* media. *Plant Cell, Tissue and Organ Culture* **22**: 153–158.

Wilson, G. (1980) Continuous culture of plant cells using the chemostat principle. *Adv. Biochem. Eng.* **16**: 1–25.

Winton, L. (1970) Shoot and tree production from aspen tissue culture. *Am. J. Bot.* **57**: 904–909.

Winton, L. (1971) Tissue culture propagation of European aspen. *For. Sci.* **17**: 348–350.

Withers, L.A. (1976) Studies on the growth in culture of plant cells. XXI. Fine structural features of *Acer pseudoplatanus* L. cells responding to 2,4-dichlorophenoxyacetic acid withdrawal in turbidostat culture. *J. Exp. Bot.* **27**: 1073–1084.

Yeoman, M.M. and Truman, D.E.S. (1982) *Differentiation In Vitro*. Cambridge University Press, Cambridge.

Zajączkowski, S. (1973) Auxin stimulation of cambial activity in *Pinus silvestris*. I. The differential cambial response. *Physiol. Plant.* **29**: 281–287.

Zakrzewski, J. (1983) Hormonal control of cambial activity and vessel differentiation in *Quercus robur*. *Physiol. Plant.* **57**: 537–542.

Zhong, Y. and Savidge, R.A. (1995a) Effects of IAA and GA_3 on *in vitro* wood formation in merchantable stems of white ash (*Fraxinus americana* (L.)). In *Proc. Plant Growth Regul. Soc. Am.*, 22nd annu. mtg. (ed. Fritz, C.), P.O. Box 12014, Research Triangle Park, NC, pp. 231–236.

Zhong, Y. and Savidge, R.A. (1995b) Manipulating wood and in particular lignin formation in one-year-old white ash cuttings using IAA and GA_3. In *Proc. Plant Growth Regul. Soc. Am.*, 22nd annu. mtg. (C. Fritz, ed.), P.O. Box 12014, Research Triangle Park, NC, pp. 160–165.

Zimmermann, U., Meinzer, F. and Bentrup, F.W. (1995) How does water ascend in tall trees and other vascular plants? *Ann. Bot.* **75**: 545–551.

Index

Plate 1. *IIF of cytoskeleton in the SVS of* Aesculus hippocastanum *seen in radial longitudinal sections of stem ((a) – (f)) and taproot ((g)). (a) Randomly oriented MTs, and (b) axially oriented MF bundles (arrows) in fusiform cambial cells; note the MF-fluorescence surrounding the nuclei (n). (c) Dense, helically oriented MTs (arrows) in fibre; note the holes which are probably sites of pit-formation. (d) Rings of MFs (darts) in a vessel element associated with bordered pit formation. (e) MT rings (darts) and polygonal MT arrays associated with bordered pit formation in a vessel element. (f) Axially oriented MF bundles (arrow) in a developing fibre; note the association of MFs with pits (darts). (g) Vessel element with MT rings (darts) associated with late stages of bordered pit formation, and MT ring (arrows) associated with perforation formation. Scale bar (in (e)) = 7 μm (a), (b), 10 μm ((c) – (g). (See Chapter 2)*

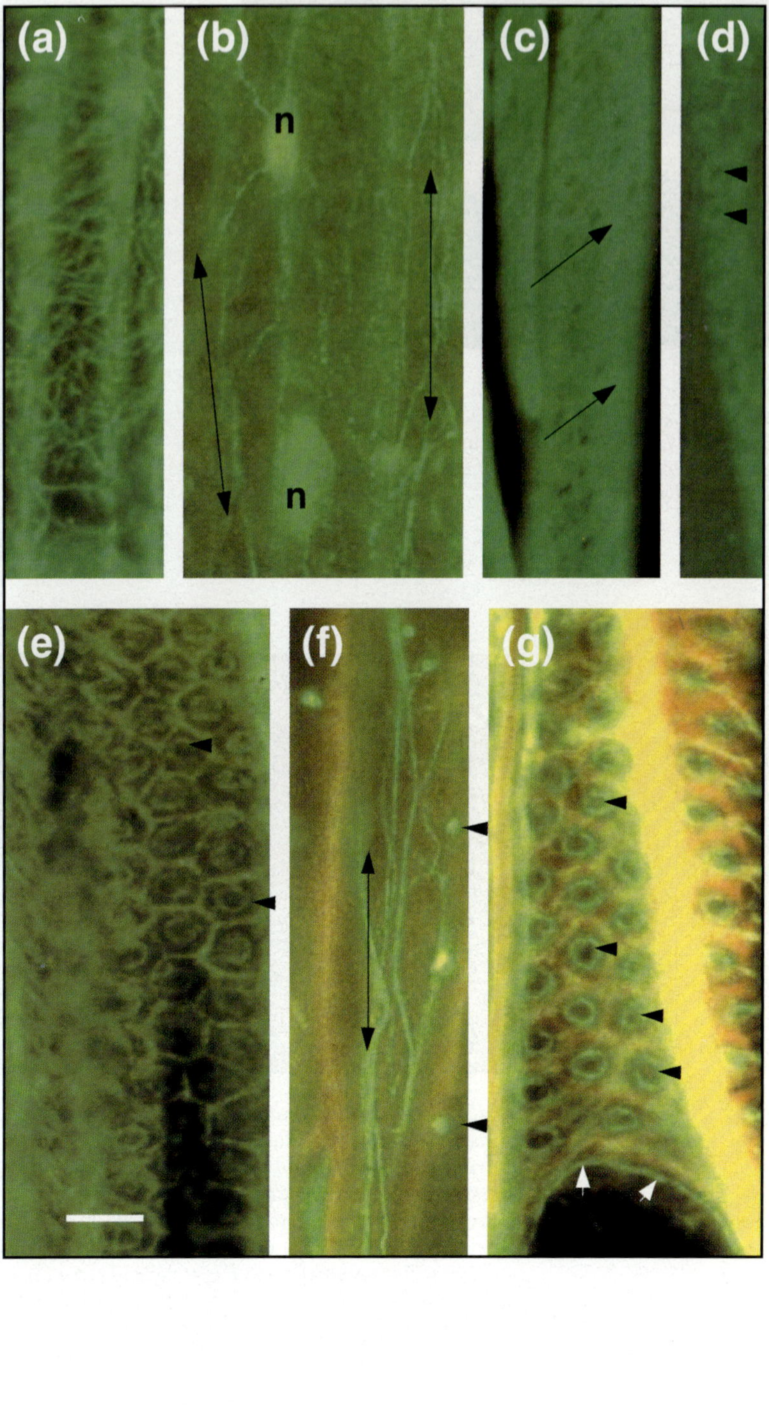
(a)
(b)
n
n
(c)
(d)
(e)
(f)
(g)

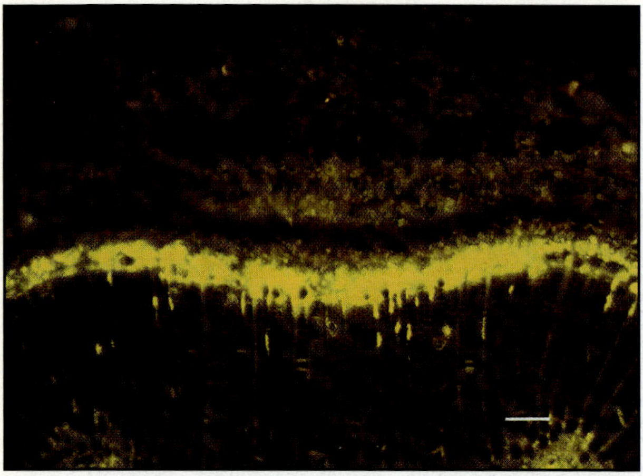

Plate 2. *Fluorescence micrograph of a transverse section of a poplar stem collected in April and treated with monoclonal antibody 46 E5 B11 F6. Bright yellow antibody-specific fluorescence occurs in almost all cells of the cambial zone (Bar: 100 μm). (See Chapter 5)*

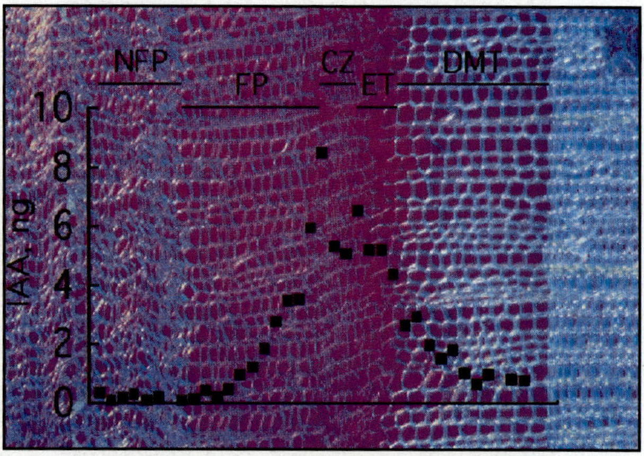

Plate 3. *Interference micrograph of a transverse section from the cambial region of* Pinus sylvestris. *The radial auxin distribution is overlain on the photograph. Each data point represent the IAA content in a cm^{-2}, 30-μm wide, tangential section. NFP = zone of non-functional phloem from previous year's growth; FP = zone of differentiating and functional phloem; CZ = cambial zone; ET = zone of expanding differentiating tracheids; DMT = zone of differentiating tracheids forming secondary walls as well as mature dead tracheids. From Uggla* et al. *(1996). (See Chapter 13)*

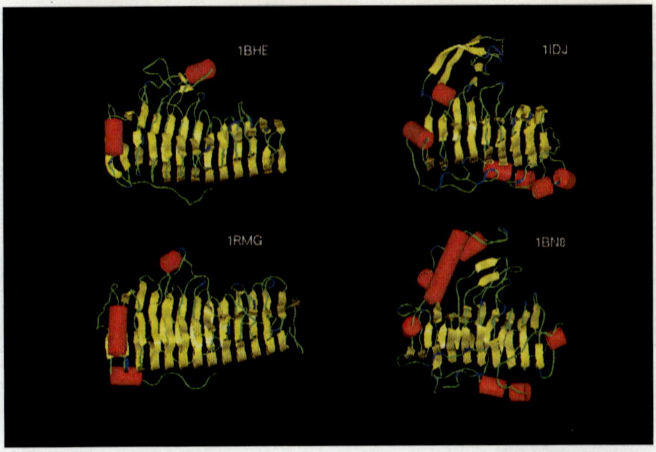

Plate 4. *A schematic drawing of* Bacillus subtilis *pectate lyase (1BN8; Pickersgill* et al., *1994), compared with pectin lyase from* Aspergillus *(1IDJ, Mayans* et al., *1997), rhamnogalacturonase from* Aspergillus *(IRMG; Peterson* et al., *1997) and polygalacturonase from* Erwinia *(1BHE; Pickersgill* et al., *1998). Software in the insight 11 programs was used to display these images. (See Chapter 24)*

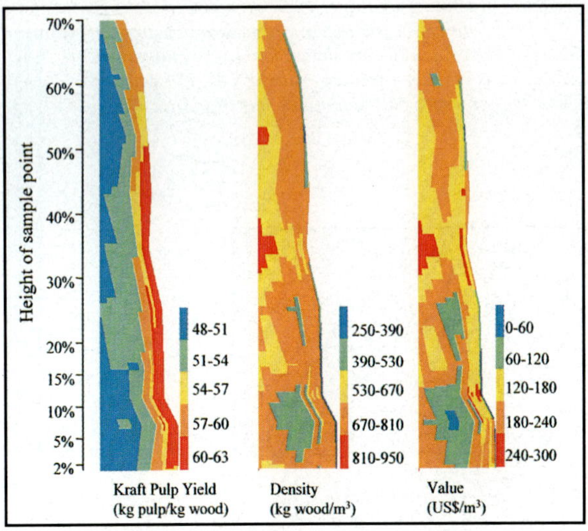

Plate 5. *Maps of within tree variation in Predicted Kraft pulp yield, wood density, and dollar value. (Pulp yield data are based on those in Schlmleck and Michell, 1998). Dollar values were determined using a price of US$420 per ton pulp. (See Chapter 26)*

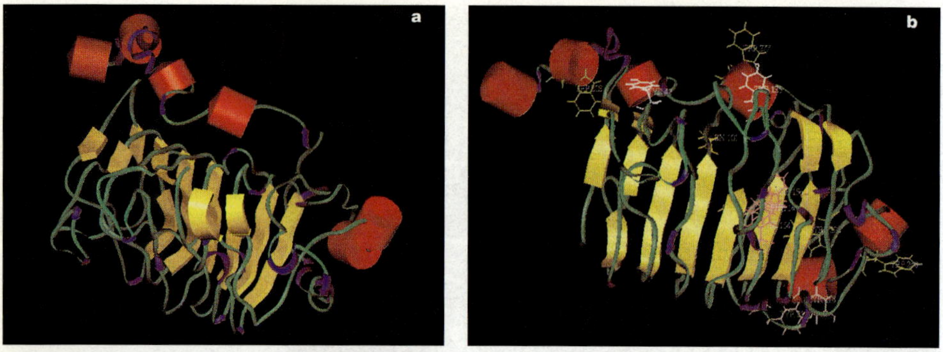

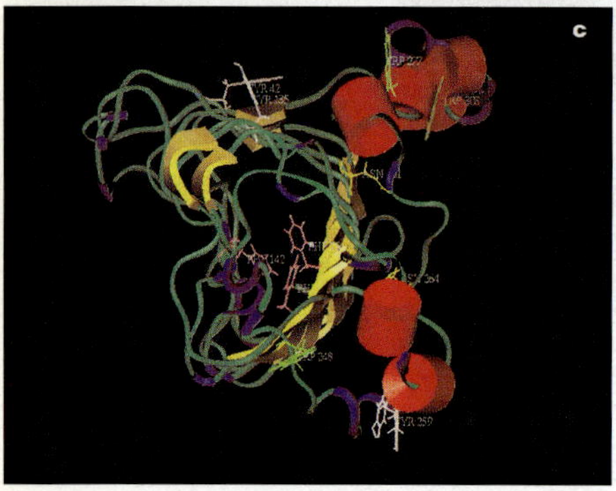

Plate 6. *Homology model of* Populus *Pectin methyl esterase. a. This shows the overall parallel b helix motif. There are seven /eight coils, each is probably made up of three sections of b sheet despite these not all being identified in the model. b. The same model but showing the conserved N101 (homologous with an Asn. in pectin lyase). Also shown are the four highly conserved tryptophanes probably involved in binding the substrate (W 223,248, 277 and 303), these side chains are shown in a situation where they are exposed to solvent as it impossible to show how they will be arranged in the final structure. Four strictly conserved tyrosines are shown (Y 42, 135, 218 and 259) and these are thought to be part of the mechanism for binding substrate. These aromatic residues extend along the active site cleft and would be able to stack against the pyranose rings of pectin. Finally three conserved phenylalanines are shown within the molecule which, it is postulated, are important in forming the hydrophobic core of the parallel b helix (F 142, 154 and 156). c shows a view of the model which indicates how the conserved phenylalanines may form hydrophobic packing in the molecule. (See Chapter 24)*

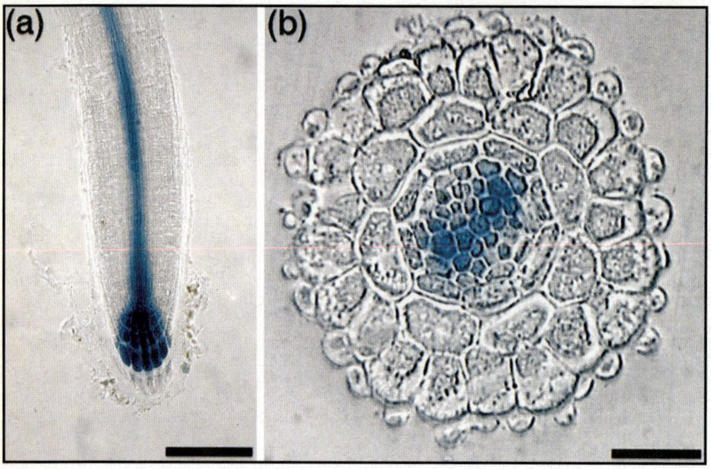

Plate 7. *ATHB-8: : GUS gene expression in the primary root of young* Arabidopsis *seedlings. (a) The GUS staining results associated with the stele and the columella region. (b) Transverse section of the elongation zone: the signal is restricted to the xylem regions. Scale bars: 50 μm (a), 20 μm (b). (See Chapter 36).*